BARRON'S

AP* CHEMISTRY

9TH EDITION

Neil D. Jespersen, Ph.D.
Professor of Chemistry
St. John's University
Jamaica, New York

Pamela K. Kerrigan, Ph.D.
Associate Professor, Biochemistry
College of Mount Saint Vincent
Riverdale, New York

About the Authors

Neil D. Jespersen is an active member of the Department of Chemistry at St. John's University (Queens, New York). He earned his B.S. degree with special attainments in chemistry at Washington and Lee University (Lexington, Virginia) and his Ph.D. at Pennsylvania State University (University Park, Pennsylvania). He specializes in analytical chemistry research and has mentored more than 100 undergraduate research students and 25 graduate students. Dr. Jespersen teaches graduate and undergraduate courses in instrumental analysis and quantitative chemical analysis. He also teaches general chemistry on a regular basis and coordinates the general chemistry laboratory program. Dr. Jespersen is a fellow of the American Chemical Society and currently serves as a councilor from the New York section. He is also the faculty advisor to the ACS Student Chapter at St. John's University.

Pamela K. Kerrigan is an Associate Professor and director of the Division of Natural Sciences at the College of Mount Saint Vincent (Bronx, New York). She earned her B.A. at Lakeland College as a double major in biology and chemistry, her M.S. at the University of Wisconsin (Milwaukee), and her Ph.D. at Arizona State University in chemistry. Over the past 23 years, Dr. Kerrigan has mentored numerous undergraduate research students. She teaches undergraduate courses in general chemistry, organic chemistry, and biochemistry. Dr. Kerrigan is the past chair of the New York section of the American Chemical Society and is a councilor. During her tenure at the college, she has been the mentor for the American Chemical Society Student Chapter. For 10 years, Dr. Kerrigan was a reader for the AP Chemistry exam.

Published by Kaplan, Inc., d/b/a Barron's Educational Series
750 Third Avenue
New York, NY 10017
www.barronseduc.com

ISBN: 978-1-4380-1066-3

9 8 7 6 5 4

Kaplan, Inc., d/b/a Barron's Educational Series print books are available at special quantity discounts to use for sales promotions, employee premiums, or educational purposes. For more information or to purchase books, please call the Simon & Schuster special sales department at 866-506-1949.

Contents

PART TWO: CHEMICAL BONDING

PART THREE: STATES OF MATTER

PART FIVE: CHEMICAL REACTIONS

PRACTICE TESTS

Note: This book comes with three additional practice tests that can be accessed online. Links to the practice tests can be found throughout this book (and by scanning the QR code below). These practice tests can be accessed on all mobile devices, including smartphones and tablets.

As you review the content in this book and work toward earning that **5** on your AP CHEMISTRY exam, here are five things that you **MUST** know above everything else.

1 **Knowing the basics** is universally important. Writing formulas and naming them, writing and balancing chemical equations, counting atoms, and determining molar masses along with proper use of significant figures are the "little things" that make a big difference between a 4 and a 5! Knowing the basics well also speeds your work so you will *seem* to have more time for the tough questions. Basic theories of chemistry require an understanding that can be applied to explaining chemical principles. These include the atomic theory, acid–base theories (Arrhenius and Brønsted-Lowry), VSEPR theory, kinetic molecular theory, collision theory, and transition state theory.

2 **Atomic and molecular structures** are fundamental to explaining many of the physical and chemical properties of substances. Atomic structure involves electron configurations and helps explain relationships within the periodic table. Molecular structure involves Lewis structures and VSEPR theory to obtain three-dimensional shapes and polarities. Polarity, or the lack of polarity, is the fundamental feature that allows the assessment of the strengths of intermolecular forces of attraction that then allows the explanation of many physical properties.

3 **Stoichiometric calculations** are used to solve many AP Chemistry problems. These problems include questions on how much of one substance reacts with another, limiting reactant calculations, titration calculations, and empirical formula calculations.

4 **Principles of chemical kinetics, chemical equilibrium, and thermodynamics** are used to explain and/or solve many questions. Chemical kinetics describes what happens as substances react and is used to deduce what happens during a chemical reaction. A dynamic equilibrium is the state that occurs after chemical change has ceased and may be used to determine the extent of reaction or the composition of the equilibrium mixture. The approaches for kinetics and equilibrium are distinctly different and must not be confused. Thermodynamics explains why chemical reactions occur in terms of changes in potential and kinetic energies. Thermodynamics also provides methods for relating the Gibbs free-energy with equilibrium constants and galvanic (voltaic) cell voltages.

5 **Representation and interpretation** of the concepts and facts of chemistry are necessary. In today's world, representations are used to explain complex issues and procedures. The new AP Chemistry course embraces this thought. You will find many common chemistry ideas translated to representations of the microscopic, molecular world. Interpretation of data, or evidence, is another skill fostered by the AP course. It is essential that students learn to correctly interpret what they are observing. Reviewing a graph, table of data, or representation of the molecular world and coming to a reasonable conclusion about its meaning is an essential skill in this complex world. These skills began to be assessed on the AP Chemistry exam starting in 2014.

Preface

You are about to embark on one of the more intellectually challenging experiences of your life, the Advanced Placement Examination in Chemistry. Fewer than 1 percent of all high school students take this exam. Whatever the outcome, you are to be congratulated as one of a select group. As a conscientious student, you can use this review book to help you increase your score. A higher score can lead to college course credit and a head start in your selected career.

The AP Examination in Chemistry is different from other exams and tests that you have taken. *Explain*, *compare*, and *predict* are three important words often used on the AP Chemistry exam. Remembered facts and calculation procedures are the basic groundwork of chemistry; however, high scores require a thorough understanding of fundamental chemical principles and relationships. Chemistry is rich in these relationships. The key to success on the exam is to think like a chemist and to apply your knowledge of one or more basic principles to provide a logical description of how chemicals behave.

This review book is designed with you, the student, in mind. It concentrates on the topics that are essential for a good score on the AP Chemistry exam. In particular, the book is designed to provide insights into the use of basic principles to answer seemingly complex questions.

The discussion in each chapter is interspersed with exercises in which subject-matter problems are presented and solved. At the end of each chapter are questions to test your understanding of the topics discussed. These, together with the three diagnostic and three practice tests, provide hundreds of questions with a range of difficulty and complexity typical of an advanced placement exam. Although many questions on the actual exam as well as the practice exams in this book will touch on two or more concepts, the diagnostic test and chapter questions in the book are written mainly as one-concept problems. In this way, the review material will help you to pinpoint weak areas on which you need more preparation, and the explained answers can be used to identify sources of error or confusion.

In May 2014, the first AP Chemistry exam developed for the new framework curriculum was offered. The revised curriculum is in place and this 9th edition of Barron's *AP Chemistry* review book is fully compliant with this program. If you feel that our review has not lived up to your expectations, feel free to contact the authors at *jespersn@stjohns.edu* and *pamela.kerrigan@mountsaintvincent.edu*. We would also be glad to hear your success stories.

Acknowledgments

First and foremost, a very special thank-you to my wife, Marilyn Zak Jespersen, who spent countless hours reading and correcting the manuscript and suggesting changes. Marilyn's contributions have made this book readable, understandable, and user-friendly. No other person could have been as dedicated to the work as she was.

It is my pleasure to work with Dr. Pamela (Pam) Kerrigan as a colleague in producing this edition. She is a well-known educator who has won many awards for working with undergraduate students. Dr. Kerrigan has considerable experience with the AP Chemistry exam and brings this background to writing questions consistent with the new curriculum. We both wish to acknowledge the many colleagues at our own institutions for their collaborations and insights.

The first iteration of the new AP Chemistry curriculum came to fruition with the 2014 exam. A tremendous effort was made to change the way in which chemistry is taught and learned. It has started successfully, and the College Board is to be praised for this effort. I also thank the College Board for keeping teachers and writers informed of the changes in a timely manner.

Finally, I thank the editors and reviewers for their suggestions and encouraging comments during the production of the book.

—N.J.

Introduction

IMPORTANT FACTS ABOUT THE ADVANCED PLACEMENT EXAMINATION IN CHEMISTRY

This examination is given in May each year at selected sites throughout the country. Exact dates, locations, and application forms are available in most guidance counselor offices. Information is also available from the following College Board Advanced Placement Program Offices.

The College Board has a specific e-mail system that assures they have essential information to answer your questions. Use the general website (*https://www.collegeboard.org/contact-us*) to enter the College Board e-mail system. See page 17 for contact details.

ESSENTIAL DETAILS

This section gives essential details about the exam itself so that nothing should be a surprise when taking the AP Chemistry exam.

The exam itself is **3 HOURS and 15 MINUTES** long. It consists of two separate sections.

Section I: Multiple Choice

Number of Questions = **60 questions**

Time Allowed = **90 minutes** (1 hour, 30 minutes)

Weight of the Multiple-Choice Questions = **50% of exam score**

A calculator is **NOT PERMITTED** on Section I

A **Periodic Table** is provided

A **Table of Useful Information** is provided (see Practice Exams for details)*

The aim of the multiple-choice questions is to pose concise questions based on learning objectives. These questions combine science practices with specific content that is covered in the AP Chemistry course.

Section II: Free Response

Number of Questions = **7 questions** (3 long-answer and 4 short-answer questions)

Time Allowed = **105 minutes** (1 hour, 45 minutes)

Weight of the Free-Response Questions = **50% of exam score**

A calculator **IS PERMITTED** on Section II

A **Periodic Table** is provided

A **Table of Useful Information** is provided (see Practice Exams for details)*

The 7 free-response questions ensure the assessment of the essential skills: experimental design, quantitative/qualitative translation, analysis of authentic lab data and observations to identify patterns or explain phenomena, creating or analyzing atomic and molecular views to explain observations, and following a logical/analytical pathway to solve a problem.

Exam free-response questions are based on learning objectives. These combine science practices with specific content that is covered in the AP Chemistry course. The free-response questions will ask you to do the following:

- Solve problems mathematically, including symbolically
- Design and describe experiments
- Perform data and error analysis
- Explain, reason, or justify answers
- Interpret and develop conceptual models

Students have a periodic table of the elements, a formula chart, and a constants chart to use on the entire exam. (See pages 528–530 for these tables and charts.) In addition, students may use a scientific or a graphing calculator on the free-response section.

Calculator Policy

Calculators are only allowed on Section II of the AP Chemistry exam. This includes scientific calculators and graphing calculators. You are allowed to use 4-function calculators, but they are not recommended. You may bring extra batteries and only one extra calculator. Be sure to check the website below for allowed and disallowed calculators. Sharing calculators is not allowed.

https://apstudent.collegeboard.org/takingtheexam/exam-policies/calculator-policy

SCORING OF THE EXAMINATION

The multiple-choice section is machine-graded. Care must be taken to be sure that there are no stray marks on your answer sheet and that all erasures are clean and complete. Also be sure that each answer has only one response marked. There is no penalty for wrong answers, and there is no penalty for leaving a question unanswered.

Highly trained high school and college chemistry teachers using a predefined scoring system grade the free-response section. Scoring is reviewed several ways to ensure consistent results. In addition, different readers grade each question to avoid carryover from one question to the next. Exams are chosen randomly for rescoring to be sure that scores do not drift during the scoring sessions.

Scores for Sections I and II are combined to obtain an overall score. These are then translated into the 1 to 5 rankings that students and colleges receive. The scoring varies little from year to year, as the following table indicates.

AP Chemistry Test Grade Distribution

Score	2014[1,2]	2015[1,2]	2016	2017
5	10.1%	8.4%	10.5%	10.1%
4	16.6%	15.2%	15.6%	16.2%
3	25.7%	28.1%	27.5%	26.1%
2	25.8%	25.5%	24.8%	26.2%
1	21.8%	22.8%	21.6%	21.4%
Mean	2.67	2.61	2.69	2.67
Number of Students	148,554	152,745	153,415	158,931

[1]2014 and 2015 are years with the new framework.
[2]©2014, 2015, 2016, 2017 The College Board

TEST CONTENT AND DISTRIBUTION

The AP Chemistry curriculum starts with the six BIG IDEAS of chemistry:

BIG IDEA 1: The chemical elements are fundamental building materials of matter. All matter can be understood in terms of the arrangements of atoms. These atoms retain their identity in chemical reactions.

BIG IDEA 2: Chemical and physical properties of materials can be explained by the structure and the arrangement of atoms, ions, or molecules and the forces among them.

BIG IDEA 3: Changes in matter involve the rearrangement and/or reorganization of atoms and/or the transfer of electrons.

BIG IDEA 4: Rates of chemical reactions are determined by details of the specific molecular collisions.

BIG IDEA 5: The laws of thermodynamics describe the essential role of energy. They explain and predict the direction of changes in matter.

BIG IDEA 6: Any bond or intermolecular attraction that can be formed can be broken. These two processes are in a dynamic competition, sensitive to initial conditions and external perturbations.

Importantly, science is an activity that is actually done, not just studied. In that process, the student of chemistry must attain certain skills to really understand the subject. These have been classified as SCIENCE PRACTICES:

SCIENCE PRACTICE 1: The student can use representations and models to communicate scientific phenomena and solve scientific problems.

SCIENCE PRACTICE 2: The student can use mathematics appropriately.

SCIENCE PRACTICE 3: The student can engage in scientific questioning to extend thinking or to guide investigations within the context of the AP course.

SCIENCE PRACTICE 4: The student can plan and implement data collection strategies in relation to a particular scientific question. (Note: Data can be collected from many different sources, e.g., investigations, scientific observations, the findings of others, historic reconstruction, and/or archived data.)

SCIENCE PRACTICE 5: The student can perform data analysis and evaluation of evidence.

SCIENCE PRACTICE 6: The student can work with scientific explanations and theories.

SCIENCE PRACTICE 7: The student is able to connect and relate knowledge across various scales, concepts, and representations in and across domains.

Inquiry-Based Investigations

Twenty-five percent of instructional time is devoted to inquiry-based laboratory investigations. Students ask questions, make observations and predictions, design experiments, analyze data, and construct arguments in a collaborative setting where they direct and monitor their progress.

After looking at this list, every item seems perfect since that is how a scientist performs his or her job. However, that is not what had been taught and tested in the past. Just memorizing facts and names will no longer suffice. Learning a set procedure to get an answer is not the focus of the test. You will have to interpret and evaluate data tables and graphs. You will have to explain representations of matter about the molecular and laboratory scale.

The test was quite different in 2014 than in the past, and there will be plenty of practice with the new format. It is quite possible that the modern student will actually find the AP exam easier. Time will tell.

It would be nice to design this review book around the six big ideas of chemistry. However, they do not fit together well to make a readable book. This 9th edition of Barron's *AP Chemistry* will have a familiar structure. It has been carefully reviewed to take out the material that will not be tested. However, we have retained some material that may not be tested directly but that the authors feel is necessary for proper context. In addition, new material on mass spectrometry and photoelectron spectroscopy has been added. The book focuses on basic information and techniques with the idea that a firm foundation in the basics will make it easier to work on the more difficult questions. To help you relate the chapter content to the six Big Ideas and the Learning Objectives, we have listed the relevant topics at the beginning of each chapter. In addition, each answer explanation in the practice tests lists the Essential Knowledge and Learning Objective numbers. For the complete wording, refer to the AP Chemistry Course Outline: *https://secure-media.collegeboard.org/digitalServices/pdf/ap/ap-chemistry-course-and-exam-description.pdf.* A listing of the topics in this edition follows.

CONTENT AND DISTRIBUTION IN THIS BOOK

The contents of this book are listed below so you can easily see how this 9th edition presents the AP Chemistry curriculum described on the previous pages.

STRUCTURE OF MATTER

- Atomic Theory and the Structure of the Atom
  - Evidence supporting the atomic theory
  - Atomic masses, atomic numbers, mass number, isotopes
  - Electronic structure of the atom
    - Energy levels, atomic spectra
    - Quantum numbers, atomic orbitals
  - Periodic trends and relationships
    - Ionization energies, electron affinities, electronegativity
    - Atomic and ionic radii, oxidation states
- Chemical Bonding
  - Inter- and intramolecular binding forces
    - Ionic and covalent bonding
    - Hydrogen bonds, dipole-dipole and van der Waals forces (including London forces)
    - Forces related to states, properties, and structure of matter
    - Bond polarity, electronegativity
  - Models of molecules
    - Lewis structures, resonance
    - VSEPR
    - Valence bond theory, hybrid orbitals, sigma and pi bonds
  - Molecular geometry, structural isomerism
    - Geometry of simple molecules, organic and inorganic
    - Coordination complexes
    - Dipole moments, molecular polarity
    - Relationship of properties to structure

STATES OF MATTER

- Gases
  - Laws of ideal gases
    - Ideal gas law (equation of state)
    - Partial pressures (Dalton's law)
  - Kinetic-molecular theory
    - Interpretation of gas laws
    - Avogadro's hypothesis
    - Relationship between kinetic energy and temperature
    - Differences between ideal gases and real gases
- Liquids and Solids
  - Kinetic-molecular theory applied to liquids and solids
  - Changes of state
  - Structure of solids; lattice energy
- Solutions
  - Types of solutions and solubility
  - Raoult's law

REACTIONS

Reaction Types

- Acid–base reactions; Arrhenius and Brønsted-Lowry theories
- Coordination complexes and amphoterism
- Precipitation reactions
- Oxidation-reduction reactions
  - Oxidation number
  - Electron transfer in oxidation and reduction
  - Electrochemistry including electrolytic cells and Faraday's laws, galvanic cells and standard reduction potentials, prediction of the direction of a reaction

Stoichiometry

- Ionic and molecular species in chemical systems, net ionic equations
- Balancing equations including redox equations
- Mass and volume relationships using the mole concept, empirical formulas, and limiting reactants

Equilibrium

- Dynamic equilibrium concept, Le Châtelier's principle and equilibrium constants
- Quantitative use of equilibrium
  - Equilibrium constants for gas-phase reactions
  - Equilibrium constants for reactions in solutions
    - Acid–base equilibrium, and pH calculations
    - Solubility product calculations including common ions
    - Buffer and hydrolysis equilibria

Kinetics

- Rates of reaction, general concepts, and factors
- Determination of rates, rate laws, reaction order, and rate constants from experimental data including graphs
- Effect of temperature on rates
- Activation energy and catalysis
- Relationship of rate-determining step to rate laws and reaction mechanisms

Thermodynamics

- State functions
- First law of thermodynamics, enthalpy change, heats of formation and reaction, Hess's law, calorimetry
- Second law of thermodynamics including the concepts of entropy, free energies of formation and reaction, the relationship between free energy, enthalpy, and entropy
- Relationships between free-energy change, equilibrium constants, and electrode potentials

DESCRIPTIVE CHEMISTRY

There are a large number of facts, principles, and concepts not previously listed that will be needed to demonstrate a knowledge of chemistry on the AP Chemistry exam. In addition, the principles, concepts, and properties of chemicals are part of the real world outside the classroom. This should be a continuing part of the AP course. Some appropriate areas are

Chemical reactivity and reactions and a knowledge of chemical nomenclature

Relationships in the periodic table that allow prediction of chemical and physical properties. These can be horizontal, vertical, or diagonal relationships.

LABORATORY

College-level chemistry courses include laboratory exercises. These involve measurement, preparation of solutions, experimental setup, and synthesis and analysis of various compounds. The AP Chemistry exam will have questions involving the laboratory experience. Most of those questions involve general laboratory procedures, including

Making and recording appropriate observations of chemical substances and reactions

Making quantitative measurements and recording data appropriately

Calculating results from quantitative data and making appropriate interpretations of these results

Communicating the results of experiments in laboratory reports

CHEMICAL CALCULATIONS AND RELATIONSHIPS

Below is a list of problem types and calculations that the test taker should have mastered after taking an advanced placement course. The student should be able to assess calculation results in regard to reasonableness, significant figures (including the results of logarithmic and exponential operations), and precision of measurements.

1. Percentage composition
2. Empirical formulas from experimental data and molecular formulas from empirical formulas
3. Molar masses from gas density measurements
4. Gas laws, including the ideal gas law and Dalton's law of partial pressures
5. Stoichiometric relations using the concept of the mole; titration calculations
6. Mole fractions and molar solutions
7. Faraday's laws of electrolysis
8. Equilibrium constants and their applications, including their use for simultaneous equilibria
9. Standard electrode potentials and their use
10. Thermodynamic and thermochemical calculations
11. Kinetics calculations

HOW TO USE THIS REVIEW BOOK

Every student taking the AP Chemistry examination is an individual with different plans and needs. Used properly, this book can be individualized to help any student maximize their score on the AP exam no matter what their previous preparation or success with learning chemistry has been. There are two important factors: (1) the time available for your preparation before the exam and (2) your current knowledge of chemistry. Look under the heading below that seems to fit you best.

I'm at the Top of My Class and Just Need a Tune-Up

This review book has the most problems and questions of any currently available. If you are at the top of your class, the first place to start is with a sample of the AP exam. There are six tests available, three at the end of this book and three electronic versions. Take these tests to sharpen your responses and learn how to manage your time.

I Do Pretty Well, But I Can Do Better

You will need a little time to review all the topics. We suggest taking one of the six practice tests to see how you do. If you do well, study the topics that show up as weaknesses. If your score is less than desired, you might want to try the diagnostic tests.

Average Describes Me, But I Need to Excel

Start with a diagnostic test. Evaluate where the problems lie and read those parts of the book thoroughly. Try a practice test and evaluate your results. If needed, take another diagnostic test to locate and correct your major problems.

Chemistry Confuses Me, Help

Even the totally confused can be helped. Start with the diagnostic tests. Review the chapters on subjects that show up as problems in the analysis. Repeat the diagnostic tests as needed. The more time you have, the better. However, the diagnostics in this book help identify the areas that can produce a better score with the time available.

ORGANIZING YOUR REVIEW

1. **PLAN TO START REVIEWING AS SOON AS POSSIBLE.** You don't go on a diet and lose 10 pounds in 1 day, you cannot exercise for a week and run a marathon, and you cannot read this book in one sitting and have a thorough review of chemistry. The sooner you start reviewing, the more leisurely pace you can use to digest information. Cramming does not allow time for concepts to gel and for relationships to become apparent.
2. **SET A SCHEDULE FOR YOUR REVIEW.** Depending on the available time before the AP exam, divide your review into reasonable study blocks. Schedule more time than you need, plan on 2-hour study sessions, and reward yourself with time off for topics you know well.
3. **ACTIVELY REVIEW.** Don't just read this book; actively study it. Use a red pen to cancel units in problems. Write out the answers to all problems; don't solve them mentally. Make notes of topics that you find confusing.

4. **ASK QUESTIONS.** AP teachers are dedicated educators who want you to succeed, but don't expect them to teach the whole course over again. Ask specific questions (write them down beforehand). If you need help with a problem, show your teacher that you've tried to solve it and indicate where you got stuck. Education is learning how to ask questions and get them answered.
5. **ASSESS YOUR PROGRESS AND REVIEW YOUR WEAKNESSES.** Use the diagnostic tests to help you concentrate on specific areas. Use the practice exams to become accustomed to test conditions.

The following schedule may be helpful in planning your review sessions. Each directive represents a 2-hour session. The schedule should be read horizontally, not vertically. For example, in the first week of review, you would take the first diagnostic test on Monday, review the first most-needed chapter on Wednesday, and review the second most-needed chapter on Friday.

The schedule requires 8 weeks to complete and leaves all weekends, as well as Tuesdays and Thursdays, free. You can change this to a Tuesday/Thursday/Saturday schedule or any other sequence that assures regular review. This schedule can be compressed into as few as 3 weeks if review is done on a daily basis.

Practice Test 3 is included for additional practice.

	Monday	Tuesday	Wednesday	Thursday	Friday
Week 1	Take first diagnostic test.		Review first most-needed chapter.		Review second most-needed chapter.
Week 2	Take second diagnostic test.		Review first most-needed chapter.		Review second most-needed chapter.
Week 3	Chapter 1		Chapter 2		Chapter 3
Week 4	Chapter 4		Chapter 5		Chapter 6
Week 5	Chapter 7		Chapter 8		Chapter 9
Week 6	Chapter 10		Chapter 11		Chapter 12
Week 7	Chapter 13		Chapter 14		Take first practice test.
Week 8	Review		Review weak areas.		Take second practice test.

(If results indicate continued weakness after the second practice test, repeat the last week of review and take the third diagnostic test.)

Reminder: The link to the online practice tests can be found on the inside front cover of this book. All three online tests can be accessed on mobile devices, including tablets and smartphones. Be sure to have a copy of your *AP Chemistry*, 9th edition on hand to complete the registration process.

WHAT TO EXPECT OF THE AP CHEMISTRY EXAMINATION

1. **EXAM DIFFICULTY.** There is no question that the Advanced Placement Examination in Chemistry is difficult, and there are at least three reasons. First, some questions will be totally unfamiliar, either because they are asked in a unique manner or the topic was not covered in class. Because of the volume of material, even college-level courses do not include all of the topics presented on the exam. Don't waste time on questions dealing with unfamiliar material. Second, the examination is long. The multiple-choice section allows an average of 1.5 minutes per question. Third, many questions combine two or more concepts.

   Another difficult part of the exam is Section II, where written answers—either mathematical calculations or essays—are required. In addition to knowing the material, you must produce a logical, well-organized, and well-written response. You often need a significant amount of practice with writing skills to do well on this section.
2. **SECTION I: MULTIPLE-CHOICE QUESTIONS. (60 QUESTIONS, 90 MINUTES, NO CALCULATORS)** This section contains two types of questions. *Conceptual questions* ask you to recall a concept and then use it correctly to answer a question or evaluate a hypothetical situation. *Estimation questions* require you to make calculations without using a calculator. These questions will use simple numbers or expect you to round data so that you can do calculations by hand or in your head. A table of symbols, constants, and equations is available for use along with a periodic table.
3. **SECTION II: FREE-RESPONSE QUESTIONS. (3 LONG QUESTIONS AND 4 SHORT QUESTIONS, 105 MINUTES)** A periodic table and a table of symbols, constants, and equations can be used. In addition, an approved calculator (see calculator policy) can be used. The exam will include 3 long questions that will be multipart and multiconcept. One of these may involve a laboratory experiment. The 4 short questions often involve a few parts and one or two concepts. All free-response questions must be answered in a well-written, logical, and mathematically correct fashion. In several questions, you will be asked to explain your reasoning or justify your results. You may be asked to draw representations of the molecular scale, draw diagrams of instruments or experiment setups, or extract and analyze data from tables and graphs.

What the AP Exam *DOES NOT* Include

TIP

***No equations are referred to by name*, such as the Arrhenius or Henderson-Hasselbalch equations.**

No data tables are provided. If needed, values for equilibrium constants and thermodynamic quantities will be given with the question. The periodic table uses only symbols and does not give names of the elements. Named equations such as Beer's law are given, but the name is not. Aids such as common oxidation numbers or ionic charges are not given. You must memorize polyatomic ion formulas and names along with methods for balancing redox reactions.

HOW TO MAXIMIZE YOUR SCORE

The Advanced Placement Examination in Chemistry is designed so that the average score will be approximately 50 percent. This is done by careful selection of the difficulty of the questions and of the length of the exam itself. There are well-known concepts and methods for ensuring that you will achieve the maximum score you deserve.

The multiple-choice section is designed to test your recall of fundamental chemical concepts and the use of these concepts to solve basic chemistry problems. The questions cover the entire AP course syllabus and are designed with various levels of difficulty. Each question has four choices, only one of which is the most appropriate answer. Provide a response for each and every question because there is no penalty for leaving a question unanswered.

The free-response section is very often the most challenging part of the AP Chemistry exam. It requires a very good knowledge of the theories, principles, concepts, and facts of chemistry, the same things that will earn a good score on the multiple choice section. In addition, this section demands well-considered, logical, concise, and readable presentations.

The first three questions are long, constructed-response questions that can include mathematical solutions to problems of approximately 4 to 8 parts. Briefly explain your approach to the problem in words. When doing the math, use good algebraic methods. You will make fewer mistakes if you do not skip steps. Use complete equations with equal signs, and explain any nonstandard symbols. Assess the result to indicate if it is reasonable or not.

Remember that partial credit is given but only if some work is shown. If a number from a previous part of the question is missing, make up a number (and tell the reader what you are doing) and then solve the problem. Correct methods that the reader can easily discern will earn a large fraction of the allotted score.

The last four questions are short, constructed-response questions. They often involve interpretation of a theory or application of a process such as writing Lewis structures and determining molecular geometry, polarity, and intermolecular forces. Answers should be well planned, concise, and technically correct. Technical errors detract from the presentation. You should include appropriate examples with sketches, structures, or graphs to be used to support statements of fact.

Strategies for Multiple-Choice Questions

SHOULD YOU GUESS?

DEFINITELY YES. There is no penalty for wrong answers. Therefore, a random guess when you are running out of time beats nothing at all: your odds of getting a correct answer are 1 out of 4.

***Always* take a guess if you don't know the answer.**

Of course, if you have time, it is best to read the question and answer it to the best of your ability. In chemistry there are many subtle clues that allow you to judge your selected answers. Below are some hints to help you read between the lines of a question *and* to do a quick evaluation of your answer.

DISTRACTERS

In the design of multiple-choice questions, the writer constructs the responses so that one choice is correct and the others are "distracters." A distracter is a response that looks good at first glance but has a serious flaw that makes it incorrect. The better the design of the test, the more distracters will be found in each question.

One popular method for constructing distracters is to use subtle changes in the wording to make a response incorrect. For example:

1. **All** chemicals become more soluble as the temperature increases.
2. **Most** chemicals become more soluble as the temperature increases.

or

1. The reaction is **exo**thermic.
2. The reaction is **endo**thermic.

Careful reading of questions and understanding of terminology are very important. The distinctions between "most" and "all" in the first set above and between "exothermic" and "endothermic" in the second are obviously significant. To ensure selecting the best answer to a nonnumerical problem, be sure to read each response before selecting one. Often a good-sounding, but incorrect, response is listed before the correct one. Another approach is to read the responses in reverse order. Pay special attention to responses that are exactly the opposite of each other as in the "exothermic"/"endothermic" example above. One response must necessarily be wrong and may also provide a clue as to whether or not both are incorrect. Also pay special attention to responses that differ by only one word, as in the first set above. Once again, they may provide a clue as to the correct way to think about the problem.

For numerical problems, some distracters provide answers in which the data are simply used in the wrong manner. For instance, for the question "What is the value of 5 divided by 2?" the answer choices may be as follows:

(A) 2.50 $\frac{5}{2}$

(B) 0.40 $\frac{2}{5}$

(C) 3.00 (5.00 − 2.00)

(D) 7.00 (5.00 + 2.00)

The parentheses show the calculation method used to obtain the answers. The question hinges on understanding the term *divided by,* and keeping in mind that 2, not 5, is the divisor, before the proper calculation can be made to obtain answer A. You must understand the proper method for using the data.

REASONABLENESS

There are, however, some common methods for increasing the probability of choosing the correct answer to a numerical chemistry problem. It is important to remember the *principle of reasonableness.* This means that answers must reflect obedience to fundamental principles, such as the conservation of matter and energy. Your personal experiences in everyday life also may provide clues as to the reasonableness of answers.

For example, if 2 grams of one reactant are mixed with 5 grams of another, it is impossible to have any more than 7 grams of product even under the best conditions (law of conservation of matter). Therefore any response that is greater than 7 grams may be eliminated very quickly. As another example, if a hot solution is added to a colder one, the final temperature must be somewhere between the low of the cold solution and the high of the hot solution. Any other responses may be eliminated as incorrect without any calculations at all.

Pointers about the reasonableness of answers will be given throughout this book.

ESTIMATING ANSWERS

The Advanced Placement Test places minimal focus on the use of calculators. Instead concepts are tested, and most mathematical problems use simple numbers that do not require calculators. In effect, these questions test your ability to set up problems rather than any ability to solve problems or do mathematical operations. In accord with this new approach, students must now understand how to quickly estimate rather than calculate answers. Below are the basic principles of estimation. Many of the problem solutions at the end of the chapters and for the sample tests indicate methods for estimating answers and also show detailed calculations.

TIP

Estimating answers to mathematical problems is an important skill.

The **first principle** of estimation is that all problems must still be set up in a rigorous and logical manner. This has not changed from the times when calculators were allowed. The **second principle** of all estimation is to round numbers to one, or at most two, significant figures. The **third principle** is to round in a manner that makes cancellation simple and to take every opportunity to cancel. The **fourth principle** is to add and subtract in groups. A few examples of these principles follow. Although the following are broken into steps, there are no rigorous steps to memorize.

EXAMPLE 1

Solve for X.

$X = 25.34 \times 1.890 \times 0.00318 \times 4.1689 \times 9.823$	Step 1: round to one or two digits
$= 25 \times 2 \times 0.003 \times 4 \times 10$	Step 2: note that $4 \times 2 \times 25 = 200$
$= 200 \times 0.003 \times 10$	Step 3: simplify the decimal
$= 0.6 \times 10 = 6$	Step 4: finish up

Notice (step 1) that you should try to round up as much as you round down. Also notice (step 2) that you DO NOT have to multiply numbers in sequence. The actual calculator answer is 6.24. This is off by 0.24, but when estimating, that is perfectly all right. You now know that the correct answer cannot be 13,248 and that it cannot be 5.2×10^{-5}.

EXAMPLE 2

Calculate Y.

Here we demonstrate how to estimate the high and low limits of a calculation.

$Y = 3.642 \times 1.102 \times 7.785 \times 178.2 \times 51.65 \times 0.00219$	Step 1: round to simple numbers, except 3.6
$= 3.6 \times 1 \times 8 \times 200 \times 50 \times 0.002$	Step 2: note that $200 \times 50 = 10{,}000$
$= 10{,}000 \times 3.6 \times 8 \times 0.002$	Step 3: $10{,}000 \times 0.002 = 20$
$= 20 \times 3.6 \times 8$	Step 4: 3.6 is between 3 and 4; do two calculations
$Y_1 = 20 \times 4 \times 8 = 640$ (maximum)	
$Y_2 = 20 \times 3 \times 8 = 480$ (minimum)	

The calculator answer is 630. or 6.30×10^2.

In this problem we chose one number that could be rounded either up or down and rounded it up to find a maximum and down to find a minimum. This is a useful technique to set limits on your answers.

EXAMPLE 3

Calculate Z.

$$Z = 21.47 \times \frac{0.000341}{0.06625} \times \frac{294.3}{771.9}$$

Step 1: round numbers, paying attention to cancelling opportunities

$$= 20 \times \frac{0.0003}{0.06} \times \frac{300}{800}$$

Step 2: cancel 2 of 20 with 8 of 800 to get

Step 3: cancel zeros in 300 and 400

$$= 10 \times \frac{0.0003}{0.06} \times \frac{300}{400}$$

Step 4: divide center terms by 3 to obtain the following

$$= 10 \times \frac{0.0003}{0.06} \times \frac{3}{4}$$

Step 5: multiply out the numerator and denominator

$$= 10 \times \frac{0.0001}{0.02} \times \frac{3}{4}$$

Step 6: multiply the numerator and denominator by 1000 each

$$= \frac{0.003}{0.08}$$

Step 7: estimate the answer as about 4 parts in 100, or 0.04

$$= 3/80$$

The calculated answer is 0.0421. Once again the estimate is in the ballpark. It is important to notice how advantage was taken of simple math operations in canceling. Good canceling saves a lot of work and reduces errors. Finally, each separate step was written out above. In real examples, these steps are done on a single equation without rewriting it.

EXAMPLE 4

Calculate A.

$$A = \frac{2.847 \times 10^{-3}}{9.113 \times 10^{6}} \times \frac{6.321 \times 10^{-8}}{14.34 \times 10^{-23}}$$

Step 1: regroup all exponential terms together and all numbers together as shown

$$= \frac{2.847 \times 6.321}{14.34 \times 9.113} \times \frac{10^{-3} \times 10^{-8}}{10^{6} \times 10^{-23}}$$

Step 2: evaluate exponents

$$= \frac{2.847 \times 6.321}{14.34 \times 9.113} \times 10^{6}$$

Step 3: evaluate numbers; simplify first

$$= \frac{3 \times 6}{14 \times 9} \times 10^{6}$$

Step 4: divide numerator and denominator by 3

$$= \frac{1 \times 6}{14 \times 3} \times 10^{6}$$

Step 5: divide numerator and denominator by 3 again

$$= \frac{1 \times 2}{14 \times 1} \times 10^{6}$$

Step 6: divide numerator and denominator by 2

$$= \frac{1 \times 1}{7 \times 1} \times 10^{6} = \frac{1}{7} \times 10^{6}$$

$$\text{or } 0.14 \times 10^{6}$$

The calculated answer is 0.1377×10^{6} or 1.377×10^{5}.

EXAMPLE 5

Calculate B.

$$B = 20.5 + 2.346 + 102.33 + 33.62 + 5.009$$

Round off to one significant figure and add:

$$B = 20 + 2 + 100 + 30 + 5$$
$$= 157$$

The calculated answer is 163.8.

EXAMPLE 6

Estimate pH from $[H^+]$, where pH is defined as $-\log [H^+]$. What is the approximate pH of a solution where $[H^+] = 4.1 \times 10^{-5}$ M?

In almost all cases the hydrogen ion concentration is expressed exponentially, for example, 4.1×10^{-5} M. It turns out that the pH for a 4.1×10^{-5} M solution is 4.39. This is between pH 4 and 5. In fact, any hydrogen ion concentration with an exponent of 10^{-5} has a pH between 4 and 5. Extending this, we can estimate the pH to within one unit by simply looking at the power of 10 for the hydrogen ion concentration. The pH is ALWAYS a maximum of the positive value of the exponent and a minimum of one pH unit less than that. If $[H^+] = 3.8 \times 10^{-6}$, we can quickly say that the pH is between 5 and 6. If $[H^+] = 3.8 \times 10^{-11}$, the pH is between 10 and 11.

We can use exactly the same principle if $[OH^-]$ is given and you want to know the pOH. For example, if $[OH^-] = 1.3 \times 10^{-2}$, the pOH will be between 1 and 2. If $[OH^-] = 3.3 \times 10^{-6}$, the pOH will be between 5 and 6.

In both cases the exponent tells us the maximum value for the pH or pOH. One less than this maximum is the minimum value that the pH can be.

EXAMPLE 7

Estimate $[H^+]$ from pH. What is the $[H^+]$ of a solution where the pH = 4.7?

If you are given a pH of 4.7, the corresponding $[H^+]$ will be $10^{-4.7}$. The decimal exponent is unusual but does not violate any rules of mathematics. At times it is convenient to work with decimal exponents.

In cases where the decimal exponent is inconvenient, use a high and a low value to establish a range. For example, the $[H^+]$ of $10^{-4.7}$ can also be a high of 10^{-4} to a low of 10^{-5}.

EXAMPLE 8

Determine additional logarithmic measurements. Estimate the pOH of a 3.3×10^{-6} M solution of OH^-.

If you want more precision in logarithms, it is necessary only to remember that log 2 = 0.3 and log 3 = 0.5.

For example, if the $[OH^-] = 3.3 \times 10^{-6}$, we can estimate the pOH as $-\log (3.3 \times 10^{-6})$, which is $-\log 3.3$ and $-\log (10^{-6})$. The former is approximately -0.5, and the latter is $+6$. The two add up to 5.5 for the pOH.

Strategies for the Free-Response Section

NUMERICAL CALCULATIONS

Any scientific calculator, including a graphing calculator, can be used for Section II of the exam. Be certain that the batteries are fresh and that you are familiar with the operation of your calculator. Write appropriate chemical reactions. Always write the fundamental equations or laws that the question requires. Identify variables. Use correct algebra in the solution and show as many algebraic steps as possible. Clearly state any assumptions you have used and verify that each assumption is valid before reporting the answer. Check also that you have used the correct number of significant figures in calculations.

For example, consider this problem: Calculate the pH of a 0.100 *M* solution of hydrofluoric acid, $K_a = 6.9 \times 10^{-4}$.

To logically solve the problem you need a chemical reaction, the equilibrium law, a simplification, and a solution, as shown in the following steps.

Reaction: $HF \rightleftharpoons H^+ + F^-$ or $HF + H_2O \rightleftharpoons H_3O^+ + F^-$

Equilibrium expression used: $K = \dfrac{[H^+][F^-]}{[HF]}$ or $\dfrac{[H_3O^+][F^-]}{[HF]}$

Simplified equation: $[H^+] = \sqrt{K_a C_a}$, where C_a is the initial HF concentration and the assumption is that $[H^+] << C_a$.

Solution:

$$
\begin{aligned}
[H^+] &= \sqrt{(6.9\times10^{-4})(0.100)} \\
&= \sqrt{(6.9\times10^{-5})} \\
&= 8.3\times10^{-3} \text{ (This agrees with the assumption.)} \\
pH &= -\log[H^+] \\
&= -\log(8.3\times10^{-3}) \\
&= 2.08
\end{aligned}
$$

PARTIAL CREDIT

Partial credit is not given on the exam. All work must be shown in order to receive credit for the question. In questions with multiple parts, the student will be given credit for later sections of the question if the calculation in the first part is incorrect but used correctly in subsequent steps. If you have not set up your calculations clearly, you will not be given any credit for the problem.

EXPLAIN AND JUSTIFY

You will probably not encounter a true essay question. Instead, you will most likely be asked to explain or justify your answer. Be sure to answer the question asked. The graders do not want you to rehash your problem-solving methods. Instead, they are looking to see how you evaluate your answer. For instance, you may discuss why your answer seems reasonable. You can also refer to other information, such as comparing acid strengths or atomic radii, that shows you are thinking beyond the calculated answer. Justifying your answer is slightly different. Now you are using evidence to help support your answer and perhaps show that alternative answers are not as good. A classic example of this is the use of formal charges to defend your choice of structure in relation to other possibilities. Even if you are not asked to explain or justify your answer, get into the habit of asking yourself to answer those questions with each practice problem you work.

FINAL PREPARATIONS FOR THE EXAM

Just as an athlete needs to prepare for the "big game," the student must prepare physically, mentally, and emotionally for the "big test." Here are some suggestions:

1. Eat well to have enough energy for the exam. A good dinner the night before and a relaxed breakfast on the day of the exam provide the energy essential to peak performance.
2. Get plenty of sleep. A full 8 hours of sleep is recommended for a rested body and a well-functioning mind. The night before the exam is no time to cram; in fact, such last-minute study may be detrimental.
3. The night before the exam, assemble the things you will need: plenty of #2 pencils with erasers, a scientific calculator with fresh batteries, a watch, and your admission card for the AP exam. You should also plan what you will wear to the test. Comfortable, loose-fitting clothes, including items such as sweaters that can be layered or removed to suit the room temperature, are best.
4. Be sure your transportation to the test center is reliable. Set your alarm so that you can leave early. Allow time to deal with the unexpected: a traffic jam, flat tire, or late-running bus.
5. Minimize distractions and worries. Leave all valuables at home so that you do not worry about them during the test. Put all unrelated matters firmly out of your mind.
6. Be confident of your ability. A positive attitude is very important in successful test taking.
7. Relax. This one test will not make or break your career. Enjoy the exam and show the world how well you can do.

THE COLLEGE BOARD

The College Board has a specific e-mail system that assures they have essential information to answer your questions. Use the general website (*https://www.collegeboard.org/contact-us*) to enter the College Board e-mail system.

MAIN OFFICES
The College Board National Office
250 Vesey Street
New York, NY 10281
Phone: 212-713-8000
https://www.collegeboard.org/contact-us/send-message

The College Board Washington D.C. Office
Phone: 202-741-4700
https://www.collegeboard.org/contact-us/send-message

The College Board Reston, Virginia Office
11955 Democracy Drive
Reston, VA 20190
Phone: 571-485-3000
https://www.collegeboard.org/contact-us/send-message

REGIONAL OFFICES
Middle States Regional Office
Phone: 866-392-3019
Fax: 610-227-2580
https://www.collegeboard.org/about/region-offices/middle-states/contact

Midwestern Regional Office
Phone: 866-392-4086
Fax: 847-653-4528
https://www.collegeboard.org/about/region-offices/midwest/contact

New England Regional Office
Phone: 866-392-4089
Fax: 781-663-2743
https://www.collegeboard.org/about/region-offices/new-england/contact

Southern Regional Office
Phone: 866-392-4088
Fax: 770-225-4062
https://www.collegeboard.org/about/region-offices/south/contact

Southwestern Regional Office
Phone: 866-392-3017
Fax: 512-721-1841
https://www.collegeboard.org/about/region-offices/southwest/contact

Western Regional Office
Phone: 866-392-4078
Fax: 213-416-2133
https://www.collegeboard.org/about/region-offices/west/contact

Oficina de Puerto Rico y América Latina
(Puerto Rico and Latin America Office)
Phone: 787-772-1200
Fax: 787-759-8629
E-mail: *PROLAA@collegeboard.com*

The College Board International
250 Vesey Street
New York, New York 10281
Phone: 212-373-8738
https://www.international.collegeboard.org/send-message

State Services Office
South Florida Office
Phone: 866-392-4088
Fax: 954-874-4341
E-mail: *flo@collegeboard.org*

The direct URL for the AP Chemistry Web pages is *https://apstudent.collegeboard.org/apcourse/apchemistry*. Please check this website for late-breaking news and exam changes that may affect you. In addition, this site has important information about AP exam registration, sites, test dates and times, score reporting, and fees.

It is very important to check the schedules on this website and register on time so that you do not miss an exam or pay late-registration fees. In most instances, your AP teacher will guide you through this process.

DIAGNOSTIC TESTS

ANSWER SHEET
Diagnostic Test 1

1. Ⓐ Ⓑ Ⓒ Ⓓ	16. Ⓐ Ⓑ Ⓒ Ⓓ	31. Ⓐ Ⓑ Ⓒ Ⓓ
2. Ⓐ Ⓑ Ⓒ Ⓓ	17. Ⓐ Ⓑ Ⓒ Ⓓ	32. Ⓐ Ⓑ Ⓒ Ⓓ
3. Ⓐ Ⓑ Ⓒ Ⓓ	18. Ⓐ Ⓑ Ⓒ Ⓓ	33. Ⓐ Ⓑ Ⓒ Ⓓ
4. Ⓐ Ⓑ Ⓒ Ⓓ	19. Ⓐ Ⓑ Ⓒ Ⓓ	34. Ⓐ Ⓑ Ⓒ Ⓓ
5. Ⓐ Ⓑ Ⓒ Ⓓ	20. Ⓐ Ⓑ Ⓒ Ⓓ	35. Ⓐ Ⓑ Ⓒ Ⓓ
6. Ⓐ Ⓑ Ⓒ Ⓓ	21. Ⓐ Ⓑ Ⓒ Ⓓ	36. Ⓐ Ⓑ Ⓒ Ⓓ
7. Ⓐ Ⓑ Ⓒ Ⓓ	22. Ⓐ Ⓑ Ⓒ Ⓓ	37. Ⓐ Ⓑ Ⓒ Ⓓ
8. Ⓐ Ⓑ Ⓒ Ⓓ	23. Ⓐ Ⓑ Ⓒ Ⓓ	38. Ⓐ Ⓑ Ⓒ Ⓓ
9. Ⓐ Ⓑ Ⓒ Ⓓ	24. Ⓐ Ⓑ Ⓒ Ⓓ	39. Ⓐ Ⓑ Ⓒ Ⓓ
10. Ⓐ Ⓑ Ⓒ Ⓓ	25. Ⓐ Ⓑ Ⓒ Ⓓ	40. Ⓐ Ⓑ Ⓒ Ⓓ
11. Ⓐ Ⓑ Ⓒ Ⓓ	26. Ⓐ Ⓑ Ⓒ Ⓓ	41. Ⓐ Ⓑ Ⓒ Ⓓ
12. Ⓐ Ⓑ Ⓒ Ⓓ	27. Ⓐ Ⓑ Ⓒ Ⓓ	42. Ⓐ Ⓑ Ⓒ Ⓓ
13. Ⓐ Ⓑ Ⓒ Ⓓ	28. Ⓐ Ⓑ Ⓒ Ⓓ	43. Ⓐ Ⓑ Ⓒ Ⓓ
14. Ⓐ Ⓑ Ⓒ Ⓓ	29. Ⓐ Ⓑ Ⓒ Ⓓ	44. Ⓐ Ⓑ Ⓒ Ⓓ
15. Ⓐ Ⓑ Ⓒ Ⓓ	30. Ⓐ Ⓑ Ⓒ Ⓓ	45. Ⓐ Ⓑ Ⓒ Ⓓ

Diagnostic Test 1

Directions: Answer the following multiple-choice questions. You may use a calculator and the periodic table on page 528, but no other information. Limit your time to 60 minutes. If you do not finish in 60 minutes, note the number of questions answered and then continue until all the remaining questions are answered. Record your total time. Score your test with the answer key at the end of the test. Also at the end of the test are tables to help diagnose your strengths and weaknesses. Review the topics that have the most errors and then continue with Diagnostic Test 2. For more of a challenge, do not use your calculator when answering these questions.

1. A certain beam of monochromatic light has a frequency of 6.00×10^{14} hertz. The wavelength of this radiation is ________, and this wavelength and frequency reside in the ____________ of the electromagnetic spectrum. (The speed of light is 3.00×10^8 m s^{-1}.)
   (A) 500. nm visible region
   (B) 200. nm ultraviolet region
   (C) 2,000,000 m. radio wave region
   (D) 500. pm X-ray region

2. Of the following oxo acids, which is predicted to be the strongest acid?
   (A) HBrO
   (B) HClO
   (C) HIO
   (D) $HClO_3$

3. What is the equilibrium expression for the reaction below?

$$C_3H_8(l) + 5O_2(g) \longrightarrow 3CO_2(g) + 4H_2O(l)$$

(A) $K = \dfrac{[CO_2]^3}{[O_2]^5}$

(B) $K = \dfrac{[C_3H_8][O_2]^5}{[CO_2]^3[H_2O]^4}$

(C) $K = \dfrac{[C_3H_8]}{[H_2O]^4}$

(D) $K = \dfrac{[O_2]^5}{[CO_2]^3}$

4. The rate law may be written using stoichiometric coefficients for which of the following?
(A) Precipitation reactions
(B) Acid–base reactions
(C) Elementary processes
(D) Solubility reactions

5. A gas in a 5.0 L container under 2.5 atm pressure is allowed to expand to fill a new container in which the pressure has decreased to 1.0 atm. What is the new volume of the gas?
(A) 12.5 L
(B) 2.0 L
(C) 0.50 L
(D) 10.0 L

6. A solution is prepared by dissolving 30.0 grams of $Ni(NO_3)_2$ in enough water to make 250 mL of solution. What is the molarity of this solution?
(A) 0.496 molar
(B) 0.657 mol/L
(C) 3.3 molar
(D) 6.3×10^{-3} molar

7. Hard materials such as silicon carbide, used for grinding wheels, are said to be examples of
(A) ionic crystals
(B) network crystals
(C) metallic crystals
(D) molecular crystals

8. The equilibrium constant, K_c, for the dissociation of HI into hydrogen gas and iodine vapor is 21 at a certain temperature. What will be the molar concentration of iodine vapor if 15 grams of HI gas is introduced into a 12.0-L flask and allowed to come to equilibrium?
   (A) 4.58 mol/L
   (B) 0.00687 mol L^{-1}
   (C) 4.4×10^{-3} M
   (D) 9.76×10^{-3} M

9. What is the molarity of a sodium hydroxide solution that requires 42.6 mL of 0.108 M HCl to neutralize 40.0 mL of the base?
   (A) 0.0641 M
   (B) 1.64 M
   (C) 0.115 M
   (D) 0.400 mol/L

10. Changing which of the following will change the numerical value of the equilibrium constant?
   (A) The pressure of reactants
   (B) The pressure of products
   (C) The temperature of the reaction
   (D) The total mass of the chemicals present

11. Of the reactions below, which is a decomposition reaction?
   (A) $C_7H_8O_2(l) + 8O_2(g) \rightarrow 7CO_2(g) + 4H_2O(l)$
   (B) $2KClO_3(s) \rightarrow 2KCl(s) + 3O_2(g)$
   (C) $2Cr(s) + 3Cl_2(g) \rightarrow 2CrCl_3(s)$
   (D) $6Li(s) + N_2(g) \rightarrow 2Li_3N(s)$

12. Helium effuses through a pinhole 5.33 times faster than an unknown gas. That gas is most likely
   (A) CO_2
   (B) CH_4
   (C) C_5H_{12}
   (D) C_8H_{18}

13. The Pauli exclusion principle states that
   (A) no two electrons can have the same energy
   (B) no two electrons can have the same four quantum numbers
   (C) no two electrons can occupy separate orbitals
   (D) no two electrons can pair up if there is an empty orbital available

14. Which of the following can form hydrogen bonds?
(A) $CH_3OCH_2CH_3$
(B) HCN
(C) CH_3OCH_2Br
(D) CH_3NH_2

15. Which experiment led to the notion that the atom contains an extremely small, positively charged nucleus?
(A) Millikan's oil drop experiment
(B) Rutherford's gold foil experiment
(C) Thomson's cathode ray experiment
(D) Dalton's atomic experiment

16. Which of the following is expected to be a polar molecule?
(A) PCl_4F
(B) BF_3
(C) CO_2
(D) $Si(CH_3)_4$

17. All of the following ions have the same electron configuration EXCEPT
(A) Rb^+
(B) Se^{2-}
(C) As^{5+}
(D) Sr^{2+}

18. In which of the following is the negative end of the bond written last?
(A) O–S
(B) Br–N
(C) N–C
(D) P–Cl

19. What is the molar mass of a monoprotic weak acid that requires 26.3 mL of 0.122 *M* KOH to neutralize 0.682 gram of the acid?
(A) 212 g mol^{-1}
(B) 4.70 g mol^{-1}
(C) 147 g mol^{-1}
(D) 682 g mol^{-1}

20. Which of the following correctly lists the individual intermolecular attractive forces from the strongest to the weakest?
(A) Induced dipole < dipole-dipole < hydrogen bond
(B) Hydrogen bond < dipole-dipole < induced dipole
(C) Induced dipole < hydrogen bond < dipole-dipole
(D) Dipole-dipole < hydrogen bond < induced dipole

21. A mechanism is a sequence of elementary reactions that add up to the overall reaction stoichiometry. A substance that is produced in one elementary reaction and consumed in another is called
(A) a catalyst
(B) an intermediate
(C) a reactant
(D) a complex

22. How many moles of propane (C_3H_8) are there in 6.2 g of propane?
(A) 14.1 mol
(B) 1.4×10^{-1} mol
(C) 71 mol
(D) 7.1 mol

23. What is the work involved if a gas in a 2.0-liter container at 2.4 atmospheres pressure is allowed to expand against a pressure of 0.80 atmosphere?
(A) −3.2 L atm
(B) −9.62 L atm
(C) −4.8 L atm
(D) +14.4 L atm

24. Monatomic ions of the representative elements are often
(A) very soluble
(B) very electronegative
(C) isoelectronic with a noble gas
(D) highly colored

25. A first-order reaction has a half-life of 34 minutes. What is the rate constant for this reaction?
(A) 3.4×10^{-4} s^{-1}
(B) 2.04×10^{-2} s^{-1}
(C) 2.9×10^{-1} min^{-1}
(D) 34 min

DIAGNOSTIC TEST 1

26. The electrolysis of molten magnesium bromide is expected to produce
   (A) magnesium at the anode and bromine at the cathode
   (B) magnesium at the cathode and bromine at the anode
   (C) magnesium at the cathode and oxygen at the anode
   (D) bromine at the anode and hydrogen at the cathode

27. The oxidation number of chlorine in the perchlorate ion, ClO_4^-, is
   (A) +2
   (B) −3
   (C) −1
   (D) +7

28. A certain reaction has a $\Delta H^\circ = -43.2$ kJ mol^{-1} and an entropy change ΔS° of +22.0 J mol^{-1} K^{-1}. What is the value of ΔG° at 800°C?
   (A) −66.8 kJ mol^{-1}
   (B) +21.2 kJ mol^{-1}
   (C) −21.2 kJ/mol
   (D) −2365 kJ mol^{-1}

29. How many electrons, neutrons, and protons are in an atom of ^{52}Cr?
   (A) 24 electrons, 24 protons, 24 neutrons
   (B) 27 electrons, 27 protons, 24 neutrons
   (C) 24 electrons, 28 protons, 24 neutrons
   (D) 24 electrons, 24 protons, 28 neutrons

30. The ideal gas law is successful for most gases because
   (A) room temperature is high
   (B) volumes are small
   (C) gas particles do not interact significantly
   (D) gases are dimers

31. Which of the following is a conjugate acid–base pair?

(A)	HCl	NaOH
(B)	HNO_2	NO_2^-
(C)	H_2SO_3	SO_3^{2-}
(D)	$Fe(OH)_3$	$Fe(OH)_2$

32. The electron configuration $1s^2\,2s^2\,2p^6\,3s^2\,3p^6\,4s^2$ signifies the ground state of the element

(A) V

(B) Ti

(C) Co

(D) Ca

33. What is the molar concentration of chloride ion when 35.0 g of $MgCl_2$ (molar mass = 95.21 g/mol) are dissolved in 250. mL of water?

(A) 1.47 *M*

(B) 0.00294 *M*

(C) 2.75 *M*

(D) 2.94 *M*

34. Which of the following reactions is expected to have the greatest decrease in entropy?

(A) $HCl(aq) + NaOH(aq) \rightarrow NaCl(aq) + H_2O(l)$

(B) $CH_4(g) + 2O_2(g) \rightarrow CO_2(g) + 2H_2O(l)$

(C) $C(s) + O_2(g) \rightarrow CO_2(g)$

(D) $2SO_2(g) + O_2(g) \rightarrow 2SO_3(g)$

35. What is the conjugate acid of the $H_2PO_4^-$ ion?

(A) HPO_4^{2-}

(B) $H_2PO_4^-$

(C) H_3PO_4

(D) PO_4^{3-}

36. A thimble of water contains 4.0×10^{21} molecules. The number of moles of H_2O is

(A) 2.4×10^{45}

(B) 6.6×10^{-23}

(C) 6.6×10^{-3}

(D) 2.4×10^{23}

37. Each resonance form of the nitrate ion, NO_3^-, has how many sigma and how many pi bonds?

(A) 1 sigma and 2 pi

(B) 2 sigma and 1 pi

(C) 1 sigma and 1 pi

(D) 3 sigma and 1 pi

38. Twenty-five milligrams of sucrose ($C_{12}H_{22}O_{11}$) are dissolved in enough water to make 1.00 liter of solution. What is the molarity of the solution?
    (A) 7.3×10^{-5}
    (B) 7.31×10^{-2}
    (C) 73.1
    (D) 1.36

39. Which of the following is considered a metalloid?
    (A) Cr
    (B) Mn
    (C) Si
    (D) S

40. A fast reaction rate for a chemical reaction depends on
    (A) having a large activation energy
    (B) being exothermic
    (C) being endothermic
    (D) having a small activation energy

41. Which of the following is a reduction half-reaction?
    1. $Cu^{2+} + e^- \rightarrow Cu^+$
    2. $Cu^+ + e^- \rightarrow Cu^0$
    3. $Fe^{2+} \rightarrow Fe^{3+} + e^-$

    (A) 1 only because copper(II) ions are reduced
    (B) 3 only because the iron is reduced
    (C) 1 and 2 because they both reduce copper ions
    (D) 1 and 3 because they do not have insoluble ions

42. Which of the following is expected to have the largest bond polarity?
    (A) S–O
    (B) P–F
    (C) C–B
    (D) C–N

43. How many electrons, neutrons, and protons are in an atom of Cr?
    (A) 24 electrons, 24 protons, 24 neutrons
    (B) 27 electrons, 27 protons, 24 neutrons
    (C) 24 electrons, 28 protons, 24 neutrons
    (D) More information is needed to answer this question.

44. Which of the following is incorrectly named?

(A) $CaCl_2$ calcium chloride

(B) $Fe(NO_3)_3$ iron(III) nitrate

(C) $AlBr_3$ aluminum tribromide

(D) $K_2Cr_2O_7$ potassium dichromate

45. As the atomic number increases within a period, what happens to the atomic radius of an atom?

(A) It increases.

(B) It remains the same.

(C) It decreases.

(D) It increases and then decreases.

ANSWER KEY

Diagnostic Test 1

1. **(A)**	10. **(C)**	19. **(A)**	28. **(A)**	37. **(D)**
2. **(D)**	11. **(B)**	20. **(B)**	29. **(D)**	38. **(A)**
3. **(A)**	12. **(D)**	21. **(B)**	30. **(C)**	39. **(C)**
4. **(C)**	13. **(B)**	22. **(B)**	31. **(B)**	40. **(D)**
5. **(A)**	14. **(D)**	23. **(A)**	32. **(D)**	41. **(C)**
6. **(B)**	15. **(B)**	24. **(C)**	33. **(D)**	42. **(B)**
7. **(B)**	16. **(A)**	25. **(A)**	34. **(B)**	43. **(D)**
8. **(C)**	17. **(C)**	26. **(B)**	35. **(C)**	44. **(C)**
9. **(C)**	18. **(D)**	27. **(D)**	36. **(C)**	45. **(C)**

EVALUATING YOUR RESULTS

Score your test using the answer key. Then complete the following tables. The first table is designed to find your general strengths and weaknesses based on four broad categories. The second table is a more specific diagnostic chart that will suggest which particular chapters you should concentrate your studies on. In combination, these two tables will help you focus your efforts on the material that needs the most study.

Question Categories

Question Type	Questions	Number Wrong
Basic Facts	2, 3, 7, 10, 11, 13, 15, 16, 21, 24, 29, 35, 37, 39, 44, 45	
Basic Concepts	4, 14, 18, 27, 28, 30, 31, 32, 36, 41, 42	
Calculations	1, 5, 6, 8, 9, 12, 19, 22, 23, 25, 27, 28, 33, 38	
Mixed Concepts	9, 17, 20, 26, 32, 34, 35, 40, 43	

Breakdown by Topics

Chapter	Questions	Number Wrong
1. Structure of the Atom	1, 13, 15, 32	
2. Periodic Table	29, 39, 43, 45	
3. Ionic Compounds and Reactions	11, 17, 24, 44	
4. Covalent Compounds	16, 18, 37, 42	
5. Stoichiometry	9, 19, 22, 36	
6. Gases	5, 12, 20	
7. Liquids and Solids	7, 14, 20	

Chapter	Questions	Number Wrong
8. Solutions	6, 33, 38	
9. Equilibrium	3, 8, 10	
10. Kinetics	4, 21, 25, 40	
11. Thermodynamics	23, 28, 34	
12. Redox and Electrochemistry	26, 27, 41	
13. Acids and Bases	2, 31, 35	

ANSWERS EXPLAINED

1. **(A)** The basic equation is $\lambda\nu = c$, which reads "The wavelength times the frequency is equal to the speed of light, which is 3.00×10^8 m s^{-1}."

$$\lambda = (3.00 \times 10^8 \text{ m s}^{-1})/(6.00 \times 10^{14} \text{ s}^{-1})$$
$$\lambda = 5.0 \times 10^{-7} \text{ m} = 500 \times 10^{-9} \text{ m} = 500 \text{ nm}$$

If we convert the units to meters (m), centimeters (cm), picometers (pm), we do not end up with any of the other choices.

2. **(D)** Among HBrO, HClO, and HIO, the more electronegative central atom indicates a stronger oxo acid. Therefore HClO is the strongest of these three. Comparing HClO and $HClO_3$, we select $HClO_3$ as the strongest because it has the larger number of unshared or double-bonded oxygen atoms that stabilize the anion, creating a stronger acid.

3. **(A)** By convention, liquids are not written as part of the equilibrium expression. Liquids always have the same number of molecules per liter, which is a constant, combined into the value of the *K*. The only substances that are not solids or pure liquids in this reaction are $O_2(g)$ and $CO_2(g)$. In this reaction, CO_2 to the third power appears in the numerator because it is a product, and O_2 to the fifth power is in the denominator.

4. **(C)** Only reactions that are known to be elementary processes (i.e., actual collisions of molecules) can be used to write a rate law. Most other reactions proceed by multistep mechanisms that have to be deduced from experimental evidence.

5. **(A)** Pressure and volume are inversely proportional when a gas is at constant temperature. The relationship is expressed as $P_1V_1 = P_2V_2$ and we solve for the answer as

$$(2.50 \text{ atm})(5.00 \text{ L}) = (1.00 \text{ atm})\ V_2$$

$$V_2 = \frac{(2.50 \text{ atm})(5.00 \text{ L})}{(1.00 \text{ atm})}$$

$$= 12.5 \text{ L}$$

6. **(B)** Molarity is the moles of solute in each liter of solution; molarity = mol solute/L solution

$$\text{mol Ni(NO}_3)_2 = 30.0\text{ g Ni(NO}_3)_2\left(\frac{1\text{ mol Ni(NO}_3)_2}{182.70\text{ g Ni(NO}_3)_2}\right)$$
$$= 0.1642\text{ mol Ni(NO}_3)_2$$

$$\text{molarity Ni(NO}_3)_2 = (0.1642\text{ mol Ni(NO}_3)_2)/0.250\text{ L}$$
$$= 0.657\text{ mol Ni(NO}_3)_2/\text{L}$$

7. **(B)** Network crystals are held together with covalent bonds. These bonds make the crystal one large, rigid, molecule. As a result, the macroscopic substance is very hard.

8. **(C)** The reaction is $2\text{ HI}(g) \rightleftharpoons H_2(g) + I_2(g)$, and the equilibrium expression is

$$K = \frac{[H_2][I_2]}{[HI]^2} = 21$$

$$\frac{?\text{ mol HI}}{\text{L}} = \frac{15\text{ g HI}}{12\text{ L}}\left[\frac{1\text{ mol HI}}{128\text{ g HI}}\right] = 9.77 \times 10^{-3}\text{ mol L}^{-1}$$

When HI reacts, $2x$ moles of HI form x moles of H_2 and x moles of I_2. The equilibrium expression is

$$\frac{(x)\,(x)}{(9.77\times10^{-3} - 2x)^2} = 21$$ Take the square root of both sides of the equation.

$$\frac{(x)}{9.77\times10^{-3} - 2x} = 4.58$$

$$x = 4.47 \times 10^{-2} - 9.16x$$
$$10.16x = 4.47 \times 10^{-2}$$
$$x = 4.4 \times 10^{-3}\text{ mol L}^{-1}$$

Because $x = [I_2]$, that is the concentration of I_2 vapor expected. We keep only two significant figures because K_c had only two significant figures.

9. **(C)** Molarity is a ratio of units (moles per liter); therefore, the setup of the calculation must start with a ratio, such as another molarity.

Setup:

$$?M_{\text{NaOH}} = 0.108\ M_{\text{HCl}}$$

Expand molarities to mol/L and insert conversion factors:

$$?\ \frac{\text{mol NaOH}}{\text{L NaOH}} = \frac{0.108\text{ mol HCl}}{\text{L HCl}} \times \frac{1\text{ mol NaOH}}{1\text{ mol HCl}} \times \frac{0.0426\text{ L HCl}}{0.0400\text{ L NaOH}}$$
$$= 0.115\ M\text{ NaOH}$$

10. **(C)** ONLY temperature changes will result in changes in the numerical value of the equilibrium constant.

11. **(B)** A decomposition reaction occurs when a large molecule breaks down into its composite elements or smaller molecules. In this case, $KClO_3$ breaks down into KCl and O_2.

12. **(D)** Graham's law of effusion is written as

$$\frac{\text{rate}_1}{\text{rate}_2} = \sqrt{\frac{\text{mass}_2}{\text{mass}_1}}$$

Defining helium as compound 1 in the equation above, we can substitute

$$5.33 = \sqrt{\frac{\text{mass}_2}{4}}$$

Square both sides:

$$28.4 = \frac{\text{mass}_2}{4}$$
$$\text{mass}_2 = 4 \times 28.4 = 114 \text{ g/mol}$$

The molecular masses of the possible compounds are (A) $CO_2 = 44$, (B) $CH_4 = 16$, (C) $C_5H_{12} = 72$, and (D) $C_8H_{18} = 114$, which is obviously the sample in this experiment.

13. **(B)** The Pauli exclusion principle states that no two electrons in an element can have the same set of four quantum numbers. This is another way of saying that no two electrons can occupy the same orbital with the same spin. Response (A) is false—electrons often have the same energy. Response (C) is wrong because although electrons can occupy separate orbitals, this has nothing to do with the question. Response (D) is a statement of Hund's rule.

14. **(D)** Hydrogen bonds can form when a compound contains a fluorine, an oxygen, or a nitrogen atom with a hydrogen bonded to it. In this question, this criterion is fulfilled only by CH_3NH_2.

15. **(B)** In Rutherford's gold foil experiment, he bombarded a piece of gold foil with heavy alpha particles. Rutherford discovered that most of the particles went directly through the foil rather than being reflected back. This led to the conclusion that the atom was mostly comprised of a small, dense, positive core surrounded by empty space with sparingly spaced electrons.

16. **(A)** The Lewis structures of all molecules except PCl_4F are symmetric. The fact that two different atoms are attached to the central P atom immediately suggests that this molecule is not symmetric. We test it first by drawing its Lewis structure. We find that structure to be a trigonal bipyramid that is not symmetrical, and we predict it will be polar.

17. **(C)** All of the ions in this question are isoelectronic with the noble gas krypton except arsenic. Arsenic would have to be a 3– ion to be isoelectronic with krypton. It is written as a 5+ ion.

18. **(D)** Generally, with a pair of elements, the one closest to fluorine in the periodic table is negative (largest electronegativity), and the atom farthest from fluorine is positive (lowest electronegativity). The only one of the five pairs where the second element is closest to fluorine is the P–Cl pair.

19. **(A)** To determine the molar mass, we need to calculate the moles of the acid and divide it into the mass of the acid used in this experiment. Since each acid molecule has one proton (it is *mono*protic), determining the moles of protons gives us the moles of the acid.

moles of protons = (0.0263 L) (0.122 M KOH) = 3.21×10^{-3} mol

molar mass = (0.682 gram)/(3.21×10^{-3} mol) = 212 g mol^{-1}

20. **(B)** Hydrogen bonding is the strongest intermolecular attractive force and is listed first. Induced dipoles are the weakest attractive forces and last for a short period of time.

21. **(B)** An intermediate is defined as a substance that is neither a reactant nor a product. Intermediates are often difficult to detect. Catalysts fit the above description, but they are substances that are added to the reaction mixture and can be isolated afterward.

22. **(B)** Mathematically, we start with ? mol C_3H_8 = 6.2 g C_3H_8. We then use the conversion factor 1 mol C_3H_8 = 44 g C_3H_8 to convert g to mol.

$$? \text{ mol } C_3H_8 = 6.2 \text{ g } C_3H_8 \left[\frac{1 \text{ mol } C_3H_8}{44 \text{ g } C_3C_8}\right]$$

$$= 1.4 \times 10^{-1} \text{ mol } C_3H_8$$

23. **(A)** Work can be determined from *PV* data. The equation is

$$\text{work } (w) = -P\,\Delta V.$$

We can use $P_1V_1 = P_2V_2$ to calculate the final volume in the problem above:

$$(2.4 \text{ atm})(2.0 \text{ L}) = (0.80 \text{ atm})\ (x \text{ liters})$$

$$x \text{ liters} = 6.0 \text{ L}$$

$$\text{work} = -(0.80 \text{ atm})(6.0 \text{ L} - 2.0 \text{ L}) = -3.2 \text{ L atm}$$

24. **(C)** This is the best answer because all group 1A, 2A, and 3A metals, as well as representative nonmetals, have ions that are isoelectronic with a noble gas.

25. **(A)** From the integrated rate law we can derive that ln(2) = $kt_{1/2}$, where $t_{1/2}$ is the half-life.

Converting 34 minutes to seconds yields $34 \times 60 = 2040$ seconds. The rate constant is calculated as

$$k = \ln(2)/2040 \text{ s}$$
$$= 0.693/2040 \text{ s}$$
$$= 3.4 \times 10^{-4} \text{ s}^{-1}$$

We round the final answer to two significant figures because 34 minutes has only two significant figures.

26. **(B)** In this molten salt the only possible products are magnesium and bromine. The question is at which electrode are these products found. Remembering that oxidation always occurs at the anode, we see that $2Br^- \rightarrow Br_2 + 2e^-$ is the oxidation process. Therefore bromine is produced at the anode, and magnesium at the cathode.

27. **(D)** To calculate the oxidation number of an element in a polyatomic ion, the charges of the atoms must add up to the charge on the ion.

$$\begin{aligned} \text{Charge of ion} &= (\text{ox. no. Cl}) + 4(\text{ox. no. O}) \\ -1 &= (\text{ox. no. Cl}) + 4(-2) \\ -1 + 8 &= \text{ox. no. Cl} \\ +7 &= \text{ox. no. Cl} \end{aligned}$$

28. **(A)** The Gibbs free-energy equation is $\Delta G° = \Delta H° - T\,\Delta S°$. To determine $\Delta G°$ at a temperature other than 298 K we need $\Delta H°$ and $\Delta S°$ so that the above equation can be solved at a temperature different from 298.

$$\begin{aligned} \Delta G° &= -43.2 \text{ kJ mol}^{-1} - (1073 \text{ K})(22.0 \text{ J mol}^{-1} \text{ K}^{-1}) \\ &= -43.2 \text{ kJ mol}^{-1} - (23.6 \times 10^3 \text{ J mol}^{-1}) \\ &= -43.2 \text{ kJ mol}^{-1} - (23.6 \text{ kJ mol}^{-1}) \\ &= -66.8 \text{ kJ mol}^{-1} \end{aligned}$$

29. **(D)** This is an uncharged isotope of chromium that has a mass number of 52. The atomic number of chromium is 24 (see periodic table). So this symbol represents an isotope that has 24 protons. Because the charge is 0, the isotope has 24 electrons. It also has 52 – 24 = 28 neutrons.

30. **(C)** This response is true because of the large distance between gas particles. The other three responses are not necessarily true about any given gas sample.

31. **(B)** This pair differs by only one H^+ between their two formulas.

32. **(D)** This electron configuration has 20 electrons in the lowest possible energy levels according to the aufbau ordering. Element 20 is calcium.

33. **(D)** To calculate the concentration of chloride ion, first calculate the number of moles of Cl^- in 35.0 g of $MgCl_2$

$$35.0 \text{ g MgCl}_2 \times \frac{1 \text{ mol MgCl}_2}{95.21 \text{ g MgCl}_2} \times \frac{2 \text{ mol Cl}^-}{1 \text{ mol MgCl}_2} = 0.735 \text{ mol Cl}^-$$

Calculate the molarity after converting the volume to L

$$\frac{0.735 \text{ mol Cl}^-}{0.250 \text{ L}} = 2.94\ M$$

34. **(B)** The most important factor in entropy change is whether or not there is a change in volume as denoted by the change in the number of gas molecules in the chemical equation (Δn_g). For the reactions given, the Δn_g is: (A) 0, (B) –2, (C) 0, (D) –1. Reaction (B) has the greatest decrease in the moles of gas and should have the largest decrease in entropy.

35. **(C)** The conjugate acid of a base is obtained by adding one hydrogen atom and one positive charge to the base. Therefore, $H_2PO_4^- + H^+ \rightarrow H_3PO_4$.

36. **(C)** One mole of any substance contains 6.02×10^{23} units of that substance whether the substance contains atoms, molecules, or formula units. This can be used as a conversion factor to determine either the number of moles or the units of a substance

$$(4.0 \times 10^{21} \text{ molecules}) \times \frac{1 \text{ mole}}{6.02 \times 10^{23} \text{ molecules}} = 6.6 \times 10^{-3} \text{ moles}$$

37. **(D)** The nitrate ion has two oxygen atoms bonded to the nitrogen with single bonds, and one oxygen is bonded to the nitrogen with a double bond. This adds up to three sigma and one pi bond.

38. **(A)** We will calculate molarity as $(g/M)/(\text{L of solution})$

$$\text{molarity} = (25 \times 10^{-3} \text{ g sucrose}/342 \text{ g mol}^{-1})/1.00 \text{ L}$$
$$= 7.31 \times 10^{-5} \text{ molar} = 7.3 \times 10^{-5}\ M \text{ (when correctly rounded)}$$

39. **(C)** Silicon is a metalloid and is better known as a semiconductor material used in transistors.

40. **(D)** When the activation energy is low there are a large number of collisions with the proper energy for the reaction to occur. So a small activation energy results in a fast reaction rate.

41. **(C)** Half-reactions 1 and 2 are reductions because the charge of the copper ions decreases in the process. The iron half-reaction is an oxidation because the oxidation number increases from +2 to +3.

42. **(B)** The difference in electronegativity is 1.9. The others are (a) 1.1, (c) 0.5, and (d) 0.6. This answer can be estimated since the P and F are more widely separated than the other pairs that are adjacent atoms in the periodic table.

43. **(D)** Because chromium has more than one isotope and this question does not specify which isotope, it is impossible to state the number of neutrons. The atomic mass in the periodic table is NOT an isotope mass and cannot be used for that purpose.

44. **(C)** The correct name is aluminum bromide. Aluminum ions are always 3+, and bromide ions are always 1–. The prefix *tri-* is used only for compounds composed of two nonmetals.

45. **(C)** As the atomic number increases across a period, the size of the atom decreases. This is due to an increase in the effective nuclear charge. Adding more protons to the nucleus from left to right across a period increases the positive charge. However, increasing the number of electrons does not cause these electrons to repel each other sufficiently to maintain or increase the size.

ANSWER SHEET
Diagnostic Test 2

1. Ⓐ Ⓑ Ⓒ Ⓓ	16. Ⓐ Ⓑ Ⓒ Ⓓ	31. Ⓐ Ⓑ Ⓒ Ⓓ
2. Ⓐ Ⓑ Ⓒ Ⓓ	17. Ⓐ Ⓑ Ⓒ Ⓓ	32. Ⓐ Ⓑ Ⓒ Ⓓ
3. Ⓐ Ⓑ Ⓒ Ⓓ	18. Ⓐ Ⓑ Ⓒ Ⓓ	33. Ⓐ Ⓑ Ⓒ Ⓓ
4. Ⓐ Ⓑ Ⓒ Ⓓ	19. Ⓐ Ⓑ Ⓒ Ⓓ	34. Ⓐ Ⓑ Ⓒ Ⓓ
5. Ⓐ Ⓑ Ⓒ Ⓓ	20. Ⓐ Ⓑ Ⓒ Ⓓ	35. Ⓐ Ⓑ Ⓒ Ⓓ
6. Ⓐ Ⓑ Ⓒ Ⓓ	21. Ⓐ Ⓑ Ⓒ Ⓓ	36. Ⓐ Ⓑ Ⓒ Ⓓ
7. Ⓐ Ⓑ Ⓒ Ⓓ	22. Ⓐ Ⓑ Ⓒ Ⓓ	37. Ⓐ Ⓑ Ⓒ Ⓓ
8. Ⓐ Ⓑ Ⓒ Ⓓ	23. Ⓐ Ⓑ Ⓒ Ⓓ	38. Ⓐ Ⓑ Ⓒ Ⓓ
9. Ⓐ Ⓑ Ⓒ Ⓓ	24. Ⓐ Ⓑ Ⓒ Ⓓ	39. Ⓐ Ⓑ Ⓒ Ⓓ
10. Ⓐ Ⓑ Ⓒ Ⓓ	25. Ⓐ Ⓑ Ⓒ Ⓓ	40. Ⓐ Ⓑ Ⓒ Ⓓ
11. Ⓐ Ⓑ Ⓒ Ⓓ	26. Ⓐ Ⓑ Ⓒ Ⓓ	41. Ⓐ Ⓑ Ⓒ Ⓓ
12. Ⓐ Ⓑ Ⓒ Ⓓ	27. Ⓐ Ⓑ Ⓒ Ⓓ	42. Ⓐ Ⓑ Ⓒ Ⓓ
13. Ⓐ Ⓑ Ⓒ Ⓓ	28. Ⓐ Ⓑ Ⓒ Ⓓ	43. Ⓐ Ⓑ Ⓒ Ⓓ
14. Ⓐ Ⓑ Ⓒ Ⓓ	29. Ⓐ Ⓑ Ⓒ Ⓓ	44. Ⓐ Ⓑ Ⓒ Ⓓ
15. Ⓐ Ⓑ Ⓒ Ⓓ	30. Ⓐ Ⓑ Ⓒ Ⓓ	45. Ⓐ Ⓑ Ⓒ Ⓓ

Diagnostic Test 2

Directions: Answer the following multiple-choice questions. You may use a calculator and the periodic table on page 528, but no other information. Limit your time to 60 minutes. If you do not finish in 60 minutes, note the number of questions answered and then continue until all the remaining questions are answered. Record your total time. Score your test with the answer key at the end of the test. Also at the end of the test are tables to help diagnose your strengths and weaknesses. Review the topics that have the most errors and then continue with Diagnostic Test 3. For more of a challenge, do not use your calculator when answering these questions.

1. ^{40}Ca, ^{39}K, and ^{41}Sc all have the same
   (A) atomic mass
   (B) atomic number
   (C) number of neutrons
   (D) number of electrons

2. Balance the following skeleton reaction in acid solution using the ion-electron method to obtain a reaction with the smallest possible whole-number coefficients. The sum of all the coefficients in the balanced equation is

$$IO_3^- + Cl^- \rightarrow ClO^- + I_2$$

   (A) 8
   (B) 14
   (C) 16
   (D) 24

3. The reaction of $Br_2(g)$ with $Cl_2(g)$ to form $BrCl(g)$ has an equilibrium constant of 15.0 at a certain temperature. If 10.0 grams of BrCl is initially present in a 15.0-liter reaction vessel, what will the concentration of BrCl be at equilibrium?
   (A) 3.8×10^{-3} mol/L
   (B) 5.77×10^{-3} mol/L
   (C) 1.97×10^{-3} M
   (D) 9.9×10^{-4} M

4. Which of the following is a balanced chemical equation?
   (A) $Na_2SO_4 + Ba(NO_3)_2 \longrightarrow BaSO_4 + NaSO_4$
   (B) $AgNO_3 + K_2CrO_4 \longrightarrow Ag_2CrO_4 + 2KNO_3$
   (C) $5FeCl_2 + 8HCl + KMnO_4 \longrightarrow 5FeCl_3 + MnCl_2 + 4H_2O + KCl$
   (D) $Al(NO_3)_3 + 3KOH \longrightarrow Al(OH)_3 + KNO_3$

5. A 50.0 mL sample of 0.0025 *M* HBr is mixed with 50.0 mL of 0.0023 *M* KOH. What is the pH of the resulting mixture?
   (A) 1.00
   (B) 4.00
   (C) 5.00
   (D) 11.00

6. How many milliliters of 0.250 *M* KOH does it take to neutralize completely 50.0 mL of 0.150 $M\,H_3PO_4$?
   (A) 30.0 mL
   (B) 27 mL
   (C) 90.0 mL
   (D) 270 mL

7. Which of the following pairs have the same electron configuration and are isotopes of each other?
   (A) $^{56}Fe^{2+}$ and $^{57}Fe^{3+}$
   (B) $^{39}K^{+}$ and $^{40}K^{+}$
   (C) $^{24}Mg^{2+}$ and ^{25}Mg
   (D) $^{40}Ca^{2+}$ and ^{40}Ar

8. In acid solution the bromate ion, BrO_3^- can react with other substances, resulting in Br_2. Balance the half-reaction for bromate ions forming bromine. The balanced half-reaction has
   (A) 6 electrons on the left
   (B) 6 electrons on the right
   (C) 3 electrons on the left
   (D) 10 electrons on the left

9. The quantum number ℓ signifies the
   (A) relative distance of the electron from the nucleus
   (B) orientation in space of a particular orbital
   (C) shape of an orbital
   (D) spin of the electron

10. When an ideal gas is allowed to expand isothermally, which one of the following is true?
   (A) $q = 0$
   (B) $w = 0$
   (C) $E = 0$
   (D) $q = -w$

11. Which pair of reactants will have no net ionic equation (that is, all the ions cancel)?
   (A) $Na_2SO_3 + FeCl_2$
   (B) $CaCl_2 + MnBr_2$
   (C) $NH_4I + Pb(NO_3)_2$
   (D) $KOH + HClO_4$

12. The rate law for the reaction of $2A + B \longrightarrow 2P$ is
   (A) impossible to determine without experimental data
   (B) $[A]^2[B]$
   (C) $k[A]^2[B]$
   (D) second order with respect to A

13. Hund's rule requires that
   (A) no two electrons can have the same four quantum numbers
   (B) no two electrons with the same spin can occupy an orbital
   (C) no two electrons can occupy separate orbitals
   (D) no two electrons can pair up if there is an empty orbital at the same energy level available

14. Which is a conjugate acid–base pair in the following equation?

$$H_3PO_4 + H_2O \rightarrow H_3O^+ + H_2PO_4^-$$

   (A) H_2O and H_3O^+
   (B) H_3O^+ and $H_2PO_4^-$
   (C) H_3PO_4 and H_2O
   (D) H_3PO_4 and H_3O^+

15. The collision theory of reaction rates does not include
   (A) the number of collisions per second
   (B) the transition state
   (C) the energy of each collision
   (D) the orientation of each collision

16. A 2.35-gram sample was dissolved in water, and the chloride ions were precipitated by adding silver nitrate ($Ag^+ + Cl^- \longrightarrow AgCl$). If 0.435 g of precipitate was obtained, what is the percentage of chlorine in the sample?
(A) 10.8%
(B) 4.60%
(C) 43.5%
(D) 18%

17. Given the two standard reduction equations and their potentials below, write the thermodynamically favored chemical reaction and its standard cell potential.

$Ag^+(aq) + e^- \longrightarrow Ag(s)$	+0.80 V
$Mg^{2+}(aq) + 2e^- \longrightarrow Mg(s)$	–2.37 V

(A)	$Ag^+(aq) + Mg^{2+}(aq)$	$\rightarrow$	$Ag(s) + Mg(s)$	–1.57 V
(B)	$Ag^+(aq) + Mg(s)$	$\rightarrow$	$Ag(s) + Mg^{2+}(aq)$	+3.17 V
(C)	$2Ag^+(aq) + Mg(s)$	$\rightarrow$	$2Ag(s) + Mg^{2+}(aq)$	+3.17 V
(D)	$2Ag^+(aq) + Mg(s)$	$\rightarrow$	$2Ag(s) + Mg^{2+}(aq)$	+3.97 V

18. Which of the following has an octet of electrons around the central atom?
(A) BF_3
(B) NH_4^+
(C) PF_5
(D) SF_6

19. What is the pH of a solution made by dissolving 0.0300 mol of sodium ethanoate in enough water to make 50 mL of solution (K_a for ethanoic acid is 1.8×10^{-5})?
(A) 7.00
(B) 9.26
(C) 4.74
(D) 11.02

20. The weight percent of sodium hydroxide dissolved in water is 50%. What is the mole fraction of sodium hydroxide?
(A) 31.0%
(B) 0.164
(C) 0.311
(D) 0.500

21. Which of the following should be named using Roman numerals after the cation?
(A) $CaCl_2$
(B) $CuCl_2$
(C) $AlBr_3$
(D) $K_2Cr_2O_7$

22. Carbon exists in various forms called allotropes. Which of the following is not an allotrope of carbon?
(A) Diamond
(B) Soot
(C) Buckminsterfullerene
(D) Graphite

23. What is the molecular mass of a gas that has a density of 2.05 g/L at 26.0°C and 722 torr?
(A) 53.0 g/mol
(B) 46.7 g/mol
(C) 4.67 g/mol
(D) 2876 g/mol

24. What is the ground state electron configuration for nickel?
(A) $1s^2 2s^2 2p^6 3s^2 3p^6 3d^{10}$
(B) $1s^2 2s^2 2p^6 3s^2 4s^2 3d^{10} 4p^4$
(C) $1s^2 2s^2 2p^6 3s^2 3p^6 4s^2 4d^8$
(D) $1s^2 2s^2 2p^6 3s^2 3p^6 4s^2 3d^8$

25. Under which conditions will a real gas most closely behave as an ideal gas?
(A) High temperature and high pressure
(B) High temperature and low pressure
(C) High volume and high temperature
(D) Low temperature and low pressure

26. The equilibrium constant of a certain reaction is 2.6×10^8 at 25°C. What is the value of $\Delta G°$?
(A) –48.0 kJ/mol
(B) 20.8 J mol^{-1}
(C) 4.68×10^{-3} kJ/mol
(D) –4.03 kJ mol^{-1}

27. Which of the following geometries corresponds to a substance that has five sigma bonds and one nonbonding pair of electrons on the central atom?
(A) Tetrahedron
(B) Square planar
(C) Octahedron
(D) Square pyramid

DIAGNOSTIC TEST 2

28. Carbon dioxide, CO_2, when in the form of dry ice is a(n)
(A) network solid
(B) ionic solid
(C) molecular solid
(D) amorphous solid

29. What is the pH of a solution prepared by dissolving 0.1665 mole of hypochlorous acid (HClO) in enough water to make 500 mL of solution? The K_a is 3.0×10^{-4}.
(A) 1.00
(B) 2.00
(C) 5.4×10^{-3}
(D) 1.76

30. Substances whose Lewis structures must be drawn with an unpaired electron are called
(A) ionic compounds
(B) free radicals
(C) resonance structures
(D) polar molecules

31. Potassium-40 is a minor isotope found in naturally occurring potassium. It is radioactive and can be detected on simple radiation counters. How many protons, neutrons, and electrons does potassium-40 have when it is part of K_2SO_4?
(A) 21 neutrons, 19 protons, 18 electrons
(B) 20 neutrons, 19 protons, 19 electrons
(C) 21 neutrons, 19 protons, 19 electrons
(D) 19 neutrons, 19 protons, 19 electrons

32. An ideal solution is a
(A) mixture where two solvents cannot be dissolved in all ratios
(B) mixture that has the same physical properties as the individual solvents
(C) mixture where the potential energy of the mixture is the same as that of the individual solvents
(D) mixture that is colorless

33. Which of the following lists the electromagnetic spectral regions in order of decreasing wavelength?
(A) Ultraviolet, visible, infrared, X-ray
(B) X-ray, visible, ultraviolet, infrared
(C) X-ray, ultraviolet, visible, infrared
(D) Infrared, visible, ultraviolet, X-ray

34. Sodium fluoride dissolves in water and undergoes hydrolysis. What is the equilibrium expression for the hydrolysis reaction?

(A) $K = \frac{[HF][OH^-]}{[F^-]}$

(B) $K = \frac{[HF][OH^-]}{[F^-][Na^+]}$

(C) $K = \frac{[F^-][OH^-]}{[HF]}$

(D) $K = \frac{[HF]}{[F^-][OH^-]}$

35. Of the following pairs of elements, which pair has the second element with the larger electronegativity based on its position in the periodic table?

(A) Oxygen, chromium
(B) Chlorine, iodine
(C) Calcium, cesium
(D) Sulfur, nitrogen

36. What is the oxidation number on phosphorous in the compound Na_3PO_4?

(A) +7
(B) +5
(C) –5
(D) –1

37. Which of the following molecules is a strong electrolyte when dissolved in water?

(A) CH_3COOH
(B) $HC_2H_3O_2$
(C) PCl_5
(D) HBr

38. When using the ideal gas law, standard conditions for temperature and pressure are

(A) 0 K and 0 torr
(B) 25°C and 1 atmosphere pressure
(C) 0°C and 760 torr
(D) 0°F and 1 atmosphere pressure

39. The dimerization of $NO_2(g)$ to $N_2O_4(g)$ is an endothermic process. Which of the following will, according to Le Châtelier's principle, increase the amount of N_2O_4 in a reaction vessel?
(A) Decreasing the temperature
(B) Increasing the size of the reaction vessel
(C) Adding a selective catalyst
(D) Making the reaction vessel smaller

40. Which of the following is expected to have the highest electronegativity?
(A) S
(B) Fe
(C) W
(D) Ag

41. When a solid melts, the entropy and enthalpy changes expected are
(A) positive enthalpy change and positive entropy change
(B) negative entropy change and negative enthalpy change
(C) negative entropy change and positive enthalpy change
(D) negative enthalpy change and positive entropy change

42. What is the mass of one molecule of cholesterol ($C_{27}H_{46}O$, molecular mass = 386)?
(A) 6.41×10^{-22} g
(B) 1.5×10^{21} g
(C) 1.38×10^{-21} g
(D) 3×10^{-23} g

43. Which of the following pairs of liquids is expected to be immiscible?
(A) H_2O and CH_3OH
(B) C_6H_6 and C_5H_{12}
(C) $C_{10}H_{22}$ and $CH_2CH_2CH_2OH$
(D) $CH_3CH_2NH_2$ and $CH_3CH_2CH_2OH$

44. All of the following changes affect the value of the rate constant for a reaction EXCEPT
(A) the addition of a catalyst
(B) decreasing the temperature
(C) decreasing the activation energy
(D) increasing or decreasing the concentration of the reactants

45. Argon can be liquefied at low temperature because of
(A) dipole-dipole attractive forces
(B) hydrogen bonding
(C) instantaneous and induced dipoles
(D) the very low temperature

ANSWER KEY

Diagnostic Test 2

1. **(C)**	10. **(D)**	19. **(B)**	28. **(C)**	37. **(D)**
2. **(C)**	11. **(B)**	20. **(C)**	29. **(B)**	38. **(C)**
3. **(A)**	12. **(A)**	21. **(B)**	30. **(B)**	39. **(D)**
4. **(C)**	13. **(D)**	22. **(B)**	31. **(A)**	40. **(A)**
5. **(B)**	14. **(A)**	23. **(A)**	32. **(C)**	41. **(A)**
6. **(C)**	15. **(B)**	24. **(D)**	33. **(D)**	42. **(A)**
7. **(B)**	16. **(B)**	25. **(B)**	34. **(A)**	43. **(C)**
8. **(D)**	17. **(C)**	26. **(A)**	35. **(D)**	44. **(D)**
9. **(C)**	18. **(B)**	27. **(D)**	36. **(B)**	45. **(C)**

EVALUATING YOUR RESULTS

Score your test using the answer key. Then complete the following tables. The first table is designed to find your general strengths and weaknesses based on four broad categories. The second table is a more specific diagnostic chart that will suggest which particular chapters you should concentrate your studies on. In combination, these two tables will help you focus your efforts on the material that needs the most study.

Question Categories

Question Type	Questions	Number Wrong
Basic Facts	1, 7, 11, 14, 21, 24, 30, 31, 33, 35, 36, 37, 38, 40	
Basic Concepts	4, 9, 10, 12, 13, 15, 18, 23, 25, 28, 32, 34, 39, 44, 45	
Calculations	2, 3, 5, 6, 8, 16, 19, 20, 23, 26, 29, 42	
Mixed Concepts	17, 22, 27, 28, 41, 43, 44	

Breakdown by Topics

Chapter	Questions	Number Wrong
1. Structure of the Atom	9, 13, 24, 33	
2. Periodic Table	1, 7, 35, 40	
3. Ionic Compounds and Reactions	4, 11, 21	
4. Covalent Compounds	18, 27, 30	
5. Stoichiometry	6, 16, 42	
6. Gases	23, 25, 38	
7. Liquids and Solids	28, 43, 45	
8. Solutions	20, 32, 37	

Chapter	Questions	Number Wrong
9. Equilibrium	3, 34, 39	
10. Kinetics	12, 15, 44	
11. Thermodynamics	10, 26, 41	
12. Redox and Electrochemistry	2, 8, 17, 36	
13. Acids and Bases	5, 14, 19, 29	

ANSWERS EXPLAINED

1. **(C)** Each of the elements listed contains 20 neutrons. They cannot contain the same number of protons since they are all different elements.

2. **(C)** The sum of the coefficients is 16. The reaction is balanced in steps. First write two half-reactions:

$$IO_3^- \rightarrow I_2$$
$$Cl^- \rightarrow ClO^-$$

Balance all atoms except H and O:

$$\mathbf{2}IO_3^- \rightarrow I_2$$
$$Cl^- \rightarrow ClO^-$$

Balance oxygens by adding water molecules:

$$2IO_3^- \rightarrow I_2 + \mathbf{6H_2O}$$
$$\mathbf{H_2O} + Cl^- \rightarrow ClO^-$$

Balance hydrogens by adding H^+:

$$\mathbf{12H^+} + 2IO_3^- \rightarrow I_2 + 6H_2O$$
$$H_2O + Cl^- \rightarrow ClO^- + \mathbf{2H^+}$$

Balance charges with electrons:

$$\mathbf{10e^-} + 12H^+ + 2IO_3^- \rightarrow I_2 + 6H_2O$$
$$H_2O + Cl^- \rightarrow ClO^- + 2H^+ + \mathbf{2e^-}$$

Multiply the second half-reaction by 5 so both equations have the same number of electrons (now the e^- will cancel):

$$10e^- + 12H^+ + 2IO_3^- \rightarrow I_2 + 6H_2O$$
$$\mathbf{5}H_2O + \mathbf{5}Cl^- \rightarrow \mathbf{5}ClO^- + \mathbf{10}H^+ + \mathbf{10}e^-$$

Add the two equations:

$$10e^- + 12H^+ + 2IO_3^- + 5H_2O + 5Cl^- \rightarrow I_2 + 6H_2O + 5ClO^- + 10H^+ + 10e^-$$

Cancel like items for the final equation:

$$2H^+ + 2IO_3^- + 5Cl^- \rightarrow I_2 + H_2O + 5ClO^-$$

Remember, I_2 and H_2O have coefficients of 1 although they are not explicitly written.

3. **(A)** The reaction is $Br_2(g) + Cl_2(g) \rightarrow 2BrCl(g)$

The equilibrium expression is $K = \dfrac{[BrCl]^2}{[Br_2][Cl_2]}$.

The initial concentration of BrCl is = $(10.0\text{ g}/115\text{ g mol}^{-1})/15\text{L} = 5.8 \times 10^{-3}$ mol/L.

If $2x$ moles of BrCl reacts, x moles each of Br_2 and Cl_2 will be formed. (You may want to set up an equilibrium table at this point.)

Substituting into the equilibrium expression gives $K = \dfrac{(0.0058 - 2x)^2}{[x][x]} = 15.0.$

Take the square root of the entire equation:

$$\begin{aligned} 3.87 &= (0.0058 - 2x)/x \\ 3.87x &= 0.0058 - 2x \\ 5.87x &= 0.0058 \\ x &= (0.0058/5.87) \\ x &= 9.9 \times 10^{-4} = [Br_2] = [Cl_2] \\ [Br\,Cl] &= 0.0058 - 2\,(9.9 \times 10^{-4}) \\ &= 0.0038\text{ mol/L} \end{aligned}$$

4. **(C)** All of the compounds are written correctly, and all the elements have the same numbers of atoms on each side of the arrow. Na is not balanced in response (A), Ag is not balanced in response (B), and the K is not balanced in response (D).

5. **(B)** This is a strong acid reacting with a strong base. We need to find out how much of either the strong acid or the strong base is left after the reaction is complete. This is a type of limiting-reactant problem. The reaction is $HBr + KOH \rightarrow KBr + H_2O$.

With solutions it is easiest to calculate the moles of each reactant as

$$?\text{ moles HBr} = 50.0\text{ mL HBr}\left[\frac{0.0025\text{ mol HBr}}{1000\text{ mL HBr}}\right] = 1.25 \times 10^{-4}\text{ mol HBr}$$

$$?\text{ moles KOH} = 50.0\text{ mL KOH}\left[\frac{0.0023\text{ mol KOH}}{1000\text{ mL KOH}}\right] = 1.15 \times 10^{-4}\text{ mol KOH}$$

Since 1 mole of HBr reacts with 1 mole of KOH, we can see that 1.15×10^{-4} mol of each will react. We can also see that ALL of the KOH is used up and must be the limiting reactant. The HBr is the excess reactant, and $(1.25 \times 10^{-4} - 1.15 \times 10^{-4}) = 1.0 \times 10^{-5}$ mol of HBr is left over. The molarity of HBr is the moles left over divided by the liters of solution, or

$$\text{molarity HBr} = \left[\frac{0.000010 \text{ mol HBr}}{0.100 \text{ L solution}}\right] = 0.00010\ M\ \text{HBr}$$

Since HBr is a strong acid, it dissociates completely, and $[H^+] = 0.00010\ M$

The $pH = -\log(0.00010) = 4.00$.

6. **(C)** The question put into mathematical terms is

? mL KOH = 50.0 mL H_3PO_4

A balanced chemical equation is needed. The complete neutralization of H_3PO_4 is

$H_3PO_4 + 3KOH \rightarrow K_3PO_4 + 3H_2O$

We change molarity units to mol/L or mol/1000 mL to use them as conversion factors:

$$0.150\ M\ H_3PO_4 = \left[\frac{0.150 \text{ mol } H_3PO_4}{1000 \text{ mL } H_3PO_4}\right] \text{ and}$$

$$0.250\ M\ \text{KOH} = \left[\frac{0.250 \text{ mol KOH}}{1000 \text{ mL KOH}}\right]$$

$$\text{mL KOH} = 50.0 \text{ mL } H_3PO_4\left[\frac{0.150 \text{ mol } H_3PO_4}{1000 \text{ mL } H_3PO_4}\right]\left[\frac{3 \text{ mol KOH}}{1 \text{ mol } H_3PO_4}\right]\left[\frac{1000 \text{ mL KOH}}{0.250 \text{ mol KOH}}\right]$$

$$= 90.0 \text{ mL KOH}$$

7. **(B)** In order for atoms or ions to be isotopes, they must have a different number of neutrons while the number of protons must remain constant. *Isoelectronic* means that the two atoms or ions have the same electron configuration.

8. **(D)** The bromate ion (by analogy with the chlorate ion ClO_3^-) is BrO_3^-, and we are told the product is Br_2. This is the basis of the half-reaction we need to balance.

$$\begin{aligned} BrO_3^- &\rightarrow Br_2 \quad \text{Now balance Br atoms.} \\ 2BrO_3^- &\rightarrow Br_2 \quad \text{Now balance O with } H_2O. \\ 2BrO_3^- &\rightarrow Br_2 + 6H_2O \quad \text{Now add } H^+ \text{ to balance H's.} \\ 12H^+ + 2BrO_3^- &\rightarrow Br_2 + 6H_2O \quad \text{Now add electrons.} \\ 10e^- + 12H^+ + 2BrO_3^- &\rightarrow Br_2 + 6H_2O \end{aligned}$$

This is the balanced half-reaction with ten electrons on the left.

9. **(C)** The ℓ quantum number designates the shape of an orbital, n designates the distance from the nucleus, and m_ℓ designates the arbitrary direction in space of the orbital.

10. **(D)** For an isothermal expansion the temperature does not change, and therefore the average kinetic energy stays the same. Because the gas is ideal, there is no change in potential energy. That means that $\Delta E = 0 = w + q$. Therefore $q = -w$, and the correct response is (D).

11. **(B)** Both $CaCl_2$ and $MnBr_2$ are soluble. The products of this reaction are $CaBr_2$ and $MnCl_2$, which are also soluble. This results in the cancellation of all ions. For the others: (A) will produce insoluble iron(II) sulfite, (C) produces PbI_2, and (D) produces water as the product of a neutralization reaction.

12. **(A)** Rate laws cannot be determined from the reaction stoichiometry. Experimental data is always necessary to determine the rate law. One exception is a reaction that is an elementary step in a mechanism.

13. **(D)** This requires electrons to fill each available orbital in an energy level before pairing. Responses (A) and (B) are parts of the Pauli exclusion principle, and (C) has no meaning at all.

14. **(A)** A conjugate acid-base pair are two chemical formulas that differ by a proton (H^+). For the reaction presented, there are two conjugate acid-base pairs. One pair is H_2O and H_3O^+ and the other is H_3PO_4 and $H_2PO_4^-$.

15. **(B)** The transition state is part of the transition state theory, not the collision theory.

16. **(B)** The percentage of chlorine in the sample is calculated as

$$\%\ \text{Cl} = \frac{\text{grams chlorine}}{\text{grams of sample}} \times 100$$

We know the grams of sample = 2.35 grams. We need to calculate the grams of chlorine from the 0.435 gram of AgCl precipitated. We start with ? g Cl = 0.435 g AgCl and add the conversion factors shown below.

$$?\ \text{g Cl} = 0.435\ \text{g AgCl}\left[\frac{1\ \text{mol AgCl}}{143\ \text{g AgCl}}\right]\left[\frac{1\ \text{mol Cl}^-}{1\ \text{mol AgCl}}\right]\left[\frac{35.5\ \text{g Cl}^-}{1\ \text{mol Cl}^-}\right]$$

$$= 0.108\ \text{g Cl}^-$$

Substitute the data into the above equation to get

$$\%\ \text{Cl} = \frac{0.108\ \text{grams chlorine}}{2.35\ \text{grams of sample}} \times 100 = 4.60\%\ \text{Cl}$$

17. **(C)** Only reactions (C) and (D) are balanced. Only reaction (C) correctly subtracts –2.37 from +0.80 to get the correct +3.17 V.

18. **(B)** The nitrogen on NH_4^+ has an octet of electrons. Boron forms compounds with only six electrons, and the remaining compounds are in period 3 or above and may utilize *d* orbitals to expand the octet on the central atom to more than eight electrons. The others all have expanded octets.

19. **(B)** $NaC_2H_3O_3 \rightarrow Na^+ + C_2H_3O_2^-$

The ethanoate ion then hydrolyzes water in the reaction

$$C_2H_3O_2^- + H_2O \rightleftharpoons HC_2H_3O_2 + OH^-$$

At this point, we can see that only (B) can possibly be correct. (Some experience will teach that hydrolysis rarely makes solutions very alkaline, so the best guess is response (B).)

$$K = K_w / K_a$$

$$K_w / K_a = \frac{[OH^-][HC_2H_3O_2]}{[C_2H_3O_2^-]}$$

The molarity of the salt is 0.300 mol/0.500 L = 0.600 mol/L.

$$\frac{(1.0 \times 10^{-14})}{(1.8 \times 10^{-5})} = \frac{(x)(x)}{[0.600 - x]}$$

Assume that 0.600 >> x, so that 0.600 – x = 0.600

$$x^2 = 3.33 \times 10^{-10}$$
$$x = 1.82 \times 10^{-5} = [OH^-]$$
$$pOH = 4.74$$
$$pH = 9.26$$

20. **(C)** Taking 100 g of solution, 50 g will be NaOH and 50 g will be H_2O. The moles of each is

$$? \text{ mol NaOH} = 50 \text{ g NaOH} \left[\frac{1 \text{ mol NaOH}}{40 \text{ g NaOH}}\right] = 1.25 \text{ mol NaOH}$$

$$? \text{ mol } H_2O = 50 \text{ g } H_2O \left[\frac{1 \text{ mol } H_2O}{18 \text{ g } H_2O}\right] = 2.78 \text{ mol } H_2O$$

$$\text{mol fraction NaOH}, \chi = \frac{\text{moles NaOH}}{\text{moles NaOH} + \text{moles } H_2O}$$

$$= \frac{1.25 \text{ moles NaOH}}{1.25 \text{ moles NaOH} + 2.78 \text{ moles } H_2O}$$

$$= 0.311$$

21. **(B)** Roman numerals are used to indicate the oxidation state of a metal in a compound where the metal may have more than one possible oxidation state. Calcium, aluminum, potassium, and ammonium ions all have one possible charge. Chromium is part of the dichromate ion. Only copper can have two oxidation states, and therefore it must be named using Roman numerals.

22. **(B)** Soot is not a pure form of carbon. It is residue from incomplete combustion. Soot contains many organic compounds and inorganic compounds in addition to fine particles of graphite. The composition of soot varies with the distance from the flame that produced it.

23. **(A)** We rearrange the ideal gas law $PV = nRT$ to $PV = \left(\frac{g}{M}\right)RT$.

Dividing by PV and multiplying by the molar mass gives:

$$M = \left(\frac{g}{V}\right)\left(\frac{RT}{P}\right)$$

We will use $R = 0.0821$ L atm mol^{-1}.

Therefore we need to convert 722 torr to atm:

$$? \text{ atm} = 722 \text{ torr}\left(\frac{1 \text{ atm}}{760 \text{ torr}}\right) = 0.950 \text{ atm}$$

$$T = {}^{\circ}\text{C} + 273 = 26 + 273 = 299 \text{ K}$$

Entering the data and solving, we get

$$M = (2.05 \text{ g L}^{-1})(0.0821 \text{ L atm mol}^{-1})(299 \text{ K})/(0.950 \text{ atm})$$
$$= 53.0 \text{ g mol}^{-1}$$

24. **(D)** The ground state for electron configuration of an atom contains the correct number of electrons as determined by the atomic number. The electrons are placed into sublevels, filling each energy level before moving onto the next one. The transition elements fill the $(n + 1)s$ orbitals before filling the $(n)d$ orbitals.

25. **(B)** At high temperature the molecules have the largest kinetic energy and pass each other rapidly to minimize interactions. The low pressure means few molecules in a given volume that increases the distance between molecules and decreases interactions. Response (C) is attractive, but high volume does not guarantee low pressure and minimal interactions.

26. **(A)** Conversion of the equilibrium constant into a free energy involves the equation

$$\Delta G^{\circ} = -RT \ln K$$
$$= -(8.314 \text{ J mol}^{-1} \text{ K}^{-1})(298 \text{ K}) \ln(2.6 \times 10^{8})$$
$$= -48{,}000 \text{ J mol}^{-1} = -48.0 \text{ kJ mol}^{-1}$$

27. **(D)** The square pyramid is the shape of the molecule considering only the location of the atoms.

28. **(C)** Carbon dioxide is a covalent compound. When in the form of dry ice, it is classified as a molecular solid.

29. **(B)** Hypochlorous acid ionizes as follows: $HClO \rightarrow H^+ + ClO^-$. The equilibrium expression is

$$K_a = \frac{[H^+][ClO^-]}{[HClO]}$$

The initial concentration of HClO is 0.1665 mole/0.500 L = 0.333 *M* HClO.

The HClO will dissociate by some amount *x*, and therefore the hydrogen ion concentration and ClO^- will increase by *x*. Entering this information into the equilibrium expression gives

$$K_a = \frac{xx}{0.333 - x} = 3.0 \times 10^{-4}$$

We will make the traditional assumption that $x << 0.333$, so that $0.333 - x = 0.333$.

Recasting the equilibrium expression, we get

$$K_a = \frac{xx}{0.333 - x} = 3.0 \times 10^{-4}$$

$$x^2 = (0.333)(3.0 \times 10^{-4}) = 1.0 \times 10^{-4}$$

$$x = 1.0 \times 10^{-2} \text{ (Note that our assumption is true.)}$$

$$\text{pH} = -\log(1.0 \times 10^{-2}) = -(-2.00) = 2.00$$

30. **(B)** Free radicals are substances with an unpaired electron that makes them highly reactive.

31. **(A)** Potassium must have 19 protons since its atomic number is 19. It must have 21 neutrons so the mass can add up to 40 as indicated by its name, potassium-40. Finally, in compounds potassium has a +1 charge, meaning it has one less electron than the number of protons, or 18.

32. **(C)** The ideal solution has the same attractive forces, or potential energy, in the mixture as in the pure solvents. This also means that mixing is neither exothermic nor endothermic.

33. **(D)** We know that energy and wavelength are inversely proportional to each other. The lowest energy radiation listed is infrared, and the highest energy is the X-ray. Therefore, their wavelengths go from high to low.

34. **(A)** Sodium fluoride dissolves in water to form $Na^+ + F^-$. The fluoride ion then reacts with water: $F^- + H_2O \rightleftharpoons HF + OH^-$. The correct equilibrium constant is (A). Water is considered a constant that is part of *K*.

35. **(D)** Generally, elements closer to fluorine in the periodic table have larger electronegativities.

36. **(B)** Na_3PO_4 has two elements, oxygen and sodium, whose oxidation numbers are known. The oxidation number of each oxygen is –2. The oxidation number of each sodium is +1. In the uncharged formula unit, the oxidation number of phosphorus must allow all oxidation numbers (ox. no.) to add up to zero.

$$0 = (\text{ox. no. Na})(3 \text{ Na atoms}) + (\text{ox. no. P})(1 \text{ atom P}) + (\text{ox. no. O})(4 \text{ atoms O})$$

$$0 = (+1)(3) + (\text{ox. no. P})(1) + (-2)(4)$$

$$0 = 3 + (\text{ox. no. P}) - 8 = -5 + (\text{ox. no. P})$$

$(\text{ox. no. P}) = +5$

37. **(D)** HBr is a strong acid when dissolved in water. This means that it is also a strong electrolyte.

38. **(C)** Standard temperature is defined as zero degrees Celsius when it comes to the gas laws. The standard pressure is one atmosphere of pressure that is also 760 torr (mm Hg).

39. **(D)** The reaction is: HEAT + $2NO_2(g) \rightleftharpoons N_2O_4(g)$

Decreasing the size of the vessel will increase the pressure of both substances. However, since there are 2 moles of gas on the left and 1 mole of gas on the right, the reaction will be forced to the right to make more product as desired.

Decreasing the temperature will favor reactants, as will increasing the vessel size. Adding argon has no effect on the process, and a catalyst will speed up both the forward and reverse reactions.

40. **(A)** In the periodic table sulfur is closest to fluorine and is expected to have the greatest electronegativity. Also, in general, metals tend to have lower electronegativities than nonmetals.

41. **(A)** We know that heat must be added to melt solids. Therefore the enthalpy must be positive. We can also readily visualize that the molecules in a solid are much more ordered and closer together than those in a liquid. Therefore the entropy change must be positive also.

42. **(A)** The question is ?g cholesterol = 1 molecule cholesterol

The important conversion factors in this problem are the number of grams per mole and the number of molecules per mole of cholesterol.

$$?\text{g } C_{27}H_{46}O = 1 \text{ molecule } C_{27}H_{46}O\left(\frac{1 \text{ mol } C_{27}H_{46}O}{6.02\times10^{23} \text{ molecules } C_{27}H_{46}O}\right)\left(\frac{386 \text{ g } C_{27}H_{46}O}{1 \text{ mol } C_{27}H_{46}O}\right)$$

$$= 6.41\times10^{-22} \text{ g } C_{27}H_{46}O$$

DIAGNOSTIC TEST 2

43. **(C)** Decane is nonpolar, and propanol is polar and forms hydrogen bonds. Their polarities are very different, and they are not expected to be miscible. All other pairs have similar polarities and therefore should mix readily.

44. **(D)** The value of the rate constant for a reaction is not affected by the concentration of the reactants. However, the presence of a catalyst or either an increase or a decrease in the temperature or activation energy will affect the rate constant.

45. **(C)** Argon has no polarity, and induced dipoles or instantaneous dipoles are the reason why it can be liquefied.

ANSWER SHEET
Diagnostic Test 3

1. Ⓐ Ⓑ Ⓒ Ⓓ	16. Ⓐ Ⓑ Ⓒ Ⓓ	31. Ⓐ Ⓑ Ⓒ Ⓓ
2. Ⓐ Ⓑ Ⓒ Ⓓ	17. Ⓐ Ⓑ Ⓒ Ⓓ	32. Ⓐ Ⓑ Ⓒ Ⓓ
3. Ⓐ Ⓑ Ⓒ Ⓓ	18. Ⓐ Ⓑ Ⓒ Ⓓ	33. Ⓐ Ⓑ Ⓒ Ⓓ
4. Ⓐ Ⓑ Ⓒ Ⓓ	19. Ⓐ Ⓑ Ⓒ Ⓓ	34. Ⓐ Ⓑ Ⓒ Ⓓ
5. Ⓐ Ⓑ Ⓒ Ⓓ	20. Ⓐ Ⓑ Ⓒ Ⓓ	35. Ⓐ Ⓑ Ⓒ Ⓓ
6. Ⓐ Ⓑ Ⓒ Ⓓ	21. Ⓐ Ⓑ Ⓒ Ⓓ	36. Ⓐ Ⓑ Ⓒ Ⓓ
7. Ⓐ Ⓑ Ⓒ Ⓓ	22. Ⓐ Ⓑ Ⓒ Ⓓ	37. Ⓐ Ⓑ Ⓒ Ⓓ
8. Ⓐ Ⓑ Ⓒ Ⓓ	23. Ⓐ Ⓑ Ⓒ Ⓓ	38. Ⓐ Ⓑ Ⓒ Ⓓ
9. Ⓐ Ⓑ Ⓒ Ⓓ	24. Ⓐ Ⓑ Ⓒ Ⓓ	39. Ⓐ Ⓑ Ⓒ Ⓓ
10. Ⓐ Ⓑ Ⓒ Ⓓ	25. Ⓐ Ⓑ Ⓒ Ⓓ	40. Ⓐ Ⓑ Ⓒ Ⓓ
11. Ⓐ Ⓑ Ⓒ Ⓓ	26. Ⓐ Ⓑ Ⓒ Ⓓ	41. Ⓐ Ⓑ Ⓒ Ⓓ
12. Ⓐ Ⓑ Ⓒ Ⓓ	27. Ⓐ Ⓑ Ⓒ Ⓓ	42. Ⓐ Ⓑ Ⓒ Ⓓ
13. Ⓐ Ⓑ Ⓒ Ⓓ	28. Ⓐ Ⓑ Ⓒ Ⓓ	43. Ⓐ Ⓑ Ⓒ Ⓓ
14. Ⓐ Ⓑ Ⓒ Ⓓ	29. Ⓐ Ⓑ Ⓒ Ⓓ	44. Ⓐ Ⓑ Ⓒ Ⓓ
15. Ⓐ Ⓑ Ⓒ Ⓓ	30. Ⓐ Ⓑ Ⓒ Ⓓ	45. Ⓐ Ⓑ Ⓒ Ⓓ

Diagnostic Test 3

Directions: Answer the following multiple-choice questions. You may use a calculator and the periodic table on page 528, but no other information. Limit your time to 60 minutes. If you do not finish in 60 minutes, note the number of questions answered and then continue until all the remaining questions are answered. Record your total time. Score your test with the answer key at the end of the test. Also at the end of the test are tables to help diagnose your strengths and weaknesses. Review the topics that have the most errors. For more of a challenge, do not use your calculator when answering these questions.

1. What is the coefficient for H_2O when the following reaction is correctly balanced?

$$\_Al(OH)_3 + \_H_2SO_4 \rightarrow \_Al_2(SO_4)_3 + \_H_2O$$

(A) 2
(B) 1
(C) 6
(D) 3

2. The molar heat of vaporization of water is +43.9 kJ. What is the entropy change for the vaporization of water?
(A) 8.49 J mol^{-1} K^{-1}
(B) 4.184 J mol^{-1} K^{-1}
(C) 2.78 J mol^{-1} K^{-1}
(D) 118 J mol^{-1} K^{-1}

3. Which of the following is a correct representation of the ground state valence *p* electrons in an atom of sulfur?

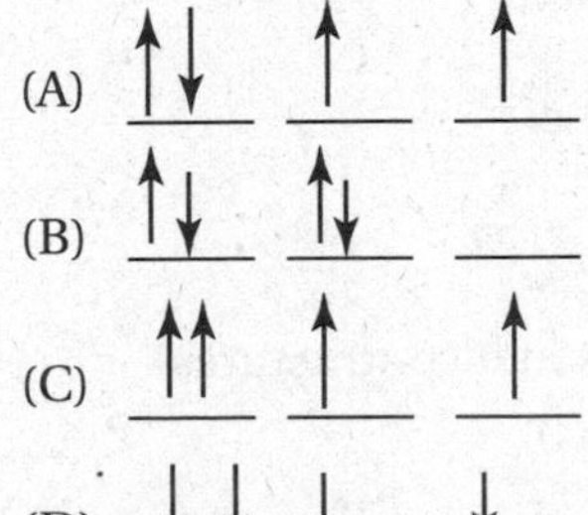

4. A liquid element that is a dark-colored, nonconducting substance at room temperature is
   (A) mercury
   (B) bromine
   (C) iodine
   (D) bismuth

5. A large positive value for the standard Gibbs free-energy change ($\Delta G°$) for a reaction means
   (A) the reaction is thermodynamically favored with virtual complete conversion of reactants to products
   (B) an extremely fast chemical reaction
   (C) a reaction with a very large increase in entropy
   (D) none of the above

6. A metal is reacted with HCl to produce hydrogen gas. If 0.0623 gram of metal produces 28.3 mL of hydrogen at STP, the mass of the metal that reacts with one mole of hydrochloric acid is
   (A) 98.6 g
   (B) 493 g
   (C) 24.7 g
   (D) 49.3 g

7. An element in its ground state
   (A) has all of its electrons in the lowest possible energy levels
   (B) is an element as found in nature
   (C) is an element that is unreactive and found free in nature
   (D) has all of its electrons paired

8. Which following pairs of substances can be used to make a buffer solution?
   (A) NaCl and HCl
   (B) $HC_2H_3O_2$ and $KC_2H_3O_2$
   (C) NaBr and KBr
   (D) HIO_3 and $KClO_3$

9. Which of the following indicates that a reaction is thermodynamically favored?
   (A) At equilibrium there are more products than reactants.
   (B) The value of $\Delta G°$ is greater than zero.
   (C) The value of $\Delta S°$ is greater than zero.
   (D) The value of K_{eq} is less than one.

10. Which of the following is expected to have two or more resonance structures?
   (A) CCl_2F_2
   (B) SO_3
   (C) PF_5
   (D) H_2O

11. Which of the following is a radioactive element?
   (A) Na
   (B) Cr
   (C) Am
   (D) Al

12. The units for the rate of a chemical reaction are
   (A) $L^2\ mol^{-2}\ s^{-1}$
   (B) $mol\ L^{-1}\ s^{-1}$
   (C) $L\ mol^{-1}\ s^{-1}$
   (D) It depends on the particular reaction.

13. Which of the following is not a good measure of relative intermolecular attractive forces?
   (A) Heat of fusion
   (B) Boiling point
   (C) Vapor pressure
   (D) Heat of vaporization

14. Which of the following is expected to be the least soluble in water?
   (A) NaBr
   (B) $NiSO_3$
   (C) $CrCl_3$
   (D) $Mn(NO_3)_2$

15. The net ionic equation expected when solutions of NH_4Br and $AgNO_3$ are mixed together is
   (A) $Ag^+(aq) + Br^-(aq) \longrightarrow AgBr(s)$
   (B) $NH_4^+(aq) + Ag^+(aq) \longrightarrow Ag(NH_4)_2^{3+}(aq)$
   (C) $Br^-(aq) + NO_3^-(aq) \longrightarrow NO_3Br(aq)$
   (D) $NH_4Br(aq) + NO_3^-(aq) \longrightarrow NH_4NO_3(aq) + Br^-(aq)$

16. Less than 1/1000 of the mass of any atom is contributed by
   (A) the electrons
   (B) the electrons and neutrons
   (C) the electrons and protons
   (D) the protons and neutrons

17. Which of the following contains the largest number of moles?
   (A) 1.0 g of aluminum
   (B) 1.0 g of sodium
   (C) 1.0 g of lithium
   (D) 1.0 g of silver

18. In determining the order for reactant A in a certain reaction, the concentrations of all other reactants are held constant while the concentration of A is tripled from one experiment to another. The reaction rate is found to triple. The appropriate exponent for A in the rate law is
   (A) 1
   (B) 2
   (C) 3
   (D) 4

19. The molecule with a tetrahedral shape is
   (A) PCl_4F
   (B) BF_3
   (C) CO_2
   (D) CBr_4

20. According to the kinetic-molecular theory of gases,
   (A) the average kinetic energy of a gas particle is directly related to the Kelvin temperature
   (B) ideal gas particles do not attract or repel each other
   (C) the atoms or molecules of an ideal gas have no volume
   (D) (A), (B), and (C) are part of the theory

21. Of the following, the most important experimental information used to deduce the structure of the atom was
   (A) the density of each element
   (B) the specific heat capacity
   (C) the emission spectrum of the elements, particularly hydrogen
   (D) the X-rays emitted from each element

22. The units for *R*, the ideal gas law equation constant, may be
   (A) L atm mol^{-1} K^{-1}
   (B) J mol^{-1} K^{-1}
   (C) volt coulomb mol^{-1} K^{-1}
   (D) (A), (B), and (C)

23. Which of the following is considered an acid anhydride?
   (A) HCl
   (B) H_2SO_3
   (C) SO_2
   (D) $Al(NO_3)_3$

24. The standard state for redox reactions includes
   (A) the temperature is 25°C
   (B) concentrations of soluble species are 1 molar
   (C) partial pressures of gases are 1 atmosphere
   (D) all of the above are true

25. What is the theoretical yield of ethyl ethanoate when 100 grams of ethanoic acid is reacted with 100 grams of ethyl alcohol?

$$CH_3COOH + CH_3CH_2OH \rightarrow CH_3COOCH_2CH_3 + H_2O$$

   (A) 337 g
   (B) 147 g
   (C) 191 g
   (D) 45 g

26. Iron(III) hydroxide has $K_{sp} = 1.6 \times 10^{-39}$. What is the molar solubility of this compound?
   (A) 1.6×10^{-18} *M*
   (B) 2.0×10^{-10} *M*
   (C) 7.4×10^{-14} mol/L
   (D) 9.4×10^{-6} mol/L

27. When the dichromate ion is reacted, one of its most common products is Cr^{3+}. What is the oxidation state (oxidation number) of chromium in the dichromate ion? Does reduction or oxidation occur when dichromate forms Cr^{3+}?
   (A) +3, reduction
   (B) +12, reduction
   (C) +6, reduction
   (D) +6, oxidation

28. Given the electronegativities below, which of the following covalent single bonds is the most polar?

Element:	H	C	N	O
Electronegativity	2.1	2.5	3.0	3.5

   (A) C–H
   (B) O–H
   (C) N–H
   (D) O–C

29. When the following reactants are mixed, what is the correct name and chemical formula for the precipitate that forms?

$$CuCl_2(aq) + Na_2CO_3(aq) \rightarrow$$

   (A) Copper(I) carbonate, Cu_2CO_3
   (B) Copper(II) carbonate, Cu_2CO_3
   (C) Copper(I) carbonate, $CuCO_3$
   (D) Copper(II) carbonate, $CuCO_3$

30. Which of the following molecules is expected to have the highest normal boiling point?
(A) $CH_3CH_2CH_2CH_3$
(B) $CH_3CH_2CH_2CH_2OH$
(C) $CH_3CH_2CH_2CH_2Cl$
(D) $CH_3CH_2CH_2CH_2Br$

31. Which is correct about the calcium atom?
(A) It contains 20 protons and neutrons.
(B) It contains 20 protons and 20 electrons.
(C) It contains 20 protons, neutrons, and electrons.
(D) All atoms of calcium have a mass of 40.078 u.

32. What is the theoretical yield of iron when 2.00 grams of carbon is reacted with 26.0 grams of Fe_2O_3?

$$2Fe_2O_3 + 3C \longrightarrow 4Fe + 3CO_2$$

(A) 5.8 g
(B) 12.4 g
(C) 74.6 g
(D) 30.6 g

33. Why do vinegar (a dilute solution of ethanoic acid in water) and vegetable oil (long-chain organic acids esterified with glycerol) not mix to form solutions?
(A) The attractive forces in vinegar are much stronger than those in vegetable oil, so the liquids always separate into two phases.
(B) Organic compounds rarely dissolve in water.
(C) Attractive forces in vinegar are mainly hydrogen bonding, while those in vegetable oil are due to instantaneous dipoles.
(D) The unfavorably large endothermic process of "separating" the molecules in the two solutes compared with the energy released when the solutes interact makes a solution thermodynamically unfavored.

34. Which of the following reactions is associated with the normal definition of K_b?
(A) $Zn(H_2O)_6^{2+}(aq) \rightleftharpoons [Zn(H_2O)_5OH]^+(aq) + H^+(aq)$
(B) $CN^-(aq) + H^+(aq) \rightleftharpoons HCN(aq)$
(C) $F^-(aq) + H_2O(l) \rightleftharpoons HF(aq) + OH^-(aq)$
(D) $Cr^{3+}(aq) + 6H_2O(l) \rightleftharpoons Cr(H_2O)_6^{3+}(aq)$

35. Which of the following salts is expected to produce an alkaline solution when one mole is dissolved in one liter of water?
(A) $NaClO_4$
(B) $CaCl_2$
(C) NH_4Br
(D) Na_2S

36. A 50.0 mL aliquot of a solution containing $Al(OH)_3$ is titrated with 0.0500 molar H_2SO_4. The end point is reached when 35.0 mL of the sulfuric acid has been added. What is the molarity of the aluminum hydroxide solution?

(A) $\frac{(0.0500)(35.0)(2)}{(50.0)(3)}$

(B) $\frac{(0.0500)(35.0)(3)}{(50.0)(2)}$

(C) $\frac{(0.0500)(50.0)(3)}{(35.0)(2)}$

(D) $\frac{(50.0)(3)}{(0.0500)(35.0)(2)}$

37. When collecting a gas over water, it is important to
(A) set the temperature at 0°C
(B) be sure the gas does not burn
(C) wait until the barometer reads 760
(D) correct for the vapor pressure of water

38. How much heat, q, is needed to raise the temperature of 35.5 g of olive oil from 25.0°C to 75.0°C? The specific heat of olive oil is 2.0 J/g°C.
(A) 5.33 kJ
(B) 3.55 kJ
(C) 0.282 kJ
(D) 0.888 kJ

39. Sulfur dioxide reacts with oxygen to form sulfur trioxide in the presence of a catalyst. The equilibrium constant, K_p, at a certain temperature is 3.0×10^{22}. A 2.0-liter flask has enough SO_3 added to it to produce a pressure of 0.789 atm. After the reaction comes to equilibrium, the expected partial pressure of O_2 will be
(A) 2.88×10^{-6} torr
(B) 3×10^{-18} mm Hg
(C) 1100 mm Hg
(D) 1.32×10^{-5} torr

40. The melting point of straight-chain hydrocarbons increases as the number of carbon atoms increase. The reason for this is the
    (A) increasing mass of the compounds
    (B) increasing polarity of the compounds
    (C) increasing number of induced dipoles per molecule
    (D) increased probability of hydrogen bonds

41. What is the empirical formula of a compound that is 51.9% carbon, 4.86% hydrogen, and 43.2% bromine?
    (A) C_7H_5Br
    (B) $C_6H_4Br_3$
    (C) C_8H_9Br
    (D) $C_{12}H_{22}Br$

42. Which of the following molecules cannot hydrogen bond with molecules identical to itself but can hydrogen bond with one of the molecules above or below it in the following responses?
    (A) CH_3CH_2OH
    (B) CH_3CH_2COOH
    (C) CH_3CH_2CHO
    (D) C_6H_5CHO

43. The standard galvanic cell voltage, $E°_{cell}$
    (A) is equal to $E°_{reduction} - E°_{oxidation}$
    (B) can be used to calculate K_{eq}
    (C) can be used to calculate $\Delta G°$
    (D) all of the above

44. When will K_p and K_c have the same numerical value?
    (A) At absolute zero for all reactions
    (B) When the concentrations are at standard state
    (C) When the concentrations are all 1.00 molar
    (D) When the reaction exhibits no change in pressure at constant volume

45. The rate of a chemical reaction is determined by
    (A) the equilibrium constant
    (B) the rate-determining or slow step of the mechanism
    (C) the reaction vessel pressure
    (D) the intermediates formed in the first step

ANSWER KEY
Diagnostic Test 3

1. **(C)**	10. **(B)**	19. **(D)**	28. **(B)**	37. **(D)**
2. **(D)**	11. **(C)**	20. **(D)**	29. **(D)**	38. **(B)**
3. **(A)**	12. **(B)**	21. **(C)**	30. **(B)**	39. **(D)**
4. **(B)**	13. **(A)**	22. **(D)**	31. **(B)**	40. **(C)**
5. **(D)**	14. **(B)**	23. **(C)**	32. **(B)**	41. **(C)**
6. **(C)**	15. **(A)**	24. **(D)**	33. **(D)**	42. **(C)**
7. **(A)**	16. **(A)**	25. **(B)**	34. **(C)**	43. **(D)**
8. **(B)**	17. **(C)**	26. **(A)**	35. **(D)**	44. **(D)**
9. **(A)**	18. **(A)**	27. **(C)**	36. **(A)**	45. **(B)**

EVALUATING YOUR RESULTS

Score your test using the answer key. Then complete the following tables. The first table is designed to find your general strengths and weaknesses based on four broad categories. The second table is a more specific diagnostic chart that will suggest which particular chapters you should concentrate your studies on. In combination, these two tables will help you focus your efforts on the material that needs the most study.

Question Categories

Question Type	Questions	Number Wrong
Basic Facts	1, 4, 11, 12, 16, 21, 23, 24, 28, 29, 31, 36	
Basic Concepts	3, 5, 7, 8, 9, 13, 18, 19, 20, 22, 30, 31, 35, 40, 42, 44, 45	
Calculations	2, 6, 17, 25, 26, 32, 33, 38, 39, 41	
Mixed Concepts	10, 12, 14, 15, 19, 20, 27, 30, 34, 35, 37, 43	

Breakdown by Topics

Chapter	Questions	Number Wrong
1. Structure of the Atom	3, 21, 31	
2. Periodic Table	4, 7, 11	
3. Ionic Compounds and Reactions	14, 29	
4. Covalent Compounds	10, 15, 19, 28	
5. Stoichiometry	1, 17, 25, 32, 41	
6. Gases	6, 20, 37	
7. Liquids and Solids	13, 22, 30, 40	

Chapter	Questions	Number Wrong
8. Solutions	33, 36, 42	
9. Equilibrium	26, 38, 44	
10. Kinetics	12, 18, 45	
11. Thermodynamics	2, 5, 9, 38	
12. Redox and Electrochemistry	24, 27, 43	
13. Acids and Bases	8, 23, 34, 35	

ANSWERS EXPLAINED

1. **(C)** When balancing a chemical reaction, the total number of each atom must be the same on both sides. The overall balanced equation is

$$2Al(OH)_3 + 3H_2SO_4 \rightarrow Al_2(SO_4)_3 + 6H_2O$$

2. **(D)** The transformation of water from $H_2O(l) \rightleftharpoons H_2O(g)$ at 100°C has $\Delta G° = 0$. Therefore $\Delta H° = T\Delta S°$. Because we know the heat of vaporization and the boiling point of water, 100°C (373 K), the entropy change can be calculated as

$$\Delta S° = (43{,}900 \text{ J mol}^{-1})/373 \text{ K}$$
$$= 118 \text{ J mol}^{-1} \text{ K}^{-1}$$

3. **(A)** Hund's rule requires that each orbital in a particular sublevel must be filled with one electron before a second electron may be added to an orbital. When the electrons become paired, they must be of opposite spins. This is usually symbolized by one up and one down arrow.

4. **(B)** There are only two elements that are liquids at room temperature. They are bromine and mercury. Mercury is a silver-colored metal, and bromine is a brown nonmetal liquid. Nonmetals do not conduct electricity.

5. **(D)** None of the three other answers fits. The positive free energy indicates a non-spontaneous system with more reactants than products. The large exothermic heat of reaction and the large increase in entropy are very unlikely with a large positive free energy.

6. **(C)** We set up the problem with the desired ratio to the left of the equal sign. To the right of the equal sign replace the given mass of metal in the numerator and the volume of H_2 in mL in the denominator. Since we already know the value of the final numerator, we just need to convert the denominator into moles of HCl.

$$\frac{\text{grams of metal}}{\text{moles of HCl}} = \frac{0.0623 \text{ g metal}}{28.3 \text{ mL } H_2}\left[\frac{22{,}400 \text{ mL } H_2}{1 \text{ mol } H_2}\right]\left[\frac{1 \text{ mol } H_2}{2 \text{ mol HCl}}\right]$$
$$= 24.7 \text{ g metal/mol HCl}$$

Notice that we used the molar volume of an ideal gas at STP to convert the H_2 from mL to moles. Then we used the mole ratio that relates H_2 and HCl for the second conversion factor.

7. **(A)** The definition of the ground state is the lowest total energy. To have the lowest total energy an atom must have its electrons in their lowest possible energy levels.

8. **(B)** Buffers are prepared from a weak acid and its conjugate base or from a weak base and its conjugate acid. (A) NaCl and HCl; this pair has HCl, which is a strong acid. (B) $HC_2H_3O_2$ and $KC_2H_3O_2$; this pair has a weak acid and its conjugate base (the $C_2H_3O_2^-$ ion). (C) NaBr and KBr; this pair has no weak acid or base. (D) HIO_3 and $KClO_3$; this pair does not have a conjugate acid–base pair.

9. **(A)** The responses for (B) and (D) indicate processes that are thermodynamically unfavorable. The response in (C) may indicate a favorable process but it is not the only thing needed for a thermodynamically favored reaction. Only response (A) is universally true for a favorable reaction.

10. **(B)** The SO_3 molecule has three resonance structures. Each has one oxygen drawn with a double bond to the sulfur and the other two oxygens drawn with a single bond to the sulfur. The remaining compounds have only one possible Lewis structure.

11. **(C)** We expect elements with atomic numbers greater than 83 to be radioactive. Americium is the only one given that fulfills that criterion.

12. **(B)** All reaction rates have the same units of moles per liter per second (mol L^{-1} s^{-1}). This may refer to the rate of appearance of a product as the reaction progresses or the disappearance of a reactant.

13. **(A)** The heat of fusion is the energy needed to disrupt the crystal lattice but not completely separate the molecules. The remaining attractive forces may be significant. (B), (C), and (D) all involve vaporizing the liquid and indicate the total attractive force.

14. **(B)** The solubility rules specify that sulfites are one group of compounds that are generally insoluble, especially if the metal ion is a transition metal.

15. **(A)** Silver bromide is insoluble, whereas all other substances in the reaction are soluble. The ammonium ions and nitrate ions are spectator ions that cancel. Silver ions will form a complex with ammonia, NH_3, not with ammonium ions, NH_4^+.

16. **(A)** The electron weighs 1/1833 of the mass of a proton and so contributes even less than 1/1000 of the mass of an atom.

17. **(C)** To determine the number of moles in 1.0 g of any substance, you must divide the mass of that substance by its atomic or molar mass. All of the choices are 1.0 g. To determine the number of moles of each, take the inverse of each atomic mass. The one with the lowest atomic mass contains the most moles.

18. **(A)** Because the reaction rate increases by the same factor as the concentration of *A*, the reaction with respect to reactant *A* must be first order.

19. **(D)** (A) is trigonal bipyramidal; (B) is triangular planar; (C) is linear.

20. **(D)** All three statements are considered to be parts of the kinetic-molecular theory of gases as stated by Clausius in 1857.

21. **(C)** The Bohr theory that preceded the quantum model of the atom relied on the atomic spectrum of hydrogen for important clues.

22. **(D)** All of the first three are units for *R*.

23. **(C)** An acid anhydride is an oxide of a nonmetal that dissolves in water to form an oxo acid. Sulfur dioxide dissolves in water to produce an acidic solution.

24. **(D)** All four of the statements define one aspect of standard state as applied to the study of redox reactions.

25. **(B)** We calculate the grams of ethanoic acid needed to react with the given amount of ethyl alcohol. Two important conversion factors are

$$1 \text{ mol } CH_3COOH = 60 \text{ g } CH_3COOH$$

and

$$1 \text{ mol } CH_3CH_2OH = 44 \text{ g } CH_3CH_2OH$$

Setup: ? g CH_3COOH = 100 g CH_3CH_2OH

Applying the above conversion factors we get

$$? \text{ g } CH_3COOH = 100 \text{ g } CH_3CH_2OH\left[\frac{1 \text{ mol } C_2H_5OH}{44 \text{ g } C_2H_5OH}\right]\left[\frac{1 \text{ mol } HC_2H_5O_2}{1 \text{ mol } C_2H_5OH}\right]\left[\frac{60 \text{ g } HC_2H_5O_2}{1 \text{ mol } HC_2H_5O_2}\right]$$
$$= 136 \text{ g } CH_3COOH$$

The problem only gave us 100 g CH_3COOH. Therefore CH_3COOH is the limiting reactant. We now calculate the mass of ethyl ethanoate formed from the GIVEN mass of 100 g CH_3COOH as follows.

Set up the question.

$$? \text{ g } CH_3COOCH_2CH_3 = 100 \text{ g } CH_3COOH$$

Apply conversion factors

$$? \text{ g } CH_3COOCH_2CH_3$$
$$= 100 \text{ g } CH_3COOH\left[\frac{1 \text{ mol } HC_2H_5O_2}{60 \text{ g } HC_2H_5O_2}\right]\left[\frac{1 \text{ mol } C_4H_8O_2}{1 \text{ mol } HC_2H_5O_2}\right]\left[\frac{88 \text{ g } C_4H_8O_2}{1 \text{ mol } C_4H_8O_2}\right]$$
$$= 147 \text{ g } C_4H_8O_2 \text{ (or } CH_3COOCH_2CH_3)$$

NOTE: We get the same result if we convert each reactant to grams of ethyl ethanoate and then choose the smaller of the two results.

26. **(A)** The dissolution reaction is $Fe(OH)_3(s) \leftrightarrows Fe^{3+} + 3\ OH^-$, and the K_{sp} equation is $K_{sp} = [Fe^{3+}][OH^-]^3$.

If s mol/L of $Fe(OH)_3$ dissolves, the solution will contain s mol/L of Fe^{3+} and $3s$ mol/L of OH^-.

The K_{sp} equation is $K_{sp} = [Fe^{3+}][OH^-]^3 = (s)(3s)^3 = 27s^4$

Calculating $s = 8.8 \times 10^{-11}$ M Fe^{3+} and $3s$ as 2.64×10^{-10} M OH^- we see there is a problem. Since pure water has an OH^- concentration of 1.0×10^{-7} M or about one thousand times larger than 2.64×10^{-10} the contribution of OH^- from water cannot be ignored. So, this is really a common-ion calculation with $[OH^-] = 1.0 \times 10^{-7}$ M. The equation to use is

$$K_{sp} = [Fe^{3+}][OH^-]^3 = (s)(3s + 1.0 \times 10^{-7})^3$$

Since $3s << 1.0 \times 10^{-7}$ the simplified equation becomes

$$= 1.6 \times 10^{-39} = (s)(1.0 \times 10^{-7})^3 = (s)(1.0 \times 10^{-21})$$

$$s = 1.6 \times 10^{-39}/1.0 \times 10^{-21} = 1.6 \times 10^{-18}\ M$$

27. **(C)** The dichromate ion is $Cr_2O_7^{2-}$. When we calculate the oxidation number for each chromium in the dichromate ion, we get +6. Because the Cr^{3+} is only +3, there is a decrease in oxidation number, and the process is called a reduction.

28. **(B)** The most polar bond is the one that has the greatest difference in electronegativities. Therefore, the O–H bond is the most polar because it has an electronegativity difference of 1.4.

29. **(D)** The reactants are copper(II) chloride and sodium carbonate. When they are mixed, copper(II) carbonate, $CuCO_3$, is the precipitate that forms.

30. **(B)** Butan-1-ol (1-butanol), $CH_3CH_2CH_2CH_2OH$, can hydrogen-bond, whereas the other compounds cannot. London forces (instantaneous dipoles) are also present, but each compound is roughly the same length, and these forces will be similar for all four molecules.

31. **(B)** All calcium atoms contain 20 protons and 20 electrons. Depending on the isotope, a calcium atom may or may not have 20 neutrons. Finally, the mass in the periodic table is a weighted average of isotopes and is NOT the mass of any calcium atom.

32. **(B)** We are asked for the theoretical yield of iron. Usually, the theoretical yield is expressed in units of grams of product, as all the responses imply. This is a limiting-reactant problem because the mass of both reactants is given. We will solve it by calculating the mass of iron that can be made from 2.00 g C assuming that Fe_2O_3 is the excess reactant. Then we will calculate the mass of iron that we can prepare from 26.0 g Fe_2O_3 assuming that carbon is the excess reactant. We set up the two equations as

$$?\ \text{g Fe} = 2.00\ \text{g C}\left[\frac{1\ \text{mol C}}{12\ \text{g C}}\right]\left[\frac{4\ \text{mol Fe}}{3\ \text{mol C}}\right]\left[\frac{55.84\ \text{g Fe}}{1\ \text{mol Fe}}\right] = 12.4\ \text{g Fe}$$

$$?\ \text{g Fe} = 26.0\ \text{g Fe}_2\text{O}_3\left[\frac{1\ \text{mol Fe}_2\text{O}_3}{159.7\ \text{g Fe}_2\text{O}_3}\right]\left[\frac{4\ \text{mol Fe}}{2\ \text{mol Fe}_2\text{O}_3}\right]\left[\frac{55.84\ \text{g Fe}}{1\ \text{mol Fe}}\right] = 18.2\ \text{g Fe}$$

Carbon produces the smaller amount of iron. Therefore, carbon is the limiting reactant, and 12.4 g of iron is the theoretical yield.

33. **(D)** Energy is always the key to chemical processes. In this case, we can use our knowledge of attractive forces to see that a solution is thermodynamically unfavored.

34. **(C)** The dissociation constant, K_b, is expressed as

$$K_b = \frac{[\text{conjugate acid}][\text{OH}^-]}{[\text{weak base}]}$$

The only equation that fits that definition is the hydrolysis of F^-.

35. **(D)** Salts made from a weak acid and a strong base will produce an anion that will hydrolyze water to form an alkaline solution. The only salt produced from a weak acid is Na_2S, and the sulfide ion hydrolyzes water according to the reaction

$$S^{2-} + H_2O \rightarrow HS^- + OH^-$$

36. **(A)** We need to write and balance the equation first:

$$2Al(OH)_3 + 3H_2SO_4 \rightarrow Al_2(SO_4)_3 + 6H_2O$$

Start with the molarity of H_2SO_4. Use dimensional analysis to convert it to the molarity of $Al(OH)_3$ as follows:

$$?\frac{\text{mol Al(OH)}_3}{\text{L Al(OH)}_3} = \frac{0.0500 \text{ mol H}_2\text{SO}_4}{\text{L H}_2\text{SO}_4} \times \frac{2 \text{ mol Al(OH)}_3}{3 \text{ mol H}_2\text{SO}_4} \times \frac{35.0 \text{ mL H}_2\text{SO}_4}{50.0 \text{ mL Al(OH)}_3}$$

When we extract the coefficients from the right side of this equation, it matches response (A).

37. **(D)** The largest error will occur if the results are not corrected for the vapor pressure of water. The temperature of the water must be measured, but it does not have to be 0°C. The barometric pressure is part of the ideal gas law and does not need to be 760 torr, but it must be measured when the experiment is performed. The use of tap water instead of distilled water and the flammability properties of the gas will not have a significant effect on the gas.

38. **(B)** To determine the amount of heat necessary to raise the temperature, you must multiply the mass × ΔT × specific heat.

$$q = (35.5 \text{ g})(75.0°\text{C} - 25.0°\text{C})(2.0 \text{ J/g°C})$$

$$q = 3{,}550 \text{ J} = 3.55 \text{ kJ}$$

39. **(D)** We need to write the chemical reaction as

$$2\ SO_2(g) + O_2(g) \rightleftharpoons 2\ SO_3(g)$$

The equilibrium expression for this reaction is written in terms of partial pressures, measured in atmospheres, and the equilibrium constant is K_p.

$$K_p = \frac{P^2_{SO_3}}{P_{O_2} P^2_{SO_2}} = 3.0\times10^{22}$$

The initial partial pressure of SO_3 is 0.789 atm. If $2x$ atm of the initial SO_3 decomposes, x atm of O_2 and $2x$ atm of SO_2 will form. This is summarized in the ICES table.

	$2\ SO_2(g)$ +	$O_2(g)$	$\rightleftharpoons$	$2\ SO_3(g)$
Initial	0	0		0.789
Change	$+2x$	$+x$		$-2x$
Equilibrium (I + C)	$2x$	x		$0.789 - 2x \approx 0.789$
Solution (sub. x into E)	3.46×10^{-8}	1.73×10^{-8}		0.789

We enter the information from the equilibrium line into the equilibrium expression. We use the approximation that $0.789 - 2x = 0.789$ because the equilibrium constant is large.

$$K_p = \frac{(0.789-2x)^2}{x(2x)^2} = \frac{(0.789)^2}{x(2x)^2}$$

$$3.0\times10^{22} = \frac{(0.789)^2}{x(2x)^2}$$

$$12.0\times10^{22}\, x^3 = 0.623$$

$$x = 1.73\times10^{-8} \text{ atm} = 1.32\times10^{-5} \text{ torr}$$

The value of x is very small, indicating that our assumption was correct. Notice how we needed to solve the problem using units of atmospheres.

40. **(C)** Straight-chain hydrocarbons are essentially nonpolar, and they interact only with London forces of attraction that include instantaneous dipoles and induced dipoles. As the chain becomes longer, each molecule has more of these forces attracting it to neighboring molecules. With long non-polar carbon chains, these instantaneous dipoles can be a strong attractive force.

41. **(C)** To determine the empirical formula we must calculate the simplest ratio of moles of each element in the formula. The percentages are converted to grams by assuming 100 g of sample and changing all percentages to grams. The moles are then calculated as

$$? \text{ mol C} = 51.9 \text{ g C}\left(\frac{1 \text{ mol C}}{12 \text{ g C}}\right) = 4.32 \text{ mol C}$$

$$? \text{ mol H} = 4.86 \text{ g H}\left(\frac{1 \text{ mol H}}{1 \text{ g H}}\right) = 4.86 \text{ mol H}$$

$$? \text{ mol Br} = 43.2 \text{ g Br}\left(\frac{1 \text{ mol Br}}{79.9 \text{ g Br}}\right) = 0.541 \text{ mol Br}$$

Divide each of the answers above by 0.541 to get 7.98 mol C, 8.98 mol H, and 1.00 mol Br. The empirical formula is C_8H_9Br.

42. **(C)** The molecules in both (A) and (B) contain the –OH group and can hydrogen bond to other identical molecules. The molecules in (C) and (D) cannot hydrogen bond to other identical molecules because they do not contain the –OH, –NH, or H–F bond. The molecule in (D) cannot hydrogen bond to the molecule in (C) and there is no molecule below (D). The molecule in (C) cannot hydrogen bond to the molecule in (D) but it can hydrogen bond to the molecule in (B).

43. **(D)** Choices (A), (B), and (C) are all correct applications of E°_{cell}.

44. **(D)** When the reaction shows no pressure changes, it indicates that $\Delta n_g = 0$. If a chemical reaction has the same number of moles of gas as reactants and products, then K_p will be equal to K_c.

45. **(B)** The rate-determining step governs the overall reaction rate. Proposed mechanisms must provide appropriate steps to include the slow step.

Note: The link to the online practice tests can be found on the inside front cover of this book. All three online tests can be accessed on mobile devices, including tablets and smartphones.

PART 1

Structure of Matter

Structure of the Atom

1

→ **ATOMIC THEORY**
→ **MODELS OF THE ATOM**
→ **STRUCTURE OF THE ATOM**
→ **PROTONS, ELECTRONS, AND NEUTRONS**
→ **ISOTOPES**
→ **ATOMIC SPECTRA**
→ **WAVE-MECHANICAL ATOM**
→ **ENERGY LEVELS**
→ **ELECTRONIC STRUCTURE**
→ **ELECTRON CONFIGURATIONS**
→ **VALENCE ELECTRONS**
→ **HUND'S RULE**
→ **ORBITAL DIAGRAMS**
→ **PAULI EXCLUSION PRINCIPLE**
→ **QUANTUM NUMBERS**

BIG IDEA 1

Learning Objectives: 1.5, 1.6, 1.7, 1.8, 1.12, 1.13, 1.14, 1.15, 1.16

For the complete list of Big Ideas and Learning Objectives, refer to the AP Chemistry Course Outline: *https://secure-media.collegeboard.org/digitalServices/pdf/ap/ap-chemistry-course-and-exam-description.pdf*

A REVIEW OF IMPORTANT DISCOVERIES ABOUT THE ATOM

Law of Conservation of Matter

Major milestones in the development of chemistry were reached in 1774, when Antoine Lavoisier performed careful experiments and measurements that led to the **law of conservation of matter**, and in 1799, when Joseph Proust made measurements on chemical reactions and compounds and developed the **law of constant composition**. The first of these important laws states that in chemical reactions matter is neither created nor destroyed. The second law states that each pure chemical compound always has the same percentage composition of each element by mass. These two laws led John Dalton to develop his **atomic theory** over the years from 1803 to 1808. Dalton's atomic theory states that:

Law of Constant Composition

Dalton's Atomic Theory

1. All matter is composed of tiny, indivisible particles, called atoms, that cannot be destroyed or created.
2. Each element has atoms that are identical to each other in all of their properties, and these properties are different from the properties of all other atoms.
3. Chemical reactions are simple rearrangements of atoms from one combination to another in small whole-number ratios.

Every scientific theory provides new predictions that may be tested by experiments to support or disprove the theory. [It is important to remember that scientists can never prove a theory to be true. Experiments may be used to support a theory, but not to prove it.]

Law of Multiple Proportions

The atomic theory led John Dalton to propose the **law of multiple proportions**:

> When two elements can be combined to make two different compounds, and if samples of these two compounds are taken so that the masses of one of the elements in the two compounds are the same in both samples, then the ratio of the masses of the other element in these compounds will be a ratio of small whole numbers.

This law was used to test Dalton's atomic theory.

Michael Faraday

In 1834 Michael Faraday showed that electric current could cause chemical reactions to occur, demonstrating the electric nature of the elements. Sir William Crookes in the 1870s developed what is known today as the **cathode ray tube**. He mistakenly thought that the cathode rays were negatively charged molecules instead of electrons. (Crookes was also the first to suggest the existence of isotopes.)

J. J. Thomson

In 1897 J. J. Thomson determined that cathode rays were a fundamental part of matter he called electrons. He also determined their **charge to mass ratio** ($e/m = -1.76 \times 10^8$ coulombs gram^{-1}) by measuring the deflection of the cathode rays in the presence of electric and magnetic fields. Twelve years later, in 1909, Robert Millikan performed his **oil drop experiment**, which allowed him to calculate the charge of the electron (-1.60×10^{-19} coulomb). Combined with Thomson's charge to mass ratio, the mass of the electron was calculated to be 9.11×10^{-28} gram. This information led to the **"plum pudding" model** of the atom, which had electrons bathed in a sea of positive charges similar to raisins in the famous English pudding.

Ernest Rutherford

At this time Ernest Rutherford was interested in radioactive materials and had identified alpha and beta particles in his research. Along with Hans Geiger and Ernest Marsden, he performed the **gold foil experiment**, in which heavy alpha particles were aimed at a thin gold foil. While most of the alpha particles went through the foil with no visible effect, a few of them were deflected from their path and some actually bounced back in the direction they came from. From these results Rutherford deduced the **nuclear model of the atom**, with an extremely small, dense, and positively charged nucleus surrounded by empty space sparsely occupied by electrons. Ten years later, in 1919, Rutherford discovered the basic unit of positive charge in the atom and named it the proton. The proton has a positive charge which is exactly equal in magnitude to the electron charge. It also has a mass of 1.67×10^{-24} gram, 1836 times heavier than the electron. In 1932 James Chadwick discovered a very penetrating form of radiation. He demonstrated that it was the third major particle that makes up the atom, and it was named the neutron since it is neutral (i.e., it has no charge). The neutron has a mass almost equal to that of the proton.

ATOMIC MODELS

Solid particle model 400 BC
Plum pudding model 1909
Nuclear model Rutherford 1910
Solar system model Bohr 1913
Wave-mechanical model Schrödinger 1927

While the fundamental particles of the atom (Table 1.1) were being discovered, other physicists were performing experiments that laid the foundation for a fundamental revolution in the way all matter is viewed. In the mid-1800s physicists were interested in the interaction of light and matter. One of the interesting things they found was that each element, when heated or sparked with electricity, gives off characteristic colors. A spectroscope was used to show that these colors consist of discrete wavelengths of light (line spectra), not the uniform rainbow observed when white light is separated by a prism. The line spectra of most elements and compounds are very complex. Hydrogen, however, has a seemingly simple series of lines. In 1885 Johann Balmer found an empirical mathematical relationship between the wavelengths of the lines he observed in the visible region of the spectrum. When similar series of lines were found in the infrared (Paschen series) and ultraviolet (Lyman series) regions, Johannes Rydberg extended Balmer's equation so that all of the wavelengths could be predicted.

Johann Balmer

Johannes Rydberg

TABLE 1.1
Fundamental Parts of the Atom

Name	Symbol	Absolute Charge (coulombs)	Absolute Mass (grams)	Relative Charge	Relative Mass (*U*)
Electron	*e* or e^-	-1.602×10^{-19}	9.109×10^{-28}	-1	5.486×10^{-4}
Proton	p	$+1.602 \times 10^{-19}$	1.673×10^{-24}	+1	1.0073
Neutron	n	0	1.675×10^{-24}	0	1.0087

In 1913 Niels Bohr completed his theory of how the hydrogen atom is constructed. He assumed that electrons move around the nucleus in circular orbits. Using a **solar system model,** he was able to duplicate the Rydberg equation from fundamental constants already known to physicists. Bohr's most important contribution was the concept that electrons exist in only certain "allowed orbits." This, along with Max Planck's description in 1900 of light as packets, or quanta, of energy, called photons, aided Bohr in developing the solar system model of the atom.

Niels Bohr

TIP

Quantum mechanics has replaced classical physics for describing events at the atomic level.

Louis de Broglie suggested in 1924 that, if light can be considered as particles, then the small particles such as electrons may have the characteristics of waves. In 1927 Erwin Schrödinger applied the equations for waves to the electrons in an atom and began the **wave-mechanical theory** of the atom. For hydrogen the results are very similar to Bohr's model except that the electron does not follow a precise orbit. The position of the electron in the wave-mechanical model is described by a probability of where it will be located. Also in the 1920s, Werner Heisenberg developed the uncertainty principle, which bears his name. It states that the position and the momentum of any particle cannot both be known exactly at the same time. As one is known more precisely, the other becomes less certain.

Max Planck

Louis de Broglie

Erwin Schrödinger

Werner Heisenberg

ATOMIC STRUCTURE

Light and the Atom

An atom (or molecule) usually exists in the lowest possible energy state, which is called the **ground state**. An atom (or molecule) that has more energy than the ground state is said to be in an **excited state**. When an atom loses energy in going from an excited state to a ground state, light may be emitted. To understand this light, we need to review the basic equations for electromagnetic radiation.

THE ELECTROMAGNETIC SPECTRUM

Visible light is only one small part of the electromagnetic spectrum. Figure 1.1 presents a representation of the electromagnetic spectrum and some of its common uses. For example, we can see that **microwaves** have wavelengths from 10^{-1} to 10^{-4} meters.

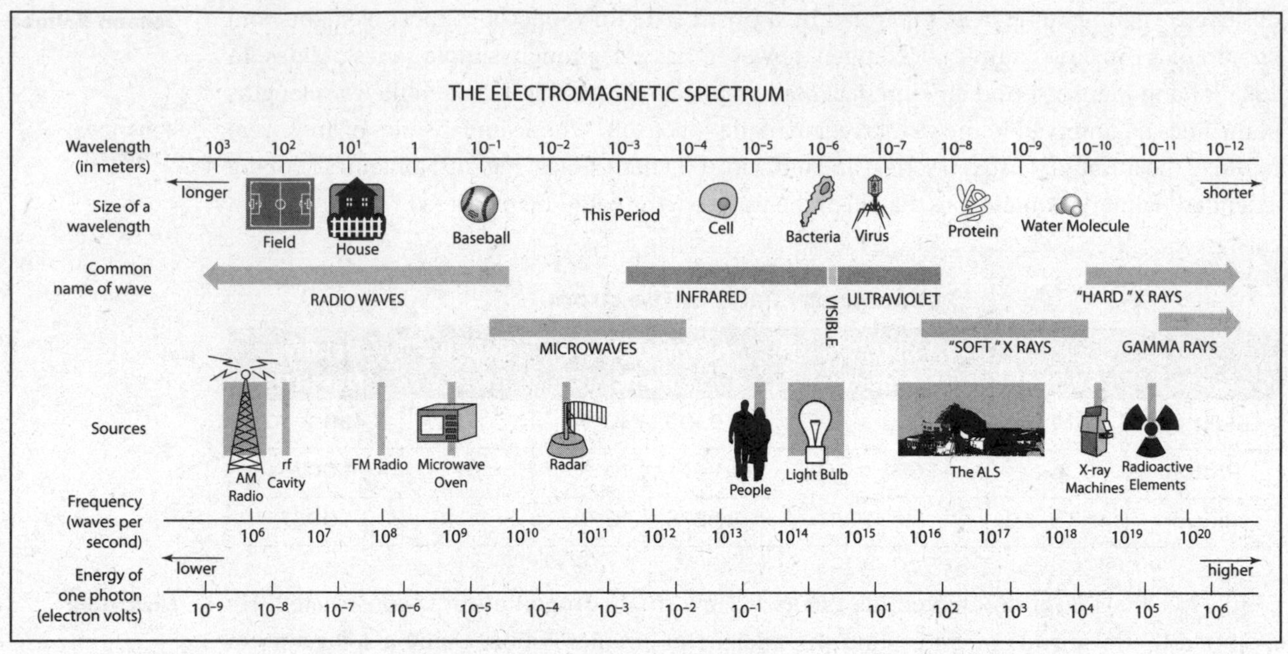

Image courtesy Lawrence Berkeley National Laboratory

FIGURE 1.1. Pictorial representation of the electromagnetic spectrum emphasizing the narrow range of visible light.

Wavelength, Frequency, and Energy of Light

All electromagnetic radiation may be considered as waves that are defined by the wavelength (λ) and frequency (ν). The wavelength (Figure 1.2) is the distance between two repeating points (either two minima or two maxima) on a sine wave. The frequency is defined as the number of waves that pass a point in space each second.

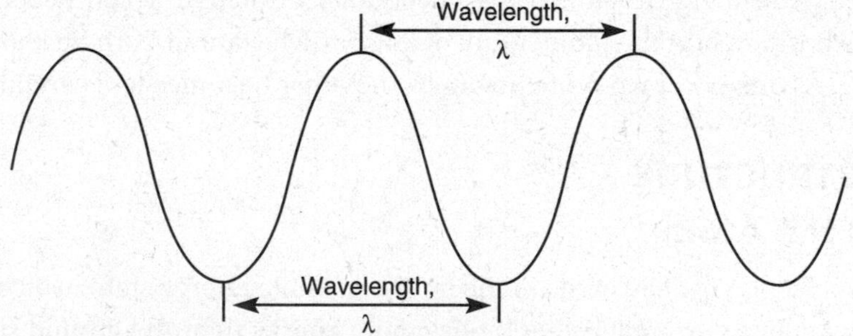

FIGURE 1.2. Definition of wavelength.

The wavelength and the frequency of light are inversely proportional to each other as shown in the graph and the equation below.

Wavelength
Frequency

$$(\text{Wavelength})(\text{Frequency}) = \text{Speed of light} \qquad (1.1)$$

$$\lambda \nu = c$$

The speed of light is 3.0×10^8 meters per second (m s^{-1}), a number worth remembering, although it is provided on the exam. Wavelength has units of meters, often with the appropriate metric prefix (cm, μm, or nm), and frequency has units of reciprocal seconds (s^{-1}), also called hertz (Hz).

Speed of light in vacuum 3.0×10^8 m s^{-1}

Max Planck found that the energy of electromagnetic waves is proportional to the frequency and inversely proportional to the wavelength as shown in the graphs below. The proportionality constant, h, is called Planck's constant; it has a value of 6.63×10^{-34} joule second. [Planck's constant need not be memorized; when required for a problem, it will be given.]

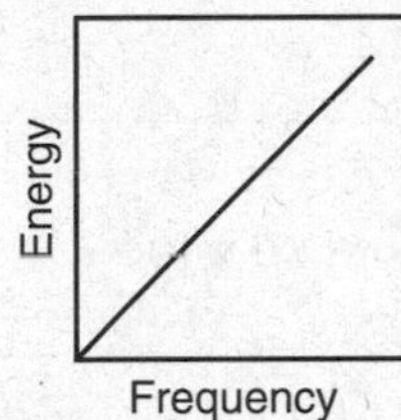

$$h\nu = E = h\frac{c}{\lambda} \qquad (1.2)$$

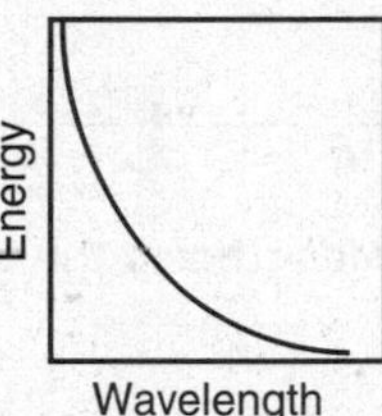

Energy, wavelength, and frequency are all related. If the speed of light and Planck's constant are known, only one of these three variables is needed to calculate the others.

EXAMPLE 1.1

(In this exercise and all others in this book it is recommended that you cancel the units with a red pencil or pen to verify the calculations.)

What are the frequency and the energy of blue light that has a wavelength of 400. nm? (Planck's constant = 6.63×10^{-34} J s)

TIP

Meters per second, m/s, and m s^{-1} all represent the same units.

Solution

Substitute the given values into the equation $\lambda\nu = c$

$$(400.\ \text{nm})(\nu) = 3.0 \times 10^8\ \text{m s}^{-1}$$

Substitute 10^{-9} for the prefix *nano-* in the wavelength

$$(400. \times 10^{-9}\ \text{m})(\nu) = 3.0 \times 10^8\ \text{m s}^{-1}$$

Rearrange this equation to solve for the frequency,

$$\nu = \frac{3.0\times10^8\ \text{m s}^{-1}}{400.\times10^{-9}\ \text{m}} = 7.5\times10^{14}\ \text{s}^{-1}$$

The energy of this light may be calculated directly from the frequency or the wavelength by using Equation 1.2.

$$E = h\nu = \left(6.63\times10^{-34}\ \text{J s}\right)\left(7.5\times10^{14}\ \text{s}^{-1}\right)$$
$$= 49.725\times10^{-20}\ \text{J}$$
$$= 5.0\times10^{-19}\ \text{J (correctly rounded)}$$

or

$$E = h\frac{c}{\lambda} = (6.63\times10^{-34}\text{ J s})\left(\frac{3.0\times10^{8}\text{ m s}^{-1}}{400.\times10^{-9}\text{ m}}\right)$$

$$= 5.0\times10^{-19}\text{ J}$$

It is very important to be sure that the units cancel properly when solving these problems. The meter units cancel each other since there is one in the numerator and another in the denominator, and the seconds units cancel because s is present in the first term and s^{-1} in the numerator of the second.

EXAMPLE 1.2

What are the wavelength and the energy of light that has a frequency of 1.50×10^{15} s^{-1}?

Solution

The relationship between the wavelength and the frequency is $\lambda\nu = c$. Substitution of the given values yields

$$\lambda(1.50 \times 10^{15}\text{ s}^{-1}) = 3.0 \times 10^{8}\text{ m s}^{-1}$$

Rearrange to solve for the wavelength, λ:

$$\lambda = \frac{3.0\times10^{8}\text{ m s}^{-1}}{1.50\times10^{-15}\text{ s}^{-1}}$$

$$= 2.0\times10^{-7}\text{ m} = 2.0 \times 10^{2} \times 10^{-9}\text{ m}$$

Use the metric prefix, 1 nm = 10^{-9} m, to simplify the answer:

$$\lambda = 2.0 \times 10^{2}\text{ nm}$$

The energy of this light is calculated from $E = h\nu$. Substituting given data yields

$$E = (6.63 \times 10^{-34}\text{ J s})(1.50 \times 10^{15}\text{ s}^{-1})$$

Cancelling units and solving gives the result (remember that $s \times s^{-1} = 1$)

$$E = 9.94 \times 10^{-19}\text{ J}$$

THE BOHR MODEL OF THE ATOM

Neils Bohr revolutionized concepts about the atom with his solar system model. This model of the atom required that the electrons be confined to specific allowed orbits. Using conventional physics the energy of an orbit with a number n is

$$E_n = \frac{-2\pi^2 me^4}{n^2h^2} = \frac{-2.178\times10^{-18}}{n^2}\text{ joule} \quad (1.3)$$

where m = mass of the electron, e = charge on the electron, h = Planck's constant, and n = orbit number. Later, the number n became known as the principal quantum number.

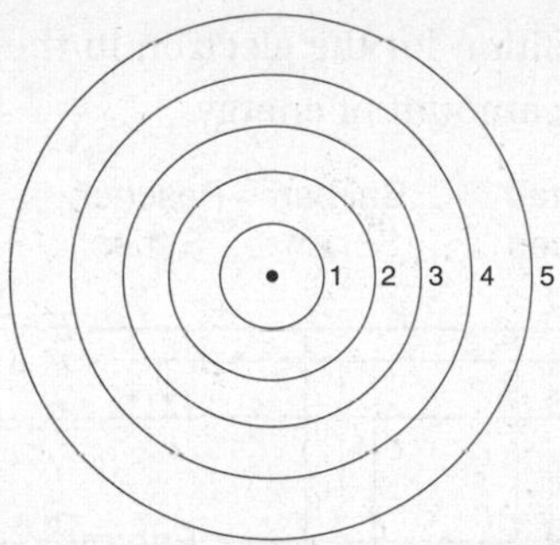

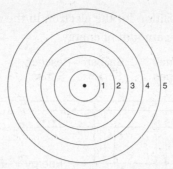

FIGURE 1.3. Diagram of Bohr orbits, showing numbering and relative sizes.

In Bohr's theory the symbol n represents the number of each orbit, starting with the one closest to the nucleus. Energy in the form of light is emitted from an atom when an electron moves from its initial orbit to an orbit with a lower value of n (Figure 1.4a). When an electron is promoted from a low orbit to a higher numbered orbit, energy must be added (Figure 1.4b). The energy difference between any two orbits is constant, and the same amount of energy needed to raise an electron from one orbit to another will be released when the electron drops back to the original orbit. Line spectra result from the emission of light by atoms and therefore represent electrons in excited atoms dropping from high orbits to lower ones.

The energy emitted can be determined by subtracting the energy of the lower orbit from the energy of the higher orbit using Equation 1.3. When energy is added to an atom, the atom moves from a lower energy state to a higher energy state as shown in Figure 1.4b. We can calculate the energy needed to excite an electron as the difference in energy between the final and initial Bohr energy.

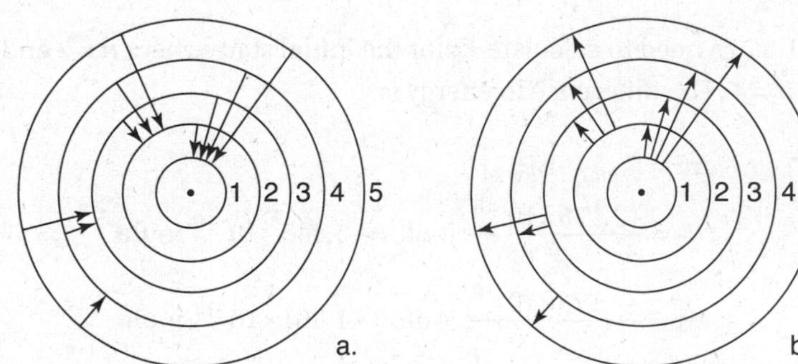

FIGURE 1.4. Movement of electrons between Bohr energy levels. (a) This represents electrons losing energy (in the form of light) as they drop to lower energy levels. (b) This represents electrons gaining energy as they are excited from a lower energy level to a higher energy level.

Another way of describing this process is the energy-level diagram, where the y-axis represents the energy of each orbit. A horizontal line, rather than the circles shown above, represents the energy level of an orbit. Arrows are drawn from one energy level to another to show where an electron starts and where it ends up. All energy-level diagrams for line spectra show electrons moving from high orbits (energy levels) to lower orbits.

The energy of each level in Figure 1.5 is calculated from Equation 1.3 as shown in Example 1.3. The energies have a negative sign since the electron loses energy as it drops toward the

$n = 1$ level. The most stable position for the electron in the atom is the first level since the electron has the lowest possible amount of energy.

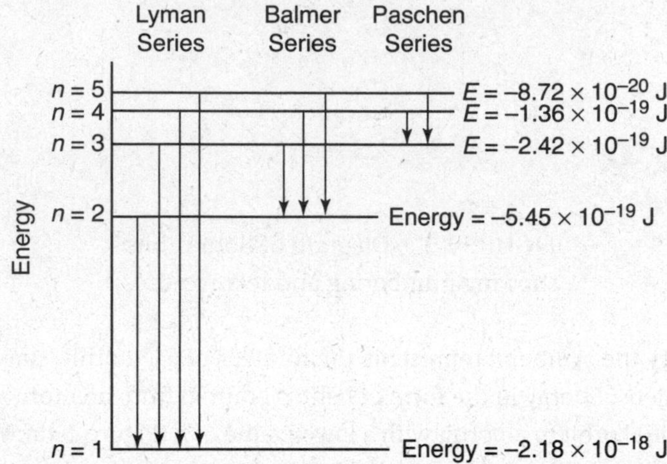

FIGURE 1.5. Energy level diagram for the Lyman, Balmer, and Paschen series of lines in the spectra of hydrogen.

EXAMPLE 1.3

Determine the energy and wavelength of light associated with an electron moving from the second to the fourth energy level in a hydrogen atom.

Solution

To use Equation 1.3, we need to calculate E_n for the initial state where $n = 2$ and again for the final state where $n = 4$. The difference in energy is

$$\Delta E = E_{\text{final}} - E_{\text{initial}}$$

$$E_2 = \frac{-2.178 \times 10^{-18}}{2^2} \text{ joule} = -5.445 \times 10^{-19} \text{ joule}$$

$$E_4 = \frac{-2.178 \times 10^{-18}}{4^2} \text{ joule} = -1.361 \times 10^{-19} \text{ joule}$$

$$\Delta E = E_4 - E_2$$

$$= \left(-1.361 \times 10^{-19} \text{ joule}\right) - \left(-5.445 \times 10^{-19} \text{ joule}\right)$$

$$= +4.084 \times 10^{-19} \text{ joule}$$

(The positive sign indicates that energy must be put into the atom to achieve this transition.)

The wavelength of this energy is calculated from

$$\Delta E = \frac{hc}{\lambda}$$

$$\text{rearranged to } \lambda = \frac{hc}{\Delta E} = \frac{(6.63 \times 10^{-34} \text{ J s})(3.00 \times 10^8 \text{ m s}^{-1})}{(4.084 \times 10^{-19} \text{ J})}$$

$$= 4.87 \times 10^{-7} \text{ m} = 487 \text{ nm}$$

The Size of the Atom

Bohr postulated that the momentum (mass × velocity) of the electron must be related to the size of the electron's orbit. The relationship he used was

Bohr Radius:

r **= 53 pm**

$$mv = \frac{nh}{2\pi r} \quad (1.4)$$

When Planck's constant, h, the electron mass, m, and the electron velocity, v, are entered into the equation for the first energy level, $n = 1$, a radius of $r = 53$ picometers (pm) is calculated. If $n = 2$, the orbit radius is 106 pm. The value 53 pm is often called the Bohr radius for the hydrogen atom. The radii of the other orbits are whole-number multiples of the **Bohr radius.** The Bohr radius gave chemists a theoretical value for the size of a hydrogen atom and confirmed that the atomic sizes determined by experiment were indeed reasonable.

The Wave-Mechanical Model of the Atom

Soon after Bohr's achievement (for which he received the Nobel Prize) Louis de Broglie suggested that the electron could behave as a wave as well as a particle. One way of looking at this duality involves the equation for the energy of a wave $\left(E = h\nu = \frac{hc}{\lambda}\right)$ and Einstein's well-known equation for the energy of a particle ($E = mc^2$).

Since the electron can have only one energy at any given moment, E in both equations must be the same, resulting in the equality

Wave-Particle Duality

$$\frac{hc}{\lambda} = h\nu = mc^2 \quad (1.5)$$

This shows the obvious relationship between the particle (mass) and wave (ν or frequency) nature of the electron. Since all particles do not move at the speed of light, c, we use the symbol v for the velocity of the particle to get $\frac{h}{\lambda} = mv$.

Describing the motion of an electron as a wave required the use of complex "wave equations." Actual use of these wave equations is left for higher level college chemistry courses. However, it is important to understand the results of using these wave equations, which can be summarized as follows:

1. The wave equations require three numbers, called quantum numbers, in order to reach a solution. They are the principal quantum number, n, the azimuthal quantum number, ℓ, and the magnetic quantum number, m_ℓ. In addition, to specify an electron completely and uniquely, a fourth quantum number, called the spin quantum number, m_s, is needed. There are specific rules for assigning the four quantum numbers to electrons, but this is not required on the AP exam.
2. The wave equations changed the picture of the atom drastically (Figure 1.6). In particular, the fixed orbits of the Bohr theory are replaced with a cloud of electrons around the nucleus. The modern orbit is the region of space in which the probability of finding the electron is highest.

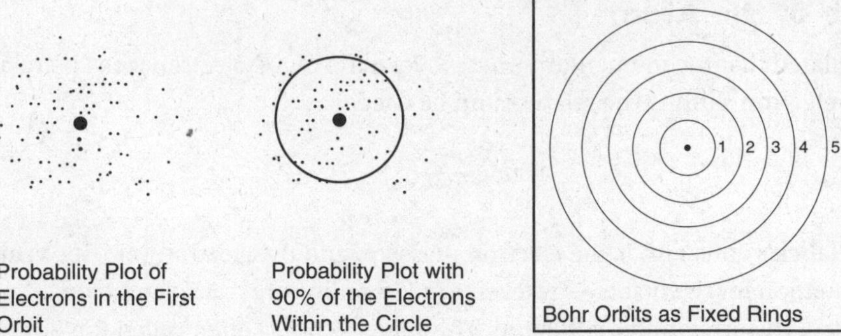

FIGURE 1.6. Comparison of the Bohr model and the wave model of the hydrogen atom.

3. The circular orbits of the Bohr theory have been replaced with spherical electron clouds. The wave equations have shown that the shapes of most electron clouds, although more complex than Bohr's orbits, are still simple geometric shapes.
4. The arrangement of electrons deduced from wave equations agrees well with the periodic table. Many physical and chemical properties of elements and compounds are more fully understood with the knowledge gained about the electronic structure and orbit shape.
5. The results of the wave equations agree completely with the Bohr model. Specifically, the energy change for an electron moving from one electron cloud to another agrees with Bohr's calculations. In addition, the identical 53 pm radius is found for the electron cloud in the wave model of the hydrogen atom.
6. The Heisenberg uncertainty principle is fundamental to the wave model of the atom. This principle states that both the position and the momentum of an electron cannot be exactly known at the same time. The more precisely that the position, *x*, of the electron is known, the more uncertainty exists as to its momentum, *mv*. Heisenberg's uncertainty equation is as follows:

Heisenberg Uncertainty Principle

$$(\Delta x)(\Delta mv) \leq \frac{h}{4\pi} \tag{1.6}$$

This equation reads, "The uncertainty in the position times the uncertainty of the momentum is equal to Planck's constant divided by four pi." The concept of this equation is important, although *no questions involving a numerical solution will be asked on the AP Chemistry exam.*

Structure of the Atom

PRINCIPAL ENERGY LEVELS (SHELLS)

In the current model of the atom the positively charged nucleus is surrounded by one or more principal energy levels or electron clouds. Principal energy levels may also be called the principal shells with reference to Bohr's atomic model. In either case, the principal energy level, or principal shell, nearest the nucleus is assigned the number 1, and each succeeding energy level is numbered with consecutive integers. The largest element known needs only seven principal energy levels to hold all of its electrons. The number of the principal energy level is given the symbol *n*, and it is the same as the principal quantum number discussed later.

Since the principal energy levels become larger the further they are from the nucleus, they can hold correspondingly more electrons. Each can hold a maximum number of electrons equal to $2n^2$, where n is the principal energy level number. From this fact we can calculate that the first four principal energy levels can hold 2, 8, 18, and 32 electrons, respectively. The last three could hold 50, 72, and 98 electrons, but they are not completely filled.

SUBLEVELS (SUBSHELLS)

Each principal energy level within an atom contains one or more sublevels. These sublevels may be called subshells in the older terminology. The number of sublevels possible in each principal energy level is equal to the value of n for that energy level. For instance, the third principal energy level ($n = 3$) may contain a maximum of three sublevels. For the 118 known elements, only four sublevels are actually used. The fifth, sixth, and seventh sublevels (for $n = 5$, 6, and 7) are theoretically possible but are not currently needed.

Sublevels are numbered with consecutive whole numbers starting with zero. These numbers are the azimuthal quantum numbers, ℓ. The value of ℓ can never be greater than $n - 1$. In addition to the numbering system, the sublevels are also given corresponding letters of *s*, *p*, *d*, and *f*. Table 1.2 shows the sublevels possible for each energy level. It is important to remember that a sublevel will not exist unless the atom has enough electrons to occupy at least part of the sublevel.

Each principal energy level has an *s* sublevel, all except the first have *p* sublevels, and the *d* and *f* sublevels also are found in more than one principal energy level. To distinguish one sublevel from another, chemists usually combine the principal energy level number with the sublevel letter in order to indicate in which principal energy level the sublevel is located. With this method, the designation $4p$ indicates a *p* sublevel in the fourth principal energy level.

TABLE 1.2
Sublevels in the Atom

Principal Level, n	Sublevel Number, ℓ	Sublevel Letter
1	0	*s*
2	0, 1	*s*, *p*
3	0, 1, 2	*s*, *p*, *d*
4	0, 1, 2, 3	*s*, *p*, *d*, *f*
5*	0, 1, 2, 3	*s*, *p*, *d*, *f*
6*	0, 1, 2	*s*, *p*, *d*
7*	0, 1	*s*, *p*

*Only sublevels used by the known elements are shown here.

TIP

The value of *n* also indicates the number of sublevels within each principal energy level.

ORBITALS

Each sublevel of the atom may contain one or more electron orbitals. An orbital is defined as a region of space that has a high electron density, and each orbital may contain a maximum of two electrons. To share an orbital, two electrons must have opposite spins. When two electrons share an orbital, they are said to be paired. Orbitals are designated as *s*, *p*, *d*, or *f* according to the sublevel they are in.

TIP

Orbitals can hold only two electrons each.

The number of orbitals that a sublevel may have depends on the azimuthal quantum number, ℓ, of the sublevel and is equal to $2\ell + 1$. From this fact we see that there is one *s* orbital in an *s* sublevel, three *p* orbitals in a *p* sublevel, five *d* orbitals in a *d* sublevel, and seven *f* orbitals in an *f* sublevel. Table 1.3 summarizes this information.

TABLE 1.3
Orbitals in the Atom

Sublevel Number, ℓ <	Sublevel Letter	Number of Orbitals, $2\ell + 1$	Number of Electrons per Sublevel
0	*s*	1	2
1	*p*	3	6
2	*d*	5	10
3	*f*	7	14

This table also shows why the principal energy levels contain 2, 8, 18, and 32 electrons, respectively. The first principal energy level has only one *s* orbital and therefore holds only 2 electrons. The second principal energy level has *s* and *p* orbitals that hold 2 and 6 electrons, respectively, or 8 for that energy level. The third principal energy level holds the sum of 2 + 6 + 10 or 18 electrons, and the fourth principal energy level holds 2 + 6 + 10 + 14 or 32 electrons. Orbitals take the same letter designation as the sublevel letter (*s, p, d,* or *f*). Each orbital is given a number, called the magnetic quantum number, m_ℓ. The possible values of m_ℓ range from $-\ell$ to $+\ell$, including zero, as Table 1.4 shows.

TABLE 1.4
Possible Values of m_ℓ

Orbital	m_ℓ Values
s	0
p	−1, 0, +1
d	−2, −1, 0, +1, +2
f	−3, −2, −1, 0, +1, +2, +3

TIP

Orbital shapes are noted as a letter or a number for the quantum number, ℓ.

The designation of a sublevel also tells the chemist the shape of the orbitals within that sublevel. In *s* sublevels the shape of the electron cloud is spherical. For the *s* sublevel in the first principal energy level, the highest electron density is found within a sphere 53 pm from the nucleus, as Bohr predicted. In the second and higher numbered principal energy levels, the *s* sublevel has a high electron density at the expected distance from the nucleus ($n \times 53$ pm). These sublevels also have an appreciable electron density at all of the lower energy level radii. Figure 1.7 illustrates this feature of the *s* sublevels.

The *p* orbitals have a dumbbell shape, with the electron density being greatest in two lobes on either side of the nucleus. There are three *p* orbitals in each *p* sublevel. Each is oriented along a different axis, as shown in Figure 1.8. These orbitals may be designated as p_x, p_y, and p_z.

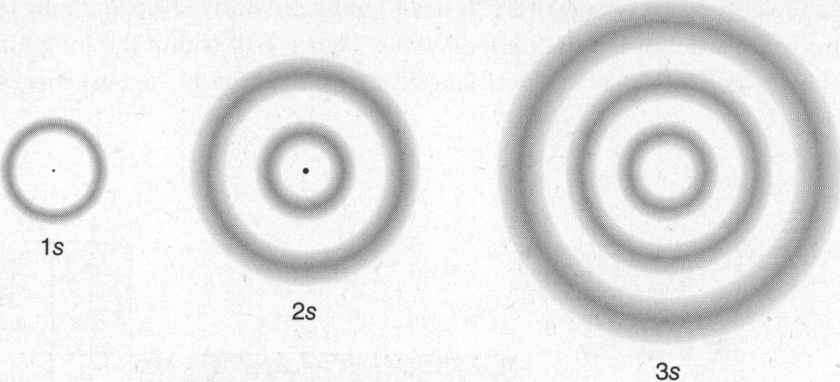

FIGURE 1.7. Diagrams of the 1***s***, 2***s***, and 3***s*** orbitals, showing that the highest radial density for the 2***s*** and 3***s*** orbitals occurs not only at the expected radius but also at intermediate levels.

TIP

Orbital orientation in space is specified by m_ℓ.

The five *d* orbitals in a *d* sublevel have the shapes shown in Figure 1.9. They often have subscripts indicating the general locations of the orbitals on the *x*-, *y*-, and *z*-axes.

The seven *f* orbitals are slightly more complex in shape than the *d* orbitals. A knowledge of the exact shapes is not needed at this time.

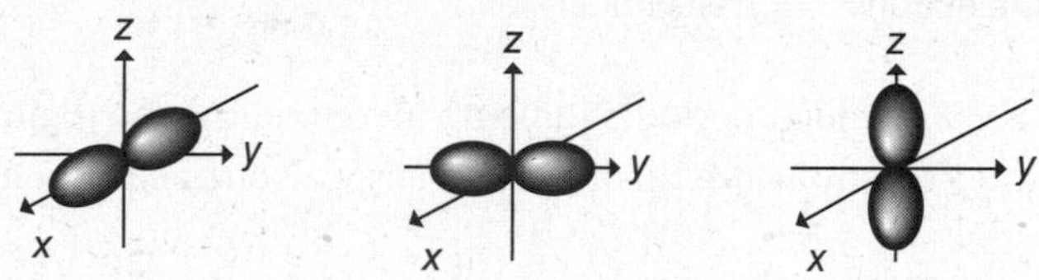

FIGURE 1.8. Diagrams of the three ***p*** orbitals aligned with the ***x***-, ***y***-, and ***z***-axes.

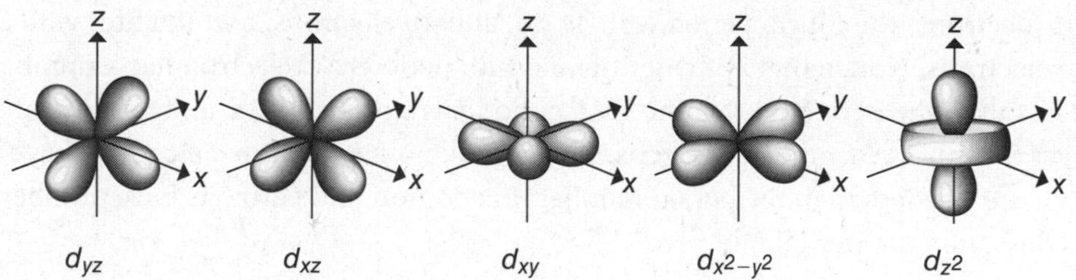

FIGURE 1.9. Shapes of the five ***d*** orbitals in the atom. Four of the shapes are identical; the fifth is very distinctive.

Electronic Structure of the Atom

At this point we need to see how the information given above is used to develop the complete picture of an atom. The underlying principle of how the electrons are arranged is based on the energy of each orbital. Electrons will fill the orbitals that have the lowest energy first, much as water always flows downhill. While the numerical values of the energies of the orbitals are not important here, the order, from lowest to highest energy, is important. That sequence

1s 2s 2p 3s 3p 4s 3d 4p 5s 4d 5p 6s 4f 5d 6p 7s 5f 6d

is known as the aufbau, or energy order. It need not be memorized since it may be quickly obtained from the structure of the periodic table. Figure 1.10 shows the long form of the periodic table divided into shaded blocks labeled *s*, *p*, *d*, and *f* for the highest energy electron in each atom.

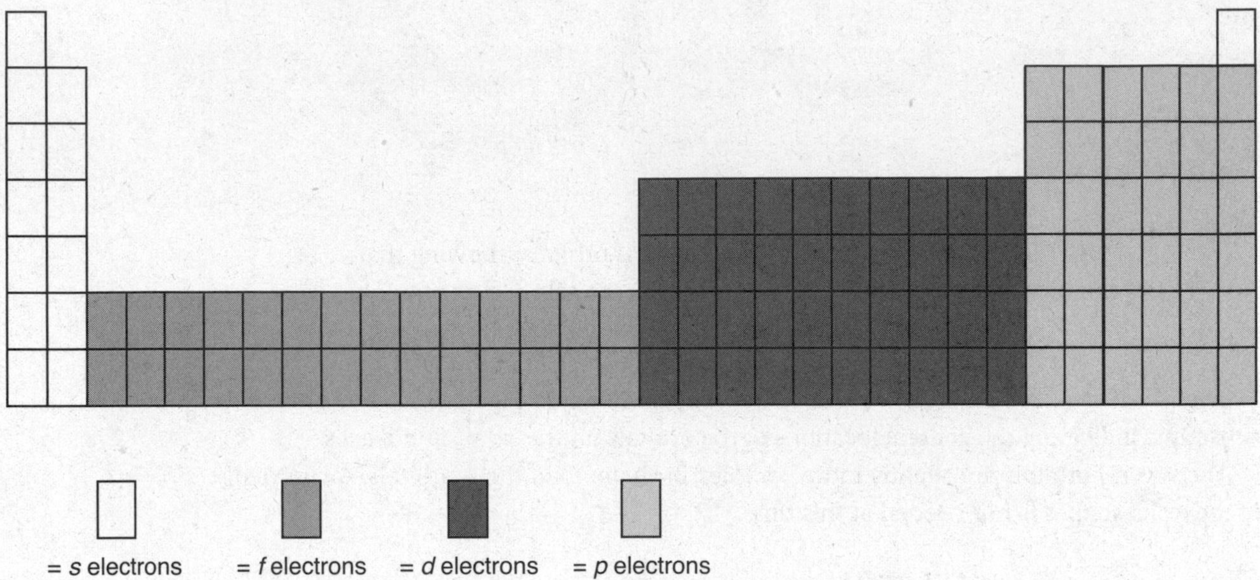

FIGURE 1.10. Expanded periodic table divided to show the regions where ***s***, ***p***, ***d***, and ***f*** electrons are the highest energy electrons in each atom.

Reading across this periodic table, we can see that the energy ordering is an integral part of this table. The first period represents the 1*s* electrons. Filling the second period adds the 2*s* electrons on the left and then 2*p* electrons on the right. The third period fills with 3*s* and then 3*p* electrons, the fourth period with 4*s*, 3*d*, and 4*p* electrons, and the fifth with 5*s*, 4*d*, and 5*p* electrons. Notice that in the fourth and fifth periods a *d* electron has a number that is one less than the period it is in. The sixth period fills with 6*s*, 4*f*, 5*d*, and 6*p* electrons, and the seventh with 7*s*, 5*f*, and 6*d* electrons. Once again, notice that the *d* electrons have numbers that are one less than the period number. In addition, the *f* orbitals have numbers that are two less than the period number.

ELECTRONIC CONFIGURATIONS

Once the energy order of the orbitals is known, the electrons in any atom may be described as the electronic configuration. Chemists use two forms of this configuration to display information. The first is the complete electronic configuration, which lists all of the electrons in the atom. The second is an abbreviated version that lumps all of the inner electrons together and lists only the highest energy electrons.

COMPLETE ELECTRONIC CONFIGURATIONS, $n\ell^x$

In complete electronic configurations the electrons present are listed by designating the principal energy level (n) by number, the sublevel (ℓ) by letter, and the number of electrons (x) in each sublevel, x. Atoms are "built up" by adding electrons to the lowest energy sublevel possible.

To obtain an electronic configuration for an element, the number of electrons is determined from the element's atomic number. Then these electrons are placed into the sublevels, completely filling one sublevel before starting to fill the next one. Several examples are shown in Table 1.5.

TABLE 1.5

Electronic Configurations of Selected Elements

Element	Complete Electronic Configuration
sodium, Na	$1s^2 2s^2 2p^6 3s^1$
lead, Pb	$1s^2 2s^2 2p^6 3s^2 3p^6 4s^2 3d^{10} 4p^6 5s^2 4d^{10} 5p^6 6s^2 4f^{14} 5d^{10} 6p^2$
radon, Rn	$1s^2 2s^2 2p^6 3s^2 3p^6 4s^2 3d^{10} 4p^6 5s^2 4d^{10} 5p^6 6s^2 4f^{14} 5d^{10} 6p^6$
antimony, Sb	$1s^2 2s^2 2p^6 3s^2 3p^6 4s^2 3d^{10} 4p^6 5s^2 4d^{10} 5p^3$
cobalt, Co	$1s^2 2s^2 2p^6 3s^2 3p^6 4s^2 3d^7$
chlorine, Cl	$1s^2 2s^2 2p^6 3s^2 3p^5$

IMPORTANT EXCEPTIONS TO AUFBAU ORDERING

The above examples follow the aufbau order in the filling of the sublevels. Many elements, however, do not follow the aufbau ordering, as shown in the table of electron configurations in Appendix 2. The only exceptions to the aufbau ordering that are important at this time occur in the electron configurations for chromium, molybdenum, copper, silver, and gold. These are listed in Table 1.6. It is apparent that the completely filled *d* sublevel of Cu, Ag, and Au confers stability to the atom at the expense of "unfilling" the previous *s* sublevel. For Cr and Mo, it appears that a half-filled *d* sublevel also confers stability, once again at the expense of a previously filled *s* sublevel. AP students do not need to memorize exceptions but do need to be able to explain them.

TABLE 1.6

Electronic Configurations of Selected Elements

Element	Electronic Configuration
copper, Cu	$1s^2 2s^2 2p^6 3s^2 3p^6 4s^1 3d^{10}$
silver, Ag	$1s^2 2s^2 2p^6 3s^2 3p^6 4s^2 3d^{10} 4p^6 5s^1 4d^{10}$
gold, Au	$1s^2 2s^2 2p^6 3s^2 3p^6 4s^2 3d^{10} 4p^6 5s^2 4d^{10} 5p^6 6s^1 4f^{14} 5d^{10}$
chromium, Cr	$1s^2 2s^2 2p^6 3s^2 3p^6 4s^1 3d^5$
molybdenum, Mo	$1s^2 2s^2 2p^6 3s^2 3p^6 4s^2 3d^{10} 4p^6 5s^1 4d^5$

ABBREVIATED ELECTRONIC CONFIGURATIONS

When atoms react with each other, their first point of contact is the outermost, highest energy electrons of each atom involved. In fact, the inner core of lower energy electrons and the nucleus play virtually no role in most chemical reactions. These inner electrons may be represented by the noble gas at the end of the period just before the period containing the element of interest. In a complete electronic configuration, all the electrons up to the last completely filled p^6 sublevel are the inner or core electrons. They may be replaced by the symbol, in brackets, for the appropriate noble gas. For iron, the complete and abbreviated electronic configurations are as follows:

$[Ar] = 1s^2\ 2s^2\ 2p^6\ 3s^2\ 3p^6$

$$Fe = 1s^2\ 2s^2\ 2p^6\ 3s^2\ 3p^6\ 4s^2\ 3d^6$$
$$Fe = [Ar]\ 4s^2\ 3d^6$$

Notice that the last completely filled *p* orbital is the 3*p* and that argon has the electronic configuration of all of the electrons up to and including the $3p^6$ electrons.

VALENCE ELECTRONS

In many instances chemists are interested in the outermost electrons in an atom. These are the valence electrons. In a practical sense, these include only *s* and *p* electrons of the atom. For any given atom the principal energy level of the outer *d* electrons will always be one less than the principal energy level of the *s* and *p* electrons. Similarly, for the *f* electrons the principal energy level is always two less than the outer *s* and *p* electrons.

How to Count Valence Electrons

Determining the number of valence electrons for any atom involves counting the groups (columns) from the left of the periodic table to the element of interest. The *d* and *f* electrons shown in Figure 1.10 are not counted. Valence electrons are often shown as dots surrounding the symbol of an atom, as shown in Figure 1.11.

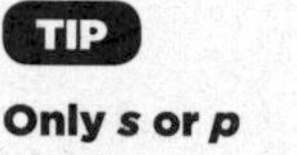

Only *s* or *p* electrons count as valence electrons.

H							He
Li	Be	B	C	N	O	F	Ne
Na	Mg	Al	Si	P	S	Cl	Ar

FIGURE 1.11. Representation of valence electrons as dots around the atomic symbol, for the first three periods in the periodic table.

You may often see the electrons unpaired in Groups IIA, IIIA, and IVA.

HUND'S RULE

Notice that the second valence electron (second column in Figure 1.11) is represented as a pair rather than placing them on opposite sides of the symbol. This is done to indicate that the electrons are paired in the completed *s* orbital. The third through eighth columns fill the *p* orbitals. Hund's rule requires that *p, d,* or *f* orbitals in a sublevel must all be filled with one electron each before a second electron is allowed to pair in any orbital. The three separate *p* orbitals fill with one electron each in boron, carbon, and nitrogen, as shown by the unpaired dots in Figure 1.11. The electrons for oxygen, fluorine, and neon then form pairs until all of the *p* orbitals are filled. Electrons will begin to pair up only if every orbital in the sublevel is first occupied with one electron.

ORBITAL DIAGRAMS

Electronic configurations and valence electrons are useful for most purposes in describing the structure of the atom. However, since all of the different orbitals in a sublevel are lumped together, some detail is lost. To see that detail, orbital diagrams, which show each of the orbitals in the valence shell of the atom as a box or circle, are often used. Arrows, representing electrons, are placed in each orbital. The second electron in each orbital has the arrow facing in the opposite direction from the first, indicating that their spins are paired. Figure 1.12 shows the three possible situations for an orbital.

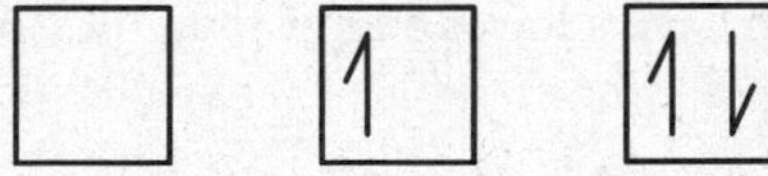

FIGURE 1.12. Orbital boxes that represent an empty orbital, an unpaired electron, and a pair of electrons, respectively.

Orbital diagrams are used mainly to describe the valence electrons since all of the inner electrons will be paired. At times, the *d* electrons are also shown in these diagrams. The orbital diagrams of the first ten elements are shown in Figure 1.13. The electrons are added to each sublevel, starting with the 1*s* sublevel. The arrow pointing upward traditionally represents the first electron in each orbital until each orbital in a sublevel contains one electron. Then the downward arrows, representing electrons of opposite spin, are added to complete the sublevel before a new sublevel starts to fill.

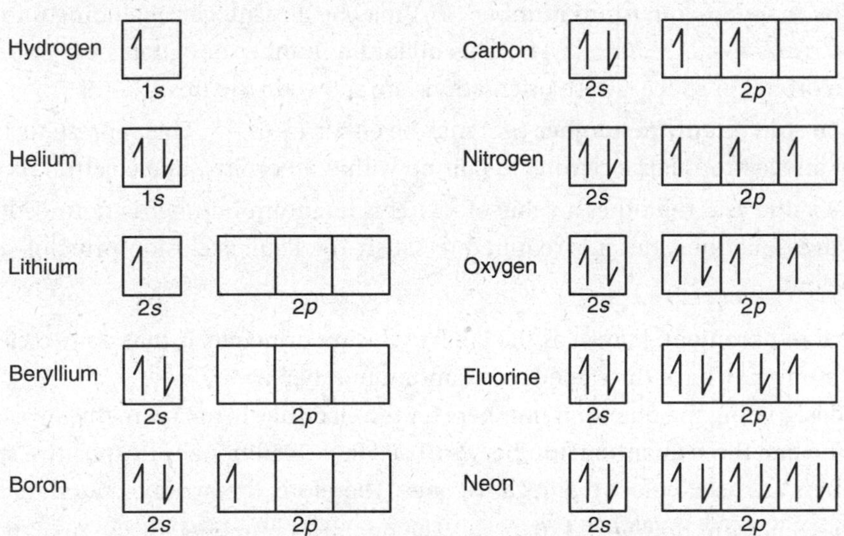

FIGURE 1.13. Orbital diagrams for the first ten elements, showing only the valence electrons.

In later chapters we will want to show energy differences between the orbitals along with the orbital diagrams. This can be done by drawing the orbitals as shown in Figure 1.14 to indicate that the $2p$ orbitals are higher in energy than the $2s$ orbitals.

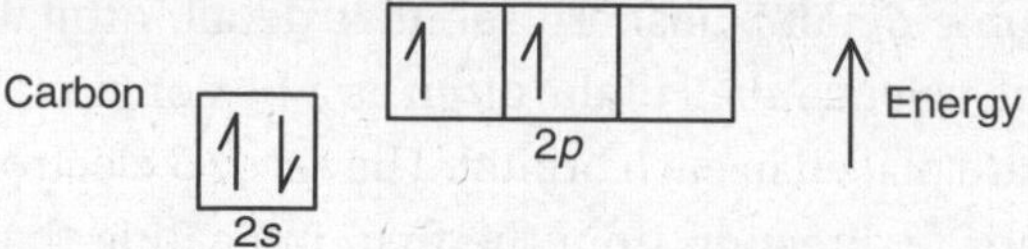

FIGURE 1.14. Orbital diagram showing that the $2s$ electrons in carbon have a lower energy than the $2p$ electrons.

Quantum Numbers

Erwin Schrödinger developed the wave equations describing the probabilities of where the electrons are located in the atom. As mentioned earlier, these equations require three integers, called quantum numbers, which describe each electron. The quantum numbers have definite rules for their possible values:

1. The **principal quantum number (n)** may have any integer value starting from 1. This represents the principal energy level of the atom in which the electron is located and is related to the average distance of the electron from the nucleus.
2. The **azimuthal quantum number (ℓ)** may have any number from 0 up to 1 less than the current value of n ($\ell = 0, \ldots, n - 1$). This designates the sublevel of the electron and also represents the shape of the orbitals in the sublevel.
3. The **magnetic quantum number (m_ℓ)** may be any integer, including 0 from $-\ell$, to $+\ell$. ($m_\ell = -\ell, \ldots, 0, \ldots, +\ell$.) This quantum number designates the orientation of an orbital in space. These orientations are shown in Figures 1.7–1.9.
4. The **spin quantum number (m_s)** may be either $+\frac{1}{2}$ or $-\frac{1}{2}$. This represents the "spin" of an electron. For electrons to pair up within an orbital, one electron must have a $+\frac{1}{2}$ value and the other a value of $-\frac{1}{2}$. This quantum number is not needed for the wave equations, but it is required to satisfy the Pauli exclusion principle described below.

The final requirement, known as the Pauli exclusion principle, is that no two electrons in the same atom may have the same four quantum numbers.

When designating the quantum numbers for the electrons in an atom, the lowest possible values of the first three quantum numbers are used first. Traditionally the positive spin quantum numbers are used before the negative ones. Therefore, the hydrogen atom has the lowest possible quantum numbers, 1, 0, 0, $+\frac{1}{2}$. The quantum numbers for the first 20 elements are listed in Table 1.7. Writing quantum numbers for a given electron is not required on the AP exam; however, it is important to know what an appropriately formulated set of quantum numbers looks like according to the above rules.

TABLE 1.7
Quantum Numbers for the Last Electron Added to Each of the First 20 Elements Using the Rules Above

Element	n	ℓ	m_ℓ	m_s
H	1	0	0	$+\frac{1}{2}$
He	1	0	0	$-\frac{1}{2}$
Li	2	0	0	$+\frac{1}{2}$
Be	2	0	0	$-\frac{1}{2}$
B	2	1	-1	$+\frac{1}{2}$
C	2	1	0	$+\frac{1}{2}$
N	2	1	+1	$+\frac{1}{2}$
O	2	1	-1	$-\frac{1}{2}$
F	2	1	0	$-\frac{1}{2}$
Ne	2	1	+1	$-\frac{1}{2}$
Na	3	0	0	$+\frac{1}{2}$
Mg	3	0	0	$-\frac{1}{2}$
Al	3	1	-1	$+\frac{1}{2}$
Si	3	1	0	$+\frac{1}{2}$
P	3	1	+1	$+\frac{1}{2}$
S	3	1	-1	$-\frac{1}{2}$
Cl	3	1	0	$-\frac{1}{2}$
Ar	3	1	+1	$-\frac{1}{2}$
K	4	0	0	$+\frac{1}{2}$
Ca	4	0	0	$-\frac{1}{2}$

Exercise 1.1

Designate each of the following sets of quantum numbers as possible or impossible according to the above rules.

(a) $1, 0, 0, \frac{1}{2}$

(b) $1, 3, 0, \frac{1}{2}$

(c) $3, 2, 0, \frac{1}{2}$

(d) $2, 2, 2, -\frac{1}{2}$

(e) $3, 2, 2, -\frac{1}{2}$

(f) $3, 1, -1, \frac{1}{2}$

(g) $4, 2, -2, -\frac{1}{2}$

(h) $4, 4, 0, \frac{1}{2}$

(i) $3, 2, 1, 0$

(j) $1, 1, 1, \frac{1}{2}$

(k) $6, 4, -4, -\frac{1}{2}$

(l) $5, 3, -2, \frac{1}{2}$

(m) $2, 0, 1, -\frac{1}{2}$

(n) $5, 0, 0, \frac{1}{2}$

(o) $3, 1, 2, -\frac{1}{2}$

Solution

(a), (c), (e), (f), (g), (k), (l), and (n) are valid. The others disobey one or more of the rules on page 98 as follows: (b), (d), (h), and (j) violate rule 2; (m) and (o) violate rule 3; (i) violates rule 4.

SIGNIFICANCE OF THE QUANTUM NUMBERS

The four quantum numbers are often looked upon as more data to be memorized along with the rules for obtaining valid sets. It is important to remember, however, that these numbers represent real physical properties of the atom that may be summarized in simplified form as follows:

1. The principal quantum number, n, represents the average distance of the electron from the nucleus, or the size of the principal energy level.
2. The azimuthal quantum number, ℓ, represents the shape(s) of the orbitals within the sublevel, as shown in Figures 1.7, 1.8, and 1.9.
3. The magnetic quantum number, m_ℓ, represents the orientation of each orbital in space.
4. The spin quantum number, m_s, represents the "spin" of the electron.

SUMMARY

This chapter has briefly reviewed the structure of the atom and the atomic theory. This includes some history about how the electron, proton, and neutron were discovered and characterized. The use of spectroscopy to discover the structure and energy levels of electrons within the atom is described along with the modern designation *s* of electron configuration. The use and meaning of the four quantum numbers is finally introduced. You should be able to discuss these discoveries and principles and work problems involving the electronic structure of the atom.

Important Concepts

Bohr model of the atom
Quantum mechanical model of the atom
Electronic configurations
Atomic theory

Important Equations

$$\lambda\nu = c$$
$$E = h\nu$$

PRACTICE EXERCISES

Multiple-Choice

1. Albert Einstein was given the Nobel Prize for the discovery and explanation of the photoelectric effect. What is this effect?
   (A) It is spectroscopy using visible light.
   (B) It is the ejection of electrons from a surface by photons.
   (C) It measures $E = mc^2$ for any element.
   (D) It is another way of stating the Heisenberg uncertainty principle.

2. Which orbital diagram represents the outer electrons of the ground state of an element?

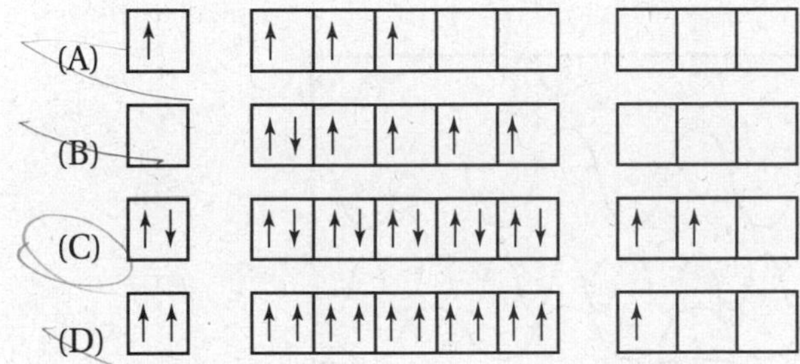

3. The wave-particle duality of nature applies to only
   (A) everything
   (B) light because of its wave properties
   (C) gamma rays since they are the most energetic electromagnetic radiation
   (D) electrons because they determine the chemical nature of the elements

4. Which of the following depicts a p orbital?

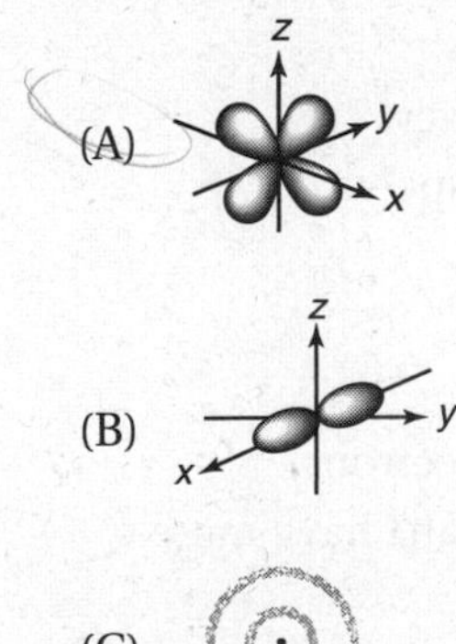

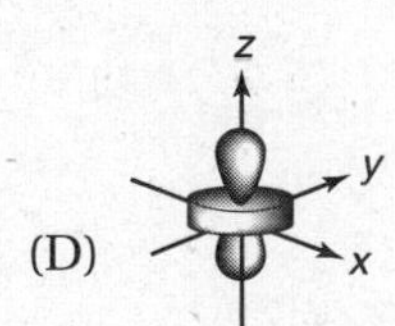

5. Calcium can have an electron in an *f* orbital
   (A) if the calcium is in the elemental state
   (B) if the calcium is in an excited state
   (C) if the calcium is a positive ion
   (D) if the calcium is a negative ion

6. What is the complete electron configuration for silicon?
   (A) [Ne] $3s^2\ 3p^2$
   (B) $1s^2\ 2s^2\ 2p^6\ 3s^2\ 3p^4$
   (C) $1s^2\ 2s^2\ 2p^6\ 3s^2\ 3p^6\ 4s^2\ 3d^{10}\ 4p^4$
   (D) $1s^2\ 2s^2\ 2p^6\ 3s^2\ 3p^2$

7. Which electromagnetic wave has the highest energy? The *x*-axis is time, and the *y*-axis is amplitude.

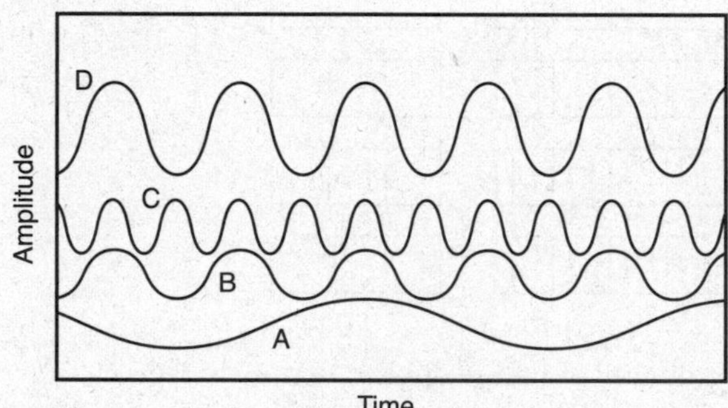

   (A) Wave *A*
   (B) Wave *B*
   (C) Wave *C*
   (D) Wave *D*

8. What is the wavelength of light that has a frequency of 4.00×10^{14} s^{-1}? (The speed of light is 3.00×10^8 m s^{-1}.)
   (A) 7.5 nm because this is the result of speed divided by frequency
   (B) 1,333 nm because this is the result of frequency divided by speed
   (C) 750. nm when the correct metric prefix is used
   (D) 1.33 cm^{-1} since the speed of light should have units of cm/s

9. Which of the following elements has the greatest number of *p* electrons?
   (A) Arsenic since it has the highest atomic number the others can't have more
   (B) Silicon since it is alphabetically last in this group
   (C) Iron since it has the most *d* electrons
   (D) Chlorine since it has 5 *p* electrons in its period

10. Which of the following shows the valence electrons of an uncharged atom of Sb?

(A) $\mathrm{X}\cdot$

(B) $\cdot\overset{\cdot}{\underset{\cdot}{\mathrm{X}}}\cdot$

(C) $\cdot\overset{\cdot\cdot}{\underset{\cdot}{\mathrm{X}}}\cdot$

(D) $\cdot\overset{\cdot\cdot}{\underset{\cdot\cdot}{\mathrm{X}}}\cdot$

11. Which region of the expanded periodic table has elements that have *d* electrons listed last in their electron configurations?

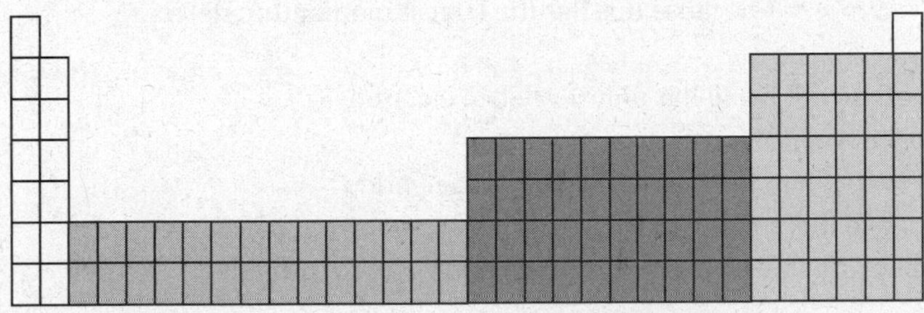

(A) □

(B) ■

(C) ■

(D) ■

12. The numbers of electrons, protons, and neutrons, respectively, in the ^{31}P isotope are

(A) 15, 31, 15

(B) 15, 15, 31

(C) 31, 15, 16

(D) 15, 15, 16

13. Which element has an electronic configuration that has the largest number of unpaired electrons?

(A) Fe because it is in the middle of the *d* block and has 4 unpaired electrons

(B) Al since it forms the Al^{3+} ion and has all unpaired electrons

(C) Ag because silver has 9 *d* electrons and 9 unpaired electrons

(D) Ni since it has 2 unpaired electrons

14. Which electronic configuration corresponds to that of a noble gas?
   (A) $1s^2\ 2s^2\ 2p^6\ 3s^2\ 3p^6\ 4s^1$ because it has the most energy levels
   (B) $1s^2\ 2s^2\ 2p^6\ 3s^2\ 3p^4$ since it ends with *s* and *p* electrons
   (C) $1s^2\ 2s^2\ 2p^6\ 3s^2\ 3p^6$ since it has the $ns^2\ np^6$ structure of a noble gas
   (D) $1s^2\ 2s^2\ 2p^6\ 3s^1$ since it has the characteristic *ns np* structure of a noble gas

15. Which electronic transition requires the addition of the most energy?
   (A) $n=1$ to $n=3$ because this has the largest energy increase
   (B) $n=5$ to $n=2$ since this has the largest size decrease
   (C) $n=2$ to $n=3$ because this has the smallest mass increase
   (D) $n=4$ to $n=1$ because this has the largest momentum decrease

16. Which of the following has only 5 valence electrons?
   (A) Rb since *n* is 5
   (B) C since it has 5 places to the end of the period
   (C) Si since it is a metalloid
   (D) P because it has 5 electrons in the outer *s* and *p* orbitals

17. Which was used to determine the charge of the electron?
   (A) The gold foil experiment
   (B) Deflection of cathode rays by electric and magnetic fields
   (C) The oil drop experiment
   (D) The mass spectrometer

18. Which of the following principles is NOT part of Dalton's atomic theory?
   (A) Chemical reactions are simple rearrangements of atoms.
   (B) Atoms follow the law of multiple proportions.
   (C) Each atom of an element is identical to every other atom of that element.
   (D) All matter is composed of atoms.

19. Which statement is correct?
   (A) Technetium, Tc, has 15 *p* electrons.
   (B) Argon has 18 valence electrons.
   (C) Oxygen has four 4 electrons.
   (D) Tungsten has no *f* electrons.

20. Which quantum number describes the shape of an orbital?
   (A) The principal quantum number, *n*
   (B) ℓ, which is the second term in an electron configuration
   (C) m_ℓ because it tells the number of directions to which the suborbitals point
   (D) m_s, which is the spin quantum number

21. You have just discovered a new, fundamental particle of nature. When measuring its mass, you obtain the following data for four samples:

4.72×10^{-34} gram
9.44×10^{-34} gram
1.18×10^{-33} gram
1.65×10^{-33} gram

Millikan found the mass of an electron by finding the common factor in a mass of data. What is the common factor in the data presented?

(A) 4.72×10^{-34} g
(B) 1.18×10^{-34} g
(C) 9.44×10^{-34} g
(D) 2.36×10^{-34} g

CHALLENGE

22. Which of the following is FALSE?
(A) The 4*d* orbitals are in the fourth period of the periodic table.
(B) The 7*s* orbitals are in the seventh period of the periodic table.
(C) The 4*f* orbitals are in the sixth period of the periodic table.
(D) The 6*s* orbitals are spherical in shape.

23. The *f* sublevel may contain a maximum of
(A) 2 electrons
(B) 14 electrons
(C) 6 electrons
(D) 10 electrons

24. The valence electrons are
(A) all electrons in an atom beyond the preceding noble gas
(B) all outermost electrons in a sublevel
(C) *s* and any *p* electrons in the highest energy level or shell
(D) electrons in the last unfilled sublevel

25. Which equation best expresses the energy of a photon?
(A) $E = \frac{1}{2}mv^2$
(B) $E = mc^2$
(C) $E = IR$
(D) $E = hv$

ANSWER KEY

1. **(B)**	6. **(D)**	11. **(C)**	16. **(D)**	21. **(D)**
2. **(C)**	7. **(C)**	12. **(D)**	17. **(C)**	22. **(A)**
3. **(A)**	8. **(C)**	13. **(A)**	18. **(B)**	23. **(B)**
4. **(B)**	9. **(A)**	14. **(C)**	19. **(D)**	24. **(C)**
5. **(B)**	10. **(C)**	15. **(A)**	20. **(B)**	25. **(D)**

See Appendix 1 for explanations of answers.

Free-Response

Answer the following question concerning the structure of the atom, the use of light as a probe of atomic structure, and the development of the modern quantum mechanical view of the atom.

(a) Describe the significant experiments that led to the discovery of the properties of the electron, proton, and neutron. Describe the results and the interpretation of results that contributed to development of the atomic model.
(b) The bond energy of a carbon-carbon bond is approximately 350 kJ/mol. Explain why electromagnetic radiation with wavelengths shorter than those in the visible region of the spectrum is often called ionizing radiation.
(c) Sketch an outline of the periodic table and discuss how the four quantum numbers n, ℓ, m_ℓ, and m_s relate to the periodic table.
(d) If an electron drops from the fifth energy level to the second energy level in a hydrogen atom, will energy be released or absorbed? Will the energy of the electron increase or decrease? Explain your answers.

ANSWERS

(a) Mention the cathode ray tube that verified the charge, particle nature, and e/m ratio of the electron (using magnetic and electric fields). The Millikan oil drop experiment was used to determine the electron charge. Mention Rutherford's alpha-particle experiment to discover the positively charged nucleus. Moseley used X-rays to determine atomic numbers, and Chadwick discovered the neutron.
(b) First we calculate the wavelength of light that has energy equivalent to the C–C bond. First calculate the energy per C–C bond by dividing 350 kJ/mol by 6.02×10^{23} to get 5.81×10^{-19} J. Use the relationship $E = \frac{hc}{\lambda}$ to determine the wavelength as

$$\lambda = \frac{hc}{E} = \frac{(6.63 \times 10^{-34}\,\text{J s})(3.00 \times 10^{8}\,\text{m s}^{-1})}{(5.81 \times 10^{-19}\,\text{J})}$$

$$= 3.42 \times 10^{-7}\,\text{m} = 342\text{ nm}$$

This calculation demonstrates that a photon with a wavelength of 342 nm has enough energy to break a C–C bond. The 342-nm wavelength is at the upper end of the ultraviolet region of the spectrum, and wavelengths below 342 nm have even more energy for breaking (ionizing) bonds.

(c) You want to point out that n represents the period of the periodic table while also noting that the d and f electrons have n quantum numbers that are 1 and 2 units less than the period in which they are found. Next, the ℓ quantum number represents various "blocks" within the periodic table similar to what is shown in Figure 1.10. Finally, the possible values of m_ℓ and m_s result in the number of elements that can occupy each of the blocks mentioned. These blocks can also be designated with the letters *s*, *p*, *d*, and *f*.

(d) The electron will emit light energy when moving from the fifth to the second energy level. In the second energy level, the electron is closer to the nucleus and has less energy than it did in the fifth level. To look at it in another way, if the electron moves from the second to the fifth level, energy must be added to overcome some of the coulombic attractive force between the negative electron and the positive nucleus. When the electron moves in the opposite direction, energy must be released.

The Periodic Table

2

- → PERIODIC TABLE
- → ATOMIC SYMBOLS
- → ATOMIC MASSES
- → ATOMIC NUMBERS
- → ISOTOPE MASSES, MASS NUMBER
- → METALS, NONMETALS, AND METALLOIDS
- → IONIZATION ENERGY
- → ELECTRON AFFINITY
- → DIAGONAL VARIATION OF PHYSICAL PROPERTIES

BIG IDEA 1

Learning Objectives: 1.9, 1.10, 1.11

For the complete list of Big Ideas and Learning Objectives, refer to the AP Chemistry Course Outline: ***https://secure-media.collegeboard.org/digitalServices/pdf/ap/ap-chemistry-course-and-exam-description.pdf***

Chemists repeatedly refer to the **periodic table** to find specific information about the elements and to understand how the properties of the various elements are related to each other. In the table we find chemical and physical similarities between elements in the same group. We also find trends where these properties vary regularly.

This chapter reviews the periodic relationships that are necessary for a complete understanding of chemistry.

THE MODERN PERIODIC TABLE

The periodic table summarizes a large amount of information useful to chemists and will be referred to many times throughout this book. A complete periodic table, updated in 2016 and similar to the one given with the AP Chemistry exam, can be found on page 528. Basic information about its construction includes the following:

1. The **symbol** for each element is shown in a separate box, in order of increasing **atomic number**, from 1 to 118. Each box shows the atomic symbol of the element with the atomic number above the symbol and the **atomic mass** below the symbol as shown here:

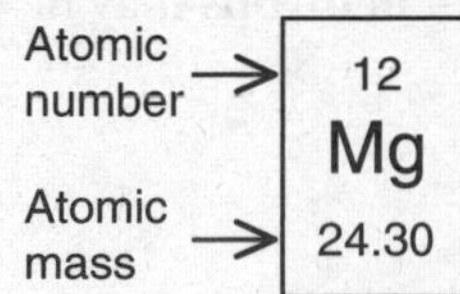

2. Each row of the periodic table is called a **period** and may contain from 2 to 32 elements. Periods with 32 elements are usually written with 14 of the elements placed below the main table.

3. Each column is called a **group,** and elements within groups have similar chemical and physical properties. Within a group, the closer elements are to each other, the more similar they are.

4. The groups of the periodic table are normally numbered; however, the groups are usually not numbered on the periodic table supplied with the AP exam.

Atomic Symbols

In the periodic table each element is designated by a one- or two-letter symbol. Most symbols for the elements are simple abbreviations of their English names. The symbols for 11 elements, however, are derived from the elements' names in other languages, mainly Latin. These elements are listed in Table 2.1, and both the symbol and the name of each element should be memorized for quick recall.

TIP

These names and symbols must be memorized.

TABLE 2.1
Non-English Chemical Symbols

sodium	Na	silver	Ag	gold	Au
potassium	K	tin	Sn	mercury	Hg
iron	Fe	antimony	Sb	lead	Pb
copper	Cu	tungsten	W		

When chemists use atomic symbols for formulas, ions, and isotopes, they add subscripts and superscripts to the four corners of the symbol to represent various features of the atom. As shown below for the element calcium, the upper left-hand corner is reserved for the mass of an isotope (mass number) of the

$$^{A}_{Z}\mathrm{Ca}^{\text{charge}}_{\text{subscript}}$$

atom designated by the symbol *A*. The lower left-hand corner is reserved for the atomic number, *Z*. In the upper right-hand corner chemists place the charge of the ion that results when the element gains or loses electrons. Subscripts used in writing chemical formulas are placed in the lower right-hand corner. These subscripts indicate how many atoms of the element are present in a formula unit of a compound. For instance, N_2 means that a molecule of nitrogen contains two atoms of nitrogen.

TIP

The four corners of an atomic symbol have specific uses.

Electrons, Protons, and Neutrons

The periodic table may be used to quickly determine the number of **electrons** and **protons** in a particular element. The number of **neutrons** may be calculated only if a specific isotopic mass of an element is known.

PROTONS

For any given element, the number of protons is *always* equal to the atomic number, *Z*, of the element:

$$\text{Number of protons} = Z \qquad (2.1)$$

ELECTRONS

For any given element, the number of electrons is equal to the atomic number:

$$\text{Number of electrons} = Z \tag{2.2}$$

For an ion of an element, the number of electrons may be calculated as:

$$\text{Number of electrons} = Z - \text{charge of the ion} \tag{2.3}$$

A positive ion (cation) has lost electrons, and a negative ion (anion) has gained electrons, compared to the neutral element.

NEUTRONS

The number of neutrons in an element depends on the specific **isotope** of the element in question. Since the atomic masses listed in the periodic table are the **weighted averages** of all the naturally occurring isotopes, the number of neutrons in an atom cannot normally be determined from the periodic table. If the specific isotope mass number is known, however, the number of neutrons can be calculated as the difference between the isotope mass (*A*) and the atomic number (*Z*):

$$\text{Number of neutrons} = A - Z \tag{2.4}$$

Exercise 2.1

Using a periodic table and Equations 2.1–2.4, fill in the blanks in the following table:

Symbol	Atomic Number	Isotope Mass	Number of Protons	Number of Electrons	Number of Neutrons
Fe		56			
	60	144			
		102	45	45	
		59			31
Al		27			

Solution

Symbol	Atomic Number	Isotope Mass*	Number of Protons	Number of Electrons	Number of Neutrons
Fe	26	56	26	26	30
Nd	60	144	60	60	84
Rh	45	102	45	45	57
Ni	28	59	28	28	31
Al	13	27	13	13	14

*Also called the mass number

Isotopes

A given element must have a certain number of protons and electrons. However, it can have varying numbers of neutrons. Atoms with the same number of protons but different numbers of neutrons are known as **isotopes**.

RADIOACTIVITY

Subatomic Particles

Name	Symbol	Mass	Charge
Beta	${}^{0}_{-1}\beta$ or ${}^{0}_{-1}e$	0	1-
Positron	${}^{0}_{+1}\beta$	0	1+
Alpha	${}^{4}_{2}He$ or ${}^{4}_{2}\alpha$	4	2+
Proton	${}^{1}_{1}H$ or ${}^{1}_{1}p$	1	1+
Neutron	${}^{1}_{0}n$	1	0

Radioactivity is a property of matter whereby an unstable nucleus spontaneously emits small particles and/or energy in order to attain a more stable nuclear state. The process is called **radioactive decay**, and an isotope that contains an unstable nucleus is termed a **radioactive isotope** or **radioisotope**. One radioactive nucleus may decay to another radioactive nucleus, and then to another. Eventually all radioactive decay results in an isotope with a stable nucleus. Some radioactive isotopes are found to exist in nature and are called natural radioactive substances. Artificial radioactive isotopes, on the other hand, are created in the laboratory in nuclear experiments.

Radioactive isotopes, both natural and artificial, emit only a few types of **subatomic particles** as they disintegrate. These include the electron (beta particle), neutron, helium nucleus (alpha particle), and positron. When these particles are emitted in a radioactive decay process, the **nuclear mass** (nuclear mass = atomic mass = A) and/or **nuclear charge** (nuclear charge = atomic number = Z) of the nucleus changes. As a result, one isotope is converted into another with a different identity. Energy may also be released in the form of X-rays or gamma rays. The energy released does not affect the identity of the isotope since these rays have neither mass nor nuclear charge. Writing and balancing nuclear reactions will not be tested on the AP Chemistry Exam. Examples of nuclear processes, though, may appear as descriptive parts of questions or with data for analysis.

Mass Spectrometry

Mass spectrometry is a very important tool for modern chemists as it was an essential instrument for understanding matter in the early days. In the high vacuum of a mass spectrometer, tiny amounts of substances are vaporized and subjected to a beam of electrons to create ions. These ions are separated and recorded based on the mass-to-charge ratio (m/z) of each ion. The pattern of mass-to-charge ratios and the intensity of each m/e ratio is called a mass spectrum. Often when observing ions with only one charge, the m/e axis can be regarded as simply mass.

The mass spectra of the compound methanol and of the element zirconium are shown in the two figures below. For the AP course, the main use of mass spectra of compounds is to determine the molar mass. The molar mass is the peak with the largest m/z ratio. The largest peak at 31 is called the base peak and is used to define 100%. For methanol, the molar mass is 32. The tiny peak at 33 is due to the deuterium isotope of the hydrogen atoms that are present in a small fraction of the molecules. The mass spectrum for the element zirconium is shown in the figure on the right. In this representation, there are five observable isotopes of Zr. Without doing any math, see if you can estimate the average atomic mass of naturally occurring zirconium (don't peek at the periodic table).

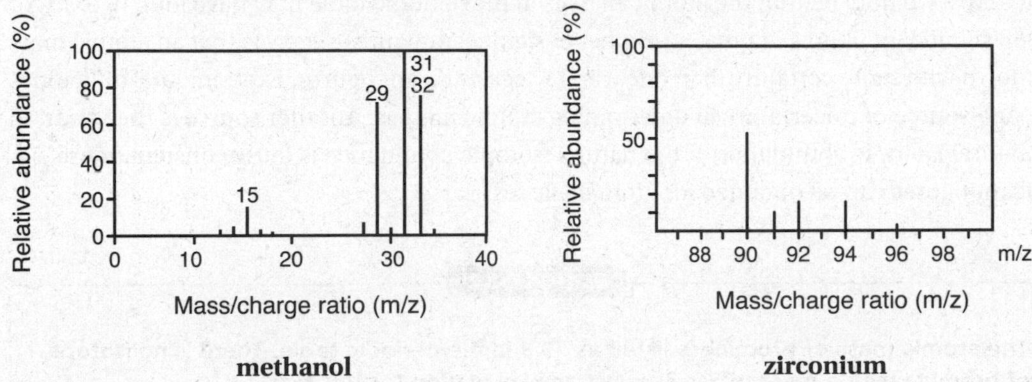

Atomic Masses

The atomic masses listed in the periodic table are based on defining the atomic mass of pure C-12 as exactly 12. All other masses are relative to that definition. Therefore Mg-24 has twice the mass of C-12. Since most elements have two or more isotopes, the masses listed in the periodic table are the weighted averages of the masses of the naturally occurring isotopes as calculated using Equation 2.5. We can see how this is done in Example 2.2.

$$\text{Weighted average} = \sum_{i=1}^{n} (\text{mass of isotope } i)(\text{abundance of isotope } i) \qquad (2.5)$$

EXAMPLE 2.1

Magnesium has three isotopes: ^{24}Mg, ^{25}Mg, and ^{26}Mg. They occur naturally with percentage abundances of 78.6%, 10.1%, and 11.3%, respectively. The exact masses of these isotopes are 23.9924, 24.9938, and 25.9898. What is the weighted average of the three isotopic masses?

Solution

The weighted average is calculated using Equation 2.5. In this equation the mass of each isotope is multiplied by the abundance of that isotope. These products are then added to obtain the weighted average:

$$\begin{aligned} \text{Weighted average} &= (\text{mass of } ^{24}\text{Mg})(\text{abundance of } ^{24}\text{Mg}) \\ &\quad + (\text{mass of } ^{25}\text{Mg})(\text{abundance of } ^{25}\text{Mg}) \\ &\quad + (\text{mass of } ^{26}\text{Mg})(\text{abundance of } ^{26}\text{Mg}) \end{aligned}$$

Since the abundances are usually given as percentages, they must be converted to fractions by dividing by 100 before using them in the equation. When the appropriate numbers are substituted, the equation becomes

$$\begin{aligned}\text{Weighted average} &= (23.9924)(0.786) + (24.9938)(0.101) + (25.9898)(0.113)\\ &= 18.86 + 2.52 + 2.94\\ &= 24.32\end{aligned}$$

You may have noticed that the atomic masses in the periodic table may have four, five, six, or seven significant figures. A greater number of significant figures suggests that an atomic mass is known with more certainty than one with fewer significant figures. Experimental difficulties are one source of uncertainty in determining atomic masses. Another source is the variation in natural isotopic abundance. If the natural isotopic abundance is fairly constant, more significant figures can be obtained for atomic masses.

EXAMPLE 2.2

The atomic mass of bromine is listed as 79.9 in the periodic table. There is no isotope of bromine with a mass of 80. Suggest an explanation for this fact.

Solution

The masses listed in the periodic table rarely represent the masses of specific isotopes. The atomic mass of bromine (79.9) can be obtained from many possible combinations of isotope masses and their relative abundances. (In this case, bromine has only two natural isotopes—^{79}Br and ^{81}Br—which occur in almost equal amounts in nature.)

PERIODIC PROPERTIES OF THE ELEMENTS

Two forms of the periodic table are shown in Figures 2.1 and 2.2. In Figure 2.1 the periodic table is arranged in the conventional manner with the **lanthanide** and **actinide** series placed below the body of the table. Figure 1.10 on page 94 places the lanthanide and actinide elements where they normally belong. However, the form in Figure 1.10 is rarely used because the boxes become too small to read.

Chemists often speak of groups of related elements such as the **alkali metals**, **alkaline earth metals**, **transition elements**, **halogens**, and **noble gases.** The locations of seven of these groupings are shown in Figure 2.1. A knowledge of the names of these groupings is important, and they will be referred to frequently in the following chapters.

Chemical reactions occur when one atom collides with another. In these collisions the outermost electrons make the first contact between the atoms. This is the main reason why elements with similar electronic structures have similar chemical properties.

A *differentiating electron* is the electron in a neutral element that makes it different from the previous element.

After the review of the electronic structure of the atom in Chapter 1, the reasons for the chemical similarities and differences of the elements become obvious. Figure 1.10 on page 94 shows the blocks of the periodic table that have *s*, *p*, *d*, and *f* electrons as the **differentiating electrons**. Each column or group has the same number and type of outermost electrons, resulting in the chemical similarities of these elements. For example, each element in Group IA has a single ns^1 valence electron. For instance, all the noble gases have completely filled *s* and *p* sublevels, which give them extraordinary stability. The halogens are missing one *p* electron; otherwise, they would be electronically the same as the noble gases. The halogens tend to enter reactions that enable them to gain that one *p* electron. Similarly, the alkali metals and the alkaline earth metals have one and two *s* electrons, respectively. They readily lose these electrons to become electronically identical to (**isoelectronic** with) a noble gas.

Isoelectronic **refers to atoms and ions that have identical electron configurations.**

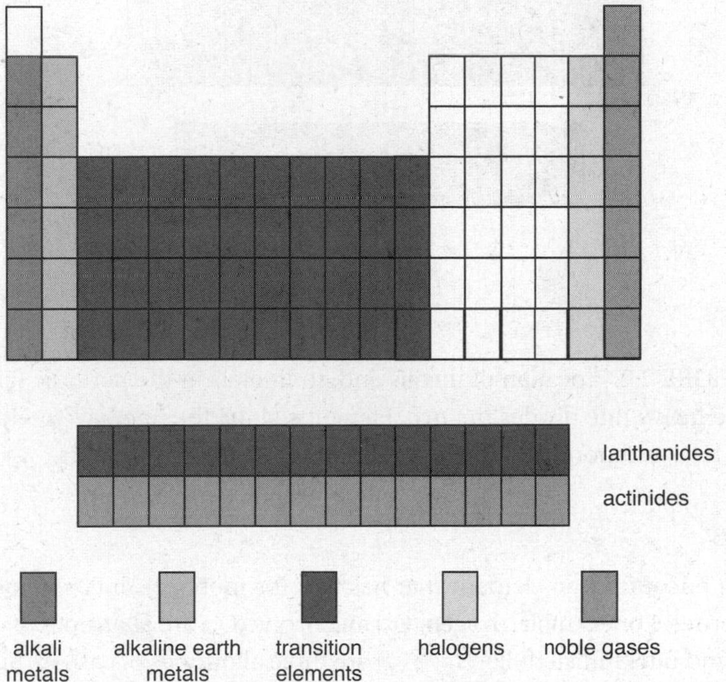

FIGURE 2.1. Common form of the periodic table, showing groups of related elements.

Physical Properties of the Elements

Of the 118 elements in the periodic table, only two, mercury and bromine, are liquids under normal conditions. The noble gases, hydrogen, nitrogen, oxygen, fluorine, and chlorine are gases at room temperature. The remaining elements are solids.

Most of the elements in the periodic table may be considered as individual atoms. A few elements, however, exist naturally as diatomic molecules. These are H_2, O_2, N_2, and the halogens. Other elements, notably sulfur and phosphorus, exist in polyatomic units such as S_8 and P_4, but are commonly represented as single atoms in chemical reactions.

METALS AND METALLOIDS

Metals dominate the elements in the periodic table. Most periodic tables show a heavy line dividing the metals from the nonmetals, as in Figure 2.2.

Elements bordering this line are often termed **metalloids** since they exhibit some properties of metals and some properties of nonmetals. The metallic character of the elements increases from the top of the periodic table to the bottom, as the group headed by nitrogen shows clearly. Nitrogen and phosphorus are nonmetals, arsenic and antimony are metalloids, and bismuth is a metal.

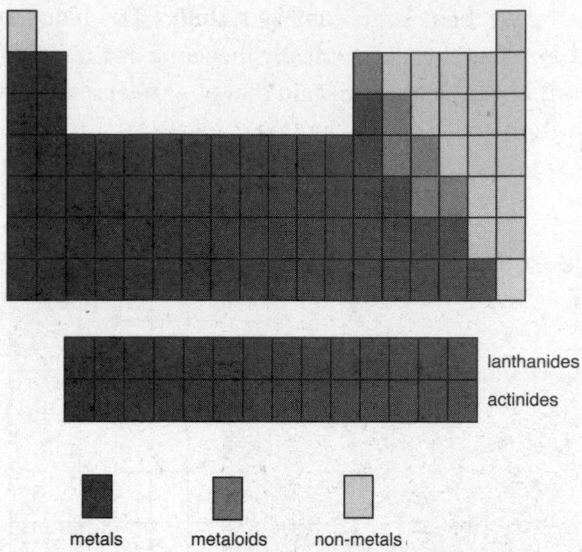

FIGURE 2.2. Location of metals and nonmetals in the periodic table. The heavy line divides the two. Elements along the line have metallic and nonmetallic properties and are called metalloids.

ALLOTROPES

Sometimes we encounter an element that has two (or more) distinct sets of chemical and physical properties. For example, oxygen (O_2) and ozone (O_3) are allotropes of oxygen. Graphite, diamond, and buckminsterfullerene (C_{60}) are three allotropes of carbon. Sulfur and phosphorous also have common allotropes. Many other elements, often metalloids, have allotropic forms.

VARIATION OF PHYSICAL PROPERTIES

As mentioned earlier, the metallic character of the elements increases from the top to the bottom of a group. Many other properties also vary regularly. The melting and boiling points of metals tend to decrease from the top to the bottom of a group. Nonmetals, on the other hand, show an increase in their melting and boiling points. Similar trends in electrical properties, densities, and specific heats are also noted within each group.

ATOMIC RADII

Textbooks often give tables of atomic radii along with graphs that illustrate the change in atomic radius with respect to an increase in atomic number. Those resources are not available on the AP exam. It is important to recall that the atomic radius increases from the top to the bottom of a group (or family) because each increase in period number involves another, larger, energy level. When moving from left to right across a period in the periodic table, there is a general decrease in atomic radius because of an increase in the **effective nuclear charge.**

EFFECTIVE NUCLEAR CHARGE

The effective nuclear charge is an important concept. We find that **core electrons (nonvalence electrons)** shield the valence electrons from an equal amount of positive nuclear charge.

At the same time, valence electrons do not shield other valence electrons from the remaining nuclear charge. The result is an increase in nuclear charge (actually unshielded nuclear charge) from left to right in any given period.

IONIZATION ENERGY

The **ionization energy** is the energy needed to remove an electron from an atom completely. The ionization energy is always endothermic (see Chapter 11). Energy must be added to remove an electron.

Removing the first electron from an atom reveals that the **first ionization energy** for most elements decreases from the top to the bottom of a group. The first ionization energy generally increases from the left to the right of a period in the periodic table. Relatively small decreases in ionization energy occur when a shell is full or half-full.

Removing more than one electron from an element reveals that the valence electrons have a relatively low ionization energy compared with the ionization energies of that same atom's core electrons. This is illustrated for some Group 1 and Group 2 elements in Table 2.2.

TABLE 2.2
Ionization Energies (kJ mol^{-1}) of Selected Elements

Metal	First Electron	Second Electron	Third Electron
Na	496	*4563*	*6913*
Mg	737	1450	*7731*
K	419	*3051*	*4411*
Ca	590	1145	*4912*

PHOTOELECTRON SPECTROSCOPY

When high-energy beams (ultraviolet, X-rays, or even visible light) are trained on the surface of the elements, electrons can be ejected. This is called the *photoelectric effect.* Einstein was awarded the Nobel Prize for elucidating this concept. Photoelectron spectrometers measure the kinetic energies of the ejected electrons. They are also set up to bathe the sample in light of very specific energy. The difference between the energy of the incoming beam and of the outgoing photoelectrons tells us how strongly the electrons are held in the atom. This result is the binding energy.

$$\text{Binding energy} = E_{\text{incoming photon}} - E_{\text{emitted photoelectron}}$$

Photoelectron spectra can be difficult to read and interpret, so idealized spectra are often presented instead. The *y*-axis (intensity) shows that peak sizes are proportional to the number of electrons in a given orbital. The filled *s*, *p*, and *d* orbitals show a 1:3:5 ratio of sizes that comes from the 2:6:10 ratios of electrons in those orbitals. The *x*-axis (binding energy in megajoules) often has high energies on the left decreasing to zero on the right side of the scale. The larger energies are due to stronger attractions of electrons near the nucleus. The lowest binding energy is, as expected, found with the valence electrons. Below is a photoelectron spectrum of nitrogen. The notation on the graph shows the ratio of electrons in their orbitals. The graph also shows the low binding energy of the valence electrons and the high binding energy of the 1*s* level.

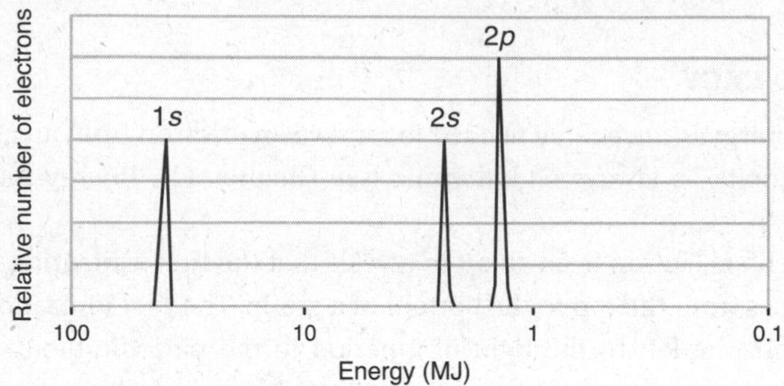

ELECTRON AFFINITY

Electron affinity is defined as the energy change that accompanies the addition of an electron to an atom. Some atoms readily attract electrons, and the electron affinity has a negative value, meaning that energy is released (see Chapter 11). Most atoms, however, do not accept additional electrons readily and the electron affinity is a positive value, indicating that energy must be used to add the electron.

Fluorine has the highest affinity for electrons and francium the lowest. Electron affinity varies diagonally across the periodic table. The atoms close to fluorine tend to accept electrons readily, and those close to francium do not.

ELECTRONEGATIVITY

The concept of **electronegativity** was developed by Linus Pauling to describe the attraction of electrons by individual atoms. Electronegativity is a combination of ionization energy, electron affinity, and other factors. Electronegativities show the same diagonal trend as do ionization energies and electron affinities. Fluorine has the highest electronegativity, and francium the lowest. The electronegativity concept is used in determining how electrons are distributed in molecules, as shown in later chapters. The periodic table in Figure 2.3 shows the electronegativities of the elements to illustrate the increasing trend from the lower left corner to the upper right corner of the table.

When we consider the trends in the periodic table, they generally change from the lower left corner to the upper right corner. Ionization energy, electron affinity, atomic radii, and electronegativity are some of these properties. If we compare an element in Period 2 with an element in Period 3 (such as lithium and magnesium), we find that they have similar physical and chemical properties. The change expected by advancing from one group to the next is canceled to a large extent by dropping from Period 2 to Period 3. In the case of Li and Mg, as we move from Group IA to Group IIA, we expect the atomic radius to decrease. However, as we move from Period 2 to Period 3, we expect the atomic radius to increase. The result is that the radius of Li is 152 pm and the radius of Mg is 160 pm. Other physical properties are similar, resulting in similar chemical properties.

H 2.1																
Li 1.0	Be 1.5											B 2.0	C 2.5	N 3.1	O 3.5	F 4.0
Na 1.0	Mg 1.3											Al 1.5	Si 1.8	P 2.1	S 2.4	Cl 2.9
K 0.8	Ca 1.1	Sc 1.2	Ti 1.3	V 1.5	Cr 1.6	Mn 1.6	Fe 1.7	Co 1.7	Ni 1.8	Cu 1.8	Zn 1.7	Ga 1.8	Ge 2.0	As 2.2	Se 2.5	Br 2.8
Rb 0.8	Sr 1.0	Y 1.1	Zr 1.2	Nb 1.3	Mo 1.3	Tc 1.4	Ru 1.4	Rh 1.5	Pd 1.4	Ag 1.4	Cd 1.5	In 1.5	Sn 1.7	Sb 1.8	Te 2.0	I 2.5
Cs 0.7	Ba 0.9	La 1.1	Hf 1.2	Ta 1.4	W 1.4	Re 1.5	Os 1.5	Ir 1.6	Pt 1.5	Au 1.4	Hg 1.5	Tl 1.5	Pb 1.6	Bi 1.7	Po 1.8	At 2.2
Fr 0.7	Ra 0.9	Ac 1.0														

FIGURE 2.3. Periodic table showing the electronegativities of the elements.

IONIC RADII

Sizes of ions:
Anions > Element
Cations < Element

There are two types of ions: cations and anions. Cations are atoms that have lost one or more electrons and carry a positive charge; anions are atoms that have gained one or more electrons and carry a negative charge.

Cations are always smaller than neutral atoms of the same element. For many cations an entire shell of electrons has been lost. In those instances, the cations are only about half the size of the neutral atoms. A further decrease in size is due to the fact that cations have more protons than electrons.

Anions are always larger than the neutral atoms; many are almost twice the size. Since the added electron(s) go into the same shell, the gain in electrons does not fully explain the increase in size observed. Chemists reason that the increase in size is due to extra repulsive forces due to the added electron. These added repulsions increase the size of the ion.

SUMMARY

Chapter 2 focused on the periodic table and how to use it effectively. This table summarizes trends of varying physical and chemical properties from group to group and also from period to period. You should be able to describe these trends. In addition, there are diagonal relationships that are best summarized as trends that vary fairly uniformly from one corner of the periodic table to the other. Once again, the ability to recall and use these diagonal trends is very useful to the chemist. This chapter also delves into the structure of the atom in terms of electrons, protons, and neutrons. Finally, the concept of atomic mass based on the carbon-12 standard is presented. The concepts of relative atomic masses and the weighted average of isotopes to get an average atomic mass should be understood.

Important Concepts

Relationship of periodic table to electronic configuration
Chemical and physical similarities within groups
Variation of properties within groups
Diagonal relationships
Electronegativity relationships
Weighted average
Diagonal trends

PRACTICE EXERCISES

Multiple-Choice

1. Trends in the periodic table show that elements become more metallic in character from the top of a group to the bottom. Which of these is an element whose properties are opposite those of the element at the top of its group?
   (A) Krypton
   (B) Strontium
   (C) Uranium
   (D) Bismuth

2. In which of the following atoms do the valence electrons feel the greatest effective nuclear charge?
   (A) Ca
   (B) K
   (C) As
   (D) Br

3. In which choice below are the elements ranked in order of increasing first ionization energy?
   (A) P, Cl, S, Al, Ar, Si
   (B) Ar, Cl, S, P, Si, Al
   (C) Al, Si, P, S, Cl, Ar
   (D) Al, Si, S, P, Cl, Ar

4. There are only two liquid elements at room temperature and atmospheric pressure. One of these is
   (A) krypton
   (B) bismuth
   (C) uranium
   (D) bromine

5. The atom with the largest radius is
   (A) krypton
   (B) strontium
   (C) tin
   (D) bromine

6. Metallic behavior is generally associated with
   (A) elements with low ionization energies
   (B) elements with very negative electron affinities
   (C) elements with small atomic radii
   (D) elements with high electronegativies

7. One way to estimate the boiling point of Pd is to
   (A) average the boiling points of Rh and Ag
   (B) average the boiling points of Ni and Pt
   (C) average the boiling points of Ir and Cu
   (D) average the boiling points of Co and Au

8. In which of the following pairs of elements is the element with the lower boiling point listed first?
   (A) Na, Cs
   (B) Te, Se
   (C) P, N
   (D) Ba, Sr

9. In which of the following pairs is the first element expected to have a higher electronegativity than the second?
   (A) O, P
   (B) Cs, Rb
   (C) I, Br
   (D) Al, P

10. Which ion has the largest radius?
   (A) Cl^-
   (B) K^+
   (C) S^{2-}
   (D) Ca^{2+}

11. The effective nuclear charge that an electron in the valence shell feels generally increases
   (A) from left to right across a period and down a group
   (B) from left to right across a period and up a group
   (C) from right to left across a period and down a group
   (D) from left to right across a period and no change down a group

12. Which one of the following groups does not contain any metals?
   (A) Xe, Hg, Ge, O
   (B) Cl, Al, Si, Ar
   (C) C, S, As, H
   (D) Cu, P, Se, Kr

13. Which of the following is expected to have the largest third ionization energy?
   (A) Be
   (B) B
   (C) C
   (D) N

14. Which pair of elements is expected to have the most similar properties?
   (A) Potassium and lithium
   (B) Sulfur and phosphorus
   (C) Silicon and carbon
   (D) Lithium and magnesium

15. Most elements in the periodic table are
   (A) metals
   (B) nonmetals
   (C) liquids
   (D) gases

CHALLENGE

16. How many protons, neutrons, and electrons are in an atom of bromine?
   (A) 35 p, 45 n, 35 e
   (B) 45 p, 35 n, 45 e
   (C) 80 p, 35 n, 80 e
   (D) Neutrons cannot be determined unless an isotope is specified.

17. Chemical properties of elements are defined by the
   (A) electrons
   (B) ionization energy
   (C) protons
   (D) electronegativity

18. The average atomic mass of Ni is 58.693 amu. There are five stable isotopes of nickel: ^{58}Ni, ^{60}Ni, ^{61}Ni, ^{62}Ni, and ^{64}Ni. Which of the following statements is correct?
   (A) Ni-58 and Ni-60 are present in equal amounts.
   (B) Ni-60 is the most abundant isotope.
   (C) Ni-58 is the most abundant isotope.
   (D) Ni-61, Ni-62, and Ni-64 are present in equal amounts.

19. Which one of the following equations correctly represents the process involved in the electron affinity of *X*?
   (A) $X(g) \rightarrow X^+(g) + e^-$
   (B) $X^+(g) \rightarrow X^+(aq)$
   (C) $X^+(g) + e^- \rightarrow X(g)$
   (D) $X(g) + e^- \rightarrow X^-(g)$

20. Which of the following elements has the most exothermic (most negative) electron affinity?
    (A) Li
    (B) F
    (C) Be
    (D) Na

CHALLENGE

21. In general, atomic radii decrease from left to right across a period. The main reason for this behavior is
    (A) the number of neutrons in the nucleus increases
    (B) the number of electrons increases
    (C) the atomic mass increases
    (D) the effective nuclear charge increases

22. Each response below is a sequential list of the first 10 ionization energies for different neutral elements. Arrange the rows in the most probable sequence they will be found in the periodic table.

W	786.5	1577.1	3231.6	4355.5	16,091	19,805	23,780	29,287	33,878	38,726
X	577.5	1816.7	2744.8	11577.0	14,842	18,379	23,326	27,465	31,853	38,473
Y	737.7	1450.7	7732.7	10542.5	13,630	18,020	21,711	25,661	31,653	35,458
Z	495.8	4562.0	6910.3	9543.0	13,354	16,613	20,117	25,496	28,932	141,362

    (A) *WXYZ*
    (B) *ZXWY*
    (C) *WYXZ*
    (D) *ZYXW*

ANSWER KEY

1. **(D)**	7. **(B)**	13. **(A)**	19. **(D)**
2. **(D)**	8. **(D)**	14. **(D)**	20. **(B)**
3. **(B)**	9. **(A)**	15. **(A)**	21. **(D)**
4. **(D)**	10. **(C)**	16. **(D)**	22. **(D)**
5. **(B)**	11. **(D)**	17. **(A)**	
6. **(A)**	12. **(C)**	18. **(C)**	

See Appendix 1 for explanations of answers.

Free-Response

PES, photoelectron spectroscopy, may be used to determine the energy of the electrons in an element. The table shows the PES information for an element.

Binding Energy (MJ)	Relative Number of Electrons
0.58	2
1.09	2
7.19	6
12.1	2
151	2

(a) Sketch a PES diagram based on the data provided. Write the electron configuration represented by each peak.

(b) Identify which element this data represent. Explain your reasoning.

(c) The first ionization energy of magnesium is higher than the first ionization energy of aluminum. Explain why this statement makes sense.

(d) The atomic radius of aluminum is 143 pm, while sulfur has a radius of 103 pm. Explain why this statement makes sense.

ANSWERS

(a) The magnitude of the binding energy determines how closely the electrons are to the nucleus. The 151 MJ value represents the electrons in the 1*s* subshell, which contains 2 electrons. The 12.1 MJ and 7.19 MJ values represent the 2*s* and 2*p* orbitals, respectively. The 1.09 MJ and 0.58 MJ values are the 3*s* and 3*p* orbitals.

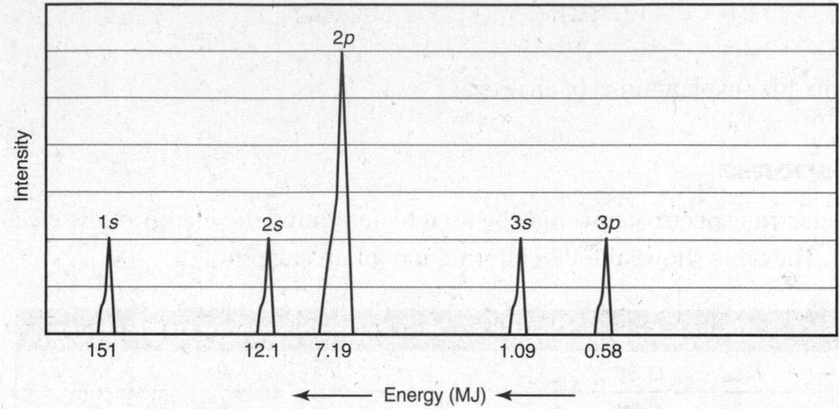

(b) The element is silicon. Each peak represents a subshell, and the intensity of the peaks represents the number of electrons. The total number of electrons is 14; therefore, the element is silicon.

(c) The electron configuration for magnesium is $1s^2\,2s^2\,2p^6\,3s^2$ and for aluminum is $1s^2\,2s^2\,2p^6\,3s^2\,3p^1$. The first electron removed from aluminum is from the 3*p* orbital, which is farther from the nucleus than the first electron removed from magnesium.

(d) Aluminum and sulfur are in the same period on the table. Aluminum is larger than sulfur since the effective nuclear charge felt by the outer electrons in aluminum is less. When going across the periodic table from left to right in the same period, the effective nuclear charge increases since the number of core-shielding electrons remains constant.

PART 2

Chemical Bonding

Ionic Compounds, Formulas, and Reactions

3

- → CHEMICAL FORMULAS
- → EMPIRICAL FORMULAS
- → STRUCTURAL FORMULAS
- → MOLECULAR FORMULAS
- → CHEMICAL REACTIONS
- → EQUATION BALANCING
- → COMBUSTION REACTIONS
- → NEUTRALIZATION REACTIONS
- → SINGLE-REPLACEMENT REACTIONS
- → DOUBLE-REPLACEMENT REACTIONS
- → FORMATION REACTIONS
- → ADDITION REACTIONS
- → DECOMPOSITION REACTIONS
- → NET IONIC REACTIONS
- → IONIC BONDING
- → ELECTRON CONFIGURATIONS
- → FORMATION OF CATIONS AND ANIONS
- → POLYATOMIC IONS
- → CONSTRUCTING IONIC FORMULAS
- → NAMING IONIC COMPOUNDS
- → SOLUTIONS OF IONIC COMPOUNDS
- → SOLUBILITY OF IONIC COMPOUNDS
- → PREDICTING CHEMICAL REACTIONS
- → CHEMICAL DRIVING FORCES
- → PREDICTING REDOX REACTIONS

BIG IDEAS 1, 2, 3

Learning Objectives: 1.1, 1.2, 1.3, 1.4, 1.5, 1.10, 1.17, 1.18, 1.19, 2.8, 2.18, 3.1, 3.2, 3.3, 3.4, 3.5, 3.6, 3.7, 3.8, 3.9, 3.10, 3.11, 3.12, 3.13

For the complete list of Big Ideas and Learning Objectives, refer to the AP Chemistry Course Outline: *https://secure-media.collegeboard.org/digitalServices/pdf/ap/ap-chemistry-course-and-exam-description.pdf*

CHEMICAL FORMULAS

Chapters 1–2 show how the atom is constructed from protons, neutrons, and, most important, electrons. A good knowledge of the electronic makeup of the atom enables us to predict formulas and reactions of many chemical compounds rather than memorizing them. However, as in learning a new language, some basics must be memorized in order to use information properly and quickly.

We start with the chemical **formula**, which is a shorthand method of describing **compounds**. It uses the **atomic symbols** in the periodic table to identify the elements in a compound. If there is more than one atom of an element in the formula, a **subscript** is used to show how many atoms are present. For example, the compound potassium permanganate contains

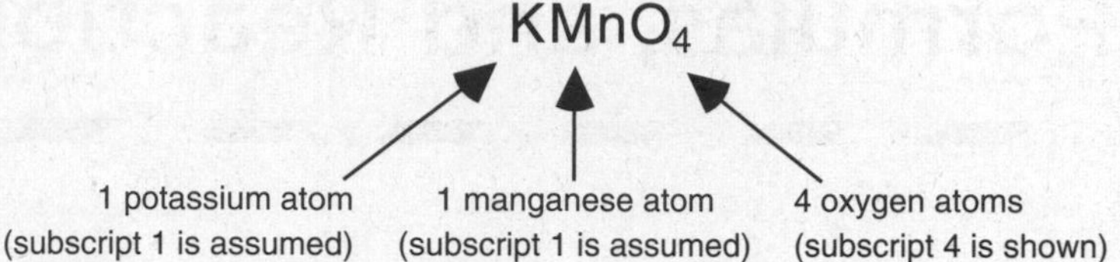

Parentheses in chemical formulas are used to clarify and to provide additional information. A subscript placed after a closing parenthesis multiplies everything within the parentheses. For example, aluminum nitrate contains

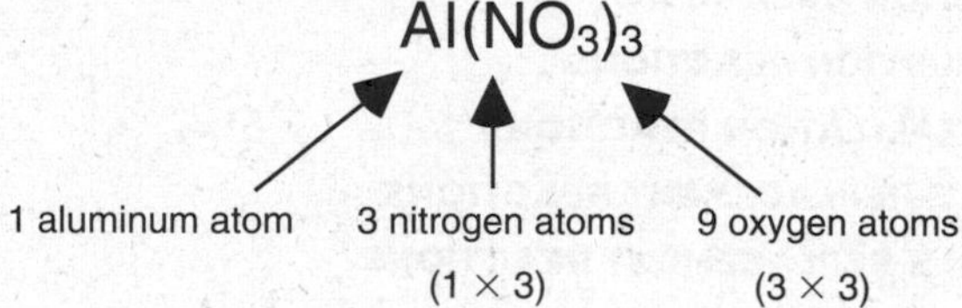

This formula could have been written as AlN_3O_9, which represents the same number of each atom as $Al(NO_3)_3$. However, the parentheses around the NO_3 gives the added information that the nitrogens and oxygens are in three groups of NO_3 units, called the nitrate group. (As will be seen later, the NO_3 should be properly written as the nitrate ion, NO_3^-, with a negative charge.)

The formula for ammonium phosphate is

$$(NH_4)_3PO_4$$

This compound contains 3 nitrogen, 12 hydrogen, 1 phosphorus, and 4 oxygen atoms. Here 3 ammonium groups (actually NH_4^+ ions) are shown in the formula by use of the parentheses.

Another type of formula is used for compounds called **hydrates**. These compounds have a fixed number of water molecules, called the **water of hydration**, in their crystal lattices. To show the water of hydration clearly in the chemical formula, it is written after a dot that is placed in the middle of the line. The dot links two separate compounds into one unit. Two examples are shown below.

The hexahydrate of cobalt(II) chloride is written as

$$CoCl_2 \cdot 6H_2O$$

it contains 1 cobalt atom, 2 chlorine atoms, 12 hydrogen and 6 oxygen atoms (6 water molecules)

The compound

$$Na_2SO_4 \cdot 10H_2O$$

contains 2 sodium, 1 sulfur, 14 oxygen, and 20 hydrogen atoms.

The names of these compounds are cobalt(II) chloride hexahydrate and sodium sulfate decahydrate. The term *hexahydrate* indicates 6 water molecules in the formula, while *decahydrate* shows 10 water molecules. Common prefixes for hydrates are listed in Table 3.1.

TABLE 3.1
Prefixes Used with Hydrates

1	mono-
2	di-
3	tri-
4	tetra-
5	penta-
6	hexa-
7	hepta-
8	octa-
9	nona-
10	deca-

The compounds discussed above are ionic, and their formulas represent the simplest ratio of the atoms in a crystal of the substance. This simplest formula is called an **empirical formula**.

For compounds that have covalent bonds **molecular formulas** are used. Benzene has a molecular formula of C_6H_6, and ethanoic acid has a formula of $HC_2H_3O_2$. These are not empirical formulas; they represent the actual number of each atom present in a single molecule of each of these compounds. A formula written this way is called a condensed formula.

Another form of the molecular formula is the **structural formula**. A structural formula shows a chemist the way the atoms are connected with each other and the covalent bonds between the atoms. The structural formula for ethanoic (acetic) acid is shown in Figure 3.1.

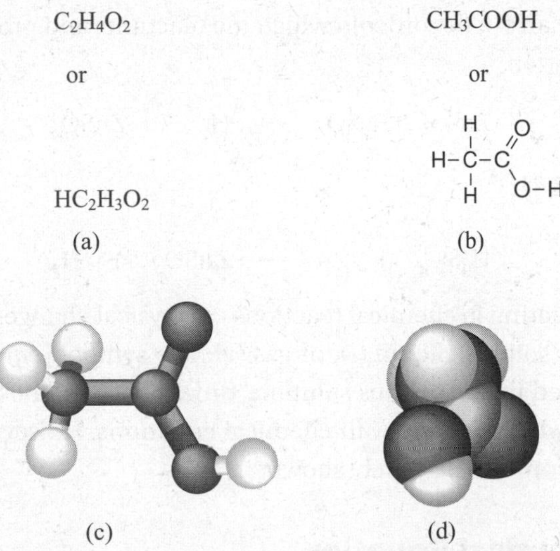

FIGURE 3.1. Representations of ethanoic (acetic) acid in several formats: (a) condensed formulas, (b) structural formulas, (c) ball-and-stick model, and (d) space-filling model.

Modern organic chemistry uses structures that attempt to show the three-dimensional shape of moleculae. These may be a ball-and-stick model or a space-filling model as shown in Figure 3.1.

Exercise 3.1

Determine the number of each different atom represented in each of the following chemical formulas:

(a) $NaClO_4$
(b) $NH_4C_2H_3O_2$
(c) LiH_2AsO_3
(d) $Ca(C_3H_5O_3)_2{\cdot}5H_2O$
(e) $Cu(NH_3)_4SO_4{\cdot}H_2O$

Solution

(a) 1 Na, 1 Cl, 4 O
(b) 1 N, 7 H, 2 C, 2 O
(c) 1 Li, 2 H, 1 As, 3 O
(d) 1 Ca, 6 C, 20 H, 11 O
(e) 1 Cu, 4 N, 14 H, 1 S, 5 O

CHEMICAL REACTIONS AND EQUATIONS

All chemical reactions are essentially the same, with **reactants** being converted into **products**. A chemical equation is written to describe the reaction process. The formulas of the reactants are placed on the left side, and the products on the right side, of an arrow that indicates that the reactants are converted into products:

$$\text{REACTANTS} \rightarrow \text{PRODUCTS}$$

On each side of the arrow, the order in which the reactants and products are written in an equation does not matter.

$$Zn + H_2SO_4 \rightarrow H_2 + ZnSO_4$$

has the same meaning as

$$H_2SO_4 + Zn \rightarrow ZnSO_4 + H_2$$

Substances participating in chemical reactions or physical changes are often in one of the three states of matter, solid (*s*), liquid (*l*), or gas (*g*). The symbol (*aq*) is used to indicate that a substance is dissolved in an aqueous solution. Unless needed, most chemists do not indicate states of matter when working with chemical equations. However, when the state is an important consideration, it is definitely shown.

Balancing a Chemical Equation

Chemical equations must be **balanced** with the same number of each atom on both sides of the arrow. (The arrow is similar to an equal sign in an ordinary mathematical equation.) A balanced chemical equation satisfies the law of conservation of matter. Equations are balanced by placing the appropriate **coefficients** in front of the formulas of the reactants and products in order to equalize the atoms on both sides of the arrow. A coefficient is a simple whole number, and it multiplies all of the atoms in the formula to which it is attached. *Neither the formulas of compounds nor their subscripts are altered to balance an equation.*

One common reaction type is a **combustion** reaction. An example is the burning of propane fuel ($CH_3CH_2CH_3$) in a barbecue grill:

$$CH_3CH_2CH_3 \ + \ O_2 \ \rightarrow \ CO_2 \ + \ H_2O \quad \text{(unbalanced)}$$

As written, this expression simply gives the reactants ($CH_3CH_2CH_3 + O_2$) and the products ($CO_2 + H_2O$). It may be balanced by using the appropriate coefficients for the two reactants and two products. There are two methods for determining these coefficients. One is the **inspection method** and the other is the **ion-electron method**, which is used for complex oxidation-reduction equations. The ion-electron method will be discussed in Chapter 12.

The first step in the inspection method of balancing equations involves counting the number of each atom in the equation on the reactant side and then on the product side. This requires care and attention to detail since the smallest mistake ruins the entire effort.

The next step is to balance one atom at a time by adding a coefficient where needed and recounting the atoms. Adding coefficients and recounting continue until the same number of atoms is present on each side of the arrow.

Chemists find the process simpler if they *balance the most complex molecule first, leaving the simple compounds and elements until last.* Also, elements that appear in more than one compound on either the reactant or product side are left to the end. Finally, it is faster to balance groups of atoms, such as the sulfate or nitrate ions discussed later, as if they were individual atoms.

Returning to the combustion of propane, we count 3 carbon, 8 hydrogen, and 2 oxygen atoms on the reactant side. On the product side we have 1 carbon, 2 hydrogen, and 3 oxygen atoms. Since propane is the most complex molecule in the reaction, it is used as the starting point. Its 3 carbon atoms can be balanced by adding a coefficient of **3** to CO_2:

$$CH_3CH_2CH_3 \ + \ O_2 \ \rightarrow \ \mathbf{3}CO_2 \ + \ H_2O \quad \text{(still unbalanced)}$$

There are now 3 carbon atoms on both sides, but the numbers of hydrogen and oxygen atoms are still not equal.

Next, the 8 hydrogen atoms in propane can be balanced by adding a coefficient of **4** to the water molecules:

$$CH_3CH_2CH_3 \ + \ O_2 \ \rightarrow \ 3CO_2 \ + \ \mathbf{4}H_2O \quad \text{(still unbalanced)}$$

Recounting the atoms, we find that there are now 3 carbon, 8 hydrogen, and 2 oxygen atoms on the reactant side and 3 carbon, 8 hydrogen, and 10 oxygen atoms on the product side. The carbon and hydrogen atoms are balanced, and only the oxygen atoms remain unequal. The equation can be balanced by using a coefficient of **5** for the O_2 molecule:

$$CH_3CH_2CH_3 \ + \ \mathbf{5}O_2 \ \rightarrow \ 3CO_2 \ + \ 4H_2O \quad \text{(balanced)}$$

The equation now has 3 carbon, 8 hydrogen, and 10 oxygen atoms on both the reactant and the product side. The coefficient for $CH_3CH_2CH_3$ is 1, but it is not written.

TIP

You may wish to make a table of the atoms to keep track of your progress.

Element	Reactant	Product
Cr	2	1
S	3	1
O	13	7
K	1	2
H	1	3

Another type of reaction is the **double-replacement** reaction. The reaction of chromium(III) sulfate with potassium hydroxide illustrates this class of reactions.

$$Cr_2(SO_4)_3 \ + \ KOH \ \rightarrow \ Cr(OH)_3 \ + \ K_2SO_4 \quad \text{(unbalanced)}$$

In this unbalanced form there are 2 chromium, 3 sulfur, 13 oxygen, 1 hydrogen, and 1 potassium atoms on the reactant side and 1 chromium, 1 sulfur, 7 oxygen, 3 hydrogen, and 2 potassium atoms on the product side. The chemist would see 2 chromium atoms, 3 sulfate ions (the SO_4^{2-} unit), 1 potassium atom, and 1 hydroxide ion (the OH^- unit) on the reactant side and 1 chromium atom, 1 sulfate ion, 2 potassium atoms, and 3 hydroxide ions on the product side.

Focusing on $Cr_2(SO_4)_3$, we can balance the chromium atoms with a 2 in front of $Cr(OH)_3$:

$$Cr_2(SO_4)_3 \ + \ KOH \ \rightarrow \ \mathbf{2}Cr(OH)_3 \ + \ K_2SO_4 \quad \text{(unbalanced)}$$

Next, the 6 hydroxide ions in $2Cr(OH)_3$ can be balanced by placing a 6 in front of KOH:

$$Cr_2(SO_4)_3 \ + \ \mathbf{6}KOH \ \rightarrow \ 2Cr(OH)_3 \ + \ K_2SO_4 \quad \text{(unbalanced)}$$

TIP

In almost all cases, balanced equations with the smallest whole-number coefficients are preferred.

Then the 6 potassium atoms in 6 KOH can be balanced with a 3 in front of K_2SO_4. This also balances the sulfate ions and the equation is balanced:

$$Cr_2(SO_4)_3 \ + \ 6KOH \ \rightarrow \ 2Cr(OH)_3 \ + \ \mathbf{3}K_2SO_4 \quad \text{(balanced)}$$

As this point, you should have noticed that balancing combustion and double-replacement reactions use slightly different approaches. It will be helpful to recall the appropriate method depending upon the type of reaction you are balancing. Later we will use a third method for balancing oxidation-reduction reactions.

SIMPLEST COEFFICIENTS

The reaction of hydrogen with oxygen to form water can be balanced as

$$2H_2 \ + \ O_2 \ \rightarrow \ 2H_2O$$

It will also be balanced if written as

$$4H_2 \ + \ 2O_2 \ \rightarrow \ 4H_2O$$
$$16H_2 \ + \ 8O_2 \ \rightarrow \ 16H_2O$$
$$H_2 \ + \ \tfrac{1}{2}O_2 \ \rightarrow \ H_2O$$

The last three reactions are technically balanced since they have the same numbers of hydrogen atoms and oxygen atoms on both sides of the arrow, but they are not in the best form. **Properly balanced equations have the smallest whole-number coefficients possible.** For the three reactions,

the first one should be divided by 2.
the second should be divided by 8.
the third should be multiplied by 2.

An equation may be multiplied or divided as necessary, but it should be remembered that all coefficients in the equation must be multiplied or divided by the same factor. Balancing reactions requires practice to develop skill and speed.

Exercise 3.2

Balance the following reactions by inspection:

C_6H_6	+	O_2	$\rightarrow$	CO_2	+	H_2O		
$MgCl_2$	+	$AgNO_3$	$\rightarrow$	$AgCl$	+	$Mg(NO_3)_2$		
Al	+	O_2	$\rightarrow$	Al_2O_3				
CaO	+	H_2SO_4	$\rightarrow$	H_2O	+	$CaSO_4$		
Al	+	Fe_3O_4	$\rightarrow$	Fe	+	Al_2O_3		
NO_2	+	O_2	$\rightarrow$	N_2O_5				
HCl	+	$CaCO_3$	$\rightarrow$	$CaCl_2$	+	H_2O	+	CO_2
O_2	+	$C_4H_9NH_2$	$\rightarrow$	CO_2	+	H_2O	+	N_2
Mg	+	HCl	$\rightarrow$	H_2	+	$MgCl_2$		
Zn	+	$Cu(NO_3)_2$	$\rightarrow$	Cu	+	$Zn(NO_3)_2$		
$CoCl_3$	+	$Ba(OH)_2$	$\rightarrow$	$BaCl_2$	+	$Co(OH)_3$		
$Ba(OH)_2$	+	H_3PO_4	$\rightarrow$	H_2O	+	$Ba_3(PO_4)_2$		
C_6H_8	+	H_2	$\rightarrow$	C_6H_{12}				
SO_2	+	O_2	$\rightarrow$	SO_3				
C_4H_{10}	+	O_2	$\rightarrow$	CO_2	+	H_2O		
H_2S	+	$AuCl_3$	$\rightarrow$	Au_2S_3	+	HCl		

Solution

The balanced equations are given in the solution to Exercise 3.3, page 138.

Reaction Types

Many chemical reactions fall into distinct groups with definite similarities. By classifying chemical reactions, it is possible to compare the properties of the reactants and products. In addition, such classification often serves as a shorthand method in place of writing a complete chemical reaction. The combustion of propane illustrates one such classification. By calling the reaction a combustion process, it is immediately known that one of the reactants is oxygen and that the products are carbon dioxide and water. Some other reaction types are described below.

COMBUSTION REACTIONS

In these reactions, an organic (carbon-containing) compound reacts with oxygen to form carbon dioxide and water. If the organic compound contains elements other than carbon, hydrogen, and oxygen, it is often assumed that those elements end up in the elemental state as products. (Sulfur is an exception. It will form SO_2.) Two typical combustion reactions are

$$C_5H_{12} + 8O_2 \rightarrow 5CO_2 + 6H_2O$$
$$2C_4H_9SH + 15O_2 \rightarrow 8CO_2 + 10H_2O + 2SO_2$$

SINGLE-REPLACEMENT REACTIONS

In some reactions, an element may react with a compound to produce a different element and a new compound. A typical reaction of this sort is

$$2AgNO_3 + Cu \rightarrow 2Ag + Cu(NO_3)_2$$

In this reaction the copper replaces the silver in the silver nitrate. This type of reaction is also known as a single-displacement reaction.

DOUBLE-REPLACEMENT REACTIONS

In these reactions, two compounds react and the cation in one compound replaces the cation in the second compound, and vice versa. A double replacement reaction is

$$MgSO_4 + Ba(NO_3)_2 \rightarrow Mg(NO_3)_2 + BaSO_4$$

In this type of reaction, the magnesium replaces the barium and the barium replaces the magnesium—thus the term *double replacement.*

NEUTRALIZATION REACTIONS

These reactions are a special type of double-replacement reaction in which one reactant is an acid and the other is a base. The products are a salt and water. A typical neutralization reaction is

$$\underset{\text{(acid)}}{HCl} + \underset{\text{(base)}}{LiOH} \rightarrow LiCl + H_2O$$

SYNTHESIS REACTIONS

Reactions of two or more elements to form a compound are often called synthesis reactions. One such reaction is the formation of rust, Fe_2O_3

$$4Fe + 3O_2 \rightarrow 2Fe_2O_3$$

FORMATION REACTIONS

A formation reaction is the same as a synthesis reaction except that the product must have a coefficient of 1. The reactants are the elements in their normal state at room temperature and atmospheric pressure. The formation reaction for $Fe(NH_4)_2(SO_4)_2$ is

$$Fe + N_2 + 4H_2 + 2S + 4O_2 \rightarrow Fe(NH_4)_2(SO_4)_2$$

If necessary, the use of fractional coefficients for the reactants is permitted in a formation reaction.

ADDITION REACTIONS

In these reactions a simple molecule or an element is added to another molecule, as in the addition of HCl to pentene

$$HCl + C_5H_{10} \rightarrow C_5H_{11}Cl$$

DECOMPOSITION REACTIONS

These reactions result when a large molecule decomposes into its elements or into smaller molecules. When sucrose is heated strongly, in the absence of O_2, this reaction occurs

$$C_{12}H_{22}O_{11} \rightarrow 12C + 11H_2O$$

NET IONIC EQUATIONS

When ionic compounds react in aqueous solution, (symbolized by (*aq*) in a reaction) usually only one ion from each compound reacts. The other ions are "spectator ions" and do not react. Writing an equation in ionic form focuses attention on the actual reaction and allows the chemist to find substitute reactants to achieve the same result. For instance, the reaction of silver nitrate with sodium chloride produces a precipitate of silver chloride in the molecular equation

$$NaCl(aq) + AgNO_3(aq) \rightarrow AgCl(s) + NaNO_3(aq)$$

Written as a net ionic equation, this becomes

$$Cl^-(aq) + Ag^+(aq) \rightarrow AgCl(s)$$

In this form the chemist now knows that any soluble chloride salt (KCl, $MgCl_2$, etc.) and any soluble silver salt ($AgClO_4$, Ag_2SO_4, etc.) will also give AgCl as the product. In ionic reactions the charges must balance as well as the atoms.

HALF-REACTION EQUATIONS

These reactions are used extensively with oxidation-reduction reactions and in describing electrochemical processes in Chapter 12. The half-reaction is a reduction reaction if electrons are on the reactant side and an oxidation reaction if the electrons are products.

$$I_2 + 2e^- \rightarrow 2I^- \quad \text{(reduction half-reaction)}$$

$$Fe^{2+} \rightarrow Fe^{3+} + e^- \quad \text{(oxidation half-reaction)}$$

Half-reactions may be combined to make a complete oxidation-reduction reaction as long as the electrons all cancel.

OXIDATION-REDUCTION REACTIONS

These reactions involve the loss of electrons by one compound or ion and the subsequent gain of the same electrons by another compound or ion. The two half-reactions above may be added (after multiplying the second reaction by 2 and canceling the electrons) to obtain the oxidation-reduction reaction

$$I_2 + 2Fe^{2+} \rightarrow 2Fe^{3+} + 2I^-$$

The combustion and single-replacement reactions discussed above are also oxidation-reduction reactions.

Exercise 3.3

Classify each of the reactions balanced in Exercise 3.2 as one of the reaction types described in this section.

Solution

$2C_6H_6$	$+ 15O_2$	$\rightarrow$	$12CO_2 + 6H_2O$		(combustion)
$MgCl_2$	$+ 2AgNO_3$	$\rightarrow$	$2AgCl + Mg(NO_3)_2$		(double replacement)
$4Al$	$+ 3O_2$	$\rightarrow$	$2Al_2O_3$		(synthesis)
CaO	$+ H_2SO_4$	$\rightarrow$	$H_2O + CaSO_4$		(double replacement)
$8Al$	$+ 3Fe_3O_4$	$\rightarrow$	$9Fe + 4Al_2O_3$		(single replacement)
$4NO_2$	$+ O_2$	$\rightarrow$	$2N_2O_5$		(combustion or addition)
$2HCl$	$+ CaCO_3$	$\rightarrow$	$CaCl_2 + H_2O$	$+ CO_2$	(double replacement)
$27O_2$	$+ 4C_4H_9NH_2$	$\rightarrow$	$16CO_2 + 22H_2O$	$+ 2N_2$	(decomposition)
Mg	$+ 2HCl$	$\rightarrow$	$H_2 + MgCl_2$		(single replacement)
Zn	$+ Cu(NO_3)_2$	$\rightarrow$	$Cu + Zn(NO_3)_2$		(single replacement)
$2CoCl_3$	$+ 3Ba(OH)_2$	$\rightarrow$	$3BaCl_2 + 2Co(OH)_3$		(double replacement)
$3Ba(OH)_2$	$+ 2H_3PO_4$	$\rightarrow$	$6H_2O + Ba_3(PO_4)_2$		(neutralization)
C_6H_8	$+ 2H_2$	$\rightarrow$	C_6H_{12}		(addition)
$2SO_2$	$+ O_2$	$\rightarrow$	$2SO_3$		(addition or combustion)
$2C_4H_{10}$	$+ 13O_2$	$\rightarrow$	$8CO_2 + 10H_2O$		(combustion)
$3H_2S$	$+ 2AuCl_3$	$\rightarrow$	$Au_2S_3 + 6HCl$		(double replacement)

BONDING

When elements combine with each other to form compounds, a chemical bond is formed. An understanding of how and why bonds are formed helps the chemist to predict many physical and chemical properties of molecules and compounds, including chemical reactivity, shape, solubility, physical state, and polarity. The key to bond formation is the behavior of the outermost, or valence, electrons. When two atoms share valence electrons to form a bond, the bond is known as a **covalent bond**. When one atom loses electrons and another gains electrons, ions are formed. The attraction between ions to form a compound is called an **ionic bond**.

The underlying principle of chemical bonding can be explained using the electronic configurations of the noble gases. Noble gases are very unreactive elements. Until 1962, when Neil Bartlett produced the first noble gas compound, they were considered inert, and most periodic tables called them "inert gases." Except for helium, all noble gases have valence shells in which the outermost *s* and *p* sublevels are completely filled (ns^2, np^6), as shown in bold type in Table 3.2. This configuration gives unusual stability to the noble gases and also to atoms that can lose, gain, or share electrons to attain the same configuration.

Ionic Substances

The basis of the ionic bond is the attraction of a positively charged ion (cation) toward a negatively charged ion (anion). It is necessary to understand which elements tend to form ions and which do not. For elements that form ions we also want to develop methods to determine what kind of ion should be expected. Finally, once the ions are known, we can use that information to predict chemical formulas and chemical reactions.

TABLE 3.2
Electronic Configurations of the Noble Gases

Noble Gas	Electronic Configuration
He	$1s^2$
Ne	$1s^2$ $2s^2$ $2p^6$
Ar	$1s^2$ $2s^2$ $2p^6$ $3s^2$ $3p^6$
Kr	$1s^2$ $2s^2$ $2p^6$ $3s^2$ $3p^6$ $4s^2$ $3d^{10}$ $4p^6$
Xe	$1s^2$ $2s^2$ $2p^6$ $3s^2$ $3p^6$ $4s^2$ $3d^{10}$ $4p^6$ $5s^2$ $4d^{10}$ $5p^6$
Rn	$1s^2$ $2s^2$ $2p^6$ $3s^2$ $3p^6$ $4s^2$ $3d^{10}$ $4p^6$ $5s^2$ $4d^{10}$ $5p^6$ $6s^2$ $4f^{14}$ $5d^{10}$ $6p^6$

MONATOMIC IONS OF THE REPRESENTATIVE ELEMENTS

The representative elements are those found within the *s* and *p* blocks of the periodic table as shown in Figure 1.10. These elements have regular properties that follow basic chemical principles with very few exceptions. In the case of ion formation, the principle is that the ions of the representative elements will have electronic configurations identical to those of the noble gases.

Representative metals will lose electrons to form positively charged ions called cations. The equation

$$\mathrm{M} \rightarrow \mathrm{M}^{n+} + ne^-$$

where n is the number of electrons lost by the metal, M, is used to represent the formation of all cations.

TIP

Cations of the representative metals are formed by removing outer *s* and *p* electrons to achieve an electronic structure the same as that of a noble gas.

The electronic configuration allows the chemist to determine the number of electrons that a representative metal will lose. For instance, sodium has the electronic configuration $1s^22s^22p^63s^1$. When the element forms the Na^+ ion, the $3s^1$ electron is lost. The electronic configuration for the Na^+ ion is then $1s^22s^22p^6$ which is the same as the electronic configuration of neon.

Barium has the electronic configuration

$$\mathrm{Ba} = 1s^22s^22p^63s^23p^64s^23d^{10}4p^65s^24d^{10}5p^6\mathbf{6s^2}$$

When the Ba^{2+} ion is formed, the two 6*s* electrons are lost, giving the barium ion the configuration

$$\mathrm{Ba}^{2+} = 1s^22s^22p^63s^23p^64s^23d^{10}4p^6\mathbf{5s^2}4d^{10}\mathbf{5p^6}$$

which is identical to the electronic configuration of xenon:

$$\mathrm{Xe} = 1s^22s^22p^63s^23p^64s^23d^{10}4p^6\mathbf{5s^2}4d^{10}\mathbf{5p^6}$$

The representative metals will lose all of their valence *s* and *p* electrons. The electronic configuration of a metal will then be identical to that of the preceding noble gas in the periodic table.

In Periods 4, 5, and 6 the metals that contain outer *s* and *p* electrons may form a second ion by losing only their *p* electrons. We find that gallium, indium, and thallium form 1+ and 3+ ions. Tin and lead form 2+ and 4+ ions, and bismuth forms 3+ and 5+ ions.

TIP

Anions are formed by adding electrons to complete an electronic shell.

The representative nonmetals gain electrons to form negatively charged ions called anions. The general reaction for this process is

$$X + ne^- \rightarrow X^{n-}$$

where *n* represents the number of electrons gained by the nonmetal represented by X.

As an example, bromine has the electronic configuration

$$\text{Br} = 1s^22s^22p^63s^23p^64s^23d^{10}\mathbf{4p^5}$$

TIP

Ions and atoms that have identical electron configurations are said to be isoelectronic.

When an electron is added to the 4*p* subshell, the bromide ion, Br^-, is formed. Its electronic configuration becomes

$$\text{Br}^- = 1s^22s^22p^63s^23p^64s^23d^{10}\mathbf{4p^6}$$

which is the same as that of the noble gas krypton:

$$\text{Kr} = 1s^22s^22p^63s^23p^6\mathbf{4s^2}3d^{10}\mathbf{4p^6}$$

All halogens will gain one electron to form anions with one negative charge. Oxygen, sulfur, and selenium gain two electrons each to form 2^- ions. We will see that these elements often participate in covalent bonding as well. The remaining nonmetals, nitrogen, phosphorus, and carbon, usually bond using covalent bonds, but they can form N^{3-}, P^{3-}, and C^{4-} ions. These are called the nitride, phosphide, and carbide ions, respectively.

TIP

Cations of nonrepresentative elements can be understood by arranging electrons by their principal quantum number.

MONATOMIC IONS OF THE NONREPRESENTATIVE ELEMENTS

The nonrepresentative elements are the remaining *d* block and *f* block metals in the periodic table. These elements are characterized by the fact that many of them may have more than one possible cation and they often form polyatomic anions. In general, it is not possible to predict with certainty the charge of the cations for these elements. However, none of these elements forms monatomic anions.

Appreciating what happens to the transition elements is not always simple, but some hints may be obtained from their electronic configurations. For these elements the electronic configurations are arranged by their principal quantum number, rather than in the aufbau order. For instance, the complete electronic configuration for lead is as follows:

$$\text{Pb} = 1s^22s^22p^63s^23p^64s^23d^{10}4p^65s^24d^{10}5p^66s^24f^{14}5d^{10}6p^2$$

Grouping the electrons by shell or principal quantum number gives

$$\text{Pb} = 1s^22s^22p^63s^23p^63d^{10}4s^24p^64d^{10}4f^{14}5s^25p^65d^{10}\mathbf{6s^26p^2}$$

and allows us to see more clearly which electrons are in the outermost shell of the atom. We can now see that the Pb^{2+} ion is formed when the two 6*p* electrons are removed and that the Pb^{4+} ion forms when all of the 6*s* and 6*p* electrons are removed.

Similarly, the 2+ and 4+ ions of titanium may be deduced from the electronic structure

$$\text{Ti} = 1s^22s^22p^63s^23p^64s^23d^2$$

Regrouping by shell gives

$$\text{Ti} = 1s^22s^22p^63s^23p^6\mathbf{3d^24s^2}$$

Removal of just the two $4s$ electrons yields the Ti^{2+} ion, and removal of the two $3d$ and two $4s$ electrons gives the Ti^{4+} ion.

Iron is another example worth considering. It forms Fe^{2+} and Fe^{3+} ions. From the electronic configuration arranged by shells we get

$$Fe \quad = \quad 1s^22s^22p^63s^23p^63d^6\mathbf{4s^2}$$

Removal of the two $4s$ electrons gives the Fe^{2+} ion. To obtain the Fe^{3+} ion one more electron must be removed. Obviously it is one of the six $3d$ electrons that are now the outermost electrons. As we saw in Chapter 1, a d subshell that contains one electron in each orbital is a stable state. Removal of one of the $3d$ electrons results in the electronic structure

$$Fe^{3+} \quad = \quad 1s^22s^22p^63s^23p^6\mathbf{3d^5}$$

which has the stable half-filled $3d$ subshell.

Not all ions in the nonrepresentative group can be rationalized without much more sophisticated reasoning. However, in the absence of additional information the logic used above provides for a reasonable first approximation of why certain ions form and others do not.

POLYATOMIC IONS

Many elements combine with oxygen (and sometimes hydrogen and nitrogen) to form a charged group of atoms called a **polyatomic ion**. (In older texts polyatomic ions are called radicals.) Polyatomic ions are unusually stable groups of atoms that tend to act as single units in many chemical reactions. The formulas, names, and charges of the common polyatomic ions are listed in Table 3.3 and **should be memorized**. Note that all of these are anions (negatively charged ions) except for the ammonium ion, NH_4^+.

The atoms in the polyatomic ions (Table 3.3) are bound to each other with covalent bonds, which will be described later. Polyatomic ions form ionic compounds by combining with other ions of opposite charge.

TIP

Polyatomic ion formulas and names must be memorized.

IONIC FORMULAS

Ionic compounds are formed when cations are attracted to anions because of their opposing charges. The formulas for ionic compounds can be deduced because no compound can have a net charge. In other words, the total positive charge of the cations must be exactly canceled by the negative charge of the anions in the chemical formula. Some chemists refer to this as the **law of electroneutrality**. In addition, every ionic compound has a formula that represents the simplest ratio of the elements needed to obey the law of electroneutrality. As mentioned earlier in this chapter, this simplest ratio is called the **empirical formula**.

When the anion and cation have the same, but opposite, charges, the compound is written with only one atom of each element. This is another way of saying that the formulas for all ionic compounds are empirical formulas.

Na^+	+	F^-	→	NaF	sodium fluoride
Mg^{2+}	+	O^{2-}	→	MgO	magnesium oxide
Fe^{2+}	+	S^{2-}	→	FeS	iron(II) sulfide
Al^{3+}	+	N^{3-}	→	AlN	aluminum nitride
La^{3+}	+	PO_4^{3-}	→	$LaPO_4$	lanthanum(III) phosphate
NH_4^+	+	NO_3^-	→	NH_4NO_3	ammonium nitrate

TABLE 3.3
Common Polyatomic Ions

Ion Formula	Ion Name
NH_4^+	ammonium ion
CO_3^{2-}	carbonate ion
HCO_3^-	bicarbonate ion (hydrogen carbonate)
ClO^-	hypochlorite ion
ClO_2^-	chlorite ion
ClO_3^-	chlorate ion
ClO_4^-	perchlorate ion
NO_2^-	nitrite ion
NO_3^-	nitrate ion
SO_3^{2-}	sulfite ion
HSO_3^-	bisulfite ion (hydrogen sulfite)
SO_4^{2-}	sulfate ion
HSO_4^-	bisulfate ion (hydrogen sulfate)
MnO_4^-	permanganate
$Cr_2O_7^{2-}$	dichromate ion
CrO_4^{2-}	chromate ion
$S_2O_3^{2-}$	thiosulfate ion
PO_4^{3-}	phosphate ion
HPO_4^{2-}	hydrogen phosphate ion
$H_2PO_4^-$	dihydrogen phosphate ion

When the charges of the anion and cation are not equal and opposite, it is necessary to adjust the numbers of the ions so that the total charge adds up to zero. The most convenient way to do this is to use the charge of the cation for the subscript of the anion and the charge of the anion (without the minus sign) as the subscript for the cation, as shown below. Remember that, when a subscript is 1, it is not written.

Ca^{2+}	+	$2Cl^-$	→	$CaCl_2$	calcium chloride
$2Al^{3+}$	+	$3S^{2-}$	→	Al_2S_3	aluminum sulfide
$3Na^+$	+	PO_4^{3-}	→	Na_3PO_4	sodium phosphate
Pb^{4+}	+	$4Cl^-$	→	$PbCl_4$	lead(IV) chloride

When a subscript must be used with a polyatomic ion, it is necessary to place parentheses around the polyatomic ion before adding the subscript, as shown in the following equations:

Ca^{2+}	+	$2NO_3^-$	→	$Ca(NO_3)_2$	calcium nitrate
$2NH_4^+$	+	SO_4^{2-}	→	$(NH_4)_2SO_4$	ammonium sulfate
$2Al^{3+}$	+	$3SO_4^{2-}$	→	$Al_2(SO_4)_3$	aluminum sulfate

NAMING IONIC COMPOUNDS

Ionic compounds contain a metal and a nonmetal. [The ammonium ion (NH_4^+) is considered a metal, and polyatomic anions are considered nonmetals for this purpose.] These compounds are also known as salts. Names are created by giving the name of the cation first and then the name of the anion.

For cations that have only one possible charge, the name is the same as that of the element. For cations that may have more than one charge, such as the lead and titanium discussed previously, the element name is followed by parentheses enclosing the charge written in roman numerals. Two examples are lead(II) and lead(IV). This method, known as the Stock system, is the preferred method for naming ionic compounds.

TIP

Use the Stock system when naming ionic compounds on the AP exam.

Naming anions depends on whether the anion is a monatomic ion or a polyatomic anion. Monatomic anions are named by taking the root or first portion of the element name and then changing the ending to *-ide*, as shown in the accompanying table of examples. Polyatomic anions have unique names, given in Table 3.3, which must be memorized.

The entire compound is named by writing the name of the cation followed by the name of the anion as a separate word; for example, MgF_2 is magnesium fluoride and FeI_3 is iron(III) iodide. Names of chemical compounds are not capitalized except at the beginning of a sentence. Examples of the names of representative compounds have been given in the preceding discussion of formula writing.

COMMON ANION NAMES

sulfur	sulfide
oxygen	oxide
chlorine	chloride
nitrogen	nitride
bromine	bromide
fluorine	fluoride

To name a compound that contains a metal that may have more than one possible charge, we must know the charge on the ion. Ionic charge is determined by "taking apart" the formula unit to find out what the charges of the ions were before the ions combined. We will know the charge of the anion, which will be either a representative monatomic anion or one of the polyatomic anions in Table 3.3. If the charge of one anion and the number of anions are known, the charge on the cation can be deduced since the formula must always have a net charge of zero. Review how formulas are determined and see how the process can be reversed.

Exercise 3.4

Name each of the following compounds:

$MgCl_2$	Na_2CrO_4	$TiBr_4$	$HgSO_4$
MnO_2	$Fe(ClO_2)_2$	Na_3PO_4	$SnCl_4$
Cr_2O_3	$Hg(NO_3)_2$	$(NH_4)_2SO_3$	BiF_3
Ca_3N_2	$Al(NO_3)_3$		

Solution

$MgCl_2$	magnesium chloride	$Al(NO_3)_3$	aluminum nitrate
MnO_2	manganese(IV) oxide	$TiBr_4$	titanium(IV) bromide
Cr_2O_3	chromium(III) oxide	Na_3PO_4	sodium phosphate
Ca_3N_2	calcium nitride	$(NH_4)_2SO_3$	ammonium sulfite
Na_2CrO_4	sodium chromate	$HgSO_4$	mercury(II) sulfate
$Fe(ClO_2)_2$	iron(II) chlorite	$SnCl_4$	tin(IV) chloride
$Hg(NO_3)_2$	mercury(II) nitrate	BiF_3	bismuth(III) fluoride

Exercise 3.5

Write the formula for each of the following compounds:

aluminum sulfate	gold(III) nitrate
magnesium oxide	lithium sulfite
vanadium(III) bromide	ammonium phosphate
barium nitrite	strontium fluoride
cobalt(II) chloride	lead(IV) carbonate

Solution

Writing formulas from names is often easier than writing names from formulas since the charges of the nonrepresentative elements are given in parentheses in the names. Remember the requirement for electric neutrality or no net charge on any chemical compound.

aluminum sulfate	$Al_2(SO_4)_3$	gold(III) nitrate	$Au(NO_3)_3$
magnesium oxide	MgO	lithium sulfite	Li_2SO_3
vanadium(III) bromide	VBr_3	ammonium phosphate	$(NH_4)_3PO_4$
barium nitrite	$Ba(NO_2)_2$	strontium fluoride	SrF_2
cobalt(II) chloride	$CoCl_2$	lead(IV) carbonate	$Pb(CO_3)_2$

Ionic Reactions

IONS IN SOLUTION

Most ionic compounds dissolve in water, and in the process the compound separates into the cations and anions. This solution process may be written as

TIP

(*s*) = solid
(*g*) = gas
(*l*) = pure liquid
(*aq*) = aqueous

$$NaBr(s) \rightarrow Na^+(aq) + Br^-(aq)$$

The symbol in parentheses designates the state of each substance in the reaction: (*s*) means that the substance is a solid, and (*aq*) that the substance is in an aqueous solution. Other symbols used are (*l*) for liquid and (*g*) for gas.

Chromium(III) nitrate dissolves according to the equation

$$Cr(NO_3)_3(s) \rightarrow Cr^{3+}(aq) + 3NO_3^-(aq)$$

One chromium(III) ion and three nitrate ions are obtained from one formula unit of chromium(III) nitrate.

The following general principles apply to the dissolution of ionic compounds.

1. Only one cation and one anion are formed. Compounds containing three or more different atoms will break apart into the appropriate polyatomic ion(s). (Exceptions are discussed in higher level chemistry courses.)
2. The charges of the ions obey the same rules as discussed above. In particular, the charges of all of the ions must add up to zero, which is the charge of any compound.
3. The subscripts of monatomic ions become coefficients for the ions. For polyatomic ions, only the subscripts after parentheses become coefficients.

Exercise 3.6

Write the ions expected when the following compounds are dissolved in water:

K_2S	$MgBr_2$	$Na_2Cr_2O_7$	K_3PO_4
$FeCl_3$	$AlCl_3$	$(NH_4)_2S$	$Ti(NO_3)_4$

Solution

$K_2S(s)$	$\rightarrow$	$2K^+(aq)$	+	$S^{2-}(aq)$
$FeCl_3(s)$	$\rightarrow$	$Fe^{3+}(aq)$	+	$3Cl^-(aq)$
$MgBr_2(s)$	$\rightarrow$	$Mg^{2+}(aq)$	+	$2Br^-(aq)$
$AlCl_3(s)$	$\rightarrow$	$Al^{3+}(aq)$	+	$3Cl^-(aq)$
$Na_2Cr_2O_7(s)$	$\rightarrow$	$2Na^+(aq)$	+	$Cr_2O_7^{2-}(aq)$
$(NH_4)_2S(s)$	$\rightarrow$	$2NH_4^+(aq)$	+	$S^{2-}(aq)$
$K_3PO_4(s)$	$\rightarrow$	$3K^+(aq)$	+	$PO_4^{3-}(aq)$
$Ti(NO_3)_4(s)$	$\rightarrow$	$Ti^{4+}(aq)$	+	$4NO_3^-(aq)$

Notice that the charges on Fe^{3+} and Ti^{4+} must be calculated from the known negative charge of the anion and the fact that the total charge must add up to zero. Also observe which subscripts have become coefficients and which remained as part of a polyatomic ion. Finally, this entire process is just the reverse of the method used to determine the formulas of ionic compounds.

Any ionic compound can be broken apart into its cations and anions in this manner. Whether or not an ionic compound will dissolve to an appreciable extent in water depends on which cations and anions make up the compound. The current AP general guidelines for predicting solubility are listed below.

SOLUBILITY RULES

1. **Compounds containing sodium or potassium alkali metal cations or the ammonium ion are soluble.**
2. **Compounds containing NO_3^- anions are soluble.**

Double-Replacement Reactions

PREDICTING PRODUCTS

If we know how to determine which ions make up an ionic compound, we can then take two ionic compounds, mix them together, and predict the possible products. As the name suggests, two replacements occur in these reactions.

> In one replacement the cation of the first salt replaces the cation of the second salt.
> In the second replacement the cation of the second salt replaces the cation of the first salt.

For example, if we mix together solutions containing $AgNO_3$ and Na_2CrO_4, what products should we predict? The first step is to determine the ions that make up the two reacting compounds. They are Ag^+, NO_3^-, $2Na^+$, and CrO_4^{2-}. The next step is to pair up these four ions in various ways to make two new ionic compounds that will be the predicted products. It is worthwhile to look at all of the possible pairs to see how we arrive at our conclusions.

Ag^+	$+ NO_3^-$	$\rightarrow$	$AgNO_3$ We already have this as a reactant.
Ag^+	$+ Na^+$	$\rightarrow$	cannot form an ionic compound from two positive ions
$2Ag^+$	$+ CrO_4^{2-}$	$\rightarrow$	Ag_2CrO_4 **This is a possibility.**
Ag^+	$+ Ag^+$	$\rightarrow$	cannot form an ionic compound from two identical ions
NO_3^-	$+ Na^+$	$\rightarrow$	$NaNO_3$ **This is a possibility.**
NO_3^-	$+ CrO_4^{2-}$	$\rightarrow$	cannot form an ionic compound from two negative ions
NO_3^-	$+ NO_3^-$	$\rightarrow$	cannot form an ionic compound from two identical ions
$2Na^+$	$+ CrO_4^{2-}$	$\rightarrow$	Na_2CrO_4 We already have this as a reactant.
Na^+	$+ Na^+$	$\rightarrow$	cannot form an ionic compound from two identical ions
CrO_4^{2-}	$+ CrO_4^{2-}$	$\rightarrow$	cannot form an ionic compound from two identical ions

All of the possible combinations of the four ions are given above, with the reasons why they are good or bad choices. Only two of the combinations give reasonable new compounds. Every other combination leads to either an impossible situation or back to the original compounds. Using the only reasonable results as the products, we may begin to construct a chemical equation:

$$AgNO_3 + Na_2CrO_4 \rightarrow Ag_2CrO_4 + NaNO_3$$

The final step is to balance the equation so it has the same number of each atom on both sides of the arrow:

$$2AgNO_3 + Na_2CrO_4 \rightarrow Ag_2CrO_4 + 2NaNO_3$$

Reviewing what was done to predict this equation, we see that only the positions of the silver and sodium ions have been switched on the reactant and product sides. In switching the positions of the metal atoms, we were careful to write the new formulas properly, based on the charges of the ions.

EXAMPLE 3.1

Predict the products obtained from the following pairs of ionic compounds. Then write a balanced chemical equation for each pair.

(a) KCl and $Pb(NO_3)_2$
(b) $CaCl_2$ and $MgSO_4$
(c) NaOH and $Fe_2(SO_4)_3$

Solution

(a) The ions involved are K^+, Cl^-, Pb^{2+}, and $2NO_3^-$. The possible new combinations are KNO_3 and $PbCl_2$. Placing these into a reaction and balancing it gives

$$2KCl + Pb(NO_3)_2 \rightarrow 2KNO_3 + PbCl_2$$

(b) The ions involved are Ca^{2+}, $2Cl^-$, Mg^{2+}, and SO_4^{2-}. The two new compounds are $CaSO_4$ and $MgCl_2$. The balanced chemical reaction is

$$CaCl_2 + MgSO_4 \rightarrow CaSO_4 + MgCl_2$$

(c) The ions are Na^+, OH^-, $2Fe^{3+}$, and $3SO_4^{2-}$. Note that the charge of the iron is determined by calculation. The new compounds are Na_2SO_4 and $Fe(OH)_3$. The balanced chemical reaction is

$$6NaOH + Fe_2(SO_4)_3 \rightarrow 3Na_2SO_4 + 2Fe(OH)_3$$

We can write these chemical reactions, but the major question is whether or not a chemical reaction will actually occur if the compounds are mixed together in the laboratory. In the next sections some ways in which the chemist can predict if an actual reaction will occur are presented.

Chemical Driving Forces

So far, we can predict the products of any mixture of two ionic compounds. However, not all such mixtures react. Chemists rely on three fundamental principles to make an educated guess about the possibility for a reaction to occur in a double-replacement reaction. These principles are sometimes described as **driving forces**.

1. The formation of water is perhaps the strongest driving force. In an ionic reaction where water is a product, it is almost a certainty that a double-replacement reaction is occurring.
2. Formation of a precipitate (insoluble compound) is another indicator of a strong driving force.
3. The formation of a nonionic (covalent) compound from ionic reactants is another driving force. Many of these nonionic compounds are organic acids (ethanoic, formic, benzoic acids) or gases such as NH_3, SO_2, and CO_2.

TIP

Formations of:

water
weak electrolyte
precipitate
gas

are the main driving forces of chemical reactions.

Some common examples of driving forces are as follows:

$HCl(aq)$	$+ NaOH(aq)$	$\rightarrow H_2O(aq)$	$+ NaCl(aq)$	(water formed)
$Na_2SO_4(aq)$	$+ Ba(NO_3)_2(aq)$	$\rightarrow BaSO_4(s)$	$+ 2NaNO_3(aq)$	(precipitate formed)
$KC_2H_3O_2(aq)$	$+ HCl(aq)$	$\rightarrow HC_2H_3O_2(aq)$	$+ KCl(aq)$	(covalent compound formed)
$K_2SO_3(aq)$	$+ 2HNO_3(aq)$	$\rightarrow 2KNO_3(aq)$	$+ H_2O + SO_2(g)$	(gas formed)

In some reactions two of these driving forces may be present. As mentioned above, the driving force to form water is especially strong and will overcome another force that may be driving the reaction in the opposite direction. In the equation below, the formation of water overcomes the fact that CaO is a solid. Since CaO is on the reactant side of the equation, it is driving the equation toward the reactants. The production of water, however, is a stronger driving force, and the net result is that this reaction actually occurs:

$$CaO(s) + 2HCl(aq) \rightarrow CaCl_2(aq) + H_2O(l)$$

Net Ionic Equations

In the process of determining the products of a reaction between ionic compounds, the ions for each substance were determined. In fact, in aqueous solution only the soluble compounds appear as ions, while insoluble compounds (precipitates), gases, and covalent compounds are written as molecules in the equation. It is possible to take a balanced double-replacement reaction and convert it into a net ionic equation that shows the actual reactants, if any, for a given reaction.

For example, a simple neutralization reaction is

$$HCl(aq) \quad + \quad NaOH(aq) \quad \rightarrow \quad NaCl(aq) + H_2O(l)$$

The ionic reaction is obtained by writing all of the soluble ionic compounds as ions.

$$H^+(aq) + Cl^-(aq) + Na^+(aq) + OH^-(aq) \rightarrow Na^+(aq) + Cl^-(aq) + H_2O(l)$$

Since the Na^+ and the Cl^- are identical on both sides of the equation, they can be canceled to give the net ionic equation:

$$H^+(aq) \quad + \quad OH^-(aq) \quad \rightarrow \quad H_2O(l)$$

This allows the chemist to show that it is the H^+ and OH^- ions that are the active components of the reaction.

A reaction does not occur if potassium chloride and sodium nitrate solutions are mixed. We can demonstrate that no reaction occurs by deducing the reaction products as sodium chloride and potassium nitrate and then writing the net ionic equation. First we write the molecular equation

$$KCl(aq) + NaNO_3(aq) \rightarrow NaCl(aq) + KNO_3(aq) \quad \text{(molecular equation)}$$

Next, the ionic equation is written by separating each of the compounds into its ions:

$$K^+(aq) + Cl^-(aq) + Na^+(aq) + NO^-{}_3(aq) \rightarrow Na^+(aq) + Cl^-(aq) + K^+(aq) + NO_3^-(aq)$$
$$\text{(ionic equation)}$$

Finally, after identical ions are canceled from both sides of this equation, nothing remains. This means that there is no net ionic equation and no reaction occurs:

NO NET IONIC EQUATION POSSIBLE

Taking a close look at the molecular equation, we see also that no driving force is present. No water, no precipitate, no covalent molecule, and no gas is formed.

A common laboratory experiment is the determination of sulfate ions by precipitation with barium ions. The precipitate is carefully collected, dried, and weighed in this experiment. The reaction between potassium sulfate and barium nitrate may be predicted to produce barium sulfate and sodium nitrate. The balanced equation is

$$K_2SO_4(aq) \quad + \quad Ba(NO_3)_2(aq) \quad \rightarrow \quad BaSO_4(s) \quad + \quad 2KNO_3(aq)$$

The ionic equation is

$$2K^+(aq) + SO_4^{2-}(aq) + Ba^{2+}(aq) + 2NO_3^-(aq) \rightarrow BaSO_4(s) + 2K^+(aq) + 2NO_3^-(aq)$$

In the ionic equation the two potassium and two nitrate ions may be canceled, resulting in

$$SO_4^{2-}(aq) \quad + \quad Ba^{2+}(aq) \quad \rightarrow \quad BaSO_4(s)$$

This balanced net ionic equation represents the reaction implied above by the words "determination of sulfate ions by precipitation with barium ions." In chemical analysis, the chemist is usually interested in a specific ion, such as the sulfate ion in this example. The net ionic equation shows us how to isolate the sulfate ion from all other ions by precipitation with barium ions.

Reactions that evolve gases are a bit more complex. Archaeologists typically carry a small bottle of hydrochloric acid on field trips. Carbonate rocks can be quickly identified since they will give off carbon dioxide (evidenced by bubbling and fizzing) when a few drops of HCl are placed on them. Most of these rocks are made of calcium carbonate. Using the techniques for a double-replacement reaction, we may predict the products to be calcium chloride and carbonic acid:

$$2HCl(aq) \ + \ CaCO_3(s) \ \rightarrow \ CaCl_2(aq) \ + \ H_2CO_3(aq)$$

The ionic equation is

$$2H^+(aq) + 2Cl^-(aq) + CaCO_3(s) \rightarrow Ca^{2+}(aq) + 2Cl^-(aq) + H_2CO_3(aq)$$

In this reaction only the Cl^- ions will cancel:

$$2H^+(aq) \ + \ CaCO_3(s) \ \rightarrow \ Ca^{2+}(aq) \ + \ H_2CO_3(aq)$$

This equation does not show any carbon dioxide gas. The key is that H_2CO_3 may also be written as $CO_2 + H_2O$. When this is substituted, the final reaction is

$$2H^+(aq) \ + \ CaCO_3(s) \ \rightarrow \ Ca^{2+}(aq) \ + \ H_2O(l) \ + \ CO_2(g)$$

Table 3.4 lists the common gases and their equivalents when dissolved in water. The gas and aqueous forms are interchangeable in reactions as needed.

TABLE 3.4
Common Gases and Their Equivalents in Aqueous Solution

Gas Name	Aqueous Form	Gas Form
carbon dioxide	$H_2CO_3(aq)$	$CO_2(g) + H_2O(l)$
sulfur dioxide	$H_2SO_3(aq)$	$SO_2(g) + H_2O(l)$
hydrogen sulfide	$H_2S(aq)$	$H_2S(g)$
ammonia	NH_4OH*(aq)	$NH_3(g) + H_2O(l)$

*NH_4OH does not actually exist, and should always be written as $NH_3(aq) + H_2O$.

Exercise 3.7

Write the balanced equation for each of the following pairs of ionic substances. Then write the ionic and net ionic equations for these reactions. Use (*aq*) to show soluble substances, (*s*) for insoluble compounds, (*l*) for liquids, and (*g*) for gases.

(a) $AgNO_3$ and $CaCl_2$
(b) Na_2CO_3 and $Fe(NO_3)_3$
(c) CaF_2 and HCl
(d) NH_4Cl and KOH

Solutions

(a) $2AgNO_3(aq) + CaCl_2(aq) \rightarrow 2AgCl(s) + Ca(NO_3)_2(aq)$

$2Ag^+(aq) + 2NO_3^-(aq) + Ca^{2+}(aq) + 2Cl^-(aq) \rightarrow 2AgCl(s) + Ca^{2+}(aq) + 2NO_3^-(aq)$

$Ag^+(aq) + Cl^-(aq) \rightarrow AgCl(s)$

(b) $3Na_2CO_3(aq) + 2Fe(NO_3)_3(aq) \rightarrow Fe_2(CO_3)_3(s) + 6NaNO_3(aq)$

$6Na^+(aq) + 3CO_3^{2-}(aq) + 2Fe^{3+}(aq) + 6NO_3^-(aq) \rightarrow Fe_2(CO_3)_3(s) + 6Na^+(aq) + 6NO_3^-(aq)$

$3CO_3^{2-}(aq) + 2Fe^{3+}(aq) \rightarrow Fe_2(CO_3)_3(s)$

(c) $CaF_2(aq) + 2HCl(aq) \rightarrow 2HF(aq) + CaCl_2(aq)$

$Ca^{2+}(aq) + 2F^-(aq) + 2H^+(aq) + 2Cl^-(aq) \rightarrow 2HF(aq) + Ca^{2+}(aq) + 2Cl^-(aq)$

$F^-(aq) + H^+(aq) \rightarrow HF(aq)$

(d) $NH_4Cl(aq) + KOH(aq) \rightarrow NH_4OH(aq) + KCl(aq)$

$NH_4^+(aq) + Cl^-(aq) + K^+(aq) + OH^-(aq) \rightarrow NH_3(g) + H_2O(l) + K^+(aq) + Cl^-(aq)$

$NH_4^+(aq) + OH^-(aq) \rightarrow NH_3(g) + H_2O(l)$

Note that the NH_4OH obtained from the double-replacement technique was replaced by $NH_3(g)$ and $H_2O(l)$ in the ionic equations since NH_4OH does not exist. It should not appear in the first equation either.

SINGLE-REPLACEMENT REACTIONS

A single-replacement reaction may be described as the reaction between an element and an ionic compound (or ions in solution) to form a different element and a new ionic compound. The products of these reactions, like those of double-replacement reactions, may be predicted. In one type of single-replacement reaction the element used as a reactant may be a metal that becomes a cation as a product. The second type of single-replacement reaction involves a nonmetal as the elemental reactant that then forms an anion. Several reactions in which the reacting metal forms a cation are shown below.

One of these reactions is

$$Cu(s) + 2AgNO_3(aq) \rightarrow 2Ag(s) + Cu(NO_3)_2(aq)$$

We can write the ionic equation by breaking the $AgNO_3$ and $Cu(NO_3)_2$ into their ions:

$$Cu(s) + 2Ag^+(aq) + 2NO_3^-(aq) \rightarrow 2Ag(s) + Cu^{2+}(aq) + 2NO_3^-(aq)$$

Cancelling the two nitrate ions from both sides gives the net ionic equation:

$$Cu(s) + 2Ag^+(aq) \rightarrow 2Ag(s) + Cu^{2+}(aq)$$

Active metals (i.e., the alkali metals) react with water. This is observed in the explosive reaction of potassium when it is placed in water.

$$2K(s) + H_2O(l) \rightarrow H_2(g) + 2KOH(aq)$$

We can write the ionic equation, which, since no ions cancel, is also the net ionic equation:

$$2K(s) + H_2O(l) \rightarrow H_2(g) + 2K^+(aq) + 2OH^-(aq)$$

Less active metals react with acids, as we observe with zinc and hydrochloric acid:

$$Zn(s) + 2HCl(aq) \rightarrow H_2(g) + ZnCl_2(aq)$$

the ionic equation is

$$Zn(s) + 2H^+(aq) + 2Cl^-(aq) \rightarrow H_2(g) + Zn^{2+}(aq) + 2Cl^-(aq)$$

and the net ionic equation is

$$Zn(s) + 2H^+(aq) \rightarrow H_2(g) + Zn^{2+}(aq)$$

In general, the metal reactant will form its ion, and the cation of the reactant will become the element. These reactions may be predicted when the metal reactant has only one possible cation. If, however, the metal can form several differently charged cations, as is true of lead, tin, or iron, the cation formed must be specified before the reaction can be completed.

Nonmetals that react to form anions are usually limited to the halogens. One of these reactions is

$$Cl_2(g) + 2KBr(aq) \rightarrow Br_2(l) + 2KCl(aq)$$

The net ionic reaction is

$$Cl_2(g) + 2Br^-(aq) \rightarrow Br_2(l) + 2Cl^-(aq)$$

In this reaction, the reactants Cl_2 and KBr are virtually colorless (Cl_2 is slightly yellow) and the product Br_2 produces a dark yellow or brown solution that allows us to directly observe that a reaction has occurred.

As we saw with double-replacement reactions, we can write equations for any mixture of an element and an ionic compound. To determine which reactions actually occur, we need to know whether a given element will displace an ion in a single-replacement reaction. Commonly this information is first given as the **activity series** of the elements. Later, when discussing oxidation-reduction reactions, we will find the same information in the table of standard reduction potentials. In this book we will use only an abbreviated table of standard reduction potentials (Table 3.5). We will describe the use of this table in more detail when discussing oxidation-reduction reactions. For the time being, it is important only to remember how to use this table effectively to predict whether or not single-replacement reactions will occur.

TABLE 3.5
Standard Reduction Potentials, 25°C

Half-Reaction							$E°$ (volts)
$F_2(g)$	+			$2e^-$	→	$2F^-$	2.87
Co^{3+}	+			e^-	→	Co^{2+}	1.82
Au^{3+}	+			$3e^-$	→	Au	1.50
$Cl_2(g)$	+			$2e^-$	→	$2Cl^-$	1.36
$O_2(g)$	+	$4H^+$	+	$4e^-$	→	$2H_2O$	1.23
$Br_2(g)$	+			$2e^-$	→	$2Br^-$	1.07
$2Hg^{2+}$	+			$2e^-$	→	Hg_2^{2+}	0.92
Ag^+	+			e^-	→	Ag	0.80
Hg_2^{2+}	+			$2e^-$	→	Hg	0.79
Fe^{3+}	+			e^-	→	Fe^{2+}	0.77
I_2	+			$2e^-$	→	$2I^-$	0.53
Cu^+	+			e^-	→	Cu	0.52
Cu^{2+}	+			$2e^-$	→	Cu	0.34
Cu^{2+}	+			e^-	→	Cu^+	0.15
Sn^{4+}	+			$2e^-$	→	Sn^{2+}	0.15
S	+	$2H^+$	+	$2e^-$	→	H_2S	0.14
$2H^+$	+			$2e^-$	→	H_2	0.00
Pb^{2+}	+			$2e^-$	→	Pb	−0.13
Sn^{2+}	+			$2e^-$	→	Sn	−0.14
Ni^{2+}	+			$2e^-$	→	Ni	−0.25
Co^{2+}	+			$2e^-$	→	Co	−0.28
Tl^+	+			e^-	→	Tl	−0.34
Cd^{2+}	+			$2e^-$	→	Cd	−0.40
Cr^{3+}	+			e^-	→	Cr^{2+}	−0.41
Fe^{2+}	+			$2e^-$	→	Fe	−0.44
Cr^{3+}	+			$3e^-$	→	Cr	−0.74
Zn^{2+}	+			$2e^-$	→	Zn	−0.76
$2H_2O$	+			$2e^-$	→	$H_2(g)+2OH^-$	−0.83
Mn^{2+}	+			$2e^-$	→	Mn	−1.18
Al^{3+}	+			$3e^-$	→	Al	−1.66
Be^{2+}	+			$2e^-$	→	Be	−1.70
Mg^{2+}	+			$2e^-$	→	Mg	−2.37
Na^+	+			e^-	→	Na	−2.71
Ca^{2+}	+			$2e^-$	→	Ca	−2.87
Sr^{2+}	+			$2e^-$	→	Sr	−2.89
Ba^{2+}	+			$2e^-$	→	Ba	−2.90
Rb^+	+			e^-	→	Rb	−2.92
K^+	+			e^-	→	K	−2.92
Cs^+	+			e^-	→	Cs	−2.92
Li^+	+			e^-	→	Li	−3.05

This table will no longer be provided with the AP exam. Necessary standard potentials and half-reactions will be given as needed.

SUMMARY

This chapter focuses on ionic compounds and their reactions. Ionic compounds have simple, often predictable formulas based on information in the periodic table. The structure of ions that make up ionic compounds is presented, and we note that ions of the representative elements are often isoelectronic with noble gases. The nature of other ions can also be understood based on electronic structures that are stable. Ionic compounds also react in a process called double replacement as long as a precipitate, a gaseous product, or a weak electrolyte is part of the chemical equation. Ionic reactions are often best represented as balanced net ionic equations. This chapter introduces a set of solubility rules and also a summary of how reactions may be classified. A logical method for naming ionic compounds is also presented.

Important Concepts

Predicting ionic charges
Balancing reactions
Writing ionic formulas and electroneutrality of compounds
Writing ionic and net ionic equations
Naming ionic compounds
Solubility rules
Polyatomic ions

PRACTICE EXERCISES

Multiple-Choice

1. What is the correct formula for aluminum sulfate?
   (A) $Al_3(SO_4)_2$ where the subscripts were chosen to match the ion charges.
   (B) $AlSO_4$ since ionic compounds are in 1 to 1 ratios of the ions.
   (C) $Al_2(SO_4)_3$ where the subscripts were chosen so the positive charge and negative charge add to zero.
   (D) $Al_2(SO_3)_3$ where the subscripts were chosen so the positive charge and negative charge add to zero.

2. What are the spectator ions when nitric acid reacts with sodium hydroxide?
   (A) $H^+(aq) + NO_3^-(aq)$ because all compounds with the hydroxide ion are insoluble
   (B) $Na^+(aq) + NO_3^-(aq)$ since these ions do not combine in an aqueous solution
   (C) $Na^+(aq) + OH^-(aq)$ because all compounds with the hydroxide ion are soluble
   (D) $H^+(aq) + OH^-(aq)$ since both ions do not appear as products.

3. Which of the following will act as a driving force for a chemical reaction?
   (A) Formation of a gas that bubbles out of solution
   (B) Formation of a weak electrolyte that removes two ions from solution
   (C) Formation of water, which is a weak electrolyte
   (D) All of these are driving forces

4. Which compound, made from the indicated pairs of ions, is expected to be the least soluble based on coulombic attractive forces?
   (A) Iron(III) and nitride ion because both ions have 3 charges of opposite sign
   (B) Bromide ion and sodium ion because both ions have single charges of opposite sign
   (C) Calcium ion and bromide ion because one ion with a +2 charge is attracting 2 ions each with a −1 charge
   (D) Bromide ion and iron(III) because one ion with a +3 charge is attracting three ions each with a −1 charge

5. When lead(II) acetate is mixed with potassium chloride, what are the formulas of the products that should be written in the molecular equation?
   (A) $PbCl_4 + K_2C_2H_3O_2$
   (B) $Pb(C_2H_3O_2)_2 + KCl$
   (C) $PbCl_2 + KC_2H_3O_2$
   (D) $PbCl + KC_2H_3O_2$

6. When ammonium oxalate, $(NH_4)_2C_2O_4$, is dissolved in water, the ions formed are
   (A) $2N^{3-}(aq) + 8H^{+}(aq) + 2C^{4+}(aq) + 4O^{2-}(aq)$
   (B) $(NH_4)^{2+}(aq) + C_2O_4^{2-}(aq)$
   (C) $2NH_4^{+}(aq) + C_2O_4^{2-}(aq)$
   (D) $NH_4^{2+}(aq) + C_2O_4^{2-}(aq)$

7. Which of the following ions are isoelectronic (have the same electron configuration as) to neon?
   (A) F^-
   (B) Ga^{3+}
   (C) P^{3-}
   (D) All of these are isoelectronic to neon.

8. Which substance is expected to have the largest distance between its nuclei?
   (A) NaF
   (B) NaCl
   (C) NaI
   (D) NaBr

9. Which of the following pairs of compounds is expected to have the compound with largest solubility listed first?
   (A) $MgCl_2$ and $AlBr_3$
   (B) CaF_2 and MgS
   (C) KBr and Na_2S
   (D) Al_2O_3 and Cr_2O_3

10. Which two atoms form ions with the same $1s^2\ 2s^2\ 2p^6$ electron configuration?
    (A) Cl and Na
    (B) Cl and F
    (C) Na and F
    (D) S and Br

11. What is the formula and name for an ionic compound formed from aluminum and chlorine?
    (A) AlCl, aluminum chloride
    (B) Al_3Cl, tri-aluminum chloride
    (C) Al_3Cl_3, aluminum(III) trichloride
    (D) $AlCl_3$, aluminum chloride

12. The electronic configuration $1s^2\ 2s^2\ 2p^6\ 3s^2\ 3p^6$ corresponds to the electronic configuration of
    (A) S^{2-}
    (B) Cl^-
    (C) K^+
    (D) all of the above

13. Which of the following reactions is a combustion reaction?
    (A) $C_7H_8O_2(l) + 8O_2(g) \rightarrow 7CO_2(g) + 4H_2O(l)$
    (B) $2KClO_3(s) \rightarrow 2KCl(s) + 3O_2(g)$
    (C) $2Cr(s) + 3Cl_2(g) \rightarrow 2CrCl_3(s)$
    (D) $6Li(s) + N_2(g) \rightarrow 2Li_3N(s)$

14. Which of the following is NOT a correct chemical formula and why?
    (A) $SrBr_2$, strontium has only ions with a +1 charge
    (B) Ca_2O_3, the charges for the ions do not add up to zero
    (C) Mg_3N_2, the charges for the ions do not add up to zero
    (D) Na_2S, sodium cannot bond with sulfur

15. Which of the following is a correct formula?
    (A) KH_2PO_4
    (B) $CaC_2H_3O_2$
    (C) Na_2ClO_4
    (D) $Ba(CO_3)_2$

16. Which of the following pairs of substances undergo a reaction often classified as single replacement?
    (A) Sodium chloride, potassium nitrate
    (B) Chlorine gas, sodium metal
    (C) Aluminum metal, hydrobromic acid
    (D) Ethanol, oxygen

17. Which of the following is NOT true of a net ionic equation?
    (A) All of the nonreacting (spectator) ions have been canceled.
    (B) It shows the actual reactants in an equation.
    (C) It allows the chemist to substitute reactants in a logical manner.
    (D) It is used to determine which compounds are insoluble.

CHALLENGE

18. What is the balanced molecular equation when $FeCl_3(aq)$ is mixed with $Ba(OH)_2(aq)$?
    (A) $Ba(OH)_2(aq) + FeCl_3(aq) \rightarrow Fe(OH)_2(aq) + BaCl_3(aq)$
    (B) $2Ba(OH)_2(aq) + 3FeCl_3(aq) \rightarrow 2BaCl_2(aq) + 3Fe(OH)_3(s)$
    (C) $3Ba(OH)_2(aq) + 2FeCl_3(aq) \rightarrow 3BaCl_2(s) + 2Fe(OH)_3(s)$
    (D) $3Ba(OH)_2(aq) + 2FeCl_3(aq) \rightarrow 3BaCl_2(aq) + 2Fe(OH)_3(s)$

19. Gaseous hydrogen and nitrogen gas are reacted to form ammonia. Which of the following molecular views shows the correct stoichiometry for this reaction?

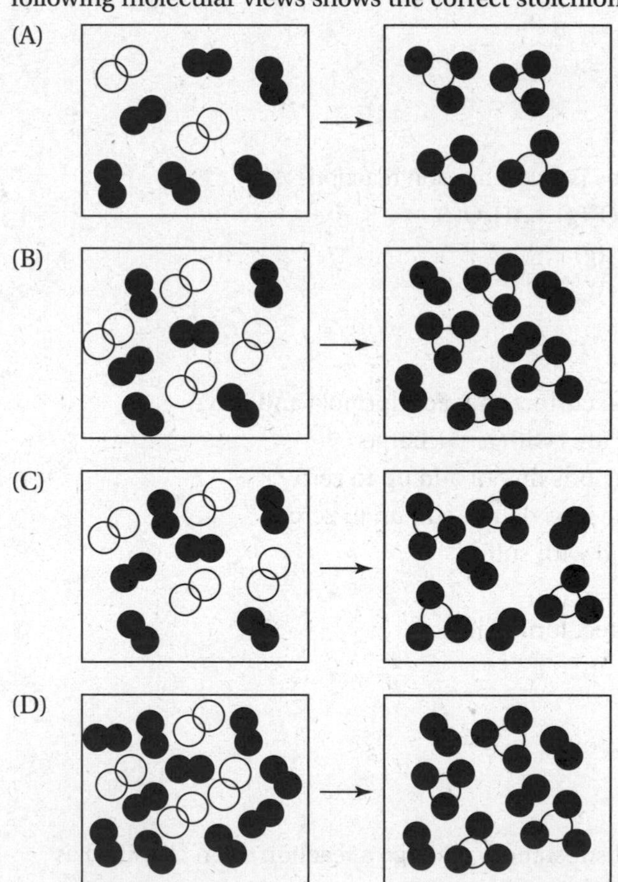

20. Which of the following molecular views does not violate the law of conservation of matter?

(A)

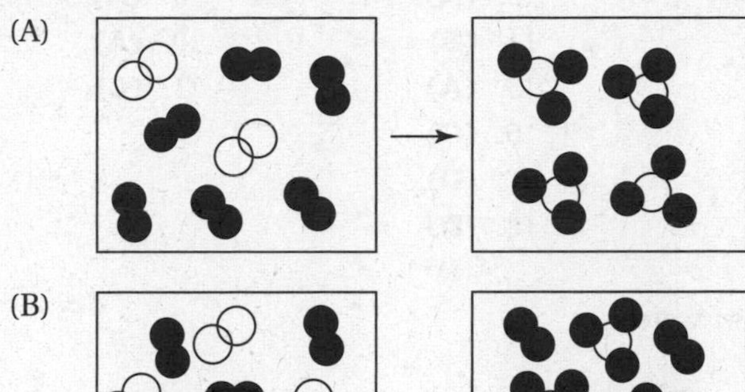

(B)

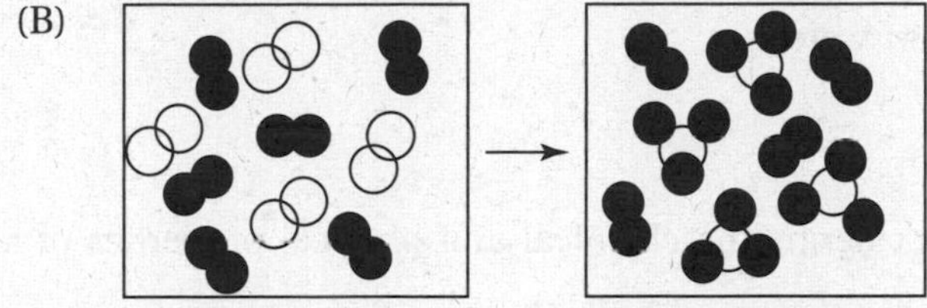

(C)

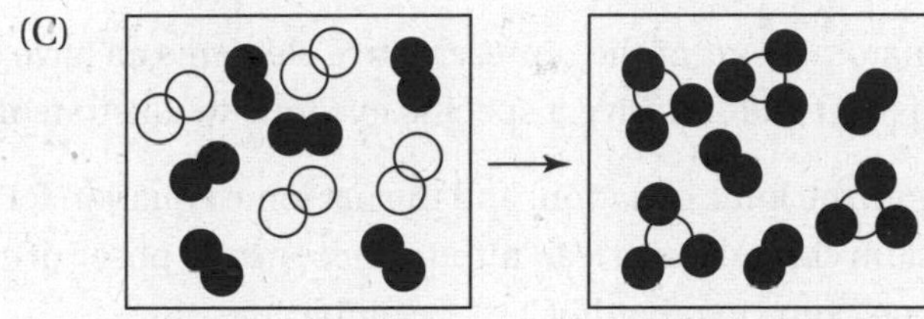

(D)

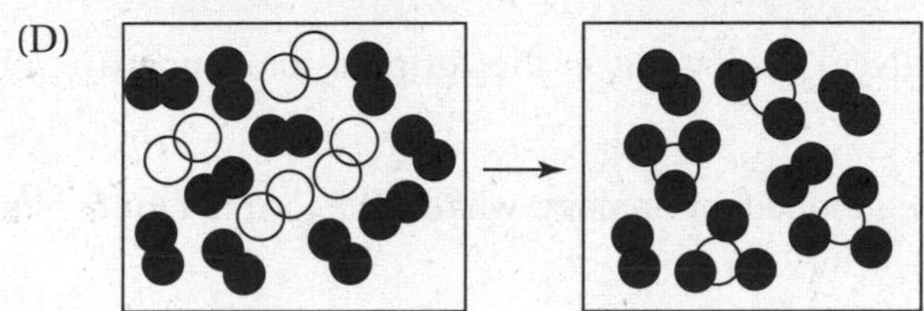

CHALLENGE

21. What is the net ionic equation for the following reaction that takes place in aqueous solution?

$$CrCl_3 + Pb(NO_3)_2 \rightarrow PbCl_2 + Cr(NO_3)_3$$

(A) $Cr^{2+}(aq) + Pb^{2+}(aq) \rightarrow CrPb(s)$
(B) $Cr^{3}(aq) + 3NO_3(aq) \rightarrow Cr(NO_3)_3(s)$
(C) $Pb^{2+}(aq) + 2Cl^{-}(aq) \rightarrow PbCl_2(s)$
(D) $Pb(NO_3)(s) + 2Cl^{-}(aq) \rightarrow PbCl_2(s) + 2NO_3(aq)$

ANSWER KEY

1. **(C)**
2. **(B)**
3. **(D)**
4. **(A)**
5. **(C)**
6. **(C)**
7. **(A)**
8. **(C)**
9. **(C)**
10. **(C)**
11. **(D)**
12. **(D)**
13. **(A)**
14. **(B)**
15. **(A)**
16. **(C)**
17. **(D)**
18. **(D)**
19. **(A)**
20. **(A)**
21. **(C)**

See Appendix 1 for explanations of answers.

Free-Response

Answer the following questions considering the chemical and physical properties of ionic substances.

(a) A large majority of the monatomic ions of the representative elements all have one feature in common. What is that feature? Give a specific example to illustrate it.

(b) Write the molecular equation, the ionic equation, and the net ionic equation for the aqueous reaction of potassium chloride with lead nitrate. Indicate the phase of each of the reactants and products, knowing that $PbCl_2$ is insoluble.

(c) What is the name for $Cr(NO_3)_3$ and what is the formula for copper(II) sulfate pentahydrate?

(d) Although it has not been mentioned, by analogy, what is the formula for the iodate ion?

(e) What is potassium permanganate used for? Does it have distinguishing physical properties?

ANSWERS

(a) Monatomic ions of the representative elements contain a single atom of an element in the *s*-block or *p*-block of the periodic table. Most of these ions are formed by gaining or losing electrons to become isoelectronic with its nearest noble gas.

(b) The molecular equation writes the complete formulas of the reactants and products as

$$2KCl(aq) + Pb(NO_3)_2(aq) \longrightarrow PbCl_2(s) + 2KNO_3(aq)$$

The ionic equation writes all soluble species as their aqueous ions:

$$2K^+(aq) + 2\,Cl^-(aq) + Pb^{2+}(aq) + 2\,NO_3^-(aq) \longrightarrow PbCl_2(s) + 2\,K^+(aq) + 2\,NO_3^-(aq)$$

Finally, the net ionic equation is obtained by canceling the potassium and nitrate ions to get

$$2Cl^-(aq) + Pb^{2+}(aq) \longrightarrow PbCl_2(s)$$

(c) $Cr(NO_3)_3$ contains Cr^{3+} and three NO_3^- ions. Chromium is a transition element (*d*-block) and has multiple oxidation states, we also identify the NO_3^- ion as the nitrate ion. Therefore we must use the Stock system of naming. We start by converting the 3+ charge to (III) and adding it to the name of the metal to get chromium(III). Because we have a polyatomic ion, all we have to do is add its name to the metal to get chromium(III) nitrate.

The second compound is copper(II) sulfate pentahydrate. This contains the copper(II) ion or Cu^{2+}. It also contains the sulfate ion, SO_4^{2-}. The charges, 2+ and 2− add to zero so the main part of the formula is simply $CuSO_4$. The name includes pentahydrate. Penta translates as 5, and hydrate indicates water molecules and we have $5H_2O$ in the formula also. Combined it is written as $CuSO_4 \bullet 5H_2O$.

(d) We have named the chlorine polyatomic ions as hypochlorite, ClO^-, chlorite, ClO_2^-, chlorate, ClO_3^-, and perchlorate, ClO_4^-. By substituting iodine for each chlorine we can obtain the formulas for the hypoiodite, iodite, iodate, and periodate ions. Therefore the iodate ion is IO_3^-.

(e) Potassium permanganate is $KMnO_4$. It is a common reactant used in many redox reactions. In fact, it is one of the strongest oxidizers. By virtue of the presence of a potassium ion we can deduce that this compound must be soluble in aqueous solution. In addition, $KMnO_4$ is highly colored with a characteristic purple color.

Covalent Compounds, Formulas, and Structures

4

- → COVALENT MOLECULES
- → COVALENT BONDS
- → LEWIS STRUCTURES
- → OCTET RULE
- → MULTIPLE BONDS
- → RESONANCE STRUCTURES
- → FORMAL CHARGES
- → BENZENE STRUCTURE
- → ELECTRONEGATIVITY
- → BOND POLARITY
- → IONIC CHARACTER
- → BOND STRENGTH
- → BOND ENERGY
- → BOND LENGTH
- → BOND ORDER
- → NAMING MOLECULES
- → VSEPR THEORY
- → MOLECULAR GEOMETRY
- → PLANAR TRIANGLE
- → TETRAHEDRON
- → TRIANGULAR BIPYRAMID
- → OCTAHEDRON
- → BOND ANGLES
- → MOLECULAR POLARITY
- → VALENCE BOND THEORY
- → HYBRID ORBITALS
- → *sp* HYBRID ORBITALS
- → sp^2 HYBRID ORBITALS
- → sp^3 HYBRID ORBITALS
- → SIGMA AND PI BONDS

BIG IDEA 3

Learning Objectives: 2.1, 2.11, 2.13, 2.17, 2.19, 2.20, 2.21, 2.22, 2.23, 2.24, 2.29, 2.30, 2.31, 2.32

For the complete list of Big Ideas and Learning Objectives, refer to the AP Chemistry Course Outline: *https://secure-media.collegeboard.org/digitalServices/pdf/ap/ap-chemistry-course-and-exam-description.pdf*

COVALENT MOLECULES

Formation of **covalent bonds** represents another way in which elements may combine to form compounds and at the same time attain a noble-gas electronic configuration. In contrast to ionic bonding, where electrons are transferred from atom to atom, covalent bonding occurs when electrons are shared between two or more atoms. In addition, whereas ionic bonding is simply electrical attraction of oppositely charged ions, in covalent compounds the atoms are physically attached to each other. Compounds composed of covalently bonded groups of atoms are called **molecules**.

To conveniently show this sharing of electrons, chemists draw structures of covalent molecules using **Lewis electron-dot structures**. In the Lewis representation the outermost *s* and *p* electrons (valence electrons) are shown as dots arranged around the atomic symbol. **Lewis structures** for 18 elements are shown in Figure 4.1.

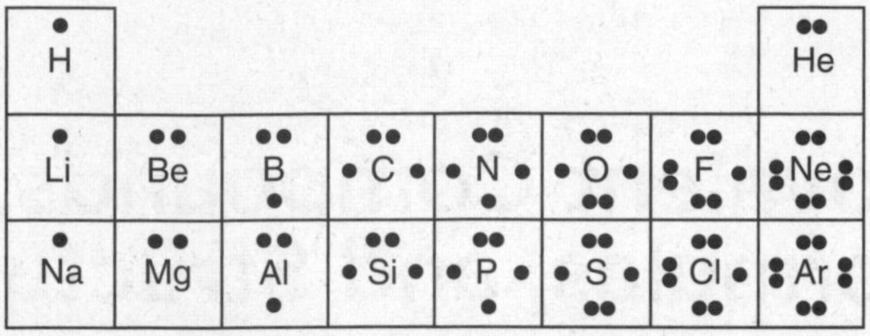

H							He
Li	Be	B	C	N	O	F	Ne
Na	Mg	Al	Si	P	S	Cl	Ar

FIGURE 4.1. Elements normally represented with Lewis electron-dot symbols. (Sometimes the paired electrons in Be, B, C, Mg, Al, and Si are separated.)

When electrons are shared between two atoms, one atom donates one electron and the other atom donates the second electron. The shared pair of electrons represents a **covalent bond**. If two pairs of electrons are shared between two atoms, a **double bond** exists. When two atoms share three pairs of electrons, a **triple bond** exists.

The sharing of electrons follows the same basic principle as prevails in the formation of ions. This fact means that the atoms are trying to attain noble-gas electronic configurations. Referring to the noble-gas configurations, we see that there are two *s* electrons and six *p* electrons in complete sublevels of each noble gas. These eight electrons represent the octet of the octet rule, which governs covalent compounds. The **octet rule** states that the noble-gas configuration will be achieved if the Lewis structure shows eight electrons around each atom. Hydrogen is an exception; its "octet" is two electrons, corresponding to the two outermost electrons in the noble gas helium.

Figure 4.2 shows the Lewis electron dot structures for the **diatomic** gases. In these structures the electrons of one atom are represented as dots and those of the other atom as circles in order to illustrate that each atom contributes the same number of electrons to the bond.

H:H N:::N O::O F:F Cl:Cl

FIGURE 4.2. Lewis structures of the diatomic elements.

In these molecules, each atom considers the shared electrons (those electrons between the two atoms) as its own. Each hydrogen in H_2 thinks it has two electrons and an electronic configuration the same as that of helium. Nitrogen molecules have triple bonds with three pairs of shared electrons. Oxygen molecules may be represented with a double bond. Fluorine and chlorine look very similar, as expected, since both are halogens.

Molecular oxygen has a simple Lewis structure, shown in Figure 4.2, which turns out to be incorrect. Experimental evidence shows that oxygen is paramagnetic and must contain unpaired electrons. More sophisticated **molecular orbital** methods must be used to obtain the correct bonding showing these unpaired electrons.

Magnetic properties of atoms, ions, and molecules arise from unpaired electrons. The electron spin generates a small magnetic field. When electrons are paired, these magnetic fields tend to cancel, and when unpaired, they impart magnetic properties to the substance. The more unpaired electrons, the stronger the magnetic effect. Substances that have unpaired electrons are said to be **paramagnetic**. They are attracted toward an external magnet, and the force of the attraction is proportional to the number of unpaired electrons. **Diamagnetic** substances have all of their electrons paired and have very small magnetic fields.

LEWIS STRUCTURES OF MOLECULES

Drawing and interpreting Lewis structures is the key to understanding many chemical and physical properties of covalent compounds.

Lewis structures of other molecules and polyatomic ions may be drawn using the basic octet rule. For larger molecules it is first necessary to determine the general arrangement of the atoms, often called the "skeleton." In most molecules this skeleton consists of a central atom with surrounding atoms bonded to it. A few general rules apply to determining the skeleton.

1. Carbon is usually a central atom in the structure. In compounds with more than one carbon atom, the carbon atoms are joined in a chain to start the skeleton.
2. Hydrogen is never a central atom because it can form only one covalent bond.
3. Halogens form only a single covalent bond when oxygen is not present, and therefore a halogen will generally not be a central atom.
4. Oxygen forms only two covalent bonds and is rarely a central atom. However, it may link two carbon atoms in a carbon chain.
5. In the simpler molecules, the atom that appears only once in the formula will often be the central atom.

Once the molecular skeleton is determined, the available valence electrons must be arranged in octets around each atom. This is done in steps as follows:

1. The valence electrons of all atoms are added together.
2. If the substance is a polyatomic ion, we must take into account the electrons used to form the ion. For anions, the charge represents additional electrons that must be added to the total of the valence electrons. For cations, the charge represents missing electrons that must be subtracted from the valence electrons.
3. A pair of electrons is placed between each two atoms in the skeletal structure to represent a covalent bond. These electrons are called the **bonding pairs**.
4. The remaining electrons are used to complete the octets of all outer atoms in the skeleton. These electrons are **nonbonding pairs** of electrons (also called **lone pairs**).
5. If any electrons are left over, they are added in pairs to the central atom. These electrons are also **nonbonding pairs** or **lone pairs**. These lone pairs are very important in determining the geometric structure of the molecule.
6. When all electrons have been placed, the outer atoms will all have octets. The central atom may have an octet, or it may have more or fewer than eight electrons.
   a. If the central atom has an octet, the structure is complete (see the section on formal charges, page 169).
   b. It is all right if the central atom has fewer than eight electrons, provided that the atom is boron. For other central atoms, double bonds must be constructed to obtain an octet. This is done by taking a nonbonding pair of electrons from an outer atom and placing them as a bonding pair to make a double bond. Enough double bonds are constructed to give the central atom an octet.
   c. The central atom may have more than eight electrons (as in PCl_5 shown in Figure 4.7) only if it is in Periods 3–7 of the periodic table. If the central atom is in Period 2, it cannot have more than an octet of electrons.

For example, we may construct the Lewis diagram for methane, CH_4, in the following manner. First we draw the skeleton with carbon in the center and the hydrogen atoms arranged symmetrically around it. Next we count the valence electrons. There are four valence electrons on the carbon atom and one on each of the four hydrogen atoms, for a total of eight electrons. In the next step we add the bonding pairs of electrons to the structure. This uses up all of the electrons, and we check to see whether each atom has an octet. (Remember that for hydrogen an "octet" is simply one pair of electrons.) Finally, to simplify the structure, we can replace the bonding pairs of electrons by a line representing a bond. These steps are shown in Figure 4.3.

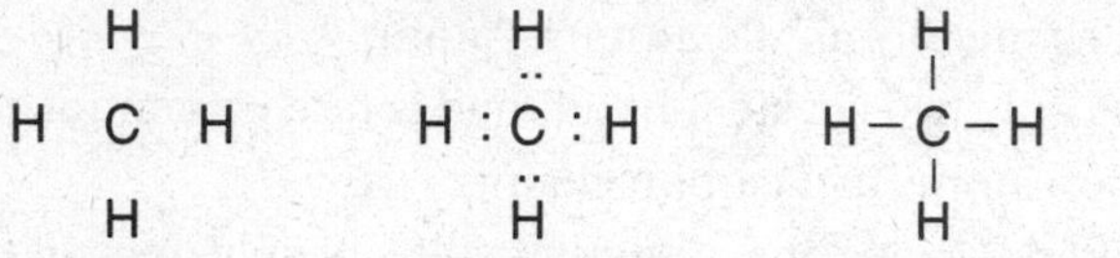

```
   H            H              H
               ..              |
 H C H       H : C : H       H-C-H
               ..              |
   H            H              H
Skeleton    Bonding Pairs   Lines Represent
Structure      Added            Bonds
```

FIGURE 4.3. Steps used to construct the Lewis structure for methane. Because of its simplicity, the structure is complete once the bonding pairs are added.

The most common **electron-deficient** molecules involve boron. The structure of boron trifluoride, BF_3, is constructed as illustrated in Figure 4.4.

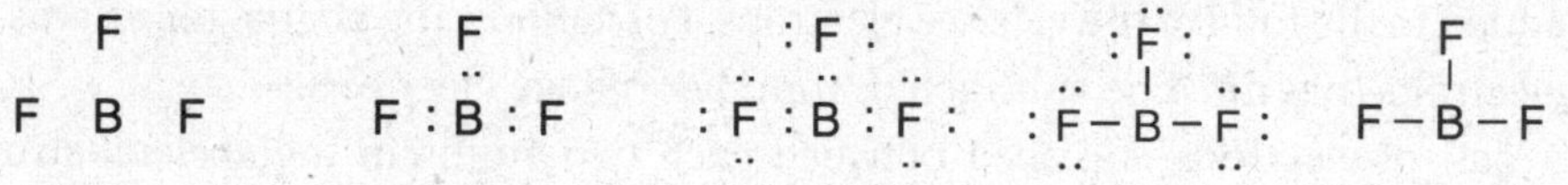

```
                            ..           ..
  F          F            : F :        : F :        F
             ..          ..  ..  ..   ..  |  ..     |
F B F      F : B : F    : F : B : F : : F-B-F :   F-B-F
                         ..      ..   ..     ..
Skeleton    Bonding      Outer Octets   Line-dot   Line Structure
            Electrons    Completed      Structure
```

FIGURE 4.4. Steps used to construct the BF_3 molecule. Compared to methane, this construction requires the additional step of constructing octets around the fluorine atoms.

The skeleton is arranged with boron as the central atom. Adding up the valence electrons, we get $3 + 7 + 7 + 7 = 24$ electrons for the boron and three fluorine atoms. Next, bonding pairs are added between each fluorine and the boron atom, using six electrons and leaving 18 to be placed. Six more electrons are placed around each fluorine atom to complete its octet. All of the remaining electrons are now utilized. Since boron is commonly found with an electron-deficient structure (i.e., with less than an octet of electrons), the structure is complete. The line structure may be drawn for simplicity. The nonbonding pairs of the outer atoms are generally not shown in a line structure.

Phosphorus forms two compounds with chlorine, PCl_3 and PCl_5. Examining their Lewis structures, we find first that PCl_3 is constructed as shown in Figure 4.5.

As before, the skeleton is drawn and the valence electrons counted. There are $5 + 7 + 7 + 7 = 26$ electrons. The bonding electron pairs and the outer octets are completed in the next two steps, leaving two electrons unused. These are placed on the central phosphorus atom as shown in Figure 4.5, completing its octet. All atoms are checked to see that each has an octet of electrons. One way to do this is to draw a circle around all of the electrons adjacent to each atom, as shown in Figure 4.6.

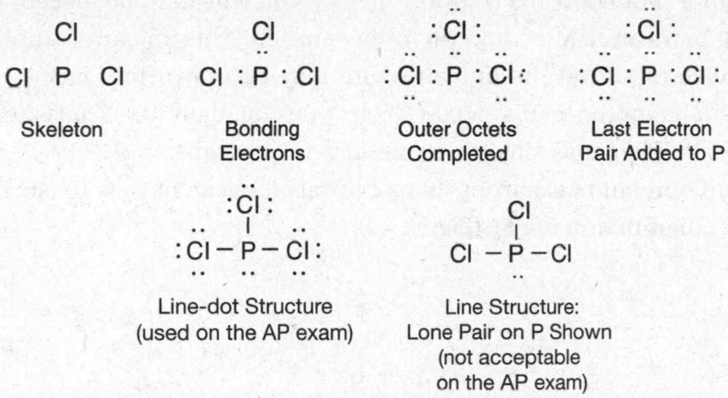

Skeleton

Bonding Electrons

Outer Octets Completed

Last Electron Pair Added to P

Line-dot Structure (used on the AP exam)

Line Structure: Lone Pair on P Shown (not acceptable on the AP exam)

FIGURE 4.5. Steps used to construct the Lewis structure for PCl_3. Compared to BCl_3, this construction requires the addition of a pair of nonbonding electrons to the central atom in step 4.

TIP

On the AP Chemistry exam, *all* electrons must be shown. Line structures with only lone pairs shown are often used by more advanced chemists. Line structures will receive zero credit on the exam.

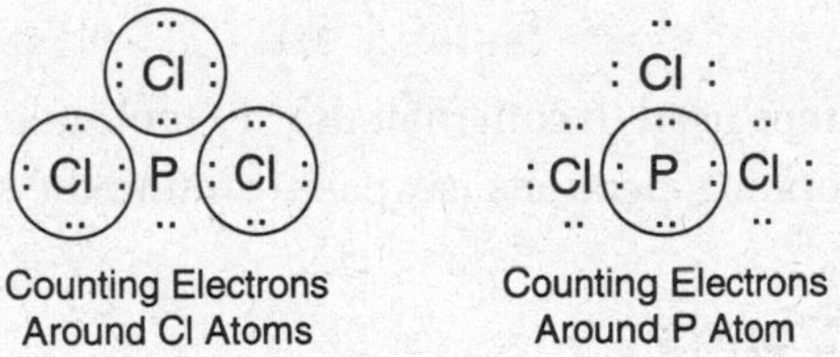

Counting Electrons Around Cl Atoms

Counting Electrons Around P Atom

FIGURE 4.6. Diagram illustrating how to count electrons around each atom by drawing a circle that includes all nonbonding electrons and all bonding electrons for that atom.

TIP

Circling electrons helps identify octets.

Note that the line structure may be used by more advanced chemists, but it is not used on the AP exam. Since nonbonding pairs on the central atom are important in determining the molecular geometry, they are shown in the line structure (see the last structure in Figure 4.5).

In constructing the PCl_5 molecule, we observe some differences, as shown in Figure 4.7.

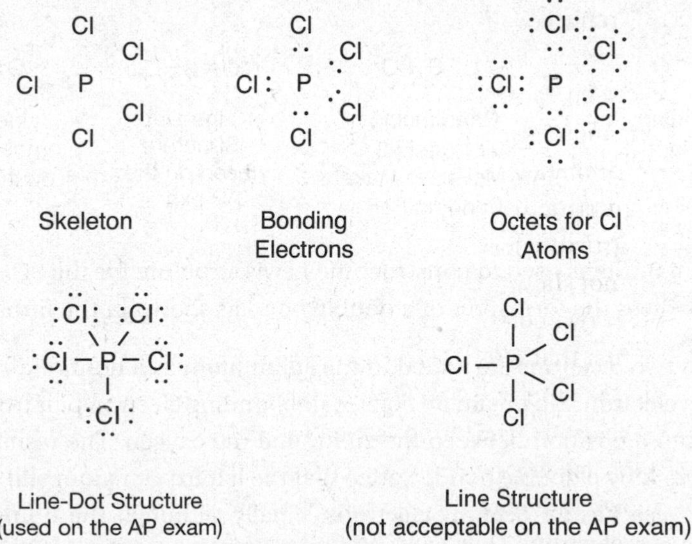

FIGURE 4.7. Steps used to construct the PCl_5 molecule. The phosphorus atom has 10 bonding electrons (five pairs), which are allowed since phosphorus is in Period 3 of the periodic table.

TIP

To avoid losing points when drawing Lewis structures for the AP exam, include ALL electrons, including lone pairs on outer atoms.

There are a total of 40 valence electrons in PCl_5, which need to be placed. The skeleton is drawn, bonding pairs are added, and then the octets for chlorine are completed. This step uses up all 40 electrons, and the line structure may be drawn for clarity. Notice that the phosphorus atom has more than an octet of electrons (actually 10). This is reasonable since phosphorus is in Period 3 and may have an excess of electrons.

The maximum number of electrons that a central atom can have is 12 (six bonding pairs), as shown in the construction of SF_6 (Figure 4.8).

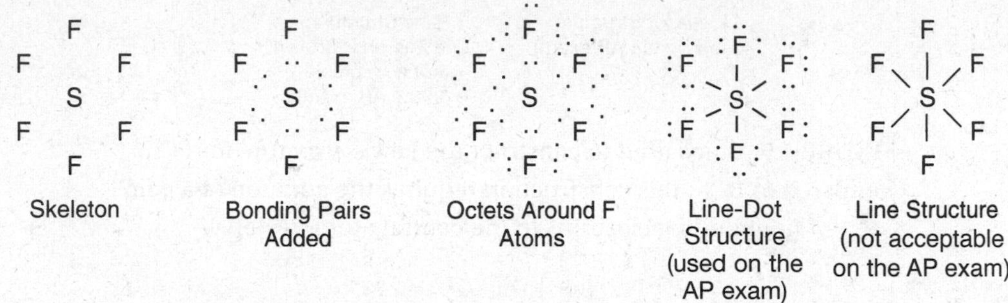

FIGURE 4.8. Steps used to construct the SF_6 molecule. This structure shows 12 bonding electrons (six pairs) around the sulfur atom.

Multiple Covalent Bonds

Some compounds require the use of double bonds, which are represented by two pairs of electrons between atoms. Sulfur dioxide, SO_2, is one of those substances. There are 18 valence electrons $(6 + 6 + 6 = 18)$ to distribute on the O S O skeleton so as to obtain octets on all atoms. The steps are illustrated in Figure 4.9.

O S O — Skeleton

O : S : O — Bonding e⁻ Added

:Ö : S : Ö: — Octets Around O Completed

:Ö : S̈ : Ö: — Nonbonding Pair Added to S

Ö : : S̈ : Ö: — Nonbonding Pair from Left O Are Moved to Make a Bonding Pair

Ö=S̈−Ö: — Line-Dot Structure (used on the AP exam)

O=S−O — Line Structure (not acceptable on the AP exam)

TIP

SO_2 and NO_2 are not linear molecules. The section on molecular geometry describes how to determine actual shapes of molecules.

FIGURE 4.9. Steps used to construct the Lewis structure for the SO_2 molecule. This involves the formation of a double bond as shown in the fifth diagram.

After the last two electrons are added to the sulfur atom as a nonbonding pair, the sulfur still has only six electrons. To obtain an octet, a nonbonding electron pair from the left oxygen atom is moved to a position between the sulfur and the oxygen. The result is an additional bonding pair, creating a double bond. Notice that the left oxygen atom still has an octet, and now the sulfur also has an octet of electrons. Finally, although the nonbonding pair was moved from the left oxygen, we could have chosen to move a pair from the oxygen atom on the right to obtain a similar structure with the double bond on the right. These two, equally probable structures (Figure 4.10) are called **resonance structures** (see page 172).

:O = S – O: ⟷ :O – S = O:

FIGURE 4.10. The two resonance structures of SO_2. These molecules differ only in the position of the double bond.

In addition, SO_2 is shown here as a linear molecule. See the Molecular Geometry section on page 180 to see how the correct shape is deduced.

Some molecules contain triple bonds. Common examples are nitrogen, N_2, acetylene, C_2H_2, and the cyanide ion, CN^-. A triple bond is formed by first constructing a double bond as described above. If the atoms still do not have octets, a second pair of nonbonding electrons may be moved to a bonding position.

Lewis Structures of Ions

In addition to the Lewis structures for molecules, we may also draw Lewis structures for covalently bonded polyatomic ions. The methods are the same as those for molecules except that we must account for the electrons that give an ion its charge. One electron is added to the valence electrons for each negative charge on an ion and one electron is subtracted from the valence electrons for each positive charge. The structure of the nitrite ion, NO_2^-, is an example. As with molecules, we count the valence electrons, $6 + 6 + 5 = 17$. The negative charge adds one electron for a total of 18. As shown in Figure 4.11, we construct the skeleton, add the bonding electrons, complete the octets around the oxygen atoms, add one nonbonding pair to nitrogen to use the remaining electrons, and finally construct a double bond so that all atoms have octets. When writing Lewis structures for ions, it is necessary to enclose the ion in brackets and indicate the charge of the ion as shown.

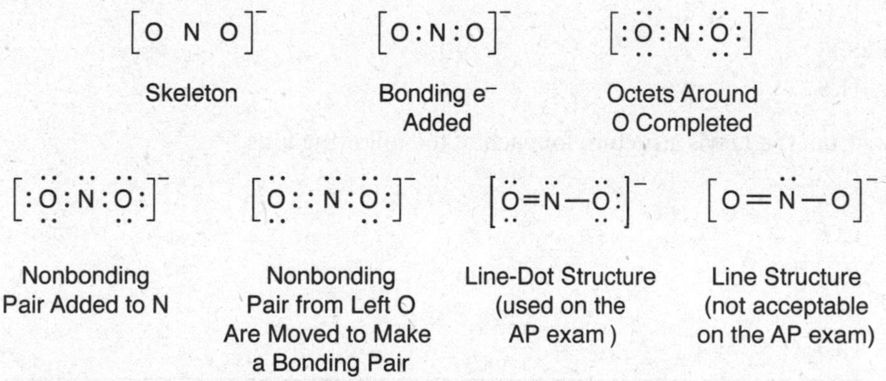

[O N O]⁻ Skeleton

[O:N:O]⁻ Bonding e⁻ Added

[:O:N:O:]⁻ Octets Around O Completed

[:O:N:O:]⁻ Nonbonding Pair Added to N

[O::N:O:]⁻ Nonbonding Pair from Left O Are Moved to Make a Bonding Pair

[O=N—O:]⁻ Line-Dot Structure (used on the AP exam)

[O=N—O]⁻ Line Structure (not acceptable on the AP exam)

FIGURE 4.11. Steps used to construct the Lewis structure of the nitrite ion.

We notice a distinct similarity in the structures of the SO_2 molecule and the NO_2^- ion. The electrons are arranged in exactly the same manner. Since the nitrogen atom has one less electron than sulfur does, the nitrite ion requires an extra electron, which results in its –1 charge. As with the SO_2, the linear shape of the NO_2^- shown here is incorrect. The next step is to determine the proper shape (see pages 180–186).

Summary: By now, we see a pattern developing for the construction of Lewis structures. Table 4.1 summarizes the logical steps involved in obtaining a Lewis structure.

TABLE 4.1
Steps Used to Construct Lewis Structures

1. Arrange atoms in a skeleton structure. Remember that H and halogens cannot be central atoms except in oxo-anions. Oxo-acids have H bonded to the oxygen atoms.
2. Count all valence electrons of all atoms. For anions, add one electron for each negative charge. For cations, subtract one electron for each positive charge. Remember to bracket ions with the overall charge of the ion.
3. Add pairs of electrons between atoms in the skeleton structure to form (sigma) bonds.
4. Then add electrons to all atoms to complete their octets (recall that an octet for H = 2 and for B = 6).
5. (a) If all atoms have octets and electrons are still left, add the extra electrons to the central atom as nonbonding (lone) pairs.
(b) If all electrons have been added and the central atom does not have an octet, move a pair of nonbonding electrons from an outer atom to make a double bond. Repeat this step as needed.

Exercise 4.1

1. Construct the Lewis structure for each of the following compounds.

(a) CH_3Cl
(b) CS_2
(c) PH_3
(d) SiF_4
(e) H_2S

2. Construct the Lewis structure for each of the following ions.

(a) NO_3^-
(b) CO_3^{2-}
(c) PO_4^{3-}
(d) SO_3^{2-}
(e) ClO_4^-

Solution

```
            H                    H
            ..   ..              ..               ..
1. (a) H : C : Cl:   (c) H : P : H   (e) H : S : H
            ..   ..              ..               ..
            H

                                   ..
                                 : F :
          ..      ..               ..
   (b) : S :: C :: S :   (d) : F : Si : F :
                           ..      ..    ..
                                 : F :
                                   ..
```

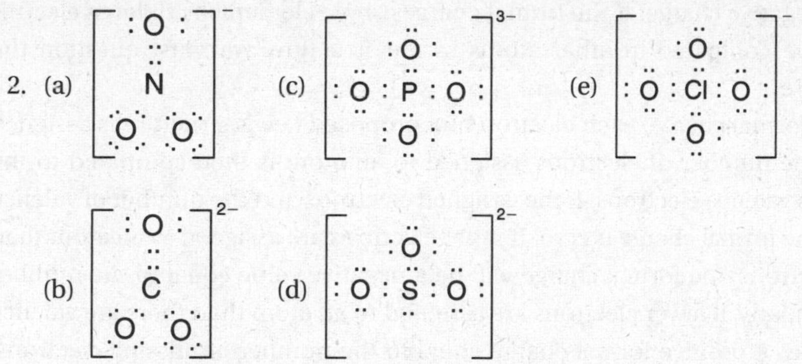

```
2. (a) ⎡  :O:  ⎤⁻     (c) ⎡    :Ö:     ⎤3–     (e) ⎡    :Ö:      ⎤⁻
       ⎢   ::  ⎥          ⎢     ::     ⎥           ⎢     ::      ⎥
       ⎢   N   ⎥          ⎢ :Ö : P : Ö:⎥           ⎢ :O : Cl : O:⎥
       ⎢  ⋰ ⋱  ⎥          ⎢     ::     ⎥           ⎢     ::      ⎥
       ⎣ :O: :O:⎦         ⎣    :O:     ⎦           ⎣    :O:      ⎦

   (b) ⎡  :O:  ⎤2–    (d) ⎡    :Ö:     ⎤2–
       ⎢   ::  ⎥          ⎢     ::     ⎥
       ⎢   C   ⎥          ⎢ :Ö : S : Ö:⎥
       ⎢  ⋰ ⋱  ⎥          ⎢            ⎥
       ⎣ :O: :O:⎦         ⎣            ⎦
```

Lewis Structures of Odd Electron Compounds

Some compounds have formulas in which the total number of valence electrons is an odd number. In such cases it is impossible to construct a Lewis structure with an octet around each atom. Nitrogen dioxide, NO_2, is one such compound. One possible Lewis structure is shown in Figure 4.12.

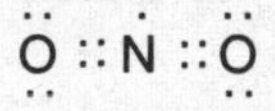

```
 ..   .   ..
 O :: N :: O
 ..        ..
```

FIGURE 4.12. Lewis structure of the NO_2 molecule, showing the unpaired electron on the nitrogen atom.

Molecules that have Lewis structures with an unpaired electron are often called **free radicals**. The unpaired electron makes the molecule unusually reactive. Free radicals have been implicated in such biological processes as aging and cancer. In an effort to pair up the single electron, free radicals may also form **dimers** or pairs of molecules. For example, the molecule NO_2 dimerizes to produce the N_2O_4 molecule in the reaction

$$2NO_2 \rightarrow N_2O_4$$

Formal Charges

How do we know whether or not a Lewis structure is reasonable? Experimental evidence, such as bond length, is the best verification of structures. Without experimental data, calculating of the **formal charge** on each atom is one technique that may be used to make this judgment. The formal charge is the difference between the number of electrons an atom has in a Lewis structure and its number of valence electrons. This calculation is described in detail below.

First it is important to understand the concept of formal charges. In a covalent compound an atom shares some of its valence electrons to form bonds, while the rest of the valence electrons remain as nonbonding electron pairs. If we count these nonbonding electrons and the electrons that an atom shares to form bonds, we should end up with a number equal to the valence of the atom. A different number means that the atom has lost or gained one or more electrons, an unlikely event since the loss or gain of electrons implies ionic behavior. Technically the formal charge is a separation of charge. In fact, some Lewis structures do require charge separation, but it is the minimum possible. In addition, we know, based on the electronegativities of the atoms, that the element with the greater electronegativity will be the atom with a negative charge. If the formal charges show elements with large electronegativities as positive compared to other atoms in the structure, we must question the validity of the structure.

In calculating the formal charge, each electron in a proposed Lewis structure is assigned to a specific atom. The number of electrons assigned to an atom is then compared to the number of that atom's valence electrons. If the assigned electrons and the number of valence electrons are equal, the formal charge is zero. If more electrons are assigned to an atom than there are valence electrons, the formal charge will be a negative value equal to the number of extra electrons. Similarly, if fewer electrons are assigned to an atom than there are valence electrons, the atom has a positive formal charge equal to the number of missing electrons. To calculate the formal charge on each atom in a Lewis structure the following steps are taken:

1. For each atom count all electrons not used for bonding by the atom.
2. Count half of the atom's bonding electrons.
3. Add steps 1 and 2 to obtain the electrons assigned to that atom.
4. Subtract the assigned electrons from the valence electrons to obtain the formal charge.

In equation form the formal charge is calculated as follows:

$$\text{Formal charge} = \text{Valence } e^- - [\text{Number of nonbonding } e^- + \tfrac{1}{2}(\text{Number of bonding } e^-)] \quad (4.1)$$

The formal charge calculations may be quickly checked since the formal charges on the atoms in a molecule must add up to zero, obeying the law of electroneutrality. For a polyatomic ion the formal charges must add up to the charge on the ion.

A molecule or polyatomic ion with the lowest possible formal charge on each atom is usually judged to be a more probable structure than one where the formal charge is larger. Consider the sulfate ion, SO_4^{2-}. Its Lewis structure may be drawn as shown in Figure 4.13.

```
 ┌    ..    ┐2−
     :O:
  ..  ..  ..
 :O : S : O:
  ..  ..  ..
     :O:
 └    ..    ┘
```

FIGURE 4.13. Lewis structure of the sulfate ion, constructed by the procedures described above.

From this structure we calculate the formal charges on sulfur and oxygen:

Formal charge on sulfur	=	$6 - 0 - \tfrac{1}{2}(8) = +2$
Formal charge on **each** oxygen	=	$6 - 6 - \tfrac{1}{2}(2) = -1$

We find that the formal charge on each oxygen is -1 and the formal charge on sulfur is +2. These charges add up to the charge of the ion, as they should $(+2 - 1 - 1 - 1 - 1 = -2)$. These formal charges are large, and an alternative structure should be sought. The structure shown in Figure 4.14 is one possible variation.

```
        ..      2−
       :O:
    ..  ..  ..
    O ::S:: O
    ..  ..  ..
       :O:
        ..
```

FIGURE 4.14. Alternative Lewis structures for the sulfate ion.

There are now two types of oxygen-sulfur bonds (two with a single bond and two with a double bond). When the formal charges are calculated, we have these results:

Formal charge on sulfur	=	$6 - 0 - \frac{1}{2}(12) = 0$
Formal charge on each oxygen with double bond	=	$6 - 4 - \frac{1}{2}(4) = 0$
Formal charge on each oxygen with single bond	=	$6 - 6 - \frac{1}{2}(2) = -1$

This is the preferred structure since it has the minimum formal charges. The formal charges add up to the charge of the ion ($-1 -1 = -2$) and cannot be any lower.

Formal charges may also be used to deduce the appropriate structure for a compound that has many possible Lewis structures. For instance, the NOCl molecule can be drawn with the four structures shown in Figure 4.15.

```
 ..   ..   ..        ..   ..   ..        ..   ..   ..        ..   ..   ..
:O :: N : Cl :      :N :: O : Cl :      :O : N :: Cl :      :N : O :: Cl :
           ..                  ..        ..                  ..
  Structure 1         Structure 2         Structure 3         Structure 4
```

FIGURE 4.15. Possible Lewis structures for the NOCl molecule.

To determine which of these structures is the most reasonable, we calculate the formal charges. Using the rules for determining these charges, we obtain the values in Table 4.2.

TABLE 4.2
Formal Charges on NOCl Structures

Element	Structure 1	Structure 2	Structure 3	Structure 4
Oxygen	0	+1	-1	+1
Nitrogen	0	-1	0	-2
Chlorine	0	0	+1	+1

From Table 4.2 we see that the first structure has the lowest formal charges on all atoms, and it is preferred. Structure 2 has larger formal charges, but also has a negative charge on nitrogen even though nitrogen is less electronegative than oxygen. This is not a reasonable situation since the more electronegative atom is expected to have the more negative formal charge. Structures 3 and 4 also have negative charges on the less electronegative elements as well as formal charges greater than zero. We therefore conclude that structure 1 is the most reasonable structure.

Exercise 4.2

In Exercise 4.1 the Lewis structures of the molecules and ions listed below were constructed. Now calculate the formal charges for each of those structures. Also, suggest which ones may have better structures, and draw them.

(a) CH_3Cl	(c) PH_3	(e) H_2S	(g) CO_3^{2-}	(i) SO_3^{2-}
(b) CS_2	(d) SiF_4	(f) NO_3^-	(h) PO_4^{2-}	(j) ClO_4^-

Solution

(a) all zero
(b) all zero
(c) all zero
(d) all zero
(e) all zero
(f) N = +1 O(single bonded) = −1 O(double bonded) = 0
(g) C = 0 O(single bonded) = −1 O(double bonded) = 0
(h) P = +1 O = −1
(i) S = +1 O = −1
(j) Cl = +3 O = −1

It is possible to find better structures for (f), (h), (i), and (j). Structures (f), (h), and (i) should each have another double-bonded oxygen; (j) should have three double-bonded oxygens.

Resonance Structures

In the discussion above we found that, when there are several possible structures, the most reasonable one can be selected by using the concept of formal charges. At times we can construct several Lewis structures for a substance that are totally equivalent, even down to the formal charges on the atoms. Chemists call these structures **resonance structures**.

Most resonance structures are very similar for a given substance, usually differing only in the geometry of the molecule or ion. It is found experimentally that none of the resonance Lewis structures properly describes the molecule. The true properties of the substance are found by blending all of the resonance structures together. For example, the resonance of the SO_3 molecule is shown in Figure 4.16 with the three possible Lewis structures.

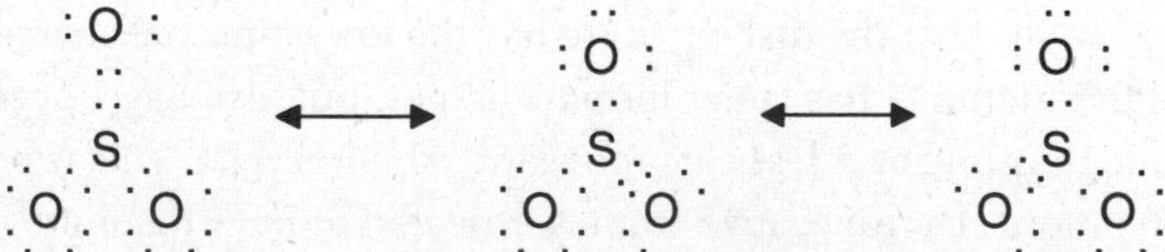

FIGURE 4.16. Three resonance structures of SO_3.
They differ only in the position of the double bond.

It is important to understand the nature of resonance. In Figure 4.16, each SO_3 has two single bonds and one double bond. In a variety of experiments it is found that all of the sulfur-oxygen bonds are identical. The measured properties of these bonds indicate that they are neither purely single bonds nor purely double bonds. Sulfur-oxygen bonds have characteristics in between those of the single bond and those of the double bond. To properly visualize the

SO_3 molecule we must think of three identical sulfur-oxygen bonds as a blend of bonds that is approximately two-thirds single bond and one-third double bond in character.

When discussing resonance, the benzene ring must be mentioned. The formula for benzene is C_6H_6. The molecule is a ring of six carbon atoms with one hydrogen atom attached to each carbon. Assigning electrons to this skeleton results in the resonance structures shown in Figure 4.17.

FIGURE 4.17. Structural formulas for the two resonance forms of benzene.

These structures are often summarized as shown in Figure 4.18. Each corner of the hexagon is assumed to represent a carbon atom, and a hydrogen atom is assumed to be attached to each carbon.

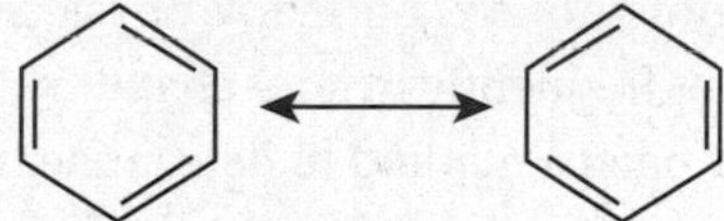

FIGURE 4.18. Abbreviated resonance structures of benzene.

The symbol

is the arrow used with resonance. The symbol

is used for chemical equilibrium.

Finally, since we know that the actual structure of the benzene molecule is not properly represented by either resonance structure, but rather involves a blending of the two, organic chemists often represent the benzene ring as shown in Figure 4.19.

FIGURE 4.19. Structure of benzene with a central circle representing the resonant nature of the molecule.

The circle within the ring reminds us that the double bonds are distributed (delocalized) over the entire molecule.

Benzene is one of many organic compounds classified as **aromatic molecules**. They definitely have a smell, but current chemical terminology recognizes the term *aromatic* as meaning that the structure contains one or more benzene rings. The benzene ring is found in such diverse compounds as aspirin, morphine, nicotine, proteins, saccharin, and many plastics. Benzene and many other aromatic compounds are considered carcinogens; however, a large number of beneficial compounds with aromatic character are not carcinogenic. Prudence dictates, however, that all aromatic compounds be treated with care in the laboratory.

Exercise 4.3

In Exercises 4.1 and 4.2 simple Lewis structures were constructed and then modified, based on the formal charges. Use those structures to draw the resonance structures of the following ions:

(a) NO_3^- (b) CO_3^{2-} (c) PO_4^{3-} (d) SO_3^{2-} (e) ClO_4^-

Solution

You should find three resonance structures each for (a), (b), and (d) and four resonance structures each for (c) and (e).

Covalent Bond Polarity and Electronegativity

Electrons are shared equally only in a covalent bond between two identical atoms (e.g., H_2, F_2, and N_2). If the electrons are not shared equally by two atoms, they will spend more time localized near one atom or the other. The result is that the atom that attracts the electrons will be relatively more negative than the other atom. When this occurs, we say that the bond between the atoms is **polar** with a positive end and a negative end. Understanding bond polarities allows the chemist to explain many physical properties of chemical compounds. We will use this concept frequently in later chapters.

To understand polarities, we need a method to determine how effectively different atoms attract electrons. Linus Pauling developed the concept of **electronegativity** to numerically represent the ability of an atom to attract electrons. Figure 4.20 shows the periodic table with the electronegativity value for each element. We see that electronegativity generally increases from left to right in each period of the periodic table. In addition, the electronegativity within any group generally increases from the bottom of the group to the top. In general, the electronegativity increases from the lower left corner of the periodic table up to the upper right corner. This is one of the important **diagonal trends** in the periodic table.

PERIODIC TREND

EN increases from lower left to upper right corner of periodic table.

H																
2.1																
Li	Be											B	C	N	O	F
1.0	1.5											2.0	2.5	3.1	3.5	4.0
Na	Mg											Al	Si	P	S	Cl
1.0	1.3											1.5	1.8	2.1	2.4	2.9
K	Ca	Sc	Ti	V	Cr	Mn	Fe	Co	Ni	Cu	Zn	Ga	Ge	As	Se	Br
0.8	1.1	1.2	1.3	1.5	1.6	1.6	1.7	1.7	1.8	1.8	1.7	1.8	2.0	2.2	2.5	2.8
Rb	Sr	Y	Zr	Nb	Mo	Tc	Ru	Rh	Pd	Ag	Cd	In	Sn	Sb	Te	I
0.8	1.0	1.1	1.2	1.3	1.3	1.4	1.4	1.5	1.4	1.4	1.5	1.5	1.7	1.8	2.0	2.5
Cs	Ba	La	Hf	Ta	W	Re	Os	Ir	Pt	Au	Hg	Tl	Pb	Bi	Po	At
0.7	0.9	1.1	1.2	1.4	1.4	1.5	1.5	1.6	1.5	1.4	1.5	1.5	1.6	1.7	1.8	2.2
Fr	Ra	Ac														
0.7	0.9	1.0														

FIGURE 4.20. Periodic table showing the electronegativities of the elements.

The values of electronegativities are generally not given on the AP exam. However, the diagonal trend of the electronegativities allows you to estimate quickly which end of a bond is negative and which end is positive. In a bond, the element closest to fluorine will be relatively negative and the element furthest from fluorine will be relatively positive. Polarities are indicated by the symbols $\delta+$ and $\delta-$ for partially positive and partially negative atoms respectively. Full positive and negative charges are used only for ions. For example, the boron-oxygen bond would be written as $^{\delta+}B–O^{\delta-}$ to show that the boron atom is more positive than the oxygen atom.

Exercise 4.4

Without using Figure 4.20, indicate the positive and negative ends of each of the following bonds by the symbols $\delta+$ and $\delta-$:

(a) S–O
(b) C–N
(c) S–P
(d) C–F
(e) Si–O
(f) H–Br
(g) H–O

Solution

Using the diagonal relationships in the periodic table, we identify the positive and negative ends as follows:

(a) $^{\delta+}S–O^{\delta-}$
(b) $^{\delta+}C–N^{\delta-}$
(c) $^{\delta-}S–P^{\delta+}$
(d) $^{\delta+}C–F^{\delta-}$
(e) $^{\delta+}Si–O^{\delta-}$
(f) $^{\delta+}H–Br^{\delta-}$
(g) $^{\delta+}H–O^{\delta-}$

The electronegativity table in Figure 4.20 gives numerical data that may be used to evaluate the magnitude of bond polarity. This is done by taking the absolute value of the difference in the electronegativities of the two atoms participating in a bond. We may call this value delta EN or, in written form, ΔEN:

$$\Delta EN = \left(\begin{array}{c}\text{Atom with largest}\\ \text{electronegativity}\end{array}\right) - \left(\begin{array}{c}\text{Atom with smallest}\\ \text{electronegativity}\end{array}\right) \tag{4.2}$$

The larger the value of ΔEN, the greater the polarity of the bond. If ΔEN is zero, the bond is considered to be nonpolar.

Exercise 4.5

For the bonds in Exercise 4.4, determine the ΔEN values, and predict which bond is the least polar and which is the most polar.

Solution

(a) 1.1
(b) 0.6
(c) 0.3
(d) 1.5
(e) 1.7
(f) 0.7
(g) 1.4

The Si–O bond is the most polar, and the S–P bond is the least polar.

Without an electronegativity table, it is possible to determine which of two bonds is the more polar if the bonds have one atom in common. The more polar bond will be the one where the second atom is furthest in the periodic table from the common atom. For instance, the nitrogen-fluorine bond is more polar than the oxygen-fluorine bond since nitrogen is further from fluorine than is oxygen.

Exercise 4.6

Using a periodic table, but not a table of electronegativities, estimate, for each of the following pairs, which bond is more polar:

(a) C–N or C–O
(b) H–Cl or H–Br
(c) S–O or S–Br
(d) H–S or H–O
(e) P–Br or S–Br

Solution

The more polar bond in each pair belongs to (a) C–O, (b) H–Cl, (c) S–O, (d) H–O, and (e) P–Br.

Dipole Moments

As shown in the previous section, we can very successfully estimate the polarity of a bond between two atoms. That is, we can tell which atom will attract electrons more easily than another. As a result, the atom that attracts electrons will have a partial negative charge, $\delta-$. The other atom will have a partial positive change, $\delta+$. From a table of electronegativities or by even just estimating from the periodic table, we can make good estimates of how polar a bond is. A better measure is the **dipole moment**. The dipole moment is a measure of the difference in charge, q, on two covalently bonded atoms and the distance, r, between the two nuclei.

$$\text{Dipole moment} = q \times r$$

The units for the dipole moment are coulomb-meters, and a common unit for dipole moments is the debye.

$$1 \text{ debye} = 3.34 \times 10^{-30} \text{ Cm}$$

You will often see dipole moments expressed as debyes. Another feature of dipole moments is that they are treated mathematically as vectors. As we will see shortly, molecules with polar bonds can be nonpolar if opposing dipole moments cancel each other.

Electronegativity and Ionic Character

At one end of the polarity scale are the completely **nonpolar** bonds between diatomic elements. We may visualize ionic compounds as being at the other end of the polarity scale since the electrons are actually transferred from one atom to another. From the table of electronegativities in Figure 4.20, we see that the largest ΔEN is 3.3 for the ionic compound FrF. The well-known ionic compounds NaF and $CaBr_2$ have ΔEN values of 3.0 and 1.7, respectively. The ΔEN of a bond has been used to estimate the percentage ionic character of a bond. Chemists say that a ΔEN of 1.7 represents a bond that is 50 percent ionic and 50 percent covalent in character. A bond with a ΔEN of 1.7 or greater is considered ionic. Bonds with ΔENs of less than 1.7 are polar covalent, and those for which ΔEN is zero are nonpolar covalent bonds.

Bond Order

Bond order is a term that refers to the average number of bonds that an atom makes in all of its bonds to other atoms. From the Lewis structures of the diatomic elements in Figure 4.2, we see that fluorine (F_2) and chlorine (Cl_2) have one bond each and a bond order of 1. Oxygen (O_2) has a double bond and a bond order of 2, and nitrogen (N_2) has a triple bond and thus a bond order of 3. In the SO_3 resonance structures there is a total of four bonds: two single bonds and one double bond. Since the sulfur is bonded to three oxygen atoms, the average number of bonds that sulfur has with its oxygen atoms is $\frac{\text{four bonds}}{\text{three O atoms}}$, and the bond order of sulfur is $^4/_3$. In benzene, the bond order for each carbon atom is $^3/_2$.

Exercise 4.7

Determine the bond order of the central atom of each of the following compounds. You may use the structures determined in Exercises 4.1 and 4.2.

(a) CH_3Cl (c) PH_3 (e) H_2S (g) CO_3^{2-} (i) SO_3^{2-}
(b) CS_2 (d) SiF_4 (f) NO_3^- (h) PO_4^{3-} (j) ClO_4^-

Solution

If we use the simple Lewis structures, we obtain

(a) 1 (c) 1 (e) 1 (g) $^4/_3$ (i) 1
(b) 2 (d) 1 (f) $^4/_3$ (h) 1 (j) 1

If we use the best formal charge structures, we obtain

(a) 1	(c) 1	(e) 1	(g) $^4/_3$	(i) $^4/_3$
(b) 2	(d) 1	(f) $^5/_3$	(h) $^5/_4$	(j) $^7/_4$

Bond Strength, Bond Energy, and Bond Length

Just as two ropes are twice as strong as one rope, a double bond is almost twice as strong as a single bond. The strength of a covalent bond is expressed as its bond energy. When two atoms are covalently bonded together, they vibrate in the same way that a spring vibrates. The frequency of this vibration is related to the two masses attached to the ends of the spring and to the strength of the spring itself. Since we know the masses of the two atoms, the frequency of vibration will be related to the strength of the bond. Frequency is related to bond energy (bond strength) by Equation 4.3, where the energy is equal to Planck's constant times the frequency of vibration.

$$E = h\nu \qquad (4.3)$$

Bond vibrations may be observed in the infrared spectral region by using an infrared spectrometer. Infrared spectra confirm that single bonds have the lowest energy, double bonds have a higher energy, and triple bonds have the highest energy.

Another method used to measure bond energies involves measurement of the energy released when organic compounds are burned. Ethane, C_2H_6, contains only single bonds; ethene, C_2H_4, contains a double bond; and ethyne, C_2H_2, contains a triple bond. When burned, ethane yields 1540 kilojoules, ethene 1387 kilojoules, and ethyne 1305 kilojoules of energy. The fact that the lowest energy is released by ethyne is taken to indicate that this compound's bonds are already in the highest energy configuration. The high energy of combustion for ethane, on the other hand, indicates that its bonds have lower energy to start with and can release more energy upon combustion.

The length of a covalent bond may be measured in several ways. One way is by X-ray crystallography, as described in Chapter 7. The positions of the atoms in a crystal can be determined based on the diffraction of X-rays by a crystal. Another method involves the fact that the frequency of vibration and the length of the vibrating medium are related. As described above, these vibrations are measured using infrared spectroscopy. Whatever method is used to measure bond lengths, consistent results are obtained. In particular, a single bond has the longest length and a triple bond has the shortest.

We have discussed resonance structures and the fact that the actual structure of a molecule is a blend of all equivalent resonance structures. In these molecules we do not have pure single, double, or triple bonds. We can, however, calculate bond order, which represents the average number of bonds per atom. Bond order is related to bond strength and length also. The general rule is that the greater the bond order, the shorter the bond length. Table 4.3 lists some typical bond lengths and energies.

TABLE 4.3
Typical Bond Lengths and Energies

Bond	Bond Order	Bond Length (pm)	Bond Energy (kJ mol^{-1})
C—C	1	154	347
C=C	2	134	612
C≡C	3	120	820
N—N	1	145	159
N=N	2	123	418
N≡N	3	110	914
C—O	1	143	351
C=O	2	120	715

Nomenclature

The naming of covalently bonded binary molecules (these are predominately molecules with two nonmetals) is quite different from the naming of ionic compounds, and the two methods should not be confused. In addition, many of these covalent substances were discovered long before the modern method of naming compounds was developed. Common covalent compounds often have trivial (older) and systematic (modern) names. Some important trivial names are shown in Table 4.4, along with their modern equivalents.

TABLE 4.4
Trivial and Systematic Names of Some Common Covalent Molecules

Molecule	Trivial Name	Systematic Name
H_2O	water	dihydrogen monoxide*
CH_4	methane	carbon tetrahydride*
N_2O	nitrous oxide	dinitrogen oxide
NO	nitric oxide	nitrogen monoxide
N_2O_3	nitrous anhydride	dinitrogen trioxide
N_2O_5	nitric anhydride	dinitrogen pentoxide
NH_3	ammonia	nitrogen trihydride*
AsH_3	arsine	arsenic trihydride
H_2O_2	hydrogen peroxide	dihydrogen dioxide*
N_2H_4	hydrazine	dinitrogen tetrahydride

*These names are almost never used.

TIP

Names of binary molecular compounds use prefixes.

In the systematic naming of covalent compounds, the prefixes listed in Table 3.1 indicate the number of each atom present in a molecular formula. Some of these prefixes are indicated in the systematic names in Table 4.4. Other examples of the use of these prefixes can also be given. The formula for carbon dioxide is CO_2, and the name indicates one carbon atom and, because of the prefix *di–*, two oxygen atoms. Sulfur forms two compounds, SO_2 and SO_3, named sulfur dioxide and sulfur trioxide, respectively. Again the prefixes *di–* and *tri–* indicate the number of oxygen atoms bound to the sulfur in the two compounds. The compound N_2O_4 has two nitrogen and four oxygen atoms and is named dinitrogen tetroxide.

Exercise 4.8

Name each of the following covalent molecules:

(a) SO_2 (c) N_2O_3 (e) PBr_5 (g) $BrCl_3$
(b) P_4O_{10} (d) SF_4 (f) XeF_4 (h) S_4N_4

Solution

(a) sulfur dioxide
(b) tetraphosphorus decaoxide
(c) dinitrogen trioxide
(d) sulfur tetrafluoride
(e) phosphorus pentabromide
(f) xenon tetrafluoride
(g) bromine trichloride
(h) tetrasulfur tetranitride

Exercise 4.9

Give the formula for each of the following compounds:

(a) diboron tetrabromide
(b) boron trifluoride
(c) carbon tetrafluoride
(d) carbon monoxide
(e) diphosphorus pentoxide
(f) carbon disulfide
(g) sulfur trioxide
(h) nitrogen triiodide

Solution

(a) B_2Br_4 (c) CF_4 (e) P_2O_5 (g) SO_3
(b) BF_3 (d) CO (f) CS_2 (h) NI_3

Additional rules and names for compounds are given in Chapter 13 on acids and bases.

MOLECULAR GEOMETRY

Once a valid Lewis structure has been determined, the overall geometry of a simple molecule with one central atom can be established. The process can be extended to very large macromolecules, such as proteins and DNA, by determining the geometries around individual atoms and then combining them to obtain the entire structure. This overall geometry is extremely important in understanding the properties of chemical compounds. The key to the discovery of DNA's double helix was the geometric structures of the four bases, which must hydrogen-bond to each other in order to hold the total structure together.

Valence-Shell Electron-Pair Repulsion Theory

The **valence-shell electron-pair repulsion theory (VSEPR theory)** allows us to determine the three-dimensional shape of covalently bonded molecules with a minimum of information. This theory simply states that electron pairs will repel each other since all electrons carry a negative charge. In fact, they will repel each other so they are as far apart as possible. Modern chemistry divides the electrons around an atom into bonding electron pairs and nonbonding electron pairs. The region in space occupied by a bonding electron pair (or pairs for double and triple bonds) is called the **bonding domain**. The region in space occupied by a nonbonding electron pair is called the **nonbonding domain**.

When answering questions concerning molecular geometry, the logic and results obtained from the VSEPR theory along with the effect of bonding and nonbonding electron domains are sufficient. In particular, hybrid orbitals, which are addressed later, are not needed unless specifically asked for in the question. In addition, the AXE system described in the next sections, although widespread, is artificial and should be used with appropriate explanations that A represents the central atom while X and E represent the number of bonding electron domains and nonbonding electron domains respectively that are associated with A.

Basic Structures

To determine the three-dimensional geometry around a central atom, A, all we need to know is the total number of bonding and nonbonding domains attached to A. We will start with structures that have no nonbonding domains. These involve the six basic geometries found in any molecule. Table 4.5 lists the possible geometric shapes found around an atom that may have one to six atoms bonded to a central atom (that is AX, AX_2, AX_3, AX_4, AX_5, and AX_6).

TABLE 4.5
Basic Structures for Six Geometries

Notation	Shape	Example	Angle(s)
AX*	Linear molecule	HBr	—
AX_2	Linear molecule	CS_2	180°
AX_3	Planar triangle	BCl_3	120°
AX_4	Tetrahedron	CCl_4	109.5°
AX_5	Trigonal bipyramid	PCl_5	120°, 90°
AX_6	Octahedron	XeF_6	90°

*This structure is trivial but is included for completeness.

The angles listed in this table are the angles between the bonds, assuming that the central atom is the vertex of the angle. For structures AX to AX_4 every bond is equidistant from every other one. For the AX_5 and AX_6 structures bond angles are measured between the nearest neighbors. For the AX_5 structure, the 120° angle is for the three equatorial atoms and the 90° angle is the angle between the axial atoms and the equatorial atoms. Figure 4.21 illustrates these shapes in diagram format. The AX structure is omitted from many texts as being trivial since any molecule that contains only two atoms must be linear.

The geometry around a central atom that does not have any nonbonding electron pairs is determined by counting the atoms bonded to it. For instance, three atoms bound to a central atom with no nonbonding pairs must be a planar triangle. Five atoms bound to a central atom must have the shape of a trigonal bipyramid.

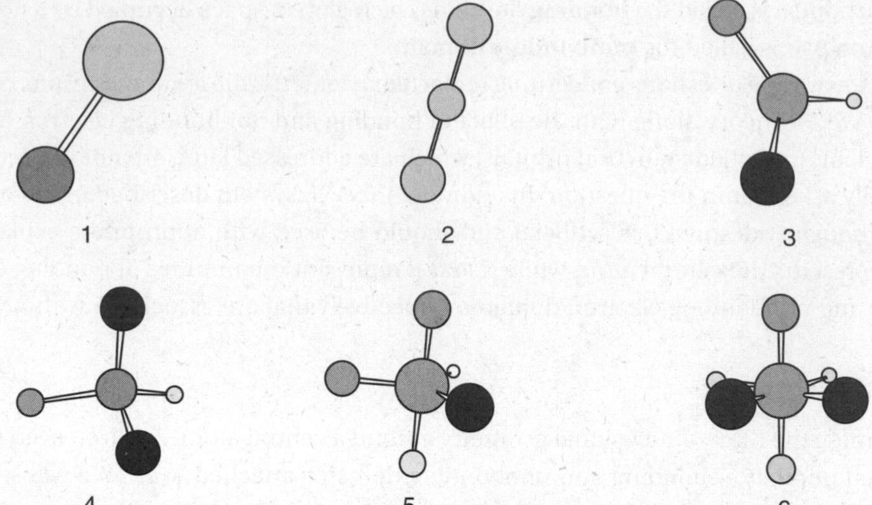

FIGURE 4.21. Perspective diagrams of the six basic geometric structures. Darkest atoms are closest to the viewer, structures are tilted to show all atoms. 1—linear diatomic; 2—linear triatomic; 3—planar triangle; 4—tetrahedron; 5—trigonal bipyramid; 6—octahedron. For diagrams 5 and 6 the axial atoms are at the top and bottom of the figures, while the equatorial atoms are in the center.

Derived Structures

TIP

Simply stated, atom and electron pairs arrange themselves so that they are as far from each other as possible.

Derived structures have one or more of the bonding domains replaced with nonbonding domains. The result is the same basic structure. However, we usually want to know the arrangement of bonding domains (these are the atoms we actually "see" in a molecule). We do this in two steps. First, we count all of the domains, bonding and nonbonding, around the central atom and then determine the basic structure. Then, as needed, we replace atoms with electron clouds and determine the geometry of the remaining atoms. Table 4.6 summarizes the possible derived structures. Figures 4.22, 4.23, 4.24, and 4.25 show diagrams of the derived structures.

Since the first two entries in Table 4.6 represent single atoms and diatomic substances, their structures need not be drawn. The geometries that correspond to the other derived structures are shown in Figures 4.22–4.25.

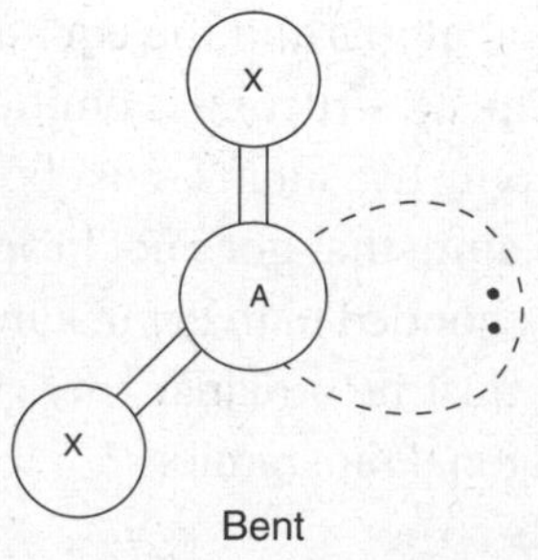

FIGURE 4.22. Bent AX_2E derived structure, showing a lone-pair electron domain occupying the space formerly occupied by an atom in the basic AX_3 (trigonal planar) structure.

TABLE 4.6
Derived Structures with Nonbonding Electron Pairs on the Central Atom

Basic Structure Notation	Derived Structure Notation	Derived Structure Shape	Example	Derived Structure Angle(s)**
A	A*	Single atom	None	None
AX_2	AXE*	Linear diatomic	CN^-	None
AX_3	AX_2E	Bent	$SnCl_2$	120°
AX_4	AX_3E	Triangular pyramid	NH_3	109.5°
AX_4	AX_2E_2	Bent	H_2O	109.5°
AX_5	AX_4E	Distorted tetrahedron	SF_4	120°, 90°
AX_5	AX_3E_2	T-shape	ICl_3	90°
AX_5	AX_2E_3	Linear	I_3^-	180°
AX_6	AX_5E	Square pyramid	IF_5	90°
AX_6	AX_4E_2	Square planar	XeF_4	90°

*These structures are trivial but are included for completeness.

**Angles listed here assume that lone pairs occupy the same space as bonding electrons.

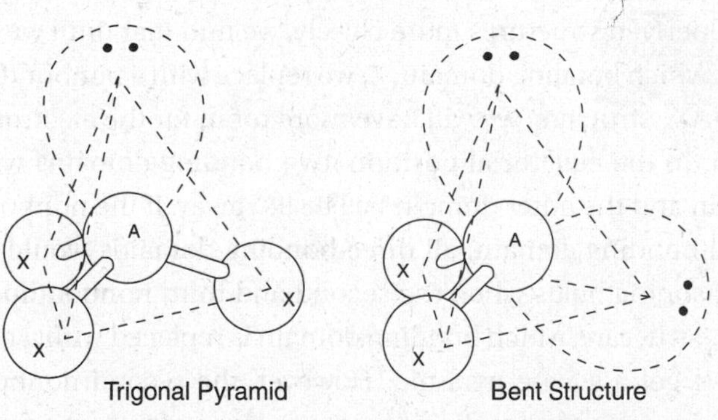

FIGURE 4.23. Pictorial representation of the AX_3E (triangular pyramid) and AX_2E_2 (bent) derived structures that are derived from the AX_4 (tetrahedral) basic structure.

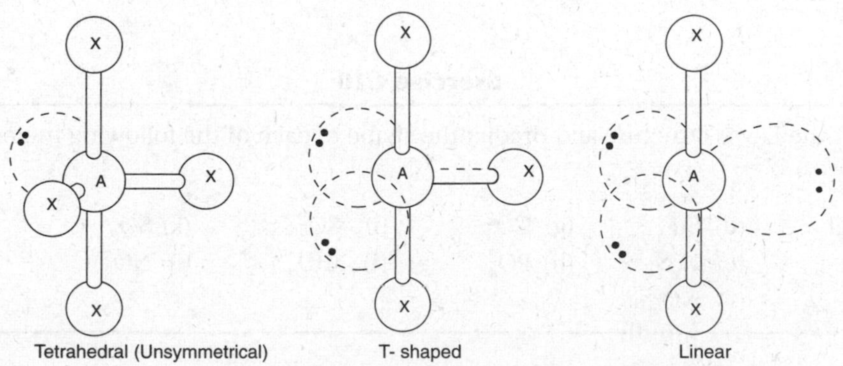

FIGURE 4.24. The three possible derived structures obtained from the AX_5 (trigonal bipyramid) basic structure. Notice that the equatorial atoms are replaced by lone-pair electron domains.

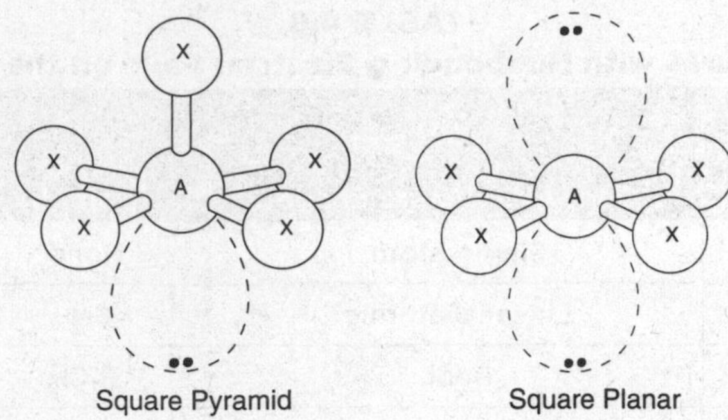

FIGURE 4.25. The two derived structures obtained from the AX_6 (octahedral) basic structure. Note that the second atom replaced by a lone-pair electron domain is on the opposite side of the molecule, so that the electron clouds have the extra space they need.

In Table 4.6 and Figures 4.22 and 4.23 we find two bent structures, AX_2E and AX_2E_2. The bond angles for these two structures will be very different, and we can distinguish the structures on the basis of their angles. Since the AX_2E bent structure is derived from the trigonal planar AX_3 structure, we expect its angle to be approximately 120°. The AX_2E_2 structure is derived from the tetrahedral AX_4 basic structure, and bond angles of approximately 109.5° are expected for structures related to the tetrahedron.

If we look at the derived structures more closely, we find that until we reach the basic AX_5 and AX_6 structures, which bonding domain, X, we replace with a nonbonding domain, E, does not matter. For the AX_5 structure, we will have more room for the electrons if they go into an equatorial position. In the equatorial position, two bonding domains will be 120° from the nonbonding domain and the other domain will be 90° away. If the nonbonding domain were to replace the axial bonding domain, all three bonding domains would be just 90° degrees away. The same reasoning holds when the second and third nonbonding domains are considered. For the AX_6 structure, which bonding domain is replaced with a nonbonding domain does not matter. We get a square pyramid. However, the second nonbonding domain will most often replace the opposite bonding domain to give a structure that is a square plane.

From this analysis, we see that proper placement of nonbonding pairs of electrons in a Lewis structure is very important. Doing so allows us to obtain the correct geometry of the molecule.

Exercise 4.10

Construct the Lewis structure and predict the shape of each of the following molecules and ions:

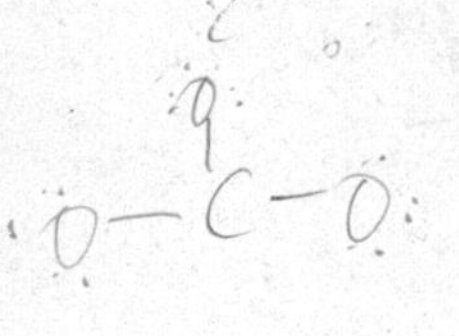

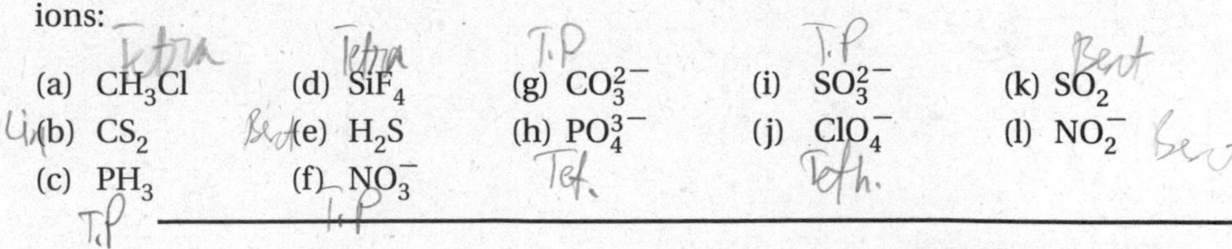

(a) CH_3Cl
(b) CS_2
(c) PH_3
(d) SiF_4
(e) H_2S
(f) NO_3^-
(g) CO_3^{2-}
(h) PO_4^{3-}
(i) SO_3^{2-}
(j) ClO_4^-
(k) SO_2
(l) NO_2^-

Solution

The shapes are as follows:

(a) tetrahedron
(b) linear
(c) triangular pyramid
(d) tetrahedron
(e) bent
(f) triangular planar
(g) triangular planar
(h) tetrahedron
(i) tetrahedron
(j) tetrahedron
(k) bent
(l) bent

These shapes are the same whether we use the simple Lewis structure or the structures optimized for the best formal charges.

Complex Structures

Geometries of more complex molecules are constructed by determining the geometry around each atom in sequence and then stringing the geometries together. Organic (carbon-based) compounds often have complex structures where these geometries are very important. The three-dimensional structures of these compounds often help define chemical, physical, and biological properties.

Carbon, with its four valence electrons, can form a maximum of four covalent bonds with four other atoms. It can also bond to three atoms as long as one of the bonds is a double bond. In bonding to two atoms, carbon will form either two double bonds, as in carbon dioxide, CO_2, or one single bond and one triple bond, as in hydrogen cyanide, HCN. In all instances, carbon never has a nonbonding pair of electrons. As a result, a carbon bonded to four atoms is tetrahedral; if bonded to three atoms, trigonal planar; and if bonded to two atoms, linear.

For instance, the molecule of ethene, CH_2CH_2, has the structure shown in Figure 4.26.

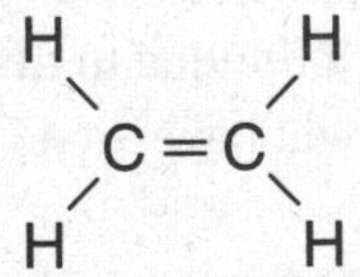

FIGURE 4.26. Ethene molecule.

Since each carbon atom is bonded to only three atoms (two hydrogen and one carbon), the carbon atoms must each have a trigonal planar geometry. We will see the reason later, but both planes are lined up so that this molecule is perfectly flat. We can also predict that the benzene ring shown in Figure 4.19 must be flat also since each of its six carbon atoms is trigonal planar.

When we have a molecule such as butane, $CH_3CH_2CH_2CH_3$, we may draw the structures shown in Figure 4.27.

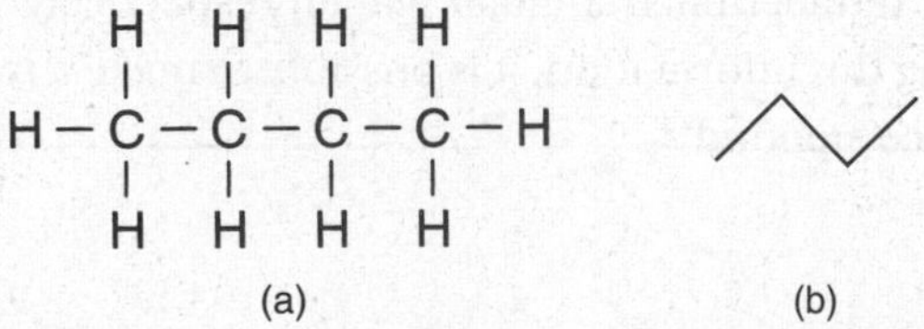

(a) (b)

FIGURE 4.27. Structure of butane, C_4H_{10} (a), with all H shown and (b) with lines only. Each end and vertex represents a carbon atom.

Each carbon atom is bonded to four other atoms; therefore, the geometries of the carbon atoms are all tetrahedral. Placing the four tetrahedral structures together, we obtain the three-dimensional structure illustrated in Figure 4.28.

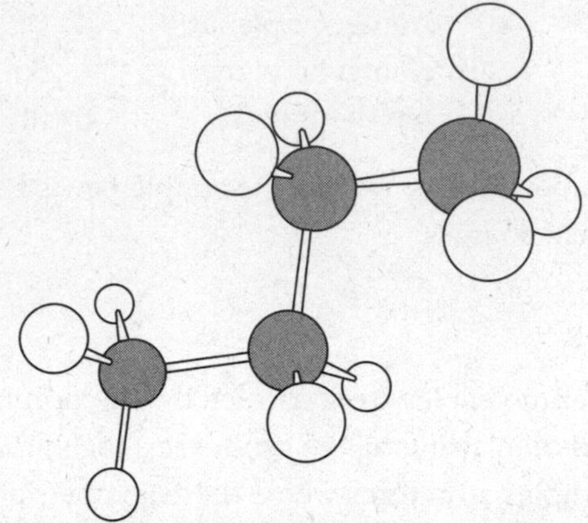

FIGURE 4.28. Three-dimensional computer-generated structure for butane, showing the tetrahedral arrangement of atoms around each carbon (shaded circles).

Since three-dimensional structures are difficult to draw on paper, organic chemists often find it convenient to build models of these structures so that they can inspect their features more easily.

Oxygen atoms in organic compounds always have two nonbonding domains. An oxygen bonded to two atoms will have an AX_2E_2 derived structure (bent), and a double-bonded oxygen with an AXE_2 derived structure (linear) will be bonded to only a single atom. Nitrogen atoms in organic compounds will have one nonbonding domain. The nitrogen atom will have an AX_3E structure (triangular pyramid) if bonded to three other atoms. When bonded to only two atoms, one with a double bond, it will have an AX_2E structure (bent).

Exercise 4.11

Predict the geometry around each of the carbon atoms in this molecule:

$$CH_3CH_2CH=\overset{\displaystyle CH_3 \atop |}{C}CH_2\underset{| \atop \displaystyle OH}{C}=O$$

Solution

From left to right along the main chain, the geometries are tetrahedral, tetrahedral, triangular planar, triangular planar, tetrahedral, triangular planar, respectively. The CH_3 above the molecule is tetrahedral. Using this information, it is possible to make a more realistic drawing, or molecular model, of the compound.

MOLECULAR POLARITY

Bond polarities depend on the electronegativities of the two elements bonded together. Very few molecules are diatomic, meaning that for most molecules more than one bond must be considered in determining the polarity of the molecule as a whole. These bonds are arranged geometrically in space as described in the preceding section. The result is that, even if a bond is polar, the molecule as a whole may or may not be polar. There are four general rules for determining whether a molecule is polar.

TIP

Molecular polarity is determined by combining bond polarity and molecular geometry.

1. A molecule that is symmetrical is nonpolar. It does not matter how polar the individual bonds are.
2. A nonsymmetrical molecule is polar if the bonds are polar.
3. A molecule with more than one type of atom attached to the central atom is often nonsymmetrical and therefore polar.
4. A central atom with nonbonding electron pairs is often nonsymmetrical and polar.

When determining the polarity of a molecule, we must remember that there is only one positive end and one negative end, directly opposite each other. We recognize the CH_3Cl and CBr_4 molecules as tetrahedral structures. As Figure 4.29 indicates, CH_3Cl is polar because it is not symmetrical, and CBr_4 is symmetrical and nonpolar. We can also calculate that the four equal dipole moments will add up to zero because of the tetrahedral geometry.

```
              H                    ..
              |    ..             :Br:
   δ+  H  —  C  — Cl:  δ−     ..    |    ..
              |    ..          :Br — C — Br:
              H                ..    |    ..
                                    :Br:
                                     ..
```

FIGURE 4.29. Structures of CH_3Cl, showing the polarity of the molecule, and of CBr_4, showing the symmetry and nonpolarity.

From our knowledge of electronegativities we may predict that the negative end of the CH_3Cl molecule is located near the chlorine atom and that the positive end is opposite the chlorine atom in a region of space between the three hydrogen atoms.

```
       δ⁻
       ↑
      :O:
     /   \
    H  ↓  H
       δ⁺
```

FIGURE 4.30. Structure of water and its polarity.

For the water molecule we may draw the structure shown in Figure 4.30. The electronegativity of the oxygen indicates that it is the negative end of the molecule and that the region of space opposite the oxygen, between the two hydrogen atoms, is the positive end.

Exercise 4.12

Construct the Lewis structure and predict the polarity of each of the following:

(a) CH_3Cl	(c) PH_3	(e) H_2S	(g) CO_3^{2-}	(i) SO_3
(b) CS_2	(d) SiF_4	(f) NO_3^-	(h) PO_4^{3-}	(j) ClO_4^-

Solution

(a) polar with Cl the negative end
(b) nonpolar
(c) polar with P the negative end
(d) nonpolar
(e) polar with S the negative end
(f–j) nonpolar because of resonance; charge on ion is distributed evenly over the ion. These ions are charged but not polar.

COVALENT BOND FORMATION

Wave Mechanics and Covalent Bond Formation

TIP

The AP exam does not ask questions on molecular orbitals.

Up to now, we have constructed molecules only according to the placement of electrons. We have determined the shapes and structures based on the very successful VSEPR theory. We now turn our attention to the wave mechanical nature of the covalent bond. Two successful theories approach the formation of the covalent bond in different ways. The **valence bond theory** (VB theory) considers a covalent bond to be the overlapping of two atomic orbitals when the electron spins are paired. The **molecular orbital theory** (MO theory) considers that a molecule is similar to an atom in that they both have distinct energy levels that can be populated by electrons. Figure 4.31 attempts to illustrate the idea that the electrons and nuclei of two hydrogen atoms are rearranged into molecular orbitals.

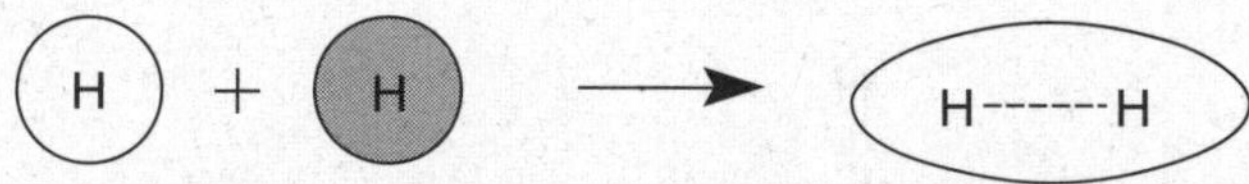

FIGURE 4.31. Combination of two hydrogen atoms to form the hydrogen molecule. The electrons in the hydrogen atoms have opposing spins so that they can pair in the molecular orbital.

In atoms these levels are called atomic orbitals, and in molecules they are molecular orbitals. Both the VB theory and the MO theory have been refined to produce similar results. The AP exam tends to focus on the valence bond approach, and it is the subject of the next section.

Valence Bond Theory

In the VB theory, illustrated in Figure 4.32, two hydrogen atoms approach and interact with an overlap of atomic orbitals with electrons of opposing spin (the opposing spins are indicated by the different shading of the hydrogen atoms).

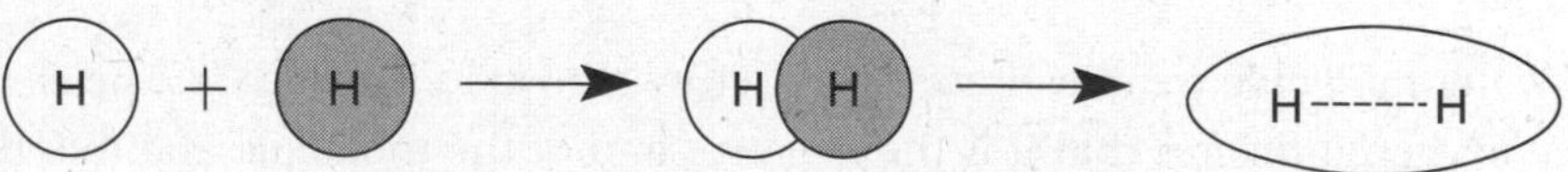

FIGURE 4.32. Overlap of two *s* atomic orbitals to form a molecular orbital in hydrogen.

As the bond is formed, the paired electrons spread out over the molecule to form the final electron cloud surrounding the nuclei. The results of the MO theory and the VB theory are the same (Figures 4.31 and 4.32), with a high electron density along the internuclear axis. The formation of H_2 and many other compounds and bonds is described as the overlap of two *s* orbitals to form a **sigma bond** (σ).

Sigma bonds may also be formed by the overlap of an *s* orbital and a *p* orbital (Figure 4.33), as in the formation of hydrogen fluoride, HF, or by the overlap of two *p* orbitals (Figure 4.34), as in F_2.

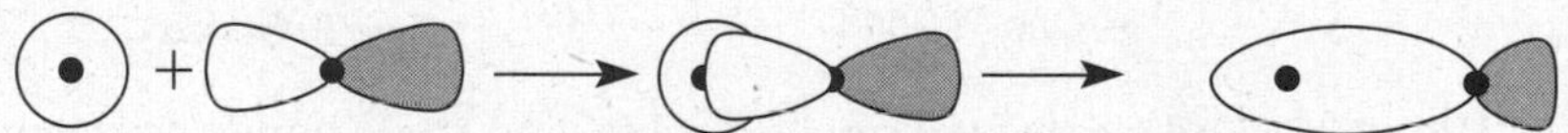

FIGURE 4.33. Overlap of an ***s*** orbital and a ***p*** orbital to form a sigma bond in a substance such as HF.

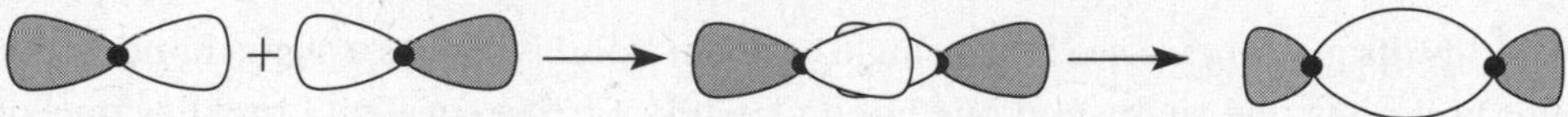

FIGURE 4.34. Overlap of two ***p*** orbitals to form a sigma bond in a molecule such as F_2.

Orbital Overlap Model (Pi Bonds)

We have described the three important ways in which sigma bonds are formed. Every covalent bond has one and only one sigma bond. If a compound has a double or triple covalent bond, additional overlap of orbitals is needed. Such a bond, called a **pi bond** (π bond), is formed by the *sideways overlap* of two *p* orbitals as shown in Figure 4.35.

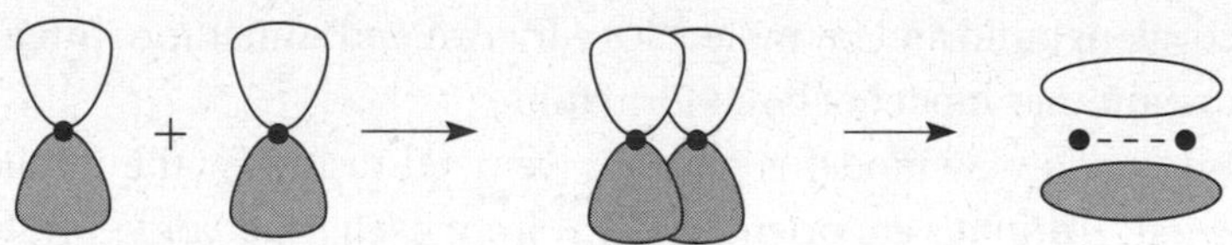

FIGURE 4.35. Sideways overlap of ***p*** orbitals to form a pi bond.

The pi bond has its electron density arranged in two electron clouds, one above and one below the **internuclear axis** (dashed line). When arranged in this manner, the electrons in the pi bond do not interfere with the electrons in the sigma bond.

A double bond involves one sigma bond and one pi bond. A triple bond between two atoms may be formed by adding a second pi bond, which has its two electron clouds centered behind and in front of the two nuclei.

If we place one atom in front of the other and look down the internuclear axis, the positions of the sigma and pi bonds are as shown in Figure 4.36.

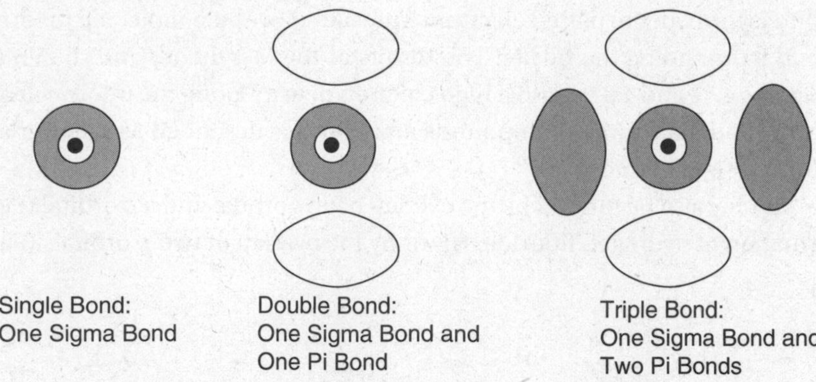

FIGURE 4.36. End view of the single, double, and triple bonds, looking down the internuclear axis. The dot and white circle represent the two nuclei. The shaded circle represents the sigma bond present in all three bonds. The two white ovals represent one of the pi bonds, and the two shaded ovals represent the second pi bond.

From this discussion, we see that a single covalent bond is always a sigma bond. A double covalent bond has one sigma and one pi bond, while a triple covalent bond has one sigma and two pi bonds. All of these bonds are arranged so that their electron clouds do not interfere with each other.

Hybrid Orbital Model

As mentioned earlier, the VSEPR theory provides all the information that we need to explain why certain molecules have the geometric shapes we measure experimentally (bond lengths and bond angles). VSEPR theory is also capable of accurately predicting the geometric shapes of molecules. This section on hybrid orbitals describes how the electrons might rearrange in a molecule to explain the observed experimental data.

The overlap of *s* and *p* orbitals to form sigma and pi bonds works well to describe some features of the covalent bond and for molecules with two and sometimes three atoms. Larger molecules require another model of bond formation.

To understand why a new model is needed, we need to review the implications of the overlap model. First, *p* orbitals are oriented at 90° from each other and *s* orbitals are spherical, having no directionality. If all covalent compounds were formed from the overlap of these orbitals, we would expect all covalent molecules to have 90° bond angles. As we have seen, however, few molecular geometries have angles of 90°. Second, even a simple molecule such as methane, CH_4, cannot be adequately explained by the overlap model. We know that methane is a tetrahedral molecule with four totally equivalent C–H bonds. Using the overlap method, we see that the carbon in methane has only two unpaired *p* electrons (the *s* electrons are paired), and we would expect the formation of the CH_2 molecule. The bonds would be oriented at a 90° angle since *p* orbitals are 90° apart. If we allowed the two *s* electrons to unpair so that they could also form bonds, we could obtain the CH_4 molecule. However, the bond angles would still not be correct, and we would expect two distinctly different C–H bond types in methane, one from the overlap of *s* orbitals and the other from the overlap of *p* orbitals. Our new model must be able to explain correctly the molecular geometries and bonding in larger molecules.

sp^3 Hybrid Orbitals

The problem posed by the CH_4 molecule requires that we develop a better model of sigma bond formation. To construct this model, we postulate the formation of hybrid orbitals. A **hybrid orbital** may be defined as a set of orbitals with identical properties formed from the combination of two or more different orbitals with different energies. The orbital diagram of carbon is presented in Figure 4.37, along with the conversion of the *s* and *p* electrons into hybrid orbitals called sp^3 orbitals.

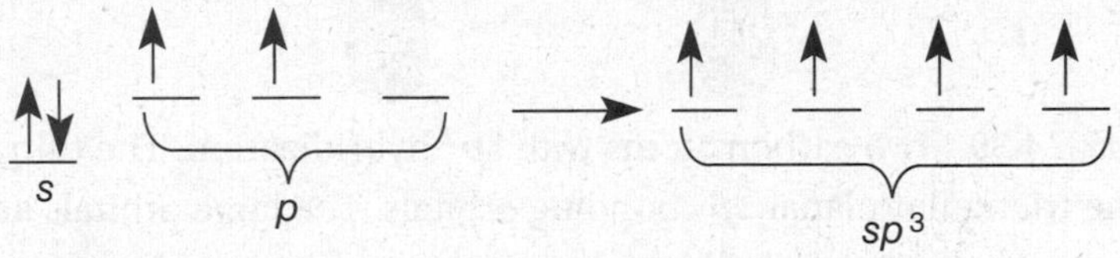

FIGURE 4.37. The ***s*** and ***p*** electrons of the carbon atom and their conversion into the $\boldsymbol{sp^3}$ hybrid orbitals used in bonding.

The designation sp^3 indicates that one *s* and three *p* orbitals have been combined to form the hybrid orbital. In this orbital diagram for carbon, the *p* electrons are shown as having a higher energy than the *s* electrons by placing the orbitals at different levels. When carbon forms methane, its electrons reorganize into the four identical sp^3 hybrid orbitals shown on the right. The energy of the electrons in the hybrid orbitals lies between the original *s* and *p* energies, as shown. When the sp^3 hybrid orbitals form, their orientation is tetrahedral. Overlap of the four identical sp^3 electrons with electrons from hydrogen atoms forms the tetrahedral methane molecule as we know it. Any molecule whose basic structure is the tetrahedron will have sp^3 hybrid orbitals. This includes the CH_4, NH_3, and H_2O molecules described previously.

sp^2 Hybrid Orbitals

Formaldehyde, CH_2O, is a carbon compound having single bonds to the hydrogen atoms and a double bond with the oxygen atom. Carbon has three sigma bonds and one pi bond in this compound. The structure is triangular planar because there are no nonbonding electron pairs on carbon. The orbitals in this compound are designated as sp^2 hybrids. The formation of these orbitals can be diagrammed as shown in Figure 4.38. Here we see that three electrons are in three identical orbitals, called sp^2 hybrids. The remaining electron stays in an unhybridized *p* orbital and overlaps with a *p* orbital on the oxygen atom to form the pi bond in the C=O double bond.

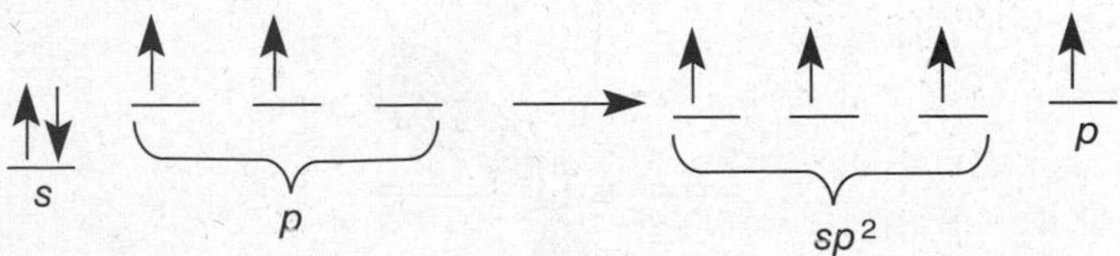

FIGURE 4.38. Formation of the $\boldsymbol{sp^2}$ hybrid orbitals for carbon.

In the earlier discussion of the ethene and benzene molecules, it was stated that these molecules are totally flat. Figure 4.39 shows why. In order for the *p* orbitals to overlap, they must be aligned as shown. This requirement fixes the remaining sp^2 bonds in one plane, resulting in the planar ethylene molecule. Proper alignment of the *p* orbitals in benzene forces this molecule also to be planar.

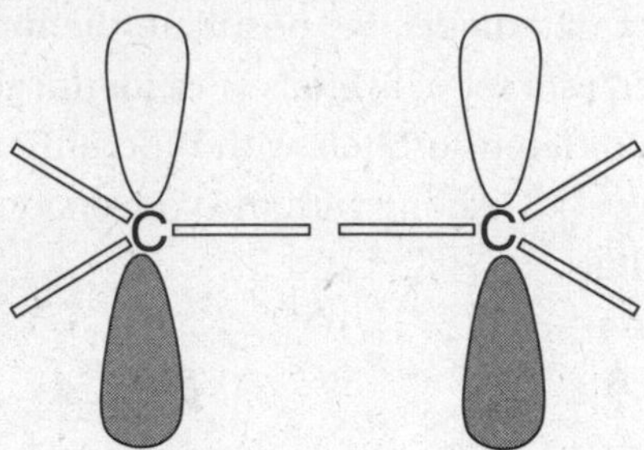

FIGURE 4.39. Two carbon atoms with ***sp*²** hybridization. The thin lines are the triangular planar ***sp*²** bonding orbitals. The large orbitals are the unhybridized ***p*** orbitals that overlap to form a pi bond.

sp Hybrid Orbitals

Carbon dioxide, O=C=O, has two sigma bonds and two pi bonds. The hybridization for this molecule is shown in Figure 4.40.

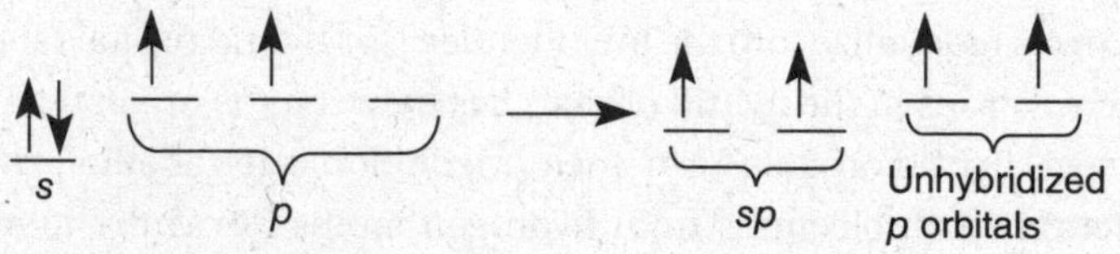

s *p* → *sp* Unhybridized *p* orbitals

FIGURE 4.40. Hybridization of carbon to produce the ***sp*** hybrid orbitals. The two unhybridized ***p*** electrons are available to form pi bonds.

In forming *sp* hybrid orbitals, we obtain two equivalent electrons that can form sigma bonds. The two remaining, unhybridized *p* electrons can overlap with *p* electrons from the oxygen atoms to form the required double bonds. The hybridized and unhybridized orbitals in carbon's sp^2 hybrid may be pictured as shown in Figure 4.41. This diagram shows the *p* orbitals available for pi bonding. Since there are two *p* orbitals, two additional pi bonds can form. These two pi bonds can be directed toward different atoms to form compounds, such as O=C=O, with two double bonds. They can also be directed toward the same atom to form a triple bond, as in the cyanide ion, $C{\equiv}N^-$.

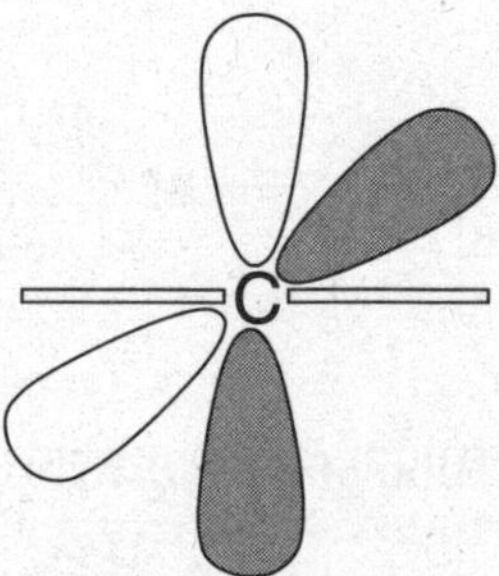

FIGURE 4.41. The ***sp*** hybrid orbitals, shown as thin lines. The remaining two ***p*** orbitals are shown as larger lobes. These ***p*** orbitals overlap with other ***p*** orbitals to form two pi bonds to the carbon.

sp^3d and sp^3d^2 Hybrid Orbitals

Over the past several years, researchers have not been able to demonstrate the existence and importance of the sp^3d and sp^3d^2 hybrid orbitals. They are omitted from the AP exam.

TABLE 4.7
Correspondence Between Hybridization and Structure

Basic Structure	Derived Structure	Hybrid	Bonding e^- Pairs	Nonbonding e^- Pairs
Linear		*sp*	2	0
Planar triangle		sp^2	3	0
Planar triangle	Bent	sp^2	2	1
Tetrahedron		sp^3	4	0
Tetrahedron	Triangular pyramid	sp^3	3	1
Tetrahedron	Bent	sp^3	2	2
Trigonal bipyramid			5	0
Trigonal bipyramid	Distorted tetrahedron		4	1
Trigonal bipyramid	T-shape		3	2
Trigonal bipyramid	Linear		2	3
Octahedron			6	0
Octahedron	Square pyramid		5	1
Octahedron	Square planar		4	

Exercise 4.13

Determine the total number of sigma and pi bonds in each of the following. Using the simple Lewis structure, also determine if the substance's shape can be explained by *sp*, sp^2, or sp^3 hybrids.

(a) CH_3Cl
(b) CS_2
(c) PH_3
(d) SiF_4
(e) H_2S
(f) NO_3^-
(g) CO_3^{2-}
(h) PO_4^{3-}
(i) SO_3^{2-}
(j) ClO_4^-

Solution

(a) 4 σ, 0 π, sp^3
(b) 2 σ, 2 π, *sp*
(c) 3 σ, 0 π, sp^3
(d) 4 σ, 0 π, sp^3
(e) 2 σ, 0 π, sp^3
(f) 3 σ, 1 π, sp^2
(g) 3 σ, 1 π, sp^2
(h) 4 σ, 0 π, sp^3
(i) 3 σ, 0 π, sp^2
(j) 4 σ, 0 π, sp^3

Before leaving the topic of hybrid orbitals, we must recognize that this model is used to explain experimental results. For a molecule that has a particular shape, the concept of hybrid orbitals may be used to explain that shape. **The reverse is not true.** We say that H_2O has sp^3 hybridization because it is a bent structure with a bond angle of 104°, which is close to the 109.5° bond angle expected for a tetrahedral structure. Experiments show, however, that the similar molecule H_2S has a bond angle of 90°. In this case the simple overlap of the *p* orbitals of sulfur with the *s* orbitals of hydrogen is sufficient to explain the structure. Hybridization is not needed, and is apparently unwarranted, in this example.

Experimental evidence always outweighs theory.

SUMMARY

After discussing ionic compounds in Chapter 3, this chapter covers molecular compounds that are characterized by covalent bonds. In order to understand molecular compounds and their structures, the chapter starts by reviewing the logical methods for drawing Lewis structures. Equivalent Lewis structures are described as resonance structures that don't resemble any one structure but are a blend of each contributing structure. Formal charges help us decide which Lewis structures may be better than others. The chapter then discusses bond order and bond strength along with electronegativity and bond polarity.

From here the chapter looks at the structures of the molecules using the VSEPR theory. There are five basic structures of matter that need to be remembered. Several additional structures are derived from the five basic structures. From the three-dimensional structures we find that molecules will be nonpolar if they are totally symmetrical (even if their bonds are polar). Nonsymmetric molecules are often polar. Because the geometry of the orbitals is different from the geometry of the molecules, hybrid orbitals are introduced to explain this observation. The relationships between hybrid orbitals and structure are also developed in this chapter. Finally, valence bond theory is used to describe the formation of sigma and pi bonds in molecular compounds.

Important Concepts

Octet rule and when it can be disobeyed
Lewis structures and formal charges
Bond polarity
Electronegativity
Dipole moment
Molecular geometry and molecular polarity
Hybrid orbitals

PRACTICE EXERCISES

Multiple-Choice

1. Which of these molecules has a shape related to a tetrahedron (or has sp^3 bonding)?
   (A) CBr_4 and NH_3
   (B) PF_5
   (C) NH_3 and HCN
   (D) SO_3

2. Which of these molecules has the largest bond angle?
   (A) HCN
   (B) PF_5
   (C) NH_3
   (D) SO_3

3. The observation that the bond lengths found in the carbonate ion are equal is due to
   (A) resonance structures
   (B) hybridization of the central atom
   (C) the presence of ionic bonds
   (D) the existence of lone pairs of electrons

4. Which of these molecules have all atoms lying in the same plane?
   (A) CBr_4 and NH_3
   (B) PF_5
   (C) NH_3 and HCN
   (D) SO_3 and HCN

5. Which of the following molecules are paired correctly with their molecular geometry?
   (A) PF_5, tetrahedron
   (B) XeF_4, square planar
   (C) CCl_4, trigonal pyramid
   (D) SF_6, square pyramid

6. In which of the following are the elements listed in order of increasing electronegativity?
   (A) Ba, Zn, C, Cl
   (B) N, O, S, Cl
   (C) N, P, As, Sb
   (D) K, Ba, Si, Ga

7. The difference in bond angles for SBr_2, PBr_3, and CBr_4 is due to
   (A) the dissimilarity in bond polarity between the central atom and bromine
   (B) the size of the central atom
   (C) the presence of several lone pairs of electrons
   (D) the variance in bond length and strength

8. For which of the following may we draw both polar and nonpolar Lewis structures?
   (A) $CHCl_3$
   (B) NH_3
   (C) BF_3
   (D) SF_2Cl_4

9. Which of the following has the fewest pi bonds and is nonpolar?
   (A) HCCH
   (B) CO_2
   (C) CO_3^{2-}
   (D) N_2

10. The following diagram is an incomplete Lewis structure.

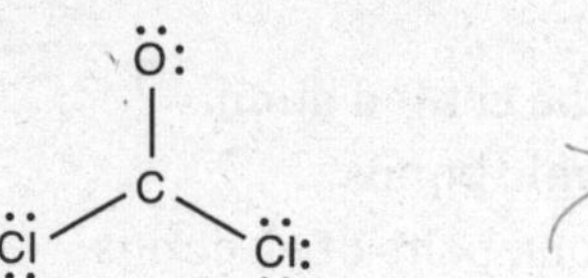

What does the complete Lewis structure for the diagram have?
(A) Both polar and nonpolar bonds
(B) At least one double bond
(C) Resonance forms
(D) At least one lone pair of electrons on each atom

11. Which of the following is NOT a linear structure?
(A) I_2
(B) I_3^-
(C) CO_2
(D) H_2S

12. The Lewis structure of the cyanide ion most closely resembles
(A) N_2
(B) O_2
(C) CO_2
(D) NO

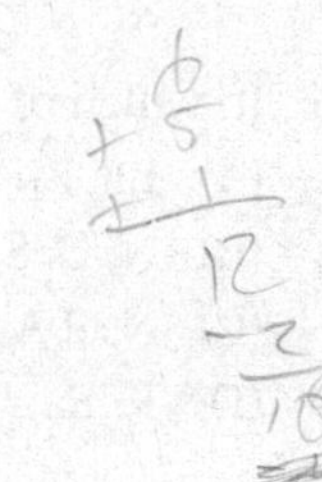

13. In which of the following pairs are the two items NOT properly related?
(A) sp^3 and 109.5°
(B) Trigonal planar shape and 120°
(C) sp and 180°
(D) Square planar shape and 120°

14. The nitrate ion, NO_3^-, can be described by the following Lewis structures:

What does this diagram mean?
(A) Two nitrogen to oxygen bonds are single bonds and a third is a double bond.
(B) The nitrate ion can exist as three equal forms in equilibrium.
(C) The nitrate ion exists in one form that is an average of all three structures shown.
(D) The electrons are rapidly exchanged among the three forms.

15. Which of the following has a nonbonding pair of electrons on the central atom?

(A) BCl_3
(B) NH_3
(C) CCl_2Br_2
(D) PF_5

16. Which of the following is true when the C=C and C≡C bonds are compared?

(A) The triple bond is shorter than the double bond.
(B) The double bond vibrates at a lower frequency than the triple bond.
(C) The double-bond energy is lower than the triple-bond energy.
(D) All of the above are true.

17. Why does a molecule of BF_3 have no overall dipole moment but a molecule of PF_3 does?

(A) The overall structure of BF_3 is linear.
(B) The atomic radius of phosphorus is larger than the atomic radius of boron.
(C) The overall structure of BF_3 is trigonal planar.
(D) BF_3 does not have a symmetrical structure but PF_3 does.

18. There are 6 sigma bonds and 3 pi bonds in the following compound.

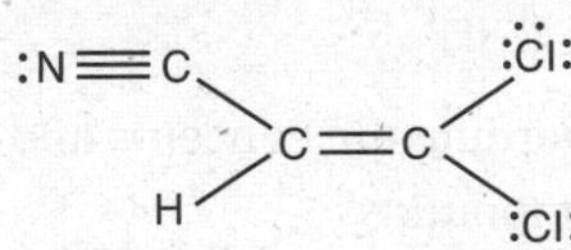

(A) 7, 2
(B) 6, 3
(C) 8, 1
(D) 9, 0

19. Which angle is NOT expected in any simple molecule?

(A) 60°
(B) 90°
(C) 109.5°
(D) 120°

CHALLENGE

20. Sulfur forms the following compounds: SO_2, SCl_2, and SO_3^{2-}. Which form of hybridization is NOT represented by these molecules?

(A) *sp*
(B) sp^2
(C) sp^3
(D) None of these

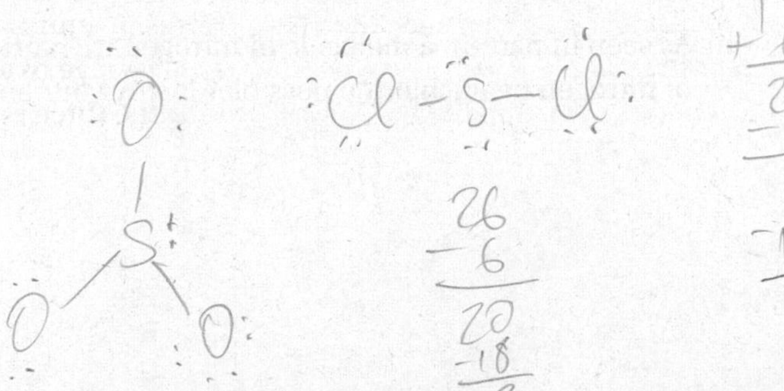

CHALLENGE

21. Which of the following is least related to the strength of a covalent bond?
    (A) vibrational frequency
    (B) bond order
    (C) bond length
    (D) bond direction

ANSWER KEY

1. **(A)**	8. **(D)**	15. **(B)**
2. **(A)**	9. **(C)**	16. **(D)**
3. **(A)**	10. **(B)**	17. **(C)**
4. **(D)**	11. **(D)**	18. **(B)**
5. **(B)**	12. **(A)**	19. **(A)**
6. **(A)**	13. **(D)**	20. **(A)**
7. **(C)**	14. **(C)**	21. **(D)**

See Appendix 1 for explanations of answers.

Free-Response

Answer the following questions regarding the concepts and properties concerning the structure and geometry of molecular compounds.

(a) Diethyl ether, $CH_3CH_2OCH_2CH_3$, is a highly flammable liquid. It is a common solvent in many laboratories.

(i) Draw the Lewis structure for this compound. Identify the electron pairs and molecular geometry of the oxygen. Explain your response.

(ii) Explain using VSEPR theory why the bond angle for the C–O–C bonds are not at 180°.

(iii) Are there any polar bonds in this molecule? If so, indicate the positive and negative poles of each bond. Is this molecule polar or nonpolar? Defend your answer.

(b) (i) Draw the Lewis structure for nitrogen trichloride, and determine its molecular geometry. Explain what hybrid orbitals are required to produce this structure.

(ii) Based on the structure of nitrogen trichloride, are the bonds in this molecule polar? If so, indicate the positive and negative poles in an appropriate manner. Explain your reasoning.

(iii) As seen in part (i), a molecule of nitrogen trichloride is possible. Is a molecule of nitrogen pentachloride possible? Justify your answer.

ANSWERS

(a) (i) Based on the structure, the electron pair geometry around the oxygen atom in diethyl ether is tetrahedral and the molecular geometry is bent or angular. This is due to the two lone pairs of electrons on the oxygen atom.

```
     H   H       H   H
     |   |       |   |
  H—C—C—Ö—C—C—H
     |   |   ..  |   |
     H   H       H   H
```

(ii) This molecule has an AX_2E_2 shape, where there are 2 atoms and 2 nonbonding electron pairs on the central oxygen atom. According to VSEPR theory, these 4 electrons repel each other in such a way as to be as far apart as possible. Nonbonding electron pairs require more space than do bonding electrons. Therefore the bond angle in this molecule is much smaller than the usual angle of 109.5° for a tetrahedron arrangement.

(iii) The two C–O bonds are polar due to the difference in electronegativity between the two atoms. The molecule is also polar since it is asymmetrical and the polarities of the individual C–O bonds do not cancel out each other.

(b) (i) The molecular geometry for this molecule is trigonal pyramid due to the nonbonding pair of electrons on the nitrogen. However, the basic structure pattern for NCl_3 is AX_3E, which is tetrahedron. In order to form the tetrahedron electron pair geometry, 4 sp^3 hybridized orbitals are necessary.

```
:C̈l—N̈—C̈l:
      |
     :C̤l:
```

(ii) The N–Cl bonds in this molecule are polar because of the significant difference in electronegativity between Cl and N. We should indicate the polarity of the bond as shown in the following diagram.

```
          δ+
δ– :C̈l—N̈—C̈l: δ–
          |
         :C̤l:
          δ–
```

The partial sign, δ, is used to symbolize a partial charge because neither a complete positive charge nor a complete negative charge is formed.

(iii) A molecule of nitrogen pentachloride is not possible since the nitrogen atom is too small to fit all 5 chlorines around it. This molecule would require an expanded octet, and the nitrogen atom does not have the ability to do so.

5 Stoichiometry

→ **CONVERSION CALCULATIONS USING THE DIMENSIONAL ANALYSIS METHOD**
→ **METRIC UNITS AND PREFIXES**
→ **CONVERTING COMPLEX UNITS**
→ **MOLE CONCEPT**
→ **AVOGADRO'S NUMBER**
→ **MOLAR MASS**
→ **MOLE RELATIONSHIPS IN FORMULAS**
→ **MOLE RELATIONSHIPS IN EQUATIONS**
→ **STOICHIOMETRY SEQUENCE**
→ **MOLE CONVERSIONS**
→ **GRAM-TO-GRAM CONVERSIONS**
→ **THEORETICAL YIELD CALCULATIONS**
→ **LIMITING-REACTANT CALCULATIONS**
→ **TITRATION METHODS**
→ **TITRATION STOICHIOMETRY**
→ **PERCENT COMPOSITION**
→ **EMPIRICAL FORMULAS**
→ **MOLECULAR FORMULAS**

BIG IDEAS 1, 3

Learning Objectives: 1.1, 1.2, 1.3, 1.4, 1.17, 1.19, 1.20, 3.1, 3.2, 3.3, 3.4, 3.5

For the complete list of Big Ideas and Learning Objectives, refer to the AP Chemistry Course Outline: ***https://secure-media.collegeboard.org/digitalServices/pdf/ap/ap-chemistry-course-and-exam-description.pdf***

Stoichiometry (measurement of the elements) is the name given to the quantitative relationships between the compounds in a chemical reaction. These quantitative relationships allow chemists to calculate the amounts of reactants needed for a reaction and to predict the quantity of product. Using stoichiometric methods, chemists can determine the formulas of compounds and can simplify procedures in chemical analysis. This chapter discusses and illustrates stoichiometric calculations and the fundamental concepts that make stoichiometry the most important topic in chemistry.

Quantitative calculations in chemistry fall into two groups. The first group involves taking a memorized equation, entering data for all but one variable, and then solving for the remaining variable. The second group involves converting information with one set of units into an answer with another set of units. Most stoichiometric calculations are the conversion type.

The **dimensional analysis method** is the predominant method taught for solving stoichiometry problems. Over the years many methods have been used to perform stoichiometric conversions. It has the advantage of minimizing memorization while applying basic chemical concepts to define the conversion process. In addition, graders look for logical, well thought out, and well-presented work. Using dimensional analysis and properly calculating units can only help.

CONVERSION CALCULATIONS USING THE DIMENSIONAL ANALYSIS METHOD

Defining Conversion Factors

In this section, the methods used for dimensional analysis are developed in detail. If you are adept at using this method, you can skip to page 209.

A conversion problem changes the units of a measurement but not its magnitude. For instance, a distance of 2.0 yards may be converted into 72 inches by multiplying 2.0 by the factor 36. This does not indicate how the units of yards became units of inches. If, instead of using just the factor 36, we multiply by the factor with its units, it is clear how the yards became inches:

$$2.0\ \cancel{\text{yards}}\left(\frac{36\ \text{inches}}{1\ \cancel{\text{yard}}}\right)=72\ \text{inches}$$

In this equation the yard units cancel, and it is clear that the remaining units are inches. The ratio $\left(\frac{36\ \text{inches}}{1\ \text{yard}}\right)$ is known as the **conversion factor**. Using labels along with the numerical factors enables us to keep track of the units and to produce an answer with the correct units. Proper use of conversion factors also tells us when to multiply and when to divide.

A conversion factor is derived from a defined relationship between two sets of units. The conversion factor above was obtained from the definition of 1 yard:

$$1\ \text{yard} = 36\ \text{inches}$$

Two conversion factors can be obtained from every defined equality. For example, dividing both sides of the above equivalence by 1 yard gives

$$\frac{1\ \text{yard}}{1\ \text{yard}}=\frac{36\ \text{inches}}{1\ \text{yard}}=1$$

Conversion Factor

Dividing both sides by 36 inches gives

$$\frac{1\ \text{yard}}{36\ \text{inches}}=\frac{36\ \text{inches}}{36\ \text{inches}}=1$$

Conversion Factor

The two possible conversion factors are the inverse of each other. In addition, they are both equal to 1. When a measurement is multiplied by a conversion factor, it is multiplied, in effect, by 1, and its true value does not change, although the units do change.

Using Conversion Factors

Every conversion problem must start with two pieces of information: (1) the number and the units that need to be converted and (2) the units of the answer. This is set up as

$$? \text{ answer units} = xxx \text{ given units}$$

On the right side is *xxx*, which represents the number given in the problem, and its units, which will be converted. The left side of the equal sign reminds us of the units needed for the solution to the problem.

After this step, it is necessary to find the equalities that can be used to produce the conversion factors needed to solve the problem. For example, we might be required to convert 35 yards into feet in part (a) and to convert 624 feet into yards in part (b). The solution to part (a) starts with the initial setup of the question with the desired units as the unknown in an equation and the given data as the starting point:

$$? \text{ feet} = 35 \text{ yards} \quad \text{(initial setup)}$$

Being familiar with the English system of measurement, we know that 1 yard is defined as being 3 feet in length:

$$1 \text{ yard} = 3 \text{ feet}$$

From this equality two conversion factors, $\left(\frac{3 \text{ feet}}{1 \text{ yard}}\right)$ and $\left(\frac{1 \text{ yard}}{3 \text{ feet}}\right)$, can be written.

The correct conversion factor allows us to cancel the yard units, leaving units of feet. This conversion factor is used to multiply the given 35 yards:

$$? \text{ feet} = 35 \text{ yards} \left(\frac{3 \text{ feet}}{1 \text{ yard}}\right)$$

After being sure that the units properly cancel, we calculate the answer as 105 feet. The following equation illustrates what happens if the wrong form of the conversion factor is chosen:

$$? \text{ feet} = 35 \text{ yards} \left(\frac{1 \text{ yard}}{3 \text{ feet}}\right) = 11.7 \text{ yards}^2 \text{ feet}^{-1} \quad \text{(wrong conversion factor chosen)}$$

This conversion will not work since the units do not cancel; in fact the final units are the meaningless yard^2 per foot. The correct answer for part (a) is

$$? \text{ feet} = 35 \text{ yards} \left(\frac{3 \text{ feet}}{1 \text{ yard}}\right) = 105 \text{ feet}$$

IN ALL OF THE FOLLOWING EXAMPLES, THE CANCELLATION OF THE UNITS IS NOT SHOWN. IT IS SUGGESTED THAT YOU TAKE A COLORED PENCIL AND PERFORM ALL OF THE CANCELLATIONS TO ASSURE YOURSELF THAT EACH CONVERSION FACTOR DOES INDEED CANCEL PROPERLY.

TIP

Conversion calculations, just like driving directions, need a starting point and an end destination.

In part (b) of the question, the reverse calculation, from feet to yards, is requested. Starting as before with the question and the data supplied, we write

$$? \text{ yards} = 624 \text{ feet} \quad \text{(initial setup)}$$

Selecting the correct conversion factor gives

$$? \text{ yards} = 624 \text{ feet} \left(\frac{1 \text{ yard}}{3 \text{ feet}}\right) = 208 \text{ yards}$$

For this conversion the same equality, but a different conversion factor, was used. One of these conversion factors converts from yards to feet, and the other converts feet to yards.

Many conversions require more than one step to reach the desired units. These problems may be solved stepwise, one conversion factor at a time, or the conversion factors may be combined in one large equation. Both methods are illustrated in Example 5.1.

At this point go back to the start of this chapter and cancel units with a red pencil to reinforce the cancellation process.

EXAMPLE 5.1

A typical school year includes 180 days of classes. How many minutes are there in those days?

TIP

Performing calculations by using a logical sequence of steps will improve your score.

Solution

We start with the question and the given information to obtain

$$? \text{ minutes} = 180 \text{ days}$$

Next, we find the conversion equalities that may be useful. These are as follows:

$$1 \text{ day} = 24 \text{ hours} \quad 1 \text{ hour} = 60 \text{ minutes}$$

The given data have units of days, and the only equality that also has units of days is the first one. The ratio needed for the conversion must have the day units in the denominator so that these units will cancel. Multiplying by this ratio yields

$$? \text{ minutes} = 180 \text{ days} \left(\frac{24 \text{ hr}}{1 \text{ day}}\right) = 4320 \text{ hr}$$

Paying attention to the units of conversion factors helps to minimize errors.

This result still does not have the desired units, so another step is needed. Starting with the result obtained above, we write

$$? \text{ minutes} = 4320 \text{ hr}$$

In our second conversion equality there is a relationship between hours and minutes, 1 hour = 60 minutes. When the next conversion factor is inserted, the hour units cancel:

$$? \text{ minutes} = 4320 \text{ hr} \left(\frac{60 \text{ min}}{1 \text{ hr}}\right) = 2.59 \times 10^5 \text{ min}$$

Since the units of this answer match the units that the question requires, the problem is solved. This answer is rounded off to 2.59×10^5 min.

Solving this same problem with one large equation involves writing all of the conversion factors needed until the units match:

$$? \text{ minutes} = 180 \text{ days} \left(\frac{24 \text{ hr}}{1 \text{ day}}\right)\left(\frac{60 \text{ min}}{1 \text{ hr}}\right) = 2.59 \times 10^5 \text{ min}$$

The step-by-step and the combined methods result in exactly the same answer, and either method is correct.

Conversion of Metric Units

Our modern version of the metric system is called the Système International, or S.I. The seven base units of the S.I. are defined in Table 5.1.

Table 5.1
Seven Base S.I. Units

Property Defined	Unit Name	Abbreviation
Mass	kilogram	kg
Length	meter	m
Time	second	s
Temperature	kelvin	K
Quantity	mole	mol
Electric current	ampere	A
Light intensity	candela	cd

These seven base units may be combined in a variety of ways to obtain other common units. For example, area can be expressed as square meters (m^2), and volume as cubic meters (m^3). The base units may be modified by the use of a metric prefix. Each **metric prefix** represents a number that multiplies the base unit. The most common metric prefixes are listed in Table 5.2.

Knowing these prefixes saves valuable time on the exam.

Table 5.2
Metric Prefixes

Prefix Name	Prefix Symbol	Exponential Value
mega-	M	10^6
kilo-	k	10^3
deci-	d	10^{-1}
centi-	c	10^{-2}
milli-	m	10^{-3}
micro-	μ	10^{-6}
nano-	n	10^{-9}
pico-	p	10^{-12}

It is essential to remember the first five of the metric base units in Table 5.1 and all of the metric prefixes in Table 5.2. Conversion factors between a unit with a prefix and the corresponding metric base unit may be quickly obtained by first writing an equality:

$$1 \text{ cm} = 1 \text{ cm}$$

and then replacing one of the prefixes with the corresponding exponent:

$$1 \text{ cm} = 1 \times 10^{-2} \text{ m}$$

This equality can be used to write the two possible conversion factors as $\left(\frac{1 \text{ cm}}{10^{-2} \text{ m}}\right)$ and $\left(\frac{10^{-2} \text{ m}}{1 \text{ cm}}\right)$. Conversion between a prefix and a base unit is a one-step calculation, but conversion between two different prefixes requires two steps.

EXAMPLE 5.2

Convert 2.38 cm to meters and millimeters.

Solution

For the first conversion one of the conversion factors above may be used:

$$? \text{ m} = 2.38 \text{ cm (setup)}$$
$$= 2.38 \text{ cm}\left(\frac{10^{-2} \text{ m}}{1 \text{ cm}}\right)$$
$$= 0.0238 \text{ m}$$

For the second conversion, centimeters to millimeters, two conversion factors are needed. The first converts the prefix to the base unit, and the second converts the base unit to the new prefix.

$$? \text{ mm} = 2.38 \text{ cm (setup)}$$
$$= 2.38 \text{ cm}\left(\frac{10^{-2} \text{ m}}{\text{cm}}\right)\left(\frac{\text{mm}}{10^{-3} \text{ m}}\right)$$
$$= 23.8 \text{ mm}$$

The answers to both parts of this exercise were written in exponential notation in order to show that the number 2.38 does not change in these conversions, but the exponent does.

Conversion of Complex Units

Many times the data used in chemistry involve complex units. There are area measurements that have squared units such as square kilometers (km^2), square meters (m^2), or square centimeters (cm^2). There are volume measurements with cubic units such as cubic centimeters (cm^3), cubic millimeters (mm^3), and cubic meters (m^3). For velocity the units may be meters per second, which is abbreviated as m/s or $m\ s^{-1}$. Acceleration has units of meters per second squared ($m\ s^{-2}$), and energy has units of kilogram meters squared per second squared ($kg\ m^2\ s^{-2}$).

The following exercises illustrate the conversion method used when the units have squared or cubed terms and when they involve a ratio.

EXAMPLE 5.3

How many square centimeters are there in 180 m²?

Solution

Setting up the problem as before, we have

$$? \text{ cm}^2 = 180 \text{ m}^2$$

The units are square centimeters and square meters. For clarity it is best to write the setup of this problem as

$$? \text{ cm} \times \text{cm} = 180 \text{ m} \times \text{m}$$

We can use the equality that says that 10^{-2} m is equal to 1 cm to write the needed conversion factor. Using the ratio that will cancel out the meter units, we obtain

$$? \text{ cm} \times \text{cm} = 180 \text{ m} \times \text{m} \left(\frac{1 \text{ cm}}{10^{-2} \text{ m}}\right)$$

By using the conversion ratio only once, we have canceled only one of the meter units, leaving the mixed units of centimeter meter. Applying the conversion ratio a second time cancels all of the meter units and leaves us with the desired square centimeters:

$$? \text{ cm} \times \text{cm} = 180 \text{ m} \times \text{m} \left(\frac{1 \text{ cm}}{10^{-2} \text{ m}}\right)\left(\frac{1 \text{ cm}}{10^{-2} \text{ m}}\right)$$

Once the units cancel properly, the answer can be calculated as 1.80×10^6 cm^2. For volumes, which have cubic units, each conversion factor is used three times.

When a ratio of units need to be converted, as illustrated in Example 5.4, the units in the numerator are converted into the units required by the problem. Then the units in the denominator are converted.

EXAMPLE 5.4

A train is moving at a speed of 25 km hr^{-1}. How fast is it moving in units of meters per minute?

Solution

Set up the problem as before, with the desired units as the question and the given speed as the starting point in the conversion.

$$?\frac{\text{m}}{\text{min}} = \frac{25\text{ km}}{\text{h}} \text{ (setup)}$$

The two equalities that must be used to construct conversion factors for this problem are

$$1 \text{ kilometer} = 10^3 \text{ meters}$$
$$1 \text{ hour} = 60 \text{ minutes}$$

Applying the factor obtained from the first equality converts the kilometers to meters:

$$?\frac{\text{m}}{\text{min}} = \frac{25\text{ km}}{\text{h}}\left(\frac{\mathbf{10^3\,m}}{\mathbf{1\,km}}\right)$$

Next, the second equality is used to convert the hour units in the denominator to the minutes required:

$$?\frac{\text{m}}{\text{min}} = \frac{25\text{ km}}{\text{h}}\left(\frac{10^3\text{ m}}{1\text{ km}}\right)\left(\frac{\mathbf{1\,h}}{\mathbf{60\,min}}\right)$$

The units cancel properly, leaving the desired meters per minute units. The answer is then calculated to be 4.2×10^2 m min^{-1}.

Examples 5.1–5.4 illustrate three important principles about the conversion factor method. First, it is necessary to know the equalities required to obtain the proper conversion factors. Second, there is often a proper sequence in which to use these conversion factors. Third, if the units for the given value are in the form of a ratio, they must be converted into units that also represent a ratio. Similarly, if the problem requests an answer where the units must be a ratio, the starting data must have units in the form of a ratio.

Exercise 5.1

Convert each of the following:

(a) 8.89 nm to mm
(b) 3.89×10^5 cm^2 to μm^2
(c) 2.43×10^2 kg m^2 s^{-2} to g cm^2 s^{-2}

Solution

(a) 8.89×10^{-6} mm
(b) 3.89×10^{13} μm^2
(c) 2.43×10^9 g cm^2 s^{-2}

CHEMICAL EQUALITIES AND RELATIONSHIPS

The Mole and Avogadro's Number

The **mole** (mol) is the central unit of measurement in chemistry. Numerically one mole represents 6.02×10^{23} units of a chemical substance. For the elements 1 mole represents 6.02×10^{23} atoms of the element. In compounds such as CH_4 or CO_2, 1 mole represents 6.02×10^{23} molecules. For ionic compounds it represents 6.02×10^{23} empirical formula units of a substance such as NaCl, or $MgBr_2$. The value 6.02×10^{23} is called **Avogadro's number** in honor of that chemist-physicist's pioneering work in stoichiometry.

In mathematical equations the number of moles is given the symbol n. This symbol is also used for many other purposes, however, and only a thorough understanding of any equation will tell whether n represents moles or some other quantity.

In general,

$$1 \text{ mole of X} = 6.02 \times 10^{23} \text{ units of X} \tag{5.1}$$

Some specific chemistry examples are as follows:

$$1 \text{ mole of argon atoms} = 6.02 \times 10^{23} \text{ Ar atoms}$$
$$1 \text{ mole of } CH_4 \text{ molecules} = 6.02 \times 10^{23}\ CH_4 \text{ molecules}$$
$$1 \text{ mole of } Mg^{2+} \text{ ions} = 6.02 \times 10^{23}\ Mg^{2+} \text{ ions}$$
$$1 \text{ mole of NaCl formula units} = 6.02 \times 10^{23} \text{ NaCl formula units}$$

Molar Mass

The periodic table used for the AP Chemistry Exam lists the **relative atomic mass** of each element directly underneath the chemical symbol. Relative atomic mass has no units and simply indicates the mass of one element as compared to that of another. In chemistry it is customary to add grams units to the atomic masses listed in the periodic table, calling them **gram-atomic masses**. One mole of an element is equal to the gram-atomic mass of that element:

$$1 \text{ mole of an element} = \text{gram-atomic mass of that element} \tag{5.2}$$

For a chemical compound the gram-molar mass is equal to the sum of the gram-atomic masses of all atoms in the chemical formula:

$$\text{gram-molar mass of a compound} = \Sigma(\text{gram-atomic masses in formula}) \tag{5.3}$$

Similar to the cases for elements, the **gram-molar mass** of a compound is equal to 1 mole of that compound. Gram-molar masses of ions are determined by the atom(s) present in the ion since the gain or loss of electrons has virtually no effect on the total mass.

$$1 \text{ mole of a compound} = \text{gram-molar mass of that compound} \tag{5.4}$$

Some specific examples of the molar-mass relationships are as follows:

$$1 \text{ mole of argon} = 39.948 \text{ grams of argon}$$
$$1 \text{ mole of uranium} = 238.029 \text{ grams of uranium}$$
$$1 \text{ mole of } CH_4 = 16.043 \text{ grams of } CH_4$$
$$1 \text{ mole of NaCl} = 58.4424 \text{ grams of NaCl}$$
$$1 \text{ mole of } Mg^{2+} = 24.3050 \text{ grams of } Mg^{2+}$$

Although it is correct to refer to the gram-atomic mass of an element and the gram-molar mass of a compound, it is common usage to refer to these as simply the **atomic mass**, *A*, and **molar mass**, respectively. Also note that the **formula mass** is the same as the molar mass, while the term **molecular mass** should be reserved for molecular substances.

The equalities defined in this section can be used to produce the appropriate conversion factors for conversion calculations. Currently more than ten million compounds are known, and these definitions will give twice as many conversion factors.

Exercise 5.2

Using the periodic table, determine the molar mass of each of the following compounds and round to two decimal places:

(a) $Cd(NO_3)_2$
(b) $CH_3(CH_2)_4Br$
(c) $(NH_4)_2SO_4$
(d) $(CH_3CH_2CH_2)_2O$
(e) $CuSO_4{\bullet}5H_2O$

Solution

(a) 236.42 g/mol
(b) 151.04 g/mol
(c) 132.14 g/mol
(d) 102.18 g/mol
(e) 249.69 g/mol

Conversion Factors from Chemical Formulas

Because of its structure, the formula for an ionic compound is an empirical formula representing the simplest ratio of atoms. Formulas for molecular compounds give the numbers and types of all the atoms that make up one molecule. All chemical formulas represent the ratio of atoms within the formula. This fact allows the chemist to write relationships between a formula as a whole and the individual atoms in that formula. For example, common table sugar is sucrose with the formula $C_{12}H_{22}O_{11}$. On the atomic scale this gives the following relationships, where the equal sign is read as "is chemically equivalent to":

$$\begin{aligned} \text{1 molecule of } C_{12}H_{22}O_{11} &= \text{12 atoms of carbon} \\ \text{1 molecule of } C_{12}H_{22}O_{11} &= \text{22 atoms of hydrogen} \\ \text{1 molecule of } C_{12}H_{22}O_{11} &= \text{11 atoms of oxygen} \\ \text{12 atoms of carbon} &= \text{22 atoms of hydrogen} \\ \text{12 atoms of carbon} &= \text{11 atoms of oxygen} \\ \text{22 atoms of hydrogen} &= \text{11 atoms of oxygen} \end{aligned}$$

In short, one sucrose molecule gives the chemist six different relationships with which to construct conversion factors. (It should be emphasized that these are not true equalities. The equal signs in these equations should be read as "is chemically equivalent to"; then the first relationship actually says that 1 molecule of $C_{12}H_{22}O_{11}$ "is chemically equivalent to" 12 atoms of carbon.)

It is rare that the chemist thinks of these relationships in terms of molecules and atoms. Rather they are viewed as moles of molecules or moles of atoms. In essence, each side of the relationships given above is multiplied by Avogadro's number to obtain the chemical equivalences:

$$\begin{aligned} \text{1 mol } C_{12}H_{22}O_{11} &= \text{12 mol C} \\ \text{1 mol } C_{12}H_{22}O_{11} &= \text{22 mol H} \\ \text{1 mol } C_{12}H_{22}O_{11} &= \text{11 mol O} \\ \text{12 mol C} &= \text{22 mol H} \\ \text{12 mol C} &= \text{11 mol O} \\ \text{22 mol H} &= \text{11 mol O} \end{aligned}$$

Finally, it is important to remember that these relationships are true only for the specified compound. They may be different for other compounds.

Exercise 5.3

How many different chemical equivalences may be written for the following formulas?

(a) NaCl
(b) $FeCl_3$
(c) $NiSO_4$
(d) $(NH_4)_3PO_4$
(e) O_3

Solution

(a) 3
(b) 3
(c) 6
(d) 10
(e) 1

Conversion Factors from Balanced Chemical Equations

While the chemical formula of a compound provides many relationships for conversion factors, these relationships are limited to that single compound. A balanced chemical equation, however, gives a group of relationships that can be used to generate additional conversion factors. Consider the balanced equation for the combustion of benzene, C_6H_6:

$$2C_6H_6 + 15O_2 \rightarrow 12CO_2 + 6H_2O$$

On a mole basis the following equivalences can be obtained:

$$2 \text{ mol } C_6H_6 = 15 \text{ mol } O_2$$
$$2 \text{ mol } C_6H_6 = 12 \text{ mol } CO_2$$
$$2 \text{ mol } C_6H_6 = 6 \text{ mol } H_2O$$
$$15 \text{ mol } O_2 = 12 \text{ mol } CO_2$$
$$15 \text{ mol } O_2 = 6 \text{ mol } H_2O$$
$$12 \text{ mol } CO_2 = 6 \text{ mol } H_2O$$

Once again, the equal sign does not represent mathematical equality but it should be read as "is equivalent to." These equalities apply only to the balanced equation from which they are derived.

Concentration and Density as Conversion Factors

The concentration of a solution can also be used as a conversion factor for conversion calculations. The most common concentration unit used in chemistry is molarity. **Molarity** (M) is the number of moles of a solute that are dissolved in 1 liter of solution. When a problem gives the concentration of a solution, such as 0.250 molar NaOH (also written as 0.250 M NaOH), this value can be made into a conversion factor by specifying its units:

$$0.250\ M \text{ NaOH} = \frac{0.250 \text{ mol NaOH}}{1 \text{ L NaOH}}$$

This ratio is a conversion factor for conversions between moles of NaOH and liters of NaOH solution. As with all conversion factors, its inverse is also a conversion factor:

$$\frac{1 \text{ L NaOH}}{0.250 \text{ mol NaOH}}$$

In many instances the volume is expressed in milliliters (mL). Then the two conversion factors can be written as

$$\frac{0.250 \text{ mol NaOH}}{1000 \text{ mL NaOH}} \text{ and } \frac{1000 \text{ mL NaOH}}{0.250 \text{ mol NaOH}}$$

since there are 1000 mL in each liter.

The density of a substance may be used as a conversion factor for conversions between volume and mass. Most densities in chemistry are given in units of grams per cubic centimeter (g cm^{-3}). If an organic liquid has a density of 0.741 g cm^{-3}, this fact may be written as the conversion factor:

$$\text{Density} = \frac{0.741\,\text{g}}{\text{cm}^3}$$

As with molarity, the inverse of this ratio is the other conversion factor obtained from the density:

$$\frac{\text{cm}^3}{0.741\,\text{g}}$$

TIP

Remember: 1.0 mL = 1.0 cm³

Since 1 cm^3 is the same as 1 mL, we can interchange the two terms as needed to obtain

$$\frac{\text{mL}}{0.741\,\text{g}} \text{ and } \frac{0.741\,\text{g}}{\text{mL}}$$

Other Conversion Factors

Another useful equality for constructing a conversion factor is the relationship between the moles of a gas and the volume of a gas at standard temperature and pressure (see Chapter 3). Standard temperature is 0°C, and standard pressure is 1 atmosphere of pressure. Under these conditions 1 mole of a gas occupies 22.4 liters. The equality and its two conversion factors are as follows:

$$1\text{ mol gas} = 22.4\text{ L gas}$$
$$1 = \frac{22.4\text{ L}}{1\text{ mol gas}}$$

and

$$1 = \frac{1\text{ mol gas}}{22.4\text{ L}}$$

While this expression specifically refers to an ideal gas, this conversion factor can be used for calculations involving most real gases with little error.

Finally, there are several equalities that are very useful to remember:

$$1\text{ cm}^3 = 1\text{ mL}$$
$$1\text{ L} = 1000\text{ mL} = 1000\text{ cm}^3$$
$$1\text{ cm}^3\ (H_2O) = 1\text{ g } H_2O$$

THE CONVERSION SEQUENCE

Once the various equalities and relationships are known, they must be used in the correct way to perform stoichiometric calculations. It is important to understand the sequence of operations required to perform any conversion successfully and efficiently.

Figure 5.1 illustrates how conversions in chemistry are related to each other. This diagram shows the sequence of conversions from any given item of chemical information to any other that may be desired. The notations along the arrows indicate the type of conversion factor needed to perform each conversion. At most, a conversion will require three steps, not including any conversions of metric prefixes.

To start analyzing this diagram, note that the left-side boxes all refer to "SUBSTANCE A," and those on the right side to "SUBSTANCE B." The examples that follow in the text will be divided into two groups. The first group involves only the conversion of units of one substance and therefore uses only the SUBSTANCE A side of the diagram. A typical question might be "How many atoms of iron are in a 2.00-gram sample of Fe?" The second group involves problems in which a given amount of substance A is converted into an equivalent amount of substance B. These conversions start on the SUBSTANCE A side of the diagram and end on the SUBSTANCE B side. A typical question might be "How many grams of carbon are there in 10.0 grams of $Fe_2(CO_3)_3$?" In most problems the given information will be found as one of the boxes on the SUBSTANCE A side of the diagram. The box representing the desired units of the answer is then located, and conversions are made step by step, using the indicated conversion factors. In some problems, the given information may not be exactly in the form shown in Figure 5.1. For instance, you may be given milligrams, mg, of a substance instead of grams, g. In those cases, you will need to convert the given units to the needed starting units.

Calculations Involving One Substance

Some stoichiometric questions involve converting from one set of units to another set for the same chemical substance. In this case we focus entirely on the left side of Figure 5.1. Some sample questions are given below. Most AP problems require answers with three significant figures, so we will keep at least four significant figures for data such as molar masses in the following questions.

EXAMPLE 5.5

How many grams of $FeCl_3$ (molar mass = 162.2 g/mol) need to be weighed to have 0.456 mol of $FeCl_3$?

Solution

TIP

Conversion factors and a logical conversion sequence show precise thinking to AP graders.

This problem gives the number of moles of $FeCl_3$ and asks for grams. This is a one-step conversion that uses the equality between the molar mass of $FeCl_3$ and the moles of $FeCl_3$ as the conversion factor. The question to be answered is set up as

$$?\ \text{g FeCl}_3 = 0.456\ \text{mol FeCl}_3$$

The conversion factor is obtained from the fact that 1 mole of any substance is equal to the molar mass in grams:

$$1\ \text{mol FeCl}_3 = 162.2\ \text{g FeCl}_3$$

The appropriate conversion factor for the conversion is $\dfrac{162.2\ \text{g FeCl}_3}{1\ \text{mol FeCl}_3}$ since it allows us to cancel the mol $FeCl_3$ units:

$$?\ \text{FeCl}_3 = 0.456\ \text{mol FeCl}_3\left(\frac{162.2\ \text{g FeCl}_3}{1\ \text{mol FeCl}_3}\right)$$

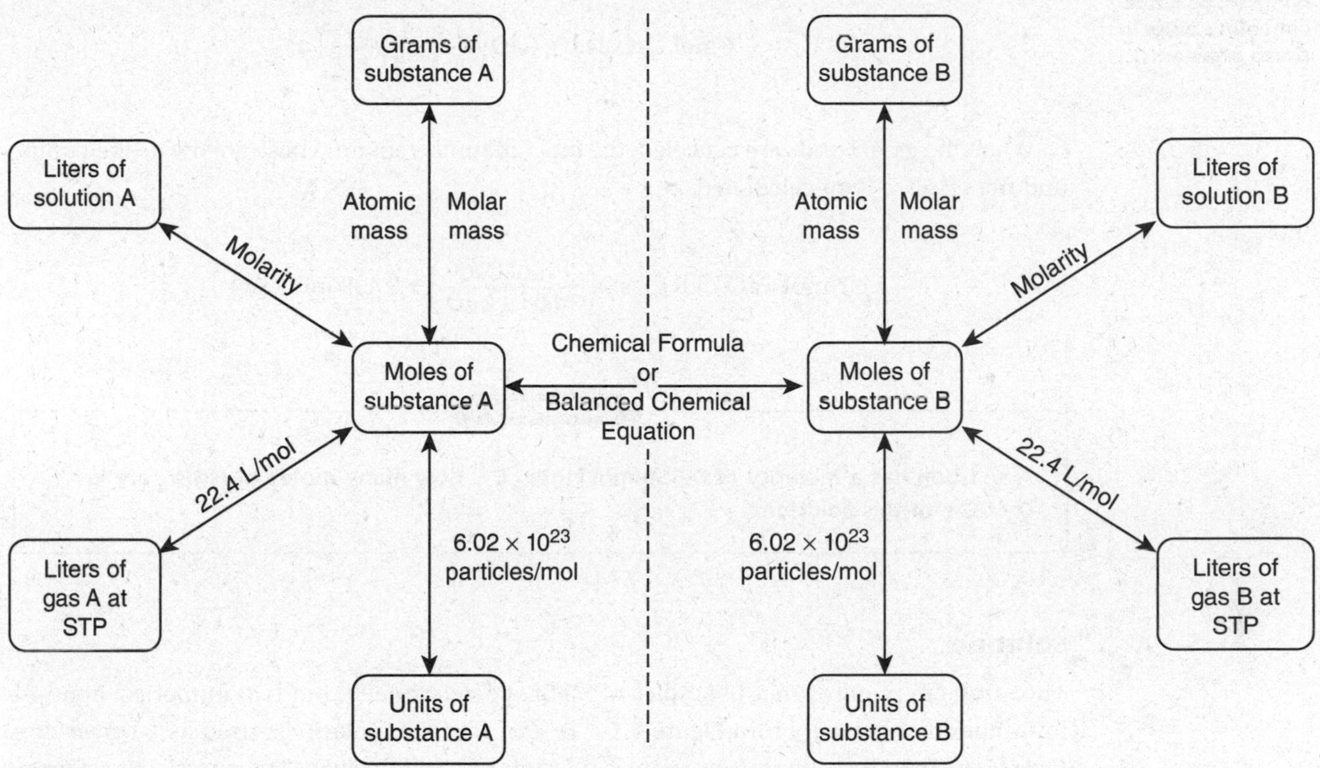

FIGURE 5.1. Diagram of the sequence of steps used in stoichiometry calculations. The box corresponding to the given data is found on the SUBSTANCE A side of the diagram. The box corresponding to the desired data is then located. The notations alongside the arrows tell what information is used to construct the required conversion factor.

The mol $FeCl_3$ units cancel, and the g $FeCl_3$ units remaining are the ones requested in the question. No other conversion factors are needed, and the answer is calculated as

$$? \text{ FeCl}_3 = 0.456 \text{ mol FeCl}_3 \left(\frac{162.2 \text{ g FeCl}_3}{1 \text{ mol FeCl}_3}\right) = 74.0 \text{ g FeCl}_3$$

EXAMPLE 5.6

A sample contains 24.6 g of CaO. How many moles of CaO (molar mass = 56.08 g/mol) are in this sample?

Solution

This problem gives the grams of sample and asks for the number of moles. In effect, it is the reverse process of the preceding calculation. The question is set up as

$$? \text{ mol CaO} = 24.6 \text{ g CaO}$$

TIP

For the rest of this chapter, practice canceling units in these problems.

The conversion equality is 1 mol CaO = 56.08 g CaO, which can be made into a conversion factor with g CaO in the denominator. Multiplying by the conversion factor gives

$$? \text{ mol CaO } 24.6 \text{ g CaO}\left(\frac{1 \text{ mol CaO}}{56.08 \text{ g CaO}}\right)$$

When the g CaO units are canceled, the mol CaO units remain. These are the desired units, and the result is then calculated as

$$? \text{ mol CaO } 24.6 \text{ g CaO}\left(\frac{1 \text{ mol CaO}}{56.08 \text{ g CaO}}\right) = 0.439 \text{ mol CaO}$$

EXAMPLE 5.7

A solution has a molarity of 0.658 mol $MgBr_2$ L^{-1}. How many moles of $MgBr_2$ are in 0.400 L of this solution?

Solution

This problem requires that the moles of $MgBr_2$ be determined, but two numerical items of information are given. From Figure 5.1 it is seen that the molarity is used as a conversion factor, and therefore the starting point is the liters of solution given. The question is set up as

$$? \text{ mol MgBr}_2 = 0.400 \text{ L MgBr}_2$$

TIP

It is often assumed that the 1 L in this type of problem is an exact number and does not affect the number of significant figures in the answer.

The molarity is already a ratio:

$$0.658\ M \text{ MgBr}_2 = \frac{0.658 \text{ mol MgBr}_2}{1 \text{ L MgBr}_2}$$

and may be used as the conversion factor:

$$? \text{ mol MgBr}_2 = 0.400 \text{ L MgBr}_2\left(\frac{0.658 \text{ mol MgBr}_2}{1 \text{ L MgBr}_2}\right)$$

Canceling the units and solving give the answer:

$$? \text{ mol MgBr}_2 = 0.263 \text{ mol MgBr}_2$$

EXAMPLE 5.8

The same $MgBr_2$ solution as in Example 5.7 must be used to obtain 0.500 mol of $MgBr_2$. How many milliliters of this solution are needed?

Solution

Now we must calculate the volume from the number of moles given, and the setup starts with

$$? \text{ mL MgBr}_2 = 0.500 \text{ mol MgBr}_2$$

Once again, the molarity is the conversion factor. However, it cannot be used directly since the units will not cancel. The molarity ratio is inverted and then used in the equation as

$$? \text{ mL MgBr}_2 = 0.500 \text{ mol MgBr}_2 \left(\frac{1 \text{ L MgBr}_2}{0.658 \text{ mol MgBr}_2}\right)$$

Although the mol units cancel properly, the answer will be calculated in liters, not the milliliters requested. We must use an additional conversion factor to change the prefix of the liter units. The appropriate conversion factor is $\left(\frac{1 \text{ mL}}{10^{-3} \text{ L}}\right)$. Multiplying by this conversion factor and canceling the L units, we obtain the desired mL units:

$$? \text{ mL MgBr}_2 = 0.500 \text{ mol MgBr}_2 \left(\frac{1 \text{ L MgBr}_2}{0.658 \text{ mol MgBr}_2}\right)\left(\frac{1 \text{ mL}}{10^{-3} \text{ L}}\right)$$

The answer is 760. mL of $MgBr_2$ solution.

Examples 5.5–5.8 demonstrated the one-step conversion of data to and from mole units. Many common calculations, however, involve two steps, as shown below.

EXAMPLE 5.9

How many grams of KCl (molar mass = 74.55 g/mol) are there in 0.250 L of a 0.300 molar solution of KCl?

Solution

From Figure 5.1 we see that to get from the given volume of the solution to the grams required for the answer involves two steps. The first step uses the molarity as a conversion factor to convert to moles, and then the second step uses the molar mass conversion factor to convert to grams. The problem starts with the volume of the solution:

$$? \text{ g KCl} = 0.250 \text{ L KCl}$$

Next the molarity is used to convert to moles:

$$? \text{ g KCl} = 0.250 \text{ L KCl}\left(\frac{\mathbf{0.300 \text{ mol KCl}}}{\mathbf{1 \text{ L KCl}}}\right)$$

Then the conversion factor for the molar mass is used:

$$? \text{ g KCl} = 0.250 \text{ L KCl}\left(\frac{0.300 \text{ mol KCl}}{1 \text{ L KCl}}\right)\left(\frac{\mathbf{74.55 \text{ g KCl}}}{\mathbf{1 \text{ mol KCl}}}\right)$$

After canceling the L KCl and mol KCl units, we have the desired g KCl units, and the calculation can be made:

$$? \text{ g KCl} = 0.250 \text{ L KCl}\left(\frac{0.300 \text{ mol KCl}}{1 \text{ L KCl}}\right)\left(\frac{74.55 \text{ g KCl}}{1 \text{ mol KCl}}\right)$$
$$= 5.59 \text{ g KCl}$$

Exercise 5.4

Perform each of the following conversions:

(a) 26.5 g of $MgCl_2$ to moles of $MgCl_2$
(b) 3.456 mol of CH_4 to grams of CH_4
(c) 1.45 mol of KCl to liters of KCl with a molarity of 0.135
(d) 23.5 mL of 0.766 *M* HF to moles of HF
(e) 1.46 L of CO_2 at STP to moles of CO_2
(f) 0.025 mol of N_2 to liters of N_2 at STP
(g) 26.5 g of $MgCl_2$ to liters of 0.200 *M* $MgCl_2$ solution
(h) 3.456 g of CH_4 to liters of CH_4 at STP
(i) 1.45 L of HCl at STP to liters of HCl with a molarity of 0.135
(j) 23.5 mL of 0.766 *M* HF to liters of HF gas at STP
(k) 0.025 g of N_2 to liters of N_2 at STP

Solution

(a) 0.278 mol $MgCl_2$
(b) 55.40 g CH_4
(c) 10.7 L KCl
(d) 0.0180 mol HF
(e) 0.0652 mol CO_2
(f) 0.56 L N_2
(g) 1.39 L $MgCl_2$
(h) 4.83 L CH_4
(i) 0.479 L HCl(*aq*)
(j) 0.403 L HF(*g*)
(k) 0.020 L N_2

Calculations Involving Two Substances

TIP

Remember that stoichiometric coefficients are exact numbers.

To this point all the sample calculations have started and ended with the same substance. In Figure 5.1, all of the conversions took place between the boxes labeled "SUBSTANCE A." When we start with one substance and end up with a different one, however, *the conversion **must always** include the central conversion from moles of substance A to moles of substance B.* There is simply no other possible way to perform the conversions. These conversions must use information obtained from a given chemical formula or from a balanced chemical reaction. The conversion factors used here are often called mole ratios.

The simplest two-substance conversions are mole-to-mole conversions, as shown in Examples 5.10 and 5.11.

EXAMPLE 5.10

How many moles of nitrogen atoms are there in 6.50 mol of ammonium phosphate, $(NH_4)_3PO_4$?

Solution

The chemical formula gives the information for the conversion factor; this compound has three nitrogen atoms for each unit of ammonium phosphate. The setup states the question as

$$? \text{ mol N} = 6.50 \text{ mol } (NH_4)_3PO_4$$

This is then multiplied by the conversion factor $\left(\frac{3 \text{ mol N}}{1 \text{ mol } (NH_4)_3PO_4}\right)$:

$$? \text{ mol N} = 6.50 \text{ mol } (NH_4)_3PO_4 \left(\frac{3 \text{ mol N}}{1 \text{ mol } (NH_4)_3PO_4}\right)$$

The mol $(NH_4)_3PO_4$ units cancel, leaving the mol N units that the problem requests. The answer is calculated as

$$? \text{ mol N} = 19.5 \text{ mol N}$$

EXAMPLE 5.11

How many moles of water will be formed in the complete combustion of 2.50 mol of methane, CH_4?

Solution

This problem starts with one substance, methane, and asks a question about a second very different substance, water. A chemical reaction will be needed to solve the problem. Every combustion reaction has oxygen as a reactant and carbon dioxide and water as the products:

$$CH_4 + 2O_2 \rightarrow CO_2 + 2H_2O$$

This equation tells us that 1 mol of methane will form 2 mol of water, and this information is used to construct the conversion factor, $\left(\frac{2 \text{ mol } H_2O}{1 \text{ mol } CH_4}\right)$. The question is written as

$$? \text{ mol } H_2O = 2.50 \text{ mol } CH_4$$

Multiplying this by the conversion factor yields

$$? \text{ mol } H_2O = 2.50 \text{ mol } CH_4 \left(\frac{2 \text{ mol } H_2O}{1 \text{ mol } CH_4}\right)$$

After cancelling the mol CH_4 units and verifying that the proper units, mol H_2O, have been obtained to satisfy the question, the answer is calculated:

$$? \text{ mol } H_2O = 2.50 \text{ mol } CH_4 \left(\frac{2 \text{ mol } H_2O}{1 \text{ mol } CH_4}\right) = 5.00 \text{ mol } H_2O$$

More complex calculations involve adding a step before the mole-to-mole conversion and a step afterwards. In all of these calculations the units for the given information in the problem are found in one of the five boxes on the left or SUBSTANCE A side of Figure 5.1. Then the units requested by the problem are found on the SUBSTANCE B side of the diagram. Proceeding from the given units to the requested units defines the sequence of conversions, and the conversion factors, that are needed to obtain the correct answer.

One of the more common calculations involves calculating the mass of a compound given the mass of another compound and the balanced chemical reaction. Figure 5.1 shows that this procedure involves three steps:

1. Convert the starting mass to moles, using the molar mass of the given compound.
2. Convert the moles of compound A to moles of compound B, using the equivalencies derived from the balanced chemical reaction.
3. Convert from moles back to grams, using the molar mass of the requested compound.

EXAMPLE 5.12

Propane, C_3H_8, is a common heating and cooking fuel in rural areas of the country. Propane is also the fuel used in outdoor grills and in small handheld torches. If 100. g of propane is burned in excess oxygen, how many grams of oxygen will be needed? In addition, how many grams of carbon dioxide and water will be formed in the reaction?

Solution

This problem asks for the calculation of three quantities, O_2, CO_2, and H_2O. These will be calculated as three separate problems. First, a balanced chemical equation is needed to determine the relationships between the reactants and products. The reactants and products are listed in Equation (a):

$$C_3H_8 + O_2 \rightarrow CO_2 + H_2O \qquad \text{(a)}$$

This is then balanced to obtain

$$C_3H_8 + 5O_2 \rightarrow 3CO_2 + 4H_2O \qquad \text{(b)}$$

To calculate the oxygen needed, the question is set up as

$$? \text{ g } O_2 = 100. \text{ g } C_3H_8$$

The first conversion uses the molar mass of the C_3H_8 molecule, which is 3(12.011) + 8(1.0080) = 44.097 rounded to 44.10 g/mol.

$$? \text{ g } O_2 = 100. \text{ g } C_3H_8 \left(\frac{\mathbf{1\ mol\ C_3H_8}}{\mathbf{44.10\ g\ C_3H_8}}\right)$$

The next conversion factor is obtained from Equation (b), which says that 1 mol of C_3H_8 is equivalent to 5 mol of O_2:

$$? \text{ g } O_2 = 100. \text{ g } C_3H_8 \left(\frac{1 \text{ mol } C_3H_8}{44.10 \text{ g } C_3H_8}\right)\left(\frac{\mathbf{5\ mol\ O_2}}{\mathbf{1\ mol\ C_3H_8}}\right)$$

The final conversion factor uses the molar mass of the O_2 molecule, 16.00 + 16.00 = 32.00:

$$? \text{ g } O_2 = 100. \text{ g } C_3H_8 \left(\frac{1 \text{ mol } C_3H_8}{44.10 \text{ g } C_3H_8}\right)\left(\frac{5 \text{ mol } O_2}{1 \text{ mol } C_3H_8}\right)\left(\frac{\mathbf{32.00\ g\ O_2}}{\mathbf{1\ mol\ O_2}}\right)$$

The units cancel to leave only units of g O_2. The answer is calculated as

$$? \text{ g } O_2 = 363 \text{ g } O_2$$

The remaining question in the statement of the problem is answered by using the following setups:

$$? \text{ g } CO_2 = 100. \text{ g } C_3H_8 \left(\frac{1 \text{ mol } C_3H_8}{44.10 \text{ g } C_3H_8}\right)\left(\frac{3 \text{ mol } CO_2}{1 \text{ mol } C_3H_8}\right)\left(\frac{44.01 \text{ g } CO_2}{1 \text{ mol } CO_2}\right)$$
$$= 299. \text{ g } CO_2$$

$$? \text{ g } H_2O = 100. \text{ g } C_3H_8 \left(\frac{1 \text{ mol } C_3H_8}{44.10 \text{ g } C_3H_8}\right)\left(\frac{4 \text{ mol } H_2O}{1 \text{ mol } C_3H_8}\right)\left(\frac{18.02 \text{ g } H_2O}{1 \text{ mol } H_2O}\right)$$
$$= 163 \text{ g } H_2O$$

Reviewing this problem, we see that it started with 100. grams of C_3H_8, which was found to require 363 grams of O_2 for complete combustion. The masses of the products were calculated as 299. grams of CO_2 and 163 grams of H_2O.

TIP

All of our conversions must obey the law of conservation of matter.

The total mass of the reactants, 100. g C_3H_8 and 364 g O_2, is 464 grams. At the end of the reaction, the total mass of the products, 299. g CO_2 and 163 g H_2O, is also 462 grams. The law of conservation of mass states that matter cannot be created or destroyed in a chemical reaction. Since this law cannot be violated, we expect that the results of our calculations will obey it. Since the total mass for the reactants and the products was within the rounding error, it indicates that the law of conservation of mass was not violated.

It is always an advantage to be able to estimate an answer to a problem to assure that no errors were made in calculations. In stoichiometry problems, only mass to mass conversions allow us to do this easily. In 95 percent of these problems, the calculated mass is between one-fifth and five times the given mass. For the problem above, this means that the answers should be between 20. and 500. grams. All of our results fell in that range, giving added confidence that the conversions were correctly done. If the results did not fit the estimates, we would be well advised to carefully recheck our calculations. The 5 percent of reactions that do not follow this general rule are those in which the molar masses of the compounds are very different, as, for example, those of H_2 and Zn.

In some instances the information given is in the form of the volume and molarity of a reactant that produces a precipitate. These calculations also involve a three-step conversion:

1. Start with the given volume and convert to moles, using the molarity of the given solution.
2. Use the balanced chemical reaction to calculate the moles of product.
3. Convert the moles of product to grams, using the molar mass.

EXAMPLE 5.13

A 45.0 mL sample of 0.300 molar $FeCl_3$ is reacted with enough NaOH solution to precipitate all the iron as $Fe(OH)_3$. How many grams of $Fe(OH)_3$ will be precipitated?

Solution

First a balanced chemical reaction must be written:

$$FeCl_3 + 3NaOH \rightarrow Fe(OH)_3 + 3NaCl \qquad \text{(c)}$$

The question is set up as

$$?\,g\ Fe(OH)_3 = 45.0\ mL\ FeCl_3$$

The first conversion factor is the molarity, written as

$$0.300\ M\ FeCl_3 = \frac{0.300\ mol\ FeCl_3}{1000\ mL\ FeCl_3}$$

The denominator of this conversion factor includes the conversion from liters to milliliters without using another conversion factor:

$$?\,g\ Fe(OH)_3 = 45.0\ mL\ FeCl_3 \left(\frac{\mathbf{0.300\ mol\ FeCl_3}}{\mathbf{1000\ mL\ FeCl_3}}\right)$$

The next conversion factor is obtained from the relationships in the chemical reaction shown in Equation (c), where 1 mol of $FeCl_3$ is equivalent to 1 mol of $Fe(OH)_3$:

$$?\,g\ Fe(OH)_3 = 45.0\ mL\ FeCl_3 \left(\frac{0.300\ mol\ FeCl_3}{1000\ mL\ FeCl_3}\right)\left(\frac{\mathbf{1\ mol\ Fe(OH)_3}}{\mathbf{1\ mol\ FeCl_3}}\right)$$

Finally, the moles of $Fe(OH)_3$ are converted to grams by using the molar mass, 106.9, for $Fe(OH)_3$:

$$?\,g\ Fe(OH)_3 = 45.0\ mL\ FeCl_3 \left(\frac{0.300\ mol\ FeCl_3}{1000\ mL\ FeCl_3}\right)\left(\frac{1\ mol\ Fe(OH)_3}{1\ mol\ FeCl_3}\right)\left(\frac{\mathbf{106.9\ g\ Fe(OH)_3}}{\mathbf{1\ mol\ Fe(OH)_3}}\right)$$

Since the units cancel, the answer is calculated as

$$? \text{ g Fe(OH)}_3 = 1.44 \text{ g Fe(OH)}_3$$

In many reactions it is important to know the volume of one reactant that will react with a given volume of a second reactant. The molarities of both reactants must be given for this type of problem to be solved.

EXAMPLE 5.14

How many milliliters of a 0.250 *M* NaOH solution are needed to completely neutralize 65.0 mL of a 0.400 *M* solution of sulfuric acid?

Solution

A balanced equation is required. Since the problem states that the sulfuric acid, H_2SO_4, is completely neutralized, both protons on the sulfuric acid react with the NaOH:

$$H_2SO_4 + 2NaOH \rightarrow Na_2SO_4 + 2H_2O \qquad \text{(d)}$$

Figure 5.1 indicates another three-step calculation:

1. Convert milliliters of H_2SO_4 to moles of H_2SO_4, using the molarity of H_2SO_4.
2. Convert moles of H_2SO_4 to moles of NaOH using the balanced chemical equation.
3. Convert moles of NaOH to milliliters, using the molarity of NaOH.

The initial setup of the question is

$$? \text{ mL NaOH} = 65.0 \text{ mL } H_2SO_4$$

Using the molarity of the H_2SO_4 as a conversion factor gives

$$? \text{ mL NaOH} = 65.0 \text{ mL } H_2SO_4 \left(\frac{\mathbf{0.400 \text{ mol } H_2SO_4}}{\mathbf{1000 \text{ mL } H_2SO_4}} \right)$$

Then Equation (d) is used to convert to moles NaOH:

$$? \text{ mL NaOH} = 65.0 \text{ mL } H_2SO_4 \left(\frac{0.400 \text{ mol } H_2SO_4}{1000 \text{ mL } H_2SO_4} \right) \left(\frac{\mathbf{2 \text{ mol NaOH}}}{\mathbf{1 \text{ mol } H_2SO_4}} \right)$$

Finally, the molarity of the NaOH is used to convert moles NaOH to milliliters NaOH:

$$? \text{ mL NaOH} = 65.0 \text{ mL } H_2SO_4 \left(\frac{0.400 \text{ mol } H_2SO_4}{1000 \text{ mL } H_2SO_4} \right) \left(\frac{2 \text{ mol NaOH}}{1 \text{ mol } H_2SO_4} \right) \left(\frac{\mathbf{1000 \text{ mL NaOH}}}{\mathbf{0.250 \text{ mol NaOH}}} \right)$$

Since the units cancel properly, the answer may be calculated as 208 mL.

In some chemical reactions a gas is evolved as one of the products. The most common cases are the reactions of active metals with mineral acids and the reactions of carbonates with acids. It is possible to calculate the volume of gas from a chemical reaction at standard temperature and pressure (STP = 0°C and 1 atm). (If the final conditions are not at STP, see

Chapter 6 on gases for further calculations using the ideal gas law.) The volume of gas evolved from a given mass or volume of reactant is calculated in Examples 5.15–5.17.

EXAMPLE 5.15

Metallic copper reacts with concentrated nitric acid to produce nitrogen dioxide. Calculate the volume of NO_2 that will form at STP when 1.25 g of copper is completely reacted according to the equation

$$Cu(s) + 4HNO_3(aq) \rightarrow Cu(NO_3)_2(aq) + 2NO_2(g) + 2H_2O(l) \qquad \text{(e)}$$

Solution

We have the chemical reaction, and need to follow three steps to convert the grams of copper to the volume of $NO_2(g)$ formed:

1. Convert grams Cu to moles Cu, using the atomic mass.
2. Convert moles Cu to moles NO_2, using the balanced equation.
3. Convert moles NO_2 to volume, using the molar volume of an ideal gas.

We start with the setup of the question:

$$?\,\text{L NO}_2 = 1.25\,\text{g Cu}$$

Using the atomic mass of copper, without rounding, as a conversion factor, we get

$$?\,\text{L NO}_2 = 1.25\,\text{g Cu}\left(\frac{1\text{ mol Cu}}{63.55\text{ g Cu}}\right)$$

The next conversion factor involves the balanced chemical equation given in Equation (e), which states that 1 mol Cu will form 2 mol NO_2. The equation becomes

$$?\,\text{L NO}_2 = 1.25\,\text{g Cu}\left(\frac{1\text{ mol Cu}}{63.55\text{ g Cu}}\right)\left(\frac{2\text{ mol NO}_2}{1\text{ mol Cu}}\right)$$

Finally, the fact that 1 mol of an ideal gas at STP occupies 22.4 L is used as a conversion factor to obtain

$$?\,\text{L NO}_2 = 1.25\,\text{g Cu}\left(\frac{1\text{ mol Cu}}{63.55\text{ g Cu}}\right)\left(\frac{2\text{ mol NO}_2}{1\text{ mol Cu}}\right)\left(\frac{22.4\text{ L NO}_2}{1\text{ mol NO}_2}\right)$$

Since all of the units cancel properly, the answer may be calculated as

$$?\,\text{L NO}_2 = 0.881\text{ L NO}_2\text{ at STP}$$

EXAMPLE 5.16

Blackboard chalk is almost 100% calcium carbonate, $CaCO_3$. What volume of carbon dioxide, CO_2, will be evolved at STP if an excess of chalk is reacted with 35.0 mL of 0.888 molar hydrochloric acid?

Solution

The chemical reaction may be obtained from the facts given in the problem. Knowing that the reactants are HCl and $CaCO_3$ and that one of the products is CO_2, we can readily deduce the other products, H_2O and $CaCl_2$:

$$2HCl(aq) + CaCO_3(s) \rightarrow CO_2(g) + H_2O(l) + CaCl_2(aq) \qquad \text{(f)}$$

The starting point for the calculation is

$$?\ \text{L } CO_2 = 35.0 \text{ mL HCl}$$

The molarity of the HCl is used to convert to moles HCl:

$$?\ \text{L } CO_2 = 35.0 \text{ mL HCl}\left(\frac{\mathbf{0.888\ mol\ HCl}}{\mathbf{1000\ mL\ HCl}}\right)$$

Next the chemical reaction shown in Equation (f) is used to obtain a conversion factor for the conversion from moles HCl to moles CO_2:

$$?\ \text{L } CO_2 = 35.0 \text{ mL HCl}\left(\frac{0.888 \text{ mol HCl}}{1000 \text{ mL HCl}}\right)\left(\frac{\mathbf{1\ mol\ CO_2}}{\mathbf{2\ mol\ HCl}}\right)$$

Finally, the molar volume of a gas is used to convert to the volume of CO_2 formed:

$$?\ \text{L } CO_2 = 35.0 \text{ mL HCl}\left(\frac{0.888 \text{ mol HCl}}{1000 \text{ mL HCl}}\right)\left(\frac{1 \text{ mol } CO_2}{2 \text{ mol HCl}}\right)\left(\frac{\mathbf{22.4\ L\ CO_2}}{\mathbf{1\ mol\ CO_2}}\right)$$

The units cancel properly, and the answer is calculated as

$$?\ \text{L } CO_2 = 0.348 \text{ L } CO_2 \text{ at STP}$$

EXAMPLE 5.17

(a) How many grams of water are obtained when 35.6 g of benzene, C_6H_6, are burned in excess oxygen? How many liters of CO_2 at STP will be produced in the same reaction?

(b) How many milliliters of 0.248 *M* HCl are needed to react with 1.36 g of zinc to produce hydrogen gas? How many milliliters of hydrogen gas are expected at STP?

Solution

(a) The balanced reaction is

$$2C_6H_6 + 15O_2 \rightarrow 12CO_2 + 6H_2O$$

$$?\ \text{g } H_2O = 35.6 \text{ g } C_6H_6\left(\frac{1 \text{ mol } C_6H_6}{78.1 \text{ g } C_6H_6}\right)\left(\frac{6 \text{ mol } H_2O}{2 \text{ mol } C_6H_6}\right)\left(\frac{18.0 \text{ g } H_2O}{1 \text{ mol } H_2O}\right)$$
$$= 24.6 \text{ g } H_2O$$

$$?\ \text{L } CO_2 = 35.6 \text{ g } C_6H_6\left(\frac{1 \text{ mol } C_6H_6}{78.1 \text{ g } C_6H_6}\right)\left(\frac{12 \text{ mol } CO_2}{2 \text{ mol } C_6H_6}\right)\left(\frac{22.4 \text{ L } CO_2}{1 \text{ mol } CO_2}\right)$$
$$= 61.3 \text{ L } CO_2$$

(b) The balanced reaction is

$$2HCl + Zn \rightarrow ZnCl_2 + H_2$$

$$? \text{ mL HCl} = 1.36 \text{ g Zn}\left(\frac{1 \text{ mol Zn}}{65.41 \text{ g Zn}}\right)\left(\frac{2 \text{ mol HCl}}{1 \text{ mol Zn}}\right)\left(\frac{1000 \text{ mL HCl}}{0.248 \text{ mol HCl}}\right)$$

$$= 168 \text{ mL HCl}$$

$$? \text{ mL } H_2 = 1.36 \text{ g Zn}\left(\frac{1 \text{ mol Zn}}{65.41 \text{ g Zn}}\right)\left(\frac{1 \text{ mol } H_2}{1 \text{ mol Zn}}\right)\left(\frac{22400 \text{ mL } H_2}{1 \text{ mol } H_2}\right)$$

$$= 466 \text{ mL } H_2$$

Limiting-Reactant Calculations

TIP

Prepare a different analogy that will help you recall the concept of a limiting reactant.

When chemicals are mixed together under the appropriate conditions, a chemical reaction is started. The reaction will stop when one of the reactants is completely used up. The reactant that is totally consumed, stopping the reaction, is called the **limiting reactant** or **limiting reagent**. The other reactant(s) are called the excess reactant(s). The amount of the limiting reactant determines how much of the other reactant(s) react and how much of each product is formed. Up to this point, only the amount of one reactant has been given in a problem, and this reactant was assumed to be the limiting reactant. When the amounts of two or more reactants are given, special procedures for limiting-reactant calculations must be used.

To understand the concept of a limiting reactant more fully, consider a vending machine that accepts quarters and dimes only and does not give any change. If an item in that machine costs 45 cents, the only way it can be purchased is with two dimes and one quarter. If you have ten dimes and ten quarters, only five 45-cent items can be obtained from this machine since the dimes will run out before the quarters do. Figure 5.2 illustrates this example.

On the new AP exam, you should be prepared to interpret and draw diagrams similar to Figure 5.2 to illustrate your ability to describe the molecular or particulate level of matter.

TIP

The limiting reactant is always completely used up.

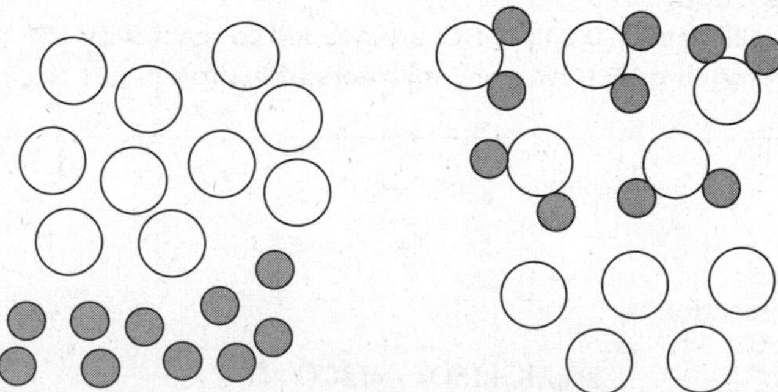

FIGURE 5.2. The vending machine example of a limiting reactant. On the left are the ten quarters (large circles) and ten dimes (small circles). On the right are five groups of one quarter and two dimes each used for 45-cent purchases. Also on the right are five leftover quarters. The dimes are the limiting reactant.

A variety of problems can be solved in the context of the limiting-reactant concept. These include the determination of which reactant is the limiting reactant, the amount of product formed, and the amount of excess reactant that does not react. Examples of these calculations are shown in Examples 5.18–5.23.

PROCEDURES FOR SOLVING LIMITING-REACTANT PROBLEMS MUST ALWAYS BE USED WHEN THE AMOUNTS OF TWO OR MORE REACTANTS ARE GIVEN IN THE STATEMENT OF THE PROBLEM.

FINDING THE LIMITING REACTANT:

EXAMPLE 5.18

For the reaction below, determine the limiting reactant if 100. g of $FeCl_3$ is reacted with 50.0 g of H_2S:

$$2FeCl_3(aq) + 3H_2S(g) \rightarrow Fe_2S_3(s) + 6HCl(aq)$$

Solution

The first thing to remember is that the compound present in the smallest amount is not necessarily the limiting reactant. To determine the limiting reactant, the amount given for one reactant is converted into the amount of the other reactant that is needed to react with it. Using the procedures above, we can convert the 50 g of H_2S into the number of grams of $FeCl_3$ needed to react with it as follows:

$$? \text{ g } FeCl_3 = 50.0 \text{ g } H_2S\left(\frac{1 \text{ mol } H_2S}{34.08 \text{ g } H_2S}\right)\left(\frac{2 \text{ mol } FeCl_3}{3 \text{ mol } H_2S}\right)\left(\frac{162.2 \text{ g } FeCl_3}{1 \text{ mol } FeCl_3}\right)$$
$$= 159 \text{ g } FeCl_3$$

From this result we see that 159 g of $FeCl_3$ is needed to react with all 50.0 g of H_2S. However, the amount of $FeCl_3$ given in the problem is only 100. g. The conclusion must be that we will run out of $FeCl_3$ before all of the H_2S can be reacted. Since the $FeCl_3$ is used up first, it is the limiting reactant.

(If the problem had stated that 200. g of $FeCl_3$, instead of 100. g, was available, it would be apparent that we had more than enough $FeCl_3$ to react all of the H_2S. Then we would have concluded that H_2S was the limiting reactant.)

ONCE A LIMITING REACTANT IS IDENTIFIED, ALL FURTHER CALCULATIONS ARE BASED ON THE AMOUNT OF THE LIMITING REACTANT GIVEN IN THE ORIGINAL STATEMENT OF THE PROBLEM.

DETERMINING THE THEORETICAL YIELD

The term **theoretical yield** refers to the maximum amount of product formed in a reaction based on the amounts of reactants used. This amount is a theoretical yield since no laboratory work is done. Many events can occur in the laboratory that result in less than the theoretical amount of product. One is that the reactants do not combine completely. Another is the possibility of side reactions that produce different products. Products can also be lost by poor lab techniques or in the purification process. In most cases, the theoretical yield refers to the maximum mass of product that can be produced.

EXAMPLE 5.19

What is the theoretical yield of Fe_2S_3 that can be obtained from 100. g of $FeCl_3$ and 50.0 g of H_2S?

Solution

These are the same data as in Example 5.18, but the question has changed. To solve the problem, we need to know the limiting reactant. In this case it has already been determined as the $FeCl_3$. We now use the 100. g of $FeCl_3$ given in the problem to calculate the mass of the Fe_2S_3.

$$\begin{aligned} ?\ \text{g Fe}_2\text{S}_3 &= 100.\ \text{g FeCl}_3\left(\frac{1\ \text{mol FeCl}_3}{162.2\ \text{g FeCl}_3}\right)\left(\frac{1\ \text{mol Fe}_2\text{S}_3}{2\ \text{mol FeCl}_3}\right)\left(\frac{207.9\ \text{g Fe}_2\text{S}_3}{1\ \text{mol Fe}_2\text{S}_3}\right) \\ &= 64.1\ \text{g Fe}_2\text{S}_3 \end{aligned}$$

(It is essential that the 100. g of $FeCl_3$ given in the problem be used for this calculation. If we had used the 159 grams calculated in Example 5.18, a totally incorrect answer would have been obtained.)

Another question that can be asked in a limiting-reactant problem is how much of the excess reactant is left over when the reaction stops. In the vending machine example in Figure 5.2 this can be done by counting the five quarters that combine with two dimes and then subtracting them from the ten quarters we started with. In the chemical calculation the amount of the excess reactant that reacts with the limiting reactant is calculated. This is then subtracted from the starting amount of the excess reactant as given in the problem. Example 5.20 illustrates the procedure.

EXAMPLE 5.20

When 100. g of $FeCl_3$ is reacted with 50.0 g of H_2S, how many grams of which reactant will be left over when the reaction is complete?

Solution

To determine which reactant is left over, the limiting reactant is identified and then all other reactants become the excess reactants. Since the data are the same for this problem as for Examples 5.18 and 5.19, we know that $FeCl_3$ is the limiting reactant and, therefore, H_2S is the excess reactant. To determine how much H_2S is left over, we first calculate the number of the grams of H_2S that react:

$$? \text{ g } H_2S = 100. \text{ g } Fe_2Cl_3 \left(\frac{1 \text{ mol } FeCl_3}{162.2 \text{ g } FeCl_3}\right)\left(\frac{3 \text{ mol } H_2S}{2 \text{ mol } FeCl_3}\right)\left(\frac{34.08 \text{ g } H_2S}{1 \text{ mol } H_2S}\right)$$
$$= 31.5 \text{ g } H_2S$$

Since the problem started with 50.0 g of H_2S and 31.4 g reacted, the remaining amount of H_2S must be

$$? \text{ g } H_2S \text{ (start)} = 50.0 \text{ g } H_2S \text{ (start)} - 31.5 \text{ g } H_2S \text{ (reacted)}$$
$$= 18.5 \text{ g } H_2S$$

To complete all of the information about this reaction, we can calculate the theoretical yield of HCl in the same way that the theoretical yield of Fe_2S_3 was calculated.

EXAMPLE 5.21

When 100. g of $FeCl_3$ is reacted with 50.0 g of H_2S, how many grams of HCl are formed?

Solution

As before, the limiting reactant must be determined. However, we already know that it is $FeCl_3$. The 100. grams of $FeCl_3$ given in the problem is used as the starting point for the calculation:

$$? \text{ g HCl} = 100. \text{ g } FeCl_3$$
$$= 100. \text{ g } FeCl_3 \left(\frac{1 \text{ mol } FeCl_3}{162.2 \text{ g } FeCl_3}\right)\left(\frac{6 \text{ mol HCl}}{2 \text{ mol } FeCl_3}\right)\left(\frac{34.46 \text{ g HCl}}{1 \text{ mol HCl}}\right)$$
$$= 67.4 \text{ g HCl}$$

This calculation follows the same principles as the calculation of the amount of Fe_2S_3 in Example 5.19.

With this calculation we have determined the masses of all reactants and products in the chemical equation. Listing the masses of the substances before and after reaction demonstrates once again that the law of conservation of matter is obeyed.

TIP

All stoichiometry calculations obey the law of conservation of mass.

REACTION:	$2FeCl_3$	+	$3H_2S$	→	Fe_2S_3	+	6HCl
Start	100 g		50.0 g		0 g		0 g
End	0 g		18.5 g		64.1 g		67.4 g

In this table we can add up the masses of all of the substances at the start of the reaction to get a total of 150 grams. At the end of the reaction the masses again add up to 150 grams.

These results show that the law of conservation of mass has been obeyed since total mass at the start and at the end of the reaction is the same.

EXAMPLE 5.22

Silver tarnishes in air because of a complex reaction with oxygen and hydrogen sulfide, H_2S, in the air, which may be written as

$$2Ag + O_2 + H_2S \rightarrow Ag_2S + 2H_2O$$

What is the theoretical yield of silver sulfide, Ag_2S, that can be produced from a mixture of 0.200 g of silver, 1.50 L of oxygen at STP, and 65.0 mL of 0.350 molar H_2S solution?

Solution

This is a limiting-reactant problem since the amounts of three different reactants are given. An added complexity is that each of these amounts has a different type of unit. To begin with, the identity of the limiting reactant must be determined. Because there are three reactants, Ag, O_2, and H_2S, a process of elimination is used. First, one pair of reactants is selected to determine which *might be* the limiting reactant, while eliminating the excess reactant. Next, the third reactant and the possible limiting reactant from the first step are used to determine the actual limiting reactant. Choosing silver and oxygen as the first pair, we have

$$\begin{aligned} ?\ \text{L O}_2 &= 0.200\ \text{g Ag}\left(\frac{1\ \text{mol Ag}}{107.9\ \text{g Ag}}\right)\left(\frac{1\ \text{mol O}_2}{2\ \text{mol Ag}}\right)\left(\frac{22.4\ \text{L O}_2}{1\ \text{mol O}_2}\right) \\ &= 0.0208\ \text{L O}_2 \end{aligned}$$

Since the problem gives the amount of oxygen as 1.5 L, there is plenty of this reactant and it cannot be the limiting reactant, but silver may be. The next step determines the amount of hydrogen sulfide needed to react with the silver given in the problem.

$$\begin{aligned} ?\ \text{mL H}_2\text{S} &= 0.200\ \text{g Ag}\left(\frac{1\ \text{mol Ag}}{107.9\ \text{g Ag}}\right)\left(\frac{1\ \text{mol H}_2\text{S}}{2\ \text{mol Ag}}\right)\left(\frac{1000\ \text{mL H}_2\text{S}}{0.350\ \text{mol H}_2\text{S}}\right) \\ &= 2.65\ \text{mL H}_2\text{S} \end{aligned}$$

Since the question states that there is 65.0 mL of H_2S solution and all that is needed is 2.65 mL, H_2S cannot be the limiting reactant. Because neither H_2S nor O_2 can be the limiting reactant, it must be the silver. Now the theoretical yield of Ag_2S can be calculated, based on the 0.200 g of silver given in the original problem:

$$\begin{aligned} ?\ \text{g Ag}_2\text{S} &= 0.200\ \text{g Ag}\left(\frac{1\ \text{mol Ag}}{107.9\ \text{g Ag}}\right)\left(\frac{1\ \text{mol Ag}_2\text{S}}{2\ \text{mol Ag}}\right)\left(\frac{247.8\ \text{g Ag}_2\text{S}}{1\ \text{mol Ag}_2\text{S}}\right) \\ &= 0.230\ \text{g Ag}_2\text{S} \end{aligned}$$

EXAMPLE 5.23

Silver nitrate, $AgNO_3$ (molar mass = 169.9 g/mol), reacts with sodium chromate, Na_2CrO_4 (molar mass = 161.9 g/mol), to form silver chromate (molar mass = 331.7 g/mol) and sodium nitrate. If 45.5 mL of 0.200 *M* $AgNO_3$ is mixed with 35.8 mL of 0.436 *M* Na_2CrO_4, what is the theoretical yield of the precipitate silver chromate? How many grams of which reactant are left over?

Solution

The reaction is

$$2AgNO_3(aq) + Na_2CrO_4(aq) \rightarrow Ag_2CrO_4(s) + 2NaNO_3(aq)$$

To determine the limiting reactant, the calculation is

$$? \text{ mL } Na_2CrO_4 = 45.5 \text{ mL } AgNO_3 \left(\frac{0.200 \text{ mol } AgNO_3}{1000 \text{ mL } AgNO_3}\right)\left(\frac{1 \text{ mol } Na_2CrO_4}{2 \text{ mol } AgNO_3}\right)\left(\frac{1000 \text{ mL } Na_2CrO_4}{0.436 \text{ mol } Na_2CrO_4}\right)$$
$$= 10.4 \text{ mL } Na_2CrO_4$$

TIP

The limiting reactant always produces fewer moles or grams of product than the excess reactant.

Since we were given 35.8 mL of Na_2CrO_4, the conclusion is that $AgNO_3$ is the limiting reactant, and all further calculations are based on the given amount of $AgNO_3$:

$$? \text{ g } Ag_2CrO_4 = 45.5 \text{ mL } AgNO_3 \left(\frac{0.200 \text{ mol } AgNO_3}{1000 \text{ mL } AgNO_3}\right)\left(\frac{1 \text{ mol } Ag_2CrO_4}{2 \text{ mol } AgNO_3}\right)\left(\frac{331.7 \text{ g } Ag_2CrO_4}{1 \text{ mol } Ag_2CrO_4}\right)$$
$$= 1.51 \text{ g } Ag_2CrO_4$$

Since $AgNO_3$ is the limiting reactant, Na_2CrO_4 is the excess reactant. The amount of Na_2CrO_4 that reacts is calculated and subtracted from the given amount:

$$? \text{ g } Na_2CrO_4 = 45.5 \text{ mL } AgNO_3 \left(\frac{0.200 \text{ mol } AgNO_3}{1000 \text{ mL } AgNO_3}\right)\left(\frac{1 \text{ mol } Na_2CrO_4}{2 \text{ mol } AgNO_3}\right)\left(\frac{161.9 \text{ g } Na_2CrO_4}{1 \text{ mol } Na_2CrO_4}\right)$$
$$= 0.737 \text{ g } Na_2CrO_4 \text{ reacts}$$

The original number of grams of Na_2CrO_4 is calculated as

$$? \text{ g } Na_2CrO_4 = 35.8 \text{ mL } Na_2CrO_4 \left(\frac{0.436 \text{ mol } Na_2CrO_4}{1000 \text{ mL } Na_2CrO_4}\right)\left(\frac{161.9 \text{ g } Na_2CrO_4}{1 \text{ mol } Na_2CrO_4}\right)$$
$$= 2.527 \text{ g } Na_2CrO_4 \text{ initially present}$$

The amount of sodium chromate left over is calculated by subtraction:

$$? \text{ g } Na_2CrO_4 \text{ left over} = 2.527 \text{ g } Na_2CrO_4 - 0.737 \text{ g } Na_2CrO_4$$
$$= 1.790 \text{ g } Na_2CrO_4$$

Titrations

The titration technique used for chemical analysis utilizes the reactions of two solutions. One reactant solution is placed in a beaker, and the other in a buret, which is a long, graduated tube with a stopcock. The stopcock is a valve that allows the chemist to add controlled amounts of solution from the buret to the beaker. An indicator, that is, a compound that changes color when the reaction is complete, is added to the solution in the beaker. The chemist reads the volume of solution in the buret at the start of the experiment and again at the point where the indicator changes color. The difference in these volumes represents the volume of reactant delivered from the buret. Figure 5.3 illustrates the experimental setup.

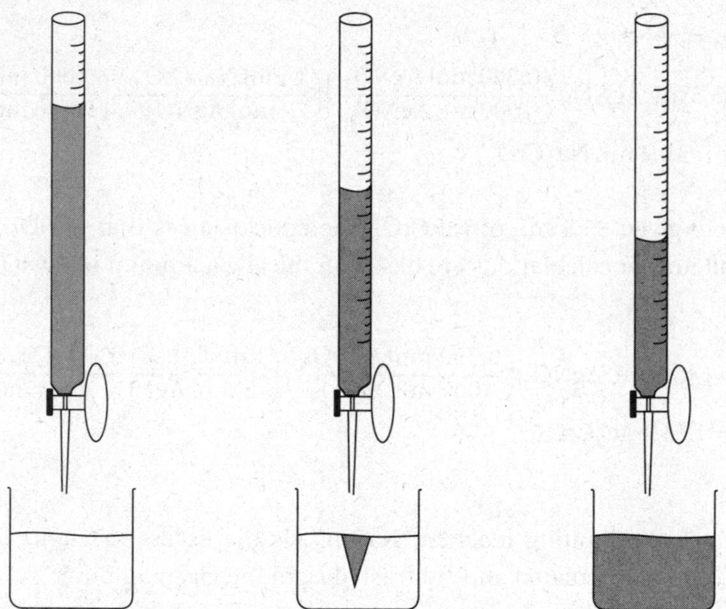

FIGURE 5.3. Titration experiment showing the initial setup, the slight color change of the indicator before the endpoint is reached, and the colored solution at the endpoint.

The crucial point about the titration experiment is that the indicator is designed to change color when the amount of reactant delivered from the buret is exactly the amount needed to react with the solution in the beaker. From this experiment, a variety of calculations may be made as shown below.

A classic chemical reaction is the one between Fe^{2+} and the permanganate ion MnO_4^-:

$$5Fe^{2+} + MnO_4^- + 8H^+ \rightarrow Mn^{2+} + 5Fe^{3+} + 4H_2O \quad (5.5)$$

In titrations, the purple permanganate ion is the indicator of the point where the correct amount has been added to completely react all of the Fe^{2+} ions in the sample. All of the calculations in Examples 5.24–5.26 refer to Equation 5.5.

EXAMPLE 5.24

It takes 34.35 mL of a 0.240 M solution of $KMnO_4$ (0.240 M MnO_4^-) to titrate an unknown sample of Fe^{2+} to its endpoint. How many grams of Fe^{2+} are in the sample?

Solution

This problem gives the amount of MnO_4^-, and the grams of Fe^{2+} are to be calculated. Figure 5.1 shows the sequence of steps, and the required conversion factors can be determined. The question is set up as

$$? \text{ g } Fe^{2+} = 34.35 \text{ mL } MnO_4^-$$

Then the necessary conversion factors are entered:

$$? \text{ g } Fe^{2+} = 34.35 \text{ mL } MnO_4^- \left(\frac{0.240 \text{ mol } MnO_4^-}{1000 \text{ mL } MnO_4^-}\right)\left(\frac{5 \text{ mol } Fe^{2+}}{1 \text{ mol } MnO_4^-}\right)\left(\frac{55.85 \text{ g } Fe^{2+}}{1 \text{ mol } Fe^{2+}}\right)$$
$$= 2.30 \text{ g } Fe^{2+}$$

EXAMPLE 5.25

How many milliliters of 0.240 M MnO_4^- solution will be needed to titrate a 1.56 g sample of pure $Fe(NO_3)_2$?

Solution

The question is set up as

$$? \text{ mL } MnO_4^- = 1.56 \text{ g } Fe(NO_3)_2$$

The grams are converted to moles by using the molar mass of $Fe(NO_3)_2$:

$$? \text{ mL } MnO_4^- = 1.56 \text{ Fe g}(NO_3)_2 \left(\frac{\mathbf{1 \text{ mol } Fe(NO_3)_2}}{\mathbf{179.9 \text{ g } Fe(NO_3)_2}}\right)$$

Using the ionization reaction

$$Fe(NO_3)_2 \rightarrow Fe^{2+} + 2NO_3^-$$

we can apply the conversion factor $\left(\frac{1 \text{ mol } Fe^{2+}}{1 \text{ mol } Fe(NO_3)_2}\right)$ to convert to the Fe^{2+} ion:

$$? \text{ mL } MnO_4^- = 1.56 \text{ g } Fe(NO_3)_2 \left(\frac{1 \text{ mol } Fe(NO_3)_2}{179.9 \text{ g } Fe(NO_3)_2}\right)\left(\frac{\mathbf{1 \text{ mol } Fe^{2+}}}{\mathbf{1 \text{ mol } Fe(NO_3)_2}}\right)$$

Equation 5.5 is used to convert to moles of MnO_4^-, and then the molarity is used to convert to milliliters of MnO_4^-.

$$? \text{ mL } MnO_4^- =$$

$$1.56 \text{ g } Fe(NO_3)_2 \left(\frac{1 \text{ mol } Fe(NO_3)_2}{179.9 \text{ g } Fe(NO_3)_2}\right)\left(\frac{1 \text{ mol } Fe^{2+}}{1 \text{ mol } Fe(NO_3)_2}\right)\left(\frac{\mathbf{1\ mol\ MnO_4^-}}{\mathbf{5\ mol\ Fe^{2+}}}\right)\left(\frac{\mathbf{1000\ mL\ MnO_4^-}}{\mathbf{0.240\ mol\ MnO_4^-}}\right)$$

$$= 7.23 \text{ mL } MnO_4^-$$

This exercise demonstrates that the stoichiometry calculations described above can be used also for calculations in titration experiments. There is one calculation, however, that the stoichiometry calculations do not address—conversion of the molarity of one solution to the molarity of another. Since the answer desired is molarity, and molarity is not one of the possible starting points in Figure 5.1, a new method for using conversion factors must be developed.

EXAMPLE 5.26

What is the molarity of an Fe^{2+} solution if 4.53 mL of a 0.687 *M* MnO_4^- solution is required to titrate 30.00 mL of the Fe^{2+}-containing solution to the endpoint?

Solution

The desired information is the molarity of the Fe^{2+} solution. The units of molarity are a ratio, and we look for a ratio of units with which to start the problem. The appropriate term is the molarity of the MnO_4^- solution. The setup of the question is as follows:

$$? \frac{\text{mol } Fe^{2+}}{\text{L } Fe^{2+}} = \frac{0.687 \text{ mol } MnO_4^-}{1 \text{ L } MnO_4^-}$$

We need conversion factors to convert the numerator from mol MnO_4^- to mol Fe^{2+} and the denominator from L MnO_4^- to L Fe^{2+}. The conversion factor for the numerator comes from Equation 5.5. We obtain the conversion factor for the denominator from the ratio of the volumes of the two solutions at the endpoint of the titration:

$$? \frac{\text{mol } Fe^{2+}}{\text{L } Fe^{2+}} = \left(\frac{0.687 \text{ mol } MnO_4^-}{1000 \text{ mL } MnO_4^-}\right)\left(\frac{5 \text{ mol } Fe^{2+}}{1 \text{ mol } MnO_4^-}\right)\left(\frac{4.53 \text{ mL } MnO_4^-}{30.00 \text{ mL } Fe^{2+}}\right)$$

$$= \frac{0.519 \text{ mol } Fe^{2+}}{1000 \text{ mL } Fe^{2+}} = 0.519\ M\ Fe^{2+}$$

EXAMPLE 5.27

A solution of HCl is titrated to the endpoint with 23.4 mL of 0.216 *M* NaOH. (a) How many grams of HCl were in the titrated sample? (b) If the volume of the HCl sample is 50.0 mL, what is the molarity of the HCl solution?

Solution

The reaction is

$$HCl + NaOH \rightarrow NaCl + H_2O$$

(a) To calculate the grams of HCl, we use the equation

$$? \text{ g HCl} = 23.4 \text{ mL NaOH}\left(\frac{0.216 \text{ mol NaOH}}{1000 \text{ mL NaOH}}\right)\left(\frac{1 \text{ mol HCl}}{1 \text{ mol NaOH}}\right)\left(\frac{35.46 \text{ g HCl}}{1 \text{ mol HCl}}\right)$$
$$= 0.184 \text{ g HCl}$$

(b) Calculating the molarity of the HCl solution involves units that are a ratio. The starting point is the molarity of the NaOH, which is also a ratio of units:

$$? \frac{\text{mol HCl}}{\text{L HCl}} = \left(\frac{0.216 \text{ mol NaOH}}{1 \text{ L NaOH}}\right)\left(\frac{1 \text{ mol HCl}}{1 \text{ mol NaOH}}\right)\left(\frac{23.4 \text{ mL NaOH}}{50.0 \text{ mL HCl}}\right)$$
$$= \frac{0.101 \text{ mol HCl}}{\text{L HCl}} = 0.101\ M \text{ HCl}$$

PERCENT COMPOSITION

The percent composition of a chemical substance tells the chemist how much of each element, or polyatomic ion, is present in a compound on a percent basis. In other words, the percent composition is the mass of an element or polyatomic ion that is present in 100 grams of a chemical compound.

EXAMPLE 5.28

What is the percentage of Fe, O, H, and the OH^- polyatomic ion in $Fe(OH)_3$?

Solution

Each of these calculations will start with 100 g of $Fe(OH)_3$, and we calculate the grams of each element in turn to obtain their percentages:

$$? \text{ g Fe} = 100 \text{ g Fe(OH)}_3 \left(\frac{1 \text{ mol Fe(OH)}_3}{106.9 \text{ g Fe(OH)}_3}\right)\left(\frac{1 \text{ mol Fe}}{1 \text{ mol Fe(OH)}_3}\right)\left(\frac{55.46 \text{ g Fe}}{1 \text{ mol Fe}}\right)$$

$$= 52.2 \text{ g Fe}$$

$$\% \text{ Fe} = \frac{52.2 \text{ g Fe}}{100 \text{ g Fe(OH)}_3} \times 100 = 52.2\% \text{ Fe}$$

$$? \text{ g O} = 100 \text{ g Fe(OH)}_3 \left(\frac{1 \text{ mol Fe(OH)}_3}{106.9 \text{ g Fe(OH)}_3}\right)\left(\frac{3 \text{ mol O}}{1 \text{ mol Fe(OH)}_3}\right)\left(\frac{16.00 \text{ g O}}{1 \text{ mol O}}\right)$$

$$= 44.9 \text{ g O}$$

$$\% \text{ O} = \frac{44.9 \text{ g O}}{100 \text{ g Fe(OH)}_3} \times 100 = 44.9\% \text{ O}$$

$$? \text{ g H} = 100 \text{ g Fe(OH)}_3 \left(\frac{1 \text{ mol Fe(OH)}_3}{106.9 \text{ g Fe(OH)}_3}\right)\left(\frac{3 \text{ mol H}}{1 \text{ mol Fe(OH)}_3}\right)\left(\frac{1.008 \text{ g H}}{1 \text{ mol H}}\right)$$

$$= 2.8 \text{ g H}$$

$$\% \text{ H} = \frac{2.8 \text{ g H}}{100 \text{ g Fe(OH)}_3} \times 100 = 2.8\% \text{ H}$$

$$? \text{ g OH}^- = 100 \text{ g Fe(OH)}_3 \left(\frac{1 \text{ mol Fe(OH)}_3}{106.9 \text{ g Fe(OH)}_3}\right)\left(\frac{3 \text{ mol OH}^-}{1 \text{ mol Fe(OH)}_3}\right)\left(\frac{17.01 \text{ g OH}^-}{1 \text{ mol OH}^-}\right)$$

$$= 47.7 \text{ g OH}^-$$

$$\% \text{ OH}^- = \frac{47.7 \text{ g OH}^-}{100 \text{ g Fe(OH)}_3} \times 100 = 47.7\% \text{ OH}^-$$

The percentages of the elements add up to 99.9%. The sum should equal 100%, but does not because the numbers were rounded to one decimal place.

Another method for calculating the percentage composition is to use the formula

$$\text{percent of element} = \frac{\begin{pmatrix}\text{atomic mass}\\ \text{of element}\end{pmatrix}\begin{pmatrix}\text{number of atoms}\\ \text{in formula}\end{pmatrix}}{\text{molar mass of compound}} \times 100 \qquad (5.6)$$

A close look at the stoichiometry equations in Example 5.30 shows that Equation 5.6 is just a summary of these stoichiometric calculations.

Exercise 5.5

What is the percentage of each element in (a) $Ca(NO_3)_2$ and (b) $CH_3CH_2NH_2$? Express your answer to four significant figures.

Solution

(a) Ca = 24.42%, N = 17.07%, O = 58.50%

(b) C = 53.28%, H = 15.65%, N = 31.07%

EMPIRICAL FORMULAS

Given a chemical formula and the atomic masses of the elements, it is possible to calculate the percent composition of a compound. More important is the fact that, with a knowledge of the percent composition and atomic masses, we can deduce the empirical formula of a compound, that is, the simplest ratio of atoms in the compound. Butyric acid has the molecular formula $HC_4H_7O_2$. Its empirical formula is C_2H_4O. Benzene, C_6H_6, has the empirical formula CH. The empirical formula of the sugar glucose, $C_6H_{12}O_6$, is CH_2O. All sugars have the same empirical formula, which is why they are also called carbohydrates (carbo- for the carbon and -hydrate for the water molecule).

To determine the empirical formula of a compound, the simplest ratio of the moles of the atoms in the compound is found by using the following steps:

1. From the given data, calculate the number of moles of each element in the compound.
2. Divide the number of moles of each element obtained in step 1 by the smallest value found to obtain whole-number subscripts for the empirical formula.
3. Note that, if step 2 does not give whole numbers (within ±0.1), the decimal portion of each number will be close to a rational fraction; for example, $0.5 = {}^1/_2$, $0.33 = {}^1/_3$, $0.67 = {}^2/_3$. In this situation, multiply each item of the data by the denominator of the rational fraction to remove the fraction and end up with a whole-number subscript for the empirical formula.

EXAMPLE 5.29

What is the empirical formula of a compound that contains 4.0 g of calcium and 7.1 g of chlorine?

Solution

The first step involves converting the grams of Ca and Cl to moles:

$$? \text{ mol Ca} = 4.0 \text{ g Ca}\left(\frac{1 \text{ mol Ca}}{40.08 \text{ g Ca}}\right) = 0.10 \text{ mol Ca}$$

$$? \text{ mol Cl} = 7.1 \text{ g Cl}\left(\frac{1 \text{ mol Cl}}{35.45 \text{ g Cl}}\right) = 0.20 \text{ mol Cl}$$

The second step is to divide both of these answers by the smaller value, 0.10:

$$\frac{0.10 \text{ mol Ca}}{0.10} = 1 \text{ mol Ca}$$

$$\frac{0.20 \text{ mol Cl}}{0.10} = 2 \text{ mol Cl}$$

This tells us that the empirical formula contains 1 mol Ca and 2 mol Cl and is written as $CaCl_2$.

EXAMPLE 5.30

A compound containing carbon, hydrogen, and oxygen is found to contain 9.1% H and 54.5% C. What is its empirical formula?

Solution

Since the percentage of an element is the number of grams per 100 g of compound, we may assume a 100 g sample of compound and convert the percent sign directly to gram units. Also, since the percent hydrogen and percent carbon do not add up to 100%, we may conclude that the missing 36.4% is oxygen. Step 1 of the procedure on page 237 is to calculate the number of moles of each element:

$$? \text{ mol C} = 54.5 \text{ g C}\left(\frac{1 \text{ mol C}}{12.01 \text{ g C}}\right) = 4.54 \text{ mol C}$$

$$? \text{ mol H} = 9.1 \text{ g H}\left(\frac{1 \text{ mol H}}{1.008 \text{ g H}}\right) = 9.0 \text{ mol H}$$

$$? \text{ mol O} = 36.4 \text{ g O}\left(\frac{1 \text{ mol O}}{16.00 \text{ g O}}\right) = 2.28 \text{ mol O}$$

Using step 2 and dividing by the smallest number, 2.28, we obtain

$$\frac{4.54 \text{ mol C}}{2.28} = 1.99 \text{ mol C}$$

$$\frac{9.0 \text{ mol H}}{2.28} = 3.95 \text{ mol H}$$

$$\frac{2.28 \text{ mol O}}{2.28} = 1.00 \text{ mol O}$$

Since these numbers are within ±0.1 of a whole number, they are rounded to 2 mol C, 4 mol H, and 1 mol O, and the empirical formula is written as C_2H_4O.

EXAMPLE 5.31

A 2.546-gram sample of a compound was burned in an excess of oxygen. Carbon dioxide was collected and weighed 4.98 g. The water collected weighed 3.56 grams. In a separate experiment, 4.738 grams of compound were reacted to form 1.79 grams of ammonia. What is the empirical formula of this compound?

Solution

We can convert the masses of CO_2, H_2O, and NH_3 to moles of N, C, and H. However, there are two problems. First, two separate experiments were performed using different starting masses. Also, we do not know if oxygen is part of the formula. To get around these roadblocks, we will calculate the percentage of C, H, and N in our compound and work from there. First we will calculate the masses of C, H, and N in our analyses.

$$? \text{ g C} = 4.98 \text{ g } CO_2 \left(\frac{1 \text{ mol } CO_2}{44.01 \text{ g } CO_2}\right)\left(\frac{1 \text{ mol C}}{1 \text{ mol } CO_2}\right)\left(\frac{12.01 \text{ g C}}{1 \text{ mol C}}\right)$$

$$= 1.36 \text{ g C}$$

$$? \text{ g H} = 3.56 \text{ g } H_2O \left(\frac{1 \text{ mol } H_2O}{18.02 \text{ g } H_2O}\right)\left(\frac{2 \text{ mol H}}{1 \text{ mol } H_2O}\right)\left(\frac{1.008 \text{ g H}}{1 \text{ mol H}}\right)$$

$$= 0.398 \text{ g H}$$

$$? \text{ g N} = 1.79 \text{ g } NH_3 \left(\frac{1 \text{ mol } NH_3}{17.03 \text{ g } NH_3}\right)\left(\frac{1 \text{ mol N}}{1 \text{ mol } NH_3}\right)\left(\frac{14.01 \text{ g N}}{1 \text{ mol N}}\right)$$

$$= 1.47 \text{ g N}$$

Now we will convert these to percentages.

$$\% \text{ C} = \frac{1.36 \text{ g C}}{2.546 \text{ g sample}} \times 100\% = 53.4\% \text{ C}$$

$$\% \text{ H} = \frac{0.398 \text{ g H}}{2.546 \text{ g sample}} \times 100\% = 15.6\% \text{ H}$$

$$\% \text{ N} = \frac{1.47 \text{ g N}}{4.738 \text{ g sample}} \times 100\% = 31.0\% \text{ N}$$

These percentages add up to 100%. That tells us that no other element is in this compound. Now we use the percentages to determine the empirical formula. As in the previous example, we will convert the percentages directly to grams by assuming we have a 100-gram sample.

$$? \text{ mol C} = 53.4 \text{ g C} \left(\frac{1 \text{ mol C}}{12.01 \text{ g C}}\right) = 4.45 \text{ mol C}$$

$$? \text{ mol H} = 15.6 \text{ g H} \left(\frac{1 \text{ mol H}}{1.008 \text{ g H}}\right) = 15.5 \text{ mol H}$$

$$? \text{ mol N} = 31.0 \text{ g N} \left(\frac{1 \text{ mol N}}{14.01 \text{ g N}}\right) = 2.21 \text{ mol N}$$

If we divide the moles of each by 2.21, the result is 2.01 mol C, 7.01 mol H, and 1 mol N. These can be rounded to integers. Therefore, the empirical formula is C_2H_7N.

Exercise 5.6

Determine the empirical formula for each of the following compounds, given its composition:

(a) A compound composed of 17.72 g Cl and 3.10 g P.
(b) A compound that is 24.74% K, 40.50% O, and 34.76% Mn.
(c) A compound containing carbon, hydrogen, and oxygen that is 40.0% C, 6.66% H, and the rest is oxygen.

Solution

(a) PCl_5
(b) $KMnO_4$
(c) CH_2O

DETERMINING AN EMPIRICAL FORMULA

1. Determine the mass of each element in the compound. You may have to use one or more of the following techniques:
   a. Convert percentages to grams.
   b. Convert the mass of compound, obtained experimentally, to the mass of an element (e.g., g CO_2 to g C).
   c. Calculate the mass of a missing element (usually oxygen).
2. Convert the mass of each element to moles.
3. Divide moles in step 2 by the smallest number of moles.
4. If the results in step 2 are integers, use them as subscripts.
5. If results in step 2 are not integers, multiply by the appropriate number to obtain integers (usually a small number such as 2, 3, 4, or 5). Use trial and error or the decimal portion of the numbers in step 2 to determine the multiplier.

MOLECULAR FORMULAS

An empirical formula, which gives the simplest ratio of atoms in a molecule, is used to represent an ionic compound. For a molecular, covalent compound, however, the actual molecular formula may be the empirical formula or some whole-number multiple of the empirical formula. Once the empirical formula has been determined as shown in the preceding section, the molar mass of the compound can be used to determine the molecular formula.

The number of empirical formula units in the molecular formula of a compound is determined by dividing the molar mass of the compound by the empirical formula mass. This will result in a small whole number:

$$\frac{\text{molar mass}}{\text{empirical formula mass}} = \text{small whole number} \qquad (5.7)$$

Each and every subscript in the empirical formula is then multiplied by this small whole number to obtain the molecular formula.

EXAMPLE 5.32

A compound has an empirical formula of CH_2O, and its molar mass is determined in a separate experiment to be 180 g mol^{-1}. What is the molecular formula of this compound?

Solution

The number of CH_2O units in the molecule is determined from Equation 5.7:

$$\frac{180\ \text{g mol}^{-1}}{30\ \text{g emp. form.}^{-1}} = 6 \text{ empirical formula units per mole}$$

Each subscript in the empirical formula is multiplied by 6 to obtain $C_6H_{12}O_6$. For the AP exam, you should be familiar with several methods to determine the molar mass.

Exercise 5.7

The following empirical formulas were determined, and their molar masses are given in parentheses after the formulas. Determine the molecular formulas.

(a) C_3H_7 (86 g mol^{-1})
(b) CH_2 (70 g mol^{-1})
(c) $C_4H_3O_2$ (166 g mol^{-1})
(d) BH_3 (27.7 g mol^{-1})
(e) CH_2ON (176 g mol^{-1})

Solution

(a) C_6H_{14}
(b) C_5H_{10}
(c) $C_8H_6O_4$
(d) B_2H_6
(e) $C_4H_8O_4N_4$

OTHER STOICHIOMETRIC EQUATIONS

Chemists become very familiar with the dimensional analysis method and see shortcuts in the calculation of the moles of a substance from a variety of units, as shown below:

$$\text{moles} = \frac{\text{grams}}{\text{molar mass}} \tag{5.8}$$

$$\text{moles} = \text{molarity} \times \text{liters of solution} \tag{5.9}$$

$$\text{moles} = \frac{\text{liters of gas}}{22.4} \tag{5.10}$$

$$\text{moles} = \frac{\text{molecules or atoms}}{6.02 \times 10^{23}} \tag{5.11}$$

These relationships can speed the calculations, but must be used with care to ensure that the proper units are chosen in all instances.

SUMMARY

This chapter summarizes approximately one half of all the calculations in chemistry. These are stoichiometric calculations, and they involve converting from one set of units to another using a logical system. The favored system is the conversion factor method, which is also called dimensional analysis. This chapter reminds you that three things are needed for a successful stoichiometric calculation. First is an understanding of both the starting point and where you need to end up. Second is a logical sequence of conversions. And third, you need the appropriate conversion factors to apply at each step. Many calculations can be performed using the three-step approach. This includes calculating the mass of an atom or molecule, mass-to-mass calculations, and limiting-reactant calculations. The chapter also shows that the conversion factor method is applicable to titration calculations.

In addition to the above calculations this chapter shows how simple mass data allowed early chemists to deduce the formulas of compounds. Methods for calculating percent composition, empirical formulas, and molecular formulas are presented here.

Important Concepts

Dimensional analysis (conversion factor) method
Equalities from: Avogadro's number, molar masses, chemical formulas, and chemical equations
Stoichiometric conversion sequence
Limiting-reactant calculations
Percent composition and empirical formulas
Chemical analysis by titration

Important Equations

moles = grams/molar mass
molarity = moles/liter
moles = molecules or atoms/6.02×10^{23}
moles = liters of gas at STP/22.4

PRACTICE EXERCISES

Multiple-Choice

1. What is the volume (in L) of a solution that contains 3.12 moles of NaCl if the concentration of this solution is 6.67 *M* NaCl?
   (A) 0.214 L
   (B) 20.8 L
   (C) 2.14 L
   (D) 0.468 L

2. The only item on this list that changes as the temperature changes is
   (A) Molarity because it uses the volume of the solution, which changes with temperature
   (B) Mass percentage because this is a ratio that changes with temperature
   (C) Molar mass because it increases only as the temperature increases
   (D) Empirical formula since it depends on the percentage composition

3. Calculate the volume (in mL) of a 2.75 *M* solution of iron(II) ammonium sulfate that must be used to make 1.25 L of a 0.150 *M* solution of $Fe(NH_4)_2(SO_4)_2$.
   (A) 0.0682 mL
   (B) 0.0330 mL
   (C) 33.0 mL
   (D) 68.2 mL

4. When calculating the moles of sodium peroxide (Na_2O_2) produced if 32.5 g of sodium react with excess oxygen, $2Na(s) + O_2(g) \rightarrow Na_2O_2(s)$, which of the following is true?
   (A) A limiting reactant calculation must be solved.
   (B) The dilution of solutions process is used.
   (C) The grams of sodium must be converted to moles of Na_2O_2 using a stoichiometric calculation.
   (D) The moles of Na_2O_2 must be converted to grams of sodium.

5. What weight of $KClO_3$ (molar mass = 122.5 g/mol) is needed to make 200 mL of a 0.150 *M* solution of this salt?
   (A) 2.73 g
   (B) 3.68 g
   (C) 27.3 g
   (D) 164 g

6. In an experiment 35.0 mL of 0.345 *M* HNO_3 is titrated with 0.130 *M* NaOH. What volume of NaOH will have been used when the indicator changes color?
   (A) 35.0 mL
   (B) 92.9 mL
   (C) 26.4 mL
   (D) 50.0 mL

7. In the reaction $CaCO_3 + 2HCl \rightarrow H_2O + CO_2 + CaCl_2$, calculating the grams of $CaCO_3$ (molar mass = 100.) needed to produce 3.00 L of CO_2 at STP requires doing all of the following EXCEPT
   (A) converting the moles of CO_2 to grams of CO_2
   (B) converting moles of CO_2 to moles of $CaCO_3$
   (C) using the conversion factor that 22.4 L of a gas is equivalent to 1 mole at STP
   (D) using the molar mass of $CaCO_3$ as a conversion factor

8. In mass spectrometry, organic compounds are deliberately fragmented in order to deduce their molecular structure. One fragment containing only carbon and hydrogen has 14.3% H. Which of the following is the fragment in question?
   (A) CH
   (B) CH_4
   (C) C_4H
   (D) CH_2

9. To two decimal places, what is the molar mass of $Al(NO_3)_3$?
   (A) 165.00 g mol^{-1}
   (B) 56.99 g mol^{-1}
   (C) 213.01 g mol^{-1}
   (D) 88.99 g mol^{-1}

10. How many milligrams of Na_2SO_4 (molar mass = 142 g/mol) are needed to prepare 100. mL of a solution that is 0.00100 *M* in Na^+ ions?
   (A) 28.4 mg
   (B) 14.2 mg
   (C) 1.00 mg
   (D) 7.10 mg

11. Which of the following compounds has the highest percentage by mass of sulfur?
   (A) Al_2S_3 since there is more mass of sulfur per gram of compound
   (B) $CaSO_4$ because the molar mass of sulfur is greater than the molar mass of calcium
   (C) Na_2S because the ratio of the molar mass of sulfur to sodium is the highest
   (D) SO_2 because the element with the highest percentage is always written first in a formula

12. In the following reaction, how many moles of aluminum will produce 1.0 mol of iron and why?

$$8Al + 3Fe_3O \rightarrow 9Fe + 4Al_2O_3$$

   (A) 1 mol Al because aluminum has the lowest atomic mass
   (B) 3/4 mol Al because 3/4 is the mole ratio of the oxides
   (C) 3/8 mol Al since this is the mole ratio of the reactants
   (D) 8/9 mol Al since this mole ratio cancels units correctly

13. In the following reaction:

$$2KOH + H_2SO_4 \rightarrow K_2SO_4 + 2H_2O$$

35.4 mL of 0.125 *M* KOH is required to neutralize completely 50.0 mL of H_2SO_4. What is the molarity of the H_2SO_4 solution?
   (A) 0.0883 *M*
   (B) 0.100 *M*
   (C) 0.0443 *M*
   (D) 0.125 *M*

14. A substance has an empirical formula of CH_2. Its molar mass is determined in a separate experiment as 83.5. What is the most probable molecular formula for this compound?
(A) C_2H_4 because it is the simplest formula after CH_2
(B) C_6H_{12} since it is the hydrocarbon with a molar mass close to the experimental molar mass
(C) C_4H_2 since it has subscripts that are simple multiples of the empirical formula
(D) $C_4H_3O_2$ because this has a molar mass closest to 83.5

15. The mass of one atom of iron is
(A) 1.66×10^{-24} g
(B) 2.11×10^{-22} g
(C) 3.15×10^{-22} g
(D) 9.28×10^{-23} g

16. How many grams of the gas SO_2 are in a 4.00 L sample of SO_2 at STP?
(A) 256.2 g since this is the mass of 4 moles of SO_2
(B) 11.4 g since this is obtained by converting L to mol using the molar volume of gases and then to mass by converting with the molar mass
(C) 358.7 g since this is obtained by multiplying by the molar volume, 22.4
(D) 2.78×10^{-3} g since this is obtained by dividing by the molar volume, 22.4

17. What is the percentage of potassium in K_3PO_4?
(A) 14.6%
(B) 29.2%
(C) 18.4%
(D) 55.3%

18. A 0.200-g sample of a compound containing only carbon, hydrogen, and oxygen is burned, and 0.357 g of CO_2 and 0.146 g of H_2O are collected. What is the percentage of carbon in the original compound and what can that value be used for?
(A) 56.0% and this can be used to obtain the empirical formula
(B) 73.0% and this can be used to name the compound
(C) 48.7% and this, with the grams of hydrogen, can be used to determine the empirical formula
(D) 24.3% and this can be used to track CO_2 in the atmosphere

19. In the following combustion reaction, when 5.50 g of glucose are burned in the presence of 2.50 L of oxygen at STP, what is the limiting reactant?

$$C_6H_{12}O_6(s) + 6O_2(g) \rightarrow 6CO_2(g) + 6H_2O(l)$$

(A) $C_6H_{12}O_6$ because this has the largest mass and therefore limits the reaction
(B) O_2 when calculated since this reacts only with a small fraction of the glucose
(C) CO_2 because the mass formed is the largest
(D) H_2O since the mass of water formed is the largest

20. In the reaction

$$2AgNO_3 + CaCl_2 \rightarrow 2AgCl + Ca(NO_3)_2$$

how many grams of AgCl (molar mass = 143.5 g/mol) will precipitate when a solution containing 20.0 g $AgNO_3$ (molar mass = 170 g/mol) is reacted with a solution containing 15.0 g $CaCl_2$ (molar mass = 111 g/mol)?
(A) 16.9 g
(B) 38.8 g
(C) 33.8 g
(D) 8.45 g

21. In the reaction

$$2AgNO_3 + CaCl_2 \rightarrow 2AgCl + Ca(NO_3)_2$$

how many grams of which reactant will remain when a solution containing 20.0 g $AgNO_3$ (molar mass = 170 g/mol) is reacted with a solution containing 15.0 g $CaCl_2$ (molar mass = 111 g/mol)?
(A) 6.53 g $CaCl_2$
(B) 6.53 g $AgNO_3$
(C) 45.9 g $CaCl_2$
(D) 8.47 g $CaCl_2$

CHALLENGE

22. Which of the following molecular views of the reaction of hydrogen and nitrogen gases to form ammonia indicates that hydrogen is the limiting reactant?

(A)

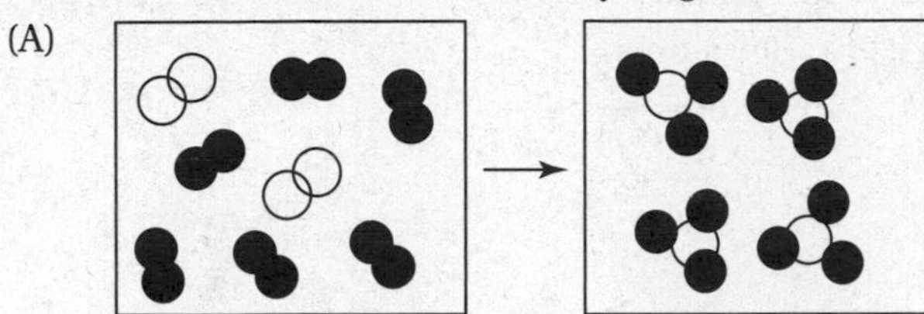

(B)

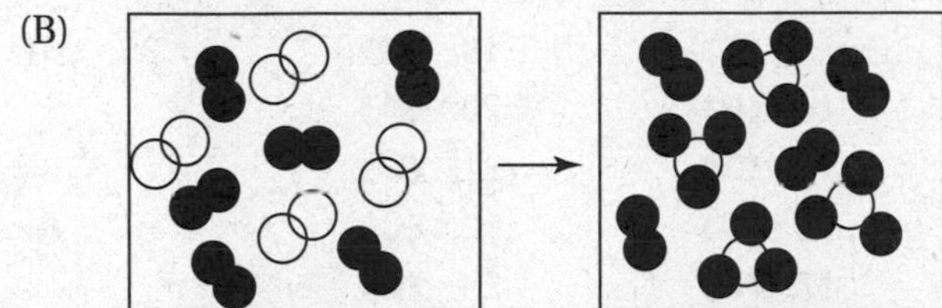

(C)

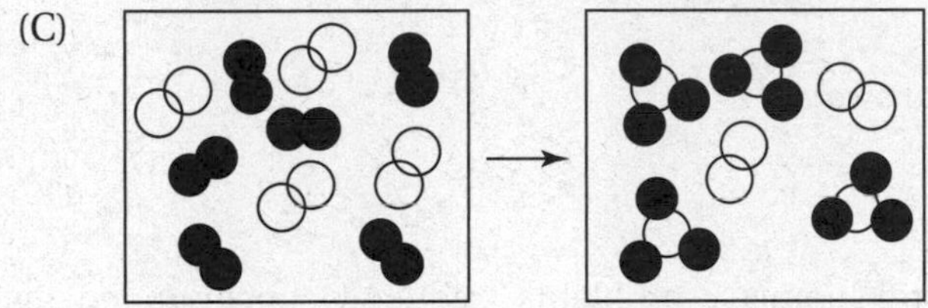

(D)

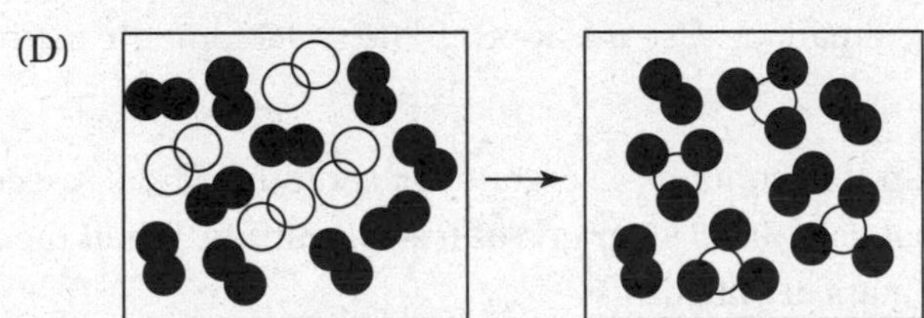

23. How many liters of air are needed to completely burn 1 mol of methane in air (20% oxygen) at STP according to the reaction

$$CH_4 + 2O_2 \rightarrow CO_2 + 2H_2O$$

(A) 22.4 L
(B) 44.8 L
(C) 11.2 L
(D) 224 L

24. Determine the empirical formula for a compound that is 25% hydrogen and 75% carbon. Is this likely to be the molecular formula?

(A) CH. No, this cannot be the molecular formula because carbon forms 4 bonds.
(B) CH_2. Yes, this can be the molecular formula because carbon often forms 2 bonds.
(C) CH_4. Yes, this is a compound with carbon having 4 bonds.
(D) C_2H_8. No, this cannot be the molecular formula since carbon atoms have more than 4 bonds in this formula.

25. One granule of sucrose ($C_{12}H_{22}O_{11}$, molar mass = 342) weighs 2.5 micrograms. How many sucrose molecules are in that granule and how many atoms are in that granule?
(A) 2.5×10^{17} molecules and 7.5×10^{17} atoms
(B) 4.4×10^{15} molecules and 2.0×10^{17} atoms
(C) 6.02×10^{17} molecules and 1.33×10^{16} atoms
(D) 4.4×10^{21} molecules and 9.8×10^{15} atoms

ANSWER KEY

1. **(D)**	7. **(A)**	13. **(C)**	19. **(B)**	25. **(B)**
2. **(A)**	8. **(D)**	14. **(B)**	20. **(A)**	
3. **(D)**	9. **(C)**	15. **(D)**	21. **(D)**	
4. **(C)**	10. **(D)**	16. **(B)**	22. **(C)**	
5. **(B)**	11. **(A)**	17. **(D)**	23. **(D)**	
6. **(B)**	12. **(D)**	18. **(C)**	24. **(C)**	

See Appendix 1 for explanations of answers.

Free-Response

The following questions involve stoichiometry problems frequently encountered by the chemist in theoretical and laboratory situations. Use the appropriate stoichiometric methods to answer the following questions.

(a) One beaker holds a solution that contains 4.65 grams of sodium sulfide. A second beaker holds a solution that contains 8.95 grams of lead(II) nitrate. When the two solutions are mixed, what mass of PbS forms?

(b) Perform the calculations to determine the empirical formula of a CHNO compound that is analyzed and found to contain 52.63 percent carbon, 7.02 percent hydrogen, and 12.28 percent nitrogen.

(c) If the compound in part (b) has a molar mass of 228, what is the molecular formula?

(d) A neutralization reaction uses a 0.125 molar solution of sodium hydroxide to titrate 50.0 mL of an unknown sulfuric acid solution. If the reaction takes 23.5 mL of the sodium hydroxide to completely neutralize the sulfuric acid, what is the molarity of the sulfuric acid solution?

ANSWERS

(a) Virtually every stoichiometry problem requires a balanced chemical equation. So we use previous knowledge of double-replacement reactions and the solubility of PbS based on

$Na_2S(aq) + Pb(NO_3)_2(aq) \rightarrow PbS(s) + 2NaNO_3(aq)$

This tells us that PbS is the precipitate and is the mass we want to determine. We start to write ? g PbS =, and we find that we have two starting points, the mass of lead(II) nitrate and the mass of sodium sulfide. That means that this is a limiting-reactant problem. We can solve it by calculating the mass of PbS obtained from each reactant and choosing the smaller mass as the limited amount that is actually produced. We write

? g PbS = 4.65 g Na_2S and ? g PbS = 8.95 g $Pb(NO_3)_2$

We then use the dimensional analysis method to perform the calculations, calculating molar masses as we go.

$$? \text{ g PbS} = 4.65 \text{ g Na}_2\text{S}\left(\frac{1\text{mol Na}_2\text{S}}{78.1\text{ g Na}_2\text{S}}\right)\left(\frac{1\text{mol PbS}}{1\text{mol Na}_2\text{S}}\right)\left(\frac{239.3\text{ g PbS}}{1\text{mol PbS}}\right)$$

? g PbS = 14.2 g PbS (from Na_2S)

? g PbS =

$$8.95 \text{ g Pb(NO}_3)_2\left(\frac{1\text{mol Pb(NO}_3)_2}{331.2\text{ g Pb(NO}_3)_2}\right)\left(\frac{1\text{mol PbS}}{1\text{mol Pb(NO}_3)_2}\right)\left(\frac{239.3\text{ g PbS}}{1\text{mol PbS}}\right)$$

? g PbS = 6.47 g PbS (from $Pb(NO_3)_2$)

The limiting reactant limits the product to the smaller of these two results. Therefore the mass of PbS is 6.47 g.

(b) First we see that the percentages do not add up to 100% and that a percentage for oxygen is not mentioned. We obtain the percentage from oxygen as (this is an application of the law of conservation of mass)

% oxygen = 100 – 52.63 – 7.02 – 12.28 = 28.07% O

Now we assume the sample size is 100 g and convert the percentages to grams. Next the number of moles of each element is calculated.

$$? \text{ mol O} = 28.07 \text{ g O}\left(\frac{1 \text{ mol O}}{16.0 \text{ g O}}\right) = 1.754 \text{ mol O}$$

$$? \text{ mol C} = 52.62 \text{ g C}\left(\frac{1 \text{ mol C}}{12.0 \text{ g C}}\right) = 4.385 \text{ mol C}$$

$$? \text{ mol H} = 7.02 \text{ g H}\left(\frac{1 \text{ mol H}}{1.00 \text{ g H}}\right) = 7.02 \text{ mol H}$$

$$? \text{ mol N} = 12.28 \text{ g N}\left(\frac{1 \text{ mol N}}{14.0 \text{ g N}}\right) = 0.8771 \text{ mol N}$$

Divide the entire list by the smallest number 0.8771 to get

? mol O = 1.754 mol O/0.8771 = 1.999

? mol C = 4.385 mol C/0.8771 = 4.999

? mol H = 7.02 mol H/0.8771 = 8.004

? mol N = 0.8771 mol N/0.8771 = 1.00

All these numbers are very close to integers, and we can use them as subscripts for the empirical formula $C_5H_8O_2N$.

(c) The mass of the empirical formula is 114. Because the molar mass is 228, we can see that 228/114 = 2, which means that two empirical formula units comprise the entire molecule. We double the subscripts to write the molecular formula as $C_{10}H_{16}O_4N_2$.

(d) We write and balance the chemical equation for the complete neutralization as

$H_2SO_4 + 2NaOH \longrightarrow Na_2SO_4 + 2H_2O$

We can start with the molarity of the sodium hydroxide solution and convert it to the molarity of the sulfuric acid.

$$?\ \frac{\text{mol } H_2SO_4}{\text{L } H_2SO_4} = \frac{0.125 \text{ mol NaOH}}{1 \text{ L NaOH}}\left(\frac{1 \text{ mol } H_2SO_4}{2 \text{ mol NaOH}}\right)\left(\frac{0.0235 \text{ L NaOH}}{0.0500 \text{ L } H_2SO_4}\right)$$

$$?\ \frac{\text{mol } H_2SO_4}{\text{L } H_2SO_4} = 0.0294\ M\ H_2SO_4$$

Notice that the last conversion factor could have been written using mL units $\left(\text{e.g., } \frac{23.5 \text{ mL NaOH}}{50.0 \text{ mL } H_2SO_4}\right)$ and would have produced the same result.

PART 3

States of Matter

Gases

6

- → IDEAL GAS LAW
- → BOYLE'S LAW
- → CHARLES'S LAW
- → GAY-LUSSAC'S LAW
- → AVOGADRO'S PRINCIPLE
- → UNIVERSAL GAS CONSTANT, *R*
- → STANDARD TEMPERATURE AND PRESSURE
- → MOLAR MASS CALCULATIONS
- → GAS DENSITY
- → MOLAR VOLUME
- → KINETIC MOLECULAR THEORY
- → GRAHAM'S LAW OF EFFUSION
- → AVERAGE KINETIC ENERGY
- → KINETIC ENERGY RELATES TO TEMPERATURE
- → REAL GASES
- → DALTON'S LAW OF PARTIAL PRESSURES
- → GAS COLLECTION

BIG IDEAS 1, 2

Learning Objectives: 1.4, 2.4, 2.5, 2.6, 2.12

For the complete list of Big Ideas and Learning Objectives, refer to the AP Chemistry Course Outline:
https://secure-media.collegeboard.org/digitalServices/pdf/ap/ap-chemistry-course-and-exam-description.pdf

While the chemical properties of gases vary widely, the physical behavior of all gases is extraordinarily similar, to the extent that one equation, the **ideal gas law**, defines the relationships among the volume, pressure, temperature, and moles of gas in any sample. Similarly, only one theory, the **kinetic molecular theory**, is commonly used to describe the behavior of all gases. A thorough understanding of the ideal gas law is necessary for numerical calculations based on gases, and a similar understanding of the kinetic molecular theory provides the basis for explaining why gases behave as they do.

THE GAS LAWS

Although we will use the ideal gas law for almost all calculations, you should be familiar with the individual gas laws. Gases have four properties: temperature, *T*; pressure, *P*; volume, *V*; and the moles of gas, *n*. Each of the gas laws holds two of these constant while measuring the change of one property as another is varied. (Note that the constants in each of these equations do not have the same value.)

As you review this chapter, pay particular attention to the graphic and pictorial representations. An increasing number of questions will involve such representations as illustrated in the practice exams at the end of this book.

Boyle's Law (1660)

Boyle's law describes the inverse pressure-volume relationship as shown below in three ways.

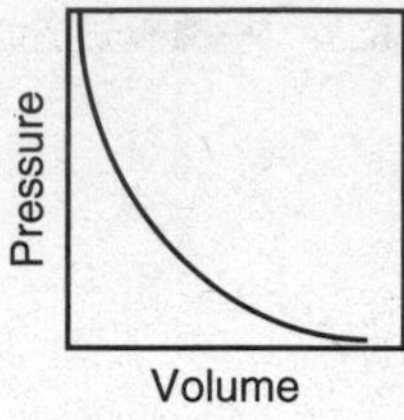

$$P \propto \frac{1}{V} \quad \text{or} \quad PV = \text{constant} \tag{6.1}$$

If a sample of gas starts with initial conditions of pressure and volume, and an experiment is done to change those conditions (without changing T or the amount of gas), then the relationship

$$P_iV_i = P_fV_f \tag{6.2}$$

is obtained. The subscripts i and f represent the initial and the final conditions, respectively.

Charles's Law (1787)

Charles's law describes the direct temperature-volume relationship as illustrated below in three ways.

$$V \propto T \quad \text{or} \quad \frac{V}{T} = \text{constant} \tag{6.3}$$

If a gas sample starts with initial conditions of volume and temperature that are changed to some final conditions (while the pressure and the amount of gas do not change), **Charles's law** may be reformulated as

$$\frac{V_i}{T_i} = \frac{V_f}{T_f} \tag{6.4}$$

The exact value for absolute zero is −273.15°C.

Absolute zero is the lowest possible temperature. It is zero on the Kelvin and −273 degrees on the Celsius temperature scales. One method of determining absolute zero is to construct a graph of the volume of a gas as its temperature is changed and to extrapolate the data to the temperature that corresponds to zero volume of the gas, as shown in Figure 6.1.

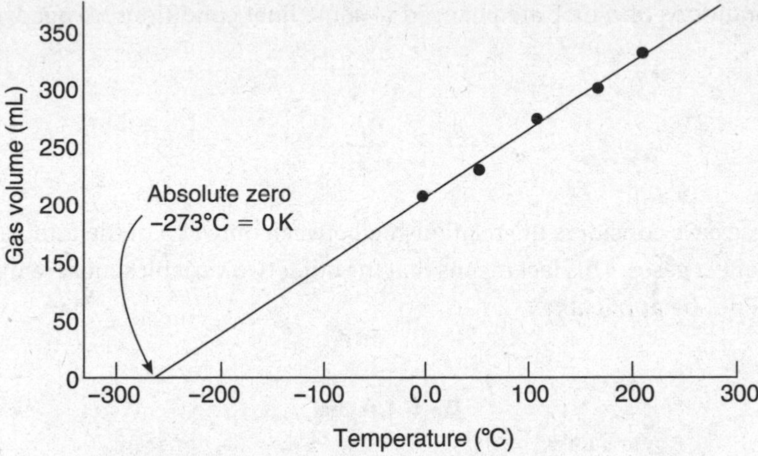

FIGURE 6.1. Using Charles's law to determine absolute zero. Each dot represents an experimental measurement. The line is the best straight line through the data, and the intercept is at −273°C.

Gay-Lussac's Law (ca. 1787)

Gay-Lussac's law describes the direct pressure-temperature relationship. It is shown in three representations below.

Pressure

Temperature (K)

$$P \propto T \quad \text{or} \quad \frac{P}{T} = \text{constant} \tag{6.5}$$

If initial conditions of P and T are changed to some final conditions, **Gay-Lussac's law** requires that

$$\frac{P_i}{T_i} = \frac{P_f}{T_f} \tag{6.6}$$

Avogadro's Principle (1811)

Finally, in 1811, Avogadro suggested the principle that equal volumes of gases contain equal numbers of molecules or atoms (i.e., moles of a gas) under identical conditions of temperature and pressure. This direct relationship between the number of moles and volume is written in equation form and shown graphically as:

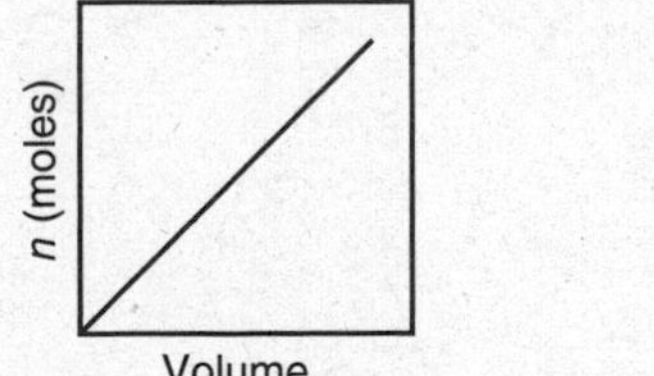

$$V \propto n \quad \text{or} \quad \frac{n}{V} = \text{constant} \tag{6.7}$$

If initial conditions of n or V are changed to some final conditions, **Avogadro's principle** requires that

$$\frac{n_i}{V_i} = \frac{n_f}{V_f} \tag{6.8}$$

Each of these laws considers the relationship between only two of the four variables, P, V, T, and n, that affect gases. This fact means that the other two variables must remain constant, or the law will not be applicable.

GAS LAWS		
Boyle's Law	PV	= constant
Charles's Law	V/T	= constant
Gay-Lussac's Law	P/T	= constant
Avogadro's Principle	n/V	= constant

IDEAL GAS LAW

The four laws of Boyle, Charles, Gay-Lussac, and Avogadro are combined into the ideal gas law:

$$PV = nRT \tag{6.9}$$

where P is the pressure, V is the volume, n is the number of moles of gas, and T is the temperature in kelvins. The constant, R, called the **universal gas constant**, is needed to make all of the relationships fit together. The numerical value of R depends upon the units used. For example, the AP exam gives three common values of R:

$$R = 0.08206 \text{ L atm mol}^{-1} \text{ K}^{-1}$$
$$R = 8.314 \text{ J mol}^{-1} \text{ K}^{-1}$$
$$R = 62.36 \text{ L torr mol}^{-1} \text{ K}^{-1}$$

Selecting the appropriate value of R and its corresponding units is an important skill for solving gas law problems. In addition, it is very important to use dimensional analysis to be sure that the units cancel properly. For instance in all cases, the temperature must be converted to the Kelvin scale. It may be necessary to convert mass to moles or volume from milliliters to liters. R appears in several other equations that chemists use. You should always pay careful attention to the units as well as the numerical value of R.

DERIVATION OF THE IDEAL GAS LAW

One method for deriving the ideal gas law uses Avogadro's principle, $\frac{n}{V} = C$, and Gay-Lussac's law, $\frac{P}{T} = C'$. Since C and C' are not the same constants, we multiply Avogadro's principle by a constant, R, so that $CR = C'$. Then

$$\frac{Rn}{V} = RC = C' \quad \text{and} \quad \frac{P}{T} = C'$$

then

$$\frac{nR}{V} = \frac{P}{T}$$

Rearranging the last two terms in this expression yields

$$PV = nRT$$

the ideal gas law, and R is the universal gas constant.

There are two ways to use the ideal gas law. First, a problem may give three of the four variables and ask that the fourth be calculated. This involves direct substitution of data into the ideal gas law equation. The second way to use the law involves taking a gas under certain initial conditions of P, V, T, and n and changing to some different, final conditions of these four variables. To solve this type of problem the ratio of the ideal gas law equations for the initial and final conditions is written as

$$\frac{P_i V_i}{P_f V_f} = \frac{n_i R T_i}{n_f R T_f} \tag{6.10}$$

Variables that the problem states (or implies) as remaining constant are canceled, along with R, and then the appropriate substitutions are made from the given data to perform the calculation. Using this approach, it can be seen that the ideal gas law becomes Boyle's law if n and T are constant (i.e., they cancel), and becomes Charles's law if n and P are held constant. If V and n are held constant, Equation 6.10 cancels to become Gay-Lussac's law, and it becomes Avogadro's principle if P and T are held constant. *You need to remember only one equation for all ideal gas law calculations!*

EXAMPLE 6.1

A gas occupies 250 mL, and its pressure is 550 mm Hg at 25°C.

(a) If the gas is expanded to 450 mL, what is the pressure of the gas now?
(b) What temperature is needed to increase the pressure of the gas to exactly 1 atmosphere and 250 mL?
(c) How many moles of gas are in this sample?
(d) The sample is an element and has a mass of 0.525 g. What is it?

Solution

(a) Use the ratio of two ideal gas law equations as shown in Equation 6.10:

$$\frac{P_iV_i}{P_fV_f}=\frac{n_iRT_i}{n_fRT_f}$$

Cancel all but the P and V terms since no change is specified for the other terms and they are assumed to be constant:

$$\frac{P_iV_i}{P_fV_f}=1$$

Assign the data to the variables: $P_i = 550$ mm Hg, $V_i = 250$ mL, and $V_f = 450$ mL, enter the data in the equation, and solve:

$$\frac{(550 \text{ mm Hg})(250 \text{ mL})}{P_f(450 \text{ mL})}=1$$

$$\frac{(550 \text{ mm Hg})(250 \text{ mL})}{450 \text{ mL}}=P_f$$

$$306 \text{ mm Hg}=P_f$$

(b) Use the same procedure as in part (a) but cancel n, R, and V since they remain constant:

$$\frac{P_iV_i}{P_fV_f}=\frac{n_iRT_i}{n_fRT_f}$$

$$\frac{P_i}{P_f}=\frac{T_i}{T_f}$$

Assign the data: $P_i = 550$ mm Hg, $T_i = 25 + 273 = 298$ K, and $P_f = 1$ atm $= 760$ mm Hg:

$$\frac{550 \text{ mm Hg}}{760 \text{ mm Hg}}=\frac{298 \text{ K}}{T_f}$$

$$T_f=298 \text{ K}\left(\frac{760 \text{ mm Hg}}{550 \text{ mm Hg}}\right)$$

$$=412 \text{ K}=139°\text{C}$$

(c) To solve this question, we use the ideal gas law equation and substitute the given values:

$$PV=nRT$$

To use the constant $R = 0.0821 \text{ L atm mol}^{-1} \text{ K}^{-1}$, the data must be converted to the proper units so that

$$P=550 \text{ mm Hg}\left(\frac{1 \text{ atm}}{760 \text{ mm Hg}}\right)=0.724 \text{ atm},$$

$$V=250 \text{ mL}\left(\frac{1 \text{ L}}{1000 \text{ mL}}\right)=0.250 \text{ L, and}$$

$$T=25°\text{C } 298 \text{ K.}$$

Enter these data and solve:

$$(0.724\ \text{atm})(0.250\ \text{L}) = n(0.0821\ \text{L atm mol}^{-1}\ \text{K}^{-1})(298\ \text{K})$$

$$n = \frac{(0.724\ \text{atm})(0.250\ \text{L})}{(0.0821\ \text{L atm mol}^{-1}\ \text{K}^{-1})(298\ \text{K})}$$

$$= 7.40 \times 10^{-3}\ \text{mol}$$

(d) We can identify a gas by its molar mass. The molar mass is

$$\text{molar mass} = \frac{\text{mass of sample}}{\text{moles of sample}}$$

$$= \frac{0.525\ \text{g}}{7.40 \times 10^{-3}\ \text{mol}}$$

$$= 70.9\ \text{g mol}^{-1}$$

The element with a molar mass closest to 70.9 is chlorine, Cl_2.

STANDARD TEMPERATURE AND PRESSURE (STP)

The ideal gas law has four variables, *P*, *V*, *n*, and *T*, along with the constant *R*. By defining the standard pressure as exactly 1 atmosphere* and the standard temperature as exactly 0 degrees Celsius, two of the variables can be stated quickly and easily. Therefore, any gas at **STP** is understood to have $P = 1.00$ atm and $T = 273$ K, and only *n* and *V* need be stated in a problem.

MOLAR MASS, DENSITY, AND MOLAR VOLUME

In the ideal gas law, *n* represents the number of moles of gas. The number of moles is calculated as $n = \frac{\text{grams of gas}}{\text{molar mass}}$. Substituting this equivalency into the ideal gas law yields

$$PV = \frac{g}{\text{molar mass}} RT \qquad (6.11)$$

This equation can be used to determine the **molar mass** of a gas if *P*, *V*, *g*, and *T* are known for a given sample.

Rearranging Equation 6.11 algebraically yields another useful relationship:

$$P(\text{molar mass}) = \frac{g}{V} RT \qquad (6.12)$$

In this equation, $\frac{g}{V}$ is the density of the gas in grams per liter. The **density** of a gas can be determined if *P*, *T*, and the molar mass are known.

---

*In 1999, the International Union of Pure and Applied Chemistry, IUPAC, decided that the standard pressure would be 100 kilopascals (kPa). The difference between the current definition of 1 atmosphere or 760 torr and the IUPAC definition is approximately 1%.

Finally, at standard temperature, 0 degrees Celsius, and pressure, 1.0 atmosphere, the ideal gas law can be solved to calculate the **molar volume**, $\frac{V}{n}$, as

$$\frac{V}{n}=\frac{RT}{P}=\frac{(0.0821\ \text{L atm mol}^{-1}\ \text{K}^{-1})(273\ \text{K})}{1.00\ \text{atm}}$$
$$=22.4\ \text{L mol}^{-1} \tag{6.13}$$

This indicates that 1 mole of an **ideal gas** at STP has a volume of 22.4 liters, a fact that is useful in stoichiometric calculations. Table 6.1 lists the molar volumes of some real gases at STP. They are all close to 22.4 liters indicating that they behave as ideal gases under these conditions.

TABLE 6.1
Molar Volumes of Some Gases at STP

Gas	Symbol	Molar (liters) Volume
argon	Ar	22.401
carbon dioxide	CO_2	22.414
helium	He	22.398
hydrogen	H_2	22.410
nitrogen	N_2	22.413
oxygen	O_2	22.414

EXAMPLE 6.2

What are the expected densities of argon, neon, and air at STP?

Solution

Each density is calculated, using Equation 6.12, as

$$\text{density}=\frac{g}{V}=\frac{(\text{molar mass})P}{RT}$$

At STP, $P = 1.00$ atm and $T = 273$ K. Substituting these data gives

$$\text{density Ar}=\frac{(39.95\ \text{g mol}^{-1})(1.00\ \text{atm})}{(0.0821\ \text{L atm mol}^{-1}\ \text{K}^{-1})(273\ \text{K})}=1.78\ \text{g L}^{-1}$$

$$\text{density Ne}=\frac{(20.18\ \text{g mol}^{-1})(1.00\ \text{atm})}{(0.0821\ \text{L atm mol}^{-1}\ \text{K}^{-1})(273\ \text{K})}=0.900\ \text{g L}^{-1}$$

$$\text{density air}=\frac{(28.8\ \text{g mol}^{-1})(1.00\ \text{atm})}{(0.0821\ \text{L atm mol}^{-1}\ \text{K}^{-1})(273\ \text{K})}=1.28\ \text{g L}^{-1}$$

The effective molar mass of air is approximated from the fact that air is 80 percent nitrogen and 20 percent oxygen:

$$\text{effective molar mass} = (0.80)\left(\frac{28 \text{ g N}_2}{\text{mol}}\right) + (0.20)\left(\frac{32 \text{ g O}_2}{\text{mol}}\right)$$
$$= 28.8 \text{ g/mol air}$$

TIP

This is similar to the calculation of the average atomic mass of an element. **See page 113.**

We may also conclude that a balloon full of neon will rise in air, whereas an argon-filled balloon will sink to the floor.

KINETIC MOLECULAR THEORY

The ideal gas law describes the relationships among *P*, *V*, *T*, and *n* for ideal gases. The **kinetic molecular theory** describes gases at the level of individual particles. This theory, developed largely by Boltzmann, Clausius, and Maxwell between 1850 and 1880, is often stated as five postulates:

1. Gases consist of molecules or atoms in continuous random motion.
2. Collisions between these molecules and/or atoms in a gas are elastic.
3. The volume occupied by the atoms and/or molecules in a gas is negligibly small.
4. The attractive or repulsive forces between the atoms and/or molecules in a gas are negligible.
5. The average kinetic energy of a molecule or atom in a gas is directly proportional to the Kelvin temperature of the gas.

The concept of gas **pressure** is important to understand since it is central to the kinetic molecular theory. Pressure is defined in physics as the force exerted per unit area. The English units for pressure, pounds per square inch, are familiar. For gases, the force is generated by collisions of the gas particles with the container walls. Each collision has a certain force, which is related to the velocity of the gas particle. The total force is the sum of the forces of all the collisions occurring each second per unit area. Thus the pressure is dependent on the velocity of the gas particles and the collision frequency. In turn, the collision frequency depends on the velocity of the gas particle and the distance to the container walls. Changing the temperature changes the force of the collisions, as well as the frequency of collision. The frequency of collision can be also changed by altering the size of the container, but the force of the collisions will not change.

Based on this understanding of pressure, the molecular meanings of the gas laws can be appreciated as follows.

Boyle's law states the inverse relationship between pressure and volume $\left(P \propto \frac{1}{V}\right)$. The kinetic molecular theory agrees with this observation. If the volume of a gas is decreased, the gas particles will strike the walls of the container more frequently. Increasing the frequency of these collisions increases the observed pressure of the gas.

Gay-Lussac's law finds a direct relationship between temperature and pressure ($P \propto T$). In this case, an increase in temperature increases the kinetic energy of the gas particles and their average velocities. This increase has two effects. First, the force of each collision is greater; second, since the average velocity of the gas particles increases with temperature, the frequency of collisions with the container walls also increases. Thus the kinetic molecular theory predicts an increase in gas pressure as temperature rises.

Charles's law predicts a direct relationship between temperature and volume ($V \propto T$). An increase in temperature increases both the force of each collision and the frequency of collisions as in Gay-Lussac's law. Both of these effects increase the pressure. If the pressure of the gas is to remain constant, the volume must increase to correspondingly decrease the frequency of collisions with the walls of the container because the gas molecules will have to travel farther to collide with a wall.

Finally, the kinetic molecular theory also explains **Graham's law of effusion**, given in Equation 6.14.

$$\sqrt{\frac{m_1}{m_2}} = \frac{v_{rms2}}{v_{rms1}} \tag{6.14}$$

When the rates at which two gases will effuse through a pinhole in a container are compared, they are found to be inversely related to the square roots of the masses of the gas particles. **Effusion** through a pinhole into a vacuum requires that a gas particle hit the pinhole just right to pass through it. The more collisions a gas has with the walls of a container, the higher the probability is that it will hit the pinhole and go through it.

Equation 6.14 is also the mathematical statement of the relative rates of **diffusion** of gases, which is the movement of one gas into a container already filled with gas. All gases will expand to fill a container, but filling it may take some time, depending on the conditions present. This equation illustrates that the root mean square velocity of the gas will be inversely proportional to the square root of the masses of the individual gas particles. Since this is a ratio and the units cancel, m_1 and m_2 may be either the actual mass or the molar mass of the gas particles. Heavier gases would be expected to diffuse more slowly than lighter gases, and they do.

EXAMPLE 6.3

Helium leaks through a very small hole into a vacuum at a rate of 3.22×10^{-5} mol s^{-1}. How fast will oxygen effuse through the same hole under the same conditions?

Solution

Graham's law of effusion is

$$\sqrt{\frac{m_1}{m_2}} = \frac{v_{rms2}}{v_{rms1}}$$

and we assign the masses as $m_1 = 4$ and $m_2 = 32$. Since the rate at which molecules effuse through a pinhole in moles per second is directly proportional to the v_{rms}, we write $v_{rms1} = 3.22 \times 10^{-5}$ mol s^{-1} while v_{rms2} is the unknown:

$$\sqrt{\frac{4}{32}} = \frac{v_{rms2}}{3.22\times10^{-5}\text{ mol s}^{-1}}$$

$$v_{rms2} = 1.14\times10^{-5}\text{ mol s}^{-1}$$

AVERAGE KINETIC ENERGIES AND VELOCITIES

Much experimentation has shown that the average kinetic energy is directly proportional to the temperature of a gas, some gas molecules will have kinetic energies above the average, and some will have kinetic energies below the average. Figure 6.2 shows the distribution of the kinetic energies of gas particles at two different temperatures. It should also be noted from Figure 6.2 that the average kinetic energy is not the same as the most probable kinetic energy. The reason is that the curves in Figure 6.2 are not symmetrical and the point at which half of the molecules have higher kinetic energies and half of the molecules have lower kinetic energies lies to the right of the peak.

Since

$$KE = \frac{1}{2}mv^2$$

Figure 6.2 can also be drawn with the *x*-axis representing the square of the molecular velocity. Some molecules would have velocities above the average, and others would have lower than average velocities.

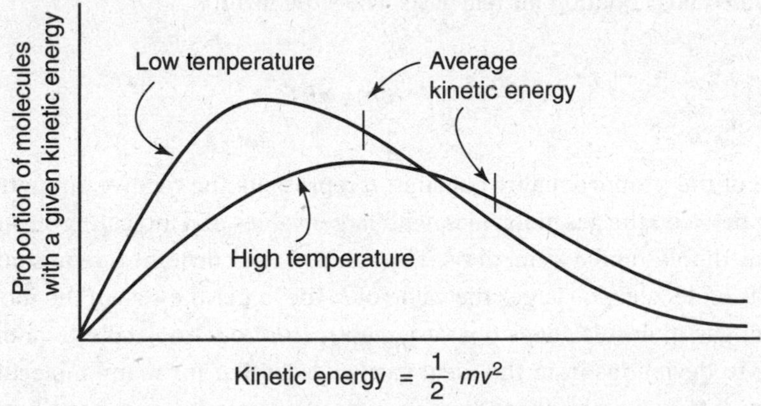

FIGURE 6.2. Kinetic energy distribution diagrams for gas particles at two different temperatures. Note that average kinetic energy is not at the curve maximum. The maximum is the most probable kinetic energy.

REAL GASES

The ideal gas law works very well for most gases; however, the law does not work well for gases under high pressures or gases at very low temperatures. These conditions are the same conditions used to condense gases, and therefore it may be generalized that any gas close to its boiling point (condensation point) will deviate significantly from the ideal gas law. The kinetic molecular theory makes two fundamental assumptions about the properties of gas particles themselves: (1) gases have no volume, and (2) they exhibit no attractive or repulsive forces. These two assumptions define an **ideal gas**. They work quite well, since gas particles are often widely separated and what little volume they have is relatively unimportant. Similarly, because of the large average distance between gas particles the attractive forces and repulsive forces are very weak. As a **real gas** is cooled and/or compressed, the distance between the particles decreases dramatically, and these real volumes and forces can no longer be ignored.

Johannes van der Waals developed a modification of the ideal gas law to deal with the nonideal behavior of real gases. He reasoned that, if the gas particles each occupied some volume, there would be a net decrease in the useful volume of the container. (Imagine a fishbowl filled with marbles. Little useful room would be left for the fish.) This decrease in volume must be proportional to the moles of gas particles present. By using the letter *b* for the proportionality constant, the volume term in the ideal gas law could be replaced by $(V - nb)$.

In evaluating the effect of intermolecular attractions, it was reasoned that, if gas particles attracted each other, they would have curved paths and therefore would take longer to collide with the container walls. The result would be a decreased collision frequency and a lower pressure for the real gas than the pressure of an ideal gas under the same conditions. The frequency of attractions between gas particles would be expected to rise with greater concentrations of the gas. The best correction factor was found to be based on the square of the concentration of the gas, $(n/V)^2$. The proportionality constant was given the symbol *a*, and the entire correction factor was added to the measured pressure $\left[P + a\left(\frac{n}{V}\right)^2\right]$ to obtain the equivalent ideal pressure.

The van der Waals equation for real gases takes the form

$$\left(P + \frac{n^2 a}{V^2}\right)(V - bn) = nRT \qquad (6.15)$$

The value of the proportionality constant *a* represents the relative strength of attractive forces acting between the gas molecules, with larger values of *a* indicating stronger attractive forces such as dipole-dipole attractions. The value of the constant *b* represents the relative size of the gas molecule. The larger the value of *b*, the larger the size of the molecule.

The size of gas molecules does not vary greatly from one molecule to another, and this contribution to deviations from the ideal gas law is similar for many molecules. Attractive forces, however, vary greatly, depending on molecular polarity, and contribute the most to deviations from the ideal gas law.

DALTON'S LAW OF PARTIAL PRESSURES

Dalton's law of partial pressures is based on the fact that, when two gases are mixed together, the gas particles tend to act independently of each other. The result is that, for a mixture of gases, the total pressure is equal to the sum of the pressures of all of the components of the mixture:

$$P_{total} = P_1 + P_2 + \cdots \qquad (6.16)$$

In this equation, the *P* stands for the **partial pressure** of each individual gas. The ellipsis (three dots) at the end of the equation indicates that, if more than two gases are mixed, the equation should be expanded to include the additional components.

EXAMPLE 6.4

A mixture of gases contains 2.00 mol of O_2, 3.00 mol of N_2, and 5.00 mol of He. The total pressure of the mixture is 850 torr. What is the partial pressure of each gas?

Solution

The ideal gas law can be interpreted to mean that the pressure is proportional to the number of moles of gas present when *T* and *V* are constant. In this problem we have a total of 10.00 mol of gases. Since there are 2.00 mol of O_2, 2.00/10.00 of all moles of gas are O_2; therefore, the same ratio applies to the partial pressure of O_2:

$$P_{O_2} = 850 \text{ torr}\left(\frac{2.00 \text{ mol } O_2}{10.00 \text{ mol total}}\right) = 170 \text{ torr}$$

Using similar calculations for N_2 and He gives

$$P_{N_2} = 850 \text{ torr}\left(\frac{3.00 \text{ mol } N_2}{10.00 \text{ mol total}}\right) = 255 \text{ torr}$$

$$P_{He} = 850 \text{ torr}\left(\frac{5.00 \text{ mol He}}{10.00 \text{ mol total}}\right) = 425 \text{ torr}$$

To check the calculations, we determine the total pressure from the three partial pressures:

$$\begin{aligned} P_{total} &= 170 \text{ torr} + 255 \text{ torr} + 425 \text{ torr} \\ &= 850 \text{ torr} \end{aligned}$$

This result agrees with the total pressure given in the problem.

EXPERIMENTS INVOLVING GASES

As a candidate for advanced placement in chemistry, you should be familiar with a variety of experiments for producing, collecting, and manipulating gases in the laboratory. A few of the more familiar reactions for producing gases are these:

$$2KClO_3(s) \rightarrow 2KCl(s) + 3O_2(g) \text{ (with heat and } MnO_2 \text{ as a catalyst)}$$

$$NH_4^+(aq) + OH^-(aq) \rightarrow NH_3(g) + H_2O(l)$$

$$CaCO_3(s) + HCl(aq) \rightarrow CO_2(g) + CaCl_2(aq) + H_2O(l)$$

$$2HCl(aq) + Zn(s) \rightarrow ZnCl_2(aq) + H_2(g)$$

In addition to knowledge of the chemical reactions, the use of a **pneumatic trough** for collecting gas samples over water should be familiar. Figure 6.4 illustrates a pneumatic trough with a bottle full of water submersed. The gas from the chemical reaction bubbles into the bottle, displacing the water as it is collected.

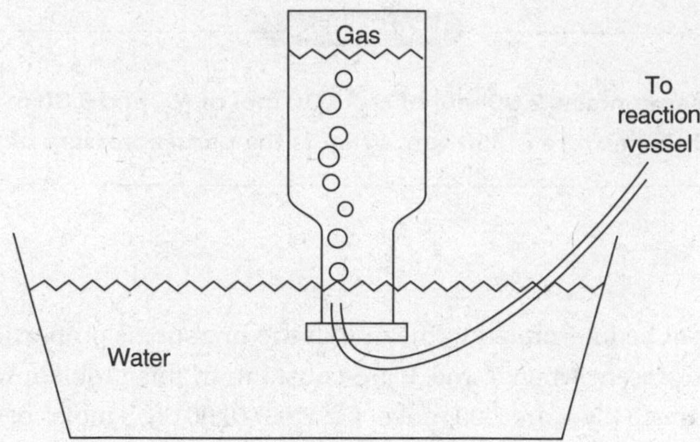

FIGURE 6.4. Collection of gases in a pneumatic trough. When the reaction is complete, the position of the bottle is adjusted so that the water level in the bottle is equal to the water level in the trough. At that point $P_{gas} = P_{atm}$.

The pressure of the gas inside the collecting bottle must be determined in order to solve the ideal gas equation. When the reaction is complete, the gas pressure is determined by moving the bottle in the water until the two liquid levels coincide. At this point the pressure inside the bottle is the same as the barometric pressure outside the bottle.

When gases are collected over water, there will always be some water vapor in the collecting bottle. The pressure of the water vapor depends on only the temperature and is obtained from the appropriate reference table. (The AP chemistry test does not include this table; the information will be given in the problem if needed.) The pressure of the gas that was generated is calculated from Dalton's law of partial pressures as

$$P_{gas} = P_{atm} - P_{water} \tag{6.17}$$

The volume of gas may be determined by marking the jar when the liquid levels are equal and then measuring the jar's volume to the mark at a later time. This is done by filling the jar to the mark with water and then carefully pouring the water into a graduated cylinder to determine the volume.

Temperature is measured with a laboratory thermometer, and atmospheric pressure is determined from a barometer for use in Equation 6.17. These measurements give P, V, and T for the ideal gas law equation, and the amount of gas produced, n, can be calculated.

The *number of moles of gas* is given the symbol "n." This is determined by measuring the mass of the gas and by using the molecular mass to convert it to moles (n = mol = g/molar mass). Measuring the mass of any gas involves evacuating a vessel to a very low pressure using a vacuum pump. The evacuated vessel is weighed and then the gas sample is introduced to the vessel and it is weighed again to determine the increase in mass due to the gas.

EXAMPLE 6.5

A 0.060-g piece of magnesium is placed in hydrochloric acid to generate hydrogen according to the equation

$$Mg(s) + 2HCl(aq) \rightarrow MgCl_2(aq) + H_2(g)$$

The gas is collected in a pneumatic trough at 25°C. A barometer reading of 755 mm Hg is made during the experiment. When bubbles of hydrogen cease forming, the bottle is adjusted to the water level in the trough and the water level is marked on the bottle. Afterwards, 65 mL of water is needed to fill the bottle to the same mark. The vapor pressure of water at 25°C is 23.8 mm Hg. How many moles of hydrogen were produced?

Solution

The volume of the gas is 0.065 L, its temperature is 298 K, and its pressure is calculated from Equation 6.17 as

$$P_{H_2} = 755 \text{ mm Hg} - 23.8 \text{ mm Hg} = 731 \text{ mm Hg}$$

This is converted to 0.962 atm. The ideal gas law is then used to calculate *n*, the number of moles of hydrogen:

$$(0.962 \text{ atm})(0.065 \text{ L}) = n(0.0821 \text{ L atm mol}^{-1} \text{ K}^{-1})(298 \text{ K})$$
$$n = 2.6 \times 10^{-3} \text{ mol } H_2(g)$$

SUMMARY

The chapter on gases introduces the start of the discussion on the states of matter and their properties. Gases all obey the ideal gas law, and various solutions to this equation are demonstrated in the chapter. An ideal gas is defined as one that has no molecular volume and no attractive forces between gas particles. The ideal gas law also allows one method for determining the molar mass of a compound. Gases are an important part of many experiments. This chapter describes in detail how to make measurements on gases and how to collect a gas in the laboratory. The kinetic molecular theory is introduced along with a discussion of how it can be used to explain all the gas laws. The chapter ends with a discussion of how a real gas differs from an ideal gas.

Important Concepts

Ideal gas law and ideal gas
Standard temperature and pressure, STP
Universal gas law constant, *R*
Dalton's law of partial pressures
Kinetic molecular theory
Average kinetic energy
Graham's law of effusion

Important Equations

$$PV = nRT$$

$$\sqrt{\frac{m_1}{m_2}} = \frac{\bar{v}_2}{\bar{v}_1}$$

$$P_{total} = P_1 + P_2 + \cdots$$

PRACTICE EXERCISES

Multiple-Choice

1. Which of the lines on the figure below is the best representation of the relationship between the pressure and volume of a gas when other factors remain constant?

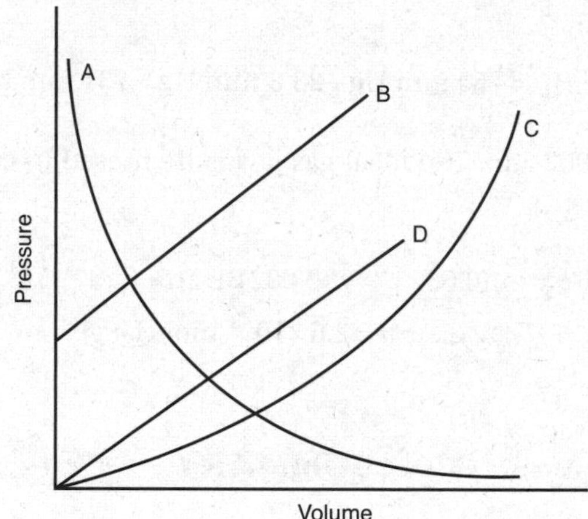

(A) Line *A*
(B) Line *B*
(C) Line *C*
(D) Line *D*

2. If methane, CH_4, rather than ethane, CH_3CH_3 is the gas used under comparable conditions, there will be an increase in
(A) volume and average kinetic energy
(B) temperature
(C) average kinetic energy and pressure
(D) effusion rate

3. The kinetic molecular theory postulates a direct relationship between
   (A) increased volume and increased average kinetic energy
   (B) increased temperature and increased average kinetic energy
   (C) increased average kinetic energy and increased pressure
   (D) increased effusion rate and increased pressure

4. The figure below represents the relative number of gas molecules in separate 2.0 L rigid containers of Cl_2, N_2, and Ne. The pressure of the container with Cl_2 is 6.0 atm. If all three gases are transferred to an evacuated 4.0 L rigid container at constant temperature, what would be the total pressure?

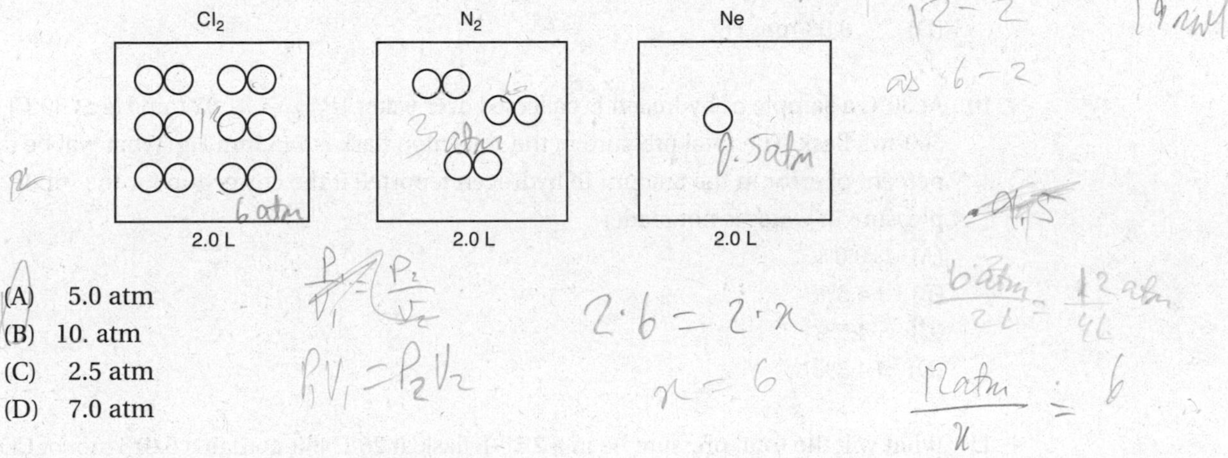

   (A) 5.0 atm
   (B) 10. atm
   (C) 2.5 atm
   (D) 7.0 atm

5. A gas will behave more like an ideal gas if we have an ________ and an ____________.
   (A) increased volume, increased temperature
   (B) increased temperature, increased average kinetic energy
   (C) increased average kinetic energy, increased pressure
   (D) increased effusion rate, increased pressure

6. The measured pressure exerted by CO_2 gas is less than that predicted by the ideal gas equation when at moderate pressures (5 atm at 298 K). This is mainly because
   (A) the attractive intermolecular forces between CO_2 molecules is now a factor
   (B) CO_2 condenses to a liquid at pressures greater than 5 atm at 298 K
   (C) the volume of the CO_2 molecules becomes significant at high pressures
   (D) the gas phase collisions prevent the CO_2 molecules from colliding with the walls of the container

7. How many moles of helium are needed to fill a balloon that has a volume of 6.45 L and a pressure of 800 mm Hg at a room temperature of 24°C? Assume ideal gas behavior.
   (A) 0.288
   (B) 214
   (C) 0.278
   (D) 2.65×10^3

8. If ideal gas behavior is assumed, what is the density of neon at STP?
(A) 1.11 g L^{-1}
(B) 448 g L^{-1}
(C) 0.009 g L^{-1}
(D) 0.901 g L^{-1}

9. A sample of CO has a pressure of 58 mm Hg and a volume of 155 mL. When the CO is quantitatively transferred to an evacuated 1.00-L flask, the pressure of the gas will be
(A) 374 mm Hg
(B) 8990 mm Hg
(C) 111 mm Hg
(D) 8.99 mm Hg

10. At 30°C a sample of hydrogen is collected over water [P_{H_2O} = 31.82 mm Hg at 30°C] in a 500-mL flask. The total pressure in the collection flask is 745 mm Hg. What will be the percent of error in the amount of hydrogen reported if the correction for the vapor pressure of water is not made?
(A) 0.0%
(B) +4.5%
(C) −4.5%
(D) +4.3%

11. What will the total pressure be in a 2.50-L flask at 25°C if it contains 0.016 mol of CO_2 and 0.035 mol of CH_4?
(A) 31.4 mm Hg
(B) 380 mm Hg
(C) 0.041 mm Hg
(D) 935 mm Hg

12. The carbon dioxide from the combustion of 1.50 g of C_2H_6 is collected over water at 25°C. The pressure of CO_2 in the collection flask is 746 mm Hg, and the volume is 2.00 L. How much of the CO_2 formed apparently dissolved in the water of the pneumatic trough?
(A) 0.0814 mol
(B) 0.86 g
(C) 1.79 g
(D) 0.100 mol

The next two questions involve the figures below. The volume of the box containing argon is twice the volume of the box containing neon. Both gases are at the same temperature.

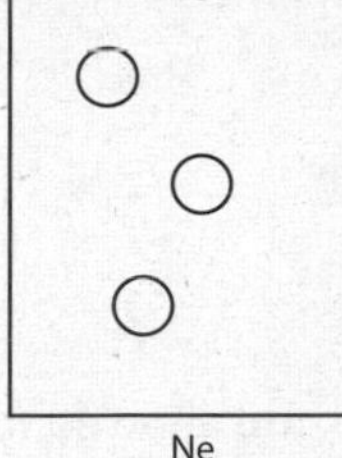

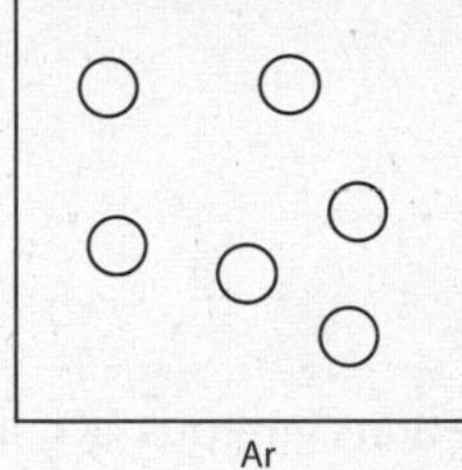

13. When the densities of the two gases were compared, what was found?
    (A) The densities of both are equal.
    (B) Ar is more dense than Ne.
    (C) Ne is more dense than Ar.
    (D) Not enough information is given to answer the question.

14. If the neon was transferred to the box containing the argon, what would be the mole fraction of the neon in the mixture?
    (A) 1/2
    (B) 2/3
    (C) 4/5
    (D) 1/3

---

15. The effect that increasing the temperature has on the pressure may be explained by the kinetic molecular theory as due to
    (A) the increase in force with which the gas molecules collide with the container walls
    (B) the increase in rotational energy of the gas molecules
    (C) the increase in average velocity of the gas molecules, which causes a corresponding increase in the frequency of collisions with the container walls
    (D) a combination of (A) and (C)

16. Hydrogen gas can be prepared by the addition of hydrochloric acid to a sample of zinc. Upon completion of the reaction, 195 mL of gas were collected by water displacement at 25°C and 753 torr. What mass of the hydrogen gas was collected? P_{water} = 24 torr at 25°C.
    (A) 0.00765 g
    (B) 0.0164 g
    (C) 0.159 g
    (D) 0.0154 g

17. Under which conditions will a real gas behave most like an ideal gas?
    (A) Low pressure and high temperature
    (B) Low pressure and low temperature
    (C) Low volume and high temperature
    (D) High pressure and high temperature

18. A compound composed of carbon, hydrogen, and chlorine is allowed to effuse through a pinhole. The rate of effusion for this compound is 0.411 times as fast as neon. What is the correct molecular formula for this compound?
   (A) CH_2Cl_2
   (B) $C_2H_2Cl_2$
   (C) C_2H_3Cl
   (D) $CHCl_3$

19. Under identical conditions gaseous CO_2 and CCl_4 are allowed to diffuse through a pinhole. If the rate of diffusion of the CO_2 is 6.3×10^{-2} mol s^{-1}, what is the rate of diffusion of the CCl_4?
   (A) 6.3×10^{-2} mol s^{-1}
   (B) 2.2×10^{-1} mol s^{-1}
   (C) 1.8×10^{-2} mol s^{-1}
   (D) 3.4×10^{-2} mol s^{-1}

20. A gas has a density, at STP, of 3.48 g L^{-1}. The most reasonable formula for this compound is
   (A) C_2H_6
   (B) C_6H_6
   (C) CCl_4
   (D) CaF_2

21. Assuming all other factors remain constant, which of the following changes will NOT affect the total pressure of a gas in a container?
   (A) Half of the molecules are replaced by an equal number of molecules of a gas with a different molar mass.
   (B) The average velocity of the molecules is lowered.
   (C) The frequency of collisions of molecules with the walls is increased.
   (D) The temperature of the sample is altered.

22. A gas mixture contains twice as many moles of O_2 as N_2. Addition of 0.200 mol of argon to this mixture increases the pressure from 0.800 atm to 1.10 atm. How many moles of O_2 are in the mixture?
   (A) 0.355 mol O_2
   (B) 0.178 mol O_2
   (C) 0.533 mol O_2
   (D) 0.200 mol O_2

23. A gas in a 1.50-L container has a pressure of 245 mm Hg. When the gas is transferred completely to a 350. mL container at the same temperature, the pressure will be
   (A) 1.05 mm Hg
   (B) 1.05 atm
   (C) 2.14 mm Hg
   (D) 1050 mm Hg

24. At STP a 5.00-L flask filled with air has a mass of 543.251 g. The air in the flask is replaced with another gas, and the mass of the flask is then determined to be 566.107 g. The density of air is 1.290 g L^{-1}. What is the gas that replaced the air?
(A) Ne
(B) O_2
(C) Ar
(D) Xe

CHALLENGE

25. What volume of hydrogen gas, at STP, will a 0.100-g sample of magnesium (molar mass = 24.31) produce when reacted with an excess of HCl?

$$Mg + 2HCl \rightarrow MgCl_2 + H_2$$

(A) 92.1 mL
(B) 46.1 mL
(C) 184 mL
(D) 9.2 mL

ANSWER KEY

1. **(A)**	6. **(A)**	11. **(B)**	16. **(D)**	21. **(A)**
2. **(D)**	7. **(C)**	12. **(B)**	17. **(A)**	22. **(A)**
3. **(B)**	8. **(D)**	13. **(B)**	18. **(D)**	23. **(D)**
4. **(A)**	9. **(D)**	14. **(D)**	19. **(D)**	24. **(D)**
5. **(A)**	10. **(B)**	15. **(D)**	20. **(B)**	25. **(A)**

See Appendix 1 for explanations of answers.

Free-Response

Answer the following questions concerning the properties of gases and the theories used to explain their behavior.

(a) Propane gas, C_3H_8, is used extensively as a fuel for gas grills. The complete combustion of this gas in the presence of oxygen produces carbon dioxide gas, water vapor, and heat.

(i) Write a balanced chemical equation for this combustion reaction.

(ii) Assuming the reaction is proceeding in the presence of excess oxygen, how many liters of carbon dioxide will be produced from a 5.0 L tank at 28°C and a pressure of 745 mm Hg?

(iii) In some cases, not enough oxygen is present to yield a complete combustion of the gas. So carbon monoxide is produced. Write a balanced equation for the incomplete combustion of propane.

(b) A student performed the decomposition of potassium chlorate, $KClO_3$, to produce oxygen and potassium chloride. In order to collect the oxygen safely, the student used water displacement.

(i) Write the balanced chemical equation for this reaction.

(ii) How much oxygen gas (in mL) did the student collect if 0.25 g of $KClO_3$ were heated at 25°C and 735 mm Hg? (The vapor pressure of water at 25°C is 23.8 mm Hg.)

(iii) The theoretical volume of oxygen gas calculated by the student for the gas was 76 mL. Is this number higher or lower than the answer for (ii)? Explain the error the student made in his/her calculation.

ANSWERS

(a) (i) The balanced chemical equation for the complete combustion is

$$C_3H_8(g) + 5O_2(g) \rightarrow 3CO_2(g) + 4H_2O(g)$$

(ii) Since the pressure and temperature are constant, the volume of carbon dioxide can be found using Avogadro's law.

$$\frac{V_1}{n_1} = \frac{V_2}{n_2}$$

$$V_2 = \frac{V_1 n_2}{n_1} = \frac{(5.0 \text{ L})(3 \text{ mol})}{(1 \text{ mol})} = 15.0 \text{ L}$$

(iii) The balanced chemical equation for the incomplete combustion is

$$2C_3H(g) + 7O_2(g) \rightarrow 6CO(g) + 8H_2O(g)$$

(b) (i) The balanced chemical equation is

$$2KClO_3(s) \rightarrow 2KCl(s) + 3O_2(g)$$

(ii) First calculate the number of moles of $KClO_3$ present:

$$n(\text{KClO})_3 = 0.25 \text{ g KCLO}_3 \times \frac{1 \text{ mol KClO}_3}{122.5 \text{ g KClO}_3} = 2.04 \times 10^{-3} \text{ mol KClO}_3$$

Then find the number of moles of oxygen using the mole ratio from the equation:

$$n(\text{O}_2) = 2.04 \times 10^{-3} \text{ mol KClO}_3 \times \frac{3 \text{ mol O}_2}{2 \text{ mol KClO}_3} = 3.1 \times 10^{-3} \text{ mol O}_2$$

Since the oxygen collected was done by water displacement, the vapor pressure of water must be subtracted from the total pressure before conversion to atm:

$$P_{O_2} = P_T - P_{H_2O} = 734 \text{ mm Hg} - 23.8 \text{ mm Hg} = 710 \text{ mm Hg} = 0.934 \text{ atm}$$

Using $PV = nRT$, the volume can be calculated:

$$V = \frac{nRT}{P} = \frac{3.1 \times 10^{-3} \text{ mol O}_2 (0.0821 \text{ L-atm/mol-K})(298 \text{ K})}{0.934 \text{ atm}} = 0.081 \text{ L} = 81 \text{ mL}$$

(iii) This value is lower than the calculated value found in (ii). The student neglected to correct for the water vapor pressure since the oxygen was collected over water.

7 Liquids and Solids

- → INTERMOLECULAR FORCES
- → DIPOLE-DIPOLE ATTRACTIONS
- → HYDROGEN BONDING
- → LONDON FORCES
- → DISPERSION FORCES
- → INSTANTANEOUS AND INDUCED DIPOLES
- → SURFACE TENSION
- → VISCOSITY
- → EVAPORATION
- → VAPOR PRESSURE
- → BOILING POINTS
- → HEATS OF VAPORIZATION
- → CRYSTAL TYPES
- → METALLIC CRYSTALS
- → IONIC CRYSTALS
- → MOLECULAR CRYSTALS
- → NETWORK CRYSTALS
- → AMORPHOUS SUBSTANCES
- → PHASE CHANGES
- → HEATING AND COOLING CURVES

BIG IDEAS 2, 3

Learning Objectives: 2.1, 2.3, 2.7, 2.10, 2.11, 2.13, 2.14, 2.16, 2.19, 2.23, 2.24, 2.25, 2.26, 2.27, 2.28, 2.31, 3.1, 3.10, 3.11

For the complete list of Big Ideas and Learning Objectives, refer to the AP Chemistry Course Outline: ***https://secure-media.collegeboard.org/digitalServices/pdf/ap/ap-chemistry-course-and-exam-description.pdf***

COMPARISON OF LIQUIDS AND SOLIDS TO GASES

Liquids and solids are distinctly different from the gases discussed in Chapter 6. First, liquids and solids are much more dense than gases. Inorganic liquids and solids have densities that range from 1 to 8 g cm^{-3}; a few have densities up to 20 g cm^{-3}. Most organic liquids and solids have densities from 0.7 to 2.0 g cm^{-3}. In contrast, gas densities at STP are generally between 10^{-2} and 10^{-4} g cm^{-3}. Second, gases expand to fill all available space and must be kept in an enclosed container, while a liquid fills any container from the bottom up to a level dictated only by the mass of liquid present. Liquids also conform to the shape of the container. Solids maintain their shape without any container. Third, and most important, is the lack of significant attractive forces in gases and the presence of significant attractive forces in liquids and solids. The obvious physical differences between the three states of matter are explained on the basis of these forces.

INTERMOLECULAR FORCES

When considering why gases can be condensed to liquids and why liquids can be solidified, the forces that attract one molecule to another must be understood. There are only a few of these forces. They are presented below.

Dipole-Dipole Attractive Forces

Molecular compounds share electrons in a covalent bond. This electron sharing is rarely equal, particularly between dissimilar elements. Consequently, the electrons may congregate at one end of the molecule, giving it polarity. Polar molecules are also called **dipoles** to remind us that there is only one positive and only one negative end to each molecule. The positive end has a partial positive charge, indicated as δ+. Similarly, the negative end of a molecule is only partially negative and is designated as δ–. Polar molecules are attracted toward each other, with the negative end of one molecule attracted to the positive end of another molecule. It may seem logical that since there are always the same number of δ+ and δ– ends to any molecule that the attractive and repulsive forces would cancel. However, dipoles move away from repulsion orientations and tend to maintain attractive orientations. This results in an overall attraction between dipoles.

One of the simplest dipoles is hydrogen chloride. In Figure 7.1 the electron clouds around the nonpolar H_2 and the polar HCl molecules are compared.

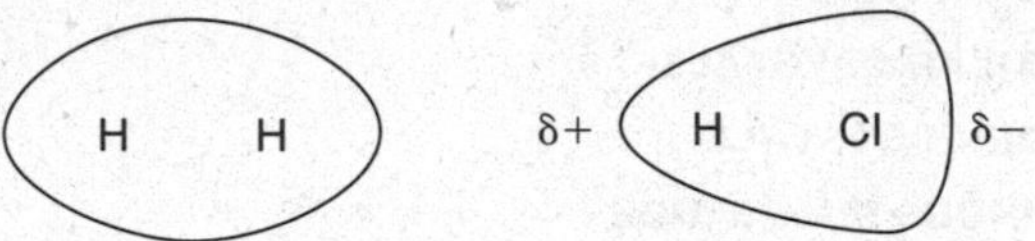

FIGURE 7.1. Representations of the electron clouds around the nonpolar H_2 and the polar HCl molecules.

In the gaseous state, polar molecules show little attraction for each other because they are so far apart (about 3000 pm). The molecules in solids and liquids, however, are approximately ten times closer (about 300 pm). Attractive forces between dipoles may be represented by Equation 7.1:

$$\text{Force} = \frac{(\delta+)(\delta-)}{r^2} \tag{7.1}$$

This equation shows that the attractive force is inversely proportional to the square of the distance, r, between two polar molecules. In gases r is so large that the attractive force is negligibly small. In liquids, where the distance between molecules is much smaller, these forces are significant.

For a gas to become a liquid, the attractive forces must overcome the kinetic energy of the moving gas molecule. Equation 7.1 indicates that decreasing the distance between molecules will increase the attractive force. Increasing the pressure on a gas forces the molecules closer together, and cooling a gas reduces its average kinetic energy. Therefore, decreasing the temperature of a gas and/or increasing the pressure on it will help condense the gas to the liquid phase. The boiling point, which is also the same as the condensation point, is an indication of the attractive forces between molecules since it is a measure of how much the kinetic energy must be increased so that it can overcome the attractive forces in a liquid. Low boiling points indicate low attractive forces, and high boiling points indicate higher attractive forces.

In the condensed state of a liquid the dipole-dipole forces define many of the observed properties. For instance, highly polar molecules have higher boiling points than molecules with lower polarities. The vapor pressure, surface tension, viscosity, and solubilities of liquids are also based on considerations involving attractive forces, as described in the following sections.

London Forces of Attraction

Dipole-dipole interactions are used to explain how and why polar molecules may be condensed to the liquid state. It remained for Fritz London, in 1928, to give a logical explanation of how nonpolar gases develop the forces necessary for condensation. He postulated that nonpolar atoms and molecules may become momentarily polar when an unsymmetrical distribution of their electrons results in the formation of instantaneous dipoles. These instantaneous dipoles provide weak attractive forces in nonpolar substances. **London forces** may also be called **dispersion forces**, **instantaneous dipole forces**, or **induced dipole forces**.

To describe how London forces develop, consider a noble gas such as argon. Previously, argon was described as an atom with 14 electrons arranged in symmetrical orbitals around the 14 protons in its nucleus. The electrons around the argon nucleus are in constant motion. This motion results in a high probability that, at any moment in time, the electrons will not be arranged symmetrically. To illustrate this, several "instantaneous snapshots" of the argon atom are shown in Figure 7.2.

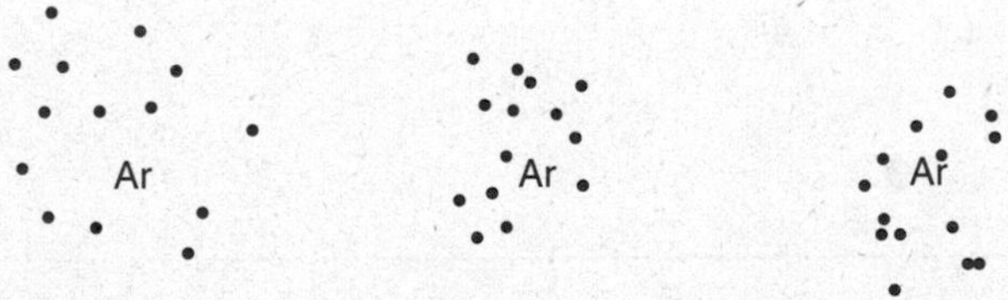

FIGURE 7.2. Random distribution of electrons around an argon nucleus. The first two obviously have more electrons on one side (the upper left quadrant). The third looks symmetrical but a close examination shows more electrons in the lower left quadrant.

When the electrons are not evenly distributed, argon will be a dipole for an instant before the electrons move to new positions. This instantaneous dipole may be attracted to another nearby instantaneous dipole, or it may induce another dipole in a neighboring atom by distorting the neighboring atom's electron cloud. The result is a very weak, attractive force, allowing argon to condense. Since such forces are very weak, argon and the other noble gases have very low boiling points.

The halogens are like the noble gases in having no permanent dipoles. Yet iodine is a solid, and bromine is a liquid, at room temperature and all of the halogens have much higher boiling points than the neighboring noble gases. The explanation for this seeming paradox lies in the **polarizability** of the electron clouds of the halogen molecules. Polarizability refers to the ease with which the electron cloud around an atom or molecule can be deformed into a dipole. Small atoms and molecules, with their electrons tightly held near the nucleus, have a low polarizability. Large atoms or molecules, with many loosely held electrons, have electron clouds with high polarizability. The difference may be visualized by comparing a small, hard golf ball and a large, soft sponge basketball. Since the large electron clouds of bromine and iodine are easily polarized, they have much higher boiling points than the neighboring noble gases.

We can explain the behavior of many molecules on the basis of London forces. For instance, methane, CH_4, is a nonpolar tetrahedral molecule. Instantaneous dipoles are used to explain why methane condenses to a liquid. Ethane, C_2H_6, has a higher boiling point than methane because the six hydrogen atoms may become instantaneous dipoles, resulting in a stronger attractive force. The related propane, C_3H_8, and butane, C_4H_{10}, molecules, have increasingly more hydrogens to form more instantaneous dipoles; therefore, they have higher boiling points. These four compounds are the first four in a series of compounds called the **normal alkanes**. All *n*-alkanes have the general formula C_nH_{2n+2}. Figure 7.3, with the number of carbon atoms on one axis and the boiling point on the other axis, shows that boiling points rise because of increased instantaneous dipole forces.

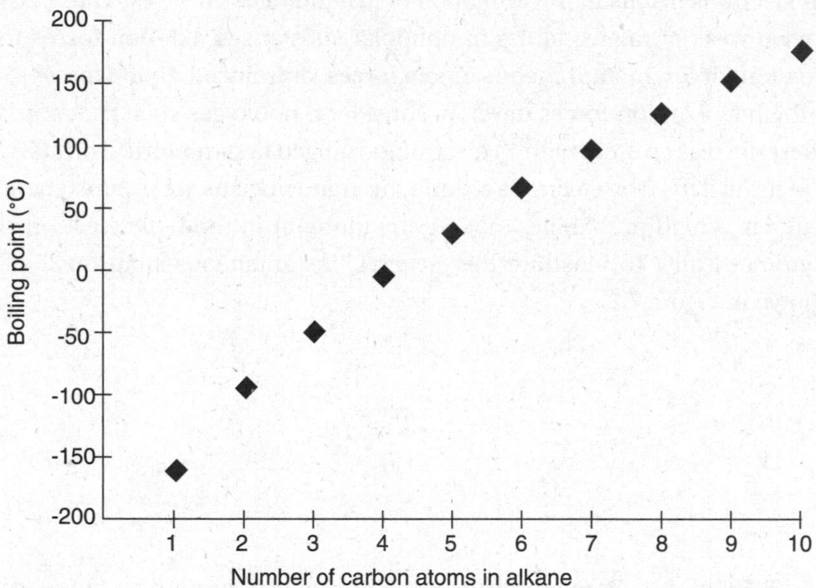

FIGURE 7.3. Plot of boiling points of the normal alkanes versus the number of carbon atoms in each ***n***-alkane. The number of hydrogen atoms is proportional to the number of carbon atoms.

In general, we may conclude that the more electrons in a molecule, the more opportunity there is to form instantaneous dipoles. The result is to increase the attractive forces and to raise the boiling point.

Hydrogen Bonding

Figure 7.3 shows a plot of the boiling points of the *n*-alkanes. These compounds are called a **homologous series** because their formulas vary in a regular fashion. For the *n*-alkanes in the graph, we added one more carbon and two more hydrogen atoms for each compound. Chemists make similar plots of other homologous series of compounds to help visualize trends in physical properties.

Figure 7.4 illustrates such a plot for the hydrogen compounds of the elements in the four groups of the periodic table headed by fluorine, oxygen, nitrogen, and carbon. We see that the compounds headed by carbon all fall on a reasonably straight line, and we conclude that they all act in a similar manner. We see also that the first compound, H_2O, NH_3, and HF, in each of the other three groups in the periodic table has a much greater boiling point than expected based on the boiling points of the other compounds in these groups.

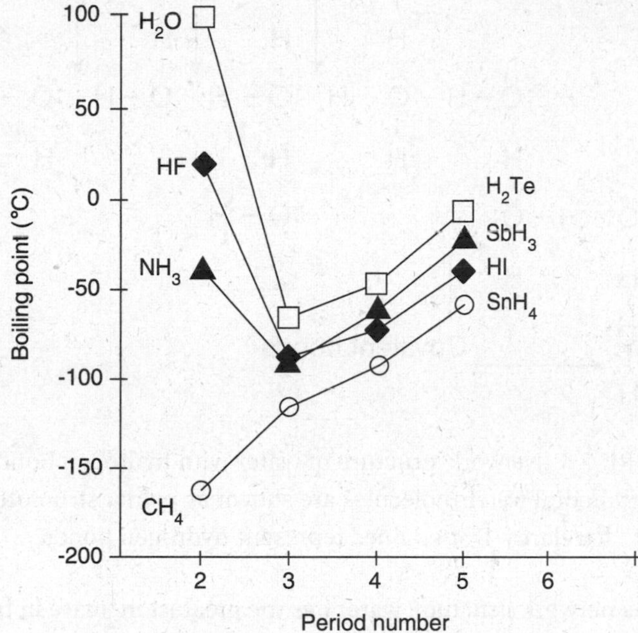

FIGURE 7.4. Plot of the boiling points of the hydrogen compounds in the groups headed by fluorine (HF, HCl, HBr, and HI), oxygen (H_2O, H_2S, H_2Se, H_2Te), nitrogen (NH_3, PH_3, AsH_3, SbH_3), and carbon (CH_4, SiH_4, GeH_4, SnH_4). Only the first and last compounds of each group are shown on the graph.

This behavior may be attributed to the large electronegativity difference (ΔEN) between hydrogen and fluorine, oxygen, and nitrogen. This large ΔEN means that H_2O, NH_3, and HF are very polar molecules with very strong dipole-dipole forces. These extraordinarily large dipole-dipole forces are given a special name, **hydrogen bonds**.

The hydrogen-bonded liquid states of HF, NH_3, and H_2O are somewhat structured. In hydrogen fluoride, the hydrogen of one HF molecule is attracted to the fluorine on another HF molecule. This attraction can extend for many HF units, creating a chainlike structure (Figure 7.5).

$$\cdots H-F\cdots H-F\cdots H-F\cdots H-F\cdots H-F\cdots H-F\cdots H-F\cdots$$

FIGURE 7.5. A structure illustrating the chain structure of the HF hydrogen bonding. The dotted lines indicate hydrogen bonds.

For ammonia also, a chainlike structure can form. Although NH_3 has three hydrogen atoms, it has only one nonbonding pair of electrons on its nitrogen, limiting it to a chainlike structure similar to that of HF. The increase in boiling point over the expected boiling point for both HF and NH_3 is similar, although it is slightly greater for HF since HF has a larger electronegativity difference.

Water is different. It has two hydrogen atoms, which can participate in two hydrogen bonds with neighboring oxygen atoms. In addition, the oxygen atoms have two lone pairs of electrons, which can hydrogen-bond with two hydrogen atoms. Water can form a large network structure, as diagrammed in Figure 7.6.

TIP

Be prepared to draw and interpret diagrams of hydrogen bonding.

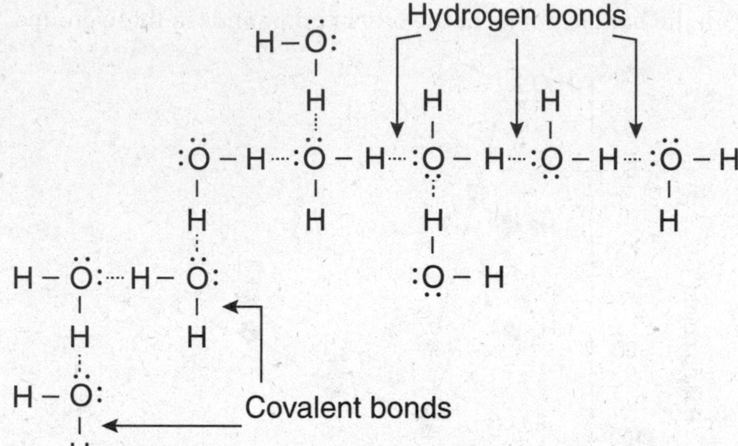

FIGURE 7.6. Network structure of water with hydrogen bonding. Tetrahedral water molecules are shown as planar structures for clarity. Dotted lines represent hydrogen bonds.

As a result of this network structure, water has the greatest increase in boiling point compared to its expected boiling point. The boiling point increase is higher in water than in HF, even though HF has a larger electronegativity difference.

Exercise 7.1

Determine the boiling points that would be expected for HF, H_2O, and NH_3 if hydrogen bonding did not exist.

Solution

Using Figure 7.4, we extrapolate each group to Period 2 and we read the temperature at that point. The approximate answers are HF = –105°C, H_2O = –95°C, NH_3 = –120°C.

Hydrogen bonding is not limited to HF, H_2O, and NH_3. This effect is observed whenever hydrogen is covalently bonded to fluorine, oxygen, or nitrogen. In the case of fluorine, there is only one compound that hydrogen-bonds: HF itself. Many compounds, however, contain the O–H bond. These include alcohols, sugars, organic acids, and phenol-type compounds. In addition to ammonia, primary and secondary amines are nitrogen containing compounds that form hydrogen bonds.

Hydrogen bonding is a phenomenon that has wide-reaching effects. It causes water to be a liquid at the temperatures normally encountered on Earth. It causes solid water, ice, to be less dense than liquid water, so that ice floats. If ice did not float on water, the entire planet would be ice-covered all year, winter, spring, summer, and fall. In addition, it is hydrogen bonding that holds the two strands of the double helix together in DNA, and gives structure to proteins such as hemoglobin and antibodies. Life as we know it is dependent on hydrogen bonding.

Exercise 7.2

Predict the type of intermolecular forces expected for each of the following compounds:

C_6H_6 (benzene), CH_3OH, CH_3NH_2, CF_4, SO_3, XeF_6, $C_{12}H_{26}$, CH_3OCH_3.

Solution

All compounds will have London attractive forces, and they should always be mentioned. In small polar molecules dipole-dipole or hydrogen-bonding attractive forces may predominate. In large organic molecules, London forces can be substantial and be the dominant force of attraction. For the compounds listed, the following are the dominant forces: C_6H_6, London forces; CH_3OH, hydrogen bonding; CH_3NH_2, hydrogen bonding; CF_4, London forces; SO_3, London forces; XeF_6, London forces; $C_{12}H_{26}$, London forces; and CH_3OCH_3, dipole-dipole forces.

SUMMARY OF ATTRACTIVE FORCES

London forces.	These forces (also called dispersion forces, instantaneous dipoles, induced dipoles, etc.) are very weak attractive forces because of the momentary unequal distribution of electrons around an atom.
Dipole-dipole forces.	The attraction between the partial positive end of one dipole and the partial negative end of another dipolar molecule. The molecules can be the same or different substances.
Hydrogen bonding.	Very strong dipole-dipole attractive forces observed exclusively in compounds that have an F, N, or O bonded directly to a hydrogen atom.

Note that some chemists combine all of the above forces under the general term van der Waals forces.

PHYSICAL PROPERTIES OF LIQUIDS

We may use the forces discussed above to describe the liquid state. Some of the physical properties of liquids are surface tension, viscosity, evaporation, vapor pressure, boiling point, and heat of vaporization.

Surface Tension

Surface tension is due to an increase in the attractive forces between molecules at the surface of a liquid compared to the forces between molecules in the center, or bulk, of the liquid. This property causes fluids to minimize their surface areas. As a result, small droplets of liquids tend to form spheres. Surface tension produces a "skin" on a liquid surface that allows small insects to literally stand on water. Also, carefully placed small iron objects such as needles and pins can float on the surface of water even though iron is almost eight times as dense as water.

When considering surface tension, we first look at a molecule of the liquid in the interior of the sample (the bulk solvent). In the interior, the solvent molecule is surrounded by other solvent molecules on all sides as shown in Figure 7.7a. At the interface between the liquid and the gas phase, some of the molecules surrounding the solvent have been removed as in Figure 7.7b. Since potential energy must be added to remove those surrounding molecules, the surface molecules will compensate by attracting neighboring molecules more strongly to reduce that added potential energy. This stronger attraction results in the molecules at the surface being closer to each other and the effect we note as surface tension.

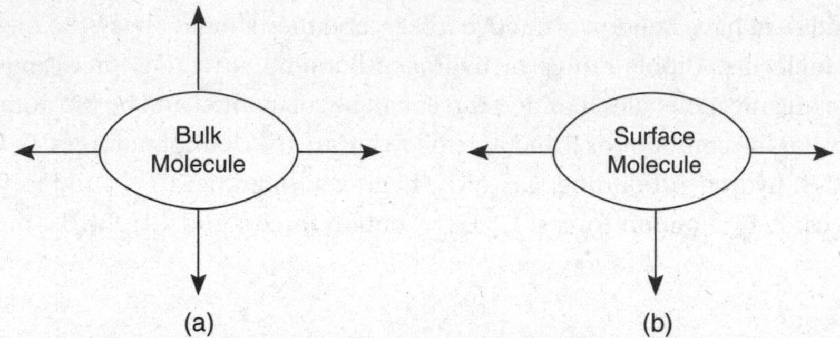

FIGURE 7.7. Diagram illustrating attractive forces for molecules (a) in the bulk of a liquid compared to those at the surface (b). Arrows indicate attractive forces; front and rear arrows are not shown.

Surface tension determines whether or not droplets of a liquid will bead up, as on a freshly waxed car, or spread out when placed on a flat surface. To describe this phenomenon, we define **cohesive forces** as attractions between identical molecules in the liquid, and **adhesive forces** as attractions between different molecules, such as those in the liquid and a flat surface. If the cohesive forces of the liquid are strong compared to the adhesive forces between that liquid and the flat surface, the liquid will retain its shape and form beads. If, however, the adhesive forces between the flat surface and the liquid are strong enough, the liquid will spread uniformly.

Very clean glass has low adhesive forces, and water beads readily on it. In dishwashers these beads of water evaporate, leaving undesired spots on glasses and dishes. For this reason dishwasher detergents contain chemicals called **surfactants**. A surfactant decreases the cohesive forces, and therefore the surface tension, of liquids. The surfactant in dishwasher detergent lowers the cohesive forces of water so that the adhesive forces between water and glass are small. Then water does not bead up on glassware, and the spotting problem is reduced.

Viscosity

Viscosity refers to a liquid's resistance to flow. A liquid such as water has a fairly low viscosity and flows easily. Pancake syrup, on the other hand, has a high viscosity, particularly when cold, and flows slowly. The attractive forces within the liquid are responsible for viscosity. In order for a liquid to flow, the molecules must move past each other. Molecules are able to move more freely in solutions that have relatively low attractive forces. The liquid alkanes have lower viscosities than water because alkanes are attracted to each other only by London forces. Water is more viscous because of hydrogen bonding. Syrup is very viscous since its bulky sugar molecules contain many –OH groups, which hydrogen-bond to the water in the mixture.

Viscosity usually decreases as the temperature of a liquid is increased. Pancake syrup flows much more easily at room temperature than it does when first taken from the refrigerator. The reason is that at the higher temperature the molecules have a higher kinetic energy since $KE = kT$. This increase in kinetic energy weakens the intermolecular forces, thus decreasing the viscosity.

Evaporation

Evaporation is a familiar process in which a liquid in an open container is slowly converted into a gas. Some liquids evaporate more rapidly than others. For example, a beaker of gasoline evaporates in a few hours, whereas a beaker of water may take a day or two. In addition, the rate at which a liquid evaporates increases as the temperature increases.

Evaporation may be explained by considering the attractive forces involved and the kinetic energy needed to overcome these forces. This process is the reverse of condensation. In order for a molecule to be converted from a liquid to a gas, it must have sufficient kinetic energy to overcome its attractive forces. In any group of molecules the average kinetic energy is proportional to the Kelvin temperature. The actual kinetic energies are distributed as shown in Figure 7.8. Some molecules have low, and others have high, kinetic energies. The escape energy is defined as the minimum kinetic energy needed for a molecule to escape from the liquid into the gas phase. All molecules with kinetic energies greater than the escape energy are capable of evaporating.

The area under the curve in Figure 7.8 represents the total number of molecules. The shaded area represents the number of molecules that have kinetic energies equal to or exceeding the escape energy. The ratio of these two areas is a constant as long as the temperature is constant. Molecules that are close enough to the surface and traveling in the correct direction will escape as gas molecules. At constant temperature, the proportion of molecules with enough kinetic energy to escape will remain constant and the liquid will evaporate at a uniform rate until all molecules have entered the gas phase.

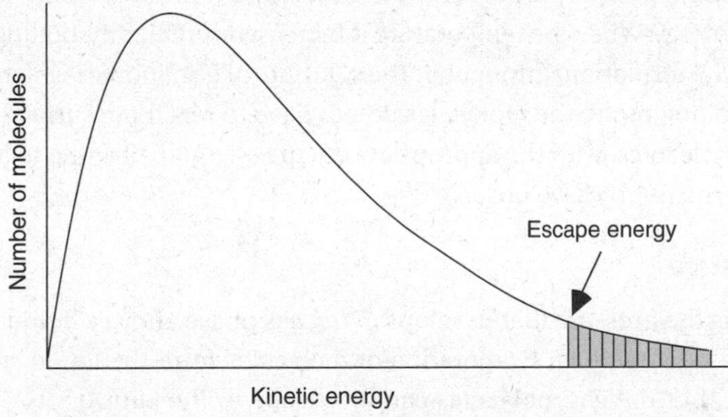

FIGURE 7.8. Distribution of kinetic energies of a group of molecules at a given temperature. Total area under the curve represents all molecules, and shaded area represents molecules with kinetic energies greater than the escape energy.

Since only molecules near the surface may escape into the gas phase, the surface area of the liquid is a factor in the rate of evaporation (Figure 7.9). Ten milliliters of a liquid in a narrow test tube will evaporate more slowly than the same 10 milliliters in a beaker. If the liquid is poured into an evaporating dish, it will evaporate even more quickly. The reason is found in the increased proportion of molecules that are close enough to the surface to escape readily.

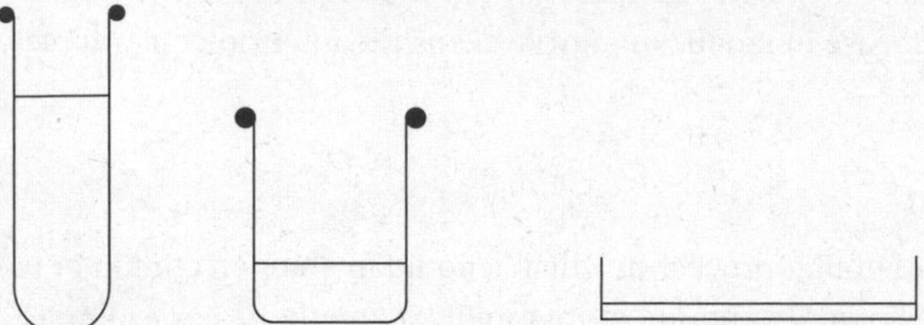

FIGURE 7.9. Surface areas of a test tube, beaker, and evaporating dish, each holding 10 mL of liquid. The increasing surface area indicates that evaporation will be slowest from the test tube and will be fastest from the evaporating dish.

Temperature is another important factor in the evaporation process. An increase in temperature increases the proportion of the molecules that have kinetic energies above the escape energy. Lowering the temperature decreases the proportion with enough kinetic energy to escape.

The temperature of a beaker of liquid evaporating on a lab bench is usually constant because it can absorb heat from its surroundings easily and quickly. If the beaker is insulated from the surroundings, however, the liquid cools and the rate of evaporation decreases. These phenomena are explained using curves similar to the one in Figure 7.8. First, when a molecule escapes from the liquid into the gas phase, the average kinetic energy of the remaining molecules decreases. A decrease in the average kinetic energy means that the temperature also decreases, explaining the cooling observed. Second, as the average kinetic energy decreases, the proportion of molecules that have kinetic energies above the escape energy decreases. Since fewer molecules have enough energy to escape, the rate of evaporation decreases.

Conversely, increasing the temperature of a liquid increases the rate of evaporation. This rise occurs because a greater proportion of all molecules now have a kinetic energy greater than the escape energy. When the temperature is increased sufficiently, boiling occurs. Boiling is recognized as the formation, throughout the solution, of gas bubbles which then rise to the surface. At the boiling point, the molecules do not have to reach the surface to enter the gas phase. Enough molecules, with the appropriate escape energy, can come together within the solution to form the bubbles we observe.

Vapor Pressure

Vapor pressure is the pressure that develops in the gas phase above a liquid when the liquid is placed in a closed container. Evaporation of molecules from the liquid still occurs in the closed container, but the gas molecules cannot escape to the surroundings. As more molecules enter the gas phase, the pressure increases, finally stopping at a level that is dependent only on the temperature. This final pressure is called the vapor pressure.

When a liquid is placed in a closed container, it starts evaporating just as it would in an open beaker. In the closed container, however, gas molecules cannot escape. As the gas molecules move, they collide with the walls of the container, the liquid in the lower part of the container being one of these "walls." When the gas molecules collide with the liquid, very few bounce off; almost all condense to the liquid state again. Initially the rate at which the molecules evaporate is much greater than the rate at which they condense. As the gas molecules increase in number, they collide with the liquid surface more frequently. Eventually the rate at which the liquid molecules evaporate is equal to the rate at which the gas molecules condense. Under these conditions the liquid and gas are said to be in **dynamic equilibrium**. Figure 7.10 illustrates this process.

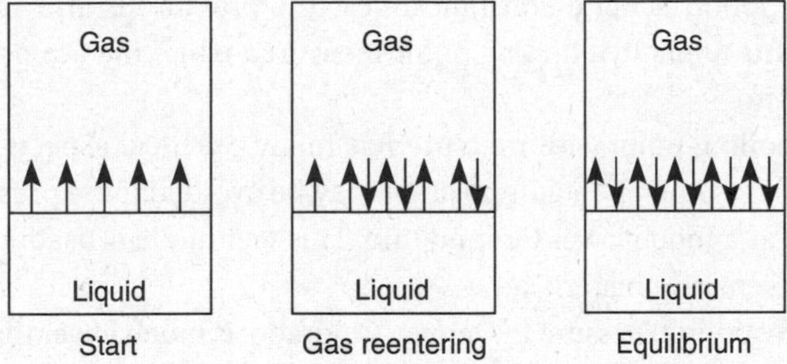

FIGURE 7.10. Diagrams illustrating the establishment of the equilibrium vapor pressure. At the start, molecules leave the liquid into empty space. The middle frame shows some gas reentering the liquid but not as rapidly as molecules leave. At equilibrium, molecules leave and enter the liquid at the same rate.

At equilibrium, the rate at which molecules leave the liquid must equal the rate at which they reenter the liquid. The rate at which molecules escape the liquid (evaporate) depends on the temperature. The rate at which they enter the liquid (condense) depends on the frequency at which the gas molecules collide with the liquid "wall" of the container. In Chapter 7 on gases, it was shown that the frequency of collision is part of the definition of gas pressure. As a result, the vapor pressure depends only on the nature of the liquid (attractive forces) and the temperature (kinetic energy). As the temperature increases, the vapor pressure increases as shown in Figure 7.11.

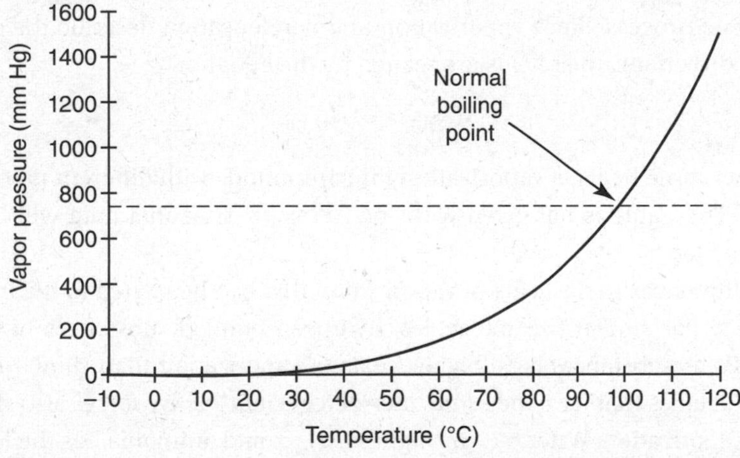

FIGURE 7.11. Vapor pressure curve for water. Dashed line at 760 mm Hg intersects the curve at 100°C, the normal boiling point of water.

A vapor pressure curve such as the one shown in Figure 7.11, or the tabular form of the data, may be used to determine the vapor pressure of a liquid at any temperature. Therefore chemists do not have to repeatedly determine the vapor pressure; it may be looked up in a convenient source.

Vapor pressure data point to some interesting facts about the boiling process.

Boiling Point

Boiling occurs when the vapor pressure of a liquid is equal to the prevailing atmospheric pressure around that liquid. The vapor pressure curve shows that the temperature at which a liquid boils can vary greatly with changes in the atmospheric pressure. Therefore the boiling point of a compound is not a constant unless the pressure is also specified. The term **normal boiling point** refers to a boiling point measured when the atmospheric pressure is 760 mm Hg (1.00 atm).

The change in boiling point with pressure has many practical aspects. Suppose a liquid decomposes instead of boiling. Decomposition may be avoided if the pressure is reduced so that boiling occurs at a much lower temperature. This technique is used in **vacuum distillation** to purify heat-sensitive materials.

The normal atmospheric pressure in Denver, Colorado, is much lower than 760 millimeters of mercury because of the mile-high altitude of the city. The result is that water boils at a lower temperature. At high elevations, longer heating times are needed to cook food properly. For this reason, many people in Denver (and elsewhere) use pressure cookers. These covered pots increase the pressure and therefore the boiling point of water. At the increased temperature, foods cook faster.

Heat of Vaporization

The **heat of vaporization** is the energy needed to convert 1 gram of liquid into 1 gram of gas at a temperature equal to the normal boiling point of the liquid. The units for the heat of vaporization are joules per gram (J g^{-1}). If 1 mole of liquid is vaporized, we call the energy the molar heat of vaporization and we use the units joules per mole (J mol^{-1}). In either case the symbol is ΔH_{vap}. Since energy must always be added to a liquid to cause it to vaporize, ΔH_{vap} is always positive. In Chapter 12 on thermodynamics, a positive ΔH_{vap} is defined as indicating an endothermic process. The reverse process, condensation, requires the gas to give off heat in an exothermic process. Since vaporization and condensation describe the same process from different directions, their heats are related by the equation

$$\Delta H_{vap} = -\Delta H_{cond} \tag{7.2}$$

Table 7.1 lists some heats of vaporization for compounds with different types of intermolecular forces. This table is not given with the AP exam. Essential data will be given with problems as needed.

There are differences in the heats of vaporization that can be related to the intermolecular attractive forces. For similar-size molecules, hydrogen-bonded substances have the largest ΔH_{vap} values. Polar substances have higher heats of vaporization than similar-size nonpolar substances. Molecules that have the same intermolecular attractive forces also show trends in their heats of vaporization. Water has the highest ΔH_{vap}, and ammonia has the lowest, among hydrogen-bonded molecules. Fluorine, chlorine, and bromine show a regular increase in ΔH_{vap}, and methane, ethane, and propane also have increasing ΔH_{vap} values because of increasing London forces.

The amount of heat required to vaporize a liquid is very large. For example, the heat energy needed to vaporize 1 gram of water could be used to raise the temperature of six times as much water from zero to 100°C. This fact explains why water can be quickly raised to its boiling point, but a long time is needed to boil away all of the water.

TABLE 7.1
Heats of Vaporization of Representative Compounds

Compound	Formula	Heat of Vaporization ($kJ\text{-}mol^{-1}$)	Attractive Force
water	H_2O	+43.9	hydrogen bonding
ammonia	NH_3	+21.7	hydrogen bonding
hydrogen fluoride	HF	+30.2	hydrogen bonding
hydrogen chloride	HCl	+15.6	dipole-dipole
hydrogen sulfide	H_2S	+18.8	dipole-dipole
fluorine	F_2	+5.9	London
chlorine	Cl_2	+10.0	London
bromine	Br_2	+15.0	London
methane	CH_4	+8.2	London
ethane	C_2H_6	+15.1	London
propane	C_3H_8	+16.9	London

Exercise 7.3

For each of the following pairs, predict which compound will have (1) the lower boiling point, (2) the higher heat of vaporization, (3) the higher evaporation rate, and (4) the lower vapor pressure.

(a) C_6H_{14} or C_8H_{18}
(b) C_6H_{14} or $C_6H_{13}OH$
(c) HF or HCl
(d) HBr or HCl
(e) $C_6H_{13}OH$ or $C_3H_7OC_3H_7$
(f) PBr_3 or PBr_5

Solution

For these pairs we have to assess the relative attractive forces. The one with the greater attractive forces will have the higher boiling point, higher heat of vaporization, lower evaporation rate, and lower vapor pressure. For the six pairs, the substances with the greater attractive forces are as follows: (a) C_8H_{18} since it has more hydrogen atoms for greater London forces; (b) $C_6H_{13}OH$ since it has the –OH group, which forms hydrogen bonds; (c) HF since it forms hydrogen bonds; (d) HBr since its electron cloud is more polarizable; (e) $C_6H_{13}OH$ since it forms hydrogen bonds; (f) PBr_5 since it has more polarizable bromine atoms (the polarity of PBr_3 is a minor factor). Once the attractive forces are determined, the answers can be obtained:

(a)
1. C_6H_{14}
2. C_8H_{18}
3. C_6H_{14}
4. C_8H_{18}

(b)
1. C_6H_{14}
2. $C_6H_{13}OH$
3. C_6H_{14}
4. $C_6H_{13}OH$

(c)
1. HCl
2. HF
3. HCl
4. HF

(d)
1. HCl
2. HBr
3. HCl
4. HBr

(e)
1. $C_3H_7OC_3H_7$
2. $C_6H_{13}OH$
3. $C_3H_7OC_3H_7$
4. $C_6H_{13}OH$

(f)
1. PBr_3
2. PBr_5
3. PBr_3
4. PBr_5

SOLIDS

At room temperature and atmospheric pressure, many substances exist as solids. In the periodic table, there are two liquids and 11 gases; the remaining 96 elements are solids. Solids have the property of retaining their shapes with or without a container. This occurs because solids have rigid crystal structures. These solid structures may be defined based on the attractive forces that hold them together or on the arrangement of the atoms in the crystals themselves.

Crystal Types Based on Attractive Forces

METALLIC CRYSTALS

All metals in the periodic table are solids at 25°C, except mercury. The **metallic crystal** is visualized as a rigid structure of metal nuclei and inner electrons. The valence electrons are thought to be very mobile in the structure, moving freely from atom to atom. These mobile electrons act to bond metal atoms together with widely varying degrees of force. Metals such as iron, chromium, cobalt, gold, platinum, and copper have melting points above 1000°C. Others, for example, mercury and gallium, have melting points near or below room temperature. Melting points are one measure of the attractive forces since melting disrupts the crystal bonding, producing a liquid. The energy needed to disrupt a crystal is often called the lattice energy.

The mobile valence electrons provide an explanation for the ability of metals to conduct electricity and heat. In both cases, the electrons can quickly carry charge (electricity) and thermal energy (heat) throughout the metal. Also, the interaction of light with these electrons is responsible for the characteristic metallic luster. Most metals have a color similar to that of silver or aluminum. A few, notably copper and gold, are yellow.

Metals such as lead, gold, sodium, and potassium are soft and can be cut with a knife. Other metals, for example, tin and zinc, are somewhat brittle. Most metals are malleable and can be formed into various shapes with a hammer or extruded into thin wires. These properties are due to the metallic crystal structure, which allows the atoms to move from one position to another without a major disruption of the crystal. The softness, hardness, and brittleness of metals can be altered by preparing solutions of one metal dissolved in another. These solutions are known as alloys.

Alloys can be classified as two types: substitutional or interstitial. When the metal atoms are of similar size, one of the metal atoms will take the place of the other in the metallic crystal. This forms a substitutional alloy. When one of the atoms is very small compared with the other, the small atom fills the spaces between the larger atoms (the interstices). The alloy is called an interstitial alloy.

Substitutional alloys, such as silver alloyed with gold, replace one set of attractive forces with an almost equal set of attractive forces with the added metal. The result is that substitutional alloys tend to have properties somewhere between the properties of the two metals. The silver-gold alloy will tend to be soft and ductile with a density that can be estimated as the average of the two constituents.

Interstitial alloys incorporate one atom into the existing structure with little change in volume. The added mass with a small change in volume results in an increase in density. This type of alloy also increases the total attractive forces in the alloy since new attractions are made to the smaller atom while the attractive forces among the larger atoms are only slightly disrupted. The result is usually an alloy that is stronger and harder than the original materials. The properties of stainless steel (iron and carbon) can be explained in this manner.

IONIC CRYSTALS

The attraction of a cation (positive ion) toward an anion (negative ion) is the strongest attractive force known in chemistry. The result is that almost all ionic compounds are solids with rigid crystalline structures (lattices). Because of these strong attractions, a large amount of energy, called the lattice energy, is required to separate the ions. The high lattice energy of ionic compounds gives them very high melting and boiling points compared to molecular compounds of similar size and molar mass. For example, sodium chloride melts at 801°C and boils at 1413°C, while butane (C_4H_{10}, molar mass = 58) melts at –135°C and boils at approximately 0°C. The melting points of ionic crystals are consistently high in contrast to the variability evident in the metals.

An **ionic crystal** has a regular structure, or lattice, of alternating positive and negative ions. Many of these crystals are cubic structures, which will be described later. Other structures may provide shapes seen in many natural minerals. It is relatively simple to describe the crystal structures of the metals since all of the atoms are the same size. Ionic crystals, however, usually have ions of different sizes, which affect the manner in which they pack. These sizes also may limit the closeness with which the ions approach each other.

Coulomb's law, $\frac{q_1 q_2}{r}$, indicates that the strength of the attractive forces in an ionic crystal will depend directly on the charges on the ions, q, and inversely on the distance between them, r. These forces tend to be rather large. Ionic bonding causes these crystals to be rigid and brittle. To understand this property, we visualize a simple ionic substance such as NaCl with alternating sodium cations and chloride anions in a crystal lattice. The strong attraction of the positive and negative charges holds the crystal rigidly together. Hitting an ionic crystal with a hammer has a very different result compared to hitting a metallic crystal with the same force. In both, the atoms can be forced to move. In a metallic crystal, the atoms shift their positions but the metallic bond is not disrupted. In an ionic crystal, however, movement of the atoms by as little as one ionic diameter will cause positive ions to be aligned with positive ions and negative ions to be aligned with negative ions. The repulsion between like-charged ions is so great that the crystal shatters, as diagrammed in Figure 7.12.

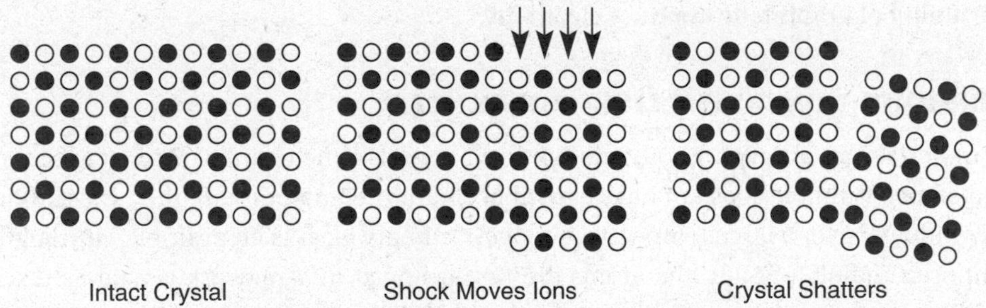

FIGURE 7.12. Illustration of why ionic crystals shatter.
Light circles are cations and dark circles are anions.

MOLECULAR CRYSTALS

Molecular crystals may be composed of either atoms of the nonmetals or of covalent molecules. These crystals are held together by London forces, dipole-dipole attractions, hydrogen bonding, or a mixture of these. All of these forces are much weaker than the attractive forces between ions in ionic crystals. As a result, molecular crystals tend to be soft, with low melting points. Some substances that form molecular crystals with London forces holding the crystal together are neon, xenon, CO_2, sulfur, fluorine, methane (CH_4), and decane ($C_{10}H_{22}$). Dipole-dipole attractive forces hold crystals of SO_2, $CHCl_3$, and other polar molecules together. Hydrogen bonding is responsible for the attractive forces in crystals of H_2O and NH_3. In many molecules there may be a combination of attractive forces. For example, *n*-decanol has the structure

$$CH_3CH_2CH_2CH_2CH_2CH_2CH_2CH_2CH_2CH_2OH$$

The long carbon chain is responsible for London forces, while the –OH at the end of the molecule provides hydrogen bonding.

NETWORK (COVALENT) CRYSTALS

A **network crystal** has a lattice structure in which the atoms are covalently bonded to each other. The result is that the crystal is one large molecule with a continuous network of covalent bonds. A diamond is pure carbon with each carbon atom covalently bonded to four other carbon atoms in a tetrahedral (sp^3) geometry. The totality of this network of covalent bonds makes the diamond the hardest natural substance known. SiO_2 is the empirical formula for sand and quartz. Silicon dioxide forms a covalent crystal with each silicon forming bonds to four oxygen atoms and each oxygen bonding to two silicon atoms with a tetrahedral geometry. Silicon carbide is another network crystal similar to diamond with alternating tetrahedral silicon and carbon atoms. It is very hard and is used as an industrial substitute for diamonds. Network crystals, like ionic substances, are represented by their empirical formulas.

Graphite is another form (allotrope) of carbon in a covalent crystal. In graphite each carbon atom is covalently bonded to three other carbon atoms in a trigonal planar (sp^2) geometry that gives graphite its structure of flat sheets. The extra *p* electron that is not used in the sp^2 bonding holds these sheets together in a manner similar to that seen in a metallic crystal. The weak bonding of the *p* electrons allows the flat sheets to slide over each other easily and is responsible for the slippery feel of graphite. In addition, these *p* electrons are responsible for the ability of graphite to conduct electricity.

AMORPHOUS (NONCRYSTALLINE) SUBSTANCES

Some materials are **amorphous** and do not form crystals. One characteristic of a noncrystalline substance is that it does not have a distinct, sharp melting point. Rather, these materials soften gradually over a large temperature range. Ordinary glass is an example. Although glass is composed mainly of SiO_2, the atoms are not arranged in a network crystal as discussed above. Glass has often been described as a supercooled liquid. Many plastics (polymers) have combined characteristics; they are partially crystalline and partially amorphous.

PHASE CHANGES

Solids, liquids, and gases can be converted from one phase to another by temperature and pressure changes, and we can make qualitative and quantitative observations about these conversions. There are two ways to represent these changes: the heating or cooling curve and the phase diagram. If the pressure is held constant, the effect of heat may be explained by a heating or cooling curve. If we are interested in the effects of both temperature and pressure, a phase diagram is used.

Heating and Cooling Curves

Starting with a solid material well below its melting point and adding heat at a constant rate will produce the following effects:

1. The temperature of the solid will increase at a constant rate until the solid starts to melt.
2. When melting begins, the temperature stops rising and remains constant until all of the solid is converted into a liquid.
3. The temperature of the liquid starts increasing at a constant rate until boiling starts.
4. When boiling begins, the temperature stops rising and remains constant until all of the liquid has been converted into gas.
5. The temperature of the gas increases at a constant rate.

This process is often summarized in a **heating curve** as shown in Figure 7.13.

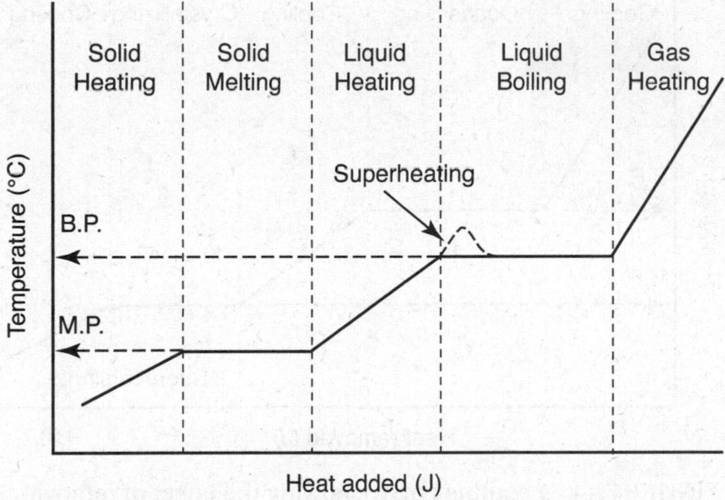

FIGURE 7.13. Typical heating curve, bringing a sample from the solid state on the left to the gaseous state on the right.

In a heating curve there are several points of interest. Adding heat to a pure solid, liquid, or gas phase increases its temperature. The **heat capacity** (J °C^{-1}) of a solid, liquid, or gas is the reciprocal of the slope of the curve (change in joules divided by the change in temperature) in those regions where the temperature increases as heat is added. The **specific heat** (J g^{-1} °C^{-1}) is the heat capacity divided by the number of grams of sample used. The plateaus represent the melting and boiling processes. The temperature does not change during melting and boiling, and two phases are present. Superheating may occur where the temperature of the liquid exceeds the boiling point. See dashed line in Figure 7.13. During melting both the solid and the liquid are in equilibrium; during boiling both the liquid and the gas phase are in equilibrium. The **molar heat of fusion** (melting) may be determined as the length of the first (solid-melting) plateau, which represents the heat added, divided by the number of moles of sample. The **molar heat of vaporization** is the length of the second (liquid-boiling) plateau divided by the number of moles of sample. Both the heat of fusion and the heat of vaporization have units of joules per mole (J mol^{-1}). These are also called the enthalpy of fusion and enthalpy of vaporization, respectively.

The **cooling curve** is the reverse of the heating curve, as shown in Figure 7.14. All of the features are the same except that we start with a gas and we end with a solid as the heat is removed from the sample. The gas condenses to a liquid at the same temperature at which the liquid boils, and the liquid crystallizes at the same temperature at which the solid melts. The terms **condensation point** and **crystallization point** are sometimes used in place of boiling point and melting point.

Some liquids exhibit an ability to be supercooled. **Supercooling** is observed when a liquid that is cooled to a temperature below its melting point remains a liquid. A supercooled liquid is described as being in a metastable state. A metastable liquid will crystallize rapidly if sufficiently disturbed by shaking or by adding a seed crystal to initiate crystallization. In Figure 7.14 the dashed line on the curve shows the alternative path taken by a substance that supercools. Gases do not supercool as some liquids can.

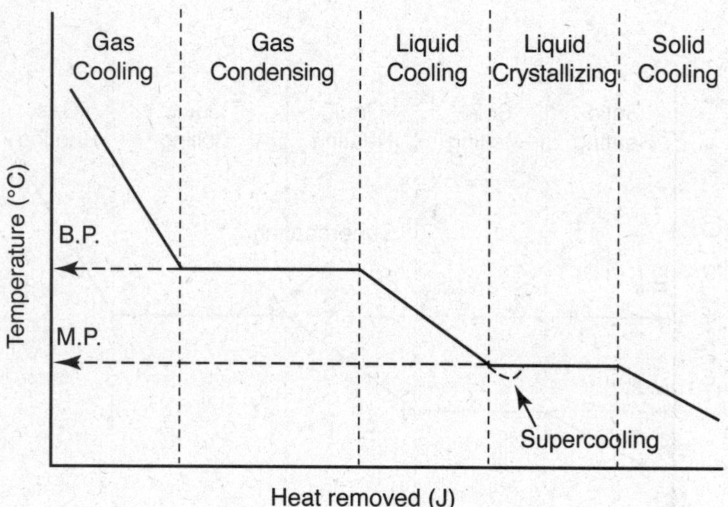

FIGURE 7.14. A cooling curve, showing the effect of removing heat from a gas to form first a liquid and then a solid. The dashed line shows an alternative path for a liquid that supercools.

Exercise 7.4

List all of the information that can be obtained from a heating curve.

Solution

We can obtain the following data from a heating curve: (1) melting point; (2) boiling point; (3) specific heats and heat capacities of the solid, liquid, and gas; (4) heat of fusion of the solid; (5) heat of vaporization of the liquid.

SUMMARY

When molecules occupy most of the available space and there are relatively strong attractive forces, we get the condensed phases, liquids and solids, described in this chapter. Intermolecular forces including hydrogen bonding, dipole-dipole forces, and the very weak London forces that arise from instantaneous dipoles are described and used to explain many observed properties of matter. Melting points, boiling points, viscosity, vapor pressure, and the behavior of liquids in contact with solids are all understandable based on knowledge of these few intermolecular forces. In turn, the forces that influence the properties above are easily understood when we know the molecular structure and polarity that were discussed in Chapter 4. Methods to describe the phases of matter include heating curves, cooling curves, and vapor pressure diagrams.

Important Concepts

Intermolecular forces, dipole-dipole, London and hydrogen bonds
Physical properties and intermolecular forces
Solid crystals and crystal structures
Heating and cooling curves

PRACTICE EXERCISES

Multiple-Choice

1. In which of the following are the intermolecular forces listed from the weakest to the strongest?
   (A) Dipole-dipole > London > hydrogen bonds
   (B) London < dipole-dipole < hydrogen bonds
   (C) Hydrogen bonds < dipole-dipole < London
   (D) London > hydrogen bonds > dipole-dipole

2. Which of the following consistently have the highest melting points and why?
   (A) Metals because the conducting electrons bind metal atoms very strongly using Coulomb forces
   (B) Salts since they have very strong attractions between ions
   (C) Molecular crystals because additional bonding between molecules takes place
   (D) Molecules having resonance structures since they can shift forms instead of breaking

3. Which of the following is expected to have the greatest surface tension?
   (A) $CH_3CH_2CH_2OH$
   (B) $CH_3CH_2CH_2CH_2CH_3$
   (C) $CH_3CH_2OCH_2CH_3$
   (D) $HOCH_2CH_2OH$

4. What is the most reasonable interpretation of the figure below?

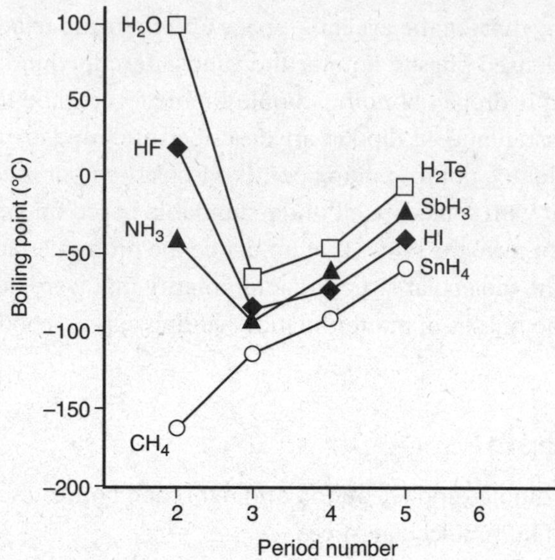

(A) Based on the numbers on the axes, this is a plot of boiling points of some compounds.
(B) The figure shows that H_2O, NH_3, and HF have unusually high boiling points because the lines on the graph of similar compounds are not extended.
(C) The figure indicates that the effect of polarizability on smaller molecules is the largest.
(D) The figure illustrates experimental error where the boiling point of CH_4 must be an error.

5. Which of the following compounds will NOT form hydrogen-bonds?
(A) CH_2F_2
(B) CH_3OH
(C) $H_2NCH_2CH_2CH_3$
(D) $HOCH_2CH_2OH$

6. When 0.100 moles of each of the following compounds are kept at the same temperature in identical open jars in a hood, what is the first compound to evaporate completely?
(A) $CH_3CH_2CH_2CH_2OCH_2CH_2CH_2CH_3$
(B) $CH_3CH_2CH_2CH_2CH_2CH_2CH_2CH_2OH$
(C) $CH_3CH_2CH_2CH_2CH_2CH_2CH_2CH_2NH_2$
(D) $CH_3CH_2CH_2CH_2CH_2CH_2CH_2CH_3$

7. What does this graph tell you?

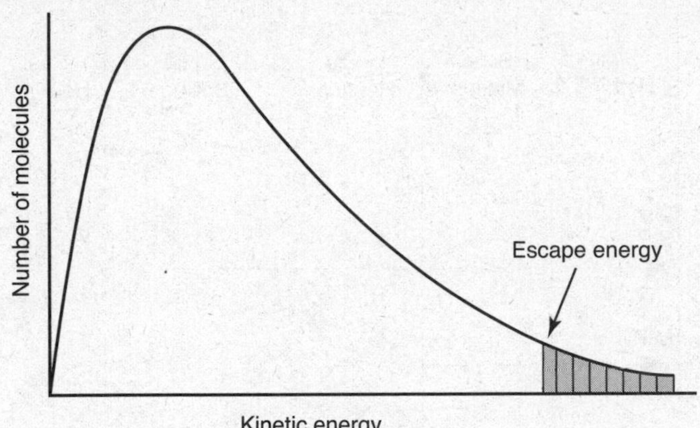

(A) Only a small fraction of all molecules of this substance has enough kinetic energy to escape into the gas phase.
(B) The kinetic energy depends on the number of molecules.
(C) A specific kinetic energy is required to escape Earth's gravitational pull.
(D) The energy needed to vaporize molecules is only a small fraction of all of the energy of the molecules based on the relative areas of the shaded and unshaded areas of this graph.

8. Which of the following statements is NOT consistent with the crystal properties of the substance?
(A) SiC is used to grind metal parts to shape.
(B) Tungsten is drawn into thin wires.
(C) Aluminum is used to cut glass.
(D) Graphite is used to lubricate locks.

9. What causes $C_{30}H_{62}$ to be a nonpolar compound that is a solid at room temperature?
(A) Ionic attractive forces
(B) Dipole-dipole attractions
(C) London forces
(D) The fact that the $C_{30}H_{62}$ is a very heavy molecule

Questions 10–11 refer to the heating curve below.

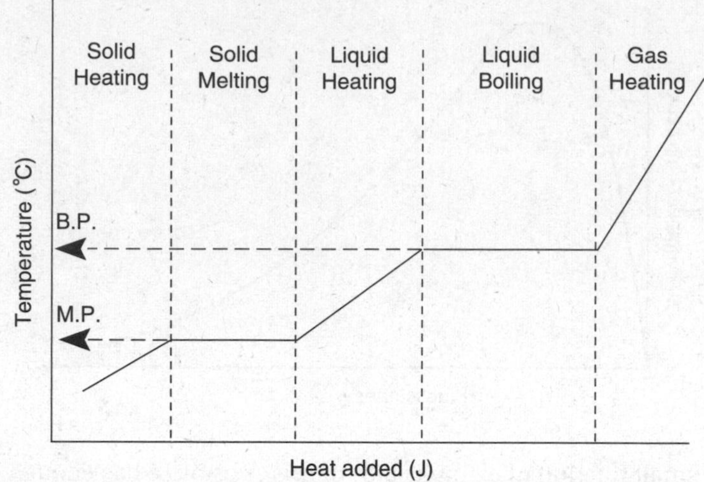

10. On the basis of this heating curve, which of the following statements is true?
    (A) The heat of fusion is less than the heat of vaporization.
    (B) The heat capacities of the solid, liquid, and gas are approximately equal.
    (C) The heat capacity of the gas is greater than that of the liquid.
    (D) The heat capacity of the gas is greater than the heat of fusion.

11. On the basis of this heating curve, which of the following statements is correct about the substance?
    (A) The substance supercools easily.
    (B) The gas is a metastable state.
    (C) The substance must be a salt that dissociates on heating.
    (D) The specific heat can be determined if the mass is known.

CHALLENGE

12. A student measured the temperature of small amount of acetone with a temperature probe interfaced to her computer and obtained a straight line that was the same as room temperature (see below). However, when the probe was removed from the acetone, the student noticed a small dip in the temperature line and that it then returned to room temperature. What is the best explanation of that observation?

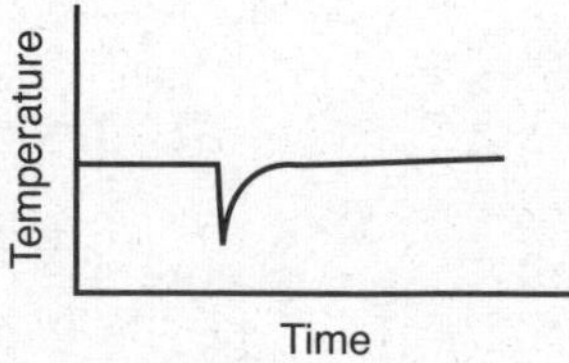

    (A) All organic compounds do this.
    (B) Acetone has a low viscosity and transfers heat quanta better than the probe.
    (C) Acetone has a high heat capacity, therefore it has more heat to lose.
    (D) The high vapor pressure of acetone results in rapid evaporation and heat loss until all the acetone evaporates, and then the temperature returns to normal.

13. When a small beaker of ethanol boils on a hot plate, small bubbles form at the bottom of the beaker and rise to the surface. An inverted test tube full of ethanol is held in the boiling ethanol to catch the bubbles by displacement. Which of the following best describes the vapor that is captured?
    (A) When the vapor is condensed, the liquid burns easily.
    (B) When a burning match approaches the test tube, the gas explodes.
    (C) When a glowing splint (thin piece of wood) is inserted in the test tube, the splint bursts into flame.
    (D) All of the above.

14. Which of the following is expected to have the lowest vapor pressure?
    (A) $C_6H_{13}OH$ because of hydrogen bonding
    (B) C_6H_{14} because of the extensive number of points of attraction via London forces
    (C) $C_3H_7OC_3H_7$ because the oxygen in the center polarizes this molecule
    (D) C_6H_{12} because of the extensive number of points of attraction via London forces

15. Which of the following representations is best for communicating the idea about dynamic equilibrium and why?

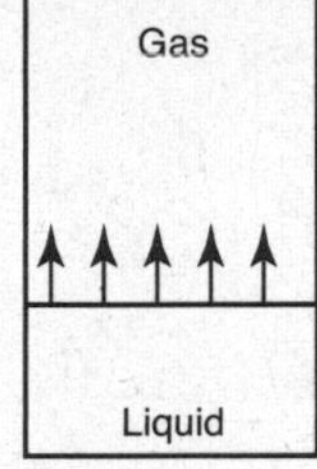

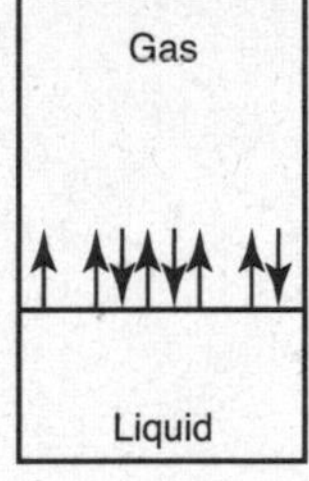

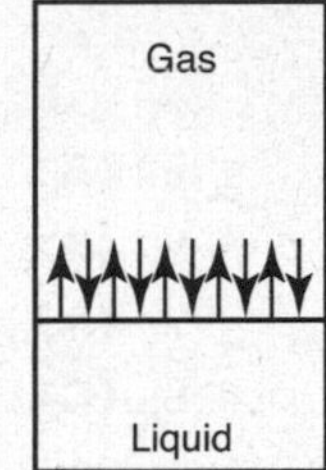

    (A) The representation on the left because liquid has to change into a gas to have an equilibrium
    (B) The center representation since it shows gas molecules coming out of the water and some coming back in
    (C) The representation on the right since there are exactly the same number of molecules condensing into the liquid as there are molecules evaporating
    (D) None of these is correct because a dynamic equilibrium occurs only in aqueous solution and most often for weak acids and bases

16. What information does this graph contain?

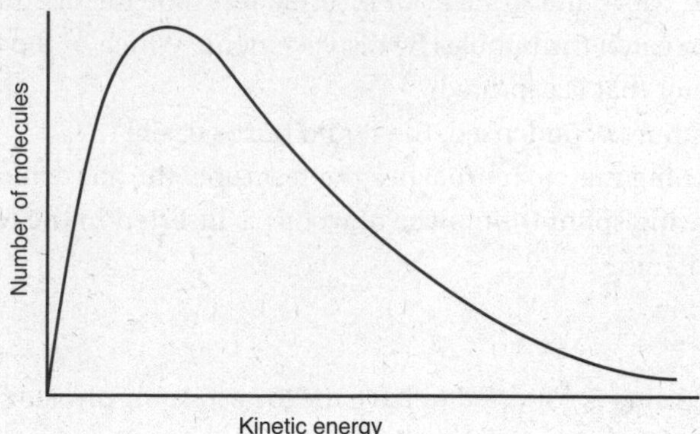

(A) Few molecules have small kinetic energies.
(B) A small number of molecules have very large kinetic energies.
(C) The average kinetic energy of a group of molecules is a little to the right of the peak of this curve.
(D) All of the above.

ANSWER KEY

1. **(B)**	5. **(A)**	9. **(C)**	13. **(A)**
2. **(B)**	6. **(D)**	10. **(A)**	14. **(A)**
3. **(D)**	7. **(A)**	11. **(D)**	15. **(C)**
4. **(B)**	8. **(C)**	12. **(D)**	16. **(D)**

See Appendix 1 for explanations of answers.

Free-Response

Use the concepts and information in this chapter and in preceding chapters to answer the following questions.

(a) Why must helium be cooled to 4 K before it will condense? Explain your answer using fundamental concepts.

(b) Use the concepts in this chapter to explain why trans fats are considered harmful whereas cis fats are not.

(c) Sketch a plot of boiling points versus period number for the hydrides of the oxygen group, the nitrogen group, and the halogen group. Explain what this demonstrates.

(d) List and describe the intermolecular forces we use to explain physical properties.

(e) Describe two methods for deciding if a substance has a predominantly ionic character or a predominantly covalent behavior.

ANSWERS

(a) Helium, the second lightest element, has the second highest average root mean square velocity at any temperature for the gases. For attractive forces to be effective in causing condensation, the molecules must move slowly enough for these attractions to occur. Helium must be cooled to extremely low temperatures so that the very weak London forces can take effect.

(b) Trans fats are essentially linear long-chain hydrocarbons and solidify readily because of multiple sites where London forces take effect. Cis acids are "V"-shaped. These molecules rarely line up effectively and do not solidify easily. Ease of solidification is thought to be related to the development of plaque and clots in arteries.

(c) For this we need to reproduce the graph in Figure 7.4. This does not have to be exact. However, it will show that the boiling points of HF, H_2O, and NH_3 are all significantly greater than expected because of hydrogen bonding.

(d) The intermolecular forces are dipole-dipole attractive forces. They arise when a molecule has a fixed dipole. The information in Chapter 4 helps us decide if a molecule will be polar or nonpolar. Molecules containing a H–O–, H–F, or H–N– bond will form hydrogen bonds, a particularly strong dipole-dipole attraction. For nonpolar molecules the presence of very weak forces of attraction due to "instantaneous dipoles" was proposed by London and these forces are often called London forces (other names are dispersion forces or van der Waals forces). These very weak attractions are present in virtually all substances. They predominate in nonpolar molecules.

(e) One method is to use the equation for dipole moment ($\mu = q \times r$) to solve for the charge, q, on 2 atoms bonded to each other. One will be negative, and the other will be positive. You will have to dig up data for the dipole moment and bond length to calculate q and then convert it to a fraction of a full electron charge. A value of q converted to e^- charges indicates the type of bonding. A nonpolar bond has 0.0 e^-. A value above 0.5 e^- indicates a more ionic character than a covalent character.

A second and easier method is to determine the difference in electronegativity. This is often written as ΔEN. If ΔEN is greater than 1.7, the bond is considered to be more than 50% ionic.

If the identity of the compound is unknown, a third, less precise, method would be to determine the melting point of the substance. The higher the melting point, the more likely it is an ionic substance and the lower melting point, the more likely it is a non-polar compound.

Solutions

8

- → SOLUTION TERMINOLOGY
- → SOLVENT/SOLUTE
- → DILUTE/CONCENTRATED SOLUTIONS
- → UNSATURATED/SATURATED AND SUPERSATURATED SOLUTIONS
- → THE SOLUTION PROCESS
- → STRONG ELECTROLYTES
- → WEAK ELECTROLYTES
- → NONELECTROLYTES
- → CONCENTRATION UNITS
- → MOLARITY
- → CONCENTRATION CALCULATIONS
- → TEMPERATURE-DEPENDENT CONCENTRATION UNITS
- → EFFECT OF TEMPERATURE ON SOLUBILITY
- → EFFECT OF PRESSURE ON SOLUBILITY

BIG IDEAS 1, 2, 3, 6

Learning Objectives: 1.4, 1.7, 2.7, 2.8, 2.9, 2.10, 2.13, 2.14, 2.15, 2.19, 3.4, 6.22, 6.24

For the complete list of Big Ideas and Learning Objectives, refer to the AP Chemistry Course Outline: ***https://secure-media.collegeboard.org/digitalServices/pdf/ap/ap-chemistry-course-and-exam-description.pdf***

INTRODUCTION AND DEFINITIONS

Chemical reactions occur only when atoms, molecules, or ions directly interact, or collide, with each other. Reactants separated by as much as a few molecular diameters may as well be separated by miles. Consequently, most reactions are carried out either in liquid solutions or in the gas phase where freely moving molecules or ions may effectively collide millions of times per second.

Every **solution** is a uniform mixture of one or more **solutes** dissolved in a **solvent**. Of all the possible mixtures of substances, only a few form solutions. These include all gas mixtures, gases, liquids, or solids dissolved in a liquid, and many **alloys**, which are solid solutions of two or more metals.

To describe a solution in **quantitative**, or numerical, terms, chemists specify the amount of solute dissolved in a specified amount of solvent. This is generally called the **concentration** of a solution. The various concentration units used by chemists are defined later in this chapter. **Solubility** is the quantitative term for the maximum amount, in units of grams per liter, of solute that can dissolve in a solvent. **Molar solubility** is the maximum amount, in moles per liter, of solute that can dissolve. Solubility depends on the temperature of the solution, as described later.

Chemists also use several **qualitative**, or nonnumerical, terms to describe solutions. In a **concentrated** solution a large amount of solute is dissolved in the solvent. A **dilute** solution has a small amount of solute dissolved in the solvent. Solutions that have the maximum amount of solute dissolved in the solvent are called **saturated solutions**. A saturated solution may be identified as one in which the solution is in equilibrium with undissolved solute. An **unsaturated solution** has less than the maximum amount of solute dissolved in the solvent. An unsaturated solution never has any undissolved solute present. In a **supersaturated solution** more than the maximum amount of solute is dissolved. With many solutes a supersaturated solution can be obtained by preparing a saturated solution at a high temperature and then carefully cooling the liquid to avoid crystallizing the excess solute. Supersaturated solutions are **metastable**, meaning that the excess solute will crystallize if the solution is shaken or if a seed crystal is added to start the crystallization process.

The terms *concentrated* and *dilute* have no relationship to the terms *saturated, supersaturated,* and *unsaturated.* Some saturated solutions may be very dilute. An example is silver chloride, which is saturated at a concentration of 1.0×10^{-5} mol AgCl per liter. On the other hand, 10 moles of sucrose dissolved in a liter of solution is a concentrated solution, but it is not saturated at 100°C.

SOLUBILITY AND THE SOLUTION PROCESS

Whether or not a substance will dissolve in a solvent is of major importance in chemistry. Separating and purifying compounds often depends on the difference in solubility between two compounds. The effectiveness of drugs depends on their solubilities in water and body fats. The ability of plants to use nutrients in the soil depends on the solubilities of the nutrients in water.

TIP

To review how to determine if a molecule is polar or nonpolar, read page 187.

Since all gases are soluble in each other in all proportions and the consideration of alloys was presented in the previous chapter, we shall focus on liquid-phase solutions. With liquids, the general rule for solubility is that solutions can be made from solutes and solvents with similar polarities, but not from solutes and solvents that have very different polarities.

Focusing on the solution process more closely, we note that the dissolution of one substance in another involves three distinct steps, as shown in Figure 8.1. First, the solute molecules are separated so that the solvent can fit in between them (step A energy is used to break attractions between solute molecules). Second, the solvent molecules are separated so that the solute can fit between them (step B energy is used to break attractions between solvent molecules). Third, the separated solute and solvent molecules are combined to form the solution (step C energy is released when new attractions are formed between solute and solvent molecules). When more energy is released in step C than is used in steps A and B, the solvent will dissolve the solute. This excess energy is released as heat when the solute and solvent are mixed, and an increase in temperature is observed.

In addition to the breaking and making of attractions among molecules, Figure 8.1 illustrates that the dispersal of molecules is another factor in the solution process. In diagram form, the solute is shown as an ordered crystal. In the solution, the solute is shown with its molecules arranged in many ways. This increase in states or arrangements that a solute can have is called an **entropy** increase and helps the solution process. In some solutions, the energy released in step C is slightly less than the energy used in steps A and B. When this occurs, the solution cools as the solvent and solute are mixed. For the solute to dissolve, however, the shortfall in energy must be compensated by the increased entropy of the solution. Solutes that are solids have the greatest increase in entropy when dissolved. Liquids have a moderate increase, and gases have virtually no increase, in entropy when dissolved.

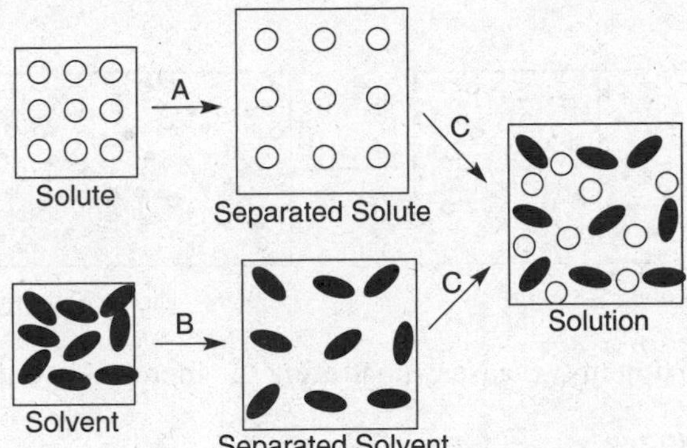

FIGURE 8.1. The three steps involved in forming a solution. Step A separates the solute, step B separates the solvent, and step C combines the separated solvent and solute into a solution.

TIP

Drawing and interpreting these diagrams will be important on the AP Chemistry exam.

Examples of the Solution Process

DISSOLVING IONIC COMPOUNDS

This example describes why ionic compounds dissolve to form aqueous solutions. Ionic compounds have the strongest attractive forces holding the ions in the crystal lattice. The energy needed to disrupt the crystal (step A in the solution process) is known as the lattice energy. Lattice energies are so high that dissolution of an ionic compound seems impossible. However, water, with its high polarity, is strongly attracted to the charged ions in the solution. As a result, many water molecules are attracted to each ion and enough energy is released (step C of the solution process) so that the dissolution is favored. The increase in entropy due to the breakup of the crystal structure also aids the solution process. Still, many ionic compounds are insoluble in water because the hydration of the ions cannot offset the lattice energy. Chapter 4 lists the solubility rules for ionic compounds.

GAS MIXTURES

In a second example we see that gases mix freely with other gases solely on the basis of the entropy increase since there are virtually no attractive forces between gas molecules. When a gas is allowed to expand into a vacuum, it does so because the increased volume means more ways for the molecules to arrange themselves, as shown in Figure 8.2.

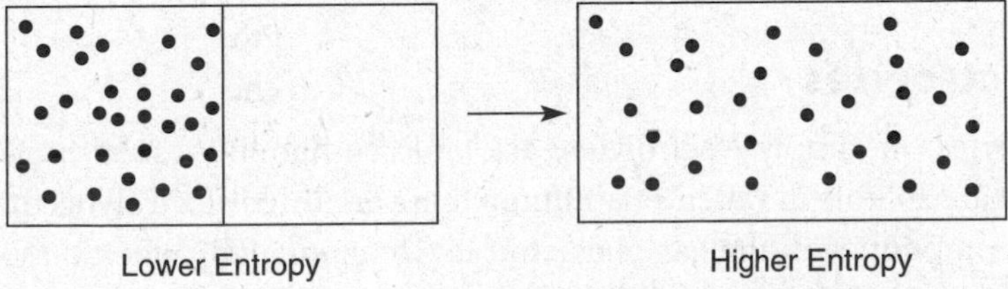

FIGURE 8.2. Increase in the entropy of a gas when the volume is increased.

When two gases are allowed to expand (Figure 8.3), each into the volume occupied by the other, the entropy of each gas increases. Since there are no attractive forces, each gas "thinks" it is expanding into a vacuum.

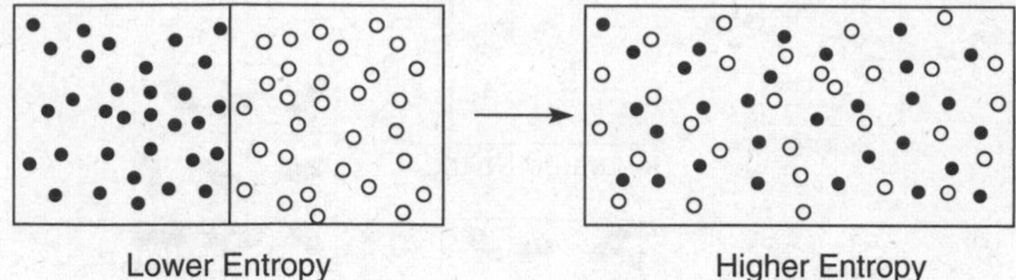

FIGURE 8.3. Mixing of two gases, illustrating the increase in entropy for each.

Rates of Dissolution

In addition to the attractive forces that determine whether or not solutes will dissolve, we are interested in the rate at which solutes dissolve. Almost intuitively, we heat solutions to increase the speed of dissolution. We also grind chemicals into small particles and stir solutions vigorously to accelerate the solution process. All three of these operations speed the rate at which solvent molecules can reach the surface of a solid and start the dissolution process. Heating accelerates the motion of the solvent molecules so that they collide with the solid more rapidly. Grinding a solute into fine particles allows more of the solute to be exposed to the solvent. Stirring a solution moves dissolved solute away from the surface of the solute and brings less concentrated solvent in contact with the solute. While heating, grinding, and stirring increase the rate of dissolution, it must be remembered that the maximum concentration achieved depends only on the temperature of the solution and the chemical identity of the solute.

AQUEOUS SOLUTIONS

Classification of Solutes

Solutions in which water is the solvent are known as **aqueous** solutions. Ionic substances **dissociate** (break apart) completely into ions when dissolved in water. Some molecular compounds, notably HCl, also break apart into ions when dissolved in water in a process called **ionization**. Their solutions conduct electricity very well, and these compounds are called **electrolytes**. Other compounds dissociate or ionize only slightly when dissolved in water; their solutions conduct electricity, but not well. These compounds are called **weak electrolytes**. The rest of the soluble compounds form absolutely no ions when dissolved in water. Their solutions do not conduct electricity, and these compounds are called **nonelectrolytes**.

Strong Electrolytes

Strong electrolytes, or simply electrolytes, are ionic compounds, such as NaCl, KBr, and $Mg(NO_3)_2$, that are soluble in water. In addition, there are three electrolytes that are molecular (covalent) compounds in the gas phase but ionize completely when dissolved in water. These compounds are HCl, HBr, and HI. A typical reaction is

$$HCl(g) + H_2O(l) \rightarrow H_3O^+(aq) + Cl^-(aq)$$

Weak Electrolytes

Weak electrolytes are soluble molecular compounds that only partially ionize into ions when dissolved in water. Most weak electrolytes ionize less than 10 percent, meaning that more than 90 percent of the molecules remain intact. The most common examples of weak electrolytes are the weak acids and weak bases. Some weak acids are listed in Table 8.1.

TABLE 8.1

Some Common Weak Acids

Name	Formula	Alternative Formula
ethanoic acid	$HC_2H_3O_2$	CH_3COOH
methanoic acid	$HCHO_2$	$HCOOH$
propanoic acid	$HC_3H_5O_2$	CH_3CH_2COOH
benzoic acid	$HC_7H_5O_2$	C_6H_5COOH
hypochlorous acid	$HClO$	$HOCl$
chlorous acid	$HClO_2$	$HOClO$
chloric acid	$HClO_3$	$HOClO_2$
hydrosulfuric acid	H_2S	None
hydrofluoric acid	HF	None
phosphoric acid	H_3PO_4	None
water	H_2O	HOH

Most weak bases are related to the weak base ammonia. Examples of some weak bases are methylamine, CH_3NH_2; diethylamine, $(CH_3CH_2)_2NH$; and ethylenediamine, $H_2NCH_2CH_2NH_2$. The reaction of a weak electrolyte with water results in an **equilibrium**, which is shown by a double arrow in a reaction:

$$NH_3(g)+H_2O(l) \rightleftharpoons NH_4^+(aq)+OH^-(aq)$$

Nonelectrolytes

Nonelectrolytes are molecular compounds that dissolve but show no tendency to form ions. Sugars such as glucose, $C_6H_{12}O_6$, and sucrose, $C_{12}H_{22}O_{11}$, and alcohols such as methanol, CH_3OH; ethanol, CH_3CH_2OH; and propanol, $CH_3CH_2CH_2OH$ are some examples of soluble nonelectrolytes.

CONCENTRATIONS OF SOLUTIONS

Definitions and Units

MOLARITY (M)

The most common unit of concentration used by chemists is **molarity**. It is defined as the number of moles of solute dissolved in 1 liter of solution and is abbreviated as the uppercase letter M:

$$\text{molarity } (M) = \frac{\text{number of moles of solute}}{\text{1 L of solution}} \tag{8.1}$$

A solution with molarity units is prepared by measuring the solute in an appropriate manner and quantitatively transferring it to a volumetric flask of the desired size. Enough solvent is then added to fill the flask about half full, and the mixture is shaken to dissolve all of the solute. Once the solute is dissolved, solvent is added exactly to the mark on the volumetric flask and mixing is repeated.

EXAMPLE 8.1

Determine the concentration of each of the following solutions:

(a) A solution prepared by dissolving 25.0 g of $MgCl_2$ in enough water to make 450 mL of solution. Calculate the molarity of the solute.

Solution

(a) The number of moles of $MgCl_2$ is calculated:

$$?\text{ mol } MgCl_2 = 25.0\text{ g } MgCl_2\left(\frac{1\text{ mol } MgCl_2}{95.21\text{ g } MgCl_2}\right) = 0.263\text{ mol } MgCl_2$$

From the definitions

$$\text{molarity } (M) = \frac{0.263\text{ mol } MgCl_2}{0.450\text{ L solution}} = 0.584\ M\ MgCl_2$$

EFFECT OF TEMPERATURE ON SOLUBILITY

Most solids are more soluble in hot solvents than in cold solvents. Figure 8.4 illustrates the changes in solubility of a few solid compounds as temperature is increased. A simple explanation is that increasing the temperature increases the disorder of all molecules. If two states are present, molecules will tend to move from the lower entropy state to a higher entropy state. A more precise explanation is left for more advanced chemistry courses.

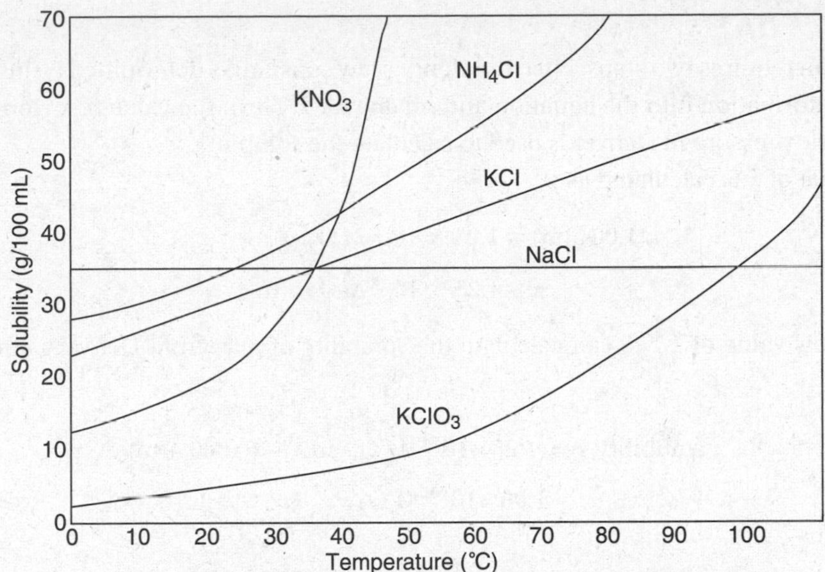

FIGURE 8.4. Temperature dependence of solubility for several representative ionic compounds.

In agreement with the general principle above, the solubility of gases in liquids, particularly water, usually decreases as the temperature is increased. You may have noticed that a warm bottle of soda fizzes more than a cold bottle of soda. In this case, the increased entropy caused by an increase in temperature favors molecules leaving the liquid phase and entering the gas phase. This occurs because molecules have more possible states in the gas phase than in the liquid phase.

EFFECT OF PRESSURE ON SOLUBILITY

External pressure has no significant effect on the solubility of solids or liquids because solids and liquids are not appreciably compressed when pressure is increased. Gases, on the other hand, are easily compressed. Compression increases the frequency with which gas molecules hit the liquid phase and enter it, thereby increasing the solubility.

HENRY'S LAW
Solubility of gases

The effect of pressure on the solubility of gases is expressed in Henry's law:

$$\text{solubility}_{gas} = kP_{gas} \tag{8.2}$$

Solubility$_{gas}$ is usually expressed as molarity or as wt/vol concentration units, k is the Henry's law proportionality constant, and P_{gas} is the partial pressure of the gas.

If the solubility of a gas is known at one pressure, Henry's law can be used to determine the solubility at another pressure as long as the temperature is the same in both cases.

EXAMPLE 8.2

The solubility of oxygen in water is 1.25×10^{-3} *M* at 25°C at sea level (1.00 atm). What will the solubility of oxygen be in Denver, where the atmospheric pressure is 0.800 atm?

Solution

The solution requires two steps. First, the Henry's law constant is determined by inserting the sea level information into the equation and solving for k. Next, the value of k along with the atmospheric pressure in Denver is used to calculate the solubility.

The value of k is calculated as

$$k(1.00 \text{ atm}) = 1.25 \times 10^{-3}\ M\ O_2$$
$$k = 1.25 \times 10^{-3}\ M\ O_2\ \text{atm}^{-1}$$

Using this value of k, we can calculate the solubility of oxygen at Denver's atmospheric pressure:

$$\text{solubility}_{O_2} = (1.25 \times 10^{-3}\ M\ O_2\ \text{atm}^{-1})(0.800 \text{ atm})$$
$$= 1.00 \times 10^{-3}\ M\ O_2$$

SUMMARY

After discussing pure matter in Chapters 6 and 7, solutions are covered in this chapter. While solutions can be prepared from almost any two substances (gas and solid solutions are possible), this chapter focuses on solutions that have at least one liquid phase. Factors that allow one substance to dissolve in another are described. Again, molecular structure and polarity give us information to understand this process. In many instances, polarity estimates help decide if two substances will form a solution.

Important Concepts

Molecular view of the solution process
Attractive forces and randomness in the solution process
Solubility related to attractive forces
Concentration units, particularly molarity

Important Equations

M = moles/liter solution
d = grams/cm^3

PRACTICE EXERCISES

Multiple-Choice

1. The solubility of cadmium chloride, $CdCl_2$, is 140 g per 100 mL of solution. What is the molar solubility (molarity) of a saturated solution of $CdCl_2$?
   (A) 0.765 *M*
   (B) 1.31 *M*
   (C) 7.64 *M*
   (D) 12.61 *M*

2. Which of the following represents the most dilute solution if ○●○ represents the solvent and ◯ represents the solute?
   (A)
   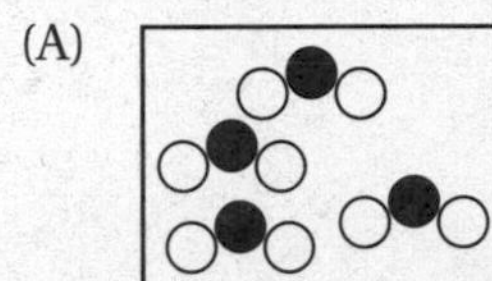
   (B)
   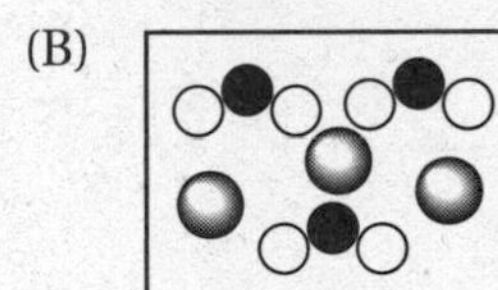
   (C)
   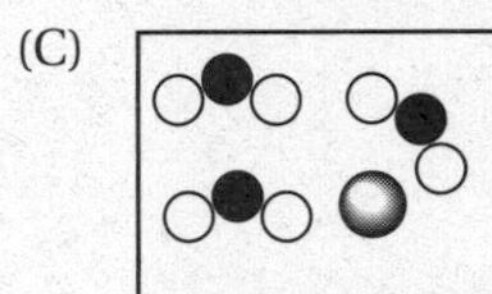
   (D)
   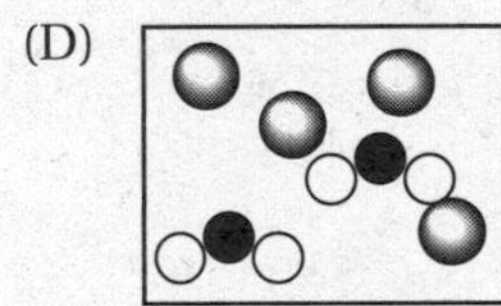

3. To make a solution, 3.45 mol of $C_6H_{13}Cl$ and 1.26 mol of C_5H_{12} are mixed. Which of the following is needed, but not readily available, to calculate the molarity of this solution?
   (A) The density of the solution
   (B) The densities of $C_6H_{13}Cl$ and C_5H_{12}
   (C) The temperature
   (D) The molar masses of $C_6H_{13}Cl$ and C_5H_{12}

4. Ethyl alcohol, C_2H_5OH, and water become noticeably warmer when mixed. This is due to
   (A) the decrease in volume when they are mixed
   (B) stronger attractive forces in the mixture than in the pure liquids
   (C) the hydrogen bonding of the two liquids
   (D) the change in vapor pressure observed

5. Which of the following concentration values changes as the temperature of a solution changes?
   (A) Mole fraction
   (B) Molarity
   (C) Molality
   (D) Mass percent

6. When algae decay in a pond, the process uses up the available oxygen. Which of the following factors will also contribute to a decrease in oxygen in a pond?
   (A) Decreasing salinity (salt concentration)
   (B) Increasing acidity due to acid rain
   (C) Increasing temperature
   (D) Increasing surface tension of the water

7. Which of the following is the best representation of a solution?
   (A)
   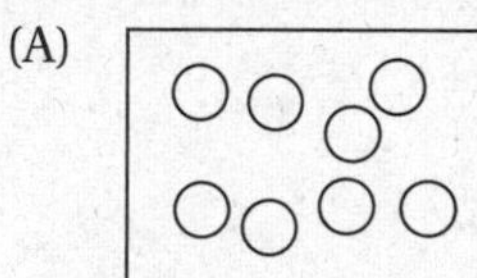
   (B)
   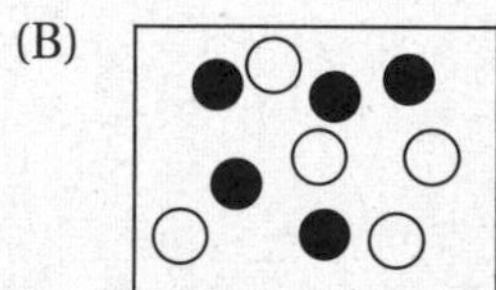
   (C)
   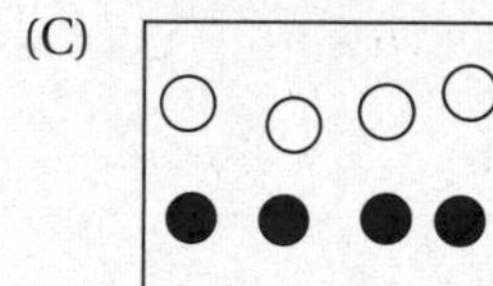
   (D)
   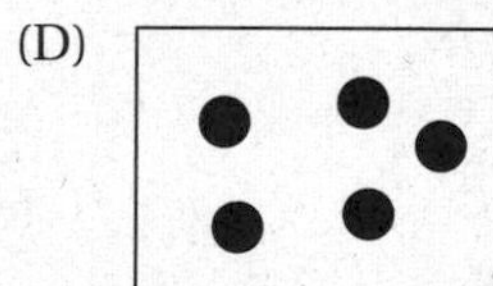

8. What is the correct process to make a 0.500 L solution of 1.0 M urea? (Molar mass = 60.06 g mol^{-1})
   (A) Dissolve 60.1 g of urea in 1.00 kg of distilled water.
   (B) Dissolve 30.0 g of urea in enough distilled water to produce 0.500 L of solution.
   (C) Dissolve 30.0 g urea in 0.500 kg of distilled water.
   (D) Dissolve 30.0 g of urea in 0.500 L of distilled water.

CHALLENGE

9. When KCl dissolves in water, the solution cools noticeably to the touch. It may be concluded that
   (A) the entropy increase overcomes the unfavorable heat of dissolution
   (B) KCl is relatively insoluble in water
   (C) the entropy decreases when KCl dissolves
   (D) the boiling point of the solution will be less than 100°C

ANSWER KEY

1. **(C)**	4. **(B)**	7. **(B)**
2. **(C)**	5. **(B)**	8. **(B)**
3. **(A)**	6. **(C)**	9. **(A)**

See Appendix 1 for explanations of answers.

Free-Response

Use the fundamental concepts of this chapter to consider solutions and their behaviors.

(a) Given the data below, explain why the solubility of alcohols decreases as the length of the carbon chain increases.

Alcohol	Solubility (g/100 mL H_2O)
CH_3CH_2OH	completely miscible
$CH_3CH_2CH_2OH$	completely miscible
$CH_3CH_2CH_2CH_2OH$	7.9
$CH_3CH_2CH_2CH_2CH_2OH$	2.7
$CH_3CH_2CH_2CH_2CH_2CH_2OH$	0.6

(b) A student is asked to make an aqueous solution of potassium chloride by dissolving 5.8 g of KCl in enough water to make 250. mL of solution.

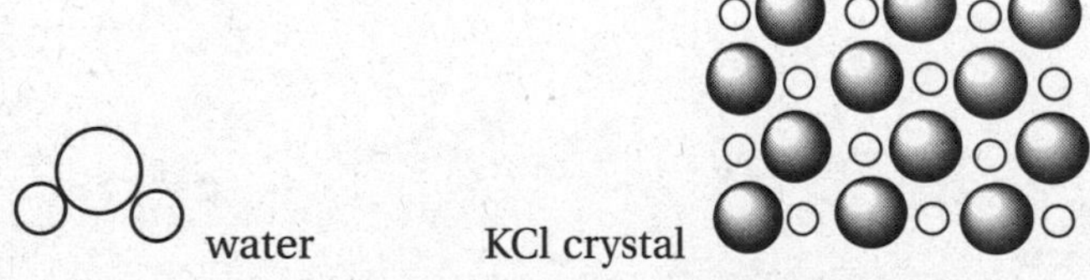

(i) Draw a diagram of the different particles existing in the solution. Base your illustration on the figures above. Use only a single formula unit and a maximum of 8 water molecules. Label all the ions and the proper orientation of the solvent.

(ii) What is the molar concentration of the aqueous potassium chloride solution?

ANSWERS

(a) All 5 compounds have an OH group that can hydrogen bond with water, thereby increasing the solubility of the compounds. However, as more carbons are added to a chain, the nonpolar portion contributes more significantly to the overall polarity of the molecule. This decreases an alcohol's water solubility.

(b) (i) The K+ ion is smaller than the Cl^- ion. Water is a polar molecule that has partial positive and partial negative regions. When the ionic solid dissolves in water, the partial positive ends of the water surround the negative chloride ions. At the same time, the partial negative ends of the water surround the positive potassium ions.

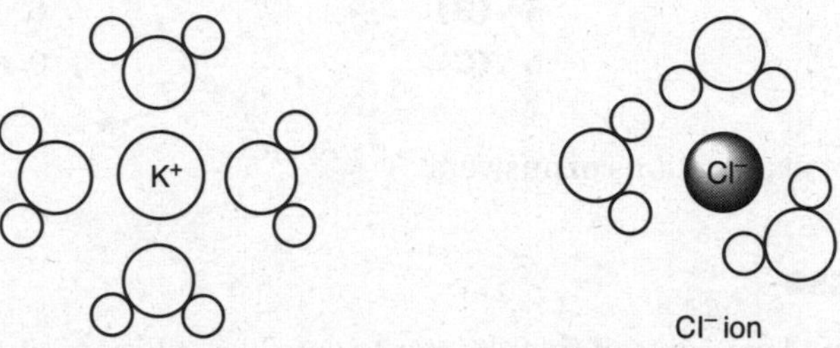

(ii) Calculate the number of moles of KCl using the molar mass. Then divide by the total volume to obtain the molar concentration.

$$5.8 \text{ g KCl}\left(\frac{1 \text{ mol KCl}}{74.55 \text{ g KCl}}\right) = 0.78 \text{ mol KCl}$$

$$\frac{0.78 \text{ mol KCl}}{0.250 \text{ L}} = 0.25 \text{ M}$$

PART 4

Physical Chemistry

Chemical Equilibrium

9

- → CHEMICAL EQUILIBRIUM
- → DYNAMIC EQUILIBRIUM
- → EQUILIBRIUM EXPRESSION
- → LAW OF MASS ACTION
- → EQUILIBRIUM CONSTANT
- → K_c, K_p, K_a, K_b, K_{sp}, K_f
- → MANIPULATING THE EQUILIBRIUM EXPRESSION
- → THERMODYNAMICALLY FAVORABLE REACTIONS
- → EXTENT OF REACTION
- → REACTION QUOTIENT, *Q*
- → EQUILIBRIUM CALCULATIONS
- → DETERMINATION OF *K*
- → SOLUBILITY PRODUCT
- → LE CHÂTELIER'S PRINCIPLE

BIG IDEAS 5, 6

Learning Objectives: 5.13, 5.14, 5.15, 5.16, 5.17, 5.18, 6.1, 6.2, 6.3, 6.4, 6.5, 6.6, 6.7, 6.8, 6.9, 6.10, 6.11, 6.21, 6.23

For the complete list of Big Ideas and Learning Objectives, refer to the AP Chemistry Course Outline: *https://secure-media.collegeboard.org/digitalServices/pdf/ap/ap-chemistry-course-and-exam-description.pdf*

The concept of a **dynamic equilibrium** is central to many aspects of chemistry. In a dynamic equilibrium, chemicals are reacting rapidly at the molecular scale, while their concentrations remain constant on the macroscopic scale.

Figure 9.1 shows that a chemical reaction has two well-defined regions in time, and these regions are studied and measured in very different ways. When compounds are first mixed in a chemical reaction, they interact to form other compounds. During the reaction process, the concentrations of the reactants decrease and the concentrations of the products increase. While the concentrations are changing, the reaction is studied using the principles of **chemical kinetics**, which are reviewed in Chapter 10. At some point in time, the concentrations of the reactants and products stop changing. Although reactions do not stop at the molecular level, at the macroscopic level the concentrations of compounds in a **dynamic equilibrium** remain constant. At this point, the compounds are in a dynamic chemical equilibrium with each other, and they are studied and described using the concepts of **chemical equilibrium**.

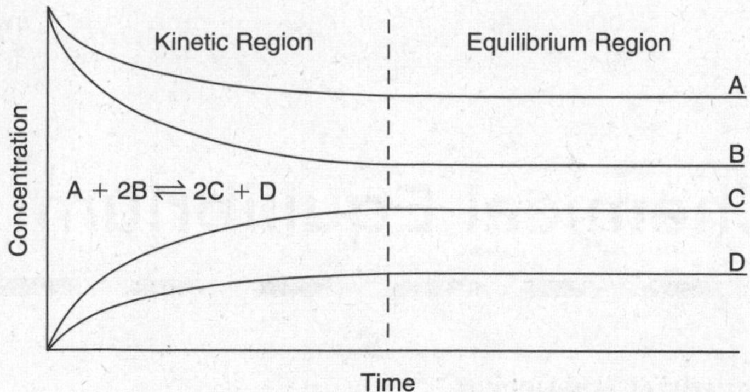

FIGURE 9.1. The two regions of chemical reactions. On the left, in the kinetic region, concentrations are changing with time. On the right, in the equilibrium region, the concentrations, on a macroscopic or laboratory scale, no longer appear to change.

THE EQUILIBRIUM EXPRESSION

TIP

The equilibrium expression is also called the law of mass action.

In a chemical reaction the actual concentrations of the reactants and products, at equilibrium, are dependent on the initial concentrations of the reacting mixture. When several experiments with different initial concentrations of reactants are performed, they result in equilibrium mixtures with different concentrations. Although these mixtures may be different, they all obey the **equilibrium expression**. This law states that the concentrations of all of the products multiplied together, divided by the concentrations of all the reactants multiplied together, will be equal to a number called the **equilibrium constant**, *K*. The value of the equilibrium constant depends only on the specific reaction and the temperature of the reaction mixture when equilibrium is reached.

TIP

The equilibrium expression is often called the equilibrium law. Scientific laws are often equations.

The uppercase letter *K* is reserved as the symbol for the equilibrium constant. To describe the type of equilibrium constant a subscript is often used after the *K*. The symbol K_c represents the equilibrium constant when concentration is expressed in molarity units (mol L^{-1}). The symbol K_p is used when the partial pressures of gases represent the amounts of reactants and products. Special symbols for the equilibrium constant are K_{sp} for the solubility product, K_a for the acid ionization constant, K_b for the base ionization constant. These special forms of *K* are described in later sections of this chapter.

The equilibrium expression depends on the chemical equation for the reaction under study. A general equilibrium reaction may be written as

$$aA + bB \rightleftharpoons pP + nN$$

and its equilibrium expression will be written as

$$K_c = \frac{[P]^p[N]^n}{[A]^a[B]^b}$$

In all cases, the chemical reaction is balanced with the smallest possible whole-number coefficients. All tabulated values for equilibrium constants refer to equations with the simplest coefficients.

A more specific example of the formulation of the equilibrium expression can be shown using the reaction between hydrogen and chlorine to form hydrogen chloride:

$$H_2(g)+Cl_2(g) \rightleftharpoons HCl(g)+HCl(g) \tag{9.1}$$

or

$$H_2(g)+Cl_2(g) \rightleftharpoons 2HCl(g) \tag{9.2}$$

The equilibrium expression for this reaction is written as

$$K_c = \frac{[HCl]\times[HCl]}{[H_2]\times[Cl_2]} = \frac{[HCl]^2}{[H_2]\times[Cl_2]}$$

Writing the chemical reaction with two separate HCl molecules as in Equation 9.1 illustrates that the concentration of HCl, written with square brackets as [HCl], should be multiplied by itself in the equilibrium expression. Writing the chemical equation in the form of Equation 9.2 illustrates that the coefficient in a chemical equation will be used as an exponent in the equilibrium expression.

For the combustion of propane, C_3H_8, the reaction is

$$C_3H_8(g)+5O_2(g) \rightleftharpoons 3CO_2(g)+4H_2O(g)$$

and the equilibrium expression for this equation is written as

$$K_c = \frac{[CO_2]^3\ [H_2O]^4}{[C_3H_8]\ [O_2]^5}$$

The coefficients of the reactants and products are written as exponents of the concentrations in the equilibrium expression.

Any substance that has a constant concentration during a reaction, including solids and all pure liquids, is not written as part of the equilibrium expression. In dilute solutions, the solvent concentration is also constant and is not written in the equilibrium expression.

Three examples are a reaction where water is a pure liquid:

$$CH_4(g)+2O_2(g) \rightleftharpoons CO_2(g)+2H_2O(l)$$

$$K_c = \frac{[CO_2]}{[CH_4]\ [O_2]^2}$$

a reaction where AgCl is a solid:

$$AgCl(s) \rightleftharpoons Ag^+(aq)+Cl^-(aq)$$

$$K_c = [Ag^+]\ [Cl^-]$$

and a reaction where water is the solvent for dilute solutions:

$$NH_3(g)+H_2O(l)\rightleftharpoons NH_4^+(aq)+OH^-(aq)$$

$$K_c=\frac{[NH_4^+]\,[OH^-]}{[NH_3]}$$

In most cases a solution is considered to be dilute when the concentration of the solute is less than 1 mole per liter.

Exercise 9.1

Write the equilibrium expression for each of the following chemical equations:

(a) $HF(aq) + H_2O(l) \rightleftharpoons F^-(aq) + H_3O^+(aq)$
(b) $2NH_3(aq) + 3I_2(s) \rightleftharpoons 2NI_3(s) + 3H_2(g)$
(c) $CO(g) + H_2O(g) \rightleftharpoons CO_2(g) + H_2(g)$
(d) $Ba^{2+}(aq) + SO_4^{2-}(aq) \rightleftharpoons BaSO_4(s)$
(e) $C_2H_4(g) + 3O_2(g) \rightleftharpoons 2CO_2(g) + 2H_2O(g)$

SOLUTION

(a) $K_c=\frac{[F^-][H_3O^+]}{[HF]}$

(b) $K_c=\frac{[H_2]^3}{[NH_3]^2}$

(c) $K_c=\frac{[CO_2][H_2]}{[CO][H_2O]}$

(d) $K_c=\frac{1}{[Ba^{2+}][SO_4^{2-}]}$

(e) $K_c=\frac{[CO_2]^2[H_2O]^2}{[C_2H_4][O_2]^3}$

MANIPULATING THE EQUILIBRIUM EXPRESSION

The equilibrium expression is written directly from the balanced chemical equation. On paper, chemical equations are easily manipulated. We can reverse the direction of the reaction by writing the reactants as products and the products as reactants. The coefficients of an equation can all be multiplied or divided by a constant factor. Equations can be added and subtracted. Each of these operations results in a different equilibrium expression and a different value for the equilibrium constant.

Reversing a Chemical Equation

The chemical equation with the corresponding K_c

$$O_2(g)+2SO_2(g) \rightleftharpoons 2SO_3(g) \qquad K_c = \frac{[SO_3]^2}{[O_2]\,[SO_2]^2} \tag{9.3}$$

can be written in the reverse direction as

$$2SO_3(g) \rightleftharpoons 2SO_2(g)+O_2(g) \qquad K_c' = \frac{[O_3]\,[SO_2]^2}{[SO_3]^2} \tag{9.4}$$

The two equilibrium constants, K_c and K'_c, are inversely related to each other:

$$K_c = \frac{1}{K_c'} \tag{9.5}$$

If a chemical reaction is reversed, the value of the new equilibrium constant will be the inverse of the original equilibrium constant.

Multiplying or Dividing Coefficients by a Constant

Taking the same reaction, Equation 9.3, of sulfur dioxide with oxygen:

$$O_2(g)+2SO_2(g) \rightleftharpoons 2SO_3(g) \qquad K_c = \frac{[SO_3]^2}{[O_2]\,[SO_2]^2}$$

we can multiply each of the coefficients by 3 to obtain another balanced equation:

$$3O_2(g)+6SO_2(g) \rightleftharpoons 6SO_3(g) \quad K_c'' = \frac{[SO_3]^6}{[O_2]^3\,[SO_2]^6}$$

The relationship between K_c and K_c'' can be shown since

$$\begin{aligned}\frac{[SO_3]^6}{[O_2]^3[SO_2]^6} &= \left(\frac{[SO_3]^2}{[O_2][SO_2]^2}\right)\left(\frac{[SO_3]^2}{[O_2][SO_2]^2}\right)\left(\frac{[SO_3]^2}{[O_2][SO_2]^2}\right)\\ &= \left(\frac{[SO_3]^2}{[O_2][SO_2]^2}\right)^3\end{aligned}$$

$$K_c'' = K_cK_cK_c = K^3$$

We see that the original equilibrium constant is raised to the power equal to the factor used in the multiplication.

Dividing an equation by 2 is the same as multiplying the equation by $^1/_2$. Therefore, when an equation is divided by 2, the new equilibrium constant, K_c''', is the square root of the original K_c:

$$K_c''' = K_c^{1/2} = \sqrt{K_c}$$

Finally, reversing a reaction may also be considered mathematically the same as multiplying it by −1. The result is that

$$K_c' = K_c^{-1} = \frac{1}{K_c}$$

which agrees with Equation 9.5.

Adding Chemical Reactions

Chemical reactions are added by adding all reactants and all products in two equations and writing them as one equation. For example:

$$2S(g) + 2O_2(g) \rightleftharpoons 2SO_2(g) \quad K_1 = \frac{[SO_2]^2}{[S]^2\ [O_2]^2}$$

$$O_2(g) + 2SO_2(g) \rightleftharpoons 2SO_3(g) \quad K_2 = \frac{[SO_3]^2}{[O_2]\ [SO_2]^2}$$

Adding these two equations and canceling the $2SO_2(g)$, which are identical on both sides, yields

$$2S(g) + 3O_2(g) \rightleftharpoons 2SO_3(g) \quad K_{\text{overall}} = \frac{[SO_3]^2}{[S]^2\ [O_2]^3}$$

Mathematically we find that

$$K_{\text{overall}} = K_1 K_2$$

When equations are added, the overall equilibrium constant will be the product of the equilibrium constants of the reactions that were added.

EXAMPLE 9.1

Given the following two reactions and their equilibrium constants:

$$Ag^+(aq) + Cl^-(aq) \rightleftharpoons AgCl(s) \qquad K_c = 5.6 \times 10^9 \qquad \text{(a)}$$

$$Ag^+(aq) + 2NH_3(aq) \rightleftharpoons Ag(NH_3)_2^+(aq) \qquad K_c = 1.6 \times 10^7 \qquad \text{(b)}$$

what is the equilibrium constant of the reaction

$$AgCl(s) + 2NH_3(aq) \rightleftharpoons Ag(NH_3)_2^+(aq) + Cl^-(aq) \qquad K_{\text{overall}} = ? \qquad \text{(c)}$$

Show how Equations (a) and (b) are added to obtain Equation (c).

Solution

To add the given equations, it is necessary to reverse Equation (a) to make AgCl(s) a reactant and Cl^- a product as required in the overall reaction:

$$AgCl(s) \rightleftharpoons Ag^+(aq) + Cl^-(aq) \quad K_c = 1/5.6 \times 10^9 = 1.8 \times 10^{-10} \qquad \text{(d)}$$

The equilibrium constant is inverted, as shown, when a reaction is reversed. Equation (d) is added to Equation (b), and the $Ag^+(aq)$ ions cancel. When reactions are added, the equilibrium constants are multiplied:

$$K_{overall} = (1.8 \times 10^{-10})(1.6 \times 10^7) = 2.9 \times 10^{-3}$$

DETERMINING THE VALUE OF THE EQUILIBRIUM CONSTANT

The equilibrium constant, K, has a numerical value that may be determined by a variety of methods. The most direct method is to measure the concentration of each reactant and product in the mixture. For example, when sulfur dioxide reacts with oxygen to produce sulfur trioxide, the reaction and equilibrium law may be written as in Equation 9.3:

$$O_2(g) + 2SO_2(g) \rightleftharpoons 2SO_3(g) \quad K_c = \frac{[SO_3]^2}{[O_2]\,[SO_2]^2}$$

If the concentrations at equilibrium are determined as $[O_2] = 2.0 \times 10^{-8}$ M, $[SO_2] = 3.4 \times 10^{-9}$ M, and $[SO_3] = 0.971$ M, they can be substituted into the equilibrium law to calculate the value of the equilibrium constant:

$$K_c = \frac{(0.971)^2}{(2.0 \times 10^{-8})\,(3.4 \times 10^{-9})^2}$$
$$= 4.2 \times 10^{24}$$

This method of determining the value of an equilibrium constant requires three measurements, and, if K_c is very large or very small, it can lead to considerable experimental error. In addition, it may be impossible to analyze the mixture for all possible components. Other ways to determine the equilibrium constant, based on chemical stoichiometry, are described later in this chapter.

USING THE EQUILIBRIUM EXPRESSION

Extent of Reaction and Thermodynamically Favorable Reactions

The value of the equilibrium constant indicates the extent to which reactants are converted into products in a chemical reaction. If the equilibrium constant is large, it indicates that the amount of products present at equilibrium is much greater than the amount of reactants. A very large equilibrium constant, greater than 10^{10}, for example, means that for all intents and purposes the reaction goes to completion. Conversely, when K is very small, less than 10^{-10}, very little product is formed and virtually no visible reaction occurs. If $K = 1$, the equilibrium mixture contains approximately equal amounts of reactants and products. These general ideas allow us to quickly estimate the composition of an equilibrium mixture.

In a **thermodynamically favorable reaction**, products are formed when reactants are mixed, without any additional assistance. Chemists generally define a thermodynamically favorable reaction as one with an equilibrium constant greater than 1.00. A reaction with an equilibrium constant less than 1.00 is called a **thermodynamically unfavorable reaction.** The fact that $K' = 1/K_c$ when a reaction is reversed tells the chemist that a reaction that is thermodynamically unfavorable in one direction is thermodynamically favorable if written in the opposite direction.

Exercise 9.2

Which of the following reactions are thermodynamically favorable in the direction written? List the reactions in order from the greatest extent of reaction to the lowest.

(a) $H_2(g) + I_2(g) \rightleftharpoons 2HI$ $\quad K_c = 49$

(b) $Br_2 + Cl_2 \rightleftharpoons 2BrCl$ $\quad K_c = 6.9$

(c) $HF(aq) + H_2O(l) \rightleftharpoons F^-(aq) + H_3O^+(aq)$ $\quad K_c = 6.8 \times 10^{-4}$

(d) $2H_2(g) + O_2(g) \rightleftharpoons 2H_2O(g)$ $\quad K_c = 9.1 \times 10^{80}$

(e) $2N_2O(g) \rightleftharpoons 2N_2(g) + O_2(g)$ $\quad K_c = 7.0 \times 10^{34}$

Solution

All the reactions except (c) have equilibrium constants greater than 1 and are favored in the direction written.

The extent of reaction follows the order (d) > (e) > (a) > (b) > (c), based on the magnitude of the equilibrium constants.

The Reaction Quotient and Predicting the Direction of a Reaction

The **reaction quotient, *Q***, is defined as the number obtained by entering all of the required concentrations into the equilibrium expression and calculating the result. For the sulfur dioxide reaction discussed previously (Equation 9.3):

$$O_2(g)+2SO_2(g)\rightleftharpoons 2SO_3(g)$$

$$K_c = \frac{[SO_3]^2}{[O_2][SO_2]^2} \text{ (equilibrium expression)}$$

$$Q = \frac{[SO_3]^2}{[O_2]\,[SO_2]^2} \text{ (reaction quotient)}$$

The equilibrium constant is the numerical value of K_c when the reaction is at equilibrium. If the chemicals in the reaction are not in equilibrium, the numerical value of the equilibrium expression is called the reaction quotient, Q. Q has exactly the same form as the equilibrium expression except that K_c has been replaced by Q.

Four principles may be ascribed to the value of Q:

1. **If Q does not change with time, the reaction is in a state of equilibrium and $Q = K_c$.**
2. **If $Q = K_c$, the reaction is in a state of equilibrium.**
3. **If $Q < K_c$, the reaction will move in the forward direction (to the right) in order to reach equilibrium.**
4. **If $Q > K_c$, the reaction will move in the reverse direction (to the left) in order to reach equilibrium.**

Exercise 9.3

Using the equilibrium constants and reactions below, determine whether or not each of the following systems is in equilibrium. If the system is not in equilibrium, predict whether it will proceed in the forward or reverse direction.

(a) $H_2(g) + I_2(g) \rightleftharpoons 2HI(g)$ $\quad K_c = 49$
$[H_2] = 0.10\ M$; $[I_2] = 0.10\ M$; $[HI] = 0.70\ M$

(b) $Br_2 + Cl_2 \rightleftharpoons BrCl$ $\quad K_c = 6.9$
$[Br_2] = 0.10\ M$; $[Cl_2] = 0.20\ M$; $[BrCl] = 0.45\ M$

(c) $HF(aq) + H_2O(l) \rightleftharpoons F^-(aq) + H_3O^+(aq)$ $\quad K_c = 6.8 \times 10^{-4}$
$[HF] = 0.20\ M$; $[F^-] = 2.0 \times 10^{-4}\ M$; $[H_3O^+] = 2.0 \times 10^{-4}\ M$

(d) $2H_2(g) + O_2(g) \rightleftharpoons 2H_2O(g)$ $\quad K_c = 9.1 \times 10^{80}$
$[H_2] = 3.0 \times 10^{-30}\ M$; $[O_2] = 4.0 \times 10^{-26}\ M$; $[H_2O] = 0.0180\ M$

(e) $2N_2O(g) \rightleftharpoons 2N_2(g) + O_2(g)$ $\quad K_c = 7.0 \times 10^{34}$
$[N_2O] = 2.4 \times 10^{-18}\ M$; $[N_2] = 0.0360\ M$; $[O_2] = 0.0090\ M$

Solution

The values of Q are calculated as follows: (a) 49; (b) 10.1; (c) 2×10^{-7}; (d) 9.1×10^{80}; (e) 2.0×10^{30}.

Reactions (a) and (d) are in equilibrium since $Q = K_c$. Reaction (b) has $Q > K_c$, indicating that the reaction must go in the reverse direction to reach equilibrium. Reactions (c) and (e) have $Q < K_c$, so the reaction must go forward to attain equilibrium.

EQUILIBRIUM CALCULATIONS

The Equilibrium Table

In solving equilibrium problems, it is convenient to organize the data in a logical format so that the proper conclusions may be drawn. To do this we construct a table of information called an **equilibrium table**. The basic table has five lines: the first line is used for the balanced chemical reaction; the next line is for the initial conditions stated in the problem; the third line represents the stoichiometric relationships, showing how the conditions on line 2 will change; the fourth line is the sum of the second and third lines and represents the concentrations at equilibrium; and the fifth will show the actual equilibrium amounts, calculated by solving the expression on line 4.

To illustrate such a table, we start with a chemical reaction, for example:

$$NO_2(g) + SO_2(g) \rightleftharpoons NO(g) + SO_3(g) \tag{9.6}$$

The table to be constructed will look like this:

Reaction	$NO_2(g)$	+	$SO_2(g)$	$\rightleftharpoons$	$NO(g)$	+	$SO_3(g)$
Initial Conc.							
Change							
Equilibrium							
Answer							

In this table, the first line is always the balanced chemical equation for the REACTION, as shown. The second and last lines will contain numerical data with units of molarity or partial pressure. All entries on these lines must have identical units. The CHANGE line represents the stoichiometric relationships inherent in the reaction. In this line the change is represented as the unknown, *x*. The coefficient for the *x* in each column is the same as the coefficient of the substance in the chemical reaction at the top of the same column. The algebraic signs of all reactants are the opposite of the signs of the products of the reaction on the third line. (Mathematically it does not matter which *x*'s are positive and which are negative as long as all reactants have the same sign and all products have the opposite sign.) The EQUILIBRIUM line is the sum of the INITIAL CONC. and CHANGE lines. Finally, the ANSWER and EQUILIBRIUM lines are mathematically equal to each other.

To illustrate a table filled out with the data given in a problem, let us assume that a 4.00-liter flask is filled with 1 mole of each of the four compounds in the given reaction. The molarity of each compound is 1.00 mole/4.00 liters = 0.250 *M*. The table will look like this:

Reaction	$NO_2(g)$	+	$SO_2(g)$	$\rightleftharpoons$	$NO(g)$	+	$SO_3(g)$
Initial Conc.	0.250 *M*		0.250 *M*		0.250 *M*		0.250 *M*
Change	$+x$		$+x$		$-x$		$-x$
Equilibrium	$0.250 + x$		$0.250 + x$		$0.250 - x$		$0.250 - x$
Answer							

Using this table, we can then perform the appropriate algebraic calculations.

Calculation of Equilibrium Constants

On page 321, the equilibrium constant was determined by measuring all of the concentrations of a mixture when equilibrium was established. Using stoichiometric relationships provides an easier way to do the same thing. To determine the equilibrium constant for the reaction in Equation 9.6:

$$NO_2(g)+SO_2(g) \rightleftharpoons NO(g)+SO_3(g)$$

an experiment can be set up where the initial amount of every compound is 0.250 *M*. The equilibrium table is the same as before:

Reaction	$NO_2(g)$	+	$SO_2(g)$	$\rightleftharpoons$	$NO(g)$	+	$SO_3(g)$
Initial Conc.	0.250 *M*		0.250 *M*		0.250 *M*		0.250 *M*
Change	$+x$		$+x$		$-x$		$-x$
Equilibrium	$0.250 + x$		$0.250 + x$		$0.250 - x$		$0.250 - x$
Answer							

When equilibrium is reached, the concentration of NO_2 is measured as 0.261 *M*. This value may be entered in the ANSWER row of the table under the NO_2 column as shown below:

Reaction	$NO_2(g)$	+	$SO_2(g)$	$\rightleftharpoons$	$NO(g)$	+	$SO_3(g)$
Initial Conc.	0.250 *M*		0.250 *M*		0.250 *M*		0.250 *M*
Change	$+x$		$+x$		$-x$		$-x$
Equilibrium	$0.250 + x$		$0.250 + x$		$0.250 - x$		$0.250 - x$
Answer	**0.261 *M***						

Since the values in the last two lines of the $NO_2(g)$ column are mathematically equal, we write

$$0.261\ M = 0.250\ M + x$$

and

$$x = 0.011\ M$$

Since the value of x is now known, the ANSWER line can be completed for all of the other compounds in the reaction by evaluating the expression on the EQUILIBRIUM line:

Reaction	$NO_2(g)$	+	$SO_2(g)$	$\rightleftharpoons$	$NO(g)$	+	$SO_3(g)$
Initial Conc.	0.250 *M*		0.250 *M*		0.250 *M*		0.250 *M*
Change	$+x$		$+x$		$-x$		$-x$
Equilibrium	$0.250 + x$		$0.250 + x$		$0.250 - x$		$0.250 - x$
Answer	**0.261 *M***		**0.261 *M***		**0.239 *M***		**0.239 *M***

These values are then entered into the equilibrium expression to determine the value of K_c:

$$\begin{aligned} K_c &= \frac{[NO][SO_3]}{[NO_2][SO_2]} \\ &= \frac{(0.239)(0.239)}{(0.261)(0.261)} \\ &= 0.8389 \end{aligned}$$

Only one measurement, the concentration of NO_2, was needed to determine the value of K_c, instead of four measurements.

DETERMINATION OF EQUILIBRIUM CONCENTRATIONS BY DIRECT ANALYSIS

When a reaction is at equilibrium, it obeys the equilibrium expression. The concentration of any compound in a reaction can be determined by measuring the concentrations of all the other compounds involved in the reaction. For instance, the reaction between H_2 and I_2 to form HI has an equilibrium constant equal to 49. At equilibrium, if $[I_2] = 0.200\ M$ and $[HI] = 0.050\ M$, we can calculate the concentration of H_2:

$$H_2 + I_2 \rightleftharpoons 2HI$$

$$K = \frac{[HI]^2}{[H_2][I_2]}$$

$$49 = \frac{(0.050)^2}{[H_2](0.200)}$$

$$[H_2] = \frac{(0.050)^2}{(49)(0.200)}$$

$$= 2.6 \times 10^{-4}$$

DETERMINATION OF EQUILIBRIUM CONCENTRATIONS FROM INITIAL CONCENTRATIONS AND STOICHIOMETRIC RELATIONSHIPS

Given initial concentrations and a known value for the equilibrium constant, we can determine the equilibrium concentrations of all compounds in a reaction. For example, the reaction

$$Br_2 + Cl_2 \rightleftharpoons 2BrCl$$

has an equilibrium constant of 6.90. If 0.100 mole of BrCl is introduced into a 500-milliliter flask, the equilibrium concentrations of Br_2, Cl_2, and BrCl can be calculated.

The solution starts with setting up the equilibrium table. The initial concentration of BrCl is 0.100 mole/0.500 liter = 0.200 *M*; and since no Br_2 or Cl_2 is added to the flask, their concentrations are entered as zero. On the CHANGE line, a positive x for Br_2 and Cl_2 is chosen because their concentrations cannot be less than zero. The change in BrCl must then be entered as $-2x$. If the signs of all the x's were reversed, the same answer would be obtained. Writing the table as suggested, however, indicates a better understanding of the equilibrium process.

Reaction	$Br_2(g)$	+	$Cl_2(g)$	$\rightleftharpoons$	$2BrCl(g)$
Initial Conc.	0 *M*		0 *M*		0.200 *M*
Change	$+x$		$+x$		$-2x$
Equilibrium	$+x$		$+x$		$0.200 - 2x$
Answer					

The algebraic expressions on the EQUILIBRIUM line are entered into the equilibrium expression:

$$K_c = \frac{[BrCl]^2}{[Br_2][Cl_2]}$$

$$6.90 = \frac{(0.200 - 2x)^2}{(x)(x)}$$

Taking the square root of both sides of this equation yields

$$2.63 = \frac{0.200 - 2x}{x}$$

$$2.63x = 0.200 - 2x$$

$$4.63x = 0.200$$

$$x = 0.0432$$

Once a value for x is determined, we return to the table and we calculate the values for the ANSWER line:

Reaction	$Br_2(g)$	+	$Cl_2(g)$	$\rightleftharpoons$	$2BrCl(g)$
Initial Conc.	0 *M*		0 *M*		0.200 *M*
Change	+x		+x		-2x
Equilibrium	+x		+x		0.200 - 2x
Answer	0.0432 *M*		0.0432 *M*		0.114 *M*

In the example above, the answer was obtained by standard mathematical methods. Such solutions are called **analytical solutions**. In other problems, the mathematics becomes very time-consuming, if not impossible, to solve analytically. Sometimes, however, as shown below, the chemist can make simplifying assumptions that result in answers that are accurate to better than ±10 percent. Other problems, which will not be described further, can be solved by making logical estimates of the answer. One way of making logical estimates is called the **method of successive approximations**. Other methods use sophisticated computer programs to make repeated estimates.

When the equilibrium constant is very small and the initial concentrations of the reactants are given, quadratic equations, or even higher order equations, may be avoided by making some assumptions based on our knowledge of the meaning of K_c. For example, the reaction in Equation 9.4:

$$2SO_3(g) \rightleftharpoons 2SO_2(g) + O_2(g)$$

has an equilibrium constant, $K_c = 2.4 \times 10^{-25}$. An equilibrium constant this small indicates that very little SO_3 will react to form the products. If 2.00 moles of SO_3 is placed in a 1.00-liter flask, we can calculate the concentrations of all three molecules when the system comes to equilibrium. First the equilibrium table is constructed:

Reaction	$2SO_3$	⇌	$2SO_2$	+	O_2
Initial Conc.	2.00 *M*		0 *M*		0 *M*
Change	$-2x$		$+2x$		$+x$
Equilibrium	$2.00 - 2x$		$+2x$		$+x$
Answer					

Entering the expressions from the EQUILIBRIUM line into the equilibrium expression gives

$$K_c = \frac{[SO_2]^2[O_2]}{[SO_3]^2} \tag{9.7}$$

$$2.4\times10^{-25} = \frac{(2x)^2(x)}{(2.00-2x)^2}$$

Equation 9.7 is a cubic equation and may be solved using advanced methods. However, because K_c is very small, it is possible to simplify this equation with an assumption based on our knowledge that this reaction produces very little product. The assumption is that $2x$ will be very small and that, when $2x$ is subtracted from 2.00 *M*, we will still have 2.00 *M* SO_3 left.

ASSUME: $2x << 2.00\ M$ so that $2.00\ M - 2x = 2.00\ M$

The assumption allows us to simplify the denominator in the equation to

$$2.4\times10^{-25} = \frac{(2x)^2(x)}{(2.00)^2}$$

A solution can now be obtained with much simpler mathematical operations. Multiplying both sides by $(2.00)^2$ and placing the unknown x on the left, we obtain

$$4x^3 = 9.6\times10^{-25}$$
$$x^3 = 2.4\times10^{-25}$$
$$x = 6.2\times10^{-9}\ M$$

Before using the calculated x to fill in the ANSWER line in the table, we check the assumption to be sure it was valid. The assumption is true since $2.00 - 6.2\times10^{-9}$ is equal to 2.00. (The actual value is 1.9999999938, which rounds off to 2.00.) Once x is shown to be reasonable, we can complete the table:

Reaction	$2SO_3$	⇌	$2SO_2$	+	O_2
Initial Conc.	2.00 *M*		0 *M*		0 *M*
Change	$-2x$		$+2x$		$+x$
Equilibrium	$2.00 - 2x$		$+2x$		$+x$
Answer	2.00 *M*		1.2×10^{-8} *M*		6.2×10^{-9} *M*

Simplifying assumptions are used only for items on the EQUILIBRIUM line of the table. In addition, these assumptions can only be used with terms that are themselves sums or differences. In the table above it is impossible to make any simplifying assumptions regarding the $+2x$ for SO_2 or the $+x$ for O_2.

In another example, the reverse of the reaction in Equation 9.4 is

$$O_2 + 2SO_2 \rightleftharpoons 2SO_3 \qquad (9.8)$$

TIP

Use simplifying assumptions only on terms with a + or − sign in them.

and its equilibrium constant will be

$$\frac{1}{2.4 \times 10^{-25}} = 4.2 \times 10^{24}$$

If the initial concentration of SO_3 is 2.00 *M*, the concentrations of O_2 and SO_2 can be determined by the same method as in the preceding example, which gives exactly the same results.

A general principle about equilibrium calculations and the assumptions used can be deduced from the two preceding examples. If the initial concentrations of reactants are given for a reaction with a very small equilibrium constant, we may assume that these concentrations will not change significantly when equilibrium is reached. Similarly, if the initial concentrations are given for the products of a reaction with a large equilibrium constant, we may assume that the concentrations of the products will not change significantly when equilibrium is reached.

The above principle indicates also, that, if the equilibrium constant is large and the initial concentrations of the reactants are given, there will be a significant change in concentration that cannot be ignored. For example, we will use the reaction in Equation 9.8 with its equilibrium constant $K_c = 4.2 \times 10^{24}$, and determine the equilibrium concentrations if 2.00 moles of O_2 and 2.00 moles of SO_2 are placed in a 1.00-liter flask.

The equilibrium table can be set up as follows:

Reaction	O_2 (*g*)	+	$2SO_2$ (*g*)	$\rightleftharpoons$	$2SO_3$ (*g*)
Initial Conc.	2.00 *M*		2.00 *M*		0 *M*
Change	$-x$		$-2x$		$+2x$
Equilibrium	$2.00 - x$		$2.00 - 2x$		$+2x$
Answer					

The equilibrium expression for this case is written as

$$4.2 \times 10^{24} = \frac{(2x)^2}{(2.00 - x)(2.00 - 2x)^2}$$

If we assume that $2x << 2.00$ and solve this equation, we find that $x = 2.9 \times 10^{12}$. This is obviously wrong, and our assumption is incorrect. This is a cubic equation and is too complex to solve analytically.

To solve this problem, we will calculate the concentration of each species that would be present if the reaction goes to completion. To do this, we set up a limiting-reactant problem. The first step will be to determine the concentration of the product and the limiting reactant's identity.

$$\text{mol } SO_3 = 2.00 \text{ mol } O_2 = \left(\frac{2 \text{ mol } SO_3}{1 \text{ mol } O_2}\right) = 4.00 \text{ mol } SO_3$$

$$\text{mol } SO_3 = 2.00 \text{ mol } SO_2 = \left(\frac{2 \text{ mol } SO_3}{2 \text{ mol } SO_2}\right) = 2.00 \text{ mol } SO_3$$

This shows that 2.00 mol of SO_3 will form. It also shows that the limiting reactant is SO_2, which is completely used up. We then calculate how much oxygen was used.

$$\text{mol } O_2 = 2.00 \text{ mol } SO_2 = \left(\frac{1 \text{ mol } O_2}{2 \text{ mol } SO_2}\right) = 1.00 \text{ mol } O_2$$

Since 1.00 moles of O_2 was consumed and we started with 2.00 moles, 1.00 moles of O_2 will be left over. To summarize, when the reaction goes to completion, we have 2.00 mol SO_3, 0.00 mol SO_2, and 1.00 mol O_2 in our 1.00 L reaction flask. Now we redraw the concentration table:

Reaction	$O_2\ (g)$	+	$2SO_2\ (g)$	⇌	$2SO_3\ (g)$
Initial Conc.	1.00 *M*		0 *M*		2.00 *M*
Change	$+x$		$+2x$		$-2x$
Equilibrium	$1.00 + x$		$+2x$		$2.00 - 2x$
Answer					

The terms on the EQUILIBRIUM line are then entered into the equilibrium expression:

$$4.2 \times 10^{24} = \frac{(2.00 - 2x)^2}{(1.00 + x)(2x)^2}$$

This equation is simplified by using two assumptions based on the fact that SO_3 is not expected to react to a large extent:

ASSUME $1.00 >> x$ so that $1.00 + x = 1.00$

ASSUME $2.00 >> 2x$ so that $2.00 - 2x = 2.00$

Then the equilibrium expression becomes

$$4.2 \times 10^{24} = \frac{(2.00)^2}{(1.00)(2x)^2}$$

Solving this equation yields

$$4x^2 = \frac{(2.00)^2}{(1.00)(4.2 \times 10^{24})}$$
$$= 9.5 \times 10^{-25}$$
$$x^2 = 2.4 \times 10^{-25}$$
$$x = 4.9 \times 10^{-13}$$

The assumption is valid, and the table with answers entered becomes as follows:

Reaction	O_2 (*g*)	+	$2SO_2$ (*g*)	⇌	$2SO_3$ (*g*)
Initial Conc.	1.00 *M*		0 *M*		2.00 *M*
Change	+*x*		+2*x*		−2*x*
Equilibrium	1.00 + *x*		+2*x*		2.00 − 2*x*
Answer	1.00 *M*		9.8×10^{-13} *M*		2.00 *M*

This problem illustrates another principle of chemical equilibrium. As long as the same number of moles of each *element* is present initially in different reacting mixtures, and these mixtures have identical volumes, the final composition of these equilibrium mixtures will always be the same.

K_p, an Equilibrium Constant for Gas-Phase Reactions

The ideal gas law

$$PV = nRT$$

can be rearranged to read

$$\frac{n}{V} = \frac{P}{RT} \qquad (9.9)$$

Since the *n*/*V* term is the concentration in moles per liter, the pressure of a gas, at a constant temperature, is directly proportional to its concentration. Therefore, the equilibrium expression may be written using the partial pressures of the gaseous reactants and products. Under these conditions the equilibrium constant is given the symbol K_p. Usually only reactions that are entirely in the gas phase are written in this manner.

One gas-phase reaction is

$$H_2(g) + I_2(g) \rightleftharpoons 2HI(g)$$

and its equilibrium expression is written as

$$K_p = \frac{P_{HI}^2}{P_{H_2}P_{I_2}}$$

where *P* represents the partial pressure of each gas. All calculations and procedures illustrated with the examples using K_c are done in exactly the same manner when the equilibrium constant is K_p.

EXAMPLE 9.2

For the reaction of gaseous sulfur with oxygen at high temperatures:

$$2S(g) + 3O_2(g) \rightleftharpoons 2SO_2(g)$$

when the system reaches equilibrium the partial pressures are measured as P_S = 0.0035 atm, P_{SO_3} = 0.0050 atm, and P_{O_2} = 0.0021 atm. What is the value of K_p under these conditions?

Solution

This problem is solved by writing the correct equilibrium expression. Since all of the chemicals in the reaction are gases and are measured in partial pressures, the equilibrium expression should be written with the constant K_p:

$$K_p = \frac{P_{SO_2}^2}{P_S^2 P_{O_2}^3}$$

With this equilibrium expression, the values for the partial pressures are entered and the solution is calculated:

$$K_p = \frac{(0.0050)^2}{(0.0021)^3(0.0035)^2}$$
$$= 2.2 \times 10^8$$

EXAMPLE 9.3

With the known value of $K_p = 2.2 \times 10^8$ from Example 9.2, determine whether each of the following systems is in equilibrium. If a system is not, determine in which direction the reaction will proceed.

System 1:	**P_{SO_3} = 1.25 atm**	**P_{O_2} = 0.256 atm**	**P_S = 0.0112 atm**
System 2:	**P_{SO_3} = 0.00677 atm**	**P_{O_2} = 0.122 atm**	**P_S = 0.212 atm**
System 3:	**P_{SO_3} = 0.123 atm**	**P_{O_2} = 0.00145 atm**	**P_S = 0.0332 atm**

Solution

The values for the pressures are entered into the equilibrium expression equation to calculate the reaction quotient, Q, which is then compared to the known value of K_p:

System 1:

$$Q = \frac{P_{SO_3}^2}{P_S^2 P_{O_2}^3} = \frac{(1.25)^2}{(0.256)^3(0.0112)^2} = 7.4 \times 10^5$$

System 2:

$$Q = \frac{P_{SO_3}^2}{P_S^2 P_{O_2}^3} = \frac{(0.00677)^2}{(1.22)^3(0.212)^2} = 5.6 \times 10^{-1}$$

System 3:

$$Q = \frac{P_{SO_3}^2}{P_S^2 P_{O_2}^3} = \frac{(0.123)^2}{(0.00145)^3(0.0332)^2} = 4.5 \times 10^9$$

None of the three systems has a value of Q equal to the known K_p of 2.2×10^8, indicating that none of the systems is in equilibrium.

In System 1 and System 2 Q is smaller than the known K_p. Q must increase as these systems approach equilibrium, and the numerator of the ratio will increase while the denominator decreases. Since the numerator contains the products and the denominator contains the reactants, the products must increase and the reactants decrease. Therefore the reaction in these systems must proceed in the forward direction.

System 3 has a value of Q that is greater than the known K_p. The ratio must decrease for the reaction to reach equilibrium. Products must be used to form more reactants, and the reaction in this system must proceed in the reverse direction.

Relationship Between K_p and K_c

The equilibrium constant for a gas-phase reaction can be written as K_p or K_c, and the two can be converted from one to another. Equation 9.9 shows that $\left(\frac{P}{RT}\right)$ can be substituted for the molar concentrations in the equilibrium expression, resulting in the relationship

$$K_p = K_c(RT)^{\Delta n_g} \qquad (9.10)$$

In this equation R is the universal gas law constant (0.0821 L atm mol^{-1} K^{-1}), T is the Kelvin temperature, and Δn_g is the change in the number of moles of gas in the balanced reaction:

$$\Delta n_g = \text{moles of gas products} - \text{moles of gas reactants}$$

In Exercise 9.2 (page 322) the reaction

$$I_2(g) + H_2(g) \rightleftharpoons 2HI(g)$$

had a $K_c = 49$. The equivalent K_p at 100°C is calculated by determining Δn_g:

$$\begin{aligned}\Delta n_g &= 2\text{ mol HI} - (1\text{ mol } I_2 + 1\text{ mol } H_2)\\ &= 0\end{aligned}$$

If $\Delta n_g = 0$, then $K_p = K_c$ for this reaction.

In the reaction of sulfur with oxygen the value of Δn_g is not zero:

$$\begin{aligned}\Delta n_g &= (2\text{ mol } SO_3) - (2\text{ mol S} + 3\text{ mol } O_2)\\ &= -3\end{aligned}$$

With $K_p = 2.2 \times 10^8$ at 573 K, we can calculate K_c as

$$2.2\times10^8 = K_c[(0.0821)(573)]^{-3}$$
$$K_c = (2.2\times10^8)[(0.0821)(573)]^3$$
$$= 2.3\times10^{13}$$

TIP

If a value of an equilibrium constant is given without specifying K_p or K_c, you can assume it is K_p unless the chemical equation requires one or more substances in solution.

Similarly, when K is calculated from $\Delta G°$ or $E°_{cell}$, it is K_p unless the chemical equation requires one or more substances in solution.

EXAMPLE 9.4

The value of K_c for the following reaction:

$$2NO(g) + O_2(g) \rightleftharpoons 2NO_2(g)$$

is 5.6×10^{12} at 290 K. What is the value of K_p?

Solution

In this problem we are required to solve Equation 9.10. Since K_p is the unknown, we need values for K_c, R, T, and Δn_g. All of these are given in the problem except Δn_g. To determine the value for Δn_g, we note that this reaction has 3 mol of gases as reactants and 2 mol of gases as products. This is a decrease of 1 mol of gas in going from reactants to products, and therefore Δn_g is −1. Entering the numbers into the equation, we get

$$K_p = 5.6\times10^{12}[(0.081)(290)]^{-1}$$

Rearranging the equation, we calculate the value of K_p as

$$K_p = \frac{5.6\times10^{12}}{(0.081)(290)} = 2.4\times10^{11}$$

Units of Equilibrium Constants

In the exact derivation of equilibrium constants there are no units, and we say that the equilibrium constants are dimensionless quantities.

SPECIAL EQUILIBRIUM CONSTANTS

Solubility Product

In Chapter 3, rules for determining solubility were given. These rules are used to determine whether an ionic compound will dissolve to an appreciable extent in water. In quantitative terms, it is loosely agreed that a salt is soluble if at least 0.1 mol of it will dissolve in 1 liter of water (0.1 *M* solution). Saturated solutions of insoluble salts have concentrations that are less than 0.1 molar. However, most insoluble salts do dissolve to a small extent.

The solution process may be written in a form similar to a chemical reaction. For solid $Fe_2(OH)_3$ we have

$$Fe(OH)_3(s) \rightleftharpoons Fe^{3+}(aq) + 3OH^-(aq)$$

The equilibrium expression for this is written as

$$K_c = [Fe^{3+}][OH^-]^3$$

In the special case of the solubility of slightly soluble compounds, the equilibrium expression always represents the product of the ions produced when the compound dissolves. The equilibrium constant is called the **solubility product** constant and is given the symbol K_{sp}:

$$K_{sp} = [Fe^{3+}][OH^-]^3$$

$Fe(OH)_3(s)$ does not appear in the equilibrium expression because it is a solid. However, in the equilibrium table, we should note in the column for the solid that initially we start with some solid. Then some small amount of solid, $-x$, will dissolve. At equilibrium, some solid must be left. This is shown in the next equilibrium table.

Questions involving K_{sp} are solved in the same way as other equilibrium problems. For example, K_{sp} has a value of 1.6×10^{-39} for $Fe(OH)_3$. The molar solubility is calculated by setting up an equilibrium table:

Reaction	$Fe(OH)_3$	$\rightleftharpoons$	Fe^{3+}	+	$3OH^-$
Initial Conc.	Solid		0 *M*		0 *M*
Change	$-x$ dissolves		$+x$		$+3x$
Equilibrium	Some solid left		$+x$		$+3x$
Answer					

Entering the information from the EQUILIBRIUM line into the K_{sp} equilibrium expression gives

$$\begin{aligned} 1.6\times10^{-39} &= (x)(3x)^3 \\ &= (x)(27x^3) \\ &= 27x^4 \\ x^4 &= 5.9\times10^{-41} \\ x &= 8.8\times10^{-11} \end{aligned}$$

The value of x is used to calculate the molar concentrations of Fe^{3+} and OH^- in the solution. We enter these data in the table:

Reaction	$Fe(OH)_3$	$\rightleftharpoons$	Fe^{3+}	+	$3OH^-$
Initial Conc.	Solid		0 *M*		0 *M*
Change	$-x$ dissolves		$+x$		$+3x$
Equilibrium	Some solid left		$+x$		$+3x$
Answer			8.8×10^{-11} *M*		2.6×10^{-10} *M*

The solubility of $Fe(OH)_3$ may be deduced from this table also. The CHANGE line indicates that $-x$ of the compound dissolves, and therefore x represents the solubility of $Fe(OH)_3$, or 8.8×10^{-11} mol L^{-1}.

If $Fe(OH)_3$ is dissolved in a solution that already contains the Fe^{3+} ion, the **common ion effect** is observed. The common ion effect is a decrease in the solubility of a compound when it is dissolved in a solution that already contains an ion in common with the salt being dissolved. As an example, we can calculate the solubility of $Fe(OH)_3$ in a solution that already has a 6.5×10^{-5} *M* concentration of Fe^{3+}. The equilibrium table is constructed as follows:

Reaction	$Fe(OH)_3$	$\rightleftharpoons$	Fe^{3+}	+	$3OH^-$
Initial Conc.	Solid		6.5×10^{-5} *M*		0 *M*
Change	$-x$ dissolves		$+x$		$+3x$
Equilibrium	Some solid left		$6.5 \times 10^{-5} + x$		$+3x$
Answer					

Entering the information into the K_{sp} equation yields

$$1.6 \times 10^{-39} = (6.5 \times 10^{-5} + x)(3x)^3$$

To avoid a very complex fifth-order equation, we can use the results of the preceding problem to assume that x will be very small compared to 6.5×10^{-5}:

$$\text{ASSUME}: x << 6.5 \times 10^{-5} \text{ so that } 6.5 \times 10^{-5} - x = 6.5 \times 10^{-5}$$

Entering this assumption into the equation gives

$$1.6 \times 10^{-39} = (6.5 \times 10^{-5})(3x)^3$$

$$27x^3 = \frac{1.6 \times 10^{-39}}{6.5 \times 10^{-5}}$$

$$x^3 = 9.1 \times 10^{-37}$$

$$= 9.7 \times 10^{-13}$$

This result satisfies the simplifying assumption and may be entered in the table:

Reaction	$Fe(OH)_3$	$\rightleftharpoons$	Fe^{3+}	+	$3OH^-$
Initial Conc.	Solid		6.5×10^{-5} *M*		0 *M*
Change	$-x$ dissolves		$+x$		$+3x$
Equilibrium	Some solid left		$6.5 \times 10^{-5} + x$		$+3x$
Answer			9.7×10^{-13} *M*		2.9×10^{-12} *M*

In addition we may conclude that the solubility of $Fe(OH)_3$ is 9.7×10^{-13} *M*.

To determine the value of the solubility product, a variety of experiments may be performed. In one of the simplest, the molar solubility of a compound is determined by measuring the amount that dissolves in 1 liter of water. For example, if the molar solubility of MgF_2 is determined to be 0.00118 mol per liter, what is its K_{sp}? When solid MgF_2 dissolves, the reaction is

$$MgF_2(s) \rightleftharpoons Mg^{2+}(aq) + 2F^-(aq)$$

and the K_{sp} expression is

$$K_{sp} = [Mg^{2+}][F^-]^2$$

The equilibrium table is set up as follows:

Reaction	MgF_2	$\rightleftharpoons$	Mg^{2+}	+	$2F^-$
Initial Conc.	Solid		0 *M*		0 *M*
Change	−*x* dissolves		+*x*		+2*x*
Equilibrium	Some solid left		+*x*		+2*x*
Answer					

The solubility given in the problem is the amount dissolved, which is also *x*. Knowing *x*, we can immediately complete the ANSWER line:

Reaction	MgF_2	$\rightleftharpoons$	Mg^{2+}	+	$2F^-$
Initial Conc.	Solid		0 *M*		0 *M*
Change	−*x* dissolves		+*x*		+2*x*
Equilibrium	Some solid left		+*x*		+2*x*
Answer			**0.00118 *M***		**0.00236 *M***

Substituting the values from the ANSWER line into the K_{sp} expression yields

$$K_{sp} = (0.00118)(0.00236)^2$$
$$= 6.6 \times 10^{-9}$$

Weak-Acid and Weak-Base Equilibria

Weak acids and weak bases are compounds that ionize only slightly when dissolved in water, as shown for ethanoic acid and ammonia, respectively, in the following equations:

TIP

Equations for K_a, K_b, and K_w are given on the AP exam.

$$CH_3COOH(aq) + H_2O(\ell) \rightleftharpoons CH_3COO^-(aq) + H_3O^+(aq)$$

$$NH_3(aq) + H_2O(\ell) \rightleftharpoons NH_4^+(aq) + OH^-(aq) \qquad (9.11)$$

The equilibrium laws for these two reactions are as follows:

$$K_c = K_a = \frac{[CH_3COO^-][H_3O^+]}{[CH_3COOH]}$$

and

$$K_c = K_b = \frac{[NH_4^+][OH^-]}{[NH_3]} \qquad (9.12)$$

For weak acids the equilibrium constant is called the **acid ionization constant** and is given the symbol K_a. Weak bases have a corresponding **base ionization constant,** K_b. Since all weak acids ionize to form an anion and the hydronium ion, H_3O^+, all of their ionization reactions have the same form as the one for ethanoic acid and their K_a expressions are similar. Weak bases all ionize the same way as ammonia, and their base ionization expressions are similar to that for ammonia. Acid and base equilibrium problems are approached in exactly the same manner as other equilibrium problems.

For example, when 0.100 mole of NH_3 is dissolved in 1 liter of solution, what are the equilibrium concentrations of all solutes? First, the equation for the ionization of ammonia is written, along with its equilibrium law, as shown in Equations 9.11 and 9.12. Next, K_b will be needed and can be found in a table of ionization constants of the bases; for ammonia $K_b = 1.8 \times 10^{-5}$. Then an equilibrium table is constructed to summarize the given information and the stoichiometric relationships:

Reaction	NH_3	+	H_2O	$\rightleftharpoons$	NH_4^+	+	OH^-
Initial Conc.	0.100 *M*		Pure Solvent		0 *M*		0 *M*
Change	$-x$		$-x$		$+x$		$+x$
Equilibrium	Not used		Not used		$+x$		$+x$
Answer							

The K_b for ammonia is 1.8×10^{-5}, and this value can be used in the equilibrium expression, along with the expressions on the EQUILIBRIUM line of the equilibrium table, to obtain

$$1.8 \times 10^{-5} = \frac{(x)(x)}{0.100 - x}$$

Since the equilibrium constant is small, it is expected that x will be much smaller than 0.100:

$$\text{ASSUME: } x << 0.100 \text{ so that } 0.100 - x = 0.100$$

Using this simplifying assumption, we have

$$1.8 \times 10^{-5} = \frac{(x)(x)}{0.100}$$

$$x^2 = 1.8 \times 10^{-6}$$

$$x = 1.3 \times 10^{-3}$$

Since x is much smaller than 0.100, the equilibrium table can be completed as follows:

Reaction	NH_3	+	H_2O	$\rightleftharpoons$	NH_4^+	+	OH^-
Initial Conc.	0.100 *M*		Pure Solvent		0 *M*		0 *M*
Change	$-x$		$-x$		$+x$		$+x$
Equilibrium	$0.100 - x$		Not used		$+x$		$+x$
Answer	0.099 *M*				0.0013 *M*		0.0013 *M*

There are many different calculations involving acids and bases that interest chemists. Chapter 13 provides the details for weak-acid and weak-base calculations, buffer calculations, and hydrolysis calculations. All of these calculations are based on the equilibrium concept developed here.

Formation Constants

Metal ions can react with anions and molecules to form chemical species called **complexes**. Ammonia complexes with copper ions in solution, turning the solution from light blue to a much darker blue. The reaction is

$$Cu^{2+}(aq) + 4NH_3(aq) \rightleftharpoons Cu(NH_3)_4^{2+}(aq)$$

$Cu(NH_3)_4^{2+}$ is a complex ion of copper and ammonia. Since this reaction is written as an equilibrium, its equilibrium expression is written as

$$K_c = K_f = \frac{[Cu(NH_3)_4^{2+}]}{[Cu^{2+}][NH_3]^4}$$

The reaction represents the formation of the complex, and the equilibrium constant is known as the **formation constant,** K_f.

When the complexation reaction is reversed, it represents the dissociation of the complex into its parts. The equilibrium constant for the dissociation is often called the **dissociation constant,** K_d. K_f and K_d are inversely proportional to each other:

$$K_f = \frac{1}{K_d}$$

Problems involving complexation equilibria are solved in exactly the same fashion as other equilibrium problems.

LE CHÂTELIER'S PRINCIPLE

In 1888 Henry Le Châtelier proposed his fundamental principle of chemical equilibrium. He observed that chemical systems react until they reach a state of equilibrium. He also observed that, if chemicals in a state of equilibrium are disturbed in some manner, they will react again until equilibrium is reestablished, if possible.

LE CHÂTELIER'S PRINCIPLE

Whenever a system in dynamic equilibrium is disrupted by changes in chemical concentrations or physical conditions, the system will respond with internal physical and chemical changes to reestablish a new equilibrium state, if possible.

Chemical changes to a system involve the addition or removal of one or more of the products or reactants. Physical changes to a system include changes in temperature, pressure, and volume. An understanding of how these factors affect chemical equilibria allows chemists to adjust experimental conditions to maximize the desired products and minimize waste.

Effect of Concentration

Changing the concentration of any reactant or product in a chemical reaction will alter the concentrations of the other chemicals present as the system reacts to reestablish equilibrium. Figure 9.2 illustrates how this process works. A two-compartment container is set up. Dividing the compartments is a very porous barrier such as a window screen. When a liquid is added to the container, the fluid flows easily to equal heights in both compartments to establish Equilibrium 1. If some more liquid is then added to the reactant (R) compartment, the equilibrium is momentarily disturbed, as shown in the middle diagram. However, the liquid flows through the barrier and reestablishes a new equilibrium condition, Equilibrium 2, in the following diagram.

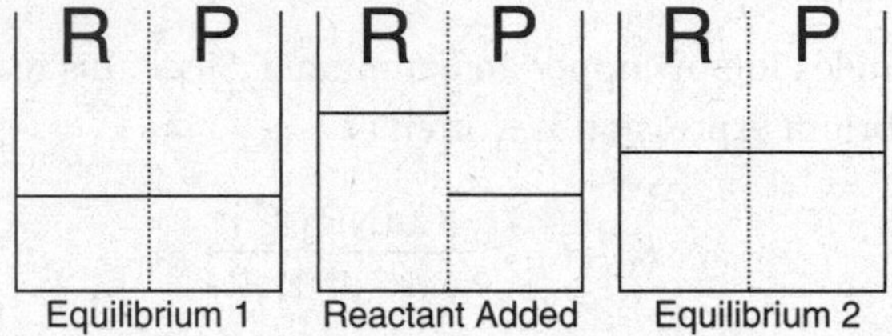

FIGURE 9.2. Diagrams illustrating an initial equilibrium of two liquids with a porous barrier, a disturbance to the equilibrium by adding liquid to one side, and finally the reestablishment of equilibrium.

Figure 9.2 clearly illustrates the action of a chemical system. Adding a reactant to an equilibrium system disturbs it by raising the reactant concentration. This disturbance causes the reaction to produce a greater amount of product and to reduce the amount of reactant in order to reestablish equilibrium. Also important is the fact that the final equilibrium has different amounts of reactants and products than were present in the initial equilibrium state.

Illustrations similar to Figure 9.2 can be used to visualize other possible concentration changes and their effects. Increasing the concentration of a product will cause an increase in reactant formation (reverse reaction). Decreasing the concentration of a reactant will cause more reactant to form (reverse reaction), and decreasing the concentration of a product will cause more product to form (forward reaction).

Increasing the concentration of a reactant or product is a simple experimental process of adding more chemicals to the reaction mixture. Decreasing the concentration of a product or reactant, however, is experimentally more difficult. Some techniques that are used to remove products or reactants from a reaction mixture are described below.

In the reaction to form ammonia:

$$N_2(g)+3H_2(g)\rightleftharpoons 2NH_3(g)$$

the ammonia gas produced is very soluble in water:

$$NH_3(g) \xrightarrow{H_2O} NH_3(aq)$$

so that a little water in the reaction system effectively removes the product NH_3. Another method used to remove a gas from a reaction is to condense it into a pure liquid.

Substances in solution can be removed by causing them to precipitate as solids. Additional chemicals may be added to a reaction mixture in order to cause precipitation.

Hydronium ions may be removed (converted to H_2O) from solution by neutralization with the base. Similarly, hydroxide ions may be removed by the addition of an acid.

Formation of a complex is an effective method for removing metal ions from solution. Complexing a metal ion changes it into a distinctly different substance, the complex. For example, the dissolution of AgCl with ammonia is a complexation reaction. The $Ag(NH_3)_2^+$ complex is very soluble, and it removes Ag^+ from the solution so that more AgCl may dissolve. The two separate steps of the reaction are as follows:

$$AgCl(s) \rightleftharpoons Ag^+(aq) + Cl^-(aq)$$

and

$$Ag^+(aq) + NH_3(aq) \rightleftharpoons Ag(NH_3)_2^+(aq)$$

which add up to

$$AgCl(s) + 2NH_3(aq) \rightleftharpoons Ag(NH_3)_2^+(aq) + Cl^-(aq)$$

The effects of changing reactant and product concentrations are summarized in Table 9.1.

TABLE 9.1
Effect of Changing Concentrations

Concentration Change	Observed Effect
Increase reactant	Favors products
Decrease reactant	Favors reactants
Increase product	Favors reactants
Decrease product	Favors products

Effect of Pressure

An increase in pressure easily compresses gases but has little effect on solids and liquids. Increasing the pressure of a gas increases its molar concentration. Changing the pressures of individual gaseous reactants by adding or removing a gas follows the same principles as changing the concentrations discussed in the preceding section. Changing the pressure of a system by adding an inert gas does nothing, however, since the gases originally present still have the same partial pressures and concentrations.

Increasing the pressure of a gaseous reaction system by decreasing its volume will have an effect on the equilibrium only if Δn_g is not zero (Δn_g is the difference between the moles of gaseous products and of gaseous reactants, used previously to make conversions between K_c and K_p). If $\Delta n_g = 0$, there will be no shift in the equilibrium since there is the same number of moles of gaseous products and of gaseous reactants. When $\Delta n_g > 0$, however, the reaction will be forced toward the reactant side because the larger number of moles of gaseous product will be compressed to a higher concentration than the reactants. Similarly, if $\Delta n_g < 0$, there will be more moles of gaseous reactant and the reaction will be forced toward producing more product. Decreasing the pressure will have the opposite effects.

These effects are summarized in Table 9.2.

TABLE 9.2
Effect of Pressure on Gaseous Reactions

Value of Δn_g	Increasing Pressure	Decreasing Pressure
Positive	Favors reactants	Favors products
Zero	No effect	No effect
Negative	Favors products	Favors reactants

TIP

Only temperature changes can cause changes in *K*.

Effect of Temperature

The only experimental variable that has any effect on the value of the equilibrium constant is temperature. For some reactions the equilibrium constant increases as the temperature increases; for others the equilibrium constant decreases. Which direction the equilibrium constant changes depends on whether the reaction is exothermic (ΔH is negative) or endothermic (ΔH is positive). An exothermic reaction gives off heat to the surroundings, and an endothermic reaction absorbs heat from the surroundings.

An exothermic reaction may be represented as a reaction in which one of the products is heat:

$$\text{reactants} \rightleftharpoons \text{products} + \text{heat}$$

Raising the temperature for an exothermic reaction is similar to increasing the concentrations of the products. As occurs with a chemical change, increasing the amounts of product will move the reaction toward the left, or reactant, side.

For an endothermic reaction, increasing the temperature is equivalent to increasing the concentrations of the reactants. The result is to move the equilibrium toward the product side:

$$\text{heat} + \text{reactants} \rightleftharpoons \text{products}$$

and represents an increase in the equilibrium constant.

TABLE 9.3
Effect of Temperature Changes

Temperature Change	Reaction Type	Effect on Reaction	Effect on *K*
Increase	Exothermic	Favors reactants	Decrease
Increase	Endothermic	Favors products	Increase
Decrease	Exothermic	Favors products	Increase
Decrease	Endothermic	Favors reactants	Decrease

The effects of temperature changes are summarized in Table 9.3. Thc last column of the table indicates that the actual effect of a change in temperature is a change in the value of the equilibrium constant. In addition to the direction of change, we will see in Chapter 11 on thermodynamics that the amount of increase or decrease in the equilibrium constant is related to the magnitude of the heat of reaction.

IMPORTANT NOTE

Le Châtelier's principle is important for evaluating what will happen to equilibrium systems. However, it does not explain why. The AP exam readers want you to use fundamental principles, laws, and theories of chemistry to explain why certain events happen. Expect to get little, if any, credit for an essay answer based only on Le Châtelier's principle.

SUMMARY

This chapter describes the essence of dynamic equilibrium in chemical systems. The equilibrium expression (law) is described in detail. The reaction quotient, *Q*, can be compared to the equilibrium constant, *K*, to predict in which direction the reaction will go or if it is at equilibrium. In many instances questions concerning equilibrium involve complex accounting for changes in concentration from initial conditions. An equilibrium table is introduced to help keep track of all the variables in a logical manner. This chapter illustrates how equilibrium constants can be determined from simple measurements. There are also calculations that allow us to calculate the composition of a mixture if the value of the equilibrium constant is known along with the initial concentrations. Techniques are described for analytical (exact) solutions to problems as well as estimation or simplification techniques. The concepts of chemical equilibrium are used to illustrate why Le Châtelier's principle works so well.

Important Concepts

Dynamic equilibrium
Equilibrium expression
Equilibrium constant and the reaction quotient
Solving equilibrium problems
Le Châtelier's principle

Important Equations

$K = [A]^a[B]^b / [C]^c[D]^d$ for reaction $cC + dD \rightleftharpoons aA + bB$

PRACTICE EXERCISES

Multiple-Choice

1. The effect of changing the temperature on a chemical system is best determined using
   (A) K_a, Q, and the enthalpy of reaction
   (B) K_{sp} and Le Châtelier's principle
   (C) Le Châtelier's principle and the enthalpy of reaction
   (D) K_c and Le Châtelier's principle

2. How can you determine if a system has come to equilibrium?
   (A) By calculating Q and the direction toward equilibrium
   (B) If K_{sp} is equal to Heisenberg's uncertainty principle
   (C) By comparing K_{eq} to Q
   (D) By comparing to see if $K_c = K_p$

3. The term(s) most useful in determining the solubility of a substance is (are)
   (A) K_a and Q
   (B) K_{sp}
   (C) Q and Le Châtelier's principle
   (D) K_c and the enthalpy of reaction

4. A chemical system in equilibrium will
   (A) have the same concentrations of all products and reactants
   (B) form more products if the temperature is increased
   (C) have a specific ratio of product to reactant concentrations
   (D) not have any precipitates

5. Using logical estimates, in which of the following expressions can the variable x be assumed to be either much smaller than or much larger than the other term(s) to simplify the mathematics?
   (A) $x[\mathrm{C}]$
   (B) $[\mathrm{C}] - x$
   (C) $[\mathrm{C}]/x$
   (D) $ax^2 - bx + c$

6. Note the following reaction:

$$\text{heat} + 2\mathrm{NO_2}(g) \rightleftharpoons \mathrm{N_2O_4}(g)$$

   Which change will not be effective in increasing the amount of $N_2O_4(g)$?
   (A) Decreasing the volume of the reaction vessel
   (B) Increasing the temperature
   (C) Adding N_2 to increase the pressure
   (D) Absorbing the $N_2O_4(g)$ with a solid absorbant

7. The reaction

$$2NO_2(g) \rightleftharpoons N_2O_4(g)$$

has an equilibrium constant of 4.5×10^3 at a certain temperature. What is the equilibrium constant of

$$2N_2O_4(g) \rightleftharpoons 4NO_2(g)?$$

(A) 4.5×10^3
(B) 9.0×10^6
(C) 2.2×10^{-4}
(D) 4.9×10^{-8}

8. The correct form of the solubility product for silver chromate, Ag_2CrO_4, is
(A) $[Ag^+]^2[CrO_4^{2-}]$
(B) $[Ag^+][CrO_4^{2-}]$
(C) $[Ag^+][CrO_4^{2-}]^2$
(D) $[Ag^+]^2[CrO^{2-}]^4$

9. In the following concentration versus time curve, at what time has this reaction reached equilibrium?

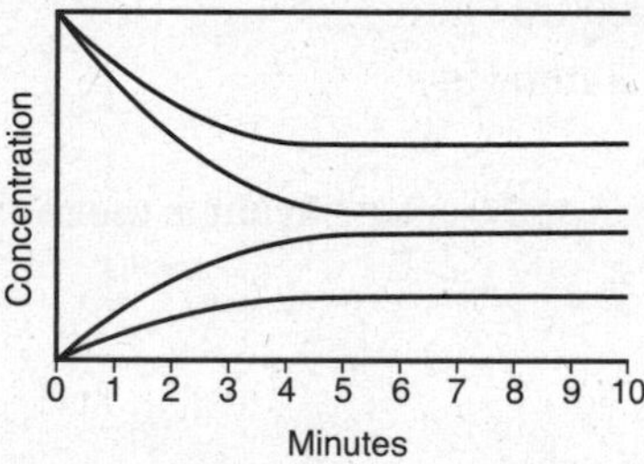

(A) Before 1 minute
(B) Between 1 and 3 minutes
(C) Around 3–4 minutes
(D) After 6 minutes

10. Which is an appropriate formulation of the equilibrium expression for the reaction

$$MgCO_3(s) + 2HCl(g) \rightleftharpoons MgCl_2(s) + CO_2(g) + H_2O(l)?$$

(A) $\frac{[CO_2]}{[HCl]}$

(B) $\frac{[MgCl_2][CO_2][H_2O]}{[HCl]^2[MgCO_3]}$

(C) $\frac{[HCl]^2[MgCO_3]}{[MgCl_2][CO_2][H_2O]}$

(D) $\frac{[CO_2]}{[HCl]^2}$

11. In the reaction

$$2HI(g) \rightleftharpoons H_2(g) + I_2(g)$$

the equilibrium constant is 0.020. If 0.200 mol of HI is placed in a 10.0-L flask, how many moles of $I_2(g)$ will be in the flask when equilibrium is reached?

(A) 0.0022 mol
(B) 0.025 mol
(C) 0.022 mol
(D) 2.2 mol

12. For the reaction

$$2NO_2(g) \rightleftharpoons N_2O_4(g)$$

$K_p = 8.8$ when pressures are measured in atmospheres. Under which of the following conditions will the reaction proceed in the forward direction?

	$NO_2(g)$	$N_2O_4(g)$
(A)	0.200 atm	0.352 atm
(B)	250 mm Hg	400 mm Hg
(C)	0.00255 atm	0.000134 atm
(D)	46.5 mm Hg	82.3 mm Hg

13. The solubility product of PbI_2 is 7.9×10^{-9}. What is the molar solubility of PbI_2 in distilled water?

(A) 2.0×10^{-3} *M*
(B) 1.25×10^{-3} *M*
(C) 5.0×10^{-4} *M*
(D) 8.9×10^{-5} *M*

14. Consider the concentration versus time curve in question 9. If we call the reactants *A* and *B* and the products *X* and *Y*, what is the most reasonable chemical equation to describe this reaction?

(A) $2A + 2B \rightleftharpoons X + Y$
(B) $2A + 3B \rightleftharpoons X + 2Y$
(C) $2A + B \rightleftharpoons 3X + 2Y$
(D) $A + 2B \rightleftharpoons X + 2Y$

15. When pure liquids and/or solids are part of a chemical equation, they are not included in the equilibrium expression for that equation. Why?

(A) The concentrations of all pure liquids and solids are defined as 1.00.
(B) Pure substances react very slowly unless ground to a fine powder.
(C) This happens for the same reason that pure substances do not need to be balanced in a chemical equation.
(D) The concentration of a solid or of a liquid at a given temperature and pressure is always a constant no matter how much of the substance is present.

16. Look at questions 9 and 14. Estimate the equilibrium constant for that reaction if the concentration at equilibrium of the bottom line in the graph represents 0.10 M.
(A) 0.037
(B) 0.27
(C) 1.0×10^{-2}
(D) 3.7

17. In which of the following cases is the reaction expected to be exothermic?
(A) Increasing the pressure increases the amount of product formed.
(B) Increasing the amount of reactants increases the amount of product formed.
(C) Increasing the temperature decreases the amount of product formed.
(D) Increasing the volume decreases the amount of product formed.

18. A reaction has a very large equilibrium constant of 3.3×10^{13}. Which statement is NOT true about this reaction?
(A) The reaction is very fast.
(B) The reaction is essentially complete.
(C) The reaction is thermodynamically favored.
(D) The equilibrium constant will change if the temperature is changed.

19. The K_{sp} of AgCl is 1.8×10^{-10}, and the K_{sp} of AgI is 8.3×10^{-17}. A solution is 0.100 *M* in I^- and Cl^-. When a silver nitrate solution is slowly added to this mixture, what is the molarity of iodide ions when AgCl just starts to precipitate?
(A) 1.0×10^{-5} *M*
(B) 9.1×10^{-9} *M*
(C) 8.3×10^{-7} *M*
(D) 4.6×10^{-8} *M*

20. A certain reaction is known to be endothermic. A temperature jump experiment is run where the system at equilibrium is rapidly heated a few degrees Celsius and then allowed to cool back to room temperature. The concentrations of the reactants and products are monitored. The following concentration versus time graphs were obtained for all species in the mixture.

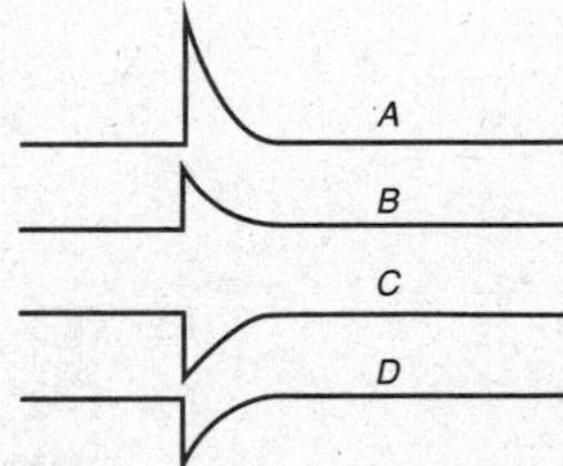

What is the chemical reaction?
(A) $A + B \rightleftharpoons C + D$
(B) $2A + B \rightleftharpoons C + D$
(C) $C + 2B \rightleftharpoons 2A + B$
(D) $C + D \rightleftharpoons 2A + B$

CHALLENGE

21. The equilibrium constant for the reaction

$$H_2(g)+I_2(g) \rightleftharpoons 2HI(g)$$

must be determined. If 1.00 g of HI is placed in a 2.00-L flask, which of the following is LEAST important in determining the equilibrium constant?

(A) The temperature must remain constant at the desired value.
(B) Several measurements must be made to assure that the reaction is at equilibrium.
(C) Only one of the three concentrations needs to be accurately determined.
(D) All three concentrations must be accurately measured.

22. In an experiment 0.0300 mol each of $SO_3(g)$, $SO_2(g)$, and $O_2(g)$ were placed in a 10.0-L flask at a certain temperature. When the reaction came to equilibrium, the concentration of $SO_2(g)$ in the flask was 3.50×10^{-5} *M*. What is K_c for the reaction

$$2SO_2(g)+O_2(g) \rightleftharpoons 2SO_3(g)$$

(A) 3.5×10^{-5}
(B) 1.9×10^{7}
(C) 5.2×10^{-8}
(D) 1.2×10^{-9}

23. The weak acid H_2A ionizes in two steps with these equilibrium constants:

$$H_2A \rightleftharpoons H^+ + HA^- \quad K_{a1} = 2.3\times10^{-4}$$
$$HA^- \rightleftharpoons H^+ + A^{2-} \quad K_{a2} = 4.5\times10^{-7}$$

What is the equilibrium constant for the reaction:

$$H_2A \rightleftharpoons 2H^+ + A^{2-}$$

(A) 6.8×10^{-11}
(B) 1.0×10^{-10}
(C) 2.3045×10^{-4}
(D) 2.0×10^{-3}

ANSWER KEY

1. **(C)**
2. **(C)**
3. **(B)**
4. **(C)**
5. **(B)**
6. **(C)**
7. **(D)**
8. **(A)**
9. **(D)**
10. **(D)**
11. **(C)**
12. **(B)**
13. **(B)**
14. **(B)**
15. **(D)**
16. **(B)**
17. **(C)**
18. **(A)**
19. **(D)**
20. **(D)**
21. **(D)**
22. **(B)**
23. **(B)**

See Appendix 1 for explanations of answers.

Free-Response

Use the principles and techniques of chemical equilibrium to answer the following questions.

1. When performing equilibrium calculations, it is possible to make simplifying assumptions. These assumptions are based in one case on the mathematical principles behind using significant figures in calculations. In a second case, the assumptions are based on an experimental principle of making a measurement.

   (a) Explain how an understanding of significant figures allows you to make a simplification and give an example.

   (b) Explain how the simplifying assumption often relates to an understanding of a measurement process.

2. (a) The solubility product of HgI_2 is 1.1×10^{-28}.

   (i) What is the molar solubility of HgI_2?
   (ii) What is the molar solubility if HgI_2 is dissolved in 0.00025 molar NaI solution?

   (b) At a certain temperature the reaction of hydrogen and chlorine to produce hydrogen chloride, all in the gas phase, has an equilibrium constant of 265. If 25.0 g of HCl is placed in a 150-L vessel and allowed to come to equilibrium, what will the concentrations of all species be?

   (c) The gas-phase reaction between oxygen and sulfur dioxide has an equilibrium constant of 8.8×10^{14} at a certain temperature. If 23.4 g of sulfur trioxide is placed in a 20.0-L vessel, calculate the concentration of all species at equilibrium.

ANSWERS

1. First of all, remember that simplifying assumptions are being made. *Assumption* is the key word. You must have criteria about whether the assumption failed or is valid. That assumption is always tested. If it fails, more complex math is needed.

   (a) The significant figure assumption states that you will be able to ignore a number because it would be ignored when considering significant figures. For example, if you are given a solution with a concentration of 0.0500 *M* and some small amount of that solution, *x*, reacts, the final concentration of the solution is 0.0500 – *x*. If it turns out that $x = 0.000036$, we can see that

$$\begin{array}{r} 0.0500 \\ -0.000036 \\ \hline 0.0500 \end{array}$$

   The 0.000036 was rounded to 0.0000 to add with the four decimal places of 0.0500. If $x = 0.00046$, it would be rounded to 0.0005 and the result would be 0.0495.

(b) If the result for x does not get rounded to zero, we now have to consider if the experiment limits the results. For instance, buffers are often precise to ±0.02 pH units. Calculating back, that is actually ±5% of the calculated concentration. Thus reading a pH meter and converting to $[H^+]$ gives a result that is precise to only ±5%. Thus values of x that do not get eliminated due to significant figures can often be used. It can be shown that a ±5% error in $[H^+]$ will not be exceeded as long as x is not greater than 10% of the initial concentration.

2. (a) The reaction is $HgI_2(s) \rightleftharpoons Hg^{2+} + 2I^-$, and $K_{sp} = [Hg^{2+}][I^-]^2$.

If the molar solubility of HgI_2 is assigned the variable s, then $[Hg^{2+}] = s$ and $[I^-] = 2s$. Making the appropriate substitutions we get

$$K_{sp} = (s)(2s)^2 = 1.1 \times 10^{-28}$$
$$4s^3 = 1.1 \times 10^{-28}$$
$$s^3 = 2.75 \times 10^{-29}$$
$$s = 3.02 \times 10^{-10}\ M\ HgI_2$$

If the HgI_2 is dissolved in a 0.000250 I^- solution, we write the iodide ion term as $[I^-] = s + 0.000250$. Because we know that the iodide concentration is on the order of 10^{-10} in distilled water, it is safe to assume that $s << 0.000250$, so that $[I^-] = 0.000250\ M$. The $K_{sp} = (s)(0.000250)^2 = 1.1 \times 10^{-28}$. Solving for s, the molar solubility of HgI_2 yields $s = 1.76 \times 10^{-22}\ M\ HgI_2$.

(b) We calculate the initial concentration of HCl as

$$\text{mol HCl} = 25.0\text{ g HCl}\left(\frac{1\text{ mol HCl}}{36.5\text{ g HCl}}\right) = 0.685\text{ mol HCl}$$

$$\text{molarity HCl} = 0.685\text{ mol HCl}/150\text{ L} = 4.57 \times 10^{-3}\ M\text{ HCl}$$

The chemical reaction is $H_2(g) + Cl_2(g) \rightleftharpoons 2HCl(g)$.

The equilibrium expression is $K_c = 265 = [HCl]^2/[H_2][Cl_2]$.

$[HCl] = (4.57 \times 10^{-3} - 2x)$ and $[H_2] = [Cl_2] = x$. Insert these into the equilibrium expression to yield

$$K_c = 265 = (4.57 \times 10^{-3} - 2x)^2/(x)(x)$$

You may want to construct an equilibrium table for these steps.

Now we take the square root of both sides to get

$$16.3 = (4.57 \times 10^{-3} - 2x)/(x)$$

and multiply through by x to get

$$16.3x = 4.57 \times 10^{-3} - 2x$$
$$18.3x = 4.57 \times 10^{-3}$$
$$x = 2.50 \times 10^{-4}\ M = [H_2] = [Cl_2] \text{ and}$$
$$[HCl] = (4.57 \times 10^{-3} - 5.0 \times 10^{-4}) = 4.07 \times 10^{-3}\ M$$

(c) The moles of SO_3 = 23.4 g/(80 g/mol) = 0.293 mol SO_3, and the molarity of SO_3 = 0.293 mol/20.0 L = 0.0146 M SO_3.

The reaction is $2SO_2 + O_2 \rightleftharpoons 2SO_3$, and we write the equilibrium expression as $K_c = 8.8 \times 10^{14} = [SO_3]^2/[O_2][SO_2]^2$. If $2x$ moles per liter of SO_3 react, then we can write (use an equilibrium table here if you need to)

$[SO_3] = 0.0146 - 2x$ and $[O_2] = x$ and $[SO_2] = 2x$.

We insert these terms into the equilibrium expression to get

$8.8 \times 10^{14} = (0.0146 - 2x)^2/((x)(2x)^2)$. Because the value of the equilibrium constant is very large, we feel safe in saying that $2x << 0.0146$, and our equation becomes

$8.8 \times 10^{14} = (0.0146)^2/((x)(2x)^2)$. This equation is easily solved as

$$4x^3 = (0.0146)^2/(8.8 \times 10^{14}) = 2.42 \times 10^{-19}$$
$$x^3 = 6.05 \times 10^{-20} \text{ and } x = 3.93 \times 10^{-7}$$

Calculating the concentrations we get

$$[SO_3] = 0.0146 - 0.00000078 = 0.0146$$
$$[O_2] = x = 3.93 \times 10^{-7} \text{ and } [SO_2] = 7.86 \times 10^{-7}.$$

Kinetics

10

- → KINETICS
- → REACTION RATES
- → FACTORS THAT AFFECT RATES
- → RATE LAWS
- → RATE CONSTANT
- → ORDER OF REACTION
- → ARRHENIUS PLOT
- → ZERO-ORDER REACTIONS
- → FIRST-ORDER REACTIONS
- → SECOND-ORDER REACTIONS
- → INTEGRATED RATE LAWS
- → COLLISION THEORY
- → KINETIC MOLECULAR THEORY
- → COLLISION FREQUENCY
- → COLLISION ORIENTATION
- → TRANSITION-STATE THEORY
- → REACTION PROFILE
- → POTENTIAL ENERGY DIAGRAM
- → ACTIVATION ENERGY
- → CATALYSTS
- → ENZYMES
- → REACTION MECHANISMS
- → ELEMENTARY REACTIONS
- → RATE-DETERMINING STEP

BIG IDEAS 2, 4

Learning Objectives: 2.4, 4.1, 4.2, 4.3, 4.4, 4.5, 4.6, 4.7, 4.8, 4.9

For the complete list of Big Ideas and Learning Objectives, refer to the AP Chemistry Course Outline: *https://secure-media.collegeboard.org/digitalServices/pdf/ap/ap-chemistry-course-and-exam-description.pdf*

REACTION RATES

The rate of a chemical reaction is the rate of change in the concentration per unit time. It is expressed as the number of moles per liter that react each second, and the units, in abbreviated form, are always mol L^{-1} s^{-1}. **Reaction rates** are determined by measuring the concentration of one or more of the chemicals involved in the reaction at different times during the course of the reaction. Figure 10.1 illustrates the results of such measurements as a **kinetic curve**, also called a **concentration versus time curve**. The kinetic curve on the left is obtained from five individual measurements of the concentration of a reaction product. These points are connected with a smooth line. The second kinetic curve is one that may be obtained using instrumentation that continuously monitors the concentration of the same product. Figure 10.1 illustrates the increase in product as a reaction occurs. If the concentration of a reactant is measured, a decrease in concentration with time will be recorded as shown in Figure 10.2.

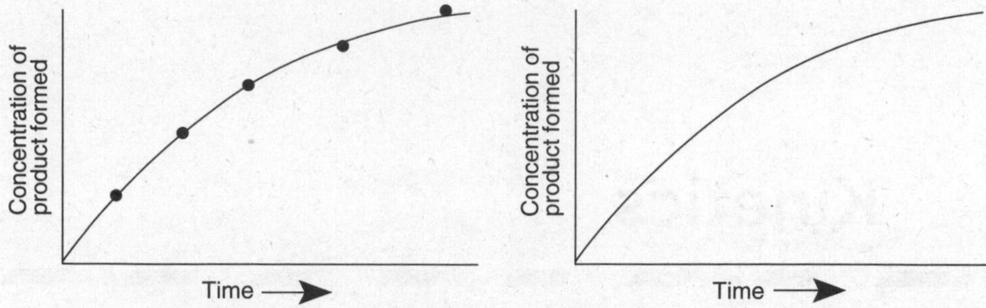

FIGURE 10.1. Kinetic curves obtained by measuring the formation of a product of a chemical reaction. The first curve is obtained from five individual measurements connected with a smooth line. The second curve is recorded automatically by an instrument designed to monitor the concentration continuously.

The rate of the production of products (or disappearance of reactants) is obtained from a kinetic curve by determining the slope of the curve at the desired point in time. The slope of a curve is determined in a three-step process:

1. Select the desired point on the curve, and draw a tangent to it.
2. Select two points on the tangent, and determine the concentrations, *c*, and times, *t*, corresponding to these points.
3. Use Equation 10.1 to calculate the slope:

$$\text{Rate} = -\left(\frac{C_2 - C_1}{t_2 - t_1}\right) = -\frac{\Delta C}{\Delta t} \tag{10.1}$$

Figure 10.2 illustrates the construction of the tangents for the determination of the initial rate, $t = 0$, and the rate at 180 seconds, $t = 180$. The small triangles on the tangents connect point 1 and point 2 with a right triangle. The necessary values for ΔC $(C_2 - C_1)$ and Δt $(t_2 - t_1)$ are the lengths of the sides of these triangles.

Another, less desirable method used to determine the rate of a chemical reaction is to measure the concentration, *C*, of one reactant or product at only two different times. The rate is then calculated using Equation 10.1. Such a two-point calculation gives the average rate of the reaction over the time interval used.

Since the reactants disappear during a chemical reaction, the rate calculated by measuring a reactant will have a negative sign. If a product is measured, the calculated rate will have a positive sign. By convention, however, chemists always use a positive value for the rate of a reaction, whether it is the positive rate of the appearance of products or the negative rate of disappearance of reactants:

$$\text{Rate} = \frac{\Delta C_{\text{products}}}{\Delta t}$$

$$\text{Rate} = \frac{-\Delta C_{\text{reactants}}}{\Delta t}$$

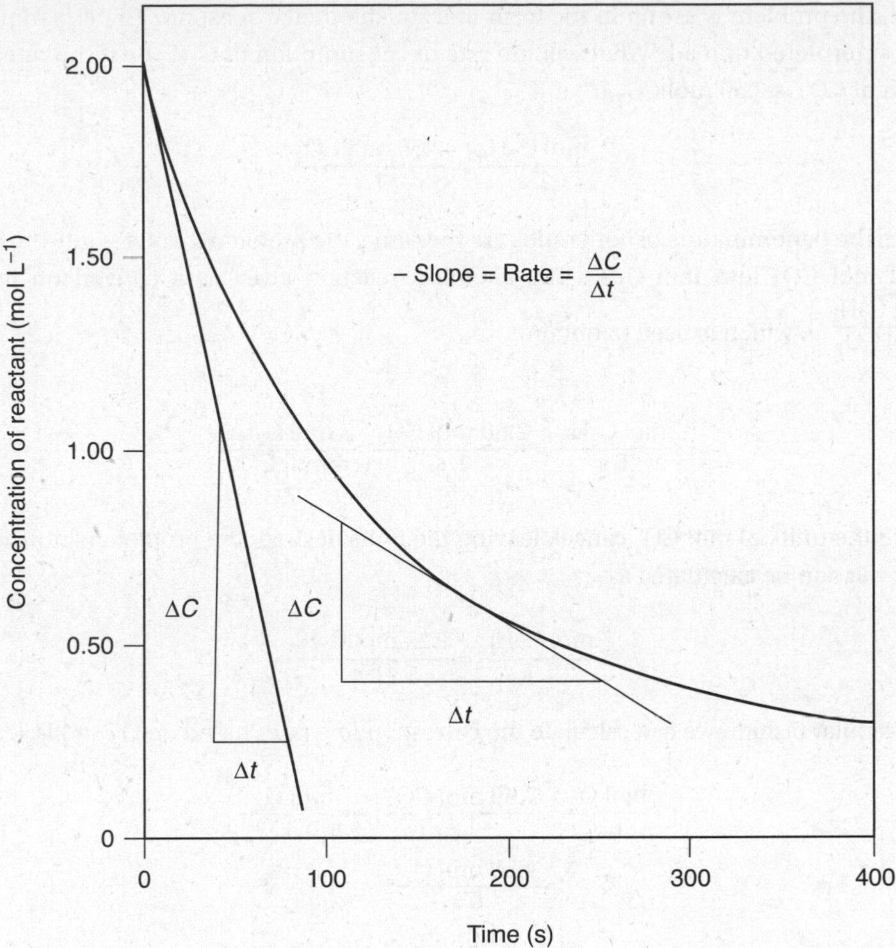

FIGURE 10.2. Tangents drawn to a kinetic curve, showing how the slope at $t = 0$ and $t = 180$ s are determined. This curve decreases with time since it was obtained by measuring the concentration of a reactant.

The sign for the rate is implied by calling the rate either the rate for the appearance (implied positive sign) of products or the rate of disappearance (implied negative sign) of reactants.

Also, the value of a reaction rate depends on which reactant or product is measured. After the rate has been measured based on one component of the reaction, the rates of change of the other components may be calculated by a stoichiometric conversion.

Consider the following reaction:

$$2C_2H_6 + 7O_2 \rightarrow 4CO_2 + 6H_2O \qquad (10.2)$$

and assume that the rate of reaction was determined by measuring the CO_2 produced and was found to be 2.50 mol CO_2 $L^{-1}s^{-1}$. This can be converted to the rates of appearance or disappearance of all other chemical species by using the stoichiometric relationships in Equation 10.2.

First, the problem is set up in the form of a stoichiometry question. The equation below can be interpreted to read "What will the rate of consumption of C_2H_6 be if the rate of production of CO_2 is 2.50 mol CO_2 L^{-1} s^{-1}?"

$$?\frac{\text{mol } C_2H_6}{\text{L s}} = \frac{2.50 \text{ mol } CO_2}{\text{L s}}$$

Since the denominators of both ratios are the same, the problem involves only the conversion of mol CO_2 into mol C_2H_6. The chemical reaction gives us a conversion factor of $\left(\frac{2\text{mol } C_2H_6}{4\text{mol } CO_2}\right)$, which is used to obtain

$$?\frac{\text{mol } C_2H_6}{\text{L s}} = \frac{2.50 \text{ mol } CO_2}{\text{L s}}\left(\frac{2 \text{ mol } C_2H_6}{4 \text{ mol } CO_2}\right)$$

Since the units of mol CO_2 cancel, leaving the units desired, the problem is finished, and the answer can be calculated as

$$?\frac{\text{mol } C_2H_6}{\text{L s}} = \frac{1.25 \text{ mol } C_2H_6}{\text{L s}}$$

In a similar fashion we can calculate the corresponding rates based on O_2 and H_2O:

$$?\frac{\text{mol } O_2}{\text{L s}} = \frac{2.50 \text{ mol } CO_2}{\text{L s}}\left(\frac{7 \text{ mol } O_2}{4 \text{ mol } CO_2}\right)$$

$$= \frac{4.38 \text{ mol } O_2}{\text{L s}}$$

and

$$?\frac{\text{mol } H_2O}{\text{L s}} = \frac{2.50 \text{ mol } CO_2}{\text{L s}}\left(\frac{6 \text{ mol } H_2O}{4 \text{ mol } CO_2}\right)$$

$$= \frac{3.75 \text{ mol } H_2O}{\text{L s}}$$

$$= 3.75 \text{ mol } H_2O \text{ L}^{-1} \text{ s}^{-1}$$

The intital reaction rates in the following problems all refer to rates as defined in the equations at the bottom of page 354.

Exercise 10.1

Using Figure 10.2, determine the rate, for the reaction illustrated, at 100 s and 200 s after the reaction starts.

Solution

When the tangents are drawn, the slopes are determined to be -6.5×10^{-3} and -2.8×10^{-3} mol L^{-1} s^{-1}. Since this curve represents the disappearance of a reactant, the rate is the negative of the slope, or 6.5×10^{-1} mol L^{-1} s^{-1} at 100 s and 2.8×10^{-3} mol L^{-1} s^{-1} at 200 s.

EXAMPLE 10.1

In the reaction

$$N_2(g) + 3H_2(g) \rightarrow 2NH_3(g)$$

the rate of disappearance of $H_2(g)$ is found to be 4.8×10^{-2} mol L^{-1} s^{-1}. What is the rate at which $N_2(g)$ is reacting, and what is the rate at which $NH_3(g)$ is being produced under the same conditions?

Solution

This problem requires the conversion of one rate into another. The initial setup is

$$? \frac{\text{mol } N_2}{\text{L s}} = \frac{4.8 \times 10^{-2} \text{ mol } H_2}{\text{L s}}$$

The next step uses the conversion factor for the conversion between $N_2(g)$ and $H_2(g)$ as

$$? \frac{\text{mol } N_2}{\text{L s}} = \frac{4.8 \times 10^{-2} \text{ mol } H_2}{\text{L s}} \left(\frac{1 \text{ mol } N_2}{3 \text{ mol } H_2} \right)$$

$$= 1.6 \times 10^{-2} \text{ mol } N_2 \text{ L}^{-1} \text{ s}^{-1}$$

The same steps are used to determine the rate for the production of $NH_3(g)$:

$$? \frac{\text{mol } NH_3}{\text{L s}} = \frac{4.8 \times 10^{-2} \text{ mol } H_2}{\text{L s}}$$

$$? \frac{\text{mol } NH_3}{\text{L s}} = \frac{4.8 \times 10^{-2} \text{ mol } H_2}{\text{L s}} \left(\frac{2 \text{ mol } NH_3}{3 \text{ mol } H_2} \right)$$

$$= 3.2 \times 10^{-2} \text{ mol } NH_3 \text{ L}^{-1} \text{ s}^{-1}$$

FACTORS THAT AFFECT REACTION RATES

FACTORS THAT AFFECT REACTION RATES

CONCENTRATION:	If a substance is part of the rate law, increasing its concentration (or pressure in the gas phase) will increase the reaction rate.
TEMPERATURE:	The Arrhenius equation shows how the rate constant (and therefore the reaction rate) is exponentially related to the Kelvin temperature.
ABILITY TO MEET:	Reactants in the gas phase or solution react most rapidly because individual molecules of ions react. Solids react more rapidly if very finely divided as in powders or dust.
CATALYST PRESENT:	A catalyst affords an alternate reaction path with a lower activation energy and therefore a larger rate constant (reaction rate).

Concentration, Temperature, and Catalysts

Most people are aware of the factors that increase reaction rates, perhaps without even realizing it. To make a fire burn more fiercely, we add wood to it. A car travels faster if we press on the accelerator pedal to give the engine more gas. In any chemical reaction, the **concentration of reactants** is an important factor in the observed rate.

In addition, to get food to cook faster, we turn up the heat on the stove. To decrease the spoilage of foods, we refrigerate or, even better, we freeze them. Obviously, **temperature** is another important factor in chemical reaction rates. As a consequence, in all rate experiments temperatures must be carefully controlled. It is important to remember that increasing temperature *always* increases reaction rates and decreasing temperature *always* decreases reaction rates.

Finally, a **catalyst** will increase the rate of reaction. A catalyst is a substance that participates in a chemical reaction but does not appear in the balanced equation. Perhaps the most familiar catalysts are those in the catalytic converters in automobiles. The platinum in a catalytic converter provides a surface on which reactants meet and react more efficiently. Although the platinum promotes the effective reaction of two other chemicals, it is not included in the balanced equation.

Other experimental factors, as long as they do not affect the concentration, temperature, or catalysts, will have no effect on the rate of a chemical reaction.

EFFECT OF CONCENTRATION ON REACTION RATES

The effect of concentration on a chemical reaction is expressed in the **rate law**. All rate laws start with the same form

$$\text{Rate} = k[A]^x[B]^y[C]^z \qquad (10.3)$$

In this equation k stands for the **rate constant,** and the square brackets indicate the concentrations of the reactants A, B, and C, which have exponents x, y, and z. These exponents are usually small whole numbers. In more complex rate laws, they may be negative numbers or rational fractions. Exponents must be determined from laboratory experiments and have no relationship to the stoichiometric coefficients of the balanced chemical equation.

However, we note that the rate law usually contains more than one substance and we might ask what rate do we use? In fact, we calculate the rate based on a coefficient of one in the balanced chemical equation. To do this, the rate of appearance (or disappearance) of a substance is divided by its stoichiometric coefficient. The general reaction

$$aA + bB \rightarrow cC + dD$$

has one rate of reaction that is

$$\text{Rate} = \frac{-dA}{a\,dt} = \frac{-dB}{b\,dt} = \frac{+dC}{c\,dt} = \frac{+dD}{d\,dt}$$

Since each rate of formation or disappearance is divided by the corresponding stoichiometric coefficient, we will get the same reaction rate no matter which reactant or product is measured.

DETERMINATION OF RATE LAWS

All rate laws must be determined by using the data from a group of well-planned experiments. By changing the concentration of one reactant while holding all other concentrations constant, we can determine whether the change we made has an effect on the rate; and, if it does, we can calculate the exponent for the changed reactant in the rate law. **There is no theoretical way to predict the exponents of a rate law.**

An example of the method used to determine rate laws can be shown using the reaction of peroxydisulfate with iodide ions according to the equation

$$S_2O_8^{2-} + 3I^- \rightarrow 2SO_4^{2-} + I_3^- \tag{10.4}$$

Table 10.1 shows three experiments using this reaction. In experiments 1 and 2 the concentration of iodide ion is the same and the concentrations of peroxydisulfate ions are different. In experiments 2 and 3 the iodide ion concentration is changed while the peroxydisulfate concentration is held constant. The measured initial rate for each experiment is given in the last column.

Table 10.1
Kinetic Data for the Peroxydisulfate Reaction at 20.0°C

Experiment Number	$[S_2O_8^{2-}]$ (mol L^{-1})	$[I^-]$ (mol L^{-1})	Initial Rate of Reaction (mol L^{-1} s^{-1})
1	0.200	0.200	2.2×10^{-3}
2	0.400	0.200	4.4×10^{-3}
3	0.400	0.400	8.8×10^{-3}

The rate law will have the form

$$\text{Rate} = k[S_2O_8^{2-}]^x[I^-]^y \tag{10.5}$$

where the exponents x and y need to be determined.

To determine the exponent for the peroxydisulfate ion, we focus attention on the change in reaction rate as the concentration of $S_2O_8^{2-}$ is changed in experiments 1 and 2. A rate law can be written for experiment 1 and another for experiment 2 by entering the data from the table into Equation 10.5:

$$\text{Rate}_1 = 2.2 \times 10^{-3} = k(0.200)^x(0.200)^y$$
$$\text{Rate}_2 = 4.4 \times 10^{-3} = k(0.400)^x(0.200)^y$$

The ratio of these two equations is then written (the calculations are usually easier if the larger numbers are used for the numerator) as

$$\frac{\text{Rate}_2}{\text{Rate}_1} = \frac{4.4 \times 10^{-3}}{2.2 \times 10^{-3}} = \frac{k(0.400)^x(0.200)^y}{k(0.200)^x(0.200)^y}$$

TIP

Write the ratio of the rate laws with the larger rate in the numerator.

The two k's and the two $(0.200)^y$ factors cancel to yield

$$\frac{\text{Rate}_2}{\text{Rate}_1} = \frac{4.4 \times 10^{-3}}{2.2 \times 10^{-3}} = \frac{(0.400)^x}{(0.200)^x}$$

Since both exponents are x, the right-hand term can be rewritten as

$$\frac{\text{Rate}_2}{\text{Rate}_1} = \frac{4.4\times10^{-3}}{2.2\times10^{-3}} = \left(\frac{0.400}{0.200}\right)^x$$

Solving this gives

$$\begin{aligned} 2.0 &= 2.00^x \\ &= 2.00^1 \end{aligned}$$

From this result we conclude that exponent $x = 1$. In a similar manner the value of exponent y is determined using experiments 2 and 3, where the concentration of $S_2O_8^{2-}$ is held constant and I^- is varied. The ratio of rate laws for experiments 2 and 3 is

$$\frac{\text{Rate}_3}{\text{Rate}_2} = \frac{8.8\times10^{-3}}{4.4\times10^{-3}} = \frac{k(0.400)^x(0.400)^y}{k(0.400)^x(0.200)^y}$$

$$\begin{aligned} 2.0 &= 2.00^y \\ &= 2.00^1 \end{aligned}$$

The value of exponent y is 1. Using the exponents determined in this manner, we write the rate law:

$$\text{Rate} = k[S_2O_8^{2-}]^1[I^-]^1 \quad \text{or} \quad k[S_2O_8^{2-}][I^-]$$

The exponents in the rate law are definitely not the same as the exponents in balanced Equation 10.4.

Once the rate law is known, the rate constant can be calculated. This is done by taking *any* one of the three experiments in Table 10.1, substituting the values into the rate equation, and solving for the rate constant k. The rate constant should be the same whether experiment 1, 2, or 3 is chosen for this calculation. To verify this, we will calculate the rate constant for each experiment and express it in the proper units:

TIP

Observe how the units cancel in these calculations.

$$\text{Experiment 1: } k = \frac{2.2\times10^{-3}\ \text{mol L}^{-1}\ \text{s}^{-1}}{(0.200\ \text{mol L}^{-1})(0.200\ \text{mol L}^{-1})} = 0.055\ \text{L mol}^{-1}\ \text{s}^{-1}$$

$$\text{Experiment 2: } k = \frac{4.4\times10^{-3}\ \text{mol L}^{-1}\ \text{s}^{-1}}{(0.400\ \text{mol L}^{-1})(0.200\ \text{mol L}^{-1})} = 0.055\ \text{L mol}^{-1}\ \text{s}^{-1}$$

$$\text{Experiment 3: } k = \frac{8.8\times10^{-3}\ \text{mol L}^{-1}\ \text{s}^{-1}}{(0.400\ \text{mol L}^{-1})(0.400\ \text{mol L}^{-1})} = 0.055\ \text{L mol}^{-1}\ \text{s}^{-1}$$

REMEMBER THE DIFFERENCE

REACTION RATE (or simply rate). This varies with the concentration of reactants and time. Rate always has units of $C\ t^{-1}$ (concentration/time) and is commonly expressed as mol L^{-1} s^{-1}.

RATE CONSTANT. This is a constant value at a fixed temperature for a given reaction. The units of a rate constant depend on the order of the reaction.

We have demonstrated that an easy way to verify the rate law is to calculate the rate constant for each experiment. The rate constants should all agree within experimental error; if they do not, something is wrong with the rate law.

Chemists use the term **order of reaction** to indicate the exponents in the rate law. The reaction we have been considering is said to be **first order** with respect to peroxydisulfate and first order with respect to iodide ions. It is also said to be **second order overall**. Individual reactants have the same orders as their exponents. The order of the overall reaction is the sum of all of the exponents.

To summarize, we can say that in this reaction:

(a) the order with respect to peroxydisulfate is 1 (first order with respect to peroxydisulfate);
(b) the order with respect to iodide ion is one (first order with respect to iodide);
(c) the overall order is $1 + 1 = 2$ (second order overall);
(d) the units for the reaction rate are always mol L^{-1} s^{-1};
(e) the units for this rate constant are (mol L^{-1} s^{-1})/(mol L^{-1})^2 = L mol^{-1} s^{-1}:

EXAMPLE 10.2

Determine the rate law and value of the rate constant, with its units, for the data in the table below.

Experiment Number	[*A*] (mol L^{-1})	[*B*] (mol L^{-1})	Initial Rate of Reaction (mol L^{-1} s^{-1})
1	0.100	0.200	1.1×10^{-6}
2	0.100	0.600	9.9×10^{-6}
3	0.400	0.600	9.9×10^{-6}

Solution

The general form of the rate law will be

$$\text{Rate} = k[A]^x [B]^y$$

Taking the ratio of the rate laws for experiments 1 and 2 yields

$$\frac{\text{Rate}_2}{\text{Rate}_1} = \frac{k(0.100)^x (0.600)^y}{k(0.100)^x (0.200)^y}$$

Cancelling the identical terms, k and $(0.100)^x$, on the right and entering the rates on the left yields

$$\frac{9.9 \times 10^{-6}}{1.1 \times 10^{-6}} = \frac{(0.600)^y}{(0.200)^y}$$

$$9 = 3^y$$

Three squared is equal to 9, and therefore $y = 2$.

To determine exponent x, the ratio of the rate laws for experiments 2 and 3 are used:

$$\frac{\text{Rate}_3}{\text{Rate}_2} = \frac{k(0.400)^x(0.600)^y}{k(0.100)^x(0.600)^y}$$

Entering the rates and cancelling as before gives

$$\frac{9.9 \times 10^{-6}}{9.9 \times 10^{-6}} = \frac{(0.400)^x}{(0.100)^x}$$

$$1.0 = 4^x$$

Any number raised to the zero power is equal to 1, and therefore $x = 0$. The rate law is

$$\text{Rate} = k[A]^0[B]^2 = k[B]^2$$

Since the concentration of A has no effect on the reaction rate, it is not part of the rate law.

The rate constant is determined by taking any of the three experiments, substituting the rate and concentrations into the rate law, and calculating k. Using the data from experiment 1 yields

$$1.1 \times 10^{-6} \text{ mol } B \text{ L}^{-1} \text{ s}^{-1} = k(0.200 \text{ mol } B \text{ L}^{-1})^2$$

$$k = \frac{1.1 \times 10^{-6} \text{ mol } B \text{ L}^{-1} \text{ s}^{-1}}{(0.200 \text{ mol } B \text{ L}^{-1})^2}$$

$$= 2.75 \times 10^{-5} \text{ L mol}^{-1} \text{ s}^{-1}$$

EXAMPLE 10.3

A certain reaction between substances X and Y was studied to determine its rate law. The balanced chemical equation shows that $2X + Y \longrightarrow$ Products. To determine the rate law, the initial reaction rates were determined for various concentrations of X and Y as shown in the following table. Determine the overall order of this reaction and the order with respect to both X and Y.

Experiment Number	$[X]$ (mol L^{-1})	$[Y]$ (mol L^{-1})	Initial Rate of Reaction (nmol L^{-1} s^{-1})
1	0.00345	0.00444	23.45
2	0.00345	0.00888	46.95
3	0.00690	0.01332	281.4
4	0.01035	0.01776	844.2

Solution

From the data in the table, we can see that in the first two experiments the concentration of X was held constant while the concentration of Y was doubled. We can see that the initial rate of reaction also doubled. From this, we can deduce without setting up the equations that the exponent for $[Y]$ must be 1.

The challenge of this problem is that no two experiments have the same concentration of Y. Let's write the rate laws for the experiments 3 and 4 using m for the exponent of $[X]$ and 1 for the exponent of $[Y]$.

$$\text{Experiment 3: } 281.4 \text{ nmol L}^{-1}\text{ s}^{-1} = k(0.00690 \text{ mol L}^{-1})^m(0.01332 \text{ mol L}^{-1})^1$$
$$\text{Experiment 4: } 844.2 \text{ nmol L}^{-1}\text{ s}^{-1} = k(0.01035 \text{ mol L}^{-1})^m(0.01776 \text{ mol L}^{-1})^1$$

Now divide both equations by their concentrations of Y to get

$$\text{Experiment 3: } (281.4 \text{ nmol L}^{-1}\text{ s}^{-1})/(0.01332 \text{ mol L}^{-1}) = k(0.00690 \text{ mol L}^{-1})^m$$
$$\text{Experiment 4: } (844.2 \text{ nmol L}^{-1}\text{ s}^{-1})/(0.01776 \text{ mol L}^{-1}) = k(0.01035 \text{ mol L}^{-1})^m$$

After dividing by their concentrations of Y, the equations are

$$\text{Experiment 3: } 2.12 \times 10^{-5}\text{ s}^{-1} = k(0.00690 \text{ mol L}^{-1})^m$$
$$\text{Experiment 4: } 4.75 \times 10^{-5}\text{ s}^{-1} = k(0.01035 \text{ mol L}^{-1})^m$$

Finally, take the ratios of the equations above. Note how the larger numbers are in the numerators

$$\frac{(4.75 \times 10^{-5}\text{ s}^{-1})}{(2.12 \times 10^{-5}\text{ s}^{-1})} = \frac{k(0.01035 \text{ mol L}^{-1})^m}{k(0.00690 \text{ mol L}^{-1})^m}$$

Cancel the k's on the right side of the equation and do the division steps to get

$$2.24 = 1.50^m$$

To find the value of m, substitute a few integers for m. When $m = 2$, $1.50^m = 2.25$, which is very close to the 2.24 shown in the previous equation. Therefore $m = 2$.

The rate law is:

$$\text{Rate} = k\,[X]^2[Y]$$

The reaction is third order overall, is second order with respect to X, and is first order with respect to Y.

Exercise 10.2

What is the overall order of each of the following rate laws, and what are the units of the rate constant, k, in each of these rate laws?

(a) Rate = $k\,[A]\,[B]\,[C]$
(b) Rate = $k\,[X]^2\,[Y]^3$
(c) Rate = $k\,[M]^2\,[N]$
(d) Rate = k
(e) Rate = $k\,[R]$

Solution

(a) Order = 3. Units are $\text{L}^2\text{ mol}^{-2}\text{ s}^{-1}$.
(b) Order = 5. Units are $\text{L}^4\text{ mol}^{-4}\text{ s}^{-1}$.
(c) Order = 3. Units are $\text{L}^2\text{ mol}^{-2}\text{ s}^{-1}$.
(d) Order = 0. Units are $\text{mol L}^{-1}\text{ s}^{-1}$.
(e) Order = 1. Units are s^{-1}.

Effect of Temperature on Reaction Rates

Almost instinctively we add heat when we wish to increase the speed of a reaction. Higher temperatures increase the rates at which foods cook and solids dissolve in water. Cold-blooded animals such as snakes and lizards are almost immobile in cold temperatures, and they sun themselves to warm up in order to hunt for food.

In 1889 Svante Arrhenius developed the equation for the relationship between the rate constant and temperature. It is called the **Arrhenius equation** in his honor.

$$k = Ae^{-E_a/RT} \tag{10.6}$$

This equation includes the rate constant, k; the activation energy, E_a; the universal gas law constant, R, which equals 8.314 J mol^{-1} K^{-1}; the Kelvin temperature, T; a proportionality constant, A; and the base of natural logarithms, e. When the natural logarithm of Equation 10.6 is taken, the result is

$$\ln k = \frac{-E_a}{RT} + \ln A \tag{10.7}$$

This equation can be utilized in two ways to eliminate the need to know the value of the constant A. First, a graph of ln k versus $1/T$ can be constructed after determining the rate constant at a variety of temperatures. This graph, shown in Figure 10.3, is often called an Arrhenius plot. The slope of the line is equal to $-E_a/R$. Arrhenius plots are used to determine the activation energy, E_a, and also the rate constant at any desired temperature.

TIP

You will not have to solve the Arrhenius equation on the AP exam. However, you do need to understand it.

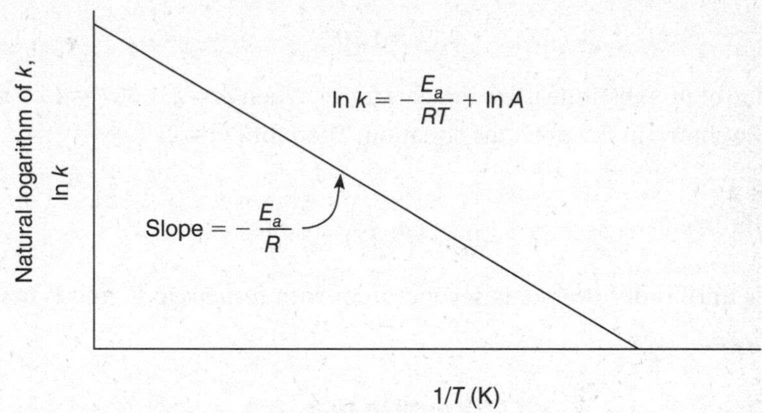

FIGURE 10.3. An Arrhenius plot of ln $\boldsymbol{k}$ versus $1/\boldsymbol{T}$. The activation energy, $\boldsymbol{E_a}$, is determined from the slope of the straight line.

The second way to apply Equation 10.7 is a two-point approach using only the rate constants determined at two different temperatures. The resulting equation is

$$\ln\left(\frac{k_1}{k_2}\right) = \frac{-E_a}{R}\left(\frac{1}{T_1} - \frac{1}{T_2}\right) \tag{10.8}$$

This equation has five variables, E_a, k_1, k_2, T_1, and T_2. Given four of these, we can determine the fifth by substitution into the equation. Equation 10.8 may also be written using the ratio of the rates rather than the rate constants:

$$\ln\left(\frac{\text{rate}_1}{\text{rate}_2}\right) = \frac{-E_a}{R}\left(\frac{1}{T_1} - \frac{1}{T_2}\right) \tag{10.9}$$

The rate and the rate constant are directly proportional to each other as long as the concentrations are held constant.

Often the signs in this equation become confused since the calculation is somewhat involved. It is always possible, however, to test the answer for reasonableness. Two principles must be remembered:

1. The larger rate constant (or rate) will *always* be associated with the higher temperature.
2. The activation energy *always* has a positive sign.

Look back at Equation 10.6. If E_a increases, it makes the exponent a larger negative number, resulting in a smaller rate constant (and rate). Also, an increase in T makes the negative exponent smaller, resulting in an increased reaction rate.

TIP

The reaction rate and rate constant are both larger at higher temperatures as long as all other factors are held constant.

APPLICATIONS OF SELECTED RATE LAWS

Zero-Order Reactions

A reaction that is **zero order** has a rate law in which the exponents of all of the reactants are zero. Since any number raised to the zero power is equal to 1 ($x^0 = 1$), the rate law is

$$\text{Rate} = k$$

This type of reaction does not depend on the concentration of any reactant. Its rate is equal to the rate constant, and the rate constant has the same units (mol L^{-1} s^{-1}) as a rate.

Catalytic reactions are often zero-order reactions. One catalyzed reaction is the decomposition of hydrogen peroxide in the presence of platinum metal:

$$2H_2O_2 \rightarrow 2H_2O + O_2$$

The concentration versus time plot for any zero-order reaction is a straight line, as shown in Figure 10.4.

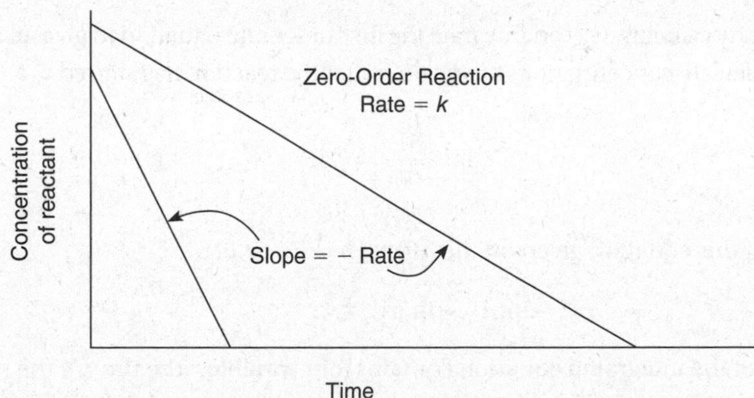

FIGURE 10.4. Kinetic curves of two different zero-order reactions, illustrating the straight-line relationship between concentration and time.

The integrated rate equation for a zero-order reaction is $[A] = [A]_0 - kt$ where $[A]_0$ is the initial concentration of the reactant. The half-life of a zero-order reaction obeys the equation $t_{1/2} = [A]_0/2k$.

First-Order Reactions

Any rate law in which the sum of the exponents is 1 is a **first-order reaction**. There are a variety of ways in which a reaction can be first order, including the case where the exponents of two reactants are 0.5 each. In the most common case the concentration of one reactant, *A*, has an exponent of 1 as in the following rate law:

$$\text{Rate} = k[A]$$

As a first-order reaction progresses, the concentration of reactant A decreases and the rate decreases. When the concentration of *A* has decreased to half its original amount, the rate will be half the initial rate. The time required for this to occur is called the **half-life**, $t_{1/2}$. In another half-life the concentration will decrease by half again, to one-fourth of the initial concentration. Figure 10.5 illustrates the shape of the kinetic curve for all first-order reactions.

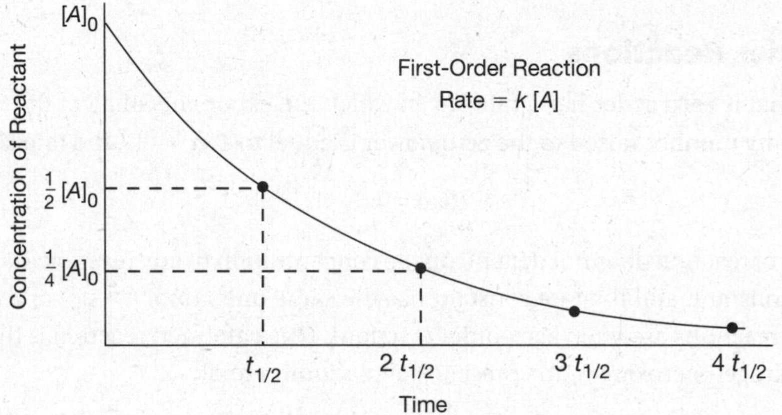

FIGURE 10.5. Kinetic curve for all first-order reactions. The time at which the concentration is one-half the original concentration is the half-life, $t_{1/2}$.

THE INTEGRATED EQUATION

Using elementary calculus, we can integrate the first-order rate equation to give an equation that allows us to calculate concentrations at any time after the reaction has started:

$$\ln\left(\frac{[A]_0}{[A]_t}\right) = kt \qquad (10.10)$$

The form of this equation given on the AP exam is

$$\ln[A]_0 - \ln[A]_t = kt \qquad (10.11)$$

Each form of the integrated equation contains four variables: the time, *t*; the rate constant, *k*; an initial concentration of *A*, $[A]_0$; and the concentration at some time after the reaction is started, $[A]_t$. Using Equation 10.11, chemists often plot the natural logarithm of the reactant concentration versus time in a graph as shown in Figure 10.6. In this type of graph, only a first-order reaction will result in a straight line, and the slope will be equal to $-k$.

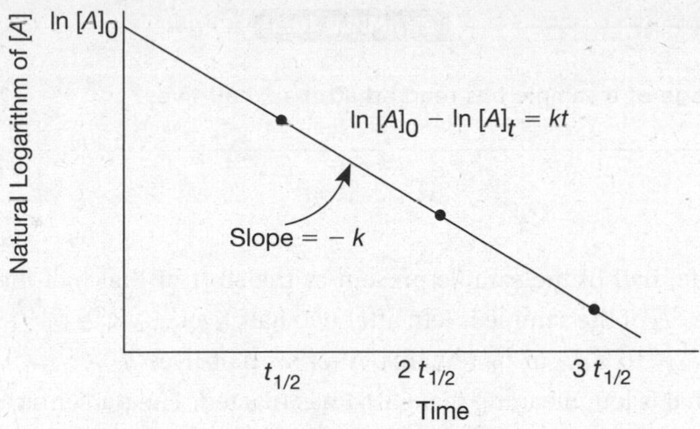

FIGURE 10.6. Logarithmic plot of a first-order reaction.

EXAMPLE 10.4

A certain first-order reaction has a rate constant of 4.5×10^{-3} s^{-1}. How much of a 50.0 millimolar (m*M*) sample will have reacted after 75.0 s?

Solution

Substitution of the data into integrated rate Equation 10.11 yields

$$\begin{aligned} \ln(0.0500) - \ln[A]_t &= (4.5\times10^{-3}\ \text{s}^{-1})(75.0\ \text{s}) \\ -2.995 - \ln[A]_t &= 0.3375 \\ \ln[A]_t &= -3.333 \\ [A]_t &= 0.0356\ M \end{aligned}$$

This indicates that the sample is now 35.6 m*M*. The question asks how much of the sample has reacted; the answer is

$$50.0\ \text{m}M - 35.6\ \text{m}M = 14.4\ \text{m}M$$

Half-Lives

As mentioned previously, a first-order reaction has a constant half-life, which is the time required for half of the reactant to be consumed. When half of the reactant is used up, $[A]_t = 0.5\,[A]_0$. Substituting into Equation 10.10 yields

$$\begin{aligned} \ln\left(\frac{0.5[A]_0}{[A]_0}\right) &= -kt_{1/2} \\ \ln(0.5) &= -kt_{1/2} \\ -0.693 &= -kt_{1/2} \\ t_{1/2} &= \frac{0.693}{k} \qquad (10.12) \end{aligned}$$

where $t_{1/2}$ is the half-life of the reaction. Half-lives may be used to quickly estimate the fraction of the starting material left after it has been allowed to react for a given number of half-lives.

TIP

All radioactive decay obeys first-order kinetics.

EXAMPLE 10.5

What percentage of a sample has reacted after six half-lives?

Solution

After each half-life, half of the sample present at the start of that half-life will be reacted. After one half-life, $1/2$ of the sample is left; after two half-lives, $1/2 \times 1/2$ or $1/4$ will remain; after three half-lives, $1/2 \times 1/2 \times 1/2$ or $1/8$ remains. After six half-lives $1/2 \times 1/2 \times 1/2 \times 1/2 \times 1/2 \times 1/2 = 1/64$ of the original is left, meaning that 63/64 has reacted. The percentage that has reacted is calculated as

$$\frac{63}{64} \times 100 = 98.4\% \text{ reacted}$$

The most prominent first-order processes in chemistry are those associated with radioactive decay of the elements. The decay of radioactive elements follows first-order kinetics.

Rate of Radioactive Decay

Radioactive decay is a random event. We cannot predict when a single isotope will spontaneously decay. For a large group of radioactive atoms, however, we can apply the well-defined laws of kinetics. Radioactive decay is a first-order reaction process. The number of nuclei that disintegrate per second depends only on the number of radioactive nuclei in the sample. The equation for the number of radioactive disintegrations per second (dps) is as follows:

$$\begin{aligned}\text{Rate (dps)} &= \text{(Constant)(Number of radioactive nuclei)} \\ &= kN \end{aligned} \qquad (10.13)$$

This rate equation may be integrated by using simple calculus to obtain the expression

$$\ln\left(\frac{N_0}{N_t}\right) = kt \qquad (10.14)$$

where N_0 is the original number of radioactive atoms, N_t is the number of radioactive atoms left after t seconds have elapsed, k is the **rate constant** with units of reciprocal seconds (s^{-1}), and t is the time in seconds from the start of the experiment. Since the number of atoms in a sample is directly proportional to the mass of the sample, we may interpret N_0 and N_t as the masses of radioactive atoms at the start and at time $= t$, respectively.

When half of the atoms have decayed, $N_0 = 2N_t$. Substituting $2N_t$ in place of N_0 in Equation 10.14 allows us to develop a relationship between the **half-life**, $t_{1/2}$, of an isotope and the specific rate constant, k:

$$\ln\left(\frac{2N_t}{N_t}\right) = kt_{1/2} \qquad (10.15)$$

N_t cancels on the left side of this equation to give

$$\ln(2) = kt_{1/2} \qquad (10.16)$$

Since the natural logarithm of 2 is 0.693,

$$t_{1/2} = \frac{0.693}{k} \tag{10.17}$$

In Equation 10.17 $t_{1/2}$ is called the half-life of the isotope.

The half-life is an important concept. It means that one-half of the radioactive isotope present at the start of an experiment will decompose in a length of time equal to the half-life. The bar graph in Figure 10.7 shows how the amount of radioactive material declines with each half-life. After one half-life half of the isotope is left; after two half-lives half of that half, or one-quarter of the original amount, is left. With the next half-life half of what remained after the second half-life decays, and only one-eighth is left. Note that each bar in Figure 10.7 is one-half the size of the one preceding it.

The half-life is another way to express the rate constant.

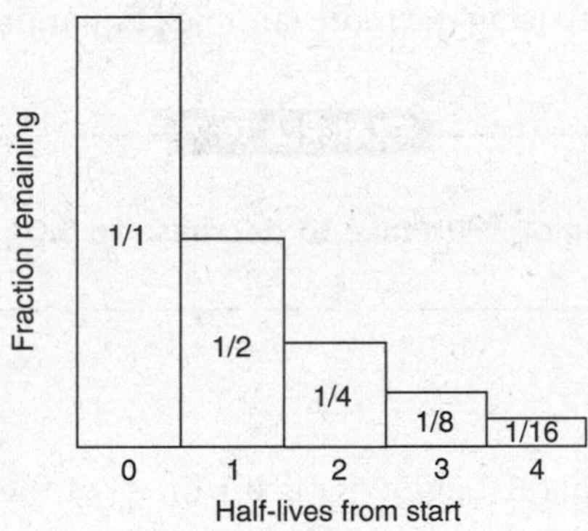

FIGURE 10.7. Bar graph illustrating how half of a radioactive material decomposes with each half-life.

We can calculate the fraction of the original sample left after a given number of half-lives by using Equation 10.18, where the fraction $\frac{1}{2}$ is raised to the power equal to the number of half-lives:

$$\text{Fraction left} = \left(\frac{1}{2}\right)^{\text{number of half-lives}} \tag{10.18}$$

The decimal value of these fractions multiplied by 100 gives us the percentage of the radioactive isotope remaining. The percentage that has decayed is obtained by subtracting the percentage remaining from 100.

Radioactive Decay Calculations

Calculations involving the disintegration of radioactive isotopes require a thorough understanding of Equations 10.13, 10.14, 10.17, and 10.18. Usually such calculations require a combination of two or more of these equations.

TIP

Half-lives can be used to estimate answers for first-order reactions.

EXAMPLE 10.6

The half-life of ^{210}Pb is 25 years. In a sample starting with 50 μg of ^{210}Pb, how much is left after 100 years?

Solution

Since 25 years is one half-life, 100 years is equal to four half-lives for ^{210}Pb. Therefore $\frac{1}{16}$ of the original amount will remain (see Figure 10.7). To calculate the amount remaining, the $\frac{1}{16}$ or its decimal equivalent (0.0625) is multiplied by the original amount given. Mathematically we may calculate:

$$? \text{ half-lives} = 100 \text{ yr}\left(\frac{1 \text{ half-life}}{25 \text{ yr}}\right) = 4 \text{ half-lives}$$

$$? \text{ fraction left} = \left(\frac{1}{2}\right)^4 = 0.0625$$

$$? \text{ amount left} = 50\ \mu\text{g}\,(0.0625) = 3.1\ \mu\text{g}$$

Whenever a problem involves a whole number of half-lives, it is usually simpler to solve it as was done here, rather than using the more complex Equation 10.14.

EXAMPLE 10.7

How many years will 50 μg of ^{210}Pb take to decrease to 5.0 μg ($t_{1/2}$ = 25 years)?

Solution

A quick calculation shows that 5.0 μg/50 μg is 0.1 or $\frac{1}{10}$ of the original amount. From the discussion above we know that after three half-lives (75 years) $\frac{1}{8}$ remains and that after four half-lives (100 years) $\frac{1}{16}$ remains. We may estimate that the answer should be between three and four half-lives (75 and 100 years). If a more precise answer is needed, Equation 10.14 must be used:

$$\ln\left(\frac{N_0}{N_t}\right) = kt$$

Here N_0 is the original number of atoms of the isotope, and N_t is the number remaining after time t has elapsed. We can use the mass of the isotope in place of the number of atoms since mass is directly proportional to number of atoms. Also, k is the rate constant for the radioactive disintegration, calculated from Equation 10.17. Then

$$t_{1/2} = \frac{0.693}{k}$$

Substitution yields:

$$k = \frac{0.693}{25 \text{ yr}} = 2.77 \times 10^{-2} \text{ yr}^{-1}$$

Entering our data into Equation 10.14 gives us

$$\ln\left(\frac{50 \text{ g}}{5.0 \text{ g}}\right) = (2.77 \times 10^{-2} \text{ yr}^{-1})t$$

$$\frac{2.30}{2.77 \times 10^{-2} \text{ yr}^{-1}} = t = 83 \text{ years}$$

As expected, the result fits the preliminary estimate of somewhere between 75 and 100 years. Effective use of good estimates cannot be overemphasized. At times an estimate allows us to select the correct answer to a multiple-choice problem. If not, the estimate serves as a check on our calculated answer.

EXAMPLE 10.8

Ninety-nine percent of a radioactive element disintegrates in 36.0 hours. What is the half-life of this isotope?

Solution

First, it must be realized that the problem states how much of the element has disintegrated. Half-life calculations using our equations refer to the amount that is left (N_t). Subtracting 99 from 100%, we find only 1% (or $\frac{1}{100}$) left after 36.0 hours. Next we can estimate the half-life by continuing the sequence depicted in Figure 10.7. We find that $\frac{1}{32}$ is left after five half-lives, $\frac{1}{64}$ after six half-lives, and $\frac{1}{128}$ after seven half-lives. Therefore our sample has undergone at least six, but not quite seven, half-lives. Six half-lives in 36.0 hours means a half-life of 6 hours. Seven half-lives in 36.0 hours is approximately 5 hours for a half-life. Our estimate of the half-life is between 5 and 6 hours.

To calculate the exact half-life (if needed), we substitute the data into Equation 10.14:

$$\ln\left(\frac{N_0}{N_t}\right) = kt$$

$$\ln\left(\frac{100}{1}\right) = k(36.0\text{ hr})$$

$$4.605 = k(36.0\text{ hr})$$

$$k = \frac{4.605}{36.0} = 0.128\text{ hr}^{-1}$$

Then we use Equation 10.17:

$$t_{1/2} = \frac{0.693}{k}$$

$$= \frac{0.693}{0.128\text{ hr}^{-1}}$$

$$= 5.41\text{ hr}$$

This answer agrees with our quick estimate. Often the estimate is sufficient to make an informed selection of the correct answer to a multiple-choice question. More important, it allows you to reject incorrect answers quickly so that you increase your chances if you must guess at an answer.

Second-Order Reactions

Two second-order rate laws are

$$\text{Rate} = k\,[A]^2$$

$$\text{Rate} = k\,[A][B]$$

In the special case where $[A] = [B]$ the two rate laws can be integrated to obtain the same equation.

$$\frac{1}{[A]_t} - \frac{1}{[A]_0} = kt \qquad (10.19)$$

The half-life of the second-order reactions specified above may be determined by substituting $0.5\,[A]_0$ for $[A]_t$ in Equation 10.19.

$$\frac{1}{0.5[A]_0} - \frac{1}{[A]_0} = kt_{1/2}$$

$$\frac{2}{[A]_0} - \frac{1}{[A]_0} = kt_{1/2}$$

$$\frac{1}{[A]_0} = kt_{1/2}$$

$$t_{1/2} = \frac{1}{k\,[A]_0}$$

The important information derived from these equations is that the half-life of a second-order reaction depends on the starting concentrations of the reactants. In addition, a straight-line graph will be obtained by plotting $1/[A]$ versus time. The slope of the line will be equal to k, as shown in Figure 10.8.

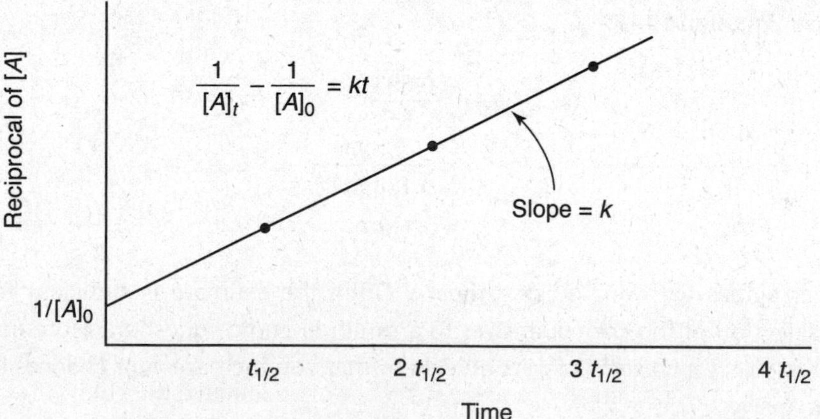

FIGURE 10.8. Plot of $1/[A]$ versus time, illustrating the straight-line relationship for a second-order reaction.

THEORY OF REACTION RATES

KINETICS THEORIES SUMMARIZED

COLLISION THEORY: Collision theory states that the rate of a chemical reaction is equal to the collision rate (a very large number) decreased by multiplying by an orientation factor and a minimum energy factor. The actual reaction is visualized as the collision of hard spheres such as billiard balls.

TRANSITION-STATE THEORY: Transition-state theory looks at the energy changes and geometric changes of molecules as they collide. During the collision process, kinetic energy is converted to potential energy. If this potential energy meets or exceeds the activation energy, the reaction can occur. In addition, there is change in the geometry of the reactants as they become products. The geometry somewhere in the middle of this conversion is called the transition state, and it occurs when the maximum kinetic energy has been converted to potential energy (i.e., at the top of the potential energy profile). The potential energy profile also indicates the heat of reaction.

Collision Theory

The **collision theory** states that the reaction rate is equal to the frequency of effective collisions between reactants. For a collision to be effective, the molecules must collide with sufficient energy and in the proper orientation so that products can form.

The minimum energy needed for a reaction is the activation energy, E_a. If two molecules collide head on, they will stop at some point and all of the kinetic energy will be converted into potential energy. If the molecules strike each other with a glancing blow, however, only part of the kinetic energy will be converted into potential energy. As long as the increase in potential energy is greater than E_a, a reaction is possible. The fraction of all collisions that have the minimum energy needed for reaction can be calculated using the **kinetic molecular theory** of gases discussed in Chapter 6. The fraction of collisions with this minimum energy increases with rising temperature since the average kinetic energy of molecules increases as temperature increases.

In addition to the energy requirement, for an effective collision the molecules must collide in the proper orientation. An example is the reaction of hydrogen iodide molecules with chlorine atoms. In this reaction the chlorine replaces the iodine in the molecule:

$$\mathrm{HI} + \mathrm{Cl} \rightarrow \mathrm{HCl} + \mathrm{I}$$

Figure 10.9 shows hydrogen iodide colliding with a chlorine atom from two directions. When, as in the top row, chlorine collides with the iodide end of HI, the reactants recoil from the collision without a reaction occurring. In the other sequence the chlorine atom collides with the hydrogen end of HI. This collision can cause the iodine atom to be released while the chlorine bonds with the hydrogen. After collision the products HCl and I are present.

The overall reaction rate predicted by the collision theory may be summarized by the equation

$$\text{reaction rate} = Nf_e f_o$$

where N represents the number of collisions per second, which depends on the temperature and concentration of the reactants; f_e is the fraction of the collisions with the minimum energy; and f_o is the fraction of collisions with the correct orientation. The fraction f_e will increase as temperature increases, while f_o remains constant for a given reaction.

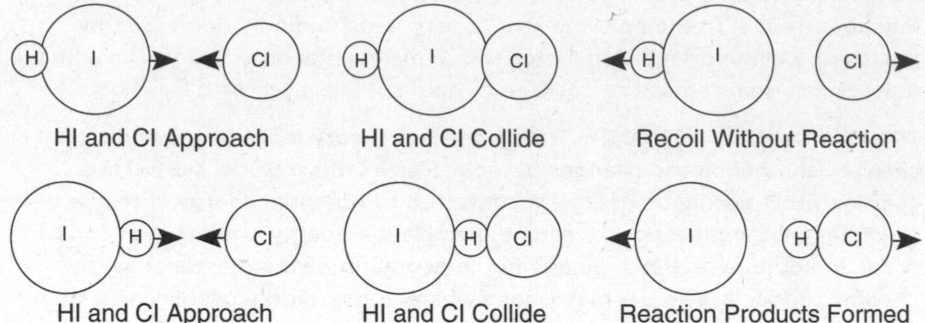

FIGURE 10.9. Orientation needed for the reaction of hydrogen iodide with chlorine. The top row shows an ineffective collision; the bottom row, an effective collision, forming products.

Transition-State Theory

The **transition-state theory** attempts to describe in detail the molecular configurations and energies as a collision of reactants occurs. This theory recognizes that, as molecules approach on a collision course, they do not act like billiard balls simply bouncing off each other. Instead, as the molecules get closer, their orbitals interact and distort each other. This distortion weakens bonds within the molecules so that at the moment of closest approach some bonds are so weak that they break and new bonds may form.

Using the diagram in Figure 10.9, we can visualize the effective collision in this sequence. First the chlorine approaches the hydrogen end of the HI molecule. As the Cl and HI get closer, the very electronegative Cl starts attracting the electrons that the hydrogen shares with the iodine atom. As a result the H–I bond is weakened and a H–Cl bond starts to form. At the moment of closest approach, the H–I bond is approximately half broken and the H–Cl bond is approximately half formed. This state is called the **activated complex**. When the atoms recoil, the activated complex breaks apart. The result may be a successful reaction giving new products, or an unsuccessful collision with the original reactants remaining intact.

THE REACTION PROFILE (POTENTIAL ENERGY DIAGRAM)

In the transition-state theory, the energies of the reactants during the collision are described by a reaction profile, which is also called a potential energy diagram. As the molecules approach, interact, and become distorted, their potential energy must increase. This potential energy increase comes from an equal decrease in the kinetic energy of the molecules. In other words, as the molecules collide, they slow down and their kinetic energy is converted to potential energy. After they reach the point of closest approach, they recoil and the potential energy is converted back to kinetic energy.

The **reaction profile** plots the increase in potential energy of the reactants as they approach, reaching a maximum at the moment of collision, and then the decrease in potential energy as the products recoil. A reaction profile is shown in Figure 10.10. In a reaction profile the minimum amount of kinetic energy that must be converted into potential energy in order to form products is called the activation energy, E_a. It is often referred to as an energy barrier between the reactants and the products.

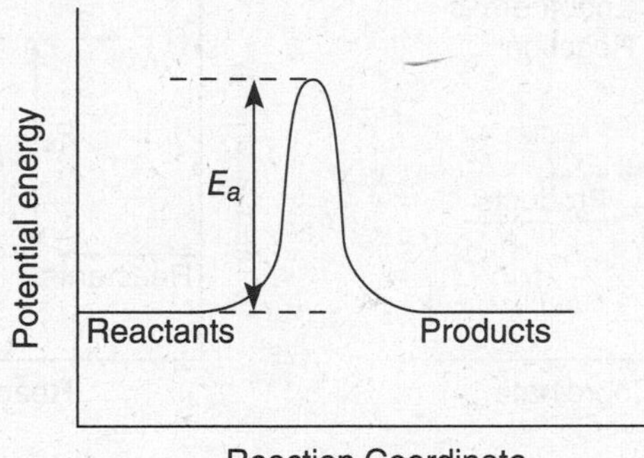

FIGURE 10.10. Reaction profile illustrating the energy barrier between reactants and products.

The transition-state theory is based on the same considerations as the collision theory. First, if reactants collide with enough energy to surmount the energy barrier, a reaction may occur. Second, if the activated complex formed at the moment of collision (top of the energy barrier) has the proper structure, it can proceed to fall apart into products. If it has the wrong structure, however, products cannot form and the molecules recoil as the original reactants. These energy and orientation factors are the same ones that are important to the collision theory.

The major difference in the two theories is that the collision theory views reactions as collisions between hard spheres, similar to the collisions between billiard balls. The transition-state theory, on the other hand, views the collisions as interactions between reactants that are deformed in the collision process. The transition-state theory involves more details about the energy and shapes of the molecules as they collide than the collision theory.

INTERPRETATION OF REACTION PROFILES

Reaction profiles provide a rich source of information about the rates of chemical reactions and are also a graphical view of the conversion of reactants into products. These graphs are often more informative than words alone in describing features of the reaction process.

In a reaction profile, the rate constant, and therefore the rate of a chemical reaction, are inversely related to the height of the energy barrier, E_a. When the activation energy is low, a large proportion of the collisions will have sufficient energy for a reaction to occur. Conversely, a high activation energy indicates that few collisions will have enough energy to convert reactants into products.

Reaction profiles can be used to determine whether a reaction is endothermic or exothermic. This is possible because the potential energy difference between the products and the reactants is equal to the heat of reaction, ΔH:

$$\Delta H = PE_{products} - PE_{reactants}$$

When heat is absorbed from the surroundings, the reaction is endothermic and ΔH has a positive sign. For **endothermic reactions** the potential energy of the products is greater than the potential energy of the reactants. The reverse is true for **exothermic reactions**. The reaction profiles for these two cases are shown in Figure 10.11.

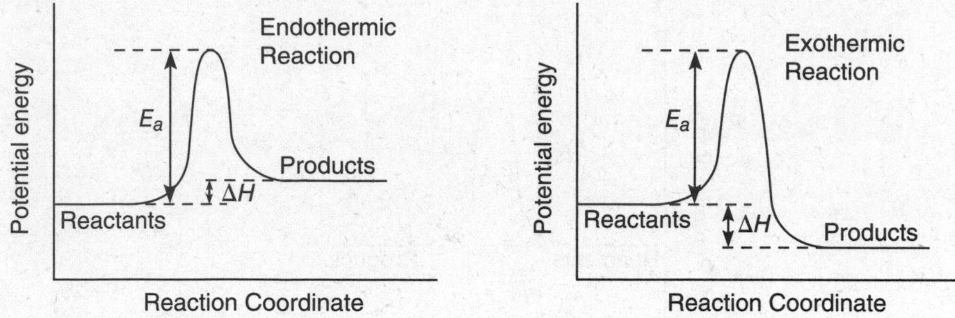

FIGURE 10.11. Reaction profiles illustrating the difference between an endothermic reaction and an exothermic reaction.

Another aspect of the reaction profile is that it allows the chemist to explain the reverse as well as the forward chemical reaction. To visualize what occurs in the reverse process, we look at the reaction profile, starting on the product side and proceeding toward the reactant side. When the products are reacting to form reactants, there is a different energy of activation and a different heat of reaction. The activation energy for the reverse reaction is the difference between the potential energy of the products and the maximum energy of the curve. For the heat of reaction, the sign of ΔH will be opposite to that for the forward reaction. Figure 10.12 is the same as Figure 10.11 except that it shows the activation energies and heats of reaction for the reverse reactions.

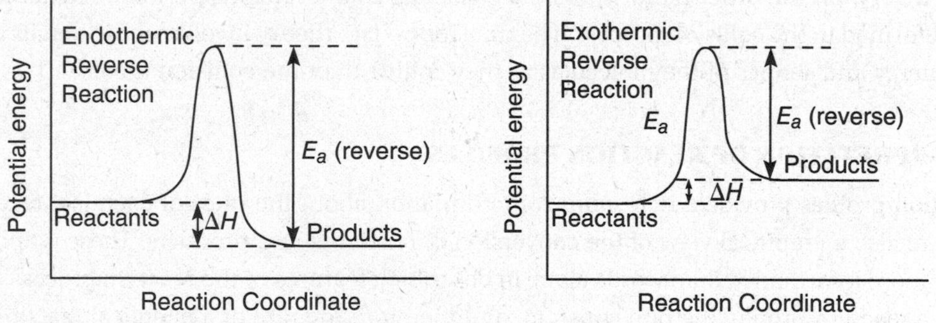

FIGURE 10.12. Reaction profiles illustrating the activation energy of reverse reactions and the fact that ΔH of the reverse reaction is opposite in sign to ΔH of the forward reaction.

Reaction profiles also allow us to explain the action of **catalysts** (Figure 10.13). As mentioned previously, a catalyst is a substance that increases the rate of a chemical reaction without itself being reacted. It speeds up a reaction by providing an alternative reaction pathway that has a lower energy barrier in the reaction profile. As a result the energies of activation of both the forward and reverse reactions are simultaneously decreased by the same amount, so that the reaction comes to chemical equilibrium more quickly. It must be kept in mind that a catalyst will not increase the amount of the product formed, nor will it alter the composition of the equilibrium mixture, as indicated by the unchanged potential energy plateaus for the reactants and products.

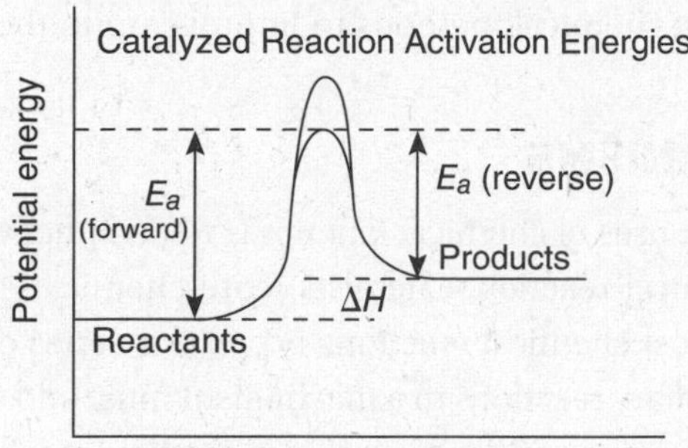

FIGURE 10.13. Reaction profile of a catalyzed reaction, illustrating that the forward and reverse activation energies are both decreased. The heat of reaction is not affected, nor is the position of equilibrium because the potential energies of the reactants and products are not affected.

There are a variety of catalysts and different ways to describe them. One scheme identifies catalysts as either homogeneous or heterogeneous. A homogeneous catalyst is in the same phase as the reactants and products. A heterogeneous catalyst is often a solid that is in contact with a liquid or gas mixture of reactants. Catalysts may also be defined by their chemical or physical composition. This classification system includes general acid–base catalysts; biochemical catalysts such as enzymes; and adsorbent catalysts, which are usually heterogeneous and include many metals and substances called zeolytes.

Enzymes are proteins made from long chains of amino acids linked together by peptide bonds. When these long chains fold into the final three-dimensional structure, reactants can be held in place to enhance reactions in a variety of ways.

General acid–base catalysts are usually conjugate acid–base systems that can provide hydrogen ions to a reaction or help remove hydrogen from reacting substances.

Adsorbent catalysts act by adsorbing reactants to the surface of a solid. While held on the catalyst, reactions can be accelerated due to changes in the shape or the electron distributions in the reactants that are caused by the catalyst. A specific example of a catalyst is the platinum metal used in the hydrogenation of ethylene. Ethylene has a double bond to which two hydrogen atoms may be added to form ethane:

$$\begin{array}{c} \text{H}\quad\text{H} \\ |\quad\;\; | \\ \text{C}=\text{C} \\ |\quad\;\; | \\ \text{H}\quad\text{H} \end{array} + H_2 \rightleftharpoons \begin{array}{c} \quad\;\;\text{H}\quad\text{H} \\ \quad\;\; |\quad\;\; | \\ \text{H}-\text{C}-\text{C}-\text{H} \\ \quad\;\; |\quad\;\; | \\ \quad\;\;\text{H}\quad\text{H} \end{array}$$

A mixture of hydrogen and ethylene at room temperature does not show appreciable reaction, but addition of a small amount of finely divided platinum catalyst causes a very rapid reaction. The apparent reason is that the hydrogen molecule is fairly stable, and the energy required to break the hydrogen atoms apart results in a high activation energy. Platinum adsorbs hydrogen readily on its surface and the hydrogen molecule separates into individual hydrogen atoms. These hydrogen atoms then readily react with the ethylene. At the end of the process, the original platinum can be recovered and used again.

The discovery of new, more effective, and more specific catalysts is a major objective of many industrial chemists. Some catalysts are naturally occurring minerals whose properties are often discovered by trial and error. Others are synthetic compounds designed by studying natural catalysts and using chemical methods to improve upon them.

REACTION MECHANISMS

One of the most important uses of chemical kinetics is to decipher the sequence of steps that lead to an observed chemical reaction. Chemists write chemical equations for reactions as a single step. However, most chemical reactions occur in a series of steps called elementary reactions. All of the elementary reactions in a mechanism must add up to the overall balanced equation. The complete sequence of steps is called a **reaction mechanism**.

Elementary Reactions

In a mechanism, the **elementary reactions** usually involve either one molecule breaking apart or the collision of only two reactant molecules. It is extremely rare that three molecules will collide simultaneously, and the simultaneous collision of more than three molecules is so infrequent that such collisions are hardly ever considered. Collisions between just two molecules, however, occur millions of times each second. Therefore, the most probable elementary reaction is one in which only two molecules collide. Another feature of the elementary reaction is that its **coefficients are the exponents** used in the rate law.

A sequence of elementary reactions will always have one step that is slower than all the rest. This slowest step determines the overall rate of reaction and is called the **rate-determining step** or the **rate-limiting step.** Since chemists can measure only the overall reaction rate in the laboratory, they are in fact measuring the rate of the slowest, rate-determining elementary reaction in the mechanism. Thus the rate law determined for a reaction is directly related to the rate-determining step.

Using these principles, chemists determine the rate law for a chemical reaction. They then postulate a series of elementary reactions based on the fact that one of the steps in the mechanism must obey the experimental rate law. When two or more mechanisms satisfy the rate law requirement, additional experiments must be done to decide which mechanism is correct.

To illustrate the procedure, we consider the reaction

$$H_2 + 2ICl \rightarrow I_2 + 2HCl \tag{10.20}$$

which has rate law

$$\text{Rate} = k[H_2][ICl]$$

A two-step mechanism that satisfies this rate law is

$$H_2 + ICl \rightarrow HI + HCl \tag{10.21}$$

$$HI + ICl \rightarrow I_2 + HCl \tag{10.22}$$

The two steps in this mechanism, Equations 10.21 and 10.22, add up to the overall reaction. They also involve only two molecules as reactants in each elementary reaction. The rate law describes the first step of the mechanism, which is presumably the slow step. Chemists can easily demonstrate that the second step is a much faster reaction by reacting HI and ICl.

The HI in the previous mechanism does not appear in the balanced equation, Equation 10.20. Any chemical species that is part of a mechanism but not part of the balanced equation is called an **intermediate**. In this example, the intermediate HI is also a known compound that made possible the experimental verification of the mechanism. In other cases intermediates are very unstable and often exotic chemical species. Demonstrating the presence of an intermediate provides evidence to support one mechanism over another.

Considering the above example, we see that there is another mechanism that might also be considered. This is a three-step process:

$$H_2 + ICl \rightarrow HI + HCl$$

$$H_2 + ICl \rightarrow HI + HCl$$

$$2HI \rightarrow H_2 + I_2 \qquad (10.23)$$

When added, these three reactions give the same overall reaction. To decide whether this mechanism is possible, scientists determined the rate of decomposition of the intermediate HI to H_2 and I_2 (the last step, Equation 10.23, in the mechanism). This reaction was found to be much slower than the reaction between H_2 and ICl. It was concluded that this is not the correct mechanism since the rate-determining step would give a completely different rate law.

From this description we can see that a kinetic study of a reaction will determine the rate law for the slowest step in the mechanism. Possible mechanisms are proposed, and the correct mechanism must have one step that obeys the observed rate law. If there are still several possible mechanisms, appropriate experiments must be devised to decide which mechanism is correct.

Determining the Rate Laws for Elementary Reactions

The coefficients of the reactants in an elementary reaction are the exponents of the reactant concentrations in the rate law. Therefore the rate law for each step of a mechanism can be predicted directly. One complication is that, after the first step, most of the elementary reactions in a mechanism will have an intermediate as a reactant. To compare an experimental rate law with the rate laws of the elementary reactions, it will be necessary to convert the rate law that has an intermediate into one that has only reactants.

For example, the reaction

$$2NO + O_2 \rightarrow 2NO_2$$

has a possible mechanism consisting of the two elementary reactions

$$NO + O_2 \rightarrow NO_3 \qquad (10.24)$$

$$NO_3 + NO \rightarrow 2NO_2$$

The rate law for the first step is

$$\text{Rate} = k[NO][O_2]$$

It contains the reactants of the experiment. The rate law for the second step includes the intermediate NO_3:

$$\text{Rate} = k[NO_3][NO]$$

To eliminate the NO_3 and to convert it into one of the reactants, we use the **steady-state assumption**. It says that if the second step is the rate-determining step, the first reaction, Equation 10.24, must be relatively fast and reversible. This means that the rate at which NO_3 is formed is equal to the rate at which it disappears:

$$NO_3 \text{ rate formation} = NO_3 \text{ rate disappearance}$$

The rate laws governing the forward and reverse reactions in the first step are

$$\text{Rate}_{\text{forward}} = k_f\,[NO][O_2]$$

$$\text{Rate}_{\text{reverse}} = k_r\,[NO_3]$$

Since the forward and reverse rates are equal, we can write

$$k_f\,[NO][O_2] = k_r\,[NO_3]$$

Solving for $[NO_3]$ gives

$$[NO]_3 = \frac{k_f}{k_r}[NO][O_2]$$

Substituting this result for the intermediate, NO_3, in the rate law gives

$$\text{Rate} = k\left(\frac{k_f}{k_r}\right)[NO][O_2][NO]$$

By combining all k, k_f, and k_r rate constants, we can write the rate law for the second step of the mechanism:

$$\text{Rate} = k\,[NO]^2\,[O_2]$$

EXAMPLE 10.9

The kinetics of the following reaction is studied:

$$2NO + O_2 \rightarrow 2NO_2$$

Two possible mechanisms are

$$2NO \rightarrow N_2O_2$$
$$N_2O_2 + O_2 \rightarrow 2NO_2$$

and

$$NO + O_2 \rightarrow NO_3$$
$$NO_3 + NO \rightarrow 2NO_2$$

Describe the method you would use to determine which mechanism is correct.

Solution

The rate laws for each of these elementary reactions can be determined. For the first mechanism they are

$$\text{Rate}_1 = k\,[\text{NO}]^2$$

$$\text{Rate}_2 = k\,[\text{N}_2\text{O}_2][\text{O}_2]$$

Using the steady-state assumption to obtain the rate law of the second elementary reaction in terms of measurable reactants, we obtain

$$k_f\,[\text{NO}]^2 = k_r\,[\text{N}_2\text{O}_2]$$

and the rate law will be

$$\text{Rate}_2 = k\,[\text{NO}]^2\,[\text{O}_2]$$

For the second mechanism the rate laws are

$$\text{Rate}_1 = k\,[\text{NO}][\text{O}_2]$$

$$\text{Rate}_2 = k\,[\text{NO}]^2\,[\text{O}_2]$$

The second rate law was derived above.

We can see that, if the first step is the slow step, the two mechanisms give two distinctly different rate laws and the decision is clear cut. If the second step is the slow step, however, both mechanisms yield the same rate law. To determine which mechanism is correct, additional experiments must be designed to identify the intermediate, NO_3 or N_2O_2, that is formed during the reaction.

SUMMARY

TIP

Data about reaction kinetics cannot be used to make conclusions about chemical equilibrium. Similarly, the value of an equilibrium constant cannot be used to draw conclusions about reaction rates.

In order to get to the equilibrium state described in Chapter 9, substances react at a finite rate. As opposed to equilibrium, where the overall concentrations do not change with time, the topic of kinetics in this chapter was all about change. The study of kinetics starts with the determination of reaction rates. Next, well-designed experiments are used to determine rate laws that express how concentrations affect reaction rates. Two special cases, first-order and second-order integrated equations, were discussed. In addition to concentration, temperature, the ability of reactants to meet, and the presence of a catalyst also affect the reaction rate. Two theories are used to explain reaction rates. One is the collision theory, and the other is the transition-state theory. These theories are part of virtually every AP exam.

Important Concepts

Reaction rates
Rate laws
Order of reaction
Half-lives
Collision theory
Transition-state theory and reaction profiles
Arrhenius equation
Reaction mechanisms
Elementary processes
Rate-determining step
Intermediate catalyst

Important Equations

Rate = $k\,[A]^x\,[B]^y \cdots$

$t_{1/2} = 0.693/k$

$\ln[A]_t - \ln[A]_0 = kt$ for a first-order reaction

PRACTICE EXERCISES

Multiple-Choice

1. Why does an increase in temperature increase the rate of a reaction?
   (A) The activation energy of the reaction increases.
   (B) The temperature acts as a catalyst in the reaction.
   (C) There are more collisions per second that will have the proper energy to exceed the activation energy.
   (D) There are a greater proportion of collisions with the correct orientation to be effective.

2. A change in temperature of 10 degrees was found to double the rate of a particular chemical reaction. How does this change in temperature affect the reacting molecules?
   (A) The average velocity of the molecules doubled.
   (B) The number of collisions per second doubled.
   (C) The number of molecules above the reaction energy threshold doubled.
   (D) The average energy of the molecules doubled.

3. In a series of chemical reactions, how do the reaction intermediates differ from the activated complexes?
   (A) The intermediates have structures with characteristics of both reactants and products.
   (B) The intermediates are unstable and can never be isolated.
   (C) The intermediate molecules contain normal bonds rather than partial bonds and can be occasionally isolated.
   (D) All reactions involve reaction intermediates, but not all have activated complexes.

4. The rate of a chemical reaction is related to
   (A) activation energy and potential energy curve
   (B) orientation
   (C) activation energy, orientation, and frequency
   (D) frequency

Questions 5 and 6 refer to the following proposed mechanism for the reaction of chlorine gas with chloroform, $CHCl_3$, to form carbon tetrachloride.

$Cl_2(g) \rightarrow 2Cl(g)$	fast
$Cl(g) + CHCl_3(g) \rightarrow CCl_3(g) + HCl(g)$	slow
$CCl_3(g) + Cl(g) \rightarrow CCl_4(g)$	fast

5. What is the overall reaction equation for this mechanism?
   (A) $Cl(g) + CHCl_3(g) \rightarrow CCl_3(g) + HCl(g)$
   (B) $Cl_2(g) + CHCl_3(g) \rightarrow HCl(g) + CCl_4(g)$
   (C) $CCl_3(g) + Cl(g) \rightarrow CCl_4(g)$
   (D) $Cl_2(g) + CHCl_3(g) + CCl_3(g) \rightarrow HCl(g) + CCl_4(g)$

6. Which of the following rate laws is consistent with the proposed mechanism?
   (A) Rate = $k\,[CHCl_3]\,[Cl]$
   (B) Rate = $k\,[CCl_3]\,[Cl]$
   (C) Rate = $k\,[Cl_2]$
   (D) Rate = $k\,[Cl_2]\,[CHCl_3]$

Questions 7–9 refer to the following diagram:

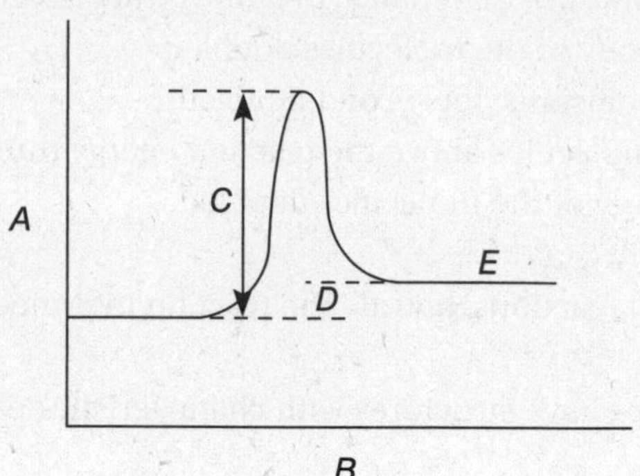

7. In the reaction profile, *A*, *B*, and *C* should be labeled as shown in

	A	*B*	*C*
(A)	potential energy	reaction coordinate	activation energy
(B)	heat of reaction	reaction coordinate	potential energy
(C)	potential energy	reaction coordinate	heat of reaction
(D)	heat of reaction	potential energy	activation energy

8. In the reaction described by the reaction profile,
   (A) forward E_a > reverse E_a and ΔH is exothermic
   (B) reverse E_a > forward E_a and ΔH is endothermic
   (C) forward E_a < reverse E_a and ΔH is exothermic
   (D) reverse E_a < forward E_a and ΔH is endothermic

9. Addition of a catalyst to the reaction mixture will affect only
   (A) *A*
   (B) *B*
   (C) *C*
   (D) *D*

10. Given the following mechanism and illustration, what is the rate of reaction for flask *B* in comparison to flask *A*?

$X_2 \rightarrow 2X$ slow

$X + Y \rightarrow XY$ fast

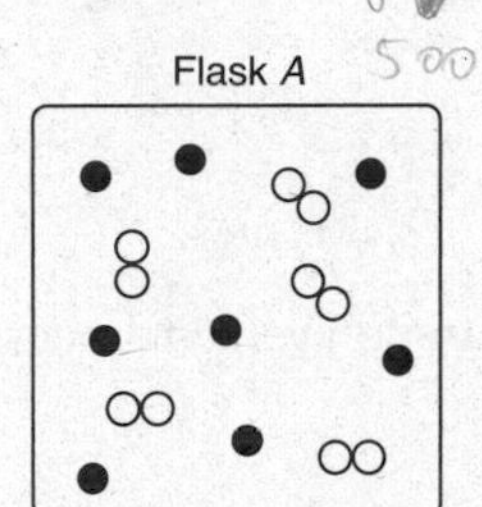

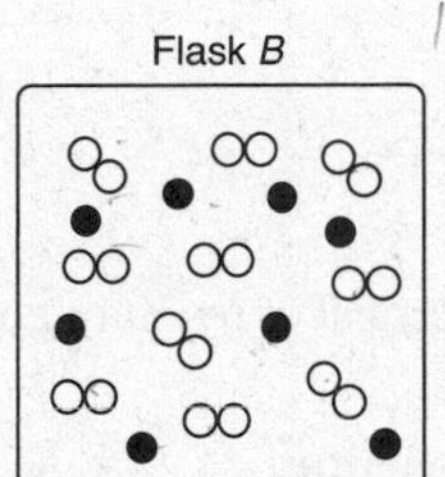

(A) The rate of reaction for flask *A* is twice as fast as the rate for flask *B*.
(B) The rate of reaction for flask *B* is the same as the rate for flask *A*.
(C) The rate of reaction for flask *A* is half as fast as the rate for flask *B*.
(D) The rate of reaction for flask *B* is half as fast as the rate for flask *A*.

11. Which of the following rate laws has a rate constant with units of $L^2\ mol^{-2}\ s^{-1}$?
(A) Rate = $k\,[A]$
(B) Rate = $k\,[A]^2$
(C) Rate = $k\,[A]\,[B]$
(D) Rate = $k\,[A]\,[B]^2$

12. Which of the following is LEAST effective in increasing the rate of a reaction?
(A) Increasing the pressure by adding an inert gas
(B) Grinding a solid reactant into small particles
(C) Increasing the temperature
(D) Eliminating reverse reactions

13. A first-order reaction has a half-life of 36 min. What is the value of the rate constant?
(A) $3.2 \times 10^{-4}\ s^{-1}$
(B) $1.9 \times 10^{-3}\ L\ mol^{-1}\ s^{-1}$
(C) $1.2\ s^{-1}$
(D) $0.028\ s^{-1}$

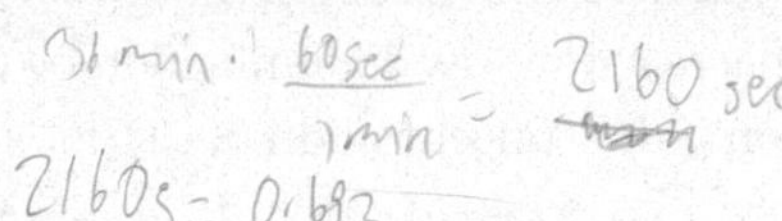

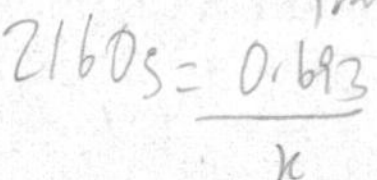

14. If a reactant's concentration is doubled and the reaction rate increases by a factor of 8, the exponent for that reactant in the rate law should be
(A) 0
(B) 1
(C) 2
(D) 3

15. In general, if the temperature of a reaction is raised from 300 K to 320 K, the reaction rate will increase by a factor of approximately

(A) $\frac{320\text{ K}}{300\text{ K}}$

(B) $\frac{22°\text{C}}{2°\text{C}}$

(C) 4

(D) 2

16. A graph of the reciprocal of reactant concentration versus time will give a straight line for

(A) a zero-order reaction

(B) a first-order reaction

(C) a second-order reaction

(D) both (A) and (C)

17. The rate at which CO_2 is produced in the following reaction:

$$2C_6H_6(g) + 15O_2(g) \longrightarrow 12CO_2(g) + 6H_2O(l)$$

is 2.2×10^{-2} mol L^{-1} s^{-1}. What is the rate at which O_2 is consumed?

(A) 2.2×10^{-2} mol L^{-1} s^{-1}

(B) 1.3×10^{-1} mol L^{-1} s^{-1}

(C) 2.8×10^{-2} mol L^{-1} s^{-1}

(D) 1.8×10^{-3} mol L^{-1} s^{-1}

CHALLENGE

18. Consider the following reaction:

$$CH_3Br(aq) + OH^-(aq) \rightarrow CH_3OH(aq) + Br^-(aq)$$

The experimental rate law was found to be first order with respect to both reactants. The rate for this reaction was determined to be 0.0432 Ms^{-1}. What is the rate constant for this reaction if the concentration of CH_3Br is 0.050 M and the OH^- is 0.025 M?

(A) 0.86 $M^{-1}s^{-1}$

(B) 0.029 $M^{-1}s^{-1}$

(C) 35 $M^{-1}s^{-1}$

(D) 69 $M^{-1}s^{-1}$

19. Which response identifies the intermediate(s) in the following mechanism?

$Cl_2(g) \rightarrow 2Cl(g)$	fast
$Cl(g) + CHCl_3 \rightarrow CCl_3(g) + HCl(g)$	slow
$CCl_3(g) + Cl(g) \rightarrow CCl_4(g)$	fast

(A) Cl

(B) CCl_3

(C) $CHCl_3$

(D) Both Cl and $CHCl_3$

20. A first-order reaction has a half-life of 85 s. What fraction of the reactant is left after 255 s?
(A) $^1/_2$
(B) $^1/_4$
(C) $^1/_8$
(D) $^1/_3$

CHALLENGE

Questions 21 and 22 refer to the following data. For the reaction

$$3A(g) + 2B(g) \rightarrow 2C(g) + 2D(g)$$

the following data were collected at constant temperature:

Trial	Initial [A] (mol/L^{-1})	Initial [B] (mol/L^{-1})	Initial rate (mol/L^{-1} min^{-1})
1	0.200	0.100	6.00×10^{-2}
2	0.100	0.100	1.50×10^{-2}
3	0.200	0.200	1.20×10^{-1}
4	0.300	0.200	2.70×10^{-1}

21. What is the correct rate law for this reaction?
(A) Rate = $k\,[A]\,[B]^2$
(B) Rate = $k\,[A]^3\,[B]^2$
(C) Rate = $k\,[A]^2\,[B]$
(D) Rate = $k\,[A]\,[B]$

22. What is the rate constant for this reaction?
(A) 30 M^{-2} min^{-1}
(B) 750 M^{-2} min^{-1}
(C) 3.0 M^{-2} min^{-1}
(D) 15 M^{-2} min^{-1}

---

23. A rate law is found to be

$$\text{Rate} = k\,[A]^2\,[B]$$

Which of the following actions will not change the initial reaction rate?
(A) Halving the concentration of *A* and quadrupling the concentration of *B*
(B) Doubling the concentration of both *A* and *B*
(C) Halving the concentration of *A* and doubling the concentration of *B*
(D) Doubling the concentration of *A* and halving the concentration of *B*

24. In an acid solution, sucrose ($C_{12}H_{22}O_{11}$) will decompose into fructose and glucose. A plot of ln[$C_{12}H_{22}O_{11}$] versus time yields a straight line with a slope of $-0.45\ hr^{-1}$. What is the rate law for this decomposition reaction?
   (A) Rate = $0.45\ hr^{-1}\ [C_{12}H_{22}O_{11}]^2$
   (B) Rate = $0.45\ hr^{-1}\ [C_{12}H_{22}O_{11}]$
   (C) Rate = $-0.45\ hr^{-1}\ [C_{12}H_{22}O_{11}]^0$
   (D) Rate = $-0.45\ hr^{-1}\ [C_{12}H_{22}O_{11}]^{-1}$

25. A catalyst will NOT
   (A) increase the forward reaction rate
   (B) shift the equilibrium to favor the products
   (C) alter the reaction pathway
   (D) increase the speed at which equilibrium will be achieved

ANSWER KEY

1. **(C)**	8. **(D)**	15. **(C)**	22. **(D)**
2. **(C)**	9. **(C)**	16. **(C)**	23. **(A)**
3. **(C)**	10. **(D)**	17. **(C)**	24. **(C)**
4. **(C)**	11. **(D)**	18. **(C)**	25. **(B)**
5. **(B)**	12. **(A)**	19. **(A)**	
6. **(A)**	13. **(A)**	20. **(C)**	
7. **(A)**	14. **(D)**	21. **(C)**	

See Appendix 1 for explanations of answers.

Free-Response

(a) In the gas phase at 500°C, cyclopropane, C_3H_6, reacts to form propene in a first-order reaction.

(i) Given the following data, plot the $\ln[C_3H_6]$ versus time on the grid provided.

Time(s)	$[C_3H_6]$ (mol/L)
0	0.100
200	0.067
400	0.050
600	0.033
800	0.027
1,000	0.017

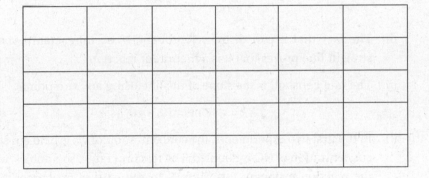

(ii) How does this plot verify that the reaction is first order? Justify your answer.

(iii) What is the rate constant for this reaction?

(b) You are given the following reaction and the table of data below.

$$2NO(g) + 2H_2(g) \rightarrow N_2(g) + 2H_2O(g)$$

Initial [NO] (mol/L)	Initial $[H_2]$ (mol/L)	Initial rate (*M*/s)
0.20	0.15	2.0×10^{-4}
0.40	0.15	8.0×10^{-4}
0.40	0.30	1.6×10^{-3}

(i) Determine the rate law for this reaction.

(ii) What is the value of the rate constant for this reaction?

ANSWERS

(a) (i) First take the natural logarithm of the concentrations. Then plot the graph.

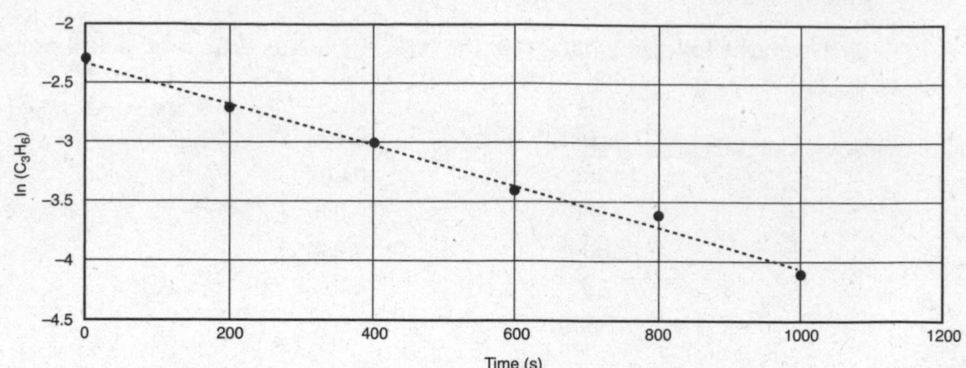

(ii) The fact that a plot of ln[cyclopropane] (i.e., ln[reactant]) versus *t* gives a straight line proves that it is a first-order reaction.

(iii) The rate constant is the slope of the line using any two points.

$$k = -\text{ slope} = 1.8 \times 10^{-5}\ s^{-1}$$

(b) (i) In the first two experiments, the concentration of H_2 is held constant while the concentration of NO is doubled. The reaction rate also doubles, indicating that the reaction exponent for NO is 1. In the case of experiments 2 and 3, the concentration of NO is held constant and the H_2 concentration is doubled. The reaction rate also doubles, so the exponent for H_2 is also 1. Therefore, the rate law is rate = k [NO] [H_2].

(ii) The rate constant is calculated as k = rate/[NO][H_2]. For all 3 experiments, we get $6.7 \times 10^{-3}\ M^{-1}s^{-1}$.

Thermodynamics 11

- → THERMODYNAMICS
- → DEFINITION OF SYSTEM
- → DEFINITION OF SURROUNDINGS
- → DEFINITION OF STATE FUNCTION
- → DEFINITION OF STANDARD STATES
- → POTENTIAL ENERGY
- → KINETIC ENERGY
- → SPECIFIC HEAT
- → FIRST LAW OF THERMODYNAMICS
- → INTERNAL ENERGY, ΔE
- → HEAT AND WORK
- → ENTHALPY, ΔH
- → HEAT OF FORMATION
- → ENTROPY, ΔS
- → SECOND LAW OF THERMODYNAMICS
- → THERMODYNAMICALLY FAVORED REACTIONS
- → FREE ENERGY, ΔG

BIG IDEAS 5, 6

Learning Objectives: 5.1, 5.2, 5.3, 5.4, 5.5, 5.6, 5.7, 5.8, 5.9, 5.10, 5.11, 5.12, 5.13, 5.14, 5.15, 5.16, 5.17, 5.18, 6.25

For the complete list of Big Ideas and Learning Objectives, refer to the AP Chemistry Course Outline: *https://secure-media.collegeboard.org/digitalServices/pdf/ap/ap-chemistry-course-and-exam-description.pdf*

ESSENTIAL DEFINITIONS

In the discussion of thermodynamics precise terminology is used to avoid confusion and to ensure that experimental results may be compared from one laboratory to another. Some of the important terms defined in this chapter are *system, state function, standard state,* and the *exo-* and *endo-* prefixes.

System

A **system** is that part of the universe that is under study. Everything else in the universe is called the **surroundings**. When hydrogen and oxygen are placed in a bomb calorimeter to study the formation of water, all of the hydrogen and oxygen atoms and the calorimeter are the system. The surrounding laboratory, the building, the city, and so on are parts of the surroundings.

There are several types of systems. An **open system** can transfer both energy and matter to and from the surroundings. An open bottle of perfume is an example of an open system. A **closed system** is one where energy can be transferred to the surroundings but matter cannot. A well-stoppered bottle of perfume is a closed system. In an **isolated system** there is no transfer of energy or matter to or from the surroundings. A thermos bottle is a close approximation of an isolated system since it minimizes heat transfer to the surroundings. Calorimeters are designed to be as close as possible to a closed system.

State Functions

State functions describe an aspect of a chemical system. Pressure, temperature, volume, and the number of atoms (or moles) are state functions that are combined in the ideal gas law. The value or magnitude of a state function does not depend on how the given state was obtained. A system that has been heated from 215 K to 228 K does not differ from one that has been cooled from 485 K to 228 K. In thermodynamics, we find other state functions, such as **internal energy**, ***E***; **enthalpy**, ***H***; **entropy**, ***S***; **and the Gibbs free energy**, ***G***. *Importantly, changes in state functions depend on only the initial and final state of a system.*

The state of a system is defined by the mass and phase (solid, liquid, or gas) of the matter in the system, as well as the temperature and pressure of the system. For the melting of 1 mole of ice, the initial state is described as 18 g $H_2O(s)$ at 1.00 atm pressure and 273 K, and the final state as 18 g $H_2O(l)$ at 1.00 atm pressure and 273 K. With this precise description all scientists should, within experimental error, obtain identical values of ΔH, ΔS, ΔG, and ΔE for the melting of ice.

Two quantities that are *not* state functions are heat (q) and work (w). The values for these quantities depend on the sequence of steps used to transform matter from the initial state to the final state.

Standard State

The thermodynamic quantities ΔH, ΔS, ΔG, and ΔE are **extensive properties** of matter, meaning that they change as the amount of sample changes. To make these quantities **intensive properties** of matter, we must define precisely the temperature, pressure, mass, and physical state of the substance. A system is in the **standard state** when the pressure is 1 atmosphere, the temperature is 25°C, and 1 mole of compound is present. When the thermodynamic quantities are determined at standard state, they are intensive properties and a superscript 0 is added to their symbols: as $\Delta H°$, $\Delta S°$, $\Delta G°$ and $\Delta E°$. For a chemical reaction, the standard state involves the number of moles designated by the stoichiometric coefficients in the simplest balanced chemical equation with the smallest possible whole-number coefficients.

Exo- and *Endo-* Prefixes and Sign Conventions

The prefix *exo-*, as in *exothermic,* indicates that energy is being lost from the system to the surroundings. Mathematically, the prefix *exo-* corresponds to a negative sign for numerical thermodynamic quantities. For an exothermic reaction the heat of reaction, ΔH, is a negative number.

The prefix *endo-* indicates that energy is gained from the surroundings. An *endothermic* reaction absorbs heat energy and appears to cool as it progresses. For an endothermic reaction the heat of reaction, ΔH, is a positive number.

TYPES OF ENERGY

Energy takes many forms. Heat and light, along with chemical, nuclear, electrical, and mechanical energy, are some of the common types. Any one of these forms of energy can be converted into any of the other forms. In addition, the **law of conservation of energy** states that energy is never created or destroyed. As a result of these properties, all forms of energy can be converted into heat energy, which we can measure in a calorimeter as described below.

Energy can also be categorized as either kinetic energy, KE, or potential energy, PE. **Kinetic energy** is the energy that matter possesses because of its motion. There is one equation to describe kinetic energy:

$$\text{KE} = \frac{1}{2}mv^2$$

When the mass, m, is expressed in kilograms and the velocity, v, in meters per second, the energy units are joules.

Potential energy is stored energy, which may be released under the appropriate conditions, as in a nuclear reaction. There are several forms of potential energy, such as **gravitational energy** and the energy of **electrostatic attraction** between oppositely charged ions. Each form of potential energy is described by its own equation, but these equations are similar to each other. They all have the forms

$$\text{PE}_{\text{grav}} = K_{\text{grav}}\left(\frac{m_1m_2}{r}\right) \quad \text{and} \quad \text{PE}_{\text{elect}} = K_{\text{elect}}\left(\frac{q_1q_2}{r}\right)$$

The two masses, m, in gravitational attraction and the two charges, q, in electrostatic attraction are separated by a distance, r. K is a proportionality constant that is different for each type of potential energy.

The total energy of a substance is the sum of its kinetic and potential energies.

$$\text{Energy (E)} = \text{potential energy (PE)} + \text{kinetic energy (KE)}$$

In chemical substances, the kinetic energy is the motion of the molecules. The potential energy of a chemical is the sum of all attractions, including all the covalent bonds, ionic bonds, or electrostatic attractions in the substance.

MEASUREMENT OF ENERGY

Heat is often called the "lowest form" of energy. In this form it is easily measurable by determining temperature changes caused by the release or absorption of heat in a chemical process.

Specific Heat

Heat energy was originally defined in terms of the calorie, which is the amount of heat needed to raise the temperature of 1 gram of pure water from 14.5 to 15.5°C. The joule is the metric unit for energy; 1 calorie is equal to exactly 4.184 joules. By virtue of these definitions, 4.184 joules of energy is needed to raise the temperature of 1 gram of water by 1 degree Celsius. This quantity is known as the **specific heat** of water:

$$\text{Specific heat of water} = 4.184 \text{ J g}^{-1}\,^{\circ}\text{C}^{-1}$$

Once the specific heat of water is defined, the specific heat of any other substance can be determined. One method is to immerse a hot object in a known quantity of water and then measure the temperature change that occurs. This method is illustrated in Example 11.1.

As an interesting sidelight, in 1818 Pierre Dulong and A.T. Petit discovered that for most metals the specific heat multiplied by the atomic mass of the metal was equal to a constant. The **Dulong and Petit law** is as follows:

$$\text{Specific heat} \times \text{molar mass} \approx 25\ \text{J mol}^{-1}\ {}^\circ\text{C}^{-1}$$

TIP

On the AP exam, this equation reads $q = mc\Delta T$.

This law helped confirm the molar masses of the elements when disagreements occurred. In addition, specific heat is an intensive physical property of all elements and compounds and can be used to identify substances. In more advanced physical chemistry courses, you will find that this law is predicted by thermodynamics.

Equation 11.1 enables chemists to determine the heat energy of any process by measuring the change in temperature of a known mass of water. The equation is as follows:

$$\begin{aligned} \text{Heat energy} &= (\text{Specific heat})(\text{Mass})(\text{Temperature change}) \\ q &= \text{sp. ht.} \times \text{g} \times \Delta T \end{aligned} \qquad (11.1)$$

In this equation q, the heat energy, is expressed in joules. The temperature change is always determined as the final temperature minus the initial temperature (${}^\circ C_{final} - {}^\circ C_{initial}$). Equation 11.1 is applicable to any substance, not just water.

There is a distinct difference between heat energy and temperature, as Equation 11.1 indicates. Temperature is a measure of the average kinetic energy of a group of atoms, and a temperature change is a change in the average kinetic energy. Heat energy is produced when both the kinetic energy and the energy of attractions between the atoms in the group change in a chemical or physical process.

EXAMPLE 11.1

An insulated cup contains 75.0 g of water at 24.00°C. A 26.00 g sample of a metal at 85.25°C is added. The final temperature of the water and metal is 28.34°C.

(a) What is the specific heat of the metal?

(b) According to the law of Dulong and Petit, what is the approximate molar mass, *MM*, of the metal?

(c) What is the apparent identity of the metal?

Solution

(a) The law of conservation of energy requires that the heat energy gained by the water be exactly equal to the heat energy lost by the metal as it cools in the water. Mathematically this is written as

$$+q_{water} = -q_{metal}$$

Using Equation 11.1, we expand this equation:

$$(\text{sp. heat}_{H_2O})(\text{g}_{H_2O})(\Delta T_{water}) = -(\text{sp. heat}_{metal})(\text{g}_{metal})(\Delta T_{metal})$$

Entering the data yields

$$4.184\ \text{J g}^{-1}\ {}^\circ\text{C}^{-1}(75.0\ \text{g})(4.34^\circ\text{C}) = -(\text{sp. heat}_{\text{metal}})(26.00\ \text{g})(-56.91^\circ\text{C})$$
$$1362\ \text{J} = 1480\ \text{g}\ {}^\circ\text{C}(\text{sp. heat}_{\text{metal}})$$
$$\text{sp. heat}_{\text{metal}} = 0.920\ \text{J g}^{-1}\ {}^\circ\text{C}^{-1}$$

(b) Using the law of Dulong and Petit (see page 394) we obtain

$$0.920\ \text{J g}^{-1}\ {}^\circ\text{C}^{-1}(\text{molar mass}) = 25\ \text{J mol}^{-1}\ {}^\circ\text{C}^{-1}$$
$$\text{molar mass} = 27\ \text{g mol}^{-1}$$

(c) Aluminum is the only metal with a molar mass of 27 g mol^{-1}.

FIRST LAW OF THERMODYNAMICS

The **first law of thermodynamics** states that energy is always conserved. In chemistry this law means that the measurable quantities heat (q) and work (w) must add up to the total energy change in a system:

$$\Delta E = q + w \tag{11.2}$$

The value of q has a positive sign if heat is added to the system. The value of w is positive if work is done on the system. Similarly, q is a negative value if heat is released from the system, and w is a negative value if work is done by the system.

If the system cools, q has a positive sign and the process is said to be endothermic. If the system heats up, it is exothermic and q has a negative sign. As we will see later, work is equal to the pressure times the change in volume, $P\,\Delta V$. If the volume increases, work is done by the system and w has a negative sign. When work is done on the system, the volume decreases and w has a positive sign.

The change in energy of a system, at constant temperature, is also the difference in potential energy between the final and initial states of the system:

$$\Delta E = \text{PE}_{\text{final}} - \text{PE}_{\text{initial}}$$

As we will see later, ΔE is mainly heat energy. Therefore a system that increases its potential energy is often said to be endothermic, while a decrease in potential energy indicates an exothermic process.

Work

In Equation 11.2 the energy change is the sum of the heat and the work. Heat is measured using a calorimeter, and work is also easily measured. **Work** is defined as the force applied to an object as it moves a certain distance:

$$\text{Work} = \text{Force} \times \text{Distance moved}$$

Force can be defined as the pressure exerted over a given area, so

$$\text{Work} = \text{Pressure} \times \text{Area} \times \text{Distance moved}$$

Multiplying the area by the distance results in volume units or an overall volume change:

$$\text{Work} = \text{Pressure} \times \text{Volume change}$$
$$w = P\,\Delta V \tag{11.3}$$

Work is the product of the pressure and the change in volume that occurs during a chemical reaction.

EXAMPLE 11.2

Demonstrate that work is not a state function by calculating the work involved in expanding a gas from an initial state of 1.00 L and 10.0 atm of pressure to (a) 10.0 L and 1.0 atm pressure, (b) 5.00 L and 2.00 atm and then to 10.0 L and 1.00 atm pressure.

Solution

(a) $w = -P\Delta V = -(1.00 \text{ atm})(10.0 \text{ L} - 1.00 \text{ L})$
$\quad = -9.00 \text{ L atm}$

(b) $w = -P\Delta V = -(2.00 \text{ atm})(5.00 \text{ L} - 1.00 \text{ L}) - (1.00 \text{ atm})(10.0 \text{ L} - 5.00 \text{ L})$
$\quad = -13.0 \text{ L atm}$

In both parts of this exercise the sample starts at the same state (1.00 L and 10.0 atm) and ends in the same state (10.0 L and 1.00 atm). However, the work in units of liter atmospheres is different. This can occur only if w is not a state function.

Use of *R* for the Conversion of Energy Units

We have different energy units—calories, joules, and the newly introduced liter-atmosphere. They are all conveniently related through the universal gas law constant, R. To make these conversions let us see how to convert L atm to joules for the last problem.

$$? \text{ joules} = -13.0 \text{ L atm}$$

We will divide by the R that has units of L atm and multiply by the R that has units of joules as follows:

$$? \text{ joules} = -13.0 \text{ L atm}\left(\frac{\text{mol K}}{0.0821 \text{ L atm}}\right)\left(\frac{8.31 \text{ J}}{\text{mol K}}\right)$$

After canceling units we end up with the desired joules and −13.0 L atm becomes -1.32×10^3 J. We use a similar approach to convert to or from other energy units such as electronvolts (EV) or calories.

Definition of q_p, q_v, ΔE, and ΔH

The first law of thermodynamics may be rewritten as

$$\Delta E = q_p - P\Delta V$$

The minus sign enters this equation because an increase in volume means that the system does work on the surroundings, and such work has been defined as a negative quantity. The heat term, q, is given the subscript p to indicate that the pressure must be constant.

When the heat energy is measured in a calorimeter that does not allow the volume to change, $P\Delta V$ must be zero. As a result $\Delta E = q_v$, where the subscript v indicates that the volume is held constant. Calorimeters that do not allow the volume to change are called **bomb calorimeters**, a name derived from the heavy stainless steel reaction vessel, which was known to explode if not used correctly.

For most real reactions chemists are interested in the heat generated at constant pressure, q_p. This variable is called the enthalpy change and is given the symbol ΔH. **Enthalpy**, H, is the heat content of a substance, and ΔH is the difference in heat content of the products and reactants:

$$\Delta H = H_{\text{products}} - H_{\text{reactants}}$$

Using Equations 11.2 and 11.3, we can write the relationship between ΔE and ΔH as

$$\Delta E = \Delta H - P\Delta V$$

For many reactions, the value of ΔH is very large and the value of $P\Delta V$ is relatively small, so that ΔE and ΔH are approximately equal.

Standard Enthalpy Changes and the Standard Heat of Reaction, $\Delta H^\circ_{\text{reaction}}$

The heat energy or enthalpy change, ΔH, produced by a chemical reaction is an **extensive property**, since reacting a larger amount of chemicals produces a larger amount of heat. For example, when propane is burned according to the equation

$$CH_3CH_2CH_3(g) + 5O_2(g) \rightarrow 3CO_2(g) + 4H_2O(g) \quad (11.4)$$

more heat is generated as more propane is burned. To make the heat produced by a reaction an **intensive property**, the amount of chemical that reacts must be specified. The standard heat of a reaction, ΔH°, is generally defined as the heat produced when the number of moles specified in the balanced chemical equation reacts. For the reaction of propane the heat of reaction, $\Delta H^\circ_{\text{react}}$, equals –2044 kJ when 1 mole of propane reacts with 5 moles of oxygen as shown in Equation 11.4. The negative sign indicates that a large amount of heat is released in this reaction, making propane an excellent fuel for cooking and heating.

As noted above, the heat of reaction, $\Delta H^\circ_{\text{reaction}}$ or simply ΔH°, has units of kJ. Only when the reaction involved is a formation reaction, with one and only one mole of product and a heat of reaction symbolized as ΔH°_f, will the units be kJ mol^{-1}. This is necessary since the heat of formation is usually multiplied by the moles of a substance resulting in the kJ units of a heat of reaction.

$$\text{kJ mol}^{-1} \times \text{mol} = \text{kJ}$$

When the $\Delta H^\circ_{\text{reaction}}$ is known for a given reaction, the heat per mole of any reactant or product can be determined by dividing the $\Delta H^\circ_{\text{reaction}}$ by the moles of substance indicated in the equation. For example, –408.8 kJ of heat will be produced per mole of O_2 when the reaction indicated by Equation 11.4 occurs.

Hess's Law

Hess's law states that, whatever mathematical operations are performed on a chemical equation, the same mathematical operations are applied also to the heat of reaction. Hess's law is summarized as follows:

1. If the coefficients of a chemical equation are all multiplied by a constant, the $\Delta H^\circ_{\text{react}}$ is multiplied by that same constant.

2. If two or more equations are added together to obtain an overall reaction, the heats of these equations are also added to give the heat of the overall reaction.

Hess's law allows the chemist to measure ΔH°_{react} for several reactions and then to combine the equations and their heats to obtain ΔH°_{react} for a completely different reaction.

For example, we saw above that the burning of propane produces a large amount of heat. The reaction in Equation 11.4 is easy to perform by igniting propane in the presence of oxygen. The reverse reaction for the synthesis of 1 mole of propane from carbon dioxide and water is impossible to perform, but it may be written as

$$3CO_2(g)+4H_2O(g) \rightarrow CH_3CH_2CH_3(g)+5O_2(g)$$

Since reversing a reaction is the same as multiplying it by −1, the heat needed for this reaction is +2044 kJ. The change from a negative to positive value is explained on the basis that ΔH°, which is a state function, depends only on the final and initial states of the system. Synthesis of 1 mole of propane has the same final and initial states as the combustion of 1 mole of propane. The only difference is the direction of the process. We must reach the conclusion that ΔH° has the same magnitude for these two reactions, but they have different signs because they go in opposite directions. Figure 11.1 illustrates this process.

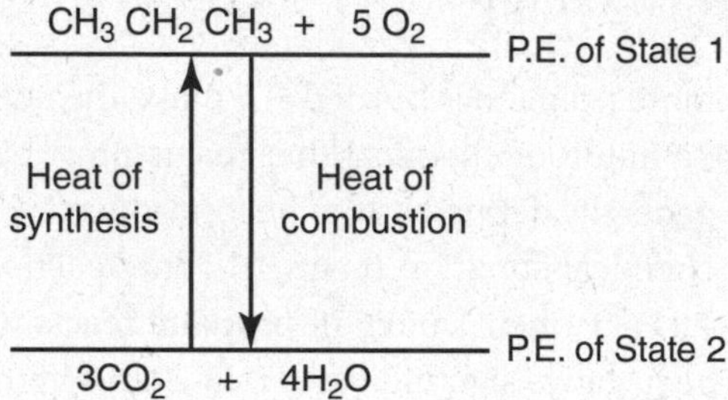

FIGURE 11.1. Diagram illustrating that the heat of a reaction has the same magnitude whether the reaction is run in the forward or reverse direction. The sign of the heat of reaction, however, is positive in one direction and negative in the other.

The general principle that $\Delta H^\circ_{forward\ react} = -\Delta H^\circ_{reverse\ react}$ is an essential part of Hess's law.

For the combustion of propane, Hess's first rule (see above) tells us that multiplying the equation by 2 will result in ΔH° also being multiplied by 2:

$$2CH_3CH_2CH_3(g) + 10O_2(g) \rightarrow 6CO_2(g) + 8H_2O(g) \quad \Delta H = -4088 \text{ kJ}$$

If the reaction is multiplied by $\frac{1}{2}$, the ΔH° will also be multiplied by $\frac{1}{2}$:

$$\tfrac{1}{2}CH_3CH_2CH_3(g) + 2.5O_2(g) \rightarrow 1.5CO_2(g) + 2H_2O(g) \quad \Delta H = -1022 \text{ kJ}$$

If the reaction is multiplied by 6, the ΔH° will also be multiplied by 6:

$$6CH_3CH_2CH_3(g) + 30O_2(g) \rightarrow 18CO_2(g) + 24H_2O(g) \quad \Delta H = -12{,}264 \text{ kJ}$$

Once ΔH° is multiplied by any coefficient, the system is no longer in standard state and the superscript zero is dropped as shown.

The second of Hess's rules concerns the addition of chemical reactions. To repeat, when chemical reactions are added, all of the reactants are written as reactants of the overall reaction. All of the products are combined as products of the overall reaction. The last step in adding reactions is the cancellation of any identical reactants and products in the overall reaction.

To add the two reactions below, the reactants and products in Equations 11.5 and 11.6 are combined in an overall reaction, Equation 11.7:

$$N_2(g) + O_2(g) \rightarrow 2NO(g) \quad (11.5)$$

$$2NO(g) + O_2(g) \rightarrow 2NO_2(g) \quad (11.6)$$

$$N_2(g) + 2NO(g) + 2O_2(g) \rightarrow 2NO(g) + 2NO_2(g) \quad (11.7)$$

Two molecules of NO(g) are canceled from both the reactant and product sides of Equation 11.7 to obtain the overall reaction:

$$N_2(g)+2O_2(g)\rightarrow 2NO_2(g)$$

Since the two reactions were added, their heats of reaction are also added:

$$\begin{array}{llll} N_2(g) & + O_2(g) \rightarrow 2NO(g) & \Delta H^\circ_1 & = +180.5\text{ kJ} \\ 2NO(g) & + O_2(g) \rightarrow 2NO_2(g) & \Delta H^\circ_2 & = -114.1\text{ kJ} \\ \hline N_2(g) & + 2O_2(g) \rightarrow 2NO_2(g) & \Delta H^\circ_{\text{overall}} & = +66.4\text{ kJ} \end{array}$$

In this example $\Delta H^\circ_1 + \Delta H^\circ_2 = \Delta H^\circ_{\text{overall}}$.

The heats of many reactions can be determined if the heats of combustion are known for each of the reactants and products. To illustrate this more complex combination of reactions, we can determine the heat of reaction for the synthesis of propane from carbon and hydrogen:

$$3C(s)+4H_2(g)\rightarrow CH_3CH_2CH_3(g) \quad (11.8)$$

from the combustion reactions of propane, hydrogen, and carbon. These reactions, along with their ΔH° values, are as follows:

$$\begin{array}{lllll} (1)\ CH_3CH_2CH_3(g) & + 5O_2(g) & \rightarrow 3CO_2(g) + 4H_2O(g) & \Delta H^\circ_1 & = -2044\text{ kJ} \\ (2)\ 2H_2(g) & + O_2(g) & \rightarrow 2H_2O(g) & \Delta H^\circ_2 & = -483.6\text{ kJ} \\ (3)\ C(s) & + O_2(g) & \rightarrow CO_2(g) & \Delta H^\circ_3 & = -393.5\text{ kJ} \end{array}$$

If it is not immediately obvious how these three equations should be combined, some general principles on how to approach the problem logically are helpful. In the list below, the first three principles tell us how to find which equation to start with. Once Equation 1 is established, the same principles are used to select and manipulate the remaining reactions. If needed, the fourth principle may be used.

1. Focus on the most complex molecules first.
2. Focus only on atoms and molecules that occur in just one reaction.
3. Focus on atoms and molecules that are in the overall equation.
4. Focus on finding atoms and molecules to cancel unneeded ones from already selected equations.

Using principles 1–3, we see that the combustion of propane in Equation 1 should be considered first. This equation contains 1 mole of propane and so does the overall equation, Equation 11.8. However, in the combustion reaction, Equation 1, propane is a reactant, and in Equation 11.8 it is a product. Consequently, Equation 1 must be reversed by multiplying it by –1. This reverses the equation and at the same time changes the sign of $\Delta H°$:

$$3CO_2(g)+4H_2O(g)\rightarrow CH_3CH_2CH_3(g)+5O_2(g) \quad \Delta H°=-2044 \text{ kJ} \times (-1)$$

Using principle 3, we now focus on the reaction in Equation 2 and note that it includes $2H_2(g)$ and that we need $4H_2(g)$ in our overall equation. The hydrogens are reactants in both Equation 2 and Equation 11.8, but Equation 2 must be multiplied by 2 so that we have the correct number of H_2 in the final reaction. We will also have to multiply $\Delta H°$ by 2:

$$4H_2(g)+2O_2(g)\rightarrow 4H_2O(g) \quad \Delta H=-483.6 \text{ kJ}\times 2$$

To demonstrate the use of principle 4, we see that the remaining reaction, Equation 3, contains one $CO_2(g)$ as a product. We need to cancel three $CO_2(g)$ molecules from the already selected equations. Since $CO_2(g)$ is a product in Equation 3 and a reactant in the rearranged Equation 3, the CO_2's will cancel when the reactions are added. However, Equation 3 must be multiplied by 3 so that all three $CO_2(g)$ molecules will cancel:

$$3C(s) + 3O_2(g) \rightarrow 3CO_2(g) \quad \Delta H=-2393.5 \text{ kJ} \times 3$$

Adding the three equations gives us

$$3C(s) + 3O_2(g) + 4H_2(g) + 2O_2(g) + 3CO_2(g) + 4H_2O(g) \rightarrow$$
$$3CO_2(g) + 4H_2O(g) + 5O_2 + CH_3CH_2CH_3(g)$$

After canceling the three $CO_2(g)$, four $H_2O(g)$, and five $O_2(g)$ molecules, the equation becomes

$$3C(s)+4H_2(g)\rightarrow CH_3CH_2CH_3(g)$$

The heat of reaction is the sum of the three $\Delta H°$ values multiplied by the operations performed:

$$\begin{aligned} \Delta H°_{overall} &= \Delta H°_1 \times (-1) + \Delta H°_2 \times 2 + \Delta H°_3 \times 3 \\ &= -2044 \text{ kJ} \times (-1) + -483.6 \text{ kJ} \times 2 + -393.5 \text{ kJ} \times 3 \\ &= +2044 \text{ kJ} - 967.2 \text{ kJ} - 1180.5 \text{ kJ} \\ &= -103.7 \text{ kJ} \end{aligned}$$

The advantage of being able to perform some simple experiments to obtain data for complex, even impossible reactions was recognized very quickly. During the energy crisis of the 1970s, the feasibility of producing alternative fuels was determined from thermochemical calculations.

Formation Reactions and Heats of Formation

In the preceding section we saw that the heats of many chemical reactions can be determined if the heats of combustion of all reactants and products are known. For other types of reactions the heat of combustion can also be tabulated. Such a table would be very long and complex, however, and finding the appropriate reactions to combine would be a monumental task. **Heats of formation** allow chemists to tabulate thermochemical data in a short, easy-to-use format.

A **formation reaction** is defined as one in which the reactants are elements in their standard state at 25°C and 1 atmosphere of pressure, and there is only 1 mole of product. Here are some examples of formation reactions:

$$Fe(s) + {}^1\!/_2\,O_2(g) \rightarrow FeO(s)$$

$$2Fe(s) + {}^3\!/_2\,O_2(g) \rightarrow Fe_2O_3(s)$$

$$2K(s) + {}^1\!/_2\,H_2(g) + 2O_2(g) + P(s) \rightarrow K_2HPO_4(s)$$

Fractional coefficients may be used in formation reactions. Since there is always 1 mole of product, the standard heats of formation, ΔH°_f, are tabulated as the heat produced per mole of product.

Since the reaction can easily be deduced if the product is known, a table of data need contain only the name or formula of the product and its corresponding ΔH°_f. A tabulation of some heats of formation is given in Appendix 3.

Heats of formation of the elements are always zero, whether they are molecules or atoms. The reason is that the formation reaction for an element such as oxygen is defined as

$$O_2(g) \rightarrow O_2(g) \qquad (11.9)$$

Since the oxygen is at 25°C and 1.00 atmosphere pressure as both the product and the reactant, the initial and final states of the oxygen in Equation 11.9 are the same, and their difference must be zero:

$$\Delta H^\circ_f \text{ (any element)} = 0$$

Heats of formation can be calculated from heats of combustion.

For the combustion of propane (Equation 11.4):

$$CH_3CH_2CH_3(g) + 5O_2(g) \rightarrow 3CO_2(g) + 4H_2O(g)$$

ΔH° can be calculated using the formation reactions and the tabulated heats of formation. Again, we will need a formation reaction for each of the reactants and products. Elements are excluded, however, since their heats of formation are always zero. In this example, the formation reactions for propane, carbon dioxide, and water are needed:

$$3C(s) + 4H_2(g) \rightarrow CH_3CH_2CH_3(g) \qquad \Delta H^\circ_f = -103.8 \text{ kJ mol}^{-1}$$

$$C(s) + O_2(g) \rightarrow CO_2(g) \qquad \Delta H^\circ_f = -393.5 \text{ kJ mol}^{-1}$$

$$H_2(g) + {}^1\!/_2\,O_2(g) \rightarrow H_2O(g) \qquad \Delta H^\circ_f = -241.8 \text{ kJ mol}^{-1}$$

The following operations are performed on these reactions so that they can be combined to yield the combustion reaction:

The propane formation reaction is reversed.

$$CH_3CH_2CH_3(g) \rightarrow 3C(s) + 4H_2(g)$$
$$\Delta H = -103.8 \text{ kJ mol}^{-1} \times (-1 \text{ mol})$$

The carbon dioxide formation reaction is multiplied by 3.

$$3C(s)+3O_2(s) \rightarrow 3CO_2(g)$$
$$\Delta H = -393.5 \text{ kJ mol}^{-1} \times 3 \text{ mol}$$

The $H_2O(g)$ formation reaction is multiplied by 4.

$$4H_2(g)+2O_2(g) \rightarrow 4H_2O(g)$$
$$\Delta H = -241.8 \text{ kJ mol}^{-1} \times 4 \text{ mol}$$

After these three operations, adding the three reactions yields the combustion reaction:

$$CH_3CH_2CH_3(g)+5O_2(g) \rightarrow 3CO_2(g)+4H_2O(g)$$

The corresponding sum of the heats is the heat of reaction:

$$\begin{aligned}\Delta H^\circ &= (-103.8 \text{ kJ mol}^{-1})(-1 \text{ mol})+(-393.5 \text{ kJ mol}^{-1})(3 \text{ mol}) \\ &\qquad +(-241.8 \text{ kJ mol}^{-1})(4 \text{ mol}) \\ &= -2044 \text{ kJ}\end{aligned}$$

After several calculations using heats of formation have been made, a pattern appears. The heat of any reaction will be the sum of ($\Delta H_f^\circ \times \text{mol}_{product}$) for all the products minus the sum of ($\Delta H_f^\circ \times \text{mol}_{reactant}$) for all the reactants. The $\text{mol}_{product}$ and $\text{mol}_{reactant}$ terms refer to the stoichiometric coefficients in the balanced equation for the reaction:

$$\Delta H^\circ = \Sigma(\Delta H_f^\circ \times \text{coeff})_{products} - \Sigma(\Delta H_f^\circ \times \text{coeff})_{reactants}$$

On the AP exam, this equation is presented as

$$\Delta H^\circ = \Sigma \Delta H_f^\circ{}_{products} - \Sigma \Delta H_f^\circ{}_{reactants}$$

EXAMPLE 11.3

Calculate the heat of combustion of CH_4. The heats of formation are as follows: ΔH_f° ($CH_4(g)$) = -74.8 kJ mol^{-1}, ΔH_f° ($CO_2(g)$) = -110.5 kJ mol^{-1}, and ΔH_f° ($H_2O(g)$) = -241.8 kJ mol^{-1}.

Solution

For the combustion of CH_4 the equation is

$$CH_4(g)+2O_2(g) \rightarrow CO_2(g)+2H_2O(g)$$

The heat of this reaction is calculated as

$$\begin{aligned}\Delta H^\circ{}_{react} &= [(-110.5 \text{ kJ mol}^{-1})+(-241.8 \text{ kJ mol}^{-1})(2 \text{ mol})] \\ &\qquad -[(-74.8 \text{ kJ mol}^{-1})(1)+(0.00 \text{ kJ mol}^{-1})(2 \text{ mol})] \\ &= -519.3 \text{ kJ}\end{aligned}$$

The value of 0.00 kJ mol^{-1} in this calculation represents the heat of formation of $O_2(g)$, which, by definition, is zero.

ENTROPY AND THE SECOND LAW OF THERMODYNAMICS

Entropy is related to the number of different ways in which a system can arrange the particles within the system. Similarly, if we divide the volume of the system into eight units, those two particles can arrange themselves in 56 different ways (8×7). (The first particle can enter any of the eight units and the second can occupy any of the remaining seven, therefore $8 \times 7 = 56$.) If the volume of the system is doubled, we will have 16 volume units the same size as before. The number of different arrangements of our particles is now 240 (16×15). That represents an increase in entropy. These examples must be extrapolated to a very large number of infinitesimally small volume units with extremely large numbers of particles (atoms and molecules) and a correspondingly large number of possible arrangements to describe real systems. However, the conclusions are relatively straightforward. An increase in the number of particles increases entropy, an increase in volume increases entropy, changing state (e.g., from liquid to gas) increases entropy, and so on.

We can visualize that water molecules in ice are constrained to the crystal lattice and have virtually no freedom to move and occupy a different volume unit. A water molecule in a liquid can move from one volume unit to another, but it does so rather slowly and is strongly influenced by other water molecules. Finally, in the gas phase the water molecule rapidly moves from one volume unit to another. Therefore we say that water in ice has the least entropy, whereas water in the gas phase has the most entropy. If a chemical reaction produces a gas such as H_2, the increase in possible positions and kinetic energies compared to that in liquids or solids leads to the conclusion that the entropy increases in such a reaction.

Entropy is assigned the symbol, S, and has units of J °C^{-1}. Standard entropy is based on the mole and is written as $S°$ with units of J °C^{-1} mol^{-1}. Values for $S°$ are tabulated in Appendix 3. It should be noted that standard entropy values for elements are not zero as they are for standard heat of formation, ΔH_f°, and standard free energy, ΔG_f°.

Unlike other thermodynamic quantities, such as energy, E, and enthalpy, H, the actual value for the entropy, S, of a substance can be determined. From fundamental principles, a perfect crystal at absolute zero (0 K or −273.16°C) has zero entropy since there is only one possible arrangement of the atoms. The Boltzmann entropy equation, $S = k \ln w$, where k is the Boltzmann constant and w is the number of microstates, describes the statistical method for determining entropy, S. Since a perfect crystal has one microstate, $k \ln (1) = 0$. Entropy can be determined experimentally since as the temperature of 1 mole of a chemical is increased from absolute zero, the entropy increases and the standard entropy is defined as

$$S° = \frac{q_{rev}}{T}$$

In this equation T represents the temperature in Kelvins, and q_{rev} is the heat added to raise the temperature very slowly from absolute zero up to T. Heat, q, is not a state function, and it seems that S should not be a state function either. However, if heat is always added in a carefully defined manner, the results will always be the same. This carefully defined path is called a **reversible process**, and the heat added is symbolized as q_{rev}. A reversible process is defined as one that occurs in infinitesimally small steps from the initial to the final state.

Entropy changes due to a chemical process are calculated in the same fashion as the heats of reaction. Just as is done for ΔH_f° values, the tables list entropies for 1 mole of substance, making entropy an intensive physical property. This can be written as

$$\Delta S° = \Sigma(\Delta S_f^\circ \times \text{coeff})_{\text{products}} - \Sigma(\Delta S_f^\circ \times \text{coeff})_{\text{reactants}}$$

On the AP exam, this equation is presented as

$$\Delta S° = \sum S°_{products} - \sum S°_{reactants}$$

For example, we can calculate the entropy change for the combustion of propane from the data in Appendix 3:

$$CH_3CH_2CH_3(g) + 5O_2(g) \rightarrow 3CO_2(g) + 4H_2O(g)$$

$$\begin{aligned}
\Delta S° &= ?\,J\,°C^{-1} \\
\Delta S° &= \left[\left(\frac{213.6\,J}{mol\,K}\right)(3\,mol) + \left(\frac{188.7\,J}{mol\,K}\right)(4\,mol)\right] - \left[\left(\frac{205.0\,J}{mol\,K}\right)(5\,mol) + \left(\frac{270.2\,J}{mol\,K}\right)(1\,mol)\right] \\
&= 1395.6\,J\,K^{-1} - 1295.2\,J\,K^{-1}. \\
&= +100.4\,J\,K^{-1} = 100.4\,J\,°C^{-1}
\end{aligned}$$

This result represents an increase in entropy. We might have predicted an increase for this reaction since there are 7 moles of gaseous products and only 6 moles of gaseous reactants. The increase in the number of moles of gas in this reaction is 1 ($\Delta n_g = 1$), indicating that the entropy change is expected to be positive.

In addition to calculating the entropy change from tabulated data, we can estimate the sign and, to some degree, the magnitude of an entropy change for a chemical process. In making such an estimate, the following principles are important:

1. Formation of a gas increases entropy greatly. The greater the position value of Δn_g, the greater the entropy increase. The reverse is true because more negative values of Δn_g represent a decrease in entropy.
2. If $\Delta n_g = 0$, changes from the solid to the liquid phase are the next leading contributors to an increase in entropy. A solid that melts or a solute that dissolves in a solvent both exhibit an increase in entropy. Conversely, the formation of solids results in a decrease in entropy.
3. An increase in temperature increases the entropy of a system, and a temperature decrease lowers the entropy.

GIBBS FREE-ENERGY, $\Delta G°$

This thermodynamic quantity was named in honor of J. Willard Gibbs, a preeminent physical chemist who developed the concept. The **free-energy change**, represented by the symbol $\Delta G°$, is the maximum amount of energy available from any chemical reaction. Two forces drive chemical reactions. The first is the enthalpy, $\Delta H°$, which represents the change in the internal potential energy of the atoms. The second is the drive toward an increase in the entropy of the system. If the enthalpy is negative, it means that the internal potential energy of the system is decreased, and this favors a spontaneous reaction. If the entropy increases, a spontaneous reaction is also favored. The combination of these two driving forces is represented as

TIP

This equation is given on the AP exam.

$$\Delta G° = \Delta H° - T\Delta S°$$

The Gibbs free-energy equation is derived directly from the **second law of thermodynamics**, which states that any physical or chemical change must result in an increase in the entropy of the universe.

Since $\Delta G°$ is a combination of $\Delta H°$ and $\Delta S°$, we can make some generalizations, shown in Table 11.1, based only on the signs of these quantities.

TABLE 11.1
Relationships of the Algebraic Signs of $\Delta H°$ and $\Delta S°$ to the Sign of $\Delta G°$

$\Delta H°$	$\Delta S°$	$\Delta G°$ as T Increases	Comment
Negative	Positive	Always negative	Always spontaneous
Positive	Negative	Always positive	Always nonspontaneous
Negative	Negative	Becomes positive	Becomes nonspontaneous as T increases
Positive	Positive	Becomes negative	Becomes spontaneous as T increases

THERMODYNAMICALLY FAVORED

What Does Thermodynamically Favored Mean?

When chemists refer to a thermodynamically favored reaction, it is one that occurs without the need for additional energy input after the reactants are mixed and the reaction initiated. Generally, they also mean that $\Delta G°$ for the process has a negative value. This implies that reactions are at standard state. It also implies that reactions are generally considered thermodynamically favored if there are more products than reactants in the equilibrium mixture.

What Does Thermodynamically Favored Not Mean?

Thermodynamically favored, or its opposite, does not tell us the direction of a reaction. ΔG tells us the direction of a reaction as it approaches equilibrium. These terms do not tell us if reactions that are not at standard state are spontaneous.

As is true of $\Delta H°$ and $\Delta S°$ calculations, we can calculate the value of $\Delta G°$ for a reaction from tabulated values of free energies of formation. These values are listed in a table in Appendix 3. Since temperature is an important variable that affects the value of the free energy, the temperature must be specified. It is most common to list $\Delta G°$ values for room temperature of 25°C or 298 K. The symbol for free energy incorporates the temperature, as in ΔG°_{298}. In using the free-energy table, we subtract the ΔG°_{298} of the reactants from the ΔG°_{298} values of the products in the equation:

$$\Delta G^\circ_{298} = \Sigma(\Delta G^\circ_{298} \times \text{coeff})_{\text{products}} - \Sigma(\Delta G^\circ_{298} \times \text{coeff})_{\text{reactants}}$$

TIP

The AP exam gives you the equation $\Delta G° = \Sigma\Delta G^\circ_f$ products − $\Sigma\Delta G^\circ_f$ reactants. The equation shown here is the same as this but reminds you to include stoichiometry coefficients.

For the combustion of propane (Equation 11.4) we obtain

$$\begin{aligned}\Delta G^\circ_{298} &= \left[\left(\frac{-394.4\text{ kJ}}{\text{mol}}\right)(3\text{ mol}) + \left(\frac{-228.6\text{ kJ}}{\text{mol}}\right)(4\text{ mol})\right] - \left[\left(\frac{-23.5\text{ kJ}}{\text{mol}}\right)(1\text{ mol})\right] \\ &= (-1183.2\text{ kJ} - 914.4\text{ kJ}) - (-23.5\text{ kJ}) \\ &= -2074.1\text{ kJ}\end{aligned}$$

The negative value indicates that the reaction is favored, as anyone who has used a barbecue grill or propane torch already knows.

We have calculated $\Delta H°$ and $\Delta S°$ for this reaction in preceding sections of this chapter. Using those values, along with a temperature of 298 K, we have a second way to determine the value of $\Delta G°_{298}$:

$$\begin{aligned}\Delta G°_{298} &= \Delta H° - T\Delta S° \\ &= -2044\text{ kJ} - (298\text{ K})(100.4\text{ J K}^{-1}) \\ &= -2044\text{ kJ} - 29.9\text{ kJ} \\ &= -2074\text{ kJ}\end{aligned}$$

Free Energy at Temperatures Other Than 298 K

Using the table of standard free energies of formation in Appendix 3, we can calculate the free-energy change at 298 K. Standard free-energy changes at other temperatures can also be calculated. For this purpose we need to know the standard heat of reaction, $\Delta H°_{react}$, and the standard entropy change, $\Delta S°_{react}$, for the reaction. These values are correct for 298 K, but we may assume that they do not change significantly with temperature. Using these values in the free-energy equation with a temperature other than 298 K gives the free-energy change at that different temperature.

EXAMPLE 11.4

For a certain reaction, $\Delta H°$ = +2.98 kJ and $\Delta S°$ = +12.3 J K^{-1}. What is $G°$ at 298 K, 200 K, and 400 K?

Solution

The equation to be solved is

$$\Delta G° = \Delta H° - T\Delta S°$$

Substituting the values given in the problem yields

$$\begin{aligned}\Delta G°_{298} &= 2.98\text{ kJ} - 298(12.3\text{ J K}^{-1}) \\ &= 2.98\text{ kJ} - 3665\text{ J}\end{aligned}$$

To complete the solution, –3665 J must be converted to –3.67 kJ. Then

$$\Delta G° = -0.69\text{ kJ (a thermodynamically favored reaction)}$$

At 200 K and 400 K the answers are

$$\Delta G°_{200} = 2.98\text{ kJ} - 200\text{ K}(12.3\text{ J K}^{-1}) = +0.52\text{ kJ}$$

and

$$\Delta G°_{400} = 2.98\text{ kJ} - 400\text{ K}(12.3\text{ J K}^{-1}) = -1.94\text{ kJ}$$

In this exercise we see that the reaction is thermodynamically unfavorable at 200 K but is favored at 298 K and 400 K.

EXAMPLE 11.5

For a certain reaction, $\Delta H° = -13.65$ kJ and $\Delta S° = -75.8$ J K^{-1}. (a) What is $\Delta G°_{298}$ at 298 K? (b) Will increasing or decreasing the temperature make the reaction thermodynamically favored? If so, at what temperature will the reaction become thermodynamically favored?

Solution

(a) At 298 K the free energy is

$$\Delta G°_{298} = -13.65\text{ kJ} - 298\text{ K}(-75.8\text{ J K}^{-1}) = +8.94\text{ kJ}$$

(b) The reaction is not thermodynamically favored at 298 K. However, since $\Delta H°$ and $\Delta S°$ both have the same sign, the free energy will change from positive to negative at some temperature. Since the number zero divides the positive numbers from the negative numbers, we may conclude that $\Delta G° = 0.00$ kJ is the dividing line between thermodynamically favored and not thermodynamically favored reactions. Consequently, the free-energy equation is set up as

$$\Delta G°_{\text{react}} = \Delta H°_{\text{react}} - T\Delta S°_{\text{react}}$$

$$0.00\text{ kJ} = -13.65\text{ kJ} - T(-75.8\text{ J K}^{-1})$$

$$T = \frac{13.65\text{ kJ}}{0.0758\text{ kJ K}^{-1}} = 180\text{ K}$$

From this result we predict that the reaction will be thermodynamically favored below 180 K and not thermodynamically favored above 180 K.

The condensation of a gas and the crystallization of a liquid are two physical processes that are thermodynamically favored at low temperatures and not thermodynamically favored at higher temperatures.

Free Energy and Equilibrium

When a system is not at standard state, the free-energy change is represented by ΔG, not $\Delta G°$. Equation 11.10 shows the relationship between ΔG and $\Delta G°$:

$$\Delta G = \Delta G° + RT\ln Q \qquad (11.10)$$

From Chapter 9 we recall that Q is the reaction quotient. If the value of Q is not equal to the equilibrium constant, further reaction occurs until the system reaches equilibrium.

Using Equation 11.10, we find that when a system is at standard state all concentrations are equal to 1 and $Q = 1$. The natural logarithm of 1 is zero ($\ln 1 = 0$), and consequently $\Delta G = \Delta G°$.

The value of ΔG (without the superscript) tells us whether the reaction will continue and, if so, in which direction it will go. When ΔG is negative, the reaction will proceed in the forward direction. When ΔG is positive, the reaction proceeds in the reverse direction. If ΔG is zero, the reaction is at equilibrium and no further reaction occurs. For the equilibrium condition we find that

TIP

The equation shown here is given on the AP exam.

$$\Delta G° = -RT \ln K$$

by setting $\Delta G = 0$, substituting the equilibrium constant, K, for the reaction quotient, Q, in Equation 11.10, and rearranging.

By using similar logic, we can state that if $K > 1$, the reaction will move in the forward direction. If $K < 1$, the reaction will move in the reverse direction.

The relationships between ΔG and $\Delta G°$ are diagrammed in Figure 11.2. The standard free energies, $\Delta G°$, are shown for the reactants on the left side and for the products on the right side of each graph. The difference between the two is $\Delta G°$ for the reaction. In each diagram the curved line connecting the two $G°$ values represents the value of G for the reaction mixture. The slope of the curved line is ΔG, and at the minimum, where the slope = 0, the reaction is in equilibrium.

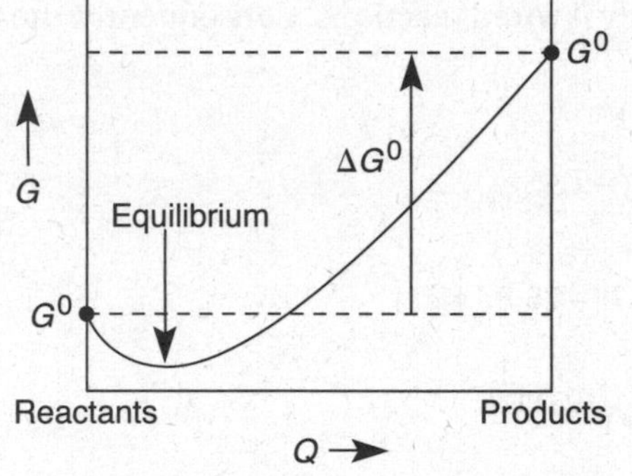

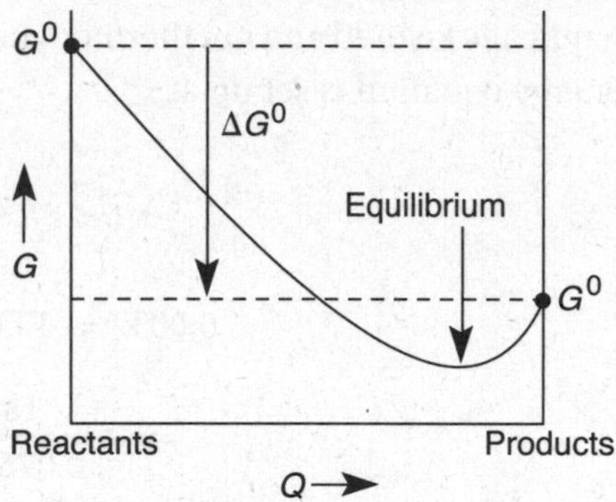

FIGURE 11.2. Free-energy diagrams for (A) a not thermodynamically favored reaction and (B) a thermodynamically favored reaction. The difference between the **$G°$** points is **$\Delta G°$**. The slope of the curved line is **ΔG**, and the equilibrium point is at the minimum of the curve.

The curves in Figure 11.2 illustrate that in a not thermodynamically favored reaction, a small amount of reactants is converted into products at equilibrium (minimum of curved line). A thermodynamically favored reaction, on the other hand, has most of the reactants converted into products because the minimum is closer toward the product side. The slope of the curved line is negative to the left of the equilibrium point, so that ΔG is negative and the reaction moves toward the products and also toward the minimum. On the right-hand side of the equilibrium point, the slope and ΔG are positive. The reaction proceeds toward the reactants and also toward the minimum.

EXAMPLE 11.6

The value of the equilibrium constant is 45 at 298 K. At the same temperature Q = 35. Determine the value of $\Delta G°$ for the reaction at 298 K, and show which direction the reaction will move to reach equilibrium.

Solution

The value of $\Delta G°$ is calculated as

$$\begin{aligned}\Delta G° &= -RT \ln K \\ &= -(8.314 \text{ J mol}^{-1} \text{ K}^{-1})(298 \text{ K})[\ln(45)] \\ &= -9.43 \text{ kJ mol}^{-1}\end{aligned}$$

The fact that Q is less than K indicates that the reaction will proceed in the forward direction.

SUMMARY

Thermodynamics reviewed in this chapter involves the consideration of energy changes in systems at equilibrium. All energy can be considered as either potential energy (energy of position, related to heat content) or kinetic energy (energy of motion, related to temperature). Absolute values for potential or kinetic energy cannot be determined, but changes in these can be measured. State functions $\Delta E°$, $\Delta H°$, $\Delta S°$, and $\Delta G°$ depend only on the initial and final states of the system. The maximum work available from a system is equal to $\Delta G° = \Delta H° - T\Delta S°$. $\Delta E°$ is the change in total internal energy and is measured with a bomb calorimeter. Enthalpy, the heat of a reaction, $\Delta E°$, is measured in a calorimeter at ambient pressure. The standard free energy change is related to the equilibrium constant, and either may be used to determine if a reaction is thermodynamically favored. A thermodynamically favored reaction is generally one where the amount of product is greater than the amount of reactants at equilibrium. This means that $\Delta E°$ is negative and $K_{eq} > 1$.

Important Concepts

Forms of energy
Law of conservation of energy
Hess's law
Standard state
Gibbs free energy, entropy, and enthalpy
Thermodynamically favored and not thermodynamically favored reactions

Important Equations

$q = mc\Delta T$
$\Delta G° = \Delta H° - T\Delta S°$
$\Delta G° = -RT \ln K$

PRACTICE EXERCISES

Multiple-Choice

1. Which of these must be negative for a reaction to be thermodynamically favored?
   (A) $\Delta G°$
   (B) $\Delta H°$
   (C) $\Delta G°$, $\Delta S°$, and ΔG
   (D) ΔG

2. Why is $\Delta H°$ a state function?
   (A) It has the required superscript zero.
   (B) Its value does not depend on the path between the initial and final states.
   (C) It requires exactly 1 molar concentrations of all reactants.
   (D) Its value is equal to the initial state minus the final state.

3. Scientists divide energy into two broad groups. They are
   (A) chemical and solar
   (B) electrical and nuclear
   (C) potential and kinetic
   (D) thermal and electromagnetic

4. How can you tell if a chemical reaction or physical process is endothermic or exothermic?
   (A) Use the $\Delta G°$ and $\Delta S°$ for the reaction at a given temperature to calculate the $\Delta H°$ of reaction or process.
   (B) Observe how increasing the temperature of the reaction or process affects the proportion of products formed.
   (C) Use Hess's law and standard heats of formation to calculate the heat of reaction.
   (D) All of the above.

5. When 0.400 g of CH_4 is burned in excess oxygen in a bomb calorimeter that has a heat capacity of 3,245 J °C^{-1}, a temperature increase of 6.795°C is observed. What is the value of q and ΔH for this combustion reaction?
   (A) 220 kJ mol^{-1} and –220 kJ mol^{-1}
   (B) –882 kJ and 882 kJ
   (C) 477 kJ and –477 kJ
   (D) –22.05 kJ and –22.05 kJ

6. In the diagram below, we have a U-shaped ramp where a stainless steel ball can roll without friction. As the ball rolls back and forth, which statement is correct?

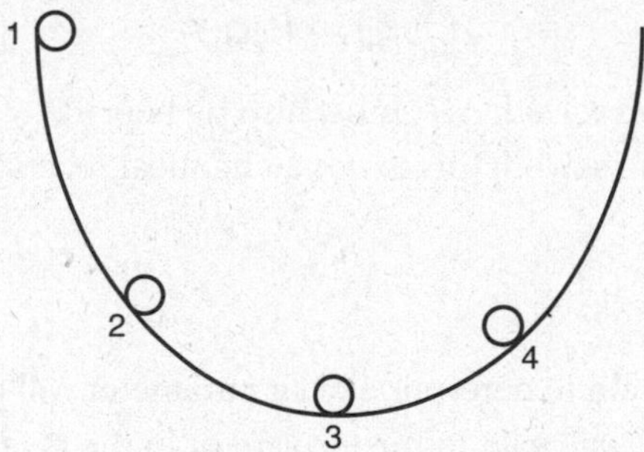

(A) Point 4 is the highest velocity, and point 3 is the lowest velocity.
(B) Point 1 is the highest potential energy, and point 3 is the highest kinetic energy.
(C) Point 1 is the highest kinetic energy, and point 3 is the lowest potential energy.
(D) Point 1 is the highest kinetic energy, and point 3 is the lowest velocity.

7. Which of the following describes a system that CANNOT be thermodynamically favored?
(A) $\Delta H°$ is positive, and $\Delta S°$ is negative.
(B) $\Delta H°$ is positive, and $\Delta S°$ is positive.
(C) $\Delta H°$ is negative, and $\Delta S°$ is negative.
(D) $\Delta H°$ is negative, and $\Delta S°$ is positive.

8. Which of the following must be true when KCl(*s*) is dissolved in water and dissociates into $K^+(aq)$ and $Cl^-(aq)$ and when you observe condensation of moisture on the outside of the glass beaker?
(A) The condensation is extra information. The dissociation equation tells us that the entropy change is positive.
(B) The condensation informs about the heat of the process. The dissociation informs about the entropy change.
(C) Only the heat of the solution process can be inferred from the condensation on the beaker.
(D) The condensation tells us that the process is exothermic. The dissolution reactions always increase the entropy.

9. The reaction with the greatest expected entropy decrease is
(A) $CH_4(g) + 2O_2(g) \rightarrow CO_2(g) + 2H_2O(g)$
(B) $CH_4(l) + 2O_2(g) \rightarrow CO_2(g) + 2H_2O(g)$
(C) $CH_4(g) + 2O_2(g) \rightarrow CO_2(g) + 2H_2O(l)$
(D) $CH_4(g) + 2O_2(g) \rightarrow CO_2(s) + 2H_2O(g)$

10. Water boils at 100°C with a molar heat of vaporization of +43.9 kJ. At 100°C what is the entropy change when water condenses?

$$H_2O(g) \rightarrow H_2O(l)$$

(A) Problem cannot be solved; $\Delta G°$ must also be known.
(B) Problem cannot be solved; this is not a chemical reaction.
(C) -439 J K^{-1}.
(D) -118 J K^{-1}.

11. Use the following diagram to determine what parameter will predict if a given mixture of products and reactants will react in the forward or in the reverse direction?

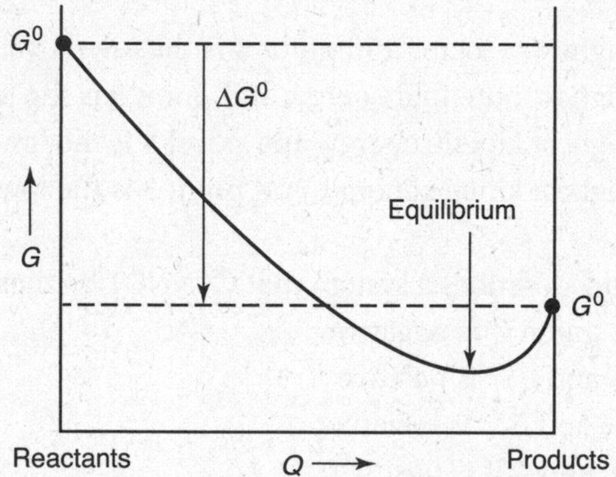

(A) ΔG because it is the slope of the line
(B) $\Delta G°$ because it indicates thermodynamic favorability
(C) G because it is the only value that can be measured experimentally
(D) $G°$ by choosing the one nearest to the product side tells the direction

12. A gas is allowed to expand from an initial volume and pressure (left piston) to a final volume and pressure (right piston). Estimate the value of w?

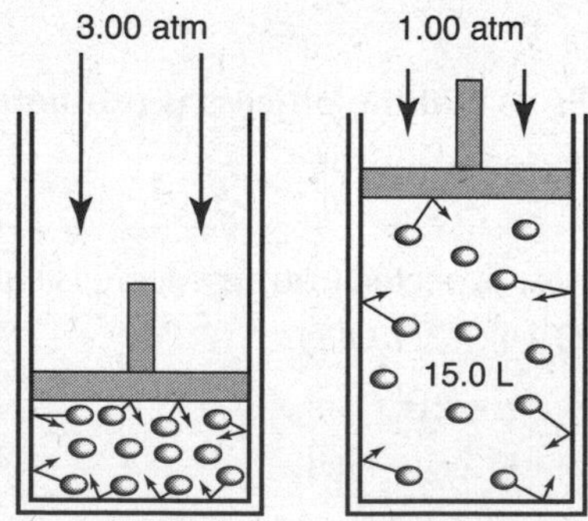

(A) +30.0 L atm
(B) −10.0 L atm
(C) +15.0 L atm
(D) The needed data are not given.

13. In question 12 the units of work are given as L atm. To convert L atm to the metric unit of joules, we need to know
   (A) Avogadro's constant and Planck's constant
   (B) the universal gas law constant in units of L atm mol^{-1} K^{-1}
   (C) the universal gas law constant in units of J mol^{-1} K^{-1}
   (D) both B and C

14. Which of the following is the LEAST probable for a combustion reaction?
   (A) $\Delta G°$ is a large negative number.
   (B) $\Delta S°$ is a large negative number.
   (C) $\Delta H°$ is a large negative number.
   (D) K_{eq} is a large positive number.

15. Of the following molecular level diagrams of argon gas, which illustrates the system with the most entropy and why?

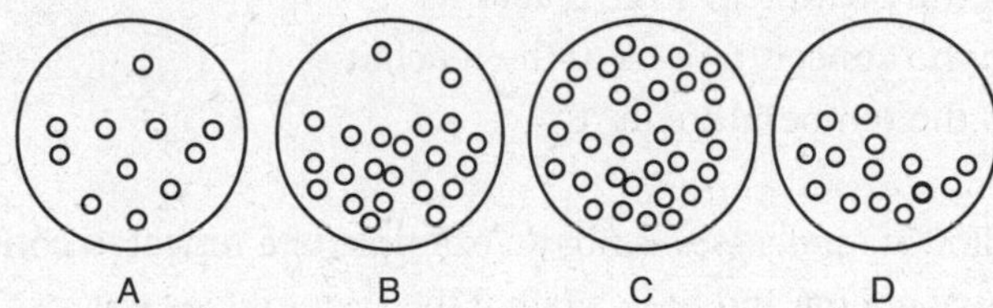

   (A) The atoms are far apart and do not interact.
   (B) The atoms are a liquid.
   (C) This diagram has the most gas atoms and states to occupy.
   (D) Argon atoms are dipoles that strongly attract.

16. The heat of formation of $CH_3OH(l)$ = –238.6 kJ mol^{-1}, of $CO_2(g)$ = –393.5 kJ mol^{-1}, and of $H_2O(g)$ = –241.8 kJ mol^{-1}. What is $\Delta H°$ for the heat of combustion of methanol to gaseous products?
   (A) –396.7 kJ
   (B) –1277 kJ
   (C) –638.5 kJ
   (D) +396.7 kJ

17. The rate of reaction will be large if
   (A) $\Delta G°$ is a large negative number.
   (B) $\Delta S°$ is a large negative number.
   (C) $\Delta H°$ is a large negative number.
   (D) None of the above can be used to estimate reaction rates.

18. Given the following thermochemical data:

$$N_2O_4(g) \rightarrow 2NO_2(g) \qquad \Delta H° = +57.93 \text{ kJ}$$
$$2NO(g) + O_2(g) \rightarrow 2NO_2(g) \qquad \Delta H° = -113.14 \text{ kJ}$$

determine the heat of the reaction

$$2NO(g)+O_2(g) \rightarrow N_2O_4(g)$$

(A) 171.07 kJ
(B) −55.21 kJ
(C) −171.07 kJ
(D) +55.21 kJ

19. Which of the following can change the value of $\Delta G°$ for a chemical reaction?
(A) Changes in the total pressure
(B) Changes in the pressures of the reactants
(C) Changes in the concentrations of the reactants
(D) Changes in the temperature in °C

20. Which of the following processes is not an example of a repeated conversion of kinetic energy to potential energy and back again if the system does not lose energy due to friction?
(A) A vibrating spring, with one end attached to a ring stand and the other to a 100 g mass
(B) A ball bouncing vertically on a lab bench
(C) A pendulum with a 100 g mass attached to a ring stand
(D) A ball bouncing down a set of stairs

21. The standard heat of formation of $SO_3(g)$ is −396 kJ mol^{-1}. The standard entropies of $S(s)$, $O_2(g)$, and $SO_3(g)$ are 31.8, 205.0, and 256 J mol^{-1} K^{-1}, respectively. Calculate the free energy for the decomposition of SO_3 in the reaction

$$2SO_3(g) \rightarrow 2S(s)+3O_2(g)$$

at 25°C.

(A) +396 kJ
(B) −446 kJ
(C) −346 kJ
(D) +742 kJ

22. To determine the standard enthalpy ($\Delta H°$) of the following combustion reaction

$$2C_6H_6(l) + 15O_2(g) \rightleftharpoons 12CO_2(g) + 6H_2O(l)$$

What do you need to know?

(A) The standard heat of combustion of carbon
(B) The standard heat of combustion of hydrogen
(C) Hess's law and its use
(D) All of the above

CHALLENGE

23. The evaporation of any liquid is expected to have
(A) a positive ΔH and a negative ΔS
(B) a negative ΔH and a negative ΔS
(C) a positive ΔH and a positive ΔS
(D) a positive ΔH and a negative ΔS

24. Which of the following is most likely to be true?
(A) Combustion of organic compounds has a negative $\Delta H°$.
(B) A positive $\Delta G°$ indicates a thermodynamically favored reaction.
(C) A positive $\Delta S°$ always means that the reaction is thermodynamically favored.
(D) A thermodynamically favored reaction always goes to completion.

ANSWER KEY

1. **(A)**	7. **(A)**	13. **(D)**	19. **(D)**
2. **(B)**	8. **(B)**	14. **(B)**	20. **(D)**
3. **(C)**	9. **(C)**	15. **(C)**	21. **(D)**
4. **(D)**	10. **(D)**	16. **(B)**	22. **(D)**
5. **(D)**	11. **(A)**	17. **(D)**	23. **(C)**
6. **(B)**	12. **(B)**	18. **(C)**	24. **(A)**

See Appendix 1 for explanations of answers.

Free-Response

Answer the following questions using the concepts of thermodynamics and equilibrium and the methods for solving problems.

1. (a) Write the balanced equation for the formation reaction of propyl amine, $CH_3CH_2CH_2NH_2$.

(b) Can propyl amine be formed in the reaction above? Explain why or why not.

(c) Can the heat of formation, $\Delta H_f°$, be determined using the reaction above? Explain why or why not.

(d) Describe the experiments needed to determine the heat of formation, $\Delta H_f°$, of propyl amine.

2. Give succinct, logical descriptions of how scientists can experimentally obtain the following information.

   (a) Describe experiments that can be used to determine the numerical value of K_{eq}.

   (b) Describe experiments that can be used to determine the numerical value of $\Delta H°$ for a substance.

   (c) Describe experiments that can be used to determine the numerical value of $\Delta G°$ for a substance.

   (d) Describe experiments that can be used to determine the numerical value of $S°$ for a substance.

3. (a) What parameters define whether or not a given reaction is thermodynamically favored? Based on those parameters, what does it mean to say a reaction is thermodynamically favored? What indicates if a mixture of chemicals will proceed in the forward direction or in the reverse direction?

   (b) What is the difference between E, ΔE, and $\Delta E°$?

   (c) Tables of thermodynamic data list heat of formation, $\Delta H_f°$, standard free energy, $\Delta G_{298}°$, and entropy, $S°$. Why don't entropy values have a delta symbol, Δ? What other difference does a table of entropy values have?

   (d) The value of K_c for the reaction $2NO + O_2 \rightleftharpoons N_2O_4$ is 36 at a certain temperature. Calculate K for the following reactions.

   (i) $6NO + 3O_2 \rightleftharpoons 3N_2O_4$

   (ii) $NO + \frac{1}{2} O_2 \rightleftharpoons \frac{1}{2} N_2O_4$

   (iii) $2N_2O_4 \rightleftharpoons 4NO + 2O_2$

FREE-RESPONSE ANSWERS

1. (a) $3C + 4.5H_2 + 0.5N_2 \rightarrow CH_3CH_2CH_2NH_2$

   (b) It is highly unlikely that mixing these three elements together will yield propyl amine. Molecular nitrogen has an extremely strong triple bond that needs to be broken first. Scientists have tried many ways to make CH_4 from carbon and hydrogen but with little success. The more complex molecule propyl amine will not form.

   (c) Since the reaction will not occur, we cannot use it to determine the heat of formation.

   (d) To obtain the heat of formation, a series of combustion reactions can be run.

$$4CH_3CH_2CH_2NH_2 + 21O_2 \rightarrow 12CO_2 + 18H_2O + 2N_2$$

$$C + O_2 \rightarrow CO_2$$

$$2H_2 + O_2 \rightarrow 2H_2O$$

Once the heats of combustion are obtained, the first reaction is reversed. The second reaction is multiplied by 12. The third equation is multiplied by 9. Adding those reacting together gives the heat of formation for 4 propyl amines. Thus one last division by 4 gives the desired results.

2. (a) K_{eq} can be determined by measuring all components of a mixture to get values for the equilibrium constant. Several determinations will assure that the system is at equilibrium.

   (b) The enthalpy is determined in a calorimetry experiment where the heat capacity of the system and the temperature change allow the calculation of heat evolved or absorbed. Normalizing to 1 mole of substance gives a molar heat of reaction.

   (c) The Gibbs free energy is not measured directly. Perhaps the easiest approach is to determine the value of the equilibrium constant or the standard galvanic cell potential. So the experiments are actually those that determine K_{eq} or $E°$. The following equation gives the standard Gibbs free energy.

$$\Delta G° = -RT \ln K_{eq} = -nFE°$$

   (d) Entropy can be determined by measuring the heat, q, needed to raise 1 mole of substance from absolute zero to 273.16 K. An easier method, since few people have access to the apparatus to reach absolute zero, is to use (b) and (c) to determine $\Delta G°$ and $\Delta H°$ and then use the following equation.

$$\Delta G° = \Delta H° - T\Delta S°$$

3. (a) At this point we have defined a thermodynamically favored reaction as one where K_{eq} is greater than 1.00. This translates into a value of $\Delta G°$ that has a negative sign. (In Chapter 12 we expand this to cases where $E°_{cell}$ has a positive sign.) For the second part of the question, this means that reactions that are symmetrical (equal moles on both sides of the arrow) will be thermodynamically favored as long as more product is formed than reactants left over. For a mixture to proceed in the forward direction, $\Delta G < 0$ or $Q < K$. The opposite conditions are required for the mixture to proceed in the reverse direction.

   (b) The difference between the three formats is that E is the total internal energy of a system. This cannot be determined exactly because all motions and attractive forces are not known. ΔE is measurable, as a change in internal energy. However, it refers to the energy change in a system that may be large or small; it is an extensive measure. Last, $\Delta E°$ represents the change in internal energy for a defined system and in an intrinsic property.

   (c) Entropies are state functions that can be precisely determined by virtue of the third law of thermodynamics. Therefore we do not need to rely on changes in the state. Another difference is that elements have values of zero for heats of formation and free energies. Elements have nonzero entropies. Finally, entropy has units of J mol^{-1} K^{-1}, whereas the free energy and enthalpy have units of J mol^{-1}.

(d) (i) This reaction has the coefficients multiplied by 3, so that $K = K_cK_cK_c = 4.66 \times 10^4$.

(ii) The coefficients are divided in half, so that $K = (K_c)^{1/2} = 6$.

(iii) In this reaction the reaction is reversed and the coefficients are multiplied by 2, so that $K = (1/(K_c)^2) = 7.7 \times 10^{-4}$.

PART 5

Chemical Reactions

Note: The link to the online practice tests can be found on the inside front cover of the book. All three online tests can be accessed on mobile devices, including tablets and smartphones.

Oxidation-Reduction Reactions and Electrochemistry

12

- → OXIDATION DEFINED
- → REDUCTION DEFINED
- → OXIDATION NUMBER/STATE
- → BALANCING REDOX REACTIONS
- → ION-ELECTRON METHOD
- → SINGLE-REPLACEMENT REACTIONS
- → PERMANGANATE REACTIONS
- → DICHROMATE REACTIONS
- → ELECTROCHEMISTRY
- → ELECTROLYSIS
- → ANODE AND OXIDATION
- → CATHODE AND REDUCTION
- → FARADAY'S CONSTANT
- → GALVANIC CELL
- → SALT BRIDGE
- → STANDARD CELL VOLTAGE, $E°_{CELL}$
- → STANDARD REDUCTION POTENTIALS
- → K_{EQ} RELATED TO, $E°_{CELL}$
- → $E°_{CELL}$ RELATED TO $\Delta G°$
- → COMBUSTION REACTIONS
- → CORROSION

BIG IDEAS 3, 6

Learning Objectives: 3.8, 3.9, 3.12, 3.13, 6.1, 6.25

For the complete list of Big Ideas and Learning Objectives, refer to the AP Chemistry Course Outline: *https://secure-media.collegeboard.org/digitalServices/pdf/ap/ap-chemistry-course-and-exam-description.pdf*

Chemical reactions in which electrons are transferred from one atom to another are called **oxidation-reduction** reactions. As a group, more reactions may be classified as oxidation-reduction reactions than as acid-base, double-replacement, or complexation reactions combined.

Oxidation is the loss of electrons, and **reduction is the gain of electrons**. When an atom of barium reacts with an atom of sulfur, the barium loses its two valence electrons and is oxidized, while the sulfur gains these two electrons and is reduced.

$$\cdot\dot{Ba} + \cdot\ddot{\underset{\cdot\cdot}{S}}\cdot \rightarrow Ba^{2+} : \ddot{\underset{\cdot\cdot}{S}}:^{2-} \rightarrow BaS$$

This reaction may be written as two **half-reactions** that show the individual oxidation and reduction steps:

$$Ba \rightarrow Ba^{2+} + 2e^- \quad \text{(oxidation)}$$

$$S + 2e^- \rightarrow S^{2-} \quad \text{(reduction)}$$

Although we may write separate half-reactions, they cannot exist without each other. Perhaps the word *redox* was coined to emphasize this point.

OXIDATION NUMBERS

Determining Oxidation Numbers

In more complex reactions it is not always obvious that electrons are transferred. To determine whether a substance has lost or gained electrons, chemists calculate the **oxidation number**, also called the **oxidation state**, of each element. If the oxidation number changes during a reaction, electrons have been transferred.

To determine the oxidation numbers of the elements in a compound or polyatomic ion, we follow a short set of rules:

TIP

Using this hierarchy removes the need to memorize peroxides and hydrides.

OXIDATION NUMBER RULE HIERARCHY

1. The oxidation numbers of all atoms add up to the charge on the atom, molecule, or ion.
2. The oxidation number of an alkali metal is +1, for an alkaline earth element it is +2, and for the metals in Group IIIA it is +3.
3. The oxidation number of hydrogen is +1, and the oxidation number of fluorine is −1.
4. The oxidation number of oxygen is −2.
5. The oxidation number of a halogen is −1.
6. The oxidation number of nonmetal in Group VIA is −2.

The oxidation number rules are a hierarchy, meaning that rule 1 is the most important and rule 6 is the least important. If two rules conflict, the rule closest to the top of the list is obeyed while the one lower on the list is ignored.

Many simple compounds follow the oxidation number rules directly. Since the charge of any element is zero, the oxidation number of any element must be zero, according to rule 1. In compounds such as $CaCl_2$, calcium is +2 according to rule 2, and each chlorine is −1 by rule 5; also, the most important rule, rule 1, is obeyed since +2 and $2 \times (-1)$ add up to zero, the charge on $CaCl_2$. Also, in potassium hydroxide, KOH, potassium is +1 according to rule 2; oxygen is −2 according to rule 4; and hydrogen is +1 according to rule 3. The oxidation numbers all obey rule 1 since $+1 + 1 - 2 = 0$, which is the charge on KOH.

Some compounds, however, do not obey one or more of the rules. Since all substances must obey rule 1, we may write rule 1 as an equation in which the total charge is equal to the sum of the elements' oxidation numbers, with each oxidation number multiplied by the number of times the element appears in the formula:

$$\text{Total charge} = (\text{ox. no. 1})(\text{subscript 1}) + (\text{ox. no. 2})(\text{subscript 2}) + \cdots$$

For example, the oxidation numbers for the elements in perchloric acid, $HClO_4$, are +1 for hydrogen, –1 for chlorine, and –2 for each oxygen if rules 3, 4, and 5 are strictly followed. These assignments violate rule 1, however, since the oxidation numbers add up to –8 and the $HClO_4$ molecule must have a charge of zero. We, therefore, ignore rule 5 in favor of rule 1 and write the equation as

$$\begin{aligned} \text{0 charge} &= (\text{ox. no. H})(1) + (\text{ox. no. Cl})(1) + (\text{ox. no. O})(4) \\ 0 &= (+1)(1) + (\text{ox. no. Cl})(1) + (-2)(4) \\ 0 &= \text{ox. no. Cl} - 7 \\ \text{ox. no. Cl} &= +7 \end{aligned}$$

In perchloric acid the oxidation number of the chlorine atom is +7.

Oxidation numbers for polyatomic ions are determined using the same method, being sure that the oxidation numbers add up to the charge on the ion. For example, we might assign all the sulfur and oxygen atoms in the sulfate ion, SO_4^{2-}, oxidation numbers of –2 according to rules 4 and 6. That would disobey rule 1, since the sum of these oxidation numbers is –2 – 8 = –10, whereas the actual charge of the sulfate ion is –2. Consequently, we abandon rule 6 and work with rules 1 and 4 to write the equation

$$\begin{aligned} -2\text{ charge} &= (\text{ox. no. S})(1) + (\text{ox. no. O})(4) \\ -2 &= \text{ox. no. S} + (-2)(4) \\ -2 &= \text{ox. no. S} - 8 \\ \text{ox. no. S} &= +6 \end{aligned}$$

Finally, the rules specify oxidation numbers for only 25 of the 111 known elements. However, by using the oxidation number rules and the formulas of the polyatomic ions, chemists can determine the oxidation numbers of most of the other elements. For example, the oxidation numbers for chromium and oxygen in the dichromate ion, $Cr_2O_7^{2-}$, can be determined using rules 1 and 4:

$$\begin{aligned} -2\text{ charge} &= (\text{ox. no. Cr})(2) + (\text{ox. no. O})(7) \\ -2 &= (\text{ox. no. Cr})(2) + (-2)(7) \\ -2 &= (\text{ox. no. Cr})(2) - 14 \\ (\text{ox. no. Cr})(2) &= +12 \\ \text{ox. no. Cr} &= +6 \end{aligned}$$

Some compounds may contain two elements that are not covered by the oxidation number rules. One such compound is nickel(II) carbonate, $NiCO_3$. In situations like this one we take advantage of knowing that the carbonate ion is CO_3^{2-} and the nickel must be a +2 ion, Ni^{2+}. For the Ni^{2+}, rule 1 applies directly, and the oxidation number is +2. For the carbonate ion we write

$$\begin{aligned} -2\text{ charge} &= (\text{ox. no. C})(1) + (\text{ox. no. O})(3) \\ -2 &= (\text{ox. no. C})(1) + (-2)(3) \\ -2 &= \text{ox. no. C} - 6 \\ \text{ox. no. C} &= +4 \end{aligned}$$

Using this procedure, we have determined that the oxidation numbers are as follows: +2 for nickel, +4 for carbon, and –2 for each oxygen.

Exercise 12.1

Determine the oxidation number of each element in the following formulas:

(a) H_2O_2 (c) K_3PO_4 (e) $Fe_2(C_2O_4)_3$ (g) SO_3^{2-} (i) NH_4^+
(b) $HClO_3$ (d) Na_2CrO_4 (f) MnO_4^- (h) N_2O_5 (j) $FeNH_4(SO_4)_2$

Solution

(a) H = +1; O = –1
(b) H = +1; Cl = +5; O = –2
(c) K = +1; P = +5; O = –2
(d) Na = +1; Cr = +6; O = –2
(e) Fe = +3; C = +3; O = –2
(f) Mn = +7; O = –2
(g) S = +4; O = –2
(h) N = +5; O = –2
(i) N = –3; H = +1
(j) Fe = +3; N = –3; H = +1; S = +6; O = –2

Using Oxidation Numbers

Oxidation numbers are used to determine whether or not a substance has been oxidized or reduced. When the permanganate ion reacts in acid solution to form Mn^{2+}, we find that the oxidation number of the manganese is +7 in the permanganate ion and +2 in the Mn^{2+} ion. The change in oxidation number indicates that an oxidation-reduction process has taken place. In addition, the fact that the oxidation number was reduced from +7 to +2 tells us that a reduction of manganese has taken place. Also, the fact that the oxidation number changes by 5 tells us that five electrons have been gained by each manganese atom in the reduction process.

Remember that oxidation numbers are artificial constructs that allow the chemist to draw some conclusions about redox reactions. However, to assign oxidation numbers, we assume that all elements in a compound are ions. This is obviously not true. Therefore, the utility of oxidation numbers is quite limited.

Exercise 12.2

Possible reactants and products for oxidation-reduction reactions are given below. In each case, determine the oxidation numbers to tell whether an oxidation or a reduction has occurred and, if so, how many electrons were lost or gained from the starting material.

(a) $Cr_2O_7^{2-}$ to CrO_4^{2-}
(b) $C_2O_4^{2-}$ to CO_2
(c) NO_3^- to NO_2
(d) I^- to IO_3^-
(e) $BaCl_2$ to $BaSO_4$
(f) Fe to Fe_2O_3
(g) C_2H_4 to C_2H_6

Solution

(a) No change occurs in any oxidation numbers; not a redox process.
(b) Each C changes from +3 to +4. Two electrons per $C_2O_4^{2-}$ are gained in this oxidation.
(c) N changes from +5 to +4. One electron per NO_3^- is used in this reduction.
(d) I changes from –1 to +5. Six electrons per I^- are used in this oxidation.
(e) Ba does not change; no redox process occurs (Cl_2 and SO_4 are ignored).
(f) Fe changes from 0 to +3. Three electrons per Fe are lost in this oxidation.
(g) Each C changes from –2 to –3. Two electrons per C_2H_4 are gained in this reduction.

In a redox equation we may now identify what is occurring. For example, Fe^{2+} is often titrated with permanganate ions. The unbalanced reaction is

$$Fe^{2+} + MnO_4^- \rightarrow Mn^{2+} + Fe^{3+}$$

From this reaction we see that Fe^{2+} is oxidized to Fe^{3+}. The manganese has an oxidation number of +7 in the permanganate ion and of +2 in the product; therefore the manganese is reduced. We can also say that the permanganate ion is reduced.

Exercise 12.3

In each of the following unbalanced equations, identify the element that is oxidized, the element that is reduced, and the changes in oxidation number.

(a) $O_2 + N_2H_4 \rightarrow H_2O_2 + N_2$
(b) $XeO_3 + I^- \rightarrow Xe + I_2$
(c) $I_2 + OCl^- \rightarrow IO_3^- + Cl^-$
(d) $NO_3^- + Cu \rightarrow NO_2 + Cu^{2+}$
(e) $PbO_2 + Cl^- \rightarrow PbCl_2 + Cl_2$

Solution

The table below lists the required information. The change in oxidation number is indicated after each element in the first two columns.

	Element Oxidized	Element Reduced
(a)	N (+2)	O (–1)
(b)	I (+1)	Xe (–6)
(c)	I (+5)	Cl (–2)
(d)	Cu (+2)	N (–1)
(e)	Cl (+1)	Pb (–2)

BALANCING REDOX REACTIONS

Redox reactions tend to be complex, and attempts to balance them by inspection often fail to achieve an answer within a reasonable length of time. Over the years chemists have devised several methods for balancing redox reactions. The simplest and most versatile is called the **ion-electron method**. To use the ion-electron method, six steps must be performed, in the order given. Steps 1–6 are for reactions that occur in acid (H^+) solution. A seventh step is added if the reaction occurs in basic (OH^-) solution, or if H^+ appears on one side of the equation and OH^- appears on the other side.

ION-ELECTRON METHOD

1. Write two half-reactions, one representing the oxidation, and the other the reduction, that occur in the reaction. It is not necessary to know which is which at this point.
2. In each half-reaction, balance all atoms except hydrogen and oxygen.
3. Balance oxygen atoms in each half-reaction by adding one H_2O molecule for each oxygen atom needed. Never use O_2, OH^-, or any other form of oxygen.
4. Balance the hydrogen atoms by adding hydrogen ions, H^+. Never use H_2, OH^-, or any other form of hydrogen.
5. Balance the charges by adding the proper number of electrons (e^-). If steps 1–4 have been done properly, electrons will be added to the left side of one half-reaction and to the right side of the other.
6. Multiply each half-reaction by the appropriate number so that the two half-reactions have the same number of electrons. Add the half-reactions, and cancel the electrons (they must cancel). Also cancel all common ions and molecules. Simplify the coefficients of the equation if possible.
7. Use only if reaction must be balanced in basic solution. Add one OH^- ion for each H^+ ion to both sides of the equation in step 6. Combine the H^+ and OH^- ions on one side of the reaction into H_2O molecules. Cancel H_2O molecules that appear on both sides of the equation, and simplify if possible.

Before using the ion-electron method, however, it is worthwhile to try to balance a reaction by inspection. For example, the unbalanced equation for the reaction of magnesium metal with acid may be written as

$$Mg + H^+ \rightarrow Mg^{2+} + H_2$$

By inspection we see that placing the coefficient 2 in front of H^+ will balance the hydrogen atoms and the charge in this equation:

$$Mg + 2H^+ \rightarrow Mg^{2+} + H_2$$

There is no need to use the longer ion-electron method.

In the following example, however, the ion-electron method must be used. Standard solutions of iodine are prepared by reacting iodide ions with the iodate ion in acid solution. The skeleton reaction is

$$I^- + IO_3^- \rightarrow I_2$$

The first step of the ion-electron method requires two half-reactions, which are written as

$$I^- \rightarrow I_2$$

$$IO_3^- \rightarrow I_2$$

Even though the skeleton reaction has only one product, there is no reason why both reactants cannot yield the same product, one by oxidation and the other by reduction. Usually two pairs of reactants can be identified to obtain the half-reactions. In addition, H^+, OH^-, or H_2O may be included in the skeleton reaction. These are generally ignored since they are added in later steps.

The second step requires that the iodine atoms be balanced by using the coefficient **2** in both half-reactions:

$$\mathbf{2}I^- \rightarrow I_2$$

$$\mathbf{2}IO_3^- \rightarrow I_2$$

The third step applies only to the second half-reaction since the first one contains no oxygen. Six water molecules are needed to balance the six oxygens in two iodate ions:

$$2I^- \rightarrow I_2$$

$$2IO_3^- \rightarrow I_2 + \mathbf{6H_2O}$$

The fourth step also involves only the second half-reaction, where 12 H^+ ions are required:

$$2I^- \rightarrow I_2$$

$$\mathbf{12H^+} + 2I_3^- \rightarrow I_2 + 6H_2O$$

At this point all the atoms are balanced, and we use step 5 to balance the charges with electrons. First we must count the charge on each side of each half-reaction. The first half-reaction has a total charge of –2 on the left and 0 on the right. The second half-reaction has a total charge of +10 on the left and 0 on the right. Two electrons are added to the right side of the first half-reaction to equalize the charges at –2. Ten electrons are added to the left side of the second half-reaction, resulting in a total charge of 0 on both sides:

$$2I^- \rightarrow I_2 + \mathbf{2e^-}$$

$$\mathbf{10e^-} + 12H^+ + 2IO_3^- \rightarrow I_2 + 6H_2O$$

In step 6 we can equalize the electrons in the two half-reactions by multiplying the entire first half-reaction by 5:

$$\mathbf{10I^-} \rightarrow \mathbf{5I_2} + \mathbf{10e^-}$$

$$10e^- + 12H^+ + 2IO_3^- \rightarrow I_2 + 6H_2O$$

Adding the equations yields

$$10I^- + 10e^- + 12H^+ + 2IO_3^- \rightarrow 5I_2 + 10e^- + I_2 + 6H_2O$$

We then cancel 10 electrons from each side, add the iodine molecules on the right to obtain $6I_2$, and divide all of the coefficients by 2 to obtain the final balanced equation:

$$5I^- + 6H^+ + IO_3^- \rightarrow 3I_2 + 3H_2O$$

All atoms must be balanced in the reaction. In addition, in ionic reactions the charges must also balance. In this equation we count a total charge of 0 on both sides of the arrow.

In a third example, the reaction between the permanganate ion and iodide ions occurs in neutral solutions. The unbalanced skeleton reaction is

$$MnO_4^- + I^- \rightarrow I_2 + MnO_2$$

The obvious pairs for the half-reactions are

$$MnO_4^- \rightarrow MnO_2$$

$$I^- \rightarrow I_2$$

In the second half-reaction the iodine atoms need to be balanced:

$$MnO_4^- \rightarrow MnO_2$$

$$\mathbf{2}I^- \rightarrow I_2$$

The first half-reaction needs two water molecules to balance the oxygen atoms:

$$MnO_4^- \rightarrow MnO_2 + \mathbf{2H_2O}$$

$$2I^- \rightarrow I_2$$

Four hydrogen ions balance the hydrogen atoms:

$$\mathbf{4H^+} + MnO_4^- \rightarrow MnO_2 + 2H_2O$$

$$2I^- \rightarrow I_2$$

The next step is the balancing of charges. We find that the first half-reaction has a total charge of +3 on the left and 0 on the right, while the second half-reaction has a total charge of –2 on the left and 0 on the right. We add three electrons to the first half-reaction on the left and two electrons to the second half-reaction on the right:

$$\mathbf{3e^-} + 4H^+ + MnO_4^- \rightarrow MnO_2 + 2H_2O$$

$$2I^- \rightarrow I_2 + \mathbf{2e^-}$$

To equalize the electrons, we multiply the first half-reaction by 2 and the second half-reaction by 3:

$$6e^- + 8H^+ + \mathbf{2}MnO_4^- \rightarrow 2MnO_2 + 4H_2O$$

$$6I^- \rightarrow 3I_2 + 6e^-$$

We add the equations:

$$6e^- + 6I^- + 8H^+ + \mathbf{2}MnO_4^- \rightarrow 2MnO_2 + 4H_2O + 3I_2 + 6e^-$$

and cancel the 6 electrons:

$$6I^- + 8H^+ + \mathbf{2}MnO_4^- \rightarrow 2MnO_2 + 4H_2O + 3I_2$$

This reaction is balanced in acid solution. To convert to a basic solution, we use step 7 and add $8OH^-$ to each side:

$$\mathbf{8OH^-} + 6I^- + 8H^+ + 2MnO_4^- \rightarrow 2MnO_2 + 4H_2O + 3I_2 + \mathbf{8OH^-}$$

The $8H^+$ and $8OH^-$ on the left are combined to make $8H_2O$:

$$\mathbf{8H_2O} + 6I^- + 2MnO_4^- \rightarrow 2MnO_2 + 4H_2O + 3I_2 + 8OH^-$$

We then cancel $4H_2O$ from each side to simplify the final equation:

$$4H_2O + 6I^- + 2MnO_4^- \rightarrow 2MnO_2 + 3I_2 + 8OH^-$$

Exercise 12.4

Balance each of the following half-reactions in acid solution:

(a) $Fe(s) \rightarrow Fe^{3+}$
(b) $Cl_2 \rightarrow Cl^-$
(c) $Cr^{3+} \rightarrow CrO_4^{2-}$
(d) $NO_3^- \rightarrow NO_2$
(e) $S_2O_3^{2-} \rightarrow SO_4^{2-}$

Solution

(a) $Fe(s) \rightarrow Fe^{3+} + 3e^-$
(b) $2e^- + Cl_2 \rightarrow 2Cl^-$
(c) $4H_2O + Cr^{3+} \rightarrow CrO_4^{2-} + 8H^+ + 3e^-$
(d) $e^- + 2H^+ + NO_3^- \rightarrow NO_2 + H_2O$
(e) $5H_2O + S_2O_3^{2-} \rightarrow 2SO_4^{2-} + 10H^+ + 8e^-$

EXAMPLE 12.1

Balance each of the following skeleton redox reactions in the solution indicated:

(a) $Cr_2O_7^{2-} + CH_3CH_2OH \rightarrow Cr^{3+} + CH_3COOH$ (acid solution)
(b) $Cu + NO_3^- \rightarrow NO_2 + Cu^{2+}$ (acid solution)
(c) $MnO_2 + ClO_3^- \rightarrow MnO_4^- + Cl^-$ (basic solution)
(d) $Al + H_2O \rightarrow Al(OH)_4^- + H_2$ (basic solution)

Solution

(a) The balanced half-reactions are

$$6e^- + 14H^+ + Cr_2O_7^{2-} \rightarrow 2Cr^{3+} + 7H_2O$$

$$H_2O + CH_3CH_2OH \rightarrow CH_3COOH + 4H^+ + 4e^-$$

The balanced equation is

$$16H^+ + \mathbf{2}Cr_2O_7^{2-} + 3CH_3CH_2OH \rightarrow 4Cr^{3+} + 3CH_3COOH + 11H_2O$$

Reduction half-reactions have electrons as reactants.

(b) The balanced half-reactions are

$$Cu \rightarrow Cu^{2+} + 2e^-$$

$$e^- + 2H^+ + NO_3^- \rightarrow NO_2 + H_2O$$

The balanced equation is

$$Cu + 2NO_3^- + 4H^+ \rightarrow 2NO_2 + Cu^{2+}$$

(c) The balanced half-reactions are

TIP

Oxidation half-reactions have electrons as products.

$$6e^- + 6H^+ + ClO_3^- \rightarrow Cl^- + 3H_2O$$

$$2H_2O + MnO_2 \rightarrow MnO_4^- + 4H^+ + 3e^-$$

The balanced equation is

$$2MnO_2 + 2OH^- + ClO_3^- \rightarrow 2MnO_4^- + Cl^- + H_2O$$

(d) The balanced half-reactions are

$$4H_2O + Al \rightarrow Al(OH)_4^- + 4H^+ + 3e^-$$

$$2e^- + 2H^+ + H_2O \rightarrow H_2 + H_2O$$

The balanced equation is

$$2Al + 2OH^- + 6H_2O \rightarrow 2Al(OH)_4^- + 3H_2$$

COMMON OXIDATION-REDUCTION REACTIONS

Single-Replacement (Displacement) Reactions

In **single-replacement** reactions an element replaces an atom in a compound, producing another element and a new compound. For example, the element zinc replaces the hydrogen atom in hydrogen chloride, forming the element H_2 and the new compound zinc chloride:

$$Zn + 2HCl \rightarrow ZnCl_2 + H_2$$

Very active metals, which have the lowest ionization energies, are Li, Na, K, Rb, Cs, Ca, Sr, and Ba. These elements all react with water in single-displacement reactions to form hydrogen. One example is

$$2Na + 2H_2O \rightarrow 2NaOH + H_2$$

Many of these reactions also produce so much heat that the hydrogen ignites.

Active metals do not react with water, but will react with acids in a single-replacement reaction. An example is

$$Mg + 2H^+ \rightarrow Mg^{2+} + H_2$$

The common active metals are Mg, Zn, Pb, Ni, Al, Ti, Cr, Fe, Cd, Sn, and Co.

Inactive metals do not undergo simple single-replacement reactions with either water or acids. The most common inactive metals are Ag, Pt, Au, and Cu. Copper and silver react with concentrated nitric acid in a reaction that produces nitrogen oxides but not hydrogen. Gold reacts with a mixture, called aqua regia, of three parts concentrated HCl and one part concentrated HNO_3. The reaction of very active, active, and inactive metals with water and acid are summarized in Table 12.1.

TABLE 12.1
Summary of Metal Reactions with Water and Acid

Type of Metal	Examples	Comments
Very active metal	Li, Na, K, Rb, Cs, Ca, Sr, Ba	React with H_2O to produce H_2, test gas by igniting a small amount
Active metal	Mg, Zn, Pb, Ni, Al, Ti, Cr, Fe, Cd, Sn, Co	React with acids to form H_2, but not with H_2O
Inactive metal	Ag, Au, Cu, Pt	DO NOT form H_2 with acids: may react with conc. oxidizing acids HNO_3 and H_2SO_4 or aqua regia

In addition to replacing hydrogen, metals will displace metal ions from their compounds. For instance, copper metal will react with silver nitrate in the single-replacement reaction

$$Cu + 2AgNO_3 \rightarrow Cu(NO_3)_2 + 2Ag$$

The net ionic equation is written as

$$Cu + 2Ag^+ \rightarrow Cu^{2+} + 2Ag$$

In these two reactions, copper is more active than silver and the reaction does occur. The reverse reaction:

$$Cu^{2+} + 2Ag \rightarrow Cu + 2Ag^+$$

does not occur, however, since silver is less active than copper.

An **activity series** is a listing of metals in the order of their strengths to cause oxidation or reduction. We can use an activity series to determine whether a certain metal will displace another metal ion from its compounds. Later in this chapter (page 437) we will introduce the concept of standard reduction potentials. A table of standard reduction potentials contains the same information as an activity series. The metal having the lower, or more negative, standard reduction potential is the more active metal.

Only the active and inactive metals will displace each other from compounds. The very active metals listed in Table 12.1 do not displace other metals. Although these metals are certainly reactive enough to result in such displacements, their high activity causes them to react preferentially with water instead.

Nonmetals such as the halogens also participate in single-displacement reactions. The activity series for the halogens, from the strongest to the weakest oxidizer, is $F_2 > Cl_2 > Br_2 > I_2$. As a result, adding chlorine to a solution of potassium bromide results in the reaction

$$Cl_2 + 2KBr \rightarrow Br_2 + 2KCl$$

The net ionic reaction is

$$Cl_2 + 2Br^- \rightarrow Br_2 + 2Cl^-$$

Industrially important single-replacement reactions use carbon to displace oxygen from metal oxides in the refining process. For example, Fe_2O_3 is refined into iron in the reaction

$$2Fe_2O_3 + 3C \rightarrow 4Fe + 3CO_2$$

Reactions of the Permanganate Ion

Permanganate solutions have a deep violet color, and the permanganate ion, MnO_4^-, is a very versatile reactant for oxidizing other substances. It reacts differently in acidic, neutral, and basic solutions.

In acid solution the half-reaction is

$$MnO_4^- + 8H^+ + 5e^- \rightarrow Mn^{2+} + 4H_2O$$

Five electrons are used in this reduction, and the soluble Mn^{2+} ion is almost colorless. In addition, this reaction is slow and the presence of Mn^{2+} ions catalyzes the process.

In neutral or slightly acid solutions the permanganate half-reaction is

$$MnO_4^- + 4H^+ + 3e^- \rightarrow MnO_2 + 2H_2O$$

The hydrogen ions in this three-electron half-reaction come from the dissociation of water molecules. The product MnO_2 is an insoluble black precipitate.

In basic solutions the reaction involves a one-electron transfer:

$$MnO_4^- + e^- \rightarrow MnO_4^{2-}$$

The reaction mixture changes color from the deep violet of the MnO_4^- ion to a green color for the MnO_4^{2-} ion.

ELECTROCHEMISTRY

The electrons in a balanced half-reaction show the direct relationship between a redox reaction and electricity. A redox reaction that normally would not occur may be forced to occur by adding electric energy in an **electrolytic cell**. A spontaneous redox reaction that occurs without added energy may be used to create a flow of electrons in a **galvanic cell**.

Electrolysis

TIP

Oxidation occurs at the anode.

In an **electrolysis** experiment, a chemical reaction that normally would not occur is forced to occur when two **electrodes** are immersed in an electrically conductive sample, and the electrical voltage applied to the two electrodes is increased until electrons flow. At the electrode supplying the electrons, reduction reactions occur. This electrode is called the **cathode**. At the other electrode, the **anode**, oxidation reactions occur.

Reduction occurs at the cathode.

One type of sample that may be electrolyzed is a **molten salt**. Salts are composed of ions; when the salt is a solid, the ions are immobile in the crystal lattice. Heating the salt until it melts frees the ions, and the mobility of the ions in the molten salt makes the salt electrically conductive. In the electrolysis of a molten salt that does not contain any polyatomic ions, the cation of the salt will be reduced at the cathode and the anion of the salt will be oxidized at the anode. For example, sodium chloride can be melted, and the electrolysis reactions will be

Cathode and *reduction* both start with consonants. *Anode* and *oxidation* both start with vowels.

Cathode reaction: $Na^+ + e^- \rightarrow Na$

Anode reaction: $2Cl^- \rightarrow Cl_2 + 2e^-$

A diagram of an electrolytic cell is shown is Figure 12.1.

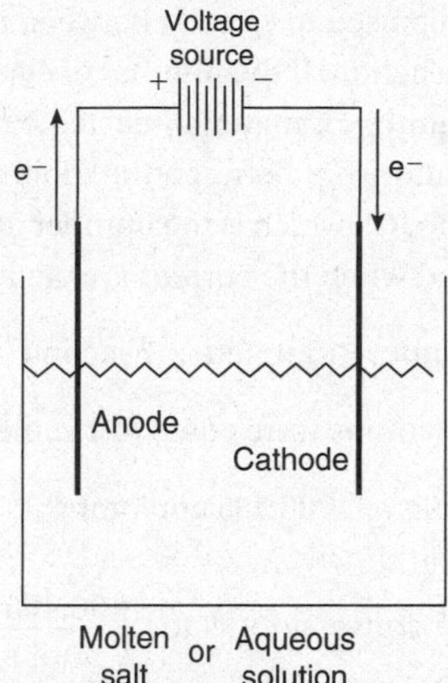

FIGURE 12.1. Setup of an electrolytic cell, described in the text.

Aqueous solutions of salts are also electrically conductive and may be electrolyzed. For solutions, two additional reactions are possible:

$$\text{Cathode: } 2H_2O + 2e^- \rightarrow H_2 + 2OH^- \qquad (12.1)$$
$$\text{Anode: } H_2O \rightarrow O_2 + 4H^+ + 4e^-$$

There are now two possible reactions at each electrode. At the cathode we will have either the reduction of water, as shown in Equation 12.1, or the reduction of a metal ion. At the anode we will have either the oxidation of water or the oxidation of the salt's anion. The following principles may be used to decide which reaction takes place at each electrode:

1. *Cathode:* Water (or the H^+ in water) will be reduced to hydrogen gas if the other cations in the solution, such as K, Na, Cs, and so on, can be reduced to very active metals (see Table 12.1). If the metal ions can be reduced to inactive metals such as Cu, Ag, Zn, and so on, they will be reduced at the cathode instead of the water.
2. *Anode:* If the anion is a polyatomic ion, it generally will not be oxidized. In particular, the sulfate, nitrate, and perchlorate polyatomic anions are not oxidized in aqueous solution. Chloride, bromide, and iodide ions will be oxidized in aqueous solution. If an anion in one salt is oxidized in an aqueous electrolysis, that same anion in any other salt will also be oxidized. For example, since a solution of NaBr results in Br^- being oxidized to Br_2, we may predict that solutions of KBr, $CaBr_2$, NH_4Br, and $AlBr_3$ will all produce Br_2 at the anode.

Electrolysis is important in the industrial production of several chemical materials. Because of the need for electricity, most operations that use electrolysis are located in areas where electricity is inexpensive. Two of these are in the Pacific Northwest and in Niagara Falls close to large hydroelectric generators.

QUANTITATIVE ELECTROCHEMISTRY

Every balanced half-reaction specifies the number of electrons lost or gained. Consequently, stoichiometric calculations can be used to convert between moles of electrons and moles of all other substances in the half-reaction. If the number of electrons flowing through the electrolysis cell is measured, the quantity of material reacted can be calculated. It was Michael Faraday who discovered the relationship between the mole and electric current. He started with the definition of the coulomb (C), which is the number of electrons that flow past a given point in a wire in exactly 1 second when the current is exactly 1 ampere:

$$1 \text{ coulomb} = 1 \text{ ampere} \times 1 \text{ second} \tag{12.2}$$

Faraday found that 96,485 coulombs were equal to 1 mole of electrons:

$$1 \text{ mole } e^- = 96{,}485 \text{ coulombs} \tag{12.3}$$

Faraday's constant, $\mathcal{F}$, is the conversion factor $\left(\frac{96{,}485 \text{ C}}{\text{mol } e^-}\right)$. Using Equations 12.2 and 12.3, we can determine the moles of electrons by measuring the current, *I*, and the time, *t*, that the current flows:

$$\text{moles of } e^- = \frac{(I \text{ C/s})(t \text{ s})}{96{,}485 \text{ C/mol}} = \frac{It}{\mathcal{F}} \tag{12.4}$$

The electric current, *I*, has units of coulombs per second (C s^{-1}), time has units of seconds (s), and Faraday's constant is 96,485 coulombs per mole of electrons.

Once the moles of electrons are calculated, the familiar stoichiometric calculations are used to determine other quantities. For example, the half-reaction for the reduction of iodine is

$$I_2 + 2e^- \rightarrow 2I^-$$

The moles of I^- produced at an electrode with a current of 0.500 ampere for 90 minutes may be calculated by first determining the moles of electrons:

$$\text{mol } e^- = \frac{(0.500 \text{ C s}^{-1})(90 \text{ min})(60 \text{ s min}^{-1})}{96{,}485 \text{ C mol}^{-1}}$$
$$= 0.0280 \text{ mol } e^-$$

Using the moles of electrons as the starting point, we can write the stoichiometric calculation:

$$? \text{ mol } I^- = 0.0280 \text{ mol } e^- \left(\frac{2 \text{ mol } I^-}{2 \text{ mol } e^-}\right) = 0.0280 \text{ mol } I^-$$

Chemists often bypass the stoichiometric step by incorporating the stoichiometry into Equation 12.4:

$$\text{moles of } X = \frac{It}{n\mathcal{F}}$$

In this equation *n* is the number of electrons per mole of substance in the balanced half-reaction.

EXAMPLE 12.2

A current of 2.34 A is delivered to an electrolytic cell for 85 min. How many grams of (a) Au from $AuCl_3$, (b) Ag from $AgNO_3$, and (c) Cu from $CuCl_2$ will be obtained?

Solution

(a) $Au^{3+} + 3e^- \rightarrow Au$

$$\text{mol Au} = \frac{(2.34\ \text{C s}^{-1})(85\ \text{min})(60\ \text{s min}^{-1})}{(3)(96{,}485\ \text{C mol}^{-1})}$$

$$= 4.12 \times 10^{-2}\ \text{mol Au}$$

$$\text{g Au} = 4.12 \times 10^{-2}\ \text{mol Au}\left(\frac{197\ \text{g Au}}{1\ \text{mol Au}}\right) = 8.12\ \text{g Au}$$

(b) $Ag^{+} + e^- \rightarrow Ag$

$$\text{mol Ag} = \frac{(2.34\ \text{C s}^{-1})(85\ \text{min})(60\ \text{s min}^{-1})}{(1)(96{,}485\ \text{C mol}^{-1})} = 0.124\ \text{mol Ag}$$

$$\text{g Ag} = 0.124\ \text{mol Ag}\left(\frac{108\ \text{g Ag}}{1\ \text{mol Ag}}\right) = 13.4\ \text{g Ag}$$

(2) $Cu^{2+} + 2e^- \rightarrow Cu$

$$\text{mol Cu} = \frac{(2.34\ \text{C s}^{-1})(85\ \text{min})(60\ \text{s min}^{-1})}{(2)(96{,}485\ \text{C mol}^{-1})}$$

$$= 6.18 \times 10^{-2}\ \text{mol Cu}$$

$$\text{g Cu} = 6.18 \times 10^{-2}\ \text{mol Cu}\left(\frac{63.5\ \text{g Cu}}{1\ \text{mol Cu}}\right) = 3.92\ \text{g Cu}$$

Galvanic Cells

Galvanic cells are used to harness the energy of spontaneous redox reactions. This is done by physically separating the chemicals in the two half-reactions so that the electrons generated by the oxidation half-reaction must flow through an electrical conductor before they can be used in the reduction half-reaction. This flow of electrons can be diverted through cell phones, iPods, laser pointers, and other devices to perform useful work before they reach their destination.

A galvanic cell is constructed as shown in Figure 12.2. All of the reactants in the oxidation half-reaction are placed in the left beaker, and all of the reactants in the reduction half-reaction in the right beaker. If a half-reaction is written with a metal, the metal serves as the electrode for that beaker; otherwise, an inert electrode made of platinum, silver, or gold is used. Electrodes are connected to each other with a metal wire, usually copper, and a device such as a meter (voltmeter or ammeter), motor, or light bulb may be inserted in the electrical circuit. If a voltmeter is used, the positive side of the voltmeter is connected to the cathode and the negative or ground of the voltmeter is connected to the anode to get the correct reading. To complete the circuit, a salt bridge is needed. The flow of charge is carried by electrons in the wires and by ions through the solutions and salt bridge.

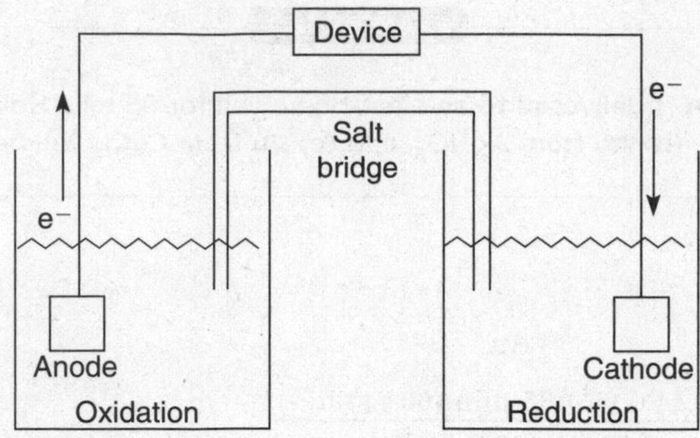

FIGURE 12.2. The galvanic cell.

In setting up a galvanic cell, we start with a balanced redox reaction. For the spontaneous reaction of permanganate with iron(II) in acid solution, the equation is

$$5Fe^{2+} + MnO_4^- + 8H^+ \rightleftharpoons 5Fe^{3+} + Mn^{2+} + 4H_2O \tag{12.5}$$

To identify where the chemicals should be placed, the half-reactions are written:

$$Fe^{2+} \rightarrow Fe^{3+} + e^- \tag{12.6}$$

$$5e^- + MnO_4^- + 8H^+ \rightarrow Mn^{2+} + H_2O \tag{12.7}$$

Equation 12.6 is the oxidation half-reaction, and Fe^{2+} and Fe^{3+} are placed in the left beaker. Since neither Fe^{2+} nor Fe^{3+} is a metal, a platinum electrode will be used. MnO_4^-, H^+, and Mn^{2+} must be placed in the right cell since they are the components of the reduction half-reaction (Equation 12.7). Once again, a platinum electrode will be used. A voltmeter will measure the tendency of electrons to flow. The complete cell is shown in Figure 12.3.

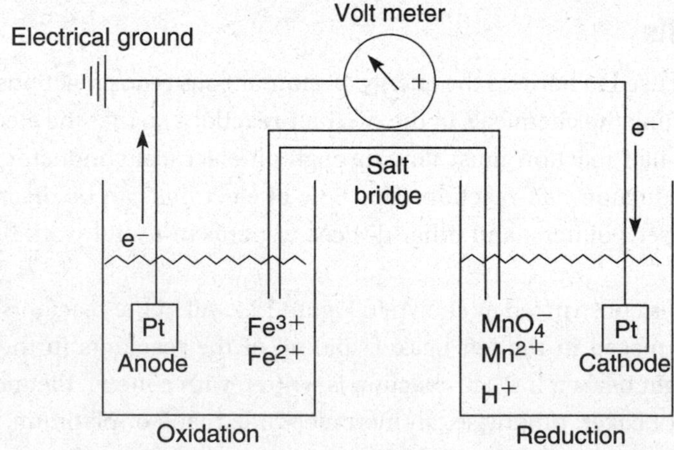

FIGURE 12.3. The galvanic cell set up to study the reaction in Equation 12.5.

To obtain consistent results from a galvanic cell, the variables of temperature, pressure, and concentration must be controlled. For this purpose electrochemists define a standard state for these experiments. Standard state for the galvanic cell is a temperature of 298 K, a pressure of 1.00 atmosphere for all gases, and concentrations of 1.00 molar for all soluble compounds. Concentrations of all solids and pure liquids are also defined as 1.00 molar.

At standard state, the galvanic cell diagram assumes that the reaction of Equation 12.5 is thermodynamically favored and that oxidation and the anode are on the left while reduction and the cathode are on the right. Since this reaction is known to be thermodynamically favored, the cell in Figure 12.3 is correct, and a positive voltmeter reading will be obtained. This voltmeter reading is called the **standard cell voltage**, E°_{cell}, or the **electromotive force**, emf, or $\mathscr{F}$.

A not thermodynamically favored reaction may be converted into a thermodynamically favored reaction by reversing the chemical equation. Then the steps outlined above are used to determine the correct setup of the galvanic cell.

In reality we often do not know whether a reaction is thermodynamically favored or not thermodynamically favored. In such a situation, a chemical equation is written and the galvanic cell is constructed for that equation. When the cell voltage is measured, it will be either positive or negative. If the cell voltage is positive, the reaction is thermodynamically favored as written. If the cell voltage turns out to be negative, the reaction is not thermodynamically favored as written and should be reversed to construct the galvanic cell.

The chemical reaction and the setup for the galvanic cell are directly related to each other. Knowing one of them allows us to determine the other. The chemicals present in the cell with the anode are written as an oxidation half-reaction; those present in the cell with the cathode, as the reduction half-reaction.

A **cell diagram** is a shorthand method of drawing a galvanic cell. The cell diagram "reads" the galvanic cell from left to right, starting with the anode and ending with the cathode. For the standard-state reaction of permanganate with iron(II) the cell diagram is as follows:

$$\text{Pt}\left|\,\text{Fe}^{2+}(1\ M), \text{Fe}^{3+}(1\ M)\right|\left|\text{MnO}_4^-(1\ M), \text{Mn}^{2+}(1\ M), \text{H}^+(1\ M)\right|\,\text{Pt}$$

The single vertical lines represent phase changes between the solid platinum electrodes and the solutions. The double vertical lines represent the salt bridge because it has a phase bounday at either end. Concentrations, if known, are shown in parentheses. The order in which the chemicals are written, within each cell, is not important.

STANDARD REDUCTION POTENTIALS

The standard cell voltage is the difference between the electric potentials (voltages) of the cathode and the anode:

$$E^\circ_{cell} = E^\circ_{cathode} - E^\circ_{anode}$$

The $E^\circ_{cathode}$ is the standard reduction potential for the reaction occurring at the cathode and represents its tendency to remove electrons from the electrode surface. E°_{anode} is the standard reduction potential for the reaction occurring at the anode and represents its tendency to remove electrons from the anode. It is impossible to independently measure the electromagnetic force values of the cathode and anode. The only measurement possible is the combined E°_{cell}.

If it were possible to measure $E^\circ_{cathode}$ or E°_{anode} for any reaction, the standard electrode potentials of all other half-reactions could be determined using the above equation. Since neither $E^\circ_{cathode}$ nor E°_{anode} can be independently measured, chemists define the reduction of hydrogen, at standard state, as having a reduction potential of exactly 0.00 volt:

$$H^+(aq) + e^- \rightarrow {}^1/_2 H_2(g) \; E^\circ = 0.00 \text{ volt}$$

This definition is used to determine the potentials of other half-reactions in relation to the reduction of hydrogen ions. For example, the reaction of

$$Zn(s) + 2H^+(aq) \rightarrow Zn^{2+}(aq) + H_2(g)$$

has a standard cell voltage of $E^\circ_{cell} = +0.76$ volt. Since the zinc oxidation occurs at the anode and the reduction of hydrogen ions occurs at the cathode, we may write

$$E^\circ_{cell} = +0.76 \text{ V} = E^\circ_{H^+} - E^\circ_{Zn}$$

The E°_{H+} has been defined as 0.00 volt, and entering that value into the preceding equation gives

$$+0.76 \text{ V} = 0.00 \text{ V} - E^\circ_{Zn}$$

Rearranging this equation yields

$$E^\circ_{Zn} = -0.76 \text{ V}$$

This is the standard reduction potential for zinc.

Once the standard reduction potential for zinc is known, we may study the reaction

$$Zn(s) + 2Ag^+(aq) \rightarrow Zn^{2+}(aq) + 2Ag(s)$$

The standard cell voltage for this reaction is measured as +1.56 volts, so

$$E^\circ_{cell} = +1.56 \text{ V} = E^\circ_{Ag^+} - E^\circ_{Zn^{2+}}$$

Substituting the value for the standard reduction potential of zinc into the above equation yields

$$+1.56 \text{ V} = E^\circ_{Ag^+} - (-0.76 \text{ V})$$

Rearranging and solving yields the standard reduction potential of silver ions:

$$E^\circ_{Ag^+} = +1.56 \text{ V} - 0.76 \text{ V} = +0.80 \text{ V}$$

In a similar manner the standard reduction potentials of all half-reactions are determined. The standard reduction potentials for most half-reactions are known, and some of these are listed in Table 12.2.

With a table of standard reduction potentials, E°_{cell} for any redox reaction can be predicted. If E°_{cell} is positive, the reaction is thermodynamically favored; if it is negative, the reaction is not thermodynamically favored.

TABLE 12.2
Standard Reduction Potentials, 25°C

Half-Reaction							$E°$ (V)
$F_2(g)$		+		$2e^-$	→	$2F^-$	2.87
Co^{3+}		+		e^-	→	Co^{2+}	1.82
Au^{3+}		+		$3e^-$	→	Au	1.50
$Cl_2(g)$		+		$2e^-$	→	$2Cl^-$	1.36
$O_2(g)$	+	$4H^+$	+	$4e^-$	→	$2H_2O$	1.23
$Br_2(g)$		+		$2e^-$	→	$2Br^-$	1.07
$2Hg^{2+}$		+		$2e^-$	→	Hg_2^{2+}	0.92
Ag^+		+		e^-	→	Ag	0.80
Hg_2^{2+}		+		$2e^-$	→	2Hg	0.79
Fe^{3+}		+		e^-	→	Fe^{2+}	0.77
I_2		+		$2e^-$	→	$2I^-$	0.53
Cu^+		+		e^-	→	Cu	0.52
Cu^{2+}		+		$2e^-$	→	Cu	0.34
Cu^{2+}		+		e^-	→	Cu^+	0.15
Sn^{4+}		+		$2e^-$	→	Sn^{2+}	0.15
S	+	$2H^+$	+	$2e^-$	→	H_2S	0.14
$2H^+$		+		$2e^-$	→	H_2	0.00
Pb^{2+}		+		$2e^-$	→	Pb	-0.13
Sn^{2+}		+		$2e^-$	→	Sn	-0.14
Ni^{2+}		+		$2e^-$	→	Ni	-0.25
Co^{2+}		+		$2e^-$	→	Co	-0.28
Tl^+		+		e^-	→	Tl	-0.34
Cd^{2+}		+		$2e^-$	→	Cd	-0.40
Cr^{3+}		+		e^-	→	Cr^{2+}	-0.41
Fe^{2+}		+		$2e^-$	→	Fe	-0.44
Cr^{3+}		+		$3e^-$	→	Cr	-0.74
Zn^{2+}		+		$2e^-$	→	Zn	-0.76
$2H_2O$		+		$2e^-$	→	$H_2(g) + 2OH^-$	-0.83
Mn^{2+}		+		$2e^-$	→	Mn	-1.18
Al^{3+}		+		$3e^-$	→	Al	-1.66
Be^{2+}		+		$2e^-$	→	Be	-1.70

Half-Reaction							$E°$ (V)
Mg^{2+}		+		$2e^-$	→	Mg	-2.37
Na^+		+		e^-	→	Na	-2.71
Ca^{2+}		+		$2e^-$	→	Ca	-2.87
Sr^{2+}		+		$2e^-$	→	Sr	-2.89
Ba^{2+}		+		$2e^-$	→	Ba	-2.90
Rb^+		+		e^-	→	Rb	-2.92
K^+		+		e^-	→	K	-2.92
Cs^+		+		e^-	→	Cs	-2.92
Li^+		+		e^-	→	Li	-3.05

EXAMPLE 12.3

Write the balanced redox reaction for each of the following, determine $E°_{cell}$, and state whether or not the reaction is thermodynamically favored:

(a) the single displacement of copper(II) by zinc metal
(b) the single displacement of H^+ by iron, forming Fe^{2+}
(c) the reduction of tin(IV) to tin(II) by Fe(II) forming Fe(III)
(d) the reduction of dichromate ions to chromium(III) ions by Mn^{2+}, forming MnO_2
(e) the oxidation of AsO_3^{3-} to $H_2AsO_4^-$ by the reduction of iodine to iodide ions

Solution

(a) $Cu^{2+} + Zn \rightarrow Zn^{2+} + Cu$
$E°_{cell} = +0.34\text{ V} - (-0.76\text{ V}) = +1.10\text{ V}$
The reaction is thermodynamically favored.

(b) $Fe + 2H^+ \rightarrow Fe^{2+} + H_2$
$E°_{cell} = 0.00\text{ V} - (-0.44\text{ V}) = +0.44\text{ V}$
The reaction is thermodynamically favored.

(c) $Sn^{4+} + 2Fe^{2+} \rightarrow 2Fe^{3+} + Sn^{2+}$
$E°_{cell} = +0.15\text{ V} - (+0.77\text{ V}) = -0.62\text{ V}$
The reaction is not thermodynamically favored.

(d) $2H^+ + Cr_2O_7^{2-} + 3Mn^{2+} \rightarrow 3MnO_2 + 2Cr^{3+} + H_2O$
$E°_{cell} = +1.33\text{ V} - (+1.23\text{ V}) = +0.10\text{ V}$
The reaction is thermodynamically favored.

(e) $H_2O + AsO_3^{3-} + I_2 \rightarrow H_2AsO_4^- + 2I^-$
$E°_{cell} = +0.54\text{ V} - (+0.58\text{ V}) = -0.04\text{ V}$
The reaction is not thermodynamically favored.

STANDARD CELL VOLTAGES AND EQUILIBRIUM

If the two electrodes of a galvanic cell are connected so that electrons flow freely, the reaction will proceed to equilibrium. At equilibrium $E_{cell} = 0$ and the mathematical rearrangement of the two Nernst equations becomes

$$E^{\circ}_{cell} = E^{\circ}_{cathode} - E^{\circ}_{anode}$$

$$= \frac{RT}{n\mathcal{F}} \ln K_{eq}$$

or

$$= \frac{0.0591}{n} \log K_{eq}$$

where n is the total number of electrons transferred in the redox reaction, and the mathematical combination of the anode reaction quotient and the cathode reaction quotient gives the equilibrium constant, K_{eq}.

The table of standard reduction potentials also provides the information necessary to calculate the equilibrium constant for a redox reaction.

EXAMPLE 12.4

Determine the equilibrium constant for each of the following reactions at 298 K:

(a) the single displacement of copper(II) by zinc metal

(b) the single displacement of H^+ by iron, forming Fe^{2+}

(c) the reduction of tin(iv) to tin(II) by the oxidation of Fe(II) being oxidized to Fe(III)

(d) the reduction of dichromate ions to chromium(III) ions Mn^{2+}, forming MnO_2

(e) the oxidation of AsO_3^{3-} to $H_2AsO_4^-$ by the reduction of iodine to iodide ions

Solution

In Exercise 12.4 the reactions and E°_{cell} values are determined. The equation $E^{\circ}_{cell} = \frac{0.0591}{n} \log K_{eq}$ is used in all cases once E°_{cell} and n are determined.

(a) $Cu^{2+} + Zn \rightarrow Zn^{2+} + Cu$

$E^{\circ}_{cell} = +0.34\ V - (-0.76\ V) = +1.10\ V$

Two electrons are transferred, and $K_{eq} = 1.7 \times 10^{37}$.

(b) $Fe + 2H^+ \rightarrow Fe^{2+} + H_2$

$E^{\circ}_{cell} = 0.00\ V - (-0.44\ V) = +0.44\ V$

Two electrons are transferred, and $K_{eq} = 7.8 \times 10^{14}$.

(c) $Sn^{4+} + 2Fe^{2+} \rightarrow 2Fe^{3+} + Sn^{2+}$

$E^{\circ}_{cell} = +0.15\ V - (+0.77\ V) = -0.62\ V$

Two electrons are transferred, and $K_{eq} = 1.0 \times 10^{-21}$.

(d) $2H^+ + Cr_2O_7^{2-} + 3Mn^{2+} \rightarrow 3MnO_2 + 2Cr^{3+} + H_2O$

$E^{\circ}_{cell} = +1.33\ V - (+1.23\ V) = +0.10\ V$

Six electrons are transferred, and $K_{eq} = 1.4 \times 10^{10}$.

(e) $H_2O + AsO_3^{3-} + I_2 \rightarrow H_2AsO_4^- + 2I^-$
$E^\circ_{cell} = +0.54\text{ V} - (+0.58\text{ V}) = -0.04\text{ V}$
Two electrons are transferred, and $K_{eq} = 4.4 \times 10^{-2}$.

In these examples we see that fairly small voltages translate into very large or very small equilibrium constants, depending on the sign of E°_{cell}.

FREE-ENERGY CHANGE, $\Delta G°$, AND STANDARD CELL VOLTAGES

As shown by the equations on the preceding page, the standard cell voltage is related to the equilibrium constant:

$$E^\circ_{cell} = \frac{RT}{n\mathcal{F}} \ln K_{eq}$$

The standard free-energy change, $\Delta G°$, is also related to the equilibrium constant:

$$\Delta G° = -RT \ln K_{eq}$$

The relationship between $\Delta G°$ and E°_{cell} is

$$\Delta G° = -n\mathcal{F}\, E^\circ_{cell} \qquad (12.8)$$

A major method for determining the standard free energy of a reaction is the measurement of the standard cell voltage for the reaction of interest, and then a mathematical conversion using Equation 12.8.

EXAMPLE 12.5

Determine the standard free-energy change for each of the following reactions at 298 K:

(a) the single displacement of copper(II) by zinc metal
(b) the single displacement of H^+ by iron, forming Fe^{2+}
(c) the reduction of tin(iv) to tin(II) by the oxidation of Fe(II) to Fe(III)
(d) the reduction of dichromate ions to chromium(III) ions Mn^{2+}, forming MnO_2
(e) the oxidation of AsO_3^{3-} to $H_2AsO_4^-$ by the reduction of iodine to iodide ions

Solution

In Example 12.3, the reactions and E°_{cell} were determined. The equation $\Delta G° = -n\mathcal{F}\, E^\circ_{cell}$ is used in all cases once n and E°_{cell} are known.

(a) $Cu^{2+} + Zn \rightarrow Zn^{2+} + Cu$
$E^\circ_{cell} = +0.34\text{ V} - (-0.76\text{ V}) = +1.10\text{ V}$
Two electrons are transferred and $\Delta G° = -2(96.485)(1.10) = -212$ kJ.

(b) $Fe + 2H^+ \rightarrow Fe^{2+} + H_2$
$E^\circ_{cell} = 0.00\text{ V} - (-0.44\text{ V}) = +0.44\text{ V}$
Two electrons are transferred, and $\Delta G° = -2(96.485)(0.44) = -84.9$ kJ.

(c) $Sn^{4+} + 2Fe^{2+} \rightarrow 2Fe^{3+} + Sn^{2+}$
$E^\circ_{cell} = +0.15\text{ V} - (+0.77\text{ V}) = -0.62\text{ V}$
Two electrons are transferred, and $\Delta G° = -2(96.485)(-0.62) = +120$ kJ.

(d) $2H^+ + Cr_2O_7^{2-} + 3Mn^{2+} \rightarrow 3MnO_2 + 2Cr^{3+} + H_2O$
$E^\circ_{cell} = +1.33\text{ V} - (+1.23\text{ V}) = +0.10\text{ V}$
Six electrons are transferred, and $\Delta G^\circ = -6(96.485)(0.10) = -57.9$ kJ.

(e) $H_2O + AsO_3^{3-} + I_2 \rightarrow H_2AsO_4^- + 2I^-$
$E^\circ_{cell} = +0.54\text{ V} - (+0.58\text{ V}) = -0.04\text{ V}$
Two electrons are transferred, and $\Delta G^\circ = -2(96.485)(-0.04) = 7.7$ kJ.

In Example 12.5, the standard cell voltages for a series of mixtures were determined. The standard state used for these values includes all solutions that contain soluble substances having concentrations of exactly 1 molar. If one of these concentrations is altered so it is not exactly 1 *M*, the *direction* of change in the cell voltage can be deduced with the information already at hand. The other important fact is that the cell voltage of any system will be zero when equilibrium is reached.

We can deduce that if the reaction moves in the direction of equilibrium, the cell voltage will decrease, eventually becoming zero when equilibrium is reached. We can also deduce that since a positive value of the cell voltage represents a thermodynamically favored reaction, the position of equilibrium lies on the product side of the arrow. We can use Le Châtelier's principle to ascertain which direction the reaction will move if simple changes in concentration are made. Let's see how this works.

EXAMPLE 12.6

Suppose the reaction of silver ions (Ag^+) with cadmium metal to give cadmium(II) ions and silver metal has a standard cell voltage of 1.20 volts. This result will be obtained if $[Ag^+]$ and $[Cd^{2+}]$ are 1 *M*. If the concentration of silver ions is changed to 0.10 *M*, how will the cell voltage change?

(a) The cell voltage will increase.
(b) The cell voltage will decrease.
(c) The cell voltage will not change.

Solution

We write the balanced chemical equation:

$$2Ag^+ + Cd(s) \rightarrow Cd^{2+} + Ag(s)$$

As demonstrated in Example 12.4, K_{eq} is very large for positive values of E°_{cell}. We expect the concentrations of products to be greater than the reactants when the reaction comes to equilibrium. Also, the positive cell voltage decreases to zero as equilibrium is approached. We can conclude that if E°_{cell} is positive and the concentration is changed from standard state (all solutes are 1 *M*) to a condition where product concentrations are greater than reactant concentrations, the cell voltage will decrease. Since $[Cd^{2+}] > [Ag^+]$, the voltage will decrease.

Exercise 12.5

Take the experiments in Example 12.5 and assume all concentrations are 1.00 *M* except for the following:

(a) $[Cu^{2+}] = 0.05000\ M$

(b) $[Fe^{2+}] = 0.00023\ M$

(c) $[Sn^{2+}] = 1.68\ M$

(d) $[Mn^{2+}] = 0.078\ M$

(e) $[I^-] = 0.003\ M$

(HINT: Be careful of reactions with negative values for the standard cell voltages.)

Solution

(a) $E°_{cell}$ is positive and $[Zn^{2+}] > [Cu^{2+}]$, so the voltage will decrease.

(b) $E°_{cell}$ is positive and $[H^+] > [Fe^{2+}]$, so the voltage will increase.

(c) $E°_{cell}$ is negative and $[Sn^{2+}] >$ [other substances], so the voltage will increase toward zero.

(d) $E°_{cell}$ is positive and [other substances] $> [Mn^{2+}]$, so the voltage will decrease.

(e) $E°_{cell}$ is negative and $[I_2] > [I^-]$, so the voltage will increase toward zero.

(NOTE: The AP exam will focus on situations in which the standard cell voltage is positive.)

IMPORTANT OXIDATION-REDUCTION REACTIONS

Combustion Reactions

Reacting organic compounds with oxygen often results in the production of a large amount of heat along with a flame that is characteristic of the combustion process. The products of **combustion** reactions are usually carbon dioxide and water, as in the combustion of glucose:

$$C_6H_{12}O_6 + 6O_2 \rightarrow 6CO_2 + 6H_2O$$

When the amount of oxygen is limited, or there is no oxygen, the products of the reaction may include carbon monoxide or elemental carbon (soot):

$$C_6H_{12}O_6 + 3O_2 \rightarrow 6CO + 6H_2O \text{ (limited } O_2)$$

$$C_6H_{12}O_6 \rightarrow 6C + 6H_2O \text{ (no } O_2)$$

The same products, CO_2 and H_2O, are formed when glucose is metabolized in the body. However, since the process is slow and does not produce a flame, it is not called a combustion reaction. The body uses the heat and energy produced in a more efficient manner in metabolism. Even when the amount of oxygen is limited, CO and C are not produced in metabolic reactions.

Oxidation of Metals

Metallic elements have differing affinities toward oxygen. The very unreactive elements, such as platinum, gold, silver, and copper, do not react with O_2. Silver tarnishes by reacting with small amounts of hydrogen sulfide in the air. Copper reacts with water and carbon dioxide to form a carbonate compound. Some metals, particularly the alkali and alkaline earth metals, however, react readily with oxygen and are completely converted into oxides if exposed long enough.

Magnesium burns with a bright white flame in oxygen. The bright light is used in magnesium flares by the military, in flashbulbs for photography, and in fireworks. The reaction is so energetic that magnesium will continue burning even in an atmosphere of carbon dioxide:

$$2Mg + CO_2 \rightarrow 2MgO + C$$

When finely divided into a powder, many metals burn in oxygen. Steel wool burns when placed in a flame, and powdered metals, such as aluminum, are classified as highly combustible.

Aluminum reacts very well with oxygen, as mentioned above. In large sheets or bars of the metal, however, the aluminum oxide formed produces an impervious coating so that complete oxidation does not occur. This oxide layer is only a few molecules thick and is not visible to the eye.

Iron and steel react poorly with gaseous oxygen. The formation of rust requires the presence of water for oxidation to occur; iron does not rust in pure oxygen or in water that contains no oxygen. In this complex electrochemical process, the iron actually acts as the anode of a chemical reaction when moisture is present and as the cathode when it is not. **Corrosion** and its prevention are major concerns of approximately 30 percent of working chemists and chemical engineers. Each year, damage due to corrosion of buildings, bridges, and even computer circuits costs billions of dollars to correct and repair.

SUMMARY

Electrochemistry and oxidation-reduction reactions are combined in this chapter. Defining and identifying oxidation and reduction is described in this chapter. Balancing of the more complex redox reactions is also described in detail. Redox reactions can occur in nature or as part of chemical reactions, including titrations. Chemical energy can be harnessed by using a galvanic or voltaic cell, better known to the consumer as a battery. Standard cell potentials, E^o_{cell}, are related to free energies, ΔG^o, and equilibrium constants, K_{eq}. Positive values for E^o_{cell} indicate a thermodynamically favored reaction. Galvanic cells can be arranged to measure concentrations of ions in solution. One of the most important, the hydrogen ion, is measured in a special galvanic cell with an instrument called a pH meter.

Galvanic cells harness chemical energy to create a flow of electrons in a circuit that can be used to power a flashlight or an MP-3 player. If electrical energy is added to a system, not thermodynamically favored reactions can be forced to occur in a process called electrolysis. Electrolysis of molten salts produces the redox products of the anions and cations of binary salts. Electrolysis of aqueous solutions produces either the redox products of the salt or hydrogen or oxygen, depending on which substances are more easily oxidized or reduced at the electrode surfaces. Electrolysis is also quantitative, and if the electric current and time are accurately measured, then the amount of substance reacted can be calculated precisely.

Important Concepts

Oxidation and reduction
Oxidation numbers
Ion electron method for balancing equations
Electrolytic and galvanic cells
Standard reduction potentials
Standard cell voltages

Important Equations

$$\text{moles} = \frac{It}{n\mathcal{F}}$$

$$E^\circ_{\text{cell}} = E^\circ_{\text{cathode}} - E^\circ_{\text{anode}}$$

$$E^\circ_{\text{cell}} = \frac{RT}{n\mathcal{F}} \ln K$$

$$\Delta G^\circ = -n\mathcal{F}E^\circ_{\text{cell}}$$

PRACTICE EXERCISES

Multiple-Choice

1. The difference between a voltaic and an electrolytic cell is that
   (A) oxidation will occur only at the cathode in a voltaic cell
   (B) the free energy change, ΔG, is negative for an electrolytic cell
   (C) the cathode is labeled as negative (–) in an electrolytic cell and as positive (+) in a voltaic cell
   (D) the electrons flow from the cathode to the anode in both types of cells through the external wire

2. Consider a voltaic cell, $E^\circ_{\text{cell}} = 0.62\ V$, composed of cobalt, copper, and their respective M^{2+} ions. If the E° for the cathode half-cell is 0.34 V, what is the E° for the anode half-cell?

$$Cu^{2+}(aq) + Co(s) \rightarrow Cu(s) + Co^{2+}(aq)$$

   (A) –0.28 V
   (B) –0.96 V
   (C) 0.28 V
   (D) 0.96 V

3. What does the line notation below indicate?

$$Pt \mid H_2(g) \mid H+(aq) \parallel Cu_2 + (aq) \mid Cu(s)$$

(A) Cu is the anode.
(B) Pt is the cathode.
(C) Cu(*s*) is a product of the cell reaction.
(D) Hydrogen gas is a product of the cell reaction.

4. Balance the following half-reaction in acid solution:

$$NO_3^- \rightarrow NH_4^+$$

When balanced with the smallest whole-number coefficients, the *sum* of all the coefficients is

(A) 13
(B) 26
(C) 15
(D) 23

5. The following reactions are known to be thermodynamically favored:

$$Cu + 2Ag^+ \rightarrow Cu^{2+} + 2Ag$$
$$Zn + 2Ag^+ \rightarrow Zn^{2+} + 2Ag$$
$$Zn + Cu^{2+} \rightarrow Zn^{2+} + Cu$$

The activity series for the three elements from the easiest to oxidize to the most difficult to oxidize is

(A) Cu > Ag > Zn
(B) Zn > Cu > Ag
(C) Ag > Cu > Zn
(D) Ag > Zn > Cu

6. The standard reduction potential for $PbO_2 \rightarrow Pb^{2+}$ is +1.46 *V*, and the standard reduction potential for $Fe^{3+} \rightarrow Fe^{2+}$ is +0.77 *V*. What is the standard cell voltage for the following reaction?

$$4H^+ + PbO_2 + 2Fe^{2+} \rightarrow 2Fe^{3+} + Pb^{2+} + 2H_2O$$

(A) −0.08 *V*
(B) +0.69 *V*
(C) +2.33 *V*
(D) −0.69 *V*

Questions 7 and 8 refer to the following half-cell reactions.

$$Cr^{3+}(aq) + 3e^- \rightarrow Cr(s) \qquad E° = -0.74\ V$$
$$Co^{2+}(aq) + 2e^- \rightarrow Co(s) \qquad E° = -0.28\ V$$

7. What would be the standard electrode potential for a voltaic cell properly constructed from these two half-cells?
   (A) 0.64 V
   (B) 1.02 V
   (C) 0.46 V
   (D) −1.02 V

8. What is the correct balanced overall equation for this voltaic cell?
   (A) $3Co^{2+}(aq) + 2Cr(s) \rightarrow 3Co(s) + 2Cr^{3+}(aq)$
   (B) $3Co(s) + 2Cr^{3+}(aq) \rightarrow 3Co^{2+}(aq) + 2Cr(s)$
   (C) $Co(s) + Cr^{3+}(aq) \rightarrow Co^{2+}(aq) + Cr(s)$
   (D) $2Co(s) + 3Cr^{3+}(aq) \rightarrow 2Co^{2+}(aq) + 3Cr(s)$

---

9. In the following pairs of compounds, which pair contains the element named with the same oxidation number?
   (A) Sulfur in $H_2S_2O_7$ and H_2SO_4
   (B) Cobalt in $Co(NH_3)_6^{3+}$ and $Co(NO_3)_2$
   (C) Mercury in $HgCl_2$ and Hg_2Cl_2
   (D) Oxygen in Na_2O_2 and H_2O

10. Sodium metal cannot be electrolyzed from an aqueous Na_2SO_4 solution because
   (A) the voltage needed is too high for any available instrument to achieve
   (B) water is reduced to O_2 before Na^+
   (C) Na^+ has a high over-potential that keeps it from being reduced
   (D) H^+ has a more favorable reduction potential than Na^+

11. Which of the following elements has the largest number of possible oxidation states?
   (A) Fe
   (B) Cl
   (C) Ca
   (D) Mn

12. For the reaction

$$2I^- + Cl_2 \rightleftharpoons 2Cl^- + I_2$$

the standard cell voltage is +0.82 V. What is the equilibrium constant for this reaction at 45 °C? ($R = 8.314\ V\ C\ mol^{-1}\ K^{-1}$, and $\mathcal{F} = 96{,}485\ C\ mol^{-1}$)
   (A) 9.8×10^{25}
   (B) 1.0×10^{-26}
   (C) 1.6×10^{5}
   (D) 6.3×10^{-6}

13. The $E°_{cell}$ = 2.04 V at 298 K for the following reaction in a lead-acid cell

$$Pb(s) + PbO_2(s) + 2H_2SO_4(aq) \rightarrow 2PbSO_4(aq) + 2H_2O(l)$$

What is the $\Delta G°$ for this reaction?
(A) -7.87×10^2 kJ
(B) -3.94×10^2 kJ
(C) -3.94×10^5 kJ
(D) -1.97×10^5 kJ

14. A metal is electrolyzed from aqueous solution by using an electrical current of 1.23 A for 2½ h, and 3.37 g of metal is deposited. In a separate experiment the number of electrons used for the reduction of the metal is 2. What is the metal?
(A) Al
(B) Ni
(C) Sn
(D) Mg

15. Using the two half-reactions listed below, a galvanic cell is constructed.

$Fe^{2+}(aq) + 2e^- \rightarrow Fe(s)$	$E° = -0.44$ V
$Cu^{2+}(aq) + 2e^- \rightarrow Cu(s)$	$E° = 0.34$ V

Once the cell is operational, which of the following species is contained in the anode compartment of the cell?
(A) $Cu^{2+}(aq)$ and $Cu(s)$
(B) $Fe^{2+}(aq)$ only
(C) $Cu^{2+}(aq)$ only
(D) $Fe^{2+}(aq)$ and $Fe(s)$

16. If the $E°$ for the reaction $X^+ + e^- \rightarrow X$ is greater than the $E°$ for $A^{2+} + 2e^- + A$, which statement would be most correct?
(A) X^+ will oxidize A^{2+}.
(B) X will oxidize A^{2+}.
(C) A will reduce X^+.
(D) X will reduce A^{2+}.

17. Which of the following compounds includes an element that has the same oxidation number as the chlorine in sodium chlorate, $NaClO_3$?
(A) $K_3Fe(CN)_6$
(B) $KMnO_4$
(C) $Al(NO_3)_3$
(D) $(NH_4)_2SO_4$

18. With a current of 1.25 A, how many minutes will be required to deposit 2.00 g of copper on a platinum electrode from a copper(II) nitrate solution? (Faraday's constant = 96,485 C mol^{-1})
(A) 4859 min
(B) 81.0 min
(C) 40.5 min
(D) 1.35 min

19. What is the minimum number of electrons needed to balance the following half-reaction with whole number coefficients?

$$IO_3^- \rightarrow I_2$$

(A) $1e^-$
(B) $2e^-$
(C) $5e^-$
(D) $10e^-$

20. Shown below are two reduction half-reactions for nicotinamide adenine dinucleotide (NAD^+) and oxaloacetate and their electrode potentials:

$$NAD^+(aq) + 2H^+(aq) + 2e^- \rightarrow NADH(aq) + H^+(aq) \qquad E° = -0.320\ V$$
$$\text{Oxaloacetate}_2^-(aq) + 2H^+(aq) + 2e^- \rightarrow \text{malate}_2^-(aq) \qquad E° = -0.166\ V$$

Determine the $E°_{cell}$ for the following reaction, and predict if the reaction is thermodynamically favored.

$$\text{Oxaloacetate}^{-2}(aq) + NADH(aq) + H^+(aq) \rightarrow NAD^+(aq) + \text{malate}^{2-}(aq)$$

(A) −0.486 *V*, not thermodynamically favored
(B) 0.154 *V*, thermodynamically favored
(C) 0.486 *V*, thermodynamically favored
(D) −0.154 *V*, not thermodynamically favored

21. Which of the following statements is FALSE?
(A) Reduction involves a gain of electrons.
(B) Batteries are galvanic cells.
(C) A thermodynamically favored reaction always has a positive $E°_{cell}$.
(D) Electrolysis reactions always produce a gas at least at one electrode.

ANSWER KEY

1. **(C)**
2. **(A)**
3. **(C)**
4. **(D)**
5. **(B)**
6. **(B)**
7. **(C)**
8. **(A)**
9. **(A)**
10.* **(D)**
11. **(B)**
12. **(A)**
13. **(B)**
14. **(B)**
15. **(D)**
16. **(C)**
17. **(C)**
18. **(B)**
19. **(D)**
20. **(D)**
21. **(D)**

See Appendix 1 for explanation of answers.

Free-Response

A galvanic cell can be constructed by using the substances in the following half-reactions:

$$Ni_2^{+}(aq) + 2e^- \rightarrow Ni(s) \qquad E° = -0.257\ V$$
$$Fe_2^{+}(aq) + 2e^- \rightarrow Fe(s) \qquad E° = -0.447\ V$$

(a) What is the $E°_{cell}$ for this reaction?

$$Ni^{2+}(aq) + Fe(s) \rightarrow Ni(s) + Fe^{2+}(aq)$$

(b) Which reaction takes place at the anode? At the cathode?

(c) Sketch a galvanic cell in which this reaction takes place. Show the direction of electron flow and the directional flow of the ions in the salt bridge.

(d) Calculate the K_{eq} for this reaction. Is this reaction thermodynamically favored?

ANSWERS

(a)

$$\begin{aligned} E°_{cell} &= E°_{reduction} - E°_{oxidation} \\ &= -0.257\ V - (-0.447\ V) \\ &= +0.190\ V \end{aligned}$$

(b) Oxidation *always* occurs at the anode:

$$Fe(s) \rightarrow Fe^{2+}(aq) + 2e^-$$

The reduction reaction occurs at the cathode.

$$Ni^{2+}(aq) + 2e^- \rightarrow Ni(s)$$

(c)

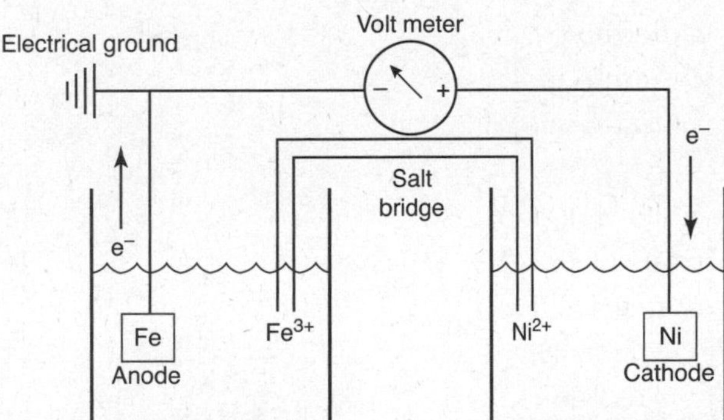

(d) Use the thermodynamic relationships that $\Delta G° = -n FE°_{cell}$ and $\Delta G° = -RT \ln K$. By combining these, we get $E°_{cell} = RT/nF \ln K_{eq}$. The standard cell voltage is $+0.190\ V$, R is 8.314 J mol^{-1}, T is 298K, $n = 2$, and $F = 96{,}495$ C/mol. Using these values the result is

$$K_{eq} = 2.62 \times 10^6$$

13 Acids and Bases

- → ACID-BASE THEORIES
- → ARRHENIUS THEORY
- → BRØNSTED-LOWRY THEORY
- → NAMING ACIDS
- → ACID STRENGTHS
- → ACID ANHYDRIDES
- → BASE ANHYDRIDES
- → NEUTRALIZATION REACTIONS
- → POLYPROTIC ACIDS
- → CONJUGATE ACID-BASE PAIRS
- → COMPLEXATION REACTIONS
- → LIGANDS
- → COORDINATE COVALENT BONDS
- → pH CALCULATIONS
- → K_W AND pK_W
- → pH OF WEAK ACIDS/BASES
- → HYDROLYSIS REACTIONS
- → pH OF SALT SOLUTIONS
- → BUFFERS
- → PREPARATION OF BUFFERS
- → pH OF POLYPROTIC ACID SOLUTIONS
- → TITRATION CURVES
- → POLYPROTIC ACID TITRATION CURVES
- → pH INDICATORS

BIG IDEAS 2, 5, 6

Learning Objectives: 2.2, 5.13, 5.14, 5.15, 5.16, 5.17, 5.18, 6.1, 6.2, 6.3, 6.4, 6.5, 6.6, 6.7, 6.8, 6.9, 6.10, 6.11, 6.12, 6.13, 6.14, 6.15, 6.16, 6.17, 6.18, 6.19, 6.20

For the complete list of Big Ideas and Learning Objectives, refer to the AP Chemistry Course Outline: *https://secure-media.collegeboard.org/digitalServices/pdf/ap/ap-chemistry-course-and-exam-description.pdf*

ACID-BASE THEORIES

ACID-BASE THEORIES

Theory Name	Brief Description
Arrhenius	An acid adds hydrogen ions to a solution, and a base adds hydroxide ions to a solution.
Brønsted-Lowry	An acid is a proton donor, and a base is a proton acceptor.

Arrhenius Theory

Svante Arrhenius considered an acid to be any substance that increases the concentration of hydrogen ions, H^+, in aqueous solution. When an acid is dissolved in water, the reaction may be written in two different forms. For example, when gaseous hydrogen chloride is dissolved in water, the equation is written as

Most acid-base reactions take place in aqueous solution. The symbol (*aq*) is often omitted for clarity and simplicity.

$$HCl(g) \xrightarrow{H_2O} H^+(aq) + Cl^-(aq)$$

Modern chemistry recognizes, however, that the hydrogen ion, H^+, is unlikely to exist in aqueous solution. Instead, the hydrogen ion is hydrated or bound to one or more water molecules. H_3O^+ is used to represent the hydration of the hydrogen ion, and the reaction of HCl with water is written as

$$HCl(g) + H_2O(l) \rightarrow H_3O^+(aq) + Cl^-(aq)$$

Common examples of acids are HCl, HBr, H_2SO_4, H_3PO_4, and $HC_2H_3O_2$. The **Arrhenius theory** considers bases to be substances that increase the hydroxide ion, OH^-, concentration when dissolved in water. A typical reaction is as follows:

$$KOH(s) \xrightarrow{H_2O} K^+(aq) + OH^-(aq)$$

Common bases are NaOH, KOH, $Ca(OH)_2$, and $Al(OH)_3$.

Brønsted-Lowry Theory

For a long time it was known that ammonia, NH_3, when dissolved in water increases the hydroxide ion concentration. In fact, ammonia solutions were often called ammonium hydroxide, with the formula NH_4OH. This fit the Arrhenius theory, but not the facts since NH_4OH does not really exist. The **Brønsted-Lowry theory** solves this problem by defining acids as proton donors, as the Arrhenius theory does, but defining bases as proton acceptors. Ammonia, NH_3, is a base since it accepts protons from water molecules in the reaction

$$H_2O(l) + NH_3(g) \rightleftharpoons NH_4^+(aq) + OH^-(aq)$$

Ethylamine, $CH_3CH_2NH_2$, and dimethylamine, $(CH_3)_2NH$, are both bases related to ammonia. In ethylamine the ethyl group, CH_3CH_2–, replaces one hydrogen in ammonia. Two methyl groups, CH_3–, in dimethylamine replace two of the hydrogen atoms in ammonia.

Modern Concept of Acids

An acid, according to both the Arrhenius and Brønsted-Lowry theories, is any compound having one or more hydrogen atoms that are weakly bound to the rest of the molecule. When dissolved in water, hydrogen ions ionize from the rest of the molecule. Most acids are written with hydrogen as the first element in the formula. Organic acids, however, can be written with hydrogen as the first element or with the –COOH unit that is the functional group for organic acids. The following three formulas all represent ethanoic acid. Ethanoic acid is often called acetic acid.

$HC_2H_3O_2$ CH_3COOH $CH_3C(=O)OH$

ACID-BASE NOMENCLATURE

Binary acids contain hydrogen and one other atom. In aqueous solutions, the name of every binary acid starts with the prefix *hydro-* and ends with the suffix *-ic* on the root of the second atom in thc formula. Table 13.1 lists several binary acid names.

TABLE 13.1
Binary Acid Names

Name	Formula
hydrofluoric acid	HF
hydrochloric acid	HCl
hydrobromic acid	HBr
hydroiodic acid	HI
hydrosulfuric acid	H_2S

Polyatomic anions can be the anions of acids. For these acids, the ending of the polyatomic anion name is changed and the word *acid* is added. If the polyatomic anion name ends in *-ate*, the ending is changed to *-ic*. If the ending is *-ite*, it is changed to *-ous*. Table 13.2 illustrates this principle for several polyatomic anions.

TABLE 13.2
Names of Acids Derived from Polyatomic Anions

Polyatomic Anion	Acid Name	Acid Formula
sulfate	sulfuric acid	H_2SO_4
sulfite	sulfurous acid	H_2SO_3
nitrate	nitric acid	HNO_3
nitrite	nitrous acid	HNO_2
hypochlorite	hypochlorous acid	HClO
chlorite	chlorous acid	$HClO_2$
chlorate	chloric acid	$HClO_3$
perchlorate	perchloric acid	$HClO_4$

Organic acids have both common and systematic names. In the systematic name for an organic acid the suffix *-oic* and the word *acid* are added to the root name of the rest of the molecule. Some organic acids and their systematic and common names are listed in Table 13.3.

There are two common types of bases, those that have hydroxide ions in their formulas and those that contain nitrogen. The **hydroxide bases** are named by using the name of the metal, with roman numerals if necessary, and then the word *hydroxide*. For example, NaOH, $Al(OH)_3$, and $Fe(OH)_3$ are called sodium hydroxide, aluminum hydroxide, and iron(III) hydroxide, respectively.

TABLE 13.3
Organic Acid Names

Systematic Name	Common Name	Formula
methanoic acid	formic acid	$HCOOH$
ethanoic acid	acetic acid	CH_3COOH
propanoic acid	propanoic acid	CH_3CH_2COOH
butanoic acid	butyric acid	$CH_3CH_2CH_2COOH$
pentanoic acid	valeric acid	$CH_3CH_2CH_2CH_2COOH$
benzoic acid	benzoic acid	C_6H_5COOH

Nitrogen bases related to ammonia are amines. Replacing a hydrogen on ammonia with a methyl, $-CH_3$, group produces methylamine. If two methyl groups replace two hydrogen atoms, the compound is called dimethylamine. If a methyl and an ethyl group, $-CH_2CH_3$, replace two hydrogen atoms, the compound is called either ethylmethylamine or methylethylamine.

The chloride salt of ammonia is known as ammonium chloride. The chloride salt of methylamine is called methyl ammonium chloride; an alternative name for this salt is methylamine hydrochloride. For chloride salts of nitrogen bases with common names, such as hydrazine, the hydrochloride ending, as in hydrazine hydrochloride, is ordinarily used.

ACIDS

Strong Acids

Strong acids are acids that ionize completely when dissolved in water.

TABLE 13.4
The Six Most Important Strong Acids

Acid	Ionization Reaction		
hydrochloric acid	HCl	→	$H^+ + Cl^-$
hydrobromic acid	HBr	→	$H^+ + Br^-$
hydroiodic acid	HI	→	$H^+ + I^-$
perchloric acid	$HClO_4$	→	$H^+ + ClO_4^-$
nitric acid	HNO_3	→	$H^+ + NO_3^-$
sulfuric acid	H_2SO_4	→	$H^+ + HSO_4^-$

For sulfuric acid, only the first hydrogen is considered strong; the second hydrogen ion ionizes only slightly. These strong acids are also called mineral acids, as opposed to organic acids.

Weak Acids

All of the remaining acids are weak, meaning that, when they are dissolved in water, only a small percentage of the molecules ionize. Most of the **weak acids** are organic acids. The most common weak mineral acids are HF, H_2CO_3, H_3PO_4, H_3AsO_4, $HClO_3$, $HClO_2$, and HClO.

Acid Strengths

The concepts of electronegativity and bonding can be used to compare the strengths of acids based on their chemical formulas. The strength of an acid is inversely proportional to the strength of the bond between the hydrogen and the remainder of the molecule. Strong acids have very weak bonds, and weak acids have stronger bonds. To determine the relative strengths of acids, an estimate of the bond strengths is needed.

As mentioned previously, a **binary acid** is composed of hydrogen and one other atom. Experimental evidence shows that acids get stronger from left to right in a period (row) of the periodic table. Thus, we find that PH_3 is weaker than H_2S, which is weaker than HBr. Across a period the acid strength parallels the electronegativity of the anion. This is reasonable because, as the anion attracts the electrons more strongly, the bond with hydrogen becomes weaker, while the size of the anions remains relatively constant.

Experimental evidence also shows that the strength of binary acids increases from the top of a group to the bottom. Thus HF is weaker than HCl, which is weaker than HBr. Electronegativity cannot explain this sequence of binary acid strengths. Instead, the size of the anion is important. An increase in anion size requires a corresponding increase in bond length. As stated in Chapter 4, longer bond lengths imply weaker bonds. In acids, a weaker bond with hydrogen means a stronger acid.

All mineral acids that are not binary acids are **oxoacids**: acids that contain hydrogen, oxygen, and another element. In all oxoacids, the oxygen atoms are bound to the central atom and the hydrogen atoms are bound to the oxygen atoms. A typical structure is represented by sulfuric acid:

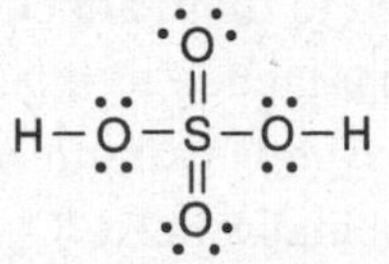

```
        O
        ‖
  H–O–S–O–H
        ‖
        O
```

The strength of an oxoacid always depends on the relative strength of the oxygen-hydrogen bond. In an oxoacid the strength of the O–H bond depends on (a) the number of oxygen atoms per hydrogen in the formula and (b) the electronegativity of the central atom in the formula.

Oxoacids that have the same central atom and the same number of hydrogen atoms increase in strength as the number of oxygen atoms increases.

Each extra oxygen withdraws more electron density from the O–H bond, thereby weakening it. The more oxygen atoms per hydrogen, the stronger the acid.

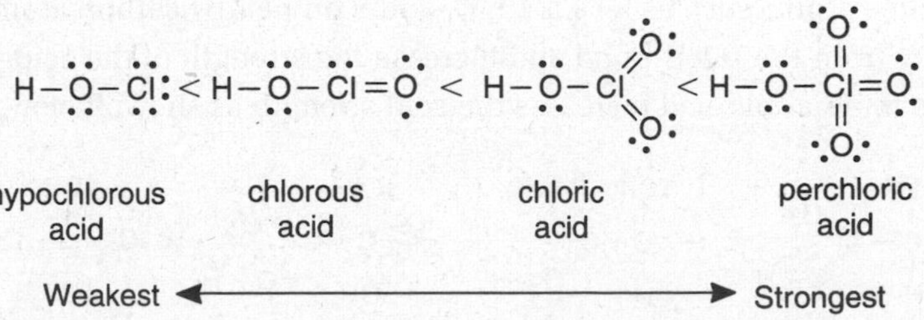

H–O–Cl < H–O–Cl=O < H–O–Cl(=O)(=O) < H–O–Cl(=O)(=O)=O

hypochlorous acid · chlorous acid · chloric acid · perchloric acid

Weakest ⟷ Strongest

We can also see that an increase in the number of oxygens helps stabilize the oxoanions by delocalizing the electrons. When there is a very stable anion, it will be less likely to react with H^+ (or hydrolyze water) to form the acid.

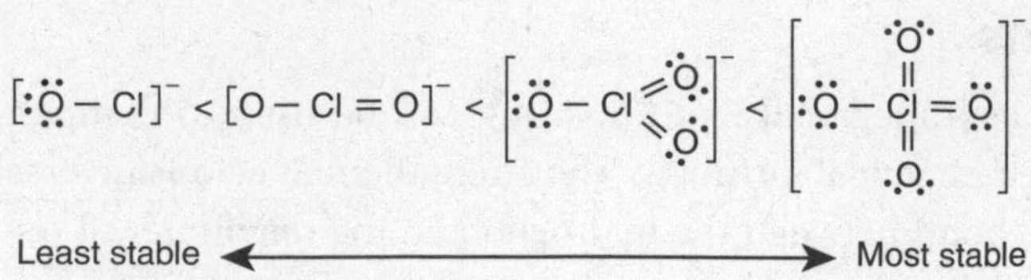

$[:\ddot{O}-Cl]^- < [O-Cl=O]^- < [:\ddot{O}-Cl(=O)(=O)]^- < [:\ddot{O}-Cl(=O)(=O)=\ddot{O}]^-$

Least stable ⟷ Most stable

The strengths of oxoacids that have the same number of hydrogen and oxygen atoms but a different central atom are affected by the electronegativity of the central atom:

$H-\ddot{O}-Te(=O)(=O)-\ddot{O}-H < H-\ddot{O}-Se(=O)(=O)-\ddot{O}-H < H-\ddot{O}-S(=O)(=O)-\ddot{O}-H$

telluric acid selenic acid sulfuric acid

Weakest ⟷ Strongest

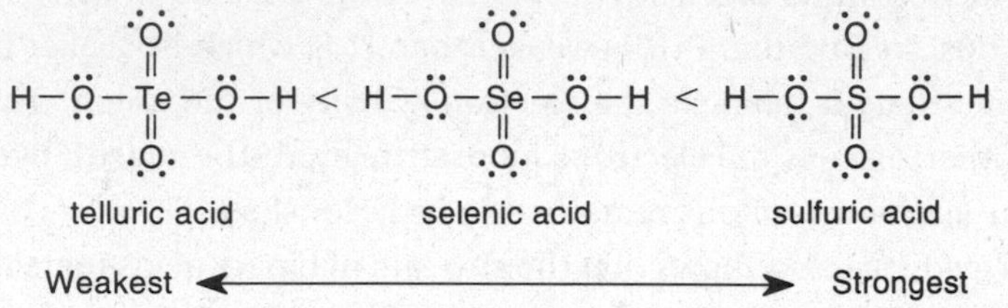

In the diagram above, the electronegativity of the central atom increases from tellurium to selenium to sulfur, weakening the O–H bond and resulting in stronger acids.

Oxoacids that have the same number of oxygen atoms but different central atoms and different numbers of hydrogens can also be compared. For instance, experiments show that

$$H_3PO_4 < H_2SO_4 < HClO_4$$

In this situation we see that the electronegativity of the central atom increases as the strength of the acid increases. In addition to the electronegativity of the central atom, the oxygen atoms that have no hydrogens attached to them affect the acid strength. H_3PO_4 has only one oxygen that does not have a hydrogen attached to it, and it is attracting electrons from three H–O bonds. In H_2SO_4 there are two oxygens without hydrogens attached, and their effect is concentrated on only two H–O bonds. Finally, in $HClO_4$ three oxygen atoms have no hydrogen attached to them, and their effect is focused on a lone H–O bond. The effect of the electronegativity of the central atom and the effect of the oxygen atoms with no hydrogens attached combine to produce the observed trend in acid strengths. To repeat: when two acids are compared, the acid with more oxygen atoms per hydrogen atom will be the stronger acid.

TIP

The weaker the O–H bond, the stronger the acid.

Every **organic acid** contains the carboxyl group, which is commonly written as

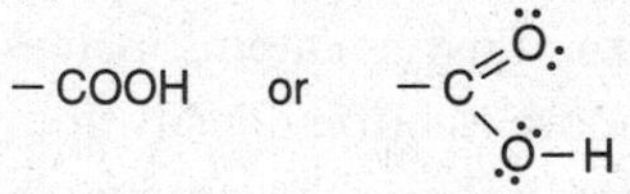

$-COOH$ or $-C(=\ddot{O}:)-\ddot{O}-H$

Electronegative atoms such as F, Cl, Br, I, O, and S on nearby carbon atoms will withdraw electron density from the –O–H bond and increase the strength of the acid. The addition of chlorine atoms to ethanoic acid increases the acid strength as shown below.

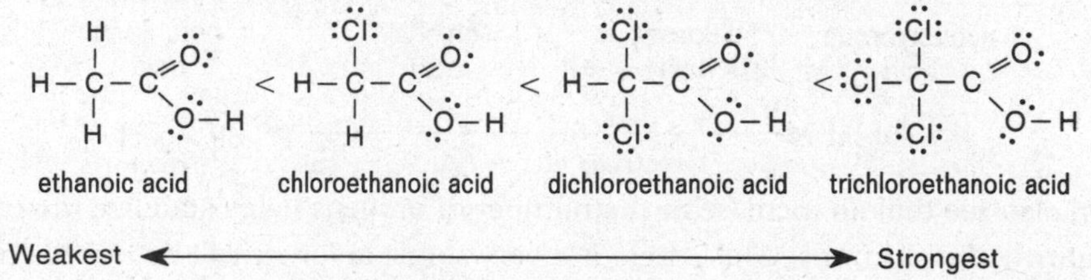

$H_3C-COOH < ClH_2C-COOH < Cl_2HC-COOH < Cl_3C-COOH$

ethanoic acid chloroethanoic acid dichloroethanoic acid trichloroethanoic acid

Weakest ⟷ Strongest

BASES

Strong Bases

All metal hydroxides are **strong bases**. However, most metal hydroxides are very slightly soluble. Only the hydroxides of Group IA metals, strontium and barium, have appreciable solubility; calcium hydroxide is moderately soluble. Soluble hydroxides may cause severe skin burns. Insoluble hydroxides are much less harmful. For instance, $Al(OH)_3$ and $Mg(OH)_2$ can safely be swallowed as an antacid to neutralize excess stomach acid.

Weak Bases

All bases related to ammonia are **weak bases**. Organic compounds related to ammonia are called amines and have carbon-containing groups replacing one or more of the hydrogen atoms of ammonia, NH_3. Two such bases, ethylamine and dimethylamine, were described previously.

Base Strengths

The relative strengths of the weak bases may be evaluated based on the electronegativities of the organic functional groups that replace the hydrogen atoms of ammonia. We saw that electronegative substituents such as chlorine increase the strength of organic acids. The reverse is true, however, of organic bases. For example, chloromethylamine is a weaker base than methylamine.

ANHYDRIDES OF ACIDS AND BASES

The word ***anhydride*** means "without water," and the acidic and basic anhydrides are compounds that, when added to water, become common acids and bases. **Acid anhydrides** are often the oxides of nonmetals. Common acid anhydrides have the following reactions with water:

$$SO_2 + H_2O \rightarrow H_2SO_3$$

$$SO_3 + H_2O \rightarrow H_2SO_4$$

$$CO_2 + H_2O \rightarrow H_2CO_3$$

$$P_2O_5 + 3H_2O \rightarrow 2H_3PO_4$$

Basic anhydrides are the oxides of metals. Some typical reactions with water are these:

$$K_2O + H_2O \rightarrow 2KOH$$

$$CaO + H_2O \rightarrow Ca(OH)_2$$

In industrial applications CaO is called lime, and $Ca(OH)_2$ is termed slaked lime. The production of lime is a major industry that involves mining naturally occurring limestone, $CaCO_3$, and heating it to very high temperatures to drive off carbon dioxide in the reaction

$$CaCO_3 \xrightarrow{\text{heat}} CaO + CO_2$$

NEUTRALIZATION REACTIONS

The reactions between acids and bases are called **neutralization reactions.** They are often double-replacement reactions, and the products can be predicted by the methods described in Chapter 3. In most cases, a neutralization reaction can be described as the mixing of an acid with a base to form a salt and water:

$$\underset{\text{(acid)}}{HBr(aq)} + \underset{\text{(base)}}{KOH(aq)} \rightarrow \underset{\text{(salt)}}{KBr(aq)} + \underset{\text{(water)}}{H_2O(l)}$$

A neutralization reaction may be written as a molecular, ionic, or net ionic equation depending on the type of information the chemist is interested in communicating. For a typical neutralization reaction with soluble acids and bases the three types of reactions would be as follows:

$$HCl + NaOH \rightarrow NaCl + H_2O \qquad \text{(molecular equation)}$$

$$H^+ + Cl^- + Na^+ + OH^- \rightarrow Na^+ + Cl^- + H_2O \qquad \text{(ionic equation)}$$

$$H^+ + OH^- \rightarrow H_2O \qquad \text{(net ionic equation)} \qquad (13.1)$$

If all of the reactants are soluble strong acids and bases, the net ionic equation of a neutralization will *always* be as shown in Equation 13.1.

Acids are often used to dissolve insoluble hydroxides or oxides of metals. For instance, lanthanum ions are used to suppress interferences in atomic absorption spectroscopy. Dissolution of lanthanum oxide, La_2O_3, is achieved by reacting it with either HCl or HNO_3:

$$6HCl + La_2O_3 \rightarrow 2LaCl_3 + 3H_2O$$

$$6HNO_3 + La_2O_3 \rightarrow 2La(NO_3)_3 + 3H_2O$$

These are reactions between an acid and a basic anhydride. In both reactions the chloride and nitrate salts of $lanthanum_3$ are soluble.

Acid anhydrides such as SO_3 and CO_2 will react with bases as in the following reactions:

$$2NaOH(aq) + SO_3(g) \rightarrow Na_2SO_4(aq) + H_2O(l)$$

$$Ca(OH)_2(aq) + CO_2(g) \rightarrow CaCO_3(s) + H_2O(l)$$

Acid and basic anhydrides may react with each other without any water present. For example, lime can react with sulfur trioxide in the reaction

$$SO_3(g) + CaO(s) \rightarrow CaSO_4(s)$$

POLYPROTIC ACIDS

Acids that contain more than one ionizing hydrogen are called **polyprotic acids**. Sulfuric acid, H_2SO_4, and phosphoric acid, H_3PO_4, are two examples. Although ethanoic acid, $HC_2H_3O_2$, has four hydrogen atoms in its formula, it is not a polyprotic acid since it has only one ionizable hydrogen.

All polyprotic acids are weak acids except sulfuric acid, which is unique in that its first proton dissociates completely but the second proton does not. One property of polyprotic acids is that the protons dissociate and react in a stepwise manner; that is, the first proton dissociates or reacts before the second proton. The stepwise dissociation of phosphoric acid is written as shown in Equations 13.2–13.4:

$$H_3PO_4 \rightleftharpoons H_2PO_4^- + H^+ \quad (13.2)$$

$$H_2PO_4^- \rightleftharpoons HPO_4^{2-} + H^+ \quad (13.3)$$

$$HPO_4^{2-} \rightleftharpoons PO_4^{3-} + H^+ \quad (13.4)$$

The product or products formed in the neutralization of a polyprotic acid depend on the amount of base used. When 1 mole of hydroxide ions per mole of phosphoric acid reacts, the equation is

$$H_3PO_4 + OH^- \rightarrow H_2PO_4^- + H_2O$$

When 2 moles of hydroxide ions per mole of phosphoric acid react, the equation is

$$H_3PO_4 + 2OH^- \rightarrow HPO_4^{2-} + 2H_2O$$

When 3 moles of hydroxide ions per mole of phosphoric acid react, the equation is

$$H_3PO_4 + 3OH^- \rightarrow PO_4^{3-} + 3H_2O$$

TIP

Remember that these are combining ratios of H_3PO_4 and OH^-. Therefore, 0.2 mol of H_3PO_4 and 0.4 mol of OH^- follows the stoichiometry written in these three equations.

Salts containing the $H_2PO_4^-$, HPO_4^{2-}, and PO_4^{3-} ions can be prepared by accurately adjusting the amount of base added to the phosphoric acid.

If exactly 1, 2, or 3 moles of hydroxide ions per mole of phosphoric acid are not reacted, a mixture of phosphate salts will form. For example, if 2.25 moles of OH^- are added per mole of phosphoric acid, we may deduce that the first 2 moles of OH^- convert the phosphoric acid to HPO_4^{2-}. The additional 0.25 mole of OH^- converts only some of the HPO_4^{2-} to PO_4^{3-}. Consequently, the final mixture will be a combination of HPO_4^{2-} and PO_4^{3-} salts.

BRØNSTED-LOWRY THEORY; CONJUGATE ACID–BASE PAIRS

The Brønsted-Lowry theory of acids and bases not only redefined these compounds but also gave us the concept of **conjugate acid–base pairs**. In the equilibrium reaction of ethanoic acid with hydroxide ions:

base 2 — acid 2: conjugate pair 2

$$HC_2H_3O_2 + OH^- \rightleftharpoons C_2H_3O_2^- + H_2O$$

acid 1 — base 1: conjugate pair 1

$HC_2H_3O_2$ is an acid that reacts with the base OH^- in the forward reaction. The products, however, are also acids and bases. $C_2H_3O_2^-$ is a base that can accept an H^+ from the water, while water is an acid since it donates a proton in the reverse reaction. $HC_2H_3O_2$ and $C_2H_3O_2^-$ are called a conjugate acid–base pair. In a similar fashion, H_2O and OH^- are another conjugate acid–base pair in this equation.

Conjugate acid–base pairs always have formulas that differ by only one H^+:

$$\text{Conjugate acid} \rightleftharpoons \text{Conjugate base} + H^+ \quad (13.5)$$

Relative Strengths of Conjugate Acids and Bases

In a conjugate acid–base pair the relative strengths of the **conjugate acid** and **conjugate base** are determined by the position of the equilibrium in Equation 13.5. If this position results in more products than reactants, the conjugate acid is stronger than the conjugate base. If there are more reactants than products at equilibrium, the conjugate base is the stronger of the pair.

Ethanoic acid dissociates slightly in the reaction

$$HC_2H_3O_2 \rightleftharpoons C_2H_3O_2^- + H^+$$

and we conclude that ethanoic acid is a weak acid and the ethanoate ion is a stronger base.

The conjugate acid–base pair for ammonia may be written as

$$NH_4^+ \rightleftharpoons NH_3 + H^+ \qquad (13.6)$$

Since an aqueous solution of ammonia has a much higher concentration of ammonia than of the ammonium ion, the equilibrium lies to the right of the double arrow in Equation 13.6. From this fact we conclude that the NH_4^+ ion is a stronger conjugate acid than NH_3 is a conjugate base.

Using similar logic, we can determine the relative strengths of conjugate acids and bases in a complete chemical reaction. For example, when equal numbers of moles of ethanoic acid and base are mixed, this reaction goes virtually to completion, leaving little $HC_2H_3O_2$ and OH^-:

$$HC_2H_3O_2 + OH^- \rightarrow C_2H_3O_2^- + H_2O$$

From the position of the equilibrium, chemists deduce that the OH^- ion is a stronger base than the ethanoate ion, $C_2H_3O_2^-$. At the same time, ethanoic acid is also a stronger acid than the water molecule. Since we already know that ethanoic acid is a weak acid, the water molecule must be a much weaker acid. Similarly, since the ethanoate ion is a relatively strong base, the hydroxide ion is an even stronger base. This information tells us that in the H_2O-OH^- conjugate acid–base pair water is a very weak acid and OH^- is a very strong base.

Another reaction takes place when ethanoic acid is dissolved in distilled water:

$$HC_2H_3O_2 + H_2O \rightleftharpoons C_2H_3O_2^- + H_3O^+$$

Again, we know that ethanoic acid is a weak acid, and by that definition most of it remains in molecular form. Since the reactants are favored in this equilibrium, we conclude that H_3O^+ is a stronger acid than $HC_2H_3O_2$. Also, the ethanoate ion is a stronger base than water. Knowing the position of equilibrium gives us another method to determine the relative strengths of conjugate acids and bases.

In aqueous solution, the strongest acid is the H_3O^+ ion (often written simply as the H^+ ion) and the strongest base is the OH^- ion. Acids that are stronger than water react with water to produce the H_3O^+ ion:

$$HCl + H_2O \rightarrow Cl^- + H_3O^+$$

Because hydrochloric acid is a very strong acid, this reaction goes all the way to completion, forming the H_3O^+ ion and leaving no HCl molecules. Bases that are stronger than water produce the OH^- ion when dissolved in water. In aqueous solutions all strong acids react completely with the weak base water. All strong bases also react completely, with water acting as a weak acid. This is known as the **leveling effect** of water.

In discussing the strengths of acids and bases, it was mentioned above that the ethanoate ion is a stronger base than water. This fact results in the concept that the anion of a weak acid may be considered as a base. Very strong acids such as HCl and HNO_3 have anions that are extremely weak conjugate bases. On the other hand, very weak acids have anions that are strong conjugate bases. Carbonic acid, H_2CO_3, is a very weak acid, and the carbonate ion, CO_3^{2-}, is a strong conjugate base. Carbonate salts such as sodium carbonate and calcium carbonate (limestone) are rather strong bases and are used in many industrial processes instead of more expensive hydroxide bases such as NaOH and KOH.

Later in this chapter we will see that the strengths of conjugate acids and bases can be expressed numerically as K_a and K_b values. These two values are related to the constant K_w in the equation

$$K_aK_b = K_w \tag{13.7}$$

Equation 13.7 indicates that K_a is inversely proportional to K_b, meaning that strong conjugate acids have weak conjugate bases and vice versa.

Amphiprotic and Amphoteric Substances

Amphiprotic substances can both gain (accept) and lose (donate) a proton. An **amphoteric** substance can act as both an acid and as a base. The difference is subtle but can also be important. Amphiprotic salts are anions, and these anions must have at least one proton so that they can act as proton donors. They must also be able to accept a proton to act as bases. The anions of partially neutralized polyprotic acids are always amphiprotic. The common amphiprotic anions are the hydrogen carbonate ion, HCO_3^-; the hydrogen sulfate ion, HSO_4^-; the hydrogen sulfite ion, HSO_3^-; the dihydrogen phosphate ion, $H_2PO_4^-$; and the monohydrogen phosphate ion, HPO_4^{2-}. Each of these ions can accept a proton and act as a base. Each ion can also lose a proton and act as an acid.

Water is an amphiprotic solvent that can act as an acid and as a base as shown in the reaction

$$H_2O + H_2O \rightleftharpoons H_3O^+ + OH^-$$

In this reaction one water molecule donates a proton and is an acid, while the other accepts a proton and is a base. In both cases water is an extremely weak acid and an extremely weak base.

Aluminum oxide is amphoteric. Its reactions as an acid and as a base are:

$$\text{with acid: } Al_2O_3 + 3H_2O + 6H_3O^+(aq) \rightarrow 2[Al(H_2O)_6]^{3+}(aq)$$

$$\text{with base: } Al_2O_3 + 3H_2O + 2OH^-(aq) \rightarrow 2[Al(OH)_4]^-(aq)$$

Aluminum oxide is not amphiprotic, since Al_2O_3 cannot donate a proton.

COMPLEXATION REACTIONS

Complexation reactions, which are not covered in the discussions of ionic reactions or covalent reactions, occur because of the formation of a coordinate covalent bond. For instance, we know that silver chloride, AgCl, is an insoluble salt. It is not a hydroxide or a basic anhydride, which we know can be dissolved in acids. However, this salt does dissolve in ammonia solutions. An analysis of this process shows that the reaction is

$$AgCl(s) + 2NH_3 \rightarrow Ag(NH_3)_2^+ + Cl^-$$

The same reaction occurs, without the Cl^- ions, when silver ions in solution react with ammonia. This reaction is a complexation reaction.

$$\begin{array}{c} \cdot\cdot \\ H : N : H \\ \cdot\cdot \\ H \end{array}$$

The silver ion started as a silver atom with the electronic configuration [Kr] $5s^1$, $4d^{10}$. In forming the Ag^+ ion, it loses the $5s^1$ electron, resulting in the [Kr] $4d^{10}$ electronic configuration. Therefore it has empty $5s$ and $5p$ orbitals that may accept pairs of electrons. In actual fact, silver ions accept only two pairs of electrons from two ammonia molecules. In a similar fashion we find that all metal ions have available orbitals that may accept electron pairs. A bond where both electrons come from one of the reactants is a coordinate covalent bond.

Ligands

Ligands, **complexing agents**, **chelates**, and **sequestering agents** are all names for substances that donate pairs of electrons. Most ligands have one pair of electrons to donate, as ammonia does. Some ligands have two pairs of electrons and some have up to six pairs. Ligands that provide more than one electron pair in forming a complex must be large, flexible molecules so that each pair of electrons can be oriented properly to form a bond. The chloride ion has four pairs of electrons but forms only one bond because the remaining six electrons cannot be aligned properly to form additional bonds.

Complexation reactions can be written generally as

$$M^{n+} + xL^{m-} \rightleftharpoons ML_x^{n-xm}$$

where M^{n+} is a metal ion with a charge of $+n$ and L^{m-} is a ligand with a charge of $-m$.

Silver tends to accept two electron pairs, and copper accepts four. The other metal ions tend to accept six electron pairs in complexes. The number of electron pairs that a metal ion will typically accept is called its coordination number. The coordination numbers of some representative ions are given in the following table.

Coordination Numbers (C.N.)

Ion	C.N.
Ag^+	2
Au^{3+}	4
Cu^{2+}	4
Zn^{2+}	4
Pt^{2+}	4
Fe^{2+}, Fe^{3+}	6
Co^{3+}	6
Ti^{4+}	6
Mn^{2+}	6
Cr^{3+}	6

The formula of a complex or a complex ion is written as any other chemical formula. For any complex ion the total charge of the ion is the sum of the charges of the ions in the complex. The complex made of Fe^{3+} and six chloride ions, $FeCl_6^{3-}$, has a charge of –3:

$$Fe^{3+} + 6Cl^- \rightleftharpoons FeCl_6^{3-}$$

Molecular complexing agents do not affect the charge since they are neutral molecules. Thus Cu^{2+} still has a +2 charge when complexed with four molecules of ammonia in $Cu(NH_3)_4^{2+}$:

$$Cu^{2+} + 4NH_3 \rightleftharpoons Cu(NH_3)_4^{2+}$$

One interesting application of complexation reactions involves gold mining. Modern gold mining involves spraying dilute cyanide solutions, along with air, on gold-bearing ore. The gold, oxidized to Au^+, complexes with two CN^- ions to form the soluble complex $Au(CN)_2^-$. The liquid containing the gold complex is then treated to recover the concentrated gold.

Coordinate Covalent Bonds

The bonds formed in complexation reactions are covalent bonds. All of the properties of covalent bonds, as well as molecular structure, discussed in Chapter 4 apply also to the bonds in complexes. However, the formation of this covalent bond is unique. Instead of each atom donating one electron to the bond, one atom donates both electrons. Covalent bonds formed in this way are called **coordinate covalent bonds.** Aside from the method of formation, these bonds are true covalent bonds.

QUANTITATIVE ACID-BASE CHEMISTRY

In describing the nature of acids and bases, reference was made to their relative strengths. These strengths can be determined experimentally as the acid and base dissociation constants, K_a and K_b. The qualitative description of acid strengths given above was developed with a knowledge of the experimental values of acid strengths.

pH and pOH: Measurements of Acidity and Basicity

The acidity of a solution is expressed either as the molar hydrogen ion concentration, $[H^+]$, or the **pH** of the solution. Soren Sorensen in 1909 developed the pH system to represent acidity data in a simplified form. The pH and the hydrogen ion concentration are related by the equation

$$pH = -\log[H^+]$$

In aqueous solutions of acids and bases the pH ranges from 0 to 14. In very concentrated solutions we may have a negative pH (extremely acidic) or a pH greater than 14 (extremely basic).

In a similar fashion, the molar hydroxide ion concentration is related to the **pOH** by

$$pOH = -\log[OH^-]$$

The pH, pOH, $[H^+]$, and $[OH^-]$ are all related to each other. Water dissociates into hydrogen ions and hydroxide ions in the reaction

$$H_2O \rightleftharpoons H^+ + OH^- \qquad (13.8)$$

TIP

K_w, the autoionization constant of water, may be called the ionization constant, the pyrolysis constant, or the ion product for water.

The equilibrium expression for Equation 13.8 is

$$K = [H^+][OH^-] = 1.0 \times 10^{-14} \tag{13.9}$$

The equilibrium constant in this case is designated as K_w, the **autoionization constant of water**. Taking the negative logarithm of Equation 13.8 gives

$$pK_w = pH + pOH = 14.00 \tag{13.10}$$

Equations 13.9 and 13.10 allow the chemist to determine the pH, pOH, $[H^+]$, or $[OH^-]$ by measuring only one of them and then finding the other three by calculation.

Exercise 13.1

Determine $[H^+]$, $[OH^-]$, pH, and pOH, given the following data. (Try to estimate the answers first.)

(a) $[H^+] = 2.3 \times 10^{-4}\ M$
(b) $[OH^-] = 6.3 \times 10^{-2}\ M$
(c) pH = 10.67
(d) pOH = 2.34

Solution

Starting with the given information, we use the equations given previously to measure pH and pOH to convert:

	$[H^+]$	$[OH^-]$	pH	pOH
(a)	2.3×10^{-4}	4.3×10^{-11}	3.64	10.36
(b)	1.6×10^{-13}	6.3×10^{-2}	12.80	1.20
(c)	2.1×10^{-11}	4.7×10^{-4}	10.67	3.33
(d)	2.2×10^{-12}	4.6×10^{-3}	11.66	2.34

TIP

Even distilled water has enough dissolved CO_2 to make it distinctly acidic.

When $[H^+]$ is greater than $[OH^-]$, the solution is considered to be acidic. A pH less than 7.0 also represents an acid solution. When $[H^+]$ is less than $[OH^-]$, the solution is basic and the pH is greater than 7.0. When $[H^+]$ is equal to the $[OH^-]$, the solution is neutral and the pH is 7.0.

Many of the common substances with which we come into contact are either acidic or basic. Many foods are acidic, and many cleaning agents are basic.

pH OF NEUTRAL SOLUTIONS

The ionization, or autopyrolysis, of water can be written two ways:

$$H_2O \rightleftharpoons [H^+] + [OH^-] \quad \text{or} \quad 2H_2O \rightleftharpoons [H_3O^+] + [OH^-]$$

The equilibrium constant for the autopyrolysis reactions is K_w, expressed as

$$K_w = [H^+]\,[OH^-] \quad \text{or} \quad K_w = [H_3O^+]\,[OH^-]$$

In a neutral solution, $[H^+] = [OH^-]$. The $[H^+]$ can be calculated as $\sqrt{K_w}$. At 25 °C the value of K_w is 1.0×10^{-14}, $[H^+] = 1.0 \times 10^{-7}$, and the pH is 7.00. However, we saw in Chapter 9 (page 342) that temperatures different from 25°C will have different values for K_w. Table 13.5 summarizes the pH of neutral solutions when the temperature is not 25°C. Normal body temperature is 36°C. A neutral solution at normal body temperature has a pH of 6.34.

TABLE 13.5
pH and K_w of Aqueous Solutions at Different Temperatures

Temperature °C	K_w	pH
10	3.0×10^{-15}	7.26
20	6.9×10^{-15}	7.08
25	1.0×10^{-14}	7.00
30	1.5×10^{-14}	6.92
36	2.1×10^{-14}	6.34
40	2.9×10^{-14}	6.77
50	5.3×10^{-14}	6.63

$[H^+]$ AND pH OF STRONG ACIDS

Except for sulfuric acid, all strong acids are monoprotic and ionize completely when dissolved in water. Consequently, the hydrogen ion concentration of a strong acid solution is equal to the molar concentration of the acid itself:

$$[H^+] = M_{\text{strong acid}}$$

Sulfuric acid is the only diprotic strong acid. When this acid is dissolved in water, only one hydrogen ion ionizes, and it acts as a monoprotic acid:

$$H_2SO_4 \rightleftharpoons H^+(aq) + HSO_4^-(aq)$$

Also, for sulfuric acid the hydrogen ion concentration is equal to the molar concentration. The pH of a strong acid solution can be written as

$$pH = -\log[H^+] = -\log M_{\text{strong acid}}$$

TIP

The ionization of the first proton of H_2SO_4 makes the solution acidic. This acidic solution suppresses the second ionization step. The result is that H_2SO_4 acts as a strong monoprotic acid when dissolved in water.

$[OH^-]$, pOH, AND pH OF STRONG BASES

Strong bases dissociate completely when dissolved in water; however, they may have more than one hydroxide ion per formula unit. The concentration of hydroxide ions is equal to the molarity of the base multiplied by the number of hydroxide ions in its chemical formula:

$$[OH^-] = M_{\text{strong base}} \times \text{number of } OH^- \text{ ions per mole}$$

For example, a 0.200 *M* solution of $Ba(OH)_2$ has a hydroxide concentration of 0.400 *M*. Once the $[OH^-]$ is known, the pOH is calculated as

$$pOH = -\log[OH^-]$$

From the pOH the pH is calculated using the relationship

$$pH = 14.00 - pOH$$

Exercise 13.2

Determine the pH of each of the following solutions. (Try to estimate the pH to ±1pH unit.)

(a) 0.020 M HNO_3
(b) 0.00043 M $Ba(OH)_2$
(c) 3.0 g of NaOH dissolved in 250 mL H_2O
(d) 0.00032 mol SO_3 dissolved in 3.4 L H_2O
(e) 0.00098 mol $Ca(OH)_2$ dissolved in 1.3 L H_2O

TIP

A very quick check of your pH calculations must have acids with pH values below 7 and bases with pH values greater than 7.

Solution

(a) $[H^+] = 0.020\ M$; pH = 1.70
(b) $[OH^-] = (0.00043\ M)\ (2) = 0.00086\ M$; pOH = 3.07 and pH = 10.93
(c) $M_{NaOH} = 0.30$; $[OH^-] = 0.30$; pOH = 0.52 and pH = 13.48
(d) $SO_3 + H_2O \rightarrow H_2SO_4$; $M_{H_2SO_4} = 9.4 \times 10^{-5}$; $[H^+] = 9.4 \times 10^{-5}$; pH = 4.03
(e) $M_{Ca(OH)_2} = 7.5 \times 10^{-4}$; $[OH^-] = (7.5 \times 10^{-4})\ (2) = 1.5 \times 10^{-3}$; pOH = 2.82 and pH = 11.18

Calculations for pH can always be checked for reasonableness since acids must have pH values less than 7 and bases must have pH values greater than 7.

pH OF WEAK ACIDS AND WEAK BASES

Hydrofluoric acid is a weak acid that dissociates in water according to the equation

$$HF + H_2O \rightleftharpoons H_3O^+ + F^- \qquad (13.11)$$

Equation 13.11 may also be written in a shorter, more convenient form by eliminating water:

$$HF \rightleftharpoons H^+ + F^- \qquad (13.12)$$

Either Equation 13.11 or Equation 13.12 can be used to write the equilibrium expression:

$$K_a = \frac{[H^+][F^-]}{[HF]} = \frac{[H_3O^+][F^-]}{[HF]}$$

The two forms are identical, and either may be used in the calculations that follow.

The constant K_a is called the **acid dissociation constant**; values for selected weak acids are tabulated in Appendixes 4 and 5. For the example above, since we know the initial concentration of HF and the value of K_a, the hydrogen ion concentration can be calculated. The technique used is similar to the equilibrium calculations in Chapter 9.

EXAMPLE 13.1

If 0.100 mol of HF is diluted with distilled water to a volume of 500 mL, what is the pH of the solution?

Solution

To solve this problem, we set up an equilibrium table with the reaction on the first line, and we enter the initial concentration on the Initial Conc. line. The initial concentration of HF is calculated as

$$\frac{0.100 \text{ mol HF}}{0.500 \text{ L solution}} = 0.200\ M \text{ HF}$$

Reaction	HF	⇌	H^+	+	F^-
Initial Conc.	0.100/0.500		0.00*		0.00
Change					
Equilibrium					
Solution					

*The hydrogen ion concentration in distilled water is 1.0×10^{-7}, but in most instances it is considered to be zero.

In the Change row we enter *x*'s to represent the stoichiometric relationships between the reactants and the products. In this table we assign a positive value to the *x*'s under the products because they start at zero and cannot possibly decrease. As a consequence any *x*'s under the reactant(s) must be negative.

Reaction	HF	⇌	H^+	+	F^-
Initial Conc.	0.200 *M*		0.00		0.00
Change	$-x$		$+x$		$+x$
Equilibrium					
Solution					

The Equilibrium row is then the sum of the Initial Conc. and Change rows of the table:

Reaction	HF	⇌	H^-	+	F^-
Initial Conc.	0.200 *M*		0.00		0.00
Change	$-x$		$+x$		$+x$
Equilibrium	$0.200 - x$		$+x$		$+x$
Solution					

At this point we enter the terms in the Equilibrium row into the equilibrium expression, along with the value of K_a:

$$K_a = \frac{[H^+][F^-]}{[HF]}$$

$$6.6 \times 10^{-4} = \frac{(x)(x)}{0.200 - x}$$

If this equation is solved for x by ordinary means, a quadratic equation results. As before, we can try a simplifying assumption to solve the problem more quickly. The only term that can be simplified is $0.200 - x$. We use the assumption that $x << 0.200$ and conclude that $0.200 - x = 0.200$. (Remember that, when we finally calculate x, we must verify that the assumption is true.) Returning to the equation, we find that the assumption yields

$$6.6 \times 10^{-4} = \frac{(x)(x)}{0.200}$$

$$(6.6 \times 10^{-4})(0.200) = x^2$$

$$x = 0.0115$$

Checking the assumption, we find that 0.0115 is less than one-tenth of 0.200 and therefore qualifies as being negligible for this type of calculation. Now we substitute 0.0115 for x in the Equilibrium row and calculate the Solution row:

Reaction	HF	$\rightleftharpoons$	H^+	+	F^-
Initial Conc.	0.200 *M*		0.00		0.00
Change	$-x$		$+x$		$+x$
Equilibrium	$0.200 - x$		$+x$		$+x$
Solution	0.189		0.0115		0.0115

With the value for $[H^+] = 0.0115$ *M*, the pH can be calculated as $-\log(0.0115) = 1.94$. Looking at this process a little more closely, we find that, if the assumption is valid, then

$$[H^+] = \sqrt{K_a C_a}$$

where C_a is the initial concentration of the weak acid. Evaluating the conditions under which the assumption holds, we find that it will always be valid as long as $C_a > 100\ K_a$.

NOTE

We use a factor of 100 here because this will keep the error in x to less than 5% compared with the value of x found when solving the quadratic equation.

Weak bases such as ammonia and methylamine, CH_3NH_2, behave similarly to weak acids. The dissociation reaction for methylamine is

$$CH_3NH_2 + H_2O \rightleftharpoons CH_3NH_3^+ + OH^-$$

Although water cannot be eliminated to simplify the chemical reaction, it does not appear in the equilibrium law since it is the solvent.

$$K_b = \frac{[CH_3NH_3^+][OH^-]}{[CH_3NH_2]}$$

The constant K_b is the **base dissociation constant**. This equilibrium expression is similar to the one for HF (page 468, top) in that it has two concentration terms in the numerator and one in the denominator. The two laws are mathematically equivalent.

Values of K_b for selected weak bases are given in Appendix 6.

EXAMPLE 13.2

What is the pH of a 0.500 *M* solution of methylamine, $K_b = 4.2 \times 10^{-4}$?

Solution

First we set up the equilibrium table in the same fashion as for the weak acid example in Example 13.1.

Reaction	CH_3NH_2	+	H_2O	$\rightleftharpoons$	$CH_3NH_3^+$	+	OH^-
Initial Conc.	0.500		55.5		0.00		0.00
Change	$-x$		$-x$		$+x$		$+x$
Equilibrium	$0.500 - x$		$55.5 - x$		$+x$		$+x$
Solution							

We entered the concentration of water:

$$\frac{1000 \text{ g H}_2\text{O L}^{-1}}{18 \text{ g H}_2\text{O mol}^{-1}} = 55.5\,M$$

for completeness although it does not appear in the equilibrium expression. The terms from the Equilibrium row are entered into the equilibrium expression

$$4.2 \times 10^{-4} = \frac{(x)(x)}{0.500 - x}$$

To simplify, we assume that x is very small since 0.500 is more than 100 times larger than K_b. The result is that the $0.500 - x$ term in the denominator becomes 0.500:

$$4.2 \times 10^{-4} = \frac{(x)(x)}{0.500}$$

Solving for x, we obtain

$$(4.2 \times 10^{-4})(0.500) = x^2$$
$$x = 0.0145$$

Once x is determined, we check the assumption. Another way to judge if our assumption worked is to see if x is less than 10% of the initial concentration. We find that it is valid since 0.500 is more than ten times larger than 0.0145. The rest of the equilibrium table is completed using the calculated value of x:

Reaction	CH_3NH_2	+	H_2O	$\rightleftharpoons$	$CH_3NH_3^+$	+	OH^-
Initial Conc.	0.500		55.5		0.00		0.00
Change	$-x$		$-x$		$+x$		$+x$
Equilibrium	$0.500 - x$		$55.5 - x$		$+x$		$+x$
Solution	0.486		55.5		0.0145		0.0145

Using the listed $[OH^-]$ we calculate pOH as 1.84, and the pH is 12.16. As with the weak acid, this entire process may be summarized in a simple equation:

$$[OH^-] = \sqrt{C_b K_b} \quad \text{as long as } C_b > 100\,K_b$$

In Exercises 13.3 and 13.4 there are two ways to check the accuracy of the results. First, the answers must be reasonable; a weak acid should have an acidic pH (below 7) and a weak base should have a basic pH (above 7). Second, the calculated values may be entered into the mass action expression, and the result should be close to the given K_a or K_b.

pH OF SALT SOLUTIONS; HYDROLYSIS REACTIONS

When an acid is neutralized with a base, a salt is formed. If the anion of a salt is the conjugate base of a weak acid, it will react with water in a **hydrolysis reaction**. For the fluoride ion the hydrolysis reaction is as follows:

$$F^-(aq) + H_2O(l) \rightleftharpoons HF(aq) + OH^-(aq) \tag{13.13}$$

The conjugate base, F^-, of the weak acid HF will produce a basic solution, as shown by the OH^- in Equation 13.13.

If the cation of a salt is the conjugate acid of a weak base, the hydrolysis reaction will result in an acid solution, as shown by the reaction of the ammonium ion:

$$NH_4^+(aq) + H_2O(l) \rightleftharpoons NH_3(aq) + H_3O^+(aq)$$

The conjugate acid of a strong base and the conjugate base of a strong acid are extremely weak acids and bases. In fact, they are so weak that they do not hydrolyze in water. For example, Cl^- from the strong acid HCl and Na^+ from the strong base NaOH do not hydrolyze in water.

ACIDITY OF SALT SOLUTIONS AND CLASSIFICATION OF SALTS

The cation of a salt will be the conjugate acid of either a strong base or a weak base and the anion of a salt will be the conjugate base of either a strong acid or a weak acid. Table 13.6 describes the different types of salts possible.

The first step in determining the pH of a salt solution is to determine the type of salt. This is done by adding H^+ to the anion and OH^- to the cation in the salt to determine the type of acid and base used to prepare the salt. For instance, ammonium chloride is composed of NH_4^+ and Cl^- ions. By adding OH^- to the cation and H^+ to the anion, NH_4OH ($NH_3 + H_2O$) and HCl are obtained. From this result we conclude that ammonium chloride is an acid salt prepared from a weak base and a strong acid. We may also quickly conclude that solutions of ammonium chloride will have pH values below 7.

TABLE 13.6

Classification of Salts Based on Acid and Base Used to Prepare Them

Acid Used	Base Used	Salt Type	Salt Solution Will Be
Strong	Strong	Neutral	Close to pH 7*
Strong	Weak	Acidic	Acidic
Weak	Strong	Basic	Basic
Weak	Weak	Mixed	Depends on K_a and K_b

*Remember that all water contains dissolved CO_2 that makes it slightly acidic.

A mixed salt has a cation that is the conjugate acid of a weak base, and an anion that is the conjugate base of a weak acid. The acidity or alkalinity of a mixed salt will be determined by whichever of these conjugates is stronger. Considering the K_a of the weak acid and the K_b of the weak base that form the salt, we conclude that the solution will be acidic if $K_b > K_a$ and basic if $K_a > K_b$.

Exercise 13.3

Determine whether the aqueous solutions of each of the following salts will be acid, neutral, or basic:

(a) $CaCl_2$	(c) Na_2SO_3	(e) SrF_2	(g) KNO_2
(b) $NaNO_3$	(d) $KC_2H_3O_2$	(f) NH_4Br	(h) Li_2CO_3

Solution

(a) Neutral	(c) Basic	(e) Basic	(g) Basic
(b) Neutral	(d) Basic	(f) Acid	(h) Basic

pH OF SALT SOLUTIONS

The pH of a neutral salt is 7.00. Although we can tell whether a mixed salt solution will be acidic or basic, calculation of the pH is left for higher level courses. We concern ourselves here with calculating the pH value of acidic and basic salts only.

Ammonium chloride was determined in the preceding section to be an acid salt. Since the chloride ion does not hydrolyze, we focus on the hydrolysis of the ammonium ion discussed above to develop the equilibrium law needed to solve problems.

NH_4^+ is the conjugate acid of ammonia, and the hydrolysis reaction is written as either

$$NH_4^+ + H_2O \rightleftharpoons NH_3 + H_3O^+$$

or

$$NH_4^+ \rightleftharpoons NH_3 + H^+$$

The equilibrium expression for this reaction is based on either of the two equivalent expressions:

$$K = \frac{[NH_3][H^+]}{[NH_4^+]} = \frac{[H_3O^+][NH_3]}{[NH_4^+]} \qquad (13.14)$$

The equilibrium constant K is equal to K_w/K_b, where K_b is the base dissociation constant for ammonia. We can demonstrate this relationship by multiplying the equilibrium expression in Equation 13.14 by $\frac{[OH^-]}{[OH^-]}$ to obtain

$$K = \left(\frac{[NH_3][H^+]}{[NH_4^+]}\right)\left(\frac{[OH^-]}{[OH^-]}\right) \qquad (13.15)$$

In the numerator the product $[H^+][OH^-] = K_w$, which we substitute into Equation 13.15:

$$K = K_w\left(\frac{[NH_3]}{[NH_4^+][OH^-]}\right)$$

The term in parentheses is the reciprocal of the equilibrium expression for the dissociation of ammonia, and its value is K_b. Performing the substitution yields

$$K = \frac{K_w}{K_b}$$

This value is sometimes called the hydrolysis constant for a salt, K_h, but a more accurate name is the acid dissociation constant of the Brønsted-Lowry conjugate acid, symbolized as K_a.

This derivation leads to a universally useful equation and concept. For a conjugate acid–base pair in aqueous solution, the K_a of the conjugate acid multiplied by the K_b of the conjugate base will always be equal to K_w:

$$K_w = K_a K_b$$

K_w is the ion product for the dissociation of water:

$$H_2O \rightleftharpoons H^+ + OH^-$$

and

$$K_w = [H^+][OH^-] = 1.00 \times 10^{-14}$$

From this derivation we see also that the salt of a weak base acts as a Brønsted-Lowry acid, and we expect its solutions to be acidic.

Once the equilibrium expression for acid and basic salts is derived, we can start solving problems. Let us calculate the pH of a 0.100 *M* solution of ammonium chloride. Since the equilibrium expression has already been described, the next step is to set up the equilibrium table:

Reaction	NH_4^+	$\rightleftharpoons$	NH_3	+	H^+
Initial Conc.	0.100		0.00		0.00
Change	$-x$		$+x$		$+x$
Equilibrium	$0.100 - x$		$+x$		$+x$
Solution					

Entering the values of K_w, K_b, and the terms from the Equilibrium line into the equilibrium expression yields

$$\frac{1.0 \times 10^{-14}}{1.8 \times 10^{-5}} = \frac{(x)(x)}{0.100 - x}$$

Assuming that $0.100 >> x$, we obtain

$$\frac{1.0 \times 10^{-14}}{1.8 \times 10^{-5}} = \frac{(x)(x)}{0.100}$$

which may be solved as

$$(5.6\times10^{-10})(0.100)=x^2$$

$$x=7.4\times10^{-6}$$

Since x is much smaller than 0.100, the assumption was valid. We now use the value of x to calculate the values in the Solution row of the table.

Reaction	NH_4^+	$\rightleftharpoons$	NH_3	+	H^+
Initial Conc.	0.100		0.00		0.00
Change	$-x$		$+x$		$+x$
Equilibrium	$0.100 - x$		$+x$		$+x$
Solution	0.100		7.4×10^{-6}		7.4×10^{-6}

The pH is then calculated to be

$$\text{pH}=-\log(7.4\times10^{-6})=5.13$$

This result is consistent with the concept that NH_4^+ is a conjugate acid of NH_3 and that the solution must be acidic.

To summarize this calculation in a simpler form, we write

$$[\text{H}^+]=\sqrt{\frac{K_w}{K_b}C_s}\ \text{provided that}\ C_s>100\frac{K_w}{K_b}$$

where K_b is the dissociation constant of the weak base used to produce the acidic salt and C_s is the molar concentration of the salt.

For basic salts, calculations are performed in a very similar manner. For example, NaF is a basic salt since it is formed by reacting the strong base NaOH with the weak acid HF. We may ignore Na^+ while writing the hydrolysis equation for F^- as follows:

$$\text{F}^-+\text{H}_2\text{O}\rightleftharpoons\text{HF}+\text{OH}^-$$

The base dissociation constant is calculated by dividing K_w by the acid dissociation constant, K_a, of HF:

$$K_b=\frac{K_w}{K_a}=\frac{[\text{HF}][\text{OH}^-]}{[\text{F}^-]}$$

Questions may then be answered by using the procedures outlined before. The general solution can be simplified to

$$[\text{OH}^-]=\sqrt{\frac{K_w}{K_a}C_s}\ \text{provided that}\ C_s>100\frac{K_w}{K_a}$$

where C_s is the initial concentration of the salt and K_a is the acid dissociation constant of HF.

The salt KNO_3 does not change the pH of the solution since it is formed from the strong acid HNO_3 and the strong base KOH. The K^+ and NO_3^- ions are such weak Brønsted-Lowry acids and bases that they have no effect on pH.

Two simple equations suffice for calculating the pH values of weak acid, weak base, acid salt, and basic salt solutions:

$$\text{weak acids and acid salts} \qquad [H^+] = \sqrt{K_a C_a} \qquad (13.16)$$

$$\text{weak bases and basic salts} \qquad [OH^-] = \sqrt{K_b C_b} \qquad (13.17)$$

For solutions of weak acids or weak bases, K_a and K_b are the acid and base dissociation constants. In salt solutions, K_a and K_b are the conjugate acid and base dissociation constants of the ions in the salt, which are calculated as K_w/K_b and K_w/K_a, respectively. Equations 13.16 and 13.17 work only when the initial concentration, *C*, is at least 100 times larger than the dissociation constant, *K*. Otherwise the problem must be solved using the quadratic equation as shown in Chapter 9.

Buffer Solutions

To this point we have considered solutions containing only pure weak acids, weak bases, or their salts dissolved in distilled water. A mixture that contains a conjugate acid–base pair is known as a **buffer solution** because the pH changes by a relatively small amount if a strong acid or base is added to it. Buffer solutions are used to control pH in many biological and chemical reactions. These solutions are governed by the same equilibrium law as are weak acids and weak bases.

For instance, a buffer can be prepared by dissolving 0.200 mole of HF and 0.100 mole of NaF in water to make 1.00 liter of solution. The dissociation reaction of HF is

$$HF(aq) \rightleftharpoons F^-(aq) + H^+(aq)$$

and its equilibrium expression is

$$K_a = \frac{[F^-][H^+]}{[HF]} = 6.8 \times 10^{-4}$$

An equilibrium table is set up using the given information. Since we have 0.100 mole NaF, the concentration of F^- is determined as 0.100 *M*. The Na^+ in NaF is a spectator ion in the dissociation reaction and is not needed.

Reaction	HF	⇌	F^-	+	H^+
Initial Conc.	0.200		0.100		0.00
Change	$-x$		$+x$		$+x$
Equilibrium	$0.200 - x$		$0.100 + x$		$+x$
Solution					

Substituting the information from the Equilibrium line into the equilibrium expression yields

$$K_a = \frac{(0.100 + x)(x)}{0.200 - x} = 6.8 \times 10^{-4}$$

Assuming that x is small compared to both 0.100 and 0.200, we simplify the equation to

$$K_a = \frac{(0.100)(x)}{0.200} = 6.8 \times 10^{-4}$$

Solving this equation for x yields

$$x = 1.36 \times 10^{-3}$$

This value of x satisfies the assumptions made, and the last line of the equilibrium table can be completed as:

Reaction	HF	$\rightleftharpoons$	F^-	+	H^+
Initial Conc.	0.200		0.100		0.00
Change	$-x$		$+x$		$+x$
Equilibrium	$0.200 - x$		$0.100 + x$		$+x$
Solution	**0.199**		**0.101**		**1.36×10^{-3}**

The last column in the table gives us $[H^+] = 1.36 \times 10^{-3}$ *M*. The pH of the buffer solution is calculated as 2.87. For almost all buffer solutions the assumptions hold true, and this type of problem is solved by simply entering the given concentrations of the conjugate acid and conjugate base directly into the equilibrium expression.

EXAMPLE 13.3

Calculate the pH of the following buffer solutions:

(a) 0.250 *M* ethanoic acid and 0.150 *M* sodium ethanoate
(b) A solution of 10.0 g each of formic acid and sodium formate dissolved in 1.00 L of H_2O
(c) 0.0345 *M* ethylamine and 0.0965 *M* ethyl ammonium chloride
(d) 0.125 *M* hydrazine and 0.321 *M* hydrazine hydrochloride

Solution

(a) $K_a = \frac{[C_2H_3O_2^-][H^+]}{[HC_2H_3O_2]}$. The given data tell us that $[HC_2H_3O_2] = 0.250$ *M* and $[C_2H_3O_2^-] =$ 0.150 *M*. Substituting these values into the equation yields

$$1.8 \times 10^{-3} = \frac{(0.150)[H^+]}{0.250}$$

$$[H^+] = 3.0 \times 10^{-3}, \text{ and pH} = 2.52$$

(b) The masses given allow us to calculate concentrations of 0.217 *M* formic acid and 0.147 *M* sodium formate. The equilibrium expression to use is $K_a = \frac{[CHO_2^-][H^+]}{[HCHO_2]}$.

Substituting data gives

$$1.8\times10^{-4} = \frac{(0.147)[H^+]}{0.217}$$

$$[H^+] = 2.7\times10^{-4}, \text{ and pH} = 3.57$$

(c) The equilibrium expression to use is $K_b = \frac{[C_2H_5NH_3^+][OH^-]}{[C_2H_5NH_2]}$. We are given

$[C_2H_5NH_2] = 0.0345\ M$ and $[C_2H_5NH_3^+] = 0.0965\ M$. Substituting these values yields

$$4.3\times10^{-4} = \frac{(0.0965)[OH^-]}{0.0345}$$

$$[OH^-] = 1.5\times10^{-4}, \text{ pOH} = 3.82, \text{ and pH} = 10.18$$

(d) The equilibrium expression to use is $K_b = \frac{[N_2H_5^+][OH^-]}{[N_2H_4]}$. We are given $[N_2H_4] =$

$0.125\ M$ and $[N_2H_5^+] = 0.321\ M$. Substituting these values yields

$$9.6\times10^{-7} = \frac{(0.321)[OH^-]}{0.125}$$

$$[OH^-] = 3.7\times10^{-7}, \text{ pOH} = 6.43, \text{ and pH} = 7.57$$

SHORTCUTS IN BUFFER CALCULATIONS

The equilibrium expression used for buffers involves a ratio of the concentrations of the conjugate acid and conjugate base. For a ratio, we may use the moles of conjugate acid and moles of conjugate base instead of the acid and base molarities, thereby eliminating a step or two in the calculations.

In addition, if the concentrations of the conjugate acid and conjugate base are equal, their ratio is exactly 1.00. The result is that, with equal molarities or equal numbers of moles of conjugate acid and conjugate base,

$$K_a = [H^+] \quad \text{and} \quad pK_a = \text{pH}$$

for a buffer made from a weak acid and its conjugate base, and

$$K_b = [OH^-] \quad \text{and} \quad pK_b = \text{pOH}$$

for a buffer made from a weak base and its conjugate acid.

PREPARATION OF BUFFERS

Preparing a buffer requires several important decisions about the buffer and the system that needs to be buffered. The following steps indicate the general process for preparing a buffer.

1. Decide what pH is required.
2. Decide what final volume is required.

3. Based on Step 1, choose a conjugate acid–base system using tables such as those in Appendices 4, 5, and 6.
4. Determine the moles per liter of acid or base that your reaction will generate.
5. Based on Step 4, the sum of the concentrations of the conjugate acid and conjugate base should be approximately 20 times the value estimated in Step 4.
6. Based on Steps 1, 3, and 5, calculate the separate concentrations of the conjugate acid and base needed.
7. Use Steps 2 and 6 to determine the masses of the conjugate acid and base to use.
8. Measure out the amounts determined in Step 7, dissolve in distilled water, and dilute to the volume determined in Step 2.

There are two alternate approaches to preparing a buffer. In the first approach, based on Steps 2 and 4, assume that all of our buffer material will be the conjugate acid. Then calculate the amount of strong base needed to neutralize enough of the acid to form the conjugate base. Mix the conjugate acid and the strong base to make the buffer.

In the second approach, based on Steps 2 and 4, assume that all of our buffer material will be the conjugate base. Then calculate the amount of strong acid needed to convert part of the conjugate base to the required amount of conjugate acid. Now mix the conjugate base with the desired amount of strong acid to prepare the buffer.

Henderson-Hasselbalch Equation

The equilibrium expression for a weak acid, HA, is written as

$$K_a = \frac{[A^-][H^+]}{[HA]} \tag{13.18}$$

Taking the logarithm of both sides of Equation 13.18 yields

$$pK_a = pH - \log\left(\frac{[A^-]}{[HA]}\right)$$

Rearrangement yields

$$pH = pK_a + \log\left(\frac{[A^-]}{[HA]}\right) = pK_a + \log\left(\frac{[\text{conjugate base}]}{[\text{conjugate acid}]}\right) \tag{13.19}$$

Equation 13.19 is known as the **Henderson-Hasselbalch equation**. A similar derivation for weak bases results in

$$pOH = pK_b + \log\left(\frac{[\text{conjugate acid}]}{[\text{conjugate base}]}\right) \tag{13.20}$$

Equations 13.19 and 13.20 illustrate that, if the pH of a buffer is known, the ratio of the conjugate acid and conjugate base concentrations can be calculated. In addition, they show clearly that, when the concentrations of conjugate acid and conjugate base are equal, $pH = pK_a$ or $pOH = pK_b$.

The symbol *A*, along with the appropriate negative charge, symbolizes an anion.

pH Values of Polyprotic Acids and Their Salts

Polyprotic acids dissociate in stepwise equilibria, each of which has its own dissociation constant. Each dissociation constant is smaller than the preceding one ($K_{a1} > K_{a2} > K_{a3}$). When a polyprotic acid such as H_2A dissolves in water, the possible reactions are as follows:

$$H_2A \rightleftharpoons HA^- + H^+$$

$$HA^- \rightleftharpoons A^{2-} + H^+$$

Since K_{a2} is always smaller than K_{a1}, the second dissociation must occur to a lesser extent than the first dissociation. In addition, the first dissociation produces hydrogen ions, and these hydrogen ions further suppress the second dissociation according to Le Châtelier's principle. These two considerations, along with experimental evidence, demonstrate that the first dissociation step is the only important one in determining the $[H^+]$ and pH of a solution of a polyprotic acid.

The equation to use is

$$[H^+] = \sqrt{K_{a1}C_a} \qquad (13.21)$$

where C_a is the initial concentration of polyprotic acid.

Exercise 13.4

What is the pH value of each of the following solutions?

(a) 0.800 M H_3PO_4
(b) 0.0200 M H_2S
(c) 0.00400 M H_2CO_3

Solution

(a) 1.12
(b) 4.35
(c) 4.37

Similar reasoning is used for the hydrolysis of the fully deprotonated anion of a polyprotic acid, such as A^{2-} in Equation 13.21. The two hydrolysis reactions are these:

$$A^{2-} + H_2O \rightleftharpoons HA^- + OH^- \qquad (13.22)$$

$$HA^- + H_2O \rightleftharpoons H_2A + OH^-$$

The hydrolysis reaction of Equation 13.22 has the largest equilibrium constant, and the OH^- formed in the first step suppresses the second step according to Le Châtelier's principle. The hydrolysis of a fully deprotonated anion of a polyprotic acid is calculated using the equation

$$[OH^-] = \sqrt{\frac{K_w}{K_{a2}}C_{A^{2-}}}$$

Exercise 13.5

What is the pH of each of the following solutions?

(a) 3.00 M Na_3PO_4
(b) 0.0500 M Na_2CO_3

Solution

(a) 13.41
(b) 11.51

Anions of polyprotic acids, such as HA^- seen recently, are amphiprotic and can act as either conjugate acids or conjugate bases. For salts that contain these anions the hydrogen ion concentration is calculated as the square root of the product of the K_a values for that anion acting as a conjugate acid and as a conjugate base:

$$[H^+] = \sqrt{K_{a1}K_{a2}}$$

Exercise 13.6

Calculate the pH of a solution of each of the following ions:

(a) HSO_3^-
(b) $H_2PO_4^-$
(c) HPO_4^{2-}
(d) HCO_3^-
(e) HS^-

Solution

(a) 4.55
(b) 4.67
(c) 9.77
(d) 8.34
(e) 10.00

Buffer solutions are often made using a conjugate acid and a conjugate base of a polyprotic acid. The pH of such a buffer is governed by the equilibrium expression, which includes both the conjugate acid and the conjugate base present in the solution. For phosphoric acid the first dissociation step involves H_3PO_4 and $H_2PO_4^-$, and buffers in the range of pH 1.1–3.1 can be prepared from various mixtures of them. The second dissociation step involves $H_2PO_4^-$ and HPO_4^{2-}, and salts with these two ions are used to prepare buffers in the range of pH 6.2–8.2. With the HPO_4^{2-} and PO_4^{3-} anions buffers in the pH range 11.3–13.3 can be prepared.

TIP

End point **is the experimentally determined end of a titration.**

Equivalence point **is the theoretical or calculated end of a titration.**

Titration Curves

The titration technique was described in Chapter 5, along with the important calculations that apply to a wide variety of titrations. Acid–base pH calculations are often used to explain what occurs as the experiment is performed. There are four major points of interest during a titration experiment:

1. The start of the titration, where the solution contains only one acid or base.
2. The region where titrant is added up to the equivalence point, and the solution now contains a mixture of unreacted sample and products.
3. The equivalence point, where all of the reactant has been converted into product.
4. The region after the equivalence point, where the solution contains product and excess titrant.

At the start of the titration the sample is a pure acid or base, and the pH is calculated as described in preceding sections. At the equivalence point, the solution is the salt of the acid or base; again, the method of calculating the pH has been described. Between the start and the equivalence point, however, we have a mixture of a conjugate acid and its conjugate base. If a weak acid or base is involved, this is a buffer solution. If only strong acids and bases are used, this region is unbuffered. After the end point, the pH depends on the excess titrant used.

The pH during a titration may be calculated or measured in an experiment. A plot of pH versus volume of titrant added is called a **titration curve**.

The general shapes of strong acid–strong base titration curves are shown in Figure 13.1. There are three important points about these curves. First, there is no buffer region since no weak acids or bases are present. Second, the end-point pH is always 7.0 because the products do not hydrolyze. Third, from the start the pH changes gradually until just before the end point where it changes sharply.

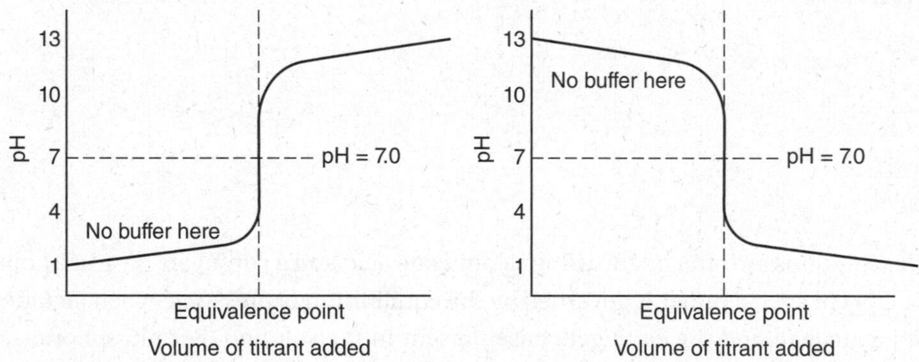

FIGURE 13.1. Titration curves for (left) a strong acid titrated with a strong base and (right) a strong base titrated with a strong acid. The region from the start to the equivalence point is not buffered. The equivalence-point pH is always 7.0.

Comparing Figure 13.1 with Figure 13.2, we find that weak acid and base titrations have a buffer region where a conjugate acid–base pair exists. In the middle of the buffer region, $pH = pK_a$. A large change in pH occurs just at the start as the buffer mixture is formed, in contrast to strong acid–base titrations. At the equivalence point the salts of weak acids and bases hydrolyze, and the pH is not 7. Salts of weak acids hydrolyze to give basic solutions, and $pH_{ep} > 7$; salts of weak bases will give acidic solutions, $pH_{ep} < 7$.

Titrations of polyprotic acids produce more than one equivalence point because of the sequence of dissociation steps. The titration of phosphoric acid with a strong base is shown in Figure 13.3. Only two end points are observed since the third occurs at a pH too high for observation. At the start, only pure H_3PO_4 is present. At each equivalence point the acid is neutralized to pure $H_2PO_4^-$ and then HPO_4^{2-}; the pH values at the equivalence points are calculated using Equation 13.18. Between the equivalence points there are mixtures of a conjugate acid and its conjugate base, as shown. These are the buffer regions. At the midpoint of each buffer region the pH is equal to the pK_a that governs that particular dissociation step.

Figures 13.2 and 13.3 illustrate why and how a buffer can be made by partial neutralization of a weak acid. The partial neutralization results in a conjugate acid–base mixture, which is a buffer solution.

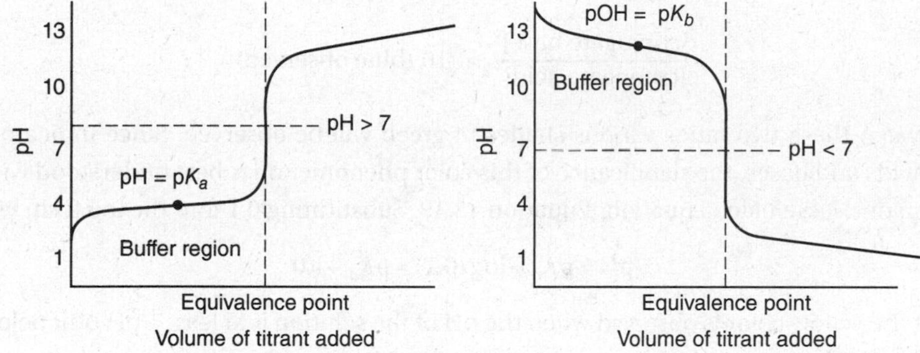

FIGURE 13.2. Titration curves for (left) a weak acid titrated with a strong base and (right) a weak base titrated with a strong acid. Between the start and the equivalence point is a buffer region due to the presence of a conjugate acid–base mixture. Because of the hydrolysis of the products, the equivalence point pH is not 7.0.

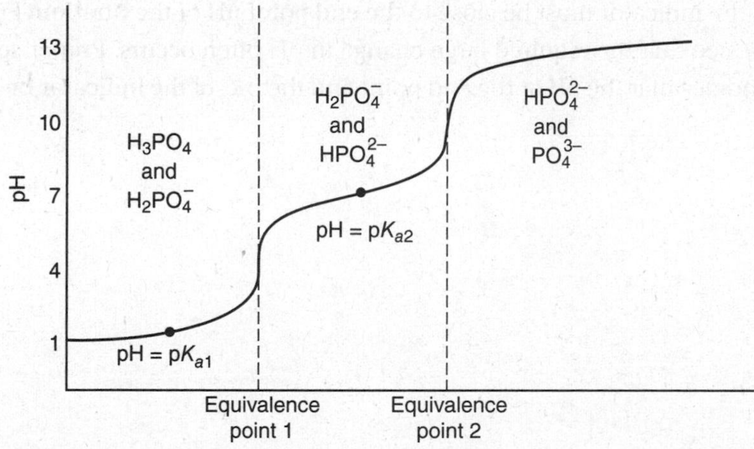

FIGURE 13.3. Titration curve for the titration of phosphoric acid with a strong base.

pH Indicators

Indicators are weak acids and weak bases whose respective conjugate bases and conjugate acids have different colors. The reason is that the loss or gain of a proton changes the energy of the electrons within their structures. This, in turn, changes the energy of light absorbed, which is observed as a change in color.

If a conjugate acid has one color, yellow for instance, and its conjugate base has another color, blue for instance, the eye will see these colors clearly only if there is ten times more of one color than of the other. We will see yellow if the conjugate acid is ten times more concentrated than the conjugate base, and blue if the conjugate base is ten times more concentrated than the conjugate acid:

$$\frac{[\text{conjugate base}]}{[\text{conjugate acid}]} < 0.1 \text{ (yellow observed)}$$

$$\frac{[\text{conjugate base}]}{[\text{conjugate acid}]} > 10 \text{ (blue observed)}$$

Between these two ratios various shades of green will be observed. Since indicators are weak acids and bases, the significance of this color phenomenon is best understood with the Henderson-Hasselbalch equation, Equation 13.19. Substituting 0.1 into the log term yields

$$\text{pH} = \text{p}K_a + \log(0.1) = \text{p}K_a - 1.0$$

so that the yellow color is observed when the pH of the solution is at least 1 pH unit below the pK_a of the indicator. For the blue color

$$\text{pH} = \text{p}K_a + \log(10) = \text{p}K_a + 1.0$$

and the pH of the solution must be 1 pH unit above the pK_a of the indicator.

This discussion tells us two important properties of indicators used in titrations: (1) the pH at the end point of a titration curve must change by at least 2 pH units very rapidly, and (2) the pK_a of the indicator must be close to the end point pH of the titration. From the titration curves we see that the required large change in pH often occurs. Proper selection of an indicator mandates that the pH at the end point and the pK_a of the indicator be close to each other.

SUMMARY

If redox reactions are considered one major class of chemical reactions, the other major group consists of acid–base chemistry. This chapter reviews the concepts of the two major theories: the Arrhenius theory and the Brønsted-Lowry theory. Each theory makes the definition of acids and bases more general, and there will always be questions concerning these theories on the AP exam. This chapter also reviews the nomenclature applied to acids and bases as well as the relationship between molecular structure, acid–base strengths, and the periodic table.

The quantitative aspects concerning the pH of acid and base solutions are covered in this chapter. The pH of salts, polyprotic acids, and highly charged cations is reviewed. In addition, the concept and quantitative aspects of buffer solutions are covered. The various problems concerning acid–base chemistry are reviewed, and methods for solving problems are presented.

Important Concepts

Arrhenius theory
Brønsted-Lowry theory
Neutralization
Titration
Buffers
Strong vs weak acids and bases
Predicting relative strengths of acids
Hydrolysis
Complexation reactions

Important Equations

$$pH = -\log [H^+] \text{ and } pOH = -\log [OH^-]$$

$$pH = pK_a + \log\left(\frac{[\text{conjugate base}]}{[\text{conjugate acid}]}\right)$$

$$K_a = \frac{[H^+][A^-]}{[HA]}$$

$$K_b = \frac{[HB^+][OH^-]}{[B]}$$

$$K_w = [H^+][OH^-]$$

PRACTICE EXERCISES

Multiple-Choice

1. Which of the following can be made by partially neutralizing an acid?
   (A) BaClOH by neutralizing $BaCl_2$
   (B) KOH by neutralizing H_2O with $K_2Cr_2O_7$
   (C) $NaHCO_3$ by neutralizing H_2CO_3 with NaOH
   (D) $MgCO_3$ by neutralizing H_2CO_3 with $Mg(OH)_2$

2. A solution containing HF is titrated with KOH. At the end point of the titration, the solution contains
   (A) Equal amounts of HF and KOH
   (B) KF and H_20
   (C) K+ and F^-
   (D) H_2O, H^+, OH^-, K+, F^-, and HF

3. A buffer at pH 5.32 is prepared from a weak acid with a $pK_a = 5.15$. If 100 mL of this buffer is diluted to 200 mL with distilled water, what is the pH of the dilute solution?
   (A) 5.32
   (B) 5.02
   (C) 5.62
   (D) The identity of the acid is needed to answer the question.

4. When an equal number of moles of each pair is mixed to make an aqueous solution, which of those solutions can be called a buffer?
   (A) $Cu(OH)_2$ and $CuCl_2$
   (B) KOH and $NaHCO_3$
   (C) $LiHCO_3$ and K_2CO_3
   (D) Na_2CO_3 and H_2SO_3

5. Which of the following has the highest pH?
   (A) 0.100 *M* HCl
   (B) 0.200 *M* $HC_2H_3O_2$
   (C) 0.100 *M* Na_2CO_3
   (D) 0.200 *M* NaCl

6. Which of the following CANNOT occur together in solution?
   (A) H_3PO_4 and $H_2PO_4^-$
   (B) HCO_3^- and CO_3^{2-}
   (C) Na^+ and SO_4^{2-}
   (D) $C_2O_4^{2-}$ and $H_2C_2O_4$

7. When 0.250 mol of NaOH is added to 1.00 L of 0.100 M H_3PO_4, the solution will contain
   (A) HPO_4^{2-}
   (B) PO_4^{3-}
   (C) A and B
   (D) NaOH

8. A buffer with a pH of 10.0 is needed. Which of the following should be used?
   (A) Ethanoic acid with a K_a of 1.8×10^{-5}
   (B) Ammonia with a K_b of 1.8×10^{-5}
   (C) Nitrous acid with a K_a of 7.1×10^{-4}
   (D) $H_2PO_4^-$ and PO_4^{3-} with a K_a of 4.5×10^{-13}

9. pH is equal to pK_a
   (A) when [conjugate acid] = [conjugate base]
   (B) at the end point of a titration
   (C) in the buffer region
   (D) in the Henderson-Hasselbalch equation

10. The pH of a 0.125 M solution of a newly synthesized weak base is 10.45. What is the K_b of this base?
   (A) 3.5×10^{-11}
   (B) 2.8×10^{-4}
   (C) 6.4×10^{-7}
   (D) 2.3×10^{-3}

11. Which of the following is the acid anhydride of a monoprotic acid?
   (A) CaO
   (B) SO_3
   (C) N_2O_5
   (D) CO_2

12. Which of the following is expected to result in a solution with the lowest pH? Assume a 0.100 M solution of the nitrate salt of each of these ions.
   (A) Sr^{2+}
   (B) Cu^{2+}
   (C) Fe^{3+}
   (D) Na^+

13. Which of the following statements is correct?
   (A) $HClO_2$ is a stronger acid than $HClO_3$.
   (B) HI is a weaker acid than HCl.
   (C) CH_3COOH is a stronger acid than $CH_2BrCOOH$.
   (D) HNO_3 is a stronger acid than HNO_2.

14. What is the pH of a 0.100 *M* solution of K_2HPO_4? (For H_3PO_4, $pK_1 = 2.15$; $pK_2 = 7.20$; $pK_3 = 12.35$.)
(A) 1.00
(B) 13.00
(C) 9.78
(D) 6.67

15. Which of the following is the correct method for preparing a buffer solution?
(A) Mix the correct amounts of a weak acid and its conjugate base.
(B) Neutralize a weak base partially with strong acid.
(C) Neutralize a weak acid partially with a strong base.
(D) All of the above methods may be used to prepare buffers.

Use the titration curve below to answer questions 16 to 21.

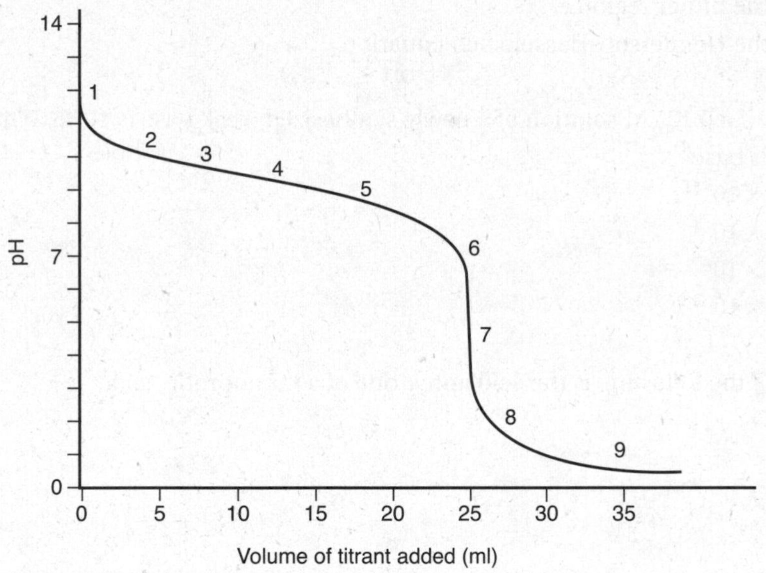

16. What is the titrant and what is the analyte in the experiment that resulted in the titration curve above?
(A) The titrant is a strong acid, and the analyte is a strong base.
(B) The titrant is a weak base, and the analyte is a weak acid.
(C) The titrant is a strong base, and the analyte is a weak acid.
(D) The titrant is a strong acid, and the analyte is a weak base.

17. What is the pH and volume of titrant at the end point?
(A) pH = 7.00; end point = 25.0 mL
(B) pH = 7.0; end point = 12.5 mL
(C) pH = 4.8; end point = 25.0 mL
(D) pH = 5; end point = 25 mL

18. At which point(s) does the analyte flask contain a buffer solution?
    (A) At points 2, 3, 4, and 5
    (B) At points 2, 3, 4, 5, and 6
    (C) Only at point 4
    (D) Only at point 7

19. How many moles of analyte were in the analyte flask if the molarity of the titrant was predetermined to be 0.0855 *M*?
    (A) 2.34 mol
    (B) 0.0021 mol
    (C) 0.002138 mol
    (D) 2.13 mol

20. What is true about the the *pK* of the analyte?
    (A) $pK_a = 9.2$
    (B) $pK_b = 9.2$
    (C) $pK_b = 4.8$
    (D) $pK_a = 11.7$

21. What is characteristic of the end point for this type of experiment?
    (A) It occurs when a color change occurs.
    (B) It occurs at the inflection point of the curve.
    (C) It occurs when the first derivative is at a minimum.
    (D) It occurs at exactly pH 7.

---

22. What is a primary standard used for?
    (A) It is used to calibrate a pH meter.
    (B) It is used to prepare calibration curves for spectroscopy.
    (C) It is used to standardize titrants.
    (D) It is the first standard that is prepared.

CHALLENGE

23. If 50.0 mL of a 0.0134 *M* HCl solution is mixed with 24.0 mL of a 0.0250 *M* NaOH solution, what is the pH of the final mixture?
    (A) 1.87
    (B) 12.40
    (C) 5.29
    (D) 3.02

CHALLENGE

24. If 50.0 g of formic acid ($HCHO_2$, $K_a = 1.8 \times 10^{-4}$) and 30.0 g of sodium formate ($NaCHO_2$) are dissolved to make 500 mL of solution, the pH of this solution is
(A) 4.76
(B) 3.76
(C) 3.35
(D) 4.12

ANSWER KEY

1. **(C)**	7. **(C)**	13. **(D)**	19. **(B)**
2. **(D)**	8. **(B)**	14. **(C)**	20. **(B)**
3. **(A)**	9. **(A)**	15. **(D)**	21. **(B)**
4. **(C)**	10. **(C)**	16. **(D)**	22. **(C)**
5. **(C)**	11. **(C)**	17. **(C)**	23. **(D)**
6. **(D)**	12. **(C)**	18. **(A)**	24. **(C)**

See Appendix 1 for explanations of answers.

Free-Response

Propanoic acid, CH_3CH_2COOH, has an acid ionization constant of $K_a = 1.3 \times 10^{-5}$. Ammonia, NH_3, has a base ionization constant of $K_b = 1.8 \times 10^{-5}$.

(a) Write the balanced equation for the reaction of propanoic acid, CH_3CH_2COOH, with ammonia.

(b) Identify the two conjugate acid–base pairs.

(c) Does the equilibrium position lie on the reactant or the product side of the equation when equal moles of propanoic acid and ammonia are mixed? Justify your choice.

(d) Of the two acids in your equation, which is stronger and why?

(e) Of the two bases in your equation, which is stronger and why?

(f) Using appropriate scientific reasoning, state why you would conclude that one binary acid is stronger than another. Give a relevant example.

(g) Using appropriate scientific reasoning, state why you would conclude that one oxoacid is stronger than another. Give a relevant example

ANSWERS

(a) $CH_3CH_2COOH(aq) + NH_3(aq) \rightleftharpoons CH_3CH_2COO^-(aq) + NH_4^+(aq)$

(b) Pair 1: $CH_3CH_2COOH(aq)$ and $CH_3CH_2COO^-(aq)$
Pair 2: $NH_3(aq)$ and $NH_4^+(aq)$

(c) This is an acid–base reaction and virtually all the propanoic acid reacts with the ammonia so the position of equilibrium lies far to the right. The solution most closely resembles a solution of ammonium propanoate ($NH_4CH_3CH_2CO_2$).

(d) Propanoic acid is a stronger acid than the ammonium ion because the K_a for the propanoic acid is larger than the K_a for the ammonium ion, which is

$$K_w/K_b = 5.56 \times 10^{-10}$$

(e) Ammonia is the stronger base since the K_b for ammonia is larger than the K_b for propanate ion, which is $K_w/K_a = 7.69 \times 10^{-12}$.

(f) Binary acids contain hydrogen ions and one other element, for example HCl and H_2S. From left to right across the periodic table, the electronegativity of the non-hydrogen atom governs the acid strength. The larger the electronegativity is, the lower the electron density is between the two atoms and the stronger the acid is.

When comparing binary acids in a periodic table group, the charges are all the same and have little effect on acid strength. However, the distance between the two atoms increases going down a period because a new shell of electrons is added for each period.

(g) Oxoacids contain hydrogen, oxygen, and one other element, usually a nonmetal. Lewis structures show that the oxygen is always bound to the nonmetal and hydrogens are always bound to the oxygen atoms. Hydrogen is not bound to the nonmetal.

If two oxoacids have the same number of oxygen atoms, such as $HClO_4$, $HBrO_4$, and H_3PO_4, the only difference in the attraction of the anion to the hydrogen ion is the central atom. The central atom with the largest electronegativity weakens the O–H bond the most, resulting in the strongest acid.

If two oxoacids have differing numbers of oxygen atoms, such as $HClO_4$, $HClO_3$, $HClO_2$, and HClO, the one with the most oxygen atoms will be the stronger acid. First, the attraction of n oxygen atoms toward the electrons in the O–H bond is greater then n-1 or n-2 oxygen atoms bound to the same central atom. Thus more oxygen atoms result in a weaker O–H bond and a stronger acid. The second reason is that the anion left after the H^+ leaves is more stable (due to delocalization of electrons) with the greater number of oxygen atoms. The more stable the anion, the less stable (and therefore stronger) the acid.

Experimental Chemistry 14

- → DATA GATHERING
- → QUALITATIVE/ QUANTITATIVE
- → ACCURACY AND PRECISION
- → SIGNIFICANT FIGURES
- → UNCERTAINTY
- → ROUNDING
- → GRAPHS
- → INDEPENDENT VARIABLE
- → DEPENDENT VARIABLE
- → READING GRADUATED SCALES
- → DETERMINING MASS
- → VOLUME MEASUREMENT
- → TEMPERATURE MEASUREMENT
- → DENSITY
- → SPECIFIC HEAT
- → SAMPLE MANIPULATION
- → SEPARATION TECHNIQUES
- → INSTRUMENTS
- → pH METERS
- → SPECTROMETERS
- → REACTIONS
- → QUALITATIVE ANALYSIS
- → CHEMICAL HAZARDS
- → SAFETY IN THE LABORATORY

BIG IDEAS 1, 2, 3, 4, 5, 6
Learning Objectives: 1.15, 1.16, 1.19, 1.20, 2.6, 2.7, 2.10, 2.22, 3.3, 3.5, 3.9, 4.1, 5.7, 6.9, 6.18

For the complete list of Big Ideas and Learning Objectives, refer to the AP Chemistry Course Outline: *https://secure-media.collegeboard.org/digitalServices/pdf/ap/ap-chemistry-course-and-exam-description.pdf*

The advanced placement program requires a laboratory component. After all, chemistry is an experimental science. ALL of the theories and concepts of chemistry, and other sciences, derive from experimental observations. The AP program lists 11 experiments that are reasonable for the AP course. You should be familiar with the experiments you did during your course. You should be able to describe the purpose of an experiment, the theory of the experiment, and the techniques and equipment used for the experiment. This section of the review guide will not rehash the actual experiments you did. Instead, it is organized to help you review the important techniques and experiences of the laboratory.

DATA GATHERING

Information gained in chemical experiments may be **quantitative**, that is, numerical, or it may not involve numbers, in which case the data are **qualitative** observations. If the concentration of an acid is determined to be 0.345 *M*, a quantitative measurement has been made. When silver nitrate is added to a solution and a white precipitate forms, the result is a qualitative observation. Both types of information are important in any experiment.

All observations and experimental details must be recorded in a **notebook**. The notebook must contain a complete description of the experiment so that any knowledgeable chemist can accurately repeat it. In particular, the description must specify the general idea of the experiment, along with the equipment and chemicals used. Diagrams are often drawn to show how an apparatus is constructed. Detailed information on the mass, volume, and source of every chemical used in the experiment is recorded. When the experiment is performed, careful observations of the reaction are noted as the experiment progresses. Finally, calculations are performed on the data gathered and conclusions are drawn. These conclusions help the chemist design the next experiment.

All information about an experiment is recorded directly in the notebook, which should be a bound, not loose-leaf or spiral, book with numbered pages. Entries are made in ink, and every entry is dated. Data should never be written on scraps of paper and then transcribed later. Erasures or removal of pages is not acceptable. Errors are crossed out in such a way that they are still readable. Although neatness is desirable, it is much more important to have a continuous record of all laboratory activity, whether a particular experiment is successful or not.

CALCULATIONS

Most calculations in chemistry involve simple algebra and two basic approaches. The first is the use of dimensional analysis to convert information from one set of units into another. The second is the use of a memorized equation or law into which data for all variables except one are inserted. The one remaining variable is the unknown for the problem. Correct use of the second method requires that all units be shown to ensure that they cancel properly to obtain the desired units for the numerical answer.

Scientific calculators simplify mathematical operations to the touch of a few keys. Understanding the principles and concepts of chemistry enables us to decide in which order to press those keys. Understanding numbers tells us how to properly interpret the answer that appears on the calculator screen.

Accuracy and Precision

The **accuracy** of a measurement refers to the closeness between the measurement obtained and the true value. Since scientists rarely know the true value, it is generally impossible to determine the accuracy with certainty. One approach to evaluating accuracy is to make a measurement by two completely independent methods. If the results from independent measurements agree, scientists have more confidence in the accuracy of their results.

Accuracy is affected by **determinate errors**, that is, errors due to poor technique or incorrectly calibrated instruments. Careful evaluation of an experiment may eliminate determinate errors.

Precision refers to the closeness of repeated measurements to each other. If the mass of an object is determined as 35.43 grams, 35.41 grams, and 35.44 grams in three measurements, the results may be considered precise. There is no guarantee that they are accurate, however, unless the balance was properly calibrated and the correct methods were used in weighing the object. When proper techniques are used, precise results infer, but do not guarantee, accurate results.

Precision is also a measure of **indeterminate errors**, that is, errors that arise in estimating the last, uncertain, digit of a measurement. Indeterminate errors are random errors and cannot be eliminated. Statistical analysis deals with the theory of random errors.

Significant Figures

Every experimental measurement is made in such a way as to obtain the most information possible from whatever instrument is used. As a result, measurements involve numbers in which the last digit is uncertain to some extent. Scientists characterize the precision of a measured number based on the number of **significant figures** it contains.

The number of significant figures in a measurement includes all digits that are not zeros. All zeros to the left of the last nonzero digit are not significant. Imbedded zeros, those between two nonzero digits, are always significant. Trailing zeros are significant only if a decimal point is somewhere in the number. For exponential numbers, the number of significant figures is determined from the digits to the left of the multiplication sign. For example, 8.32×10^3 has three significant figures.

SIGNIFICANT FIGURES

Digits ARE significant if:

1. the digit is not zero.
2. the zero is embedded.
3. the trailing zero is in a number that has a decimal point.

Digits ARE NOT significant if the zeros are to the left of all nonzero digits.

Sometimes there are trailing zeros on the right side of a number. If the number contains a decimal point, trailing zeros are always significant. Trailing zeros that are used to complete a number, however, may or may not be significant. Scientists avoid writing a number such as 12,000 since it does not definitely indicate the number of significant figures. Scientific notation is used instead. Twelve thousand can be written as 1.2×10^4, 1.20×10^4, 1.200×10^4, or 1.2000×10^4, indicating two, three, four, or five significant figures, respectively. It is the responsibility of the experimenter to write numbers in such a way that there is no ambiguity.

Exercise 14.1

Determine the number of significant figures in each of the following measured values:

(a) 23.46 mL
(b) 0.0036 s
(c) 854.236 g
(d) 6.02×10^{23} molecules
(e) 0.98 mol
(f) 0023 m
(g) 1.00026×10^{-3} cm
(h) 2.0000 J
(i) 824 mg

Solution

The numbers are repeated with the significant figures in bold type:

(a) **23.46** mL
(b) 0.00**36** s
(c) **854.236** g
(d) **6.02** $\times 10^{23}$ molecules
(e) 0.**98** mol
(f) 00**23** m
(g) **1.00026** $\times 10^{-3}$ cm
(h) **2.0000** J
(i) **824** mg

Some numbers are **exact numbers**, which involve no uncertainty. The number of plates set on a table for dinner may be determined exactly. If five plates are observed and counted, that measurement is exactly 5. In chemistry many defined equalities are exact. For instance, there are exactly 4.184 joules in each calorie. Other exact numbers are stoichiometric coefficients and subscripts in chemical formulas.

If you divide 1 by 3 on your calculator, the answer shown will be 0.333333333. How can dividing two numbers with one significant figure each give us a result with nine significant figures? The answer is that the result is faulty. To understand how to obtain and write correct answers, we need to count the numbers of significant figures, as we did above, and then apply the two simplified statistical rules described below.

1. The number with the **fewest significant figures** in a multiplication or division problem determines the number of significant figures in the answer. In these calculations an exact number is considered to have an infinite number of significant figures.
2. The number with the **fewest decimal places** in an addition or subtraction problem determines the number of decimal places in the answer. Numbers expressed in scientific notation must all be converted to the same power of 10 before determining which decimal places can be retained.

Uncertainty

There are two types of uncertainty, **absolute uncertainty** and **relative uncertainty**.

The absolute uncertainty is the uncertainty of the last digit of a measurement. For example, 45.47 mL is a measurement of volume, and the last digit is uncertain. The absolute uncertainty is ±0.01 mL. The measurement should be regarded as somewhere between 45.46 and 45.48 mL. The absolute uncertainty should have only one digit. The last digit of the number should be the first digit of the uncertainty; for example, 35.38 ± 0.02 mL.

The relative uncertainty of a number is the absolute uncertainty divided by the number itself. For the above example, the relative uncertainty is

$$\frac{0.01\text{ mL}}{45.47\text{ mL}} = 2\times10^{-4}$$

The absolute uncertainty governs the principles used for addition and subtraction. The relative uncertainty governs the principles used for multiplication and division.

Rounding

Calculations, especially those done using an electronic calculator, often generate more, and sometimes fewer, significant figures or decimal places than are required by rules 1 and 2 given above. These answers must be rounded to the proper number of significant figures or decimal places. To do this, four steps are followed:

1. The number of digits to be kept in a calculation is determined using rules 1 and/or 2 above.
2. If the digit just after the kept digits is less than 5, the remaining digits are dropped. For example, rounding 6.23499 to three significant figures yields 6.23 because 4 is less than 5.
3. If the digit just after the kept digit is 5 or more, the last kept digit is increased by 1. For example, rounding 34.2**5589** to three significant figures yields 34.3. Similarly, rounding 8.44**5000** to three significant figures yields 8.45.

Exercise 14.2

Perform each of the following calculations, and report the answer with the correct number of significant figures:

(a) $23.456 + 16.0094 + 9.21$

(b) $14.98 \times 0.00234 \times 1.5$

(c) $1.46 \times 10^3 - 5.83 \times 10^4$

(d) $(8.236 \times 10^2)\ (5.55 \times 10^{-3})$

(e) $(23.45 - 16.12)/6.233$

(f) $(6.02 \times 10^{23})\ (1.00 \times 10^{-6})/(18.23)$

(g) $(44.23)/(2.33 \times 10^2 - 2.25 \times 10^2)$

Solution

(a) 48.68

(b) 0.053

(c) -5.68×10^{-4}

(d) 4.57

(e) 1.18

(f) 3.30×10^{16}

(g) 6

Significant Figures in Atomic and Molar Masses

A quick look at a list of atomic masses of the elements reveals that only seven elements have atomic masses with four significant figures and that lead is the only element with one decimal place in its atomic mass. This means that six elements and most compounds containing lead will have atomic/molar masses with four significant figures. The remaining elements and some 50 million compounds will have atomic/molar masses with more than five significant figures. In other words, only in rare cases will atomic or molar masses affect the number of significant figures in a calculation.

When working problems with atomic or molar masses, you should first determine the number of significant figures the answer must have. (Remember that the factor with the fewest significant figures defines the number of significant figures in the answer.) *Be sure your answer has the correct number of significant figures.*

GRAPHS

A **graph** is used to illustrate the relationship between two variables. Graphs are often a more effective method of communication than tables of data. A graph has two axes: a horizontal axis, usually called the x-axis (abscissa), and a vertical axis, called the y-axis (ordinate).

It is customary to use the x-axis for the **independent variable**, and the y-axis for the **dependent variable**, in an experiment. An independent variable is one that the experimenter selects. For instance, concentrations of standard solutions that a chemist prepares are independent variables since any concentrations may be chosen. The dependent variable is a measured property of the independent variable. For instance, the dependent variable may be the amount of light that each of the standard solutions prepared by the chemist absorbs, since the absorbed light is dependent on the concentration. Each data point is an x, y pair representing the value of the independent variable and the value of the dependent variable as determined in the experiment.

The first step in constructing a graph is to label the *x*- and *y*-axes to indicate the identity of the independent and dependent variables. Next, the axes are numbered, usually from zero to the largest value expected for each variable. Finally, each data point is plotted by drawing a horizontal line at the value of the dependent variable and a vertical line upward from the value of the independent variable. The intersection of these two lines determines where that data point belongs on the graph.

Most graphs show a linear relationship between two variables. Other graphs, such as those showing kinetic curves, have curved lines. In both cases, data points are plotted on the graph and then the best smooth line is drawn through the points. Lines are never drawn by connecting the data points with straight lines. In very accurate work a statistical analysis called the "method of least squares" is used to determine the best straight line for the data. In most cases, however, the line is drawn by eye, attempting to have all data points as close as possible to the line. The usual result is a line that has the same number of data points above and below it.

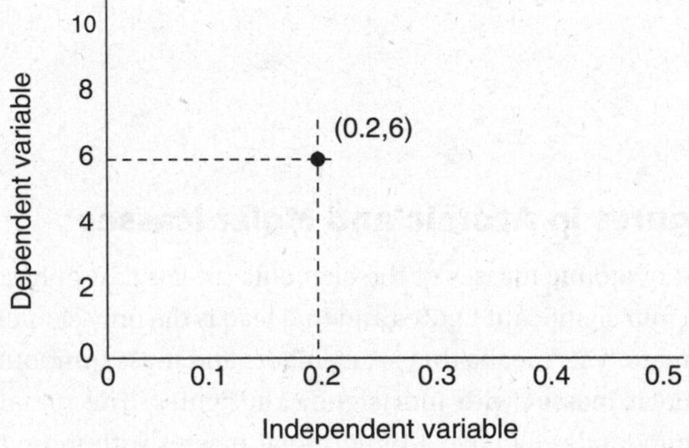

FIGURE 14.1. The positions of the independent and dependent variables in a graph. Dashed lines show the placement of a point representing a value of 0.2 for the independent variable and of 6 for the dependent variable.

Figure 14.2 illustrates that it is incorrect to draw any line beyond the measured data points. The reason is that anything beyond the measured data is unknown. Extending the line implies information that is not verified by experimental data.

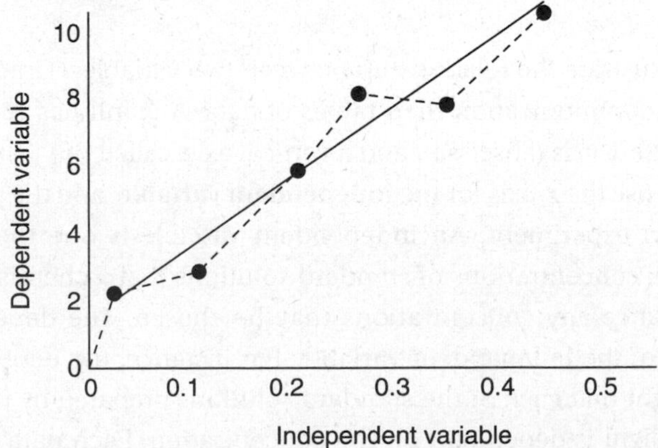

FIGURE 14.2. The correct way to draw a line using data in a graph. The solid line is the correct line; the dashed line is incorrect.

The slope of a curve or line is often needed as an experimental result. To determine the slope of a line, two points on the line are chosen. The left-hand point has coordinates (x_1,y_1), and the right-hand point has coordinates (x_2,y_2). The values of x and y at these points are determined from the graph, and Equation 14.1 is used to determine the slope:

$$\text{Slope} = \frac{y_2 - y_1}{x_2 - x_1} \qquad (14.1)$$

In a graph with a curved line the slope is determined by drawing a tangent to the curve and then determining the slope of the tangent, as is done for a straight line.

DETERMINATION OF PHYSICAL PROPERTIES

Scale Reading

Many measurements are made by comparing the level of a liquid in a container to a scale etched on the outside of the container or by observing the position of a meter pointer with reference to an adjacent scale. Such a scale is usually a series of lines in which every tenth line is numbered and is usually distinguished also by being longer than the others. The fluid level or meter pointer is never directly in contact with the scale; consequently, incorrect reading techniques can result in **parallax errors**.

The surface tension of a liquid in any container causes the liquid to have a curved surface called a **meniscus**. All glassware is calibrated on the basis that the liquid level corresponds to the bottom of the meniscus. Figure 14.3 demonstrates that, to avoid parallax errors, the eye must be at the same level as the meniscus when measuring liquids.

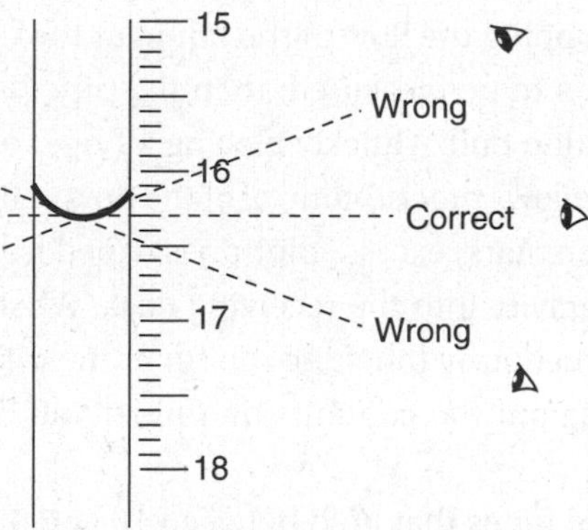

FIGURE 14.3. The correct way to read liquid levels. Parallax error results when the eye, scale, and meniscus are not lined up horizontally.

A modern meter usually has a mirror adjacent to the scale. Correct meter readings, with no parallax error, are obtained when the meter pointer and its reflection in the mirror coincide.

Determination of Mass by Weighing

The mass of a chemical substance is obtained by determining the weight of the substance. The weight of a sample is equal to the force of gravity times the mass of the sample:

$$\text{Weight} = (\text{Gravitational constant})(\text{Mass})$$

If the gravitational constant is known, the mass can be calculated from the measured weight. To avoid this calculation, the mass of a sample may be directly compared to a known mass on a double-pan balance. Modern balances are single-pan balances that are calibrated with standard masses.

Proper use of a balance requires that the balance be calibrated. A sample is always weighed in an appropriate container, never directly on a balance pan. The mass of the empty container is referred to as the tare mass. A sample's mass is the difference between the tare mass and the mass of the sample and the container. A typical notebook entry for a mass determination should look like this:

24.345 g total mass
–3.862 g tare mass

20.483 g sample mass

Liquid Volume Measurement

Graduated cylinders and graduated beakers are used to measure liquid volumes with an accuracy of ±10 percent. More accurate liquid measurements are made with **pipets** or **burets**. Glassware of this type is usually labeled with the letters "**TD**," which stand for "to deliver" and indicate that the amount of liquid poured or delivered from the pipet or buret is the volume stated. **Volumetric flasks** have the label "**TC**," which stands for "to contain." Although a 100-mL volumetric flask will contain 100 mL, it will not deliver 100 mL if the liquid is poured out. The difference is the film of solution left on the inner surface of the flask itself.

Pipets come in two types, transfer and measuring. A transfer pipet has a single mark and is used for the most accurate measurements. Measuring pipets are graduated, and any volume may be delivered by stopping the flow of the liquid at the desired point. Before filling, a pipet is rinsed in the solution to be measured; then the pipet is filled above the calibration line with suction from a suction bulb. Quickly placing a finger over the top of the pipet after the suction bulb is removed allows precise control of the flow of liquid. The solution is allowed to drain to the first calibration mark, excess solution is wiped from the tip, and then the solution is allowed to drain by gravity into the receiving flask. When draining is completed, the last drop is removed by momentarily touching the tip of the pipet to the solution surface or the side of the flask. Blowing out the contents of a pipet will deliver the wrong amount of solution.

Burets are long, graduated tubes that hold between 10 and 50 mL of solution. The flow of liquid is controlled by a valve called a **stopcock**. A buret is rinsed with solution, not distilled water, before filling. When filling is completed, the stopcock is opened fully to expel any air from the tip; and, if needed, additional solution is added so that the liquid is near the top graduation mark. In use, the starting volume is recorded, solution is delivered, and, when finished, the final volume is recorded. The difference between the final and initial volumes is the volume delivered. Here is a sample notebook entry:

23.86 mL at the end
–0.23 mL at the start

23.63 mL solution delivered

Temperature Measurement

Thermometers are used for temperature measurement. Each thermometer has a line etched near the mercury bulb. This line is the immersion depth. When the thermometer is immersed in a liquid to this etched line, the temperature reading will be the most accurate. The calibration of a thermometer is checked by immersing it in ice water, 0°C, and then in boiling water, 100°C.

Broken mercury thermometers must be disposed of carefully because of the hazards posed by elemental mercury.

Determination of Melting and Boiling Points

With sufficient liquid, the boiling point is determined by measuring the temperature as the liquid boils. Accurate determination of the normal boiling point requires that this experiment be done at 1 atmosphere of pressure.

Several instruments are available for determining melting points. For the experiment, a sample is packed into a closed-end capillary tube and inserted in the instrument, which is set to heat the sample slowly. The temperature at which the first crystals start melting is the melting point. The same result can be obtained by attaching the capillary tube to a thermometer so that the sample is next to the mercury bulb. The thermometer and attached sample are immersed in an oil with a high boiling point, and heat is applied until the sample just starts to melt.

Determination of Density

Density is an intrinsic physical property that can be used to identify unknown materials. The general equation for determining density is as follows:

$$\text{Density} = \frac{\text{Mass}}{\text{Volume}}$$

Equations 14.2 and 14.3 show the units used for mass and volume for substances in various phases:

$$\text{Density of solids or liquids} = \frac{\text{grams of material}}{\text{cubic centimeters of material}} \qquad (14.2)$$

$$\text{Density of gases} = \frac{\text{grams of gas}}{\text{liters of gas}} \qquad (14.3)$$

Since density varies somewhat with temperature, we often see the symbol d_{20}, where the subscript indicates the Celsius temperature at which the density was determined.

The density of a solid is determined by measuring first its mass and then its volume. The mass measurement was described in a preceding section. There are two methods to determine the volume of a solid. The first is to measure its dimensions and use trigonometry to calculate the volume. This method works best with a regularly shaped object such as a cylinder or a rectangular solid. The second method is to immerse the solid in a liquid and measure the volume displaced. This is commonly done by placing water in a graduated cylinder and measuring its volume. Then the solid is added and a second volume is determined. The difference in volumes is the volume of the object:

$$\text{Volume of object} = V_{\text{final}} - V_{\text{initial}}$$

Dividing the mass by the volume yields the density.

The density of a liquid is determined using a device called a **pycnometer**, which is a small flask with a volume of approximately 5–25 mL. The exact volume is obtained by determining the mass of water needed to completely fill the pycnometer. Since the density of water is known, the volume of the pycnometer can be calculated. Next, the mass of unknown liquid needed to fill the pycnometer is determined. The density is then calculated using Equation 14.2.

The density of a gas is determined by evacuating a large flask with a vacuum pump so that there is virtually no gas inside the flask. The evacuated flask is weighed, and the gas is introduced until its pressure is equal to atmospheric pressure. The flask is then weighed again to obtain the mass of the gas. The volume of the flask is determined from the amount of water it can hold or by some similar technique, and the density is determined using Equation 14.3.

Determination of Specific Heat

The specific heat of a solid, usually a metal, is determined by heating a known mass of the substance to a predetermined temperature and then submerging it in a known quantity of water in an insulated container. The final temperature of the water indicates the temperature increase of the water and the temperature decrease of the metal. The specific heat equation is as follows:

$$q = (\text{Mass})(\text{Specific heat})(\Delta T)$$

where ΔT is the temperature change.

Since the heat, q, gained by the water must be equal to the heat lost by the metal, the equality becomes

$$(\text{Mass})_{\text{water}}(\text{Specific heat})_{\text{water}}(\Delta T)_{\text{water}} = -(\text{Mass})_{\text{metal}}(\text{Specific heat})_{\text{metal}}(\Delta T)_{\text{metal}}$$

An example of specific heat calculation is given in Chapter 11.

SAMPLE MANIPULATIONS

Heating

Heating can be done in several ways, and the method chosen depends on the equipment on hand and safety factors. Water and aqueous solutions are usually heated with a Bunsen burner. Precautions should be taken to avoid **bumping**, which is a violent burst of boiling that may spatter hot liquid. The best way to avoid bumping is to add boiling chips to the mixture. Liquids heated in test tubes are very likely to bump, and care should be exercised that test tubes are not pointed toward others. Burners with open flames should be avoided when any flammable substances are in use in the laboratory.

Bumping occurs because a burner flame superheats one portion of a liquid. Hot-water baths, steam baths, sand baths, and electric heating mantles are effective heating methods that minimize bumping by spreading the applied heat. The fact that the limit for steam and hot water is approximately 100°C may be a safety feature.

Bumping also occurs when solids, particularly powders, are added to very hot liquids. For this reason solids should be added to cool liquids before heating.

Cooling

Hot solutions or objects are cooled with water or ice. A bath made of crushed ice and water is most effective. Ice by itself is not efficient since much of the flask containing the hot solution is not in contact with the ice. Only heat-resistant laboratory glassware should be used since ordinary glass will shatter with rapid temperature changes.

Lower temperatures (approx. –50°C) may be obtained using dry ice mixtures, and very low temperatures (–196°C) with liquid nitrogen. Dry ice and liquid nitrogen can cause injury, however, because of their extremely low temperatures. They also present a smaller, although real, hazard of possible suffocation.

Mixing

Preparation of solutions is the most common mixing operation in chemistry. A solid is dissolved in a solvent by adding the solid slowly, with stirring, to about half of the liquid. When dissolution is complete, the correct amount of liquid is added and mixed well. Grinding the solid to a powder and warming the mixture both speed dissolution.

Preparation of solutions with molar concentration units requires an exact total volume of solution. These solutions are prepared in **volumetric flasks**. A volumetric flask is calibrated with an etched line on its neck to hold a specified volume at a given temperature. If a solution is warmed to speed dissolution, it must be cooled to room temperature before the final volume adjustment.

When a concentrated acid is mixed with water, the acid is always added to the water. Most acids, particularly sulfuric acid, are denser than water and generate a large amount of heat when mixed with water. If water is added to sulfuric acid, it does not mix because of the density of the acid, and the high heat of mixing causes the water to boil and spatter sulfuric acid.

Drying

Drying chemical reagents before use and drying products of reactions are very common operations in the laboratory. Since the drying process involves removing water from a substance, the temperature must be above 100°C. To avoid decomposition, however, the temperature should also be as low as possible. An oven set between 105°C and 110°C is recommended.

Dilution

A common practice in chemistry laboratories is to prepare a concentrated stock solution of some solute. To make other solutions, the stock solution is diluted with water in the appropriate volume ratio. The dilution law is

$$(C_{\text{initial}})(V_{\text{initial}}) = (C_{\text{final}})(V_{\text{final}})$$

If we have a stock solution of some concentration, $C_{initial}$, we can calculate the volume needed to prepare a given volume at any other concentration.

Exercise 14.3

Calculate the volume of a stock 6.00 *M* HCl solution needed to make 1.00 L of a 0.100 *M* HCl solution.

Solution

We can define as follows: $C_{initial} = 6.00\ M$ HCl, $C_{final} = 0.100\ M$ HCl, $V_{final} = 1.00$ L. Entering these values into the dilution law equation yields

$$(6.00\ M\ \text{HCl})(V_{initial}) = (0.100\ M\ \text{HCl})(1.00\ \text{L})$$
$$V_{initial} = \frac{(0.100\ M\ \text{HCl})(1.00\ \text{L})}{6.00\ M\ \text{HCl}}$$
$$= 0.0166\ \text{L} = 16.6\ \text{mL}$$

To prepare the desired solution, 16.6 mL of the stock solution must be diluted to 1.00 L.

Exercise 14.4

How many milliliters of distilled water must be added to 100 mL of 0.250 *M* KCl to prepare a 0.100 *M* KCl solution?

Solution

The given values are as follows: $C_{initial} = 0.250\ M$ KCl, $V_{initial} = 100$ mL, $C_{final} = 0.100\ M$ KCl. From these data we calculate V_{final}:

$$(0.250\ M\ \text{KCl})(100\ \text{mL}) = (0.100\ M\ \text{KCl})(V_{final})$$
$$V_{final} = 250\ \text{mL}$$

Since the final volume of solution is the sum of the initial volume and the added distilled water:

$$V_{final} = V_{initial} + V_{water}$$
$$250\ \text{mL} = 100\ \text{mL} + V_{water}$$
$$V_{water} = 150\ \text{mL}$$

Gas Collection

When a gas is generated in a chemical reaction, it may be collected in a **pneumatic trough**, shown in Figure 14.4. A gas-collecting bottle is filled with water, and a tube from the sealed experiment leads under water to the bottle. Any gas evolved displaces the water in the bottle. Details on how to measure the amount of gas collected are described in Chapter 6.

Gases that react with water, however, cannot be collected by displacement of water. Instead, air in a container is displaced by the gas. A gas that has a density greater than the density of air will displace air from the bottom to the top of an upright gas bottle. This process is called upward displacement. When the gas density is less than the density of air, the collecting bottle is inverted and the air is displaced downward as the gas fills the bottle from the top to the bottom. Figure 14.5 illustrates the upward and downward displacement of gas.

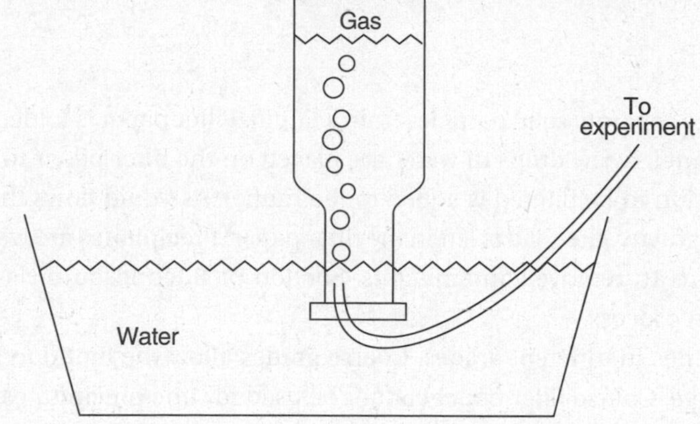

FIGURE 14.4. Diagram of a pneumatic trough.

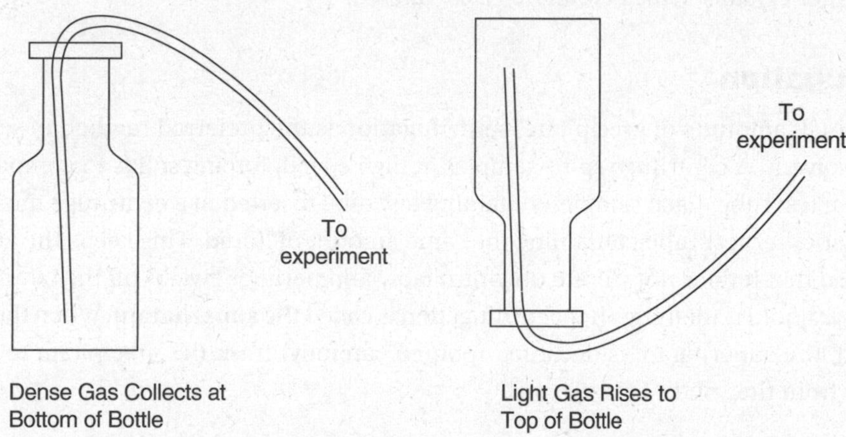

FIGURE 14.5. Experimental setups for (left) upward displacement of gas and (right) downward displacement of gas.

The density of a gas is directly proportional to its molar mass. The average molar mass of air, which is 80 percent nitrogen and 20 percent oxygen, is 29. Gases with molar masses greater than 29 are collected by upward displacement of air; gases with molar masses less than 29, by downward displacement of air.

SEPARATION TECHNIQUES

Precipitation

Solids can be precipitated by chemical reactions such as using Ba^{2+} to **precipitate** sulfate ions as barium sulfate:

$$Ba^{2+}(aq) + SO_4^{2-}(aq) \rightarrow BaSO_4(s)$$

To obtain a pure product the barium solution should be added slowly with rapid stirring. This procedure keeps the barium ion concentration low and the formation of barium sulfate crystals is slowed, thereby preventing the barium sulfate from entrapping impurities during rapid crystal growth. Once precipitation is complete, the mixture is heated to coagulate the small crystals into larger crystals for easy filtration.

Filtration

Filtration is used to separate solid particles from a liquid. Filter paper is folded into a cone and inserted into a funnel. A few drops of water are placed on the filter paper to hold it in place, and then the solution to be filtered is added to the funnel. As liquid flows through the paper under the force of gravity, the solid is left in the filter paper. Precipitates are washed with dilute electrolyte solutions to remove contaminants. Suction or Buchner funnels may be used to speed the filtration process.

Filter paper comes in different grades. Coarse grades allow the liquid to flow faster since the pore size is large. Coarse filter paper cannot be used for fine precipitates, which will pass through the pores. Fine precipitates require filter paper with smaller pores and consequently take longer to filter. Fine precipitates are often heated in a process called digestion in order to form larger crystals, which are more easily filtered.

Centrifugation

For very small amounts of precipitate, **centrifugation** is the preferred method of separation from the solvent. A centrifuge spins samples at high speed, forcing solids to compact at the bottom of a test tube. Each sample-containing test tube inserted in a centrifuge must be balanced by another test tube containing the same amount of liquid. This keeps the centrifuge balanced so that it does not vibrate uncontrollably and perhaps "walk" off the bench.

The clear liquid remaining after centrifugation is called the **supernatant**. When the process is finished, the supernatant is decanted (poured carefully) from the precipitate to separate the liquid from the solid.

Distillation

Distillation is a separation technique whereby a substance with a high vapor pressure (low boiling point) is separated from other substances with lower vapor pressures (higher boiling points). Distillation is performed by boiling a solution and passing the vapor formed through a condenser to recover the vaporized liquid. When a mixture of methyl alcohol (b.p. = 65°C) and ethyl alcohol (b.p. = 79°C) is heated to boiling, the vapor formed contains mostly methyl alcohol. When this vapor is condensed, the condensate is enriched in methyl alcohol while the residual in the boiling flask is enriched in ethyl alcohol. Two volatile substances may be purified but not completely separated.

A mixture of a salt in water is an example of a mixture containing a volatile and a nonvolatile substance. This type of mixture allows complete separation of the volatile substance, as in the distillation of seawater to produce pure water.

INSTRUMENTAL TECHNIQUES

pH Determination

A **pH meter** or **pH paper** may be used to determine pH. A pH meter is used when a precise pH value is needed. pH paper serves to estimate the pH of a solution.

A pH meter, with a glass and reference electrode, is standardized with a standard buffer. Then the electrodes are rinsed to remove any buffer and immersed in the sample. The pH is read directly from the meter scale. More accurate pH measurements are made by standardizing the pH meter with two buffers, one with a pH slightly below that of the sample and the other with a pH slightly above that of the sample.

Litmus is a type of pH paper that tells only whether a solution is acid or basic. A clean stirring rod is dipped into the sample, and a drop is transferred to the **litmus paper**. A pink color indicates an acid solution, and a blue color a basic solution. Newer pH papers turn different colors depending on the pH. A drop of sample is placed on the pH paper with a stirring rod, and the resulting color is compared to a color chart to determine the approximate pH.

Spectroscopy

Colored solutions absorb visible light, and this absorption can be measured with a **spectrophotometer**. A spectrophotometer can be used to determine the spectrum of a compound or to determine the concentration of an unknown sample.

The **absorbance** of a sample is the quantity determined in a spectrophotometer. In the first step the wavelength of light to be used for the measurement is selected by adjusting the wavelength dial. The second step is to zero the meter with a **reagent blank**, which contains the same amounts of all reagents and solvents that were used to prepare the sample. In the last step the reagent blank is replaced with the actual sample and the meter is read. More samples can be measured as long as the wavelength is not changed. If the wavelength is changed, the instrument must be zeroed again with the reagent blank.

A spectrum is a graph of the light absorbed by a sample, that is, the absorbance, A, of the sample, versus the wavelength of light. Typically the absorbance of a sample is measured at 10-nm intervals from approximately 400 nm to 700 nm. An absorbance spectrum in the visible-wavelength region is shown in Figure 14.6.

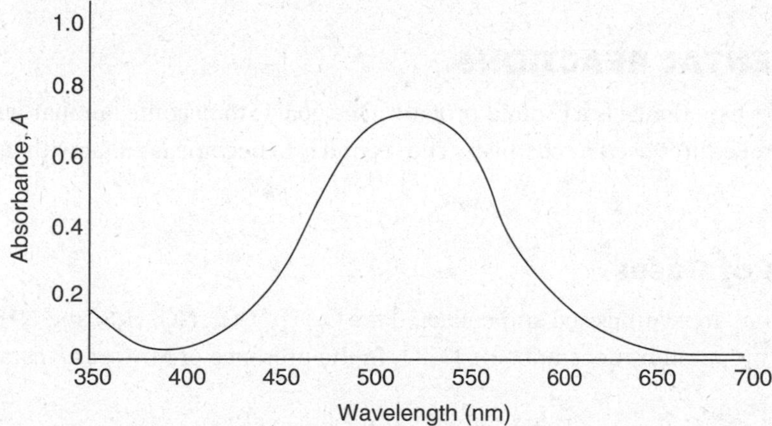

FIGURE 14.6. Absorbance spectrum for a compound that has its maximum absorbance at 515 nm. The sample has a violet color because green light is absorbed and all other wavelengths are observed by eye.

The concentration of a solute is determined at the wavelength of the largest peak in the spectrum. The absorbance of each solution of a series of concentration standards is determined, and a graph called a calibration curve is constructed. After the absorbance of the unknown sample is determined, the corresponding concentration is read from the calibration curve by drawing a horizontal line from the absorbance reading to the curve and then drawing a vertical line from the curve down to the concentration, as shown in Figure 14.7.

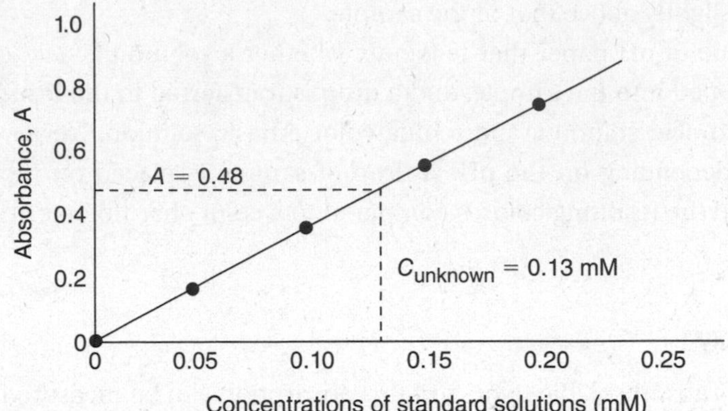

FIGURE 14.7. Calibration curve used to determine concentrations. Dots represent standard concentrations used to prepare the curve. Dashed line shows that the unknown has an absorbance of 0.48 and that the corresponding concentration is 0.13 millimolar.

The straight-line graph in Figure 14.7 is a consequence of **Beer's law**, which states that

$$\text{Absorbance (A)} = abc$$

In Beer's law, *a* is a constant called the **absorptivity**, which is characteristic of the compound and the wavelength at which the absorbance is measured. The concentration is represented by *c*, and the thickness of the sample, called the **optical path length**, by *b*. At a given wavelength, *a* and *b* are constant and there is a direct proportionality between the absorbance and the sample concentration. The slope of the line in Figure 14.7 is *ab*.

EXPERIMENTAL REACTIONS

The laboratory experience is a twofold process. One goal is to become familiar with the techniques and procedures used by chemists. The second is to become familiar with useful chemical reactions.

Synthesis of Gases

Some gases that are synthesized and collected are O_2, H_2, CO_2, NO, H_2S, and NH_3.

Oxygen is prepared by decomposing $KClO_3$ in the presence of MnO_2 as a catalyst:

$$2KClO_3(s) \xrightarrow[MnO_2]{heat} 2KCl(s) + 3O_2(g)$$

Hydrogen is prepared by reacting an active metal, usually magnesium, with acid:

$$Mg(s) + 2HCl(aq) \rightarrow MgCl_2(aq) + 2H_2(g)$$

Carbon dioxide is prepared by reacting a carbonate salt with an acid. The reaction of $CaCO_3$ with HCl is one example:

$$CaCO_3(s)+2HCl(aq) \rightarrow CaCl_2(aq)+H_2O(l)+CO_2(g) \tag{14.4}$$

Nitric oxide is produced from the reaction of copper with dilute HNO_3:

$$3Cu(s)+8HNO_3(dil.\ aq) \rightarrow 3Cu(NO_3)_2(aq)+4H_2O+2NO(g)$$

Hydrogen sulfide is produced by reacting an acid with a sulfide, for example:

$$FeS(s)+2HCl(aq) \rightarrow FeCl_2(aq)+H_2S(g)$$

Ammonia is produced from the reaction of an ammonium salt with a base:

$$NH_4Cl(aq)+NaOH(aq) \rightarrow NaCl(aq)+H_2O(l)+NH_3(g)$$

As stated in a preceding section, gases that do not dissolve in, or react with, water may be collected by displacement of water in a pneumatic trough. Other gases must be collected in the absence of water. Carbon dioxide is 1.5 times as dense as air and can be collected by upward displacement of air from a collecting bottle. Ammonia, which is much lighter than air, can be collected by downward displacement of air from an inverted collecting bottle.

Synthesis of Insoluble Salts

Insoluble salts are often precipitated in double-replacement reactions or in reactions of gases with soluble substances. Some common compounds prepared by double replacement are shown in the following net ionic equations:

$$Ag^+(aq)+Cl^-(aq) \rightarrow AgCl(s) \tag{14.5}$$

$$Ba^{2+}(aq)+SO_4^{2-}(aq) \rightarrow BaSO_4(s)$$

$$Fe^{2+}(aq)+SO^{2-}(aq) \rightarrow FeS(s) \tag{14.6}$$

In Equation 14.5, the chloride ion may be replaced by an iodide or a bromide ion. In Equation 14.6 Fe^{2+} can be replaced by almost any metal ion except the alkali metals.

Insoluble carbonates and sulfites are formed by bubbling $CO_2(g)$ and $SO_2(g)$ through solutions containing aqueous metal ions, for example:

$$Ca^{2+}(aq)+CO_2(g)+H_2O(l) \rightarrow CaCO_3(aq)+2H^+(aq)$$

$$Ni^{2+}(aq)+SO_2(g)+H_2O(l) \rightarrow NiSO_3(s)+2H^+(aq)$$

Preparation of Soluble Salts

Soluble salts are isolated from aqueous solution by removing water by evaporation with or without applying heat. To obtain a pure sample of salt, the solution must contain only the ions of that salt.

Magnesium nitrate can be prepared by reacting magnesium metal with HNO_3 until no more hydrogen is evolved. HNO_3 is the limiting reactant, and excess magnesium is removed by filtration. The filtrate can be boiled to dryness to recover $Mg(NO_3)_2(s)$:

$$Mg(s, \text{excess}) + HNO_3(aq) \rightarrow Mg(NO_3)_2(aq) + H_2(g)$$

The preparation of calcium chloride uses the same reaction as the preparation of $CO_2(g)$ in Equation 14.4. Care is taken to avoid reacting all of the $CaCO_3(s)$ with HCl, and the reaction mixture is filtered to remove excess $CaCO_3(s)$. The solution that is left contains only $Ca^{2+}(aq)$ and $Cl^-(aq)$ ions, which can be boiled to dryness to recover the $CaCl_2(s)$.

Addition of an acid to an excess of an insoluble base anhydride such as $Fe_2O_3(s)$ can be used to produce soluble salts. Adding H_2SO_4 to an excess of $Fe_2O_3(s)$ will produce $Fe_2(SO_4)_3$:

$$Fe_2O_3(s) + 3H_2SO_4(aq) \rightarrow Fe_2(SO_4)_3(aq) + 3H_2O(l)$$

Reacting a base with an excess of an acid anhydride can also be used to prepare a soluble salt, as in the reaction

$$2NaOH(aq) + O_2(g) \rightarrow Na_2SO_3(aq) + H_2O(l)$$

Soluble salts can be prepared by neutralization reactions as long as neither an excess of acid nor an excess of base is used in the reaction. For instance, LiBr may be prepared by reacting exactly 1 mole of LiOH for every mole of HBr in the reaction

$$LiOH(aq) + HBr(aq) \rightarrow LiBr(aq) + H_2O(l)$$

This method can be used to prepare three different sodium salts from phosphoric acid. NaH_2PO_4 is formed when 1 mole of NaOH is reacted with 1 mole of H_3PO_4. Two moles of NaOH will produce Na_2HPO_4, and 3 moles of NaOH per mole of H_3PO_4 produces Na_3PO_4.

Synthesis of Organic Compounds

A common synthetic organic reaction is the formation of esters by the acid-catalyzed reaction of an alcohol and an organic acid. The reaction of ethanoic acid with 1-pentanol produces pentyl ethanoate, which has the odor of bananas:

$$CH_3COOH + CH_3CH_2CH_2CH_2CH_2OH \xrightarrow{H^+} CH_3CO_2CH_2CH_2CH_2CH_2CH_3 + H_2O$$

Another common reaction is the formation of aspirin, acctylsalicylic acid, from salicylic acid and ethanoic acid. Although the anhydride of ethanoic acid is used in this reaction, the equation can be written as

$$C_6H_4(OH)COOH + CH_3COOH \xrightarrow[H^+]{heat} C_6H_4(O{-}C(=O){-}CH_3)COOH + H_2O$$

QUALITATIVE ANALYSIS OF INORGANIC IONS

Qualitative analysis techniques are used to determine whether or not a sample contains a certain ion. In qualitative analysis an unknown and a reagent are mixed, and the result of the reaction allows us to draw a logical conclusion about the presence or absence of ions in the unknown. Many ions react in a similar manner; and although the addition of one reagent to an unknown may not identify the ion, it limits the possibilities. A sequence of reactions used to analyze a sample is called the qualitative analysis scheme—qual-scheme for short.

The qual-scheme uses selective precipitation to separate the ions in a sample into groups on the basis of their chemical characteristics listed below. When separation is by filtration, the liquid is called the **filtrate**. When the precipitate is separated by centrifugation, the liquid phase is called the **supernatant**.

Table 14.1 lists the main groups in the qual-scheme, starting with the unknown solution. The analysis starts with the sample, called the unknown, and group 1. Precipitates of the ions in each group are removed by filtration or centrifugation before the next group is precipitated.

TABLE 14.1
The Qualitative Analysis Scheme

Group Number	Solution Tested	Precipitating Agent	Precipitated Compounds
1	Unknown	0.1 *M* HCl	$PbCl_2$, Hg_2Cl_2, AgCl
2	Filtrate or supernatant from group 1	H_2S at pH 1	HgS, PbS, CuS, CdS, Bi_2S_3, As_2S_3, Sb_2S_3, SnS_2
3	Filtrate or supernatant from group 2	H_2S at pH 10	MnS, FeS, NiS, CoS, ZnS, $Fe(OH)_3$, $Al(OH)_3$, $Cr(OH)_3$
4	Filtrate or supernatant from group 3	CO_3^{2-} at pH 10	$MgCO_3$, $CaCO_3$, $SrCO_3$, $BaCO_3$
5	Unknown	None	Soluble ions Na^+, K^+, NH_4^+

Separation of chlorides is done first since there are only a few insoluble chlorides, which make a convenient group. The carbonates are separated last since all of the metal ions would precipitate as carbonates, and thus too many ions would precipitate, if this step were done earlier in the scheme. Groups 2 and 3 are mainly sulfides. In acid solution only sulfides with very low K_{sp} values precipitate because the concentration of S^{2-} is low in acid solution. In basic solution the sulfide concentration is much higher, and the more soluble sulfides, with relatively high K_{sp} values, precipitate along with some hydroxides.

After each group has precipitated, additional separations and confirmation tests are run to identify the individual ions. For example, in group 1 the precipitate is washed with hot water to dissolve only $PbCl_2$, which is then confirmed by another precipitation with the chromate ion, CrO_4^{2-}. If a precipitate remains after washing with hot water, ammonia is added to dissolve the silver as the complex ion $Ag(NH_3)_2^+$. Neutralization of the ammonia solution with HCl reprecipitates AgCl and confirms its presence. Any precipitate that is still left when the Ag^+ is dissolved in ammonia is most likely Hg_2Cl_2. In the presence of ammonia Hg_2Cl_2 undergoes a redox reaction to form elemental mercury, causing the precipitate to turn dark gray in the reaction

$$Hg_2Cl_2(s) + 2NH_3(aq) \rightarrow Hg(l) + HgNH_2Cl(s) + NH_4^+(aq) + Cl^-(aq)$$

The reactions used and the decision-making process for qualitative analysis are often summarized in a flowchart, as shown in Figure 14.8.

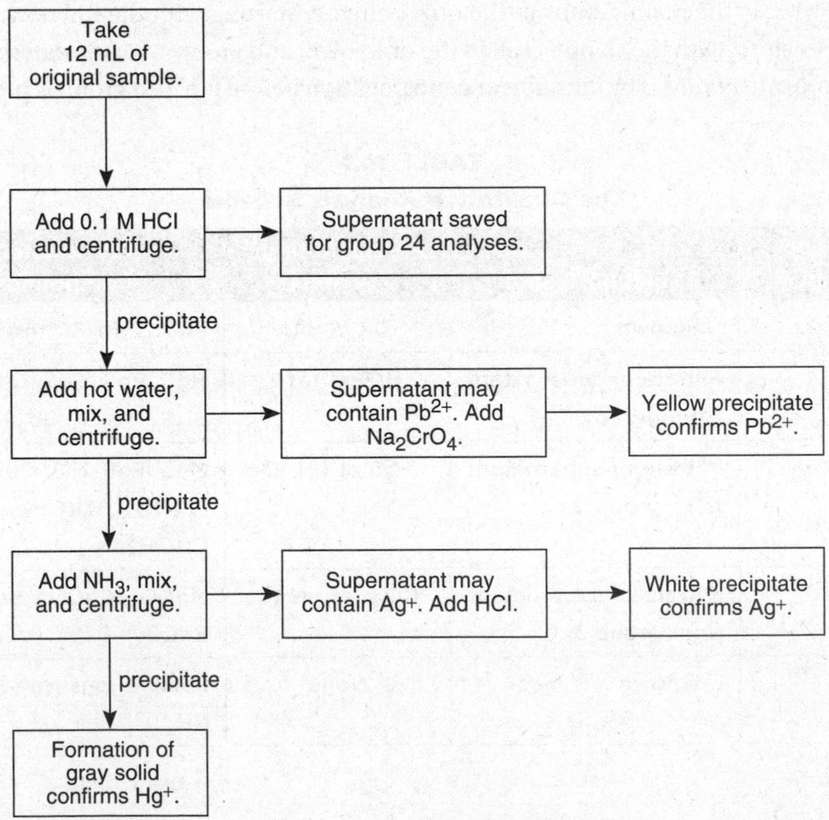

FIGURE 14.8. Flowchart for group 1 qualitative analysis of cations. Hg^+ is a dimer and is often written as Hg_2^{2+}.

Confirmation of the metal sulfides, hydroxides, and carbonates is done in a series of logical steps similar to that for the group 1 ions. A confirmation test is also performed for each metal. Some of these tests are listed in Table 14.2.

TABLE 14.2
Confirmation Spot Tests for Some Metal Ions

Cation	Test Reagent or Method	Observation
Fe^{2+}	$K_3Fe(CN)_6$	Dark blue precipitate
Fe^{3+}	$K_4Fe(CN)_6$	Dark blue precipitate
Cu^{2+}	NH_3	Dark blue solution
Ni^{2+}	Dimethylglyoxime	Red precipitate
Pb^{2+}	CrO_4^{2-}	Orange precipitate
Zn^{2+}	H_2S	White precipitate
NH_4^+	NaOH	Ammonia odor
H^+	Blue litmus paper	Paper turns red.
Na^+	Flame test	Orange flame
K^+	Flame test	Violet flame
Li^+	Flame test	Crimson red flame
Sr^{2+}	Flame test	Bright red flame

Analysis of anions is accomplished by spot tests using selected reagents. Some of these spot tests are listed in Table 14.3.

TABLE 14.3
Spot Tests for Selected Anions

Anion	Test Method	Observation
Carbonate	Add acid	Nonflammable gas evolved
Sulfide	Add acid	Rotten egg odor
Thiocyanate (SCN^-)	Add Fe^{3+}	Blood red complex
Sulfate	Add Ba^{2+}	White precipitate
Chloride	Add Ag^+	White precipitate
Bromide	Add Cl_2, water, and C_6H_{12}	Brown Br_2 in C_6H_{12} layer
Iodide	Add Cl_2, water, and C_6H_{12}	Violet I_2 in C_6H_{12} layer
Hydroxide	Red litmus paper	Paper turns blue.

CHEMICAL HAZARDS

Any human endeavor has some risk associated with it. It is much more likely, however, that you will be injured in an automobile accident or, if you smoke cigarettes, that you will develop cancer than that you will come to harm in a chemistry laboratory. While chemicals can be hazardous, one aim of chemistry education is to train students in the proper handling of these substances. Elimination of all hazards in any field of endeavor is unrealistic. The elimination of all potentially hazardous materials and techniques from the chemistry laboratory would eventually result in a society where no one understood how to handle or use chemicals properly. There are, however, certain hazards of which you should be aware so that you can take sensible precautions.

Highly Flammable Compounds

Organic compounds, particularly low-molar-mass compounds (methane, butane, etc.) and ethers, are highly flammable. They should be kept away from all sparks and flames.

Explosive Compounds

Over time, ethers react to produce explosive peroxides. Empty containers of solvents may contain explosive vapor residues. Dry picric acid is explosive. Nitrogen triiodide and nitroglycerine are shock-sensitive.

Strong Oxidizers

Concentrated perchloric acid, sulfuric acid, nitric acid, and hydrogen peroxide cause almost immediate skin injury. Perchloric acid in contact with organic material causes spontaneous combustion. White phosphorus burns spontaneously in air and emits hazardous fumes.

Compounds Incompatible with Water

The very active metals, Li, Na, K, Rb, Cs, Ca, Ba, and Sr, may react explosively with water.

Compounds with High Heats of Solution

Some substances evolve large amounts of heat when dissolved in water. If unexpected, the heat can cause spattering or can cause you to drop a hot beaker. The very active metals, their soluble oxides and hydroxides, and concentrated sulfuric and nitric acids generate heat. Calcium oxide and phosphorous pentoxide generate large amounts of heat with water. Mixing concentrated bases with concentrated acids is particularly hazardous.

Compounds with Possible Health Hazards

Benzene, chloroform, and carbon tetrachloride are suspected carcinogens and should be used only in a hood. Chlorinated organic compounds, in general, are suspected of being health hazards and should be handled carefully.

SAFETY PRINCIPLES AND EQUIPMENT

The first principle of safe experimentation in a laboratory is as follows: Never design an experiment that risks life or health. An essential part of experimental design is to assess possible safety hazards from all sources. Safety glasses are mandatory at all times; many major chemical corporations will fire a worker on the spot for failing to wear safety glasses. Protective lab clothing is worn when needed.

In experiments with hazardous materials the minimum quantities possible are used, and the work is done in a ventilated hood. When an explosion is even remotely possible, impact-resistant shields are used to protect workers. In a chemistry laboratory fire is always a possibility, and familiarity with exit routes and the location of fire extinguishers, fire blankets, and showers is essential.

SUMMARY

Laboratory work is at the core of chemistry. All the theoretical work must agree with experimental results because theories can be altered, but experimental data cannot. This chapter covers the basics of all experimental chemistry required for this course. You will draw from personal experience from labs performed but the techniques used are described in this chapter. These techniques are divided into data gathering, determination of physical properties, manipulating samples, separating and purifying products, and using simple instrumentation. Also reviewed in this chapter are reactions usually used in the laboratory along with a short description of the qualitative analysis scheme. Many calculations that you have done in the lab are also parts of other chapters in this book. Reviewing your own laboratory notebook to recall general methods, calculations, error analysis, and the purpose of your experiments will be helpful.

Important Concepts

Significant figures and calculations
Graphs
Meter reading
Common experimental reactions
Safety and chemical hazards

PRACTICE EXERCISES

Multiple-Choice

1. This ion gives a bright-orange color in a flame test.
   (A) Sodium ion
   (B) Silver ion
   (C) Bromide ion
   (D) Sulfide ion

2. This ion is usually oxidized before it is identified.
   (A) Sodium ion
   (B) Silver ion
   (C) Bromide ion
   (D) Sulfide ion

3. Which ions can be identified by a characteristic odor?
   (A) Sodium ion
   (B) Silver ion and bromide ion
   (C) Bromide ion
   (D) Sulfide ion

4. Which ion is most commonly identified as a white precipitate that dissolves in ammonia?
   (A) Sodium ion
   (B) Silver ion
   (C) Bromide ion
   (D) Sulfide ion

5. A 35.25-mL sample is needed. The best piece of glassware to use is
   (A) a buret
   (B) a graduated cylinder
   (C) a volumetric flask
   (D) a volumetric pipet

6. Methane is collected by
   (A) upward displacement of air
   (B) displacement of water
   (C) downward displacement of air
   (D) displacement of mercury

7. A very fine precipitate is best isolated by
   (A) distillation
   (B) filtration
   (C) vacuum filtration
   (D) centrifugation

8. Two students each calibrated the same balance using a standard mass of 0.1000 g. Each repeated the measurement three times, yielding the following data.

Reading	Student 1	Student 2
1	0.1004 g	0.0996 g
2	0.0998 g	0.0994 g
3	0.0992 g	0.0995 g

If we compare the two sets of data,
(A) student 2 is more accurate but student 1 is more precise
(B) student 1 is more accurate but student 2 is more precise
(C) student 1 is more precise and more accurate than student 2
(D) student 2 is more precise and more accurate than student 1

9. A 1.00 *M* solution of NaOH is prepared by weighing exactly 40.0 g of NaOH and adding it to exactly 1000 mL of distilled water at room temperature. Which of the following is most likely to be the largest source of error using the above procedure?
(A) Water should be added to the solute until the desired volume of solution is reached.
(B) The room temperature is not 25°C.
(C) The glassware is incorrectly calibrated.
(D) NaOH absorbs atmospheric water.

10. The vapor pressure of a liquid is measured at several temperatures. When a graph of the data is made,
(A) temperature is the x-axis since it is the dependent variable
(B) pressure is the y-axis since it is the dependent variable
(C) $1/T$ is the x-axis because of Raoult's law
(D) mole fraction is the x-axis and is the independent variable

11. If a student wished to transfer a coarsely powdered solid from a stock bottle to a small test tube, what would be the best method?
(A) Use an evaporating dish
(B) Transfer it from the stock bottle directly
(C) Use a thin-stemmed funnel
(D) Use a creased square of paper

12. Which measurement is expressed to the appropriate number of significant figures?
(A) 37 seconds from a timer that is graduated in tenths of a second
(B) 0.985 cm from a ruler that is graduated in millimeters
(C) 24.68 mL from a buret that is graduated in tenths of a millimeter
(D) 7.56°C from a thermometer that is graduated in degrees

13. Which of the following dissolves in both acids and bases?
(A) CaO
(B) CO_2
(C) $Al(OH)_3$
(D) AgCl

CHALLENGE

14. A 25.0-mL sample of a monoprotic acid was titrated to the end point with 20.0 mL of 0.200 *M* NaOH, and the molarity of the acid was calculated as 0.160 *M*. After the titration was complete, it was noticed that the buret was not clean, and droplets of solution were seen on the inside of the buret. What can be deduced about the molarity of the unknown?
    (A) The calculation is wrong, and the molarity is really 0.250 *M*.
    (B) The recorded volume of NaOH is low, and the molarity too high.
    (C) The recorded volume of NaOH is high, and the molarity is too high.
    (D) The recorded volume of NaOH is low, and the molarity is too low.

CHALLENGE

15. A sample is brought into a laboratory and mixed with an equal volume of a preservative solution. For analysis a 5.00-mL sample is diluted to 100 mL, and the concentration of chloride ions in the diluted solution is found to be 3.0×10^{-3} *M*. What is the chloride concentration of the sample?
    (A) 6.0×10^{-2} *M*
    (B) 1.2×10^{-1} *M*
    (C) 1.5×10^{-4} *M*
    (D) 7.5×10^{-5} *M*

16. Qualitative analysis is performed on a colored solution. Addition of HCl results in a white precipitate that dissolves completely in hot water. The solution probably contained
    (A) a transition metal ion and Ag^+
    (B) Pb^{2+} and no other cations
    (C) Pb^{2+} and possibly an alkali metal ion
    (D) Pb^{2+} and a transition metal ion

17. A solution is acidified, and a noticeable odor is observed. Which of the following is the most likely source of the odor?
    (A) CH_4
    (B) NH_3
    (C) CO_2
    (D) H_2S

18. Flame tests in which a small amount of solution is heated in a flame and the color of the flame is observed are routinely used to confirm the presence of which ion?
    (A) Ca
    (B) K
    (C) Na
    (D) All of these

19. The salt $CrCl_3$ may be prepared in pure form by
 (A) reacting an excess of Cr metal with HCl gas
 (B) reacting excess NaCl with Cr_2O_3
 (C) reacting 3 moles of HCl with 1 mole of $Cr(OH)_3$
 (D) reacting 1 mole of Na_2CrO_4 with 3 moles of HCl

20. The most common method for determining the molarity of a solution of an acid is
 (A) gravimetric analysis (weighing a precipitate)
 (B) titration with a standard base
 (C) determination of the specific gravity of the acid
 (D) determination of the volume of gas evolved when the solution is reacted with Mg metal

21. A student was performing an experiment to determine the molar mass of an unknown solid acid. The acid was dissolved in water and titrated with a standard solution of NaOH. Which of the following errors in procedure would result in the calculated molar mass being too low?

 I. Rinsing the buret with distilled water rather than the standard NaOH solution before titrating.
 II. Some of the solid acid stuck to the side of the flask and not completely dissolved.
 III. Using a NaOH solution that had absorbed carbon dioxide.

 (A) I only
 (B) III only
 (C) I and III only
 (D) I and II only

22. After a buret is filled for a titration, the bubble of air in the tip is not dislodged. What will be the effect?
 (A) No error will result if the bubble does not come out during the titration.
 (B) The volume recorded will be high if the bubble comes out.
 (C) The mass of sample will be low if the bubble comes out.
 (D) All of the above will be true.

23. If an error is made during an experiment, the appropriate action is to
 (A) stop the experiment and start over again
 (B) make a note of the error in the notebook and finish the experiment
 (C) tear the page(s) out of the notebook and start over again
 (D) adjust the results to correct for the error

24. Which of the following is appropriate when constructing a graph of experimental data?
 (A) A line is drawn connecting each data point.
 (B) The axes are scaled so that the data fill the graph as completely as possible.
 (C) A straight line is always drawn through the data points.
 (D) The line is extended beyond the last data point to the edge of the graph.

ANSWERS

1. **(A)**	7. **(D)**	13. **(C)**	19. **(C)**
2. **(C)**	8. **(B)**	14. **(C)**	20. **(B)**
3. **(D)**	9. **(A)**	15. **(B)**	21. **(C)**
4. **(B)**	10. **(B)**	16. **(D)**	22. **(D)**
5. **(A)**	11. **(D)**	17. **(D)**	23. **(A)**
6. **(B)**	12. **(C)**	18. **(D)**	24. **(B)**

See Appendix 1 for explanations of answers.

Free-Response

The following questions relate to laboratory work. All of the previous material in this book may be of use in answering these questions.

(a) Explain whether the desired result will be too high, too low, or unaffected in the following situations. Use appropriate equations to support your conclusions.

(i) You pipet 10.0 mL of monoprotic acid and titrate with standard base to calculate the molarity of the acid. You mistakenly blow the last drop of solution from the volumetric pipet each time.

(ii) You determine the molar mass of a gas by determining its density but forget to correct the pressure for the vapor pressure of the water that the gas was collected over.

(iii) In an experiment you dissolve some lead(II) nitrate but do not shake the volumetric flask well. What error results if you pipet from the top of the flask and what will happen if you pipet from the bottom of the flask?

(iv) When you prepare a solid compound, you mistakenly wash the sample with hot water instead of ice water.

(b) Separation methods include decanting, extraction, filtering, distilling, and chromatography. Give an appropriate example of using each method.

(c) Describe how you can detect and eliminate determinate errors.

(d) Flame tests are used to identify some cations, particularly those that do not tend to form complexes or precipitates. List three elements that can be easily determined using a flame test. Explain why flame tests work based on your knowledge of the structure of the atom.

ANSWERS

(a)

(i) Because you added a drop or two of extra acid, it will result in a concentration that is higher than it should be.

(ii) $M = mRT/PV$. Because P is too large (the vapor pressure of water is always subtracted), the result will be smaller than it should be.

(iii) If not well mixed, the heavy solute will stratify. The results pipeted from the top layers will have lead concentrations lower than expected. Solution pipeted from lower layers will be more concentrated than expected.

(iv) Hot water generally dissolves substances better than cold water. It is expected that more product will be lost and the results will be low.

(b) Decanting is usually done with a large amount of solid that has large crystals, allowing liquid to be poured off easily without losing solid. Extraction usually involves two immiscible solvents. The solute of interest transfers from one solvent to another because of solubility factors. Filtering can capture small particles and is preferred in many instances. Distillation is best used to separate substances that have similar boiling points. If the boiling points are very far apart, other techniques may be better, and if they are too close to each other, chromatography may be needed. Chromatography is excellent for small samples that have very similar, but not identical, physical properties. Chromatography is similar to extraction. However, one of the liquids is held stationary while the other flows by, allowing the extraction process to occur.

(c) Determinate errors are those that can be found and eliminated. Several ways to detect these errors include comparison with standards of known materials and comparison with other workers or labs with different instruments.

(d) Sodium (orange), potassium (violet), and copper (green) are three of the more common flame tests. There are other equally valid methods.

PRACTICE TESTS

ONLINE

Want more test-taking practice?

Visit *barronsbooks.com/ap/ap-chemistry/* to access the three additional online tests or scan the QR code below:

*Be sure to have a copy of your *AP Chemistry*, 9th edition, on hand to complete the registration process.

ANSWER SHEET
Practice Test 1

1. Ⓐ Ⓑ Ⓒ Ⓓ	21. Ⓐ Ⓑ Ⓒ Ⓓ	41. Ⓐ Ⓑ Ⓒ Ⓓ
2. Ⓐ Ⓑ Ⓒ Ⓓ	22. Ⓐ Ⓑ Ⓒ Ⓓ	42. Ⓐ Ⓑ Ⓒ Ⓓ
3. Ⓐ Ⓑ Ⓒ Ⓓ	23. Ⓐ Ⓑ Ⓒ Ⓓ	43. Ⓐ Ⓑ Ⓒ Ⓓ
4. Ⓐ Ⓑ Ⓒ Ⓓ	24. Ⓐ Ⓑ Ⓒ Ⓓ	44. Ⓐ Ⓑ Ⓒ Ⓓ
5. Ⓐ Ⓑ Ⓒ Ⓓ	25. Ⓐ Ⓑ Ⓒ Ⓓ	45. Ⓐ Ⓑ Ⓒ Ⓓ
6. Ⓐ Ⓑ Ⓒ Ⓓ	26. Ⓐ Ⓑ Ⓒ Ⓓ	46. Ⓐ Ⓑ Ⓒ Ⓓ
7. Ⓐ Ⓑ Ⓒ Ⓓ	27. Ⓐ Ⓑ Ⓒ Ⓓ	47. Ⓐ Ⓑ Ⓒ Ⓓ
8. Ⓐ Ⓑ Ⓒ Ⓓ	28. Ⓐ Ⓑ Ⓒ Ⓓ	48. Ⓐ Ⓑ Ⓒ Ⓓ
9. Ⓐ Ⓑ Ⓒ Ⓓ	29. Ⓐ Ⓑ Ⓒ Ⓓ	49. Ⓐ Ⓑ Ⓒ Ⓓ
10. Ⓐ Ⓑ Ⓒ Ⓓ	30. Ⓐ Ⓑ Ⓒ Ⓓ	50. Ⓐ Ⓑ Ⓒ Ⓓ
11. Ⓐ Ⓑ Ⓒ Ⓓ	31. Ⓐ Ⓑ Ⓒ Ⓓ	51. Ⓐ Ⓑ Ⓒ Ⓓ
12. Ⓐ Ⓑ Ⓒ Ⓓ	32. Ⓐ Ⓑ Ⓒ Ⓓ	52. Ⓐ Ⓑ Ⓒ Ⓓ
13. Ⓐ Ⓑ Ⓒ Ⓓ	33. Ⓐ Ⓑ Ⓒ Ⓓ	53. Ⓐ Ⓑ Ⓒ Ⓓ
14. Ⓐ Ⓑ Ⓒ Ⓓ	34. Ⓐ Ⓑ Ⓒ Ⓓ	54. Ⓐ Ⓑ Ⓒ Ⓓ
15. Ⓐ Ⓑ Ⓒ Ⓓ	35. Ⓐ Ⓑ Ⓒ Ⓓ	55. Ⓐ Ⓑ Ⓒ Ⓓ
16. Ⓐ Ⓑ Ⓒ Ⓓ	36. Ⓐ Ⓑ Ⓒ Ⓓ	56. Ⓐ Ⓑ Ⓒ Ⓓ
17. Ⓐ Ⓑ Ⓒ Ⓓ	37. Ⓐ Ⓑ Ⓒ Ⓓ	57. Ⓐ Ⓑ Ⓒ Ⓓ
18. Ⓐ Ⓑ Ⓒ Ⓓ	38. Ⓐ Ⓑ Ⓒ Ⓓ	58. Ⓐ Ⓑ Ⓒ Ⓓ
19. Ⓐ Ⓑ Ⓒ Ⓓ	39. Ⓐ Ⓑ Ⓒ Ⓓ	59. Ⓐ Ⓑ Ⓒ Ⓓ
20. Ⓐ Ⓑ Ⓒ Ⓓ	40. Ⓐ Ⓑ Ⓒ Ⓓ	60. Ⓐ Ⓑ Ⓒ Ⓓ

Practice Test 1

SECTION I: Multiple-Choice

START ONLY WHEN YOU HAVE 90 MINUTES TO COMPLETE THE WHOLE SECTION.

Time:	1 hour, 30 minutes
Number of Questions:	60
Percent of Total Score:	50%
Calculator?	None allowed
Pencil required	

Instructions

There are 60 multiple-choice questions for this part of the exam. Enter your answers on the answer sheet provided. On the actual exam, no credit will be given for answers marked on the test itself. For this test, mark your selected answer next to the question and then transfer it to the scoring sheet. (This will allow you to check that all answers are properly transferred to the scoring sheet.) Each question has only one answer. When changing answers, be sure to erase completely.

Useful Hints

Not everyone will know the answers to all of the questions. However, it is to your advantage to provide an answer to all questions. Do not waste time on difficult questions. Answer the easier ones first, and return to the difficult ones you have not answered if time remains.

Your total score is simply the number of questions answered correctly. Wrong answers or blanks on the scoring sheet do not count against you.

You will be allowed to use the following periodic table and the table of equations and constants on this part of the test.

Periodic Table of the Elements

1	2	3	4	5	6	7	8	9	10	11	12	13	14	15	16	17	18
1 **H** 1.008																	2 **He** 4.00
3 **Li** 6.94	4 **Be** 9.01											5 **B** 10.81	6 **C** 12.01	7 **N** 14.01	8 **O** 16.00	9 **F** 19.00	10 **Ne** 20.18
11 **Na** 22.99	12 **Mg** 24.30											13 **Al** 26.98	14 **Si** 28.09	15 **P** 30.97	16 **S** 32.06	17 **Cl** 35.45	18 **Ar** 39.95
19 **K** 39.10	20 **Ca** 40.08	21 **Sc** 44.96	22 **Ti** 47.87	23 **V** 50.94	24 **Cr** 52.00	25 **Mn** 54.94	26 **Fe** 55.85	27 **Co** 58.93	28 **Ni** 58.69	29 **Cu** 63.55	30 **Zn** 65.38	31 **Ga** 69.72	32 **Ge** 72.63	33 **As** 74.92	34 **Se** 78.97	35 **Br** 79.90	36 **Kr** 83.80
37 **Rb** 85.47	38 **Sr** 87.62	39 **Y** 88.91	40 **Zr** 91.22	41 **Nb** 92.91	42 **Mo** 95.95	43 **Tc** (97)	44 **Ru** 101.1	45 **Rh** 102.91	46 **Pd** 106.42	47 **Ag** 107.87	48 **Cd** 112.41	49 **In** 114.82	50 **Sn** 118.71	51 **Sb** 121.76	52 **Te** 127.60	53 **I** 126.90	54 **Xe** 131.29
55 **Cs** 132.91	56 **Ba** 137.33	57 ***La** 138.91	72 **Hf** 178.49	73 **Ta** 180.95	74 **W** 183.84	75 **Re** 186.21	76 **Os** 190.2	77 **Ir** 192.2	78 **Pt** 195.08	79 **Au** 196.97	80 **Hg** 200.59	81 **Tl** 204.38	82 **Pb** 207.2	83 **Bi** 208.98	84 **Po** (209)	85 **At** (210)	86 **Rn** (222)
87 **Fr** (223)	88 **Ra** (226)	89 **†Ac** (227)	104 **Rf** (267)	105 **Db** (270)	106 **Sg** (271)	107 **Bh** (270)	108 **Hs** (277)	109 **Mt** (276)	110 **Ds** (281)	111 **Rg** (282)	112 **Cn** (285)	113 **Nh** (286)	114 **Fl** (289)	115 **Mc** (289)	116 **Lv** (293)	117 **Ts** (294)	118 **Og** (294)

*Lanthanoid Series	58 **Ce** 140.12	59 **Pr** 140.91	60 **Nd** 144.24	61 **Pm** (145)	62 **Sm** 150.4	63 **Eu** 151.97	64 **Gd** 157.25	65 **Tb** 158.93	66 **Dy** 162.50	67 **Ho** 164.93	68 **Er** 167.26	69 **Tm** 168.93	70 **Yb** 173.05	71 **Lu** 174.97
†Actinoid Series	90 **Th** 232.04	91 **Pa** 231.04	92 **U** 238.03	93 **Np** (237)	94 **Pu** (244)	95 **Am** (243)	96 **Cm** (247)	97 **Bk** (247)	98 **Cf** (251)	99 **Es** (252)	100 **Fm** (257)	101 **Md** (258)	102 **No** (259)	103 **Lr** (262)

EQUATIONS AND CONSTANTS

GENERAL INFORMATION

L, mL	= liter(s), milliliter(s)	mm Hg	= millimeters of mercury
g	= gram(s)	J, kJ	= joule(s), kilojoule(s)
nm	= nanometer(s)	V	= volt(s)
atm	= atmosphere(s)	mol	= mole(s)

ATOMIC STRUCTURE

$E = h\nu$

$c = \lambda\nu$

E = energy

ν = frequency

λ = wavelength

Speed of light, $c = 2.998 \times 10^{8}$ m s^{-1}

Planck's constant, $h = 6.626 \times 10^{-34}$ J s

Avogadro's number = 6.022×10^{23} mol^{-1}

Electron charge, $e = -1.602 \times 10^{-19}$ coulomb

EQUILIBRIUM

$$K_a = \frac{[H^+][A^-]}{[HA]}$$

$$K_b = \frac{[OH^-][HB^+]}{[B]}$$

$$K_c = \frac{[C]^c[D]^d}{[A]^a[B]^b}, \text{ where } a\,A + b\,B \rightleftarrows c\,C + d\,D$$

$$K_p = \frac{(P_C)^c(P_D)^d}{(P_A)^a(P_B)^b}$$

$$K_w = [H^+][OH^-] = 1.0 \times 10^{-14} \text{ at } 25°C$$

$$= K_a \times K_b$$

$$pK_a = -\log K_a,\ pK_b = -\log K_b$$

$$pH = -\log[H^+],\ pOH = -\log[OH^-]$$

$$14 = pH + pOH$$

$$pH = pK_a + \log\frac{[A^-]}{[HA]}$$

Equilibrium Constants

K_a (weak acid)

K_b (weak base)

K_c (molar concentrations)

K_p (gas pressures)

K_w (water)

THERMOCHEMISTRY/ELECTROCHEMISTRY AND KINETICS

$\Delta S^\circ = \Sigma S^\circ$ products $- \Sigma S^\circ$ reactants
$\Delta H^\circ = \Sigma \Delta H^\circ_f$ products $- \Sigma \Delta H^\circ_f$ reactants
$\Delta G^\circ = \Sigma \Delta G^\circ_f$ products $- \Sigma \Delta G^\circ_f$ reactants

$\Delta G^\circ = \Delta H^\circ - T\Delta S^\circ$
$= -RT \ln K$
$= -n\mathcal{F}E^\circ$

$q = mc\Delta T$

$I = \dfrac{q}{t}$

$\ln[A]_t - \ln[A]_0 = -kt$

$\dfrac{1}{[A]_t} - \dfrac{1}{[A]_0} = kt$

$t_{1/2} = \dfrac{0.693}{k}$

m = mass
n = number of moles
q = heat
k = rate constant
c = specific heat capacity
S° = standard entropy
H° = standard enthalpy
G° = standard free energy
E° = standard reduction potential
T = temperature
I = current (amperes)
q = charge (coulombs)
t = time (seconds)
$t_{1/2}$ = half-life

Faraday's constant, $\mathcal{F}$ = 96,485 coulombs per mole of electrons

1 volt $= \dfrac{\text{1 joule}}{\text{1 coulomb}}$

GASES, LIQUIDS, AND SOLUTIONS

$PV = nRT$

$P_A = P_{total} \times X_A$, where $X_A = \dfrac{\text{moles A}}{\text{total moles}}$

$P_{total} = P_A + P_B + P_C + \ldots$

$n = \dfrac{m}{M}$

K = °C + 273

$D = \dfrac{m}{V}$

KE per molecule $= \dfrac{1}{2}mv^2$
$A = abc$

Molarity, $M = \dfrac{\text{mol solute}}{\text{liters of solution}}$

D = density
P = pressure
T = temperature
m = mass
n = number of moles
v = velocity
V = volume
A = absorbance
a = molar absorptivity
b = path length
c = concentration
KE = kinetic energy
M = molar mass

Gas constant, $R = 8.314\ \text{J mol}^{-1}\ \text{K}^{-1}$
$= 0.08206\ \text{L atm mol}^{-1}\ \text{K}^{-1}$
$= 62.36\ \text{L torr mol}^{-1}\ \text{K}^{-1}$

1 atm = 760 mm Hg = 760 torr

STP = 0.00°C and 1.000 atm

Molar volume of ideal gas = 22.4 L at STP

SECTION I

60 Multiple-Choice Questions
(Time: 90 minutes)
CALCULATORS ARE NOT ALLOWED FOR SECTION I

Note: For all questions, assume that $T = 298$ K, $P = 1.00$ atmosphere, and H_2O is the solvent for all solutions unless the problem indicates a different solvent.

Directions: The questions or incomplete statements that follow are each followed by four suggested answers or completions. Choose the response that best answers the question or completes the statement. Fill in the corresponding circle on the answer sheet.

1. Using fundamental trends in electronegativity and bond strength, which of the following should be the strongest acid?
   (A) H_2S because it is the heaviest
   (B) HI because the bond length is longest
   (C) HBr because Br has the highest electronegativity
   (D) H_2O because water forms the strongest hydrogen bonds

Compound	Formula	Normal Boiling Point, K
Ethane	CH_3CH_3	185
Ethanal	CH_3COH	294
Ethanol	CH_3CH_2OH	348
Ethanoic acid	CH_3CO_2H	391

2. Based on the data in the table above, which of the following substances has the lowest surface tension at −120°C?
   (A) Ethane since it doesn't form hydrogen bonds
   (B) Ethanal since its hydrogen bonds are the weakest
   (C) Ethanol since strong hydrogen bonds reduce surface tension
   (D) Ethanoic acid because it is the heaviest molecule

Ion	Ionic Radius, pm
O^{2-}	66
S^{2-}	104
Se^{2-}	117

3. Based on the data in the table above, which of the following correctly predicts the strength of binary acids from weakest to strongest?
   (A) $H_2S < H_2O < H_2Se$
   (B) $H_2Se < H_2O < H_2S$
   (C) $H_2S < H_2Se < H_2O$
   (D) $H_2O < H_2S < H_2Se$

PRACTICE TEST 1

4. Zinc metal is often used to reduce substances in aqueous solutions. If zinc metal is not available, which of the following would be a reasonable and safe substitute for zinc?
   (A) S
   (B) F_2
   (C) K
   (D) Cd

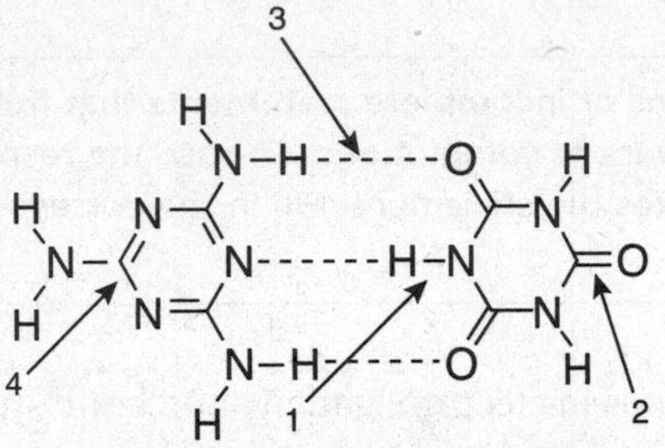

5. In the diagram above, which labeled arrow is pointing toward a hydrogen bond?
   (A) 1
   (B) 2
   (C) 3
   (D) 4

6. Which would be the easiest way to burn an iron nail?
   (A) Hold an iron nail with crucible tongs, and heat strongly in the flame of a Bunsen burner.
   (B) Use the method in (A), but use an oxyacetylene torch to reach a higher temperature.
   (C) Grind the nail into very small, dust-sized particles, and spray them into a flame.
   (D) Dissolve the nail in acid to make the oxide.

$$S(s) + O_2(g) \rightarrow SO_2(g)$$

7. The reaction of sulfur with oxygen is written in equation form above. This equation can be interpreted in all of the following ways EXCEPT
   (A) either $S(s)$ or $O_2(g)$ will be completely used up
   (B) Q must be close to 1.0 since there is one mole of gas on each side of the equation
   (C) this reaction goes to completion
   (D) adding O_2 will not change the equilibrium constant

8. An experiment was performed to determine the moles of hydrogen gas formed (collected over water) when an acid reacts with magnesium metal. To do this, a piece of dry magnesium was weighed. Then 50 mL of hydrogen was collected. Next the Mg was dried to remove approximately 0.1 mL of water and weighed again to see how much Mg had reacted. The atmospheric pressure was measured as 755 torr. The volume of hydrogen was measured and converted into moles of hydrogen. Which mistake will give the smallest error in the result?
   (A) Forgetting to dry the magnesium before weighing the second time
   (B) Failing to take the vapor pressure of water (23 torr at 25°C) into account
   (C) Failing to convert °C to K
   (D) Reading the gas-collecting container to ±20 mL

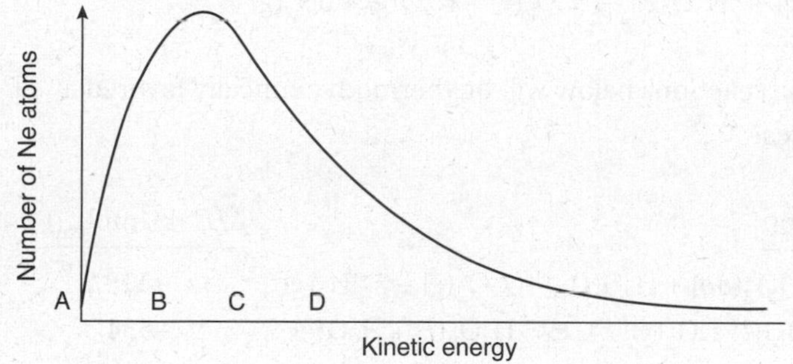

9. The graph above shows the distribution of kinetic energies of a large number of Ne atoms at 500 K. Which letter shows the average kinetic energy of this system?
   (A) A
   (B) B
   (C) C
   (D) D

10. A 25 g sample of a liquid was heated to 100°C and then quickly transferred to an insulated container holding 100 g of water at 22°C. The temperature of the mixture rose to reach a final temperature of 35°C. Which of the following can be concluded?
   (A) The sample temperature changed more than the water temperature did; therefore the sample lost more heat energy than the water gained.
   (B) The sample temperature changed more than the water temperature did, but the sample lost the same amount of heat energy as the water gained.
   (C) The sample temperature changed more than the water temperature did; therefore the C_p of the sample must be greater than the C_p of the water.
   (D) The final temperature is less than the average starting temperature of the sample and the water; therefore the total energy of the sample and water decreased.

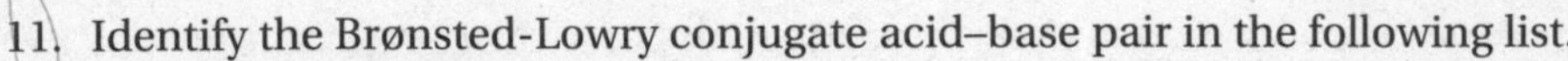

11. Identify the Brønsted-Lowry conjugate acid–base pair in the following list.
    (A) H_3O^+ and OH^-
    (B) H_3PO_4 and H_3PO_3
    (C) CO_3^{2-} and HCO_3^-
    (D) SO_3^{2-} and SO_2^{2-}

12. Chemical reactions can be classified as either heterogeneous or homogeneous. Which of the following equations below is best classified as a heterogeneous reaction?
    (A) $2C_2H_2(g) + 5O_2(g) \rightarrow 4CO_2(g) + 2H_2O(g)$
    (B) $C_2H_5OH(g) + O_2(g) \rightarrow HC_2H_3O_2(g) + H_2O(g)$
    (C) $2Fe(s) + 3CO_2(g) \rightarrow Fe_2O_3(s) + 3CO(g)$
    (D) $C_2H_2(g) + 5N_2O(g) \rightarrow CO_2(g) + H_2O(g) + 5N_2(g)$

13. Which of the reactions below will be thermodynamically favored at all temperatures?

Reaction	$\Delta H°$ (kJ/mol_{rxn})	$\Delta S°$ (J/mol_{rxn} • K)
(A) $CO(NH_2)_2(aq) + H_2O(l) \rightarrow CO_2(g) + 2NH_3(g)$	+119.2	+354.8
(B) $C_2H_5OH(l) + O_2(g) \rightarrow HC_2H_3O_2(l) + H_2O(g)$	–534.3	–131
(C) $2Fe(s) + 3CO_2(g) \rightarrow Fe_2O_3(s) + 3CO(g)$	+26.7	–15.7
(D) $C_2H_2(g) + 5N_2O(g) \rightarrow CO_2(g) + H_2O(g) + 5N_2(g)$	–437.4	+272.6

$$2SO_2(g) + O_2(g) \rightarrow 2SO_3(g)$$

14. Consider the following possible mechanism for the reaction above:

$2SO_2(g) \rightleftharpoons SO_3(g) + SO(g)$ *(fast reversible)*
$O_2(g) + SO(g) \rightarrow SO_3(g)$ *(slow)*

Which of the following statements is true?
(A) This mechanism has no free radicals.
(B) $O_2(g)$ is an intermediate.
(C) $SO(g)$ is a catalyst for this reaction.
(D) The mechanism does not add up to the overall reaction.

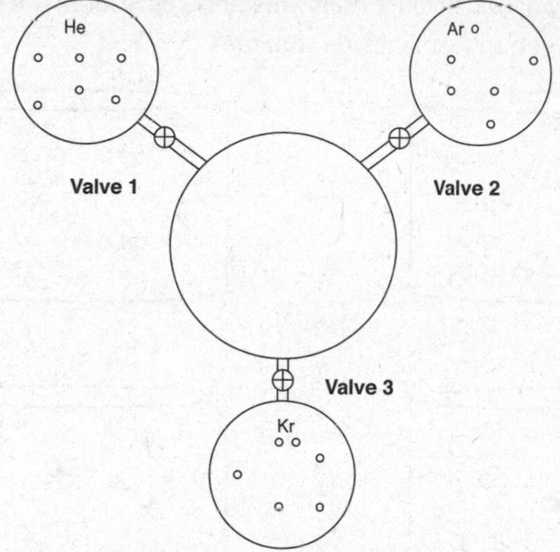

15. Three 1-liter flasks are connected to a 3-liter flask by valves. The 3-liter flask is evacuated to start and the entire system is at 298 K. The first flask contains helium, the second argon, and the third krypton. The pressure of the argon is 633 torr. The amounts of gas are proportional to their representations in the flasks. If all the valves to the center flask are opened, what will the pressure of the system be? Assume the connections have negligible volume.
(A) 633 torr
(B) 316 torr
(C) 1266 torr
(D) 211 torr

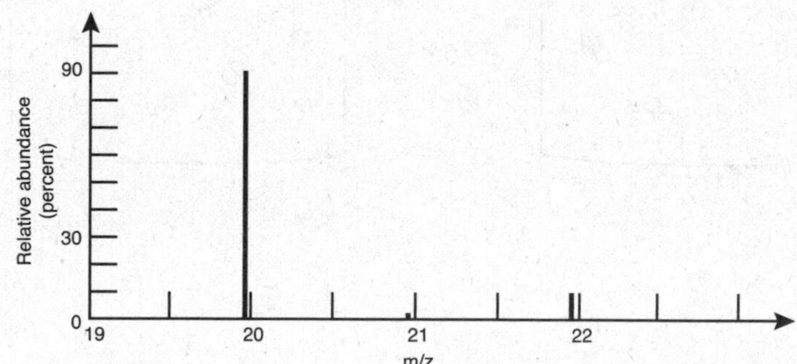

16. A portion of a mass spectrum of neon is presented above. Estimate the average mass of naturally occurring atoms of neon, assuming that the height of each line represents the relative amount of each mass.
(A) 20.3
(B) 20.0
(C) 21.0
(D) Not enough information is provided.

17. Which of the following particulate diagrams best represents the reaction of carbon monoxide with oxygen to form carbon dioxide?

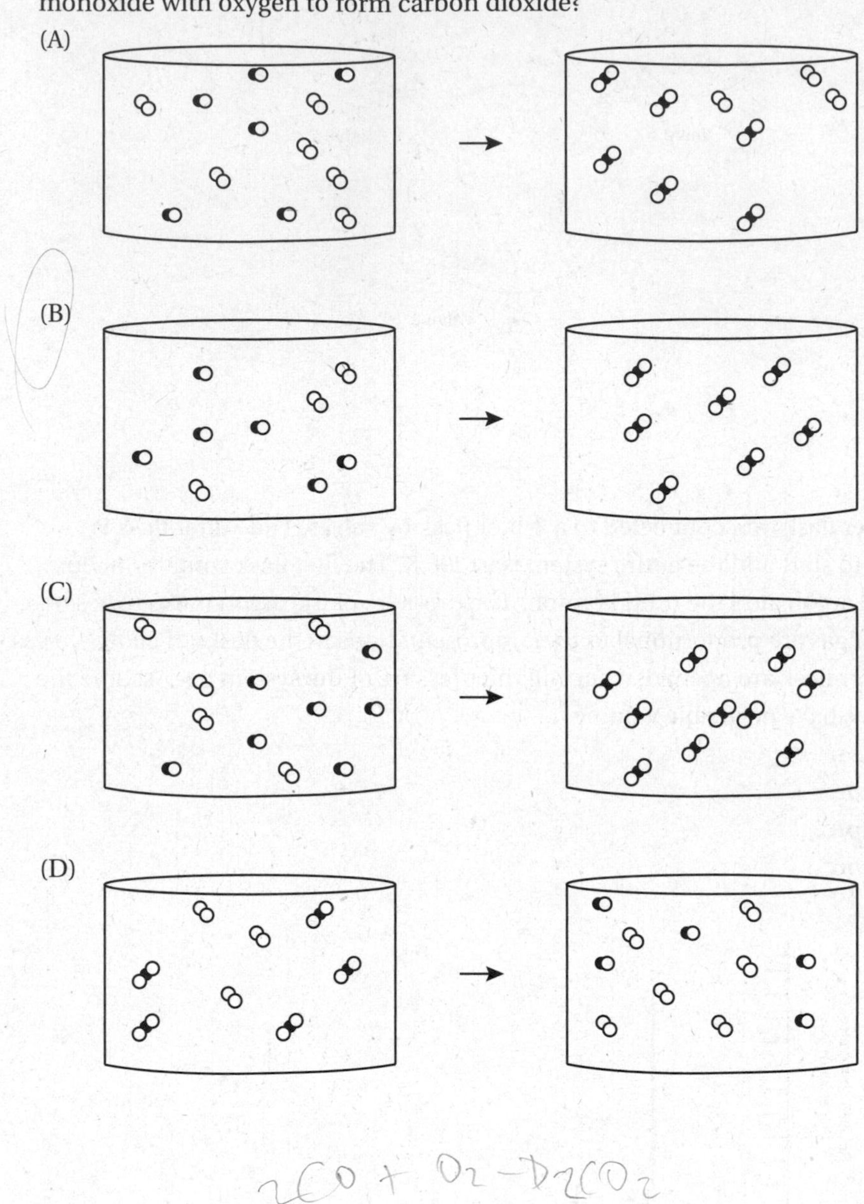

$2CO + O_2 \rightarrow 2CO_2$

Questions 18–22 refer to the following diagram and data.

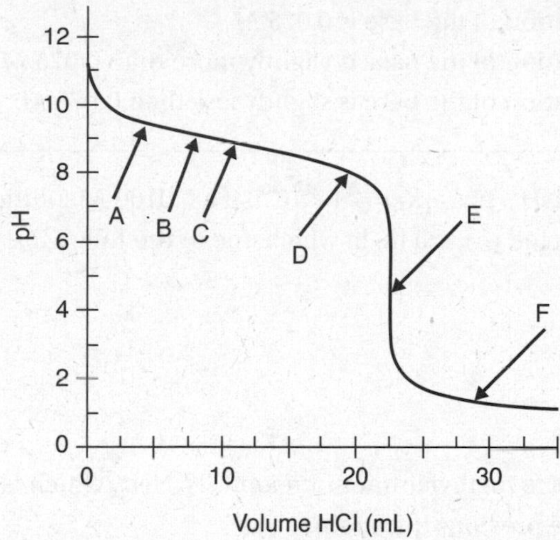

The above graph shows a titration curve of a weak base titrated with a strong acid. The pH was measured with a pH meter after small volumes of 0.075 *M* HCl were added to 25.0 mL of a weak base. Data from that experiment are shown in the graph.

18. Which arrow points to the end point of this titration?
    (A) A
    (B) E
    (C) C
    (D) F

19. Which arrow points to the place on the curve where the pH is equal to 14 – pK_b?
    (A) A
    (B) E
    (C) C
    (D) F

20. If the student used a pH indicator that changes color from pH 5 to 7, which statement best characterizes the expected observations?
    (A) The observed color change will be distinct, and the calculated molarity of the base will be accurate.
    (B) The color will change slowly, and the end point volume will be low.
    (C) The color will change slowly, and the calculated molarity of the base will be low.
    (D) The observed color change will be distinct, but the calculated molarity of the base will be high.

21. Find the two points on the curve that indicate the region where the solution can be described as a buffer. What is the change in pH from the first point to the second?
    (A) 0.5 pH unit
    (B) 1 pH unit
    (C) 2 pH units
    (D) 3 pH units

22. Which of the following describes the base that is being titrated?
    (A) The base is dibasic since two end points are observed.
    (B) The concentration of the base is 0.075 *M*.
    (C) The concentration of the base is slightly more than 0.075 *M*.
    (D) The concentration of the base is slightly less than 0.075 *M*.

---

23. Methylamine, CH_3NH_2, has a $K_b = 4.4 \times 10^{-4}$. If a 0.0100 *M* solution of methylamine is prepared, the expected pH will be in which one of the following pH ranges?
    (A) 2 to 4
    (B) 4 to 6
    (C) 8 to 10
    (D) 10 to 12

24. What is the pK_a of the methylammonium ion, $CH_3NH_3^+$, which is the conjugate acid of methylamine in the previous question?
    (A) $\log(4.4 \times 10^{-4})$
    (B) $-\log(4.4 \times 10^{-4})$
    (C) $-\log(4.4 \times 10^{-4}/1.0 \times 10^{-14})$
    (D) $\log(4.4 \times 10^{-4}/1.0 \times 10^{-14})$

Questions 25–28 refer to the table below.

Vessel	A	B	C
Gas	Bromomethane	Bromoethane	1-Bromopropane
Formula	CH_3Br	CH_3CH_2Br	$CH_3CH_2CH_2Br$
Molar mass	95 g/mol	109 g/mol	123 g/mol
Pressure	0.4 atm	0.3 atm	0.2 atm
Temperature	398 K	398 K	398 K

Note that in addition to the data in the table, the gases are held in separate, identical, rigid vessels.

25. The sample with the lowest density is
    (A) the same in all three samples since the temperatures are all the same
    (B) in vessel A
    (C) in vessel B
    (D) in vessel C

26. The average kinetic energy
    (A) is greatest in vessel A
    (B) is greatest in vessel B
    (C) is greatest in vessel C
    (D) is the same in all vessels

27. Which of these gases will condense at the lowest pressure, assuming that the temperature is held constant?
(A) Bromomethane
(B) Bromoethane
(C) 1-Bromopropane
(D) They all condense at the same pressure.

28. Which attractive force is least likely to be the cause of condensation?
(A) Hydrogen bonding
(B) London forces
(C) Dipole-dipole attractive forces
(D) Dipole-induced dipole attractive forces

Questions 29–30 refer to the diagrams below.

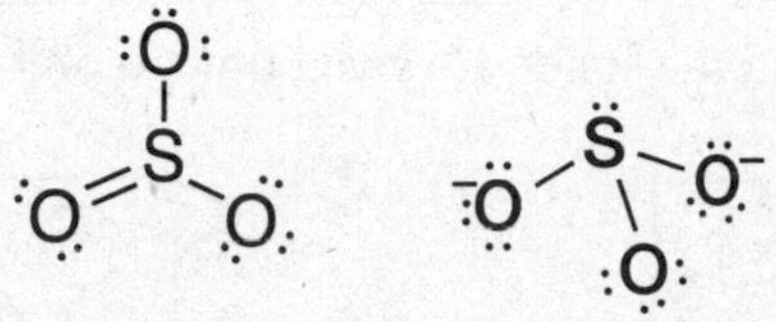

29. The Lewis electron-dot structures, ignoring formal charges, for sulfur trioxide and for the sulfite ion are given above. Which statement best describes the geometry of these two substances?
(A) The sulfite ion has a triangular pyramid shape, and the sulfur trioxide is a planar triangle.
(B) Both substances are flat, triangular, planar shapes.
(C) Both substances are triangular pyramids.
(D) SO_3 is a triangular pyramid, and SO_3^{2-} is a planar triangle.

30. Which statement below describes the charge, polarity, and resonance characteristics of the sulfite ion and of the sulfur trioxide species shown above?
(A) The sulfite ion has two negative charges along with a shape that makes it a dipole. Sulfur trioxide is symmetrical and nonpolar. In addition, as written, sulfur trioxide has three resonance structures.
(B) The sulfite ion and sulfur trioxide are both polar with the only difference being that one is an ion and the other is not.
(C) Sulfur trioxide has three resonance structures, and the sulfite ion has no resonance structures.
(D) Sulfite ions have a nonbonding electron pair domain, while sulfur trioxide has all electron domains as bonding domains.

Questions 31–33 refer to the following equation.

$$Pb^{2+}(aq) + 2Cl^{-}(aq) \rightarrow PbCl_2(s)$$

A chemist mixes a dilute solution of lead(II) perchlorate with an excess of potassium chloride to precipitate lead(II) chloride ($K_{sp} = 1.7 \times 10^{-5}$). The net ionic equation is given above for the reaction.

31. Which of the following is a molecular equation for this reaction?
    (A) $PbClO_3(aq) + KCl(aq) \rightarrow PbCl_2(s) + KClO_3(aq)$
    (B) $Pb(ClO_4)_2(aq) + 2KCl(aq) \rightarrow PbCl_2(s) + 2KClO_4(aq)$
    (C) $Pb(ClO_4)_2(aq) + KCl(aq) \rightarrow PbCl_2(s) + KClO_4(aq)$
    (D) $Pb(ClO_4)_4(aq) + 4KCl(aq) \rightarrow PbCl_2(s) + 4KClO_4(aq)$

32. Which of the following particulate views represents the experiment after the reactants are mixed thoroughly and the solids are given time to precipitate?

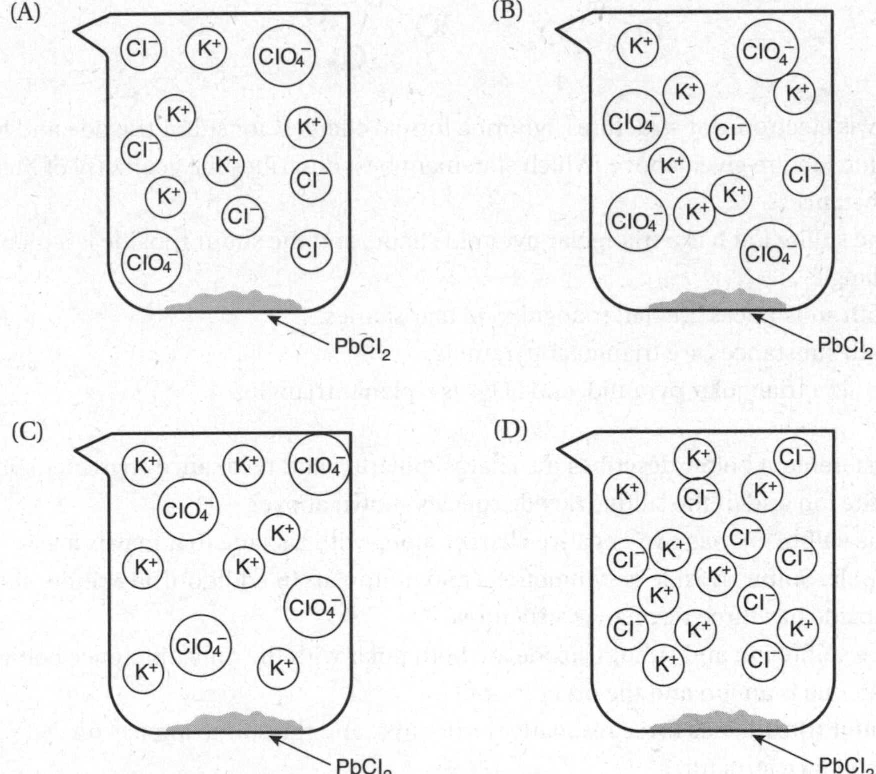

33. Within a factor of ten, what is the approximate molar solubility of lead(II) chloride in distilled water?
    (A) 1.7×10^{-5} mol/L
    (B) 10^{-7} mol/L
    (C) 10^{-2} mol/L
    (D) 10^{-15} mol/L

Questions 34–38 refer to the following equation.

$$N_2O_4(g) \rightleftharpoons 2NO_2(g)$$

$N_2O_4(g)$ decomposes into $NO_2(g)$ according to the equation above. A pure sample of $N_2O_4(g)$ is placed into a rigid, evacuated, 0.500 L container. The initial pressure of the $N_2O_4(g)$ is 760 torr. The temperature is held constant until the $N_2O_4(g)$ reaches equilibrium with its decomposition products. The figure below shows how the pressure of the system changes while reaching equilibrium.

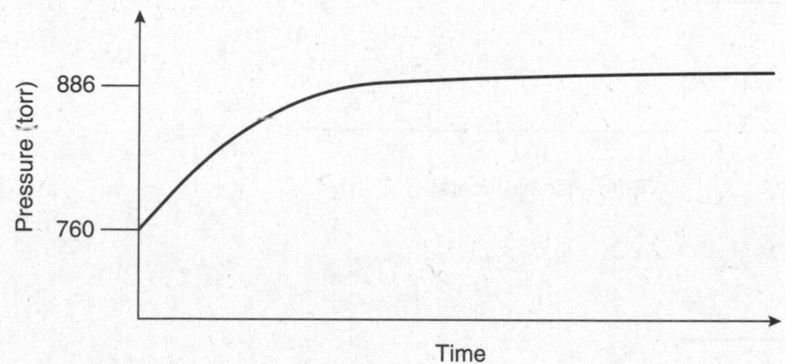

34. Why does the pressure rise in this experiment?
    (A) The intermolecular attractions inside the container decrease, so the molecules strike the walls more frequently.
    (B) The intermolecular attractions inside the container increase, which increases the force as molecules collide with the walls of the container.
    (C) The average kinetic energy increases as the reaction continues.
    (D) The number of particles striking the container walls per unit time increases.

35. The figure above gives us information about all the following EXCEPT
    (A) the activation energy
    (B) the reaction rate
    (C) the position of equilibrium
    (D) the order of reaction

36. By how much will the equilibrium pressure increase if this reaction goes to completion?
    (A) 760 torr
    (B) 900 torr
    (C) 634 torr
    (D) 886 torr

PRACTICE TEST 1

37. What can be said about the equilibrium constant, K_p, for this reaction?
    (A) $K_p > 1$
    (B) $K_p < 1$
    (C) $K_p = 1$
    (D) The data do not allow us to estimate the value of K_p.

38. Which diagram is an appropriate description of the system if more $NO_2(g)$ is rapidly injected, at time = 0, into the container after equilibrium is established?

(A)

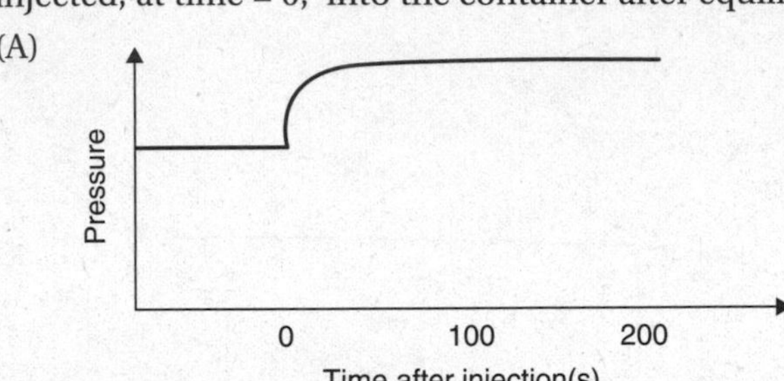

(B)

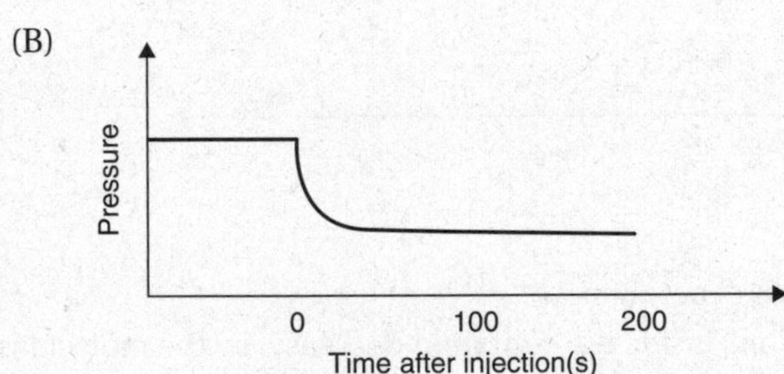

(C)

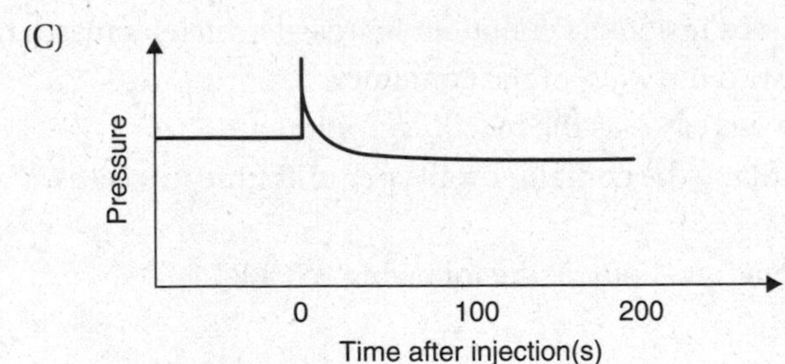

(D)

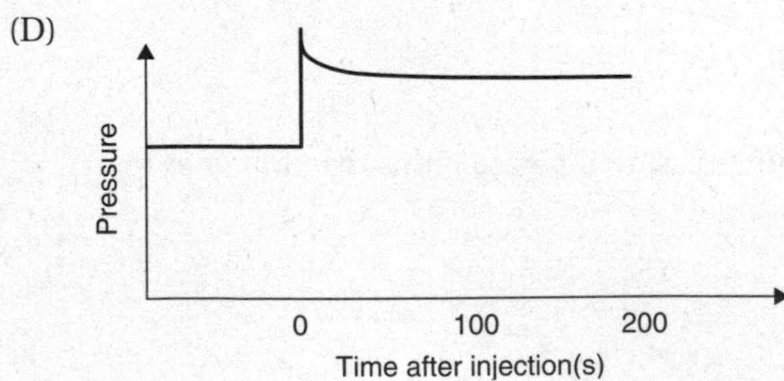

Questions 39–43 refer to the following information.

This formation equation for the reaction synthesizing RbBr(s) can be separated into a series of steps.

$Rb(s) + \frac{1}{2} Br_2(l) \rightarrow RbBr(s)$	$\Delta H° = -389\ kJ/mol_{rxn}$
$Rb(s) \rightarrow Rb(g)$	$\Delta H°(1)$
$Rb(g) \rightarrow Rb^+(g) + e^-$	$\Delta H°(2)$
$Br_2(l) \rightarrow 2Br(g)$	$\Delta H°(3)$
$Br(g) + e^- \rightarrow Br^-(g)$	$\Delta H°(4)$
$Rb^+(g) + Br^-(g) \rightarrow RbBr(s)$	$\Delta H°(5)$

39. Which of the steps in the table above are endothermic?
    (A) $\Delta H°(1)$ and $\Delta H°(2)$ only
    (B) $\Delta H°(1)$, $\Delta H°(2)$, and $\Delta H°(3)$ only
    (C) $\Delta H°(3)$ only
    (D) $\Delta H°(3)$ and $\Delta H°(4)$ only

40. If this reaction goes to completion, producing RbBr from the reactants, and if you use the overall chemical equation to estimate the entropy change for this process, which of the following statements is correct?
    (A) The reaction is favorable and driven by the enthalpy change since the entropy decreases in this process.
    (B) The reaction is unfavorable since the entropy change is a large negative value.
    (C) The reaction is favorable and driven by both enthalpy and entropy changes.
    (D) The reaction is unfavorable because of the enthalpy and entropy changes.

41. Using Hess's law, which of the following combinations will give the enthalpy for the following reaction?

$$Br_2(l) + 2e^- \rightarrow 2Br^-(g)$$

    (A) $\Delta H° = \Delta H°(3) + \Delta H°(4)$
    (B) $\Delta H° = 2\Delta H°(3) + \Delta H°(4)$
    (C) $\Delta H° = \Delta H°(3)/2 + \Delta H°(4)$
    (D) $\Delta H° = \Delta H°(3) + 2\Delta H°(4)$

42. When 0.100 mol of $Br_2(l)$ is formed from RbBr in the reaction above, how much heat is released or absorbed?
(A) 77.8 kJ of heat is released.
(B) 77.8 kJ of heat is absorbed.
(C) 38.9 kJ of heat is released.
(D) 38.9 kJ of heat is absorbed.

43. If 1.0 g of rubidium and 1.0 g of bromine are reacted, what will be left in measurable amounts (more than 0.10 mg) in the reaction vessel?
(A) RbBr only
(B) RbBr and Rb only
(C) RbBr and Br_2 only
(D) RbBr, Rb, and Br_2

Question 44 refers to the following information.

Formation of a solution can often be visualized as a three-step process.

Step 1. Solvent molecules are separated from each other to make space for the solute.
Step 2. Solute molecules are separated so they fit into spaces in the solvent.
Step 3. Separated solute and solvent are brought together, filling in the spaces.

44. Which of the following statements about the energy involved in the above is correct?
(A) Steps 1 and 2 are exothermic, and step 3 is endothermic.
(B) Steps 1 and 2 are endothermic, and step 3 is exothermic.
(C) All three steps are exothermic.
(D) All three steps are endothermic.

---

45. Nickel ($Z = 28$, $A = 59$) has a first ionization energy of 737 kJ/mol and a boiling point of 2913°C, while Pt ($Z = 78$, $A = 195$) has a first ionization energy of 870 kJ/mol and a boiling point of 3825°C. Which of the following are the most reasonable values of ionization energy and boiling point for palladium?
(A) –200 kJ/mol and 3524°C
(B) 795 kJ/mol and 2436°C
(C) 804 kJ/mol and 2963°C
(D) 1284 kJ/mol and 3416°C

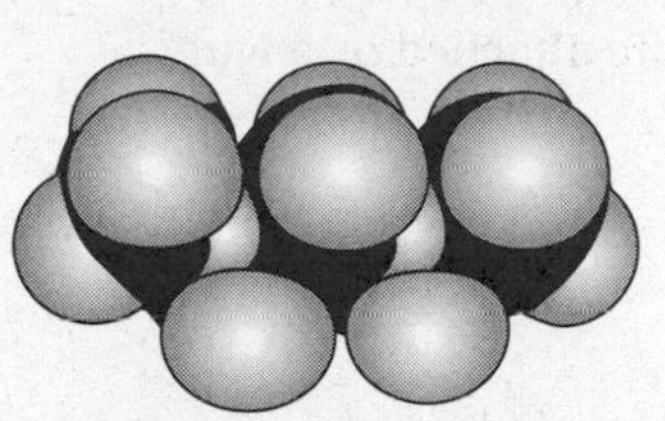

46. n-pentane (C_5H_{12} or $CH_3CH_2CH_2CH_2CH_3$) and 2,2-dimethylpropane (C_5H_{12} or $(CH_3)_4C$), shown above as space-filling models, each have the same number of carbon and hydrogen atoms but the atoms are arranged differently. n-pentane boils at 36.1°C, and 2,2-dimethylpropane boils at 9.5°C. Which statement best explains these data?
    (A) The long chains of n-pentane make it more difficult for them to reach the liquid surface and vaporize.
    (B) The long chains of n-pentane provide more sites for London attractive forces for neighboring molecules to have an effect on each other.
    (C) The compact structure of the 2,2-dimethylpropane directs the attractive forces internally in the molecule.
    (D) The bonds in the n-pentane are weaker, allowing small parts of the molecule to vaporize easily.

Use the following for questions 47–49.

A galvanic (voltaic) cell is described by the cell diagram below.

$$\text{Cu} \mid \text{Cu}^{2+} \parallel \text{Ag}^+ \mid \text{Ag} \qquad E^\circ_{\text{cell}} = +0.46\ V$$

A molecular-level view of the silver electrode is given.

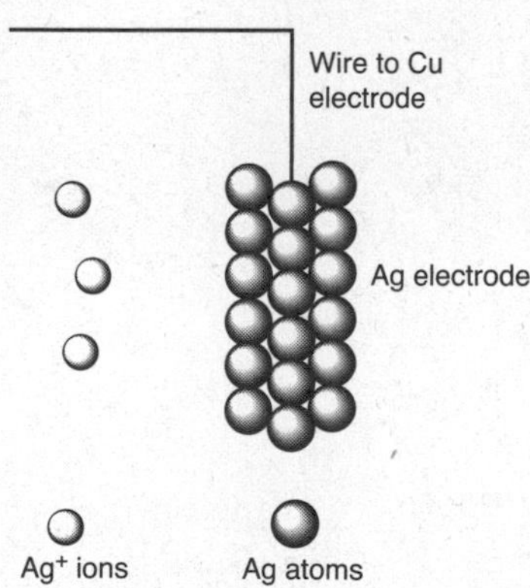

47. Which of the following best represents what happens to the silver metal electrode when the copper and silver electrodes are attached by a wire?

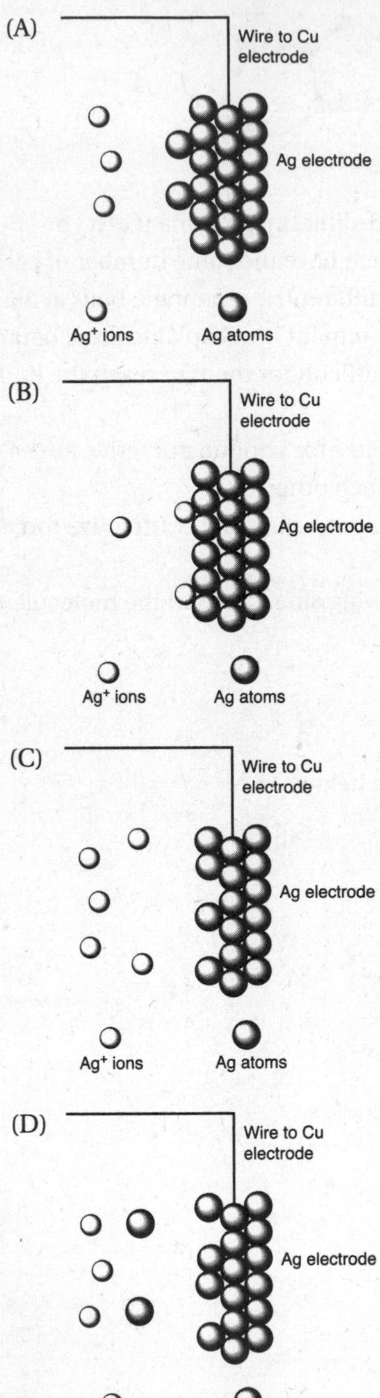

48. Which particulate view indicates that silver ions are being reduced?

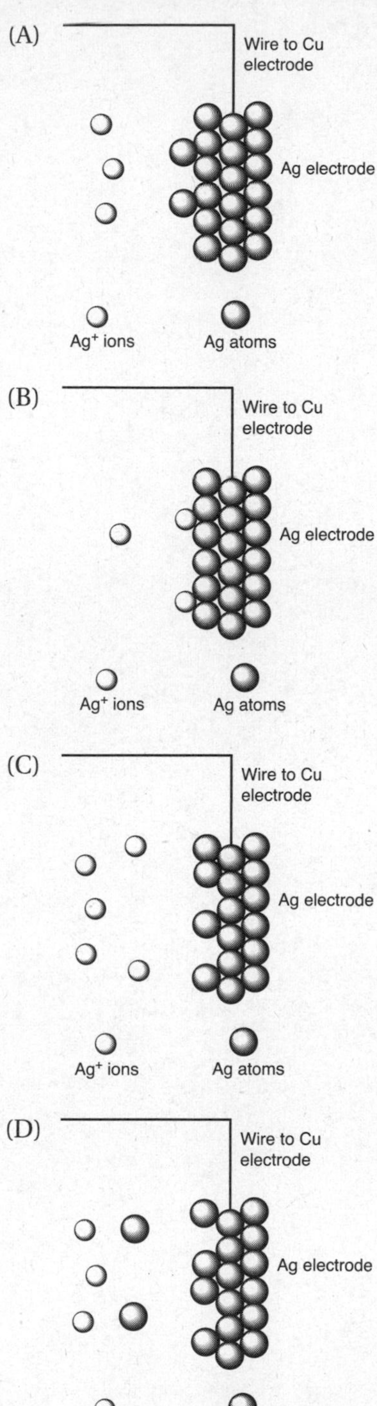

PRACTICE TEST 1

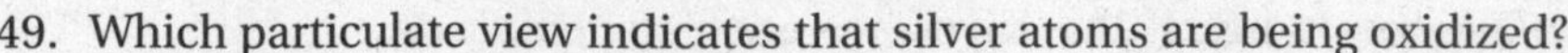

49. Which particulate view indicates that silver atoms are being oxidized?

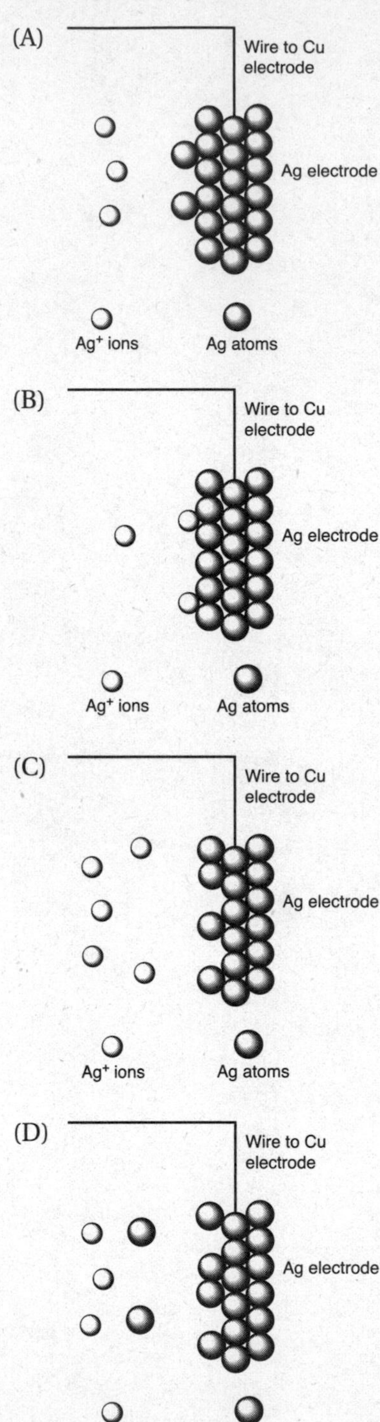

50. What is the purpose of a salt bridge in a galvanic cell?
   (A) To separate the two cells
   (B) To equalize the charge between the two cells as the redox process occurs
   (C) To allow the reducing substance to flow to the other cell to react with the oxidizing material
   (D) To allow electrons to flow from one cell to another to complete the circuit

Magnesium has an atomic radius of 160 pm and a first ionization energy of 737 kJ.

51. Based on periodic trends and the data given above, what are the most probable values for the ionization energy and atomic radius of sodium?
   (A) 186 pm, 496 kJ/mol
   (B) 186 pm, 898 kJ/mol
   (C) 135 pm, 523 kJ/mol
   (D) 147 pm, 523 kJ/mol

Iron is slowly oxidized by the permanganate in the reaction

$$5Fe^{2+} + 8H^{+} + MnO_4^{-} \rightarrow 5Fe^{3+} + Mn^{2+} + 4H_2O$$

The decrease in absorbance of a dilute solution of permanganate is followed using a spectrometer set to the appropriate wavelength in the visible region of the spectrum. The table below provides the data collected in this experiment.

Time (h)	Absorbance (A)	ln A	1/A
0	0.75	-0.29	1.3
1	0.38	-0.97	2.6
2	0.19	-1.7	5.3
3	0.095	-2.4	11

52. Which of the following is the best interpretation of the data?
   (A) The data support a first-order reaction.
   (B) The data support a second-order reaction.
   (C) The data support a zero-order reaction.
   (D) The overall order is 14.

53. 0.0025 mol of a weak, monoprotic acid is dissolved in 0.250 L of distilled water. The pH was then measured as 4.26. What is the pK_a of this weak acid?
    (A) 4.26
    (B) 8.52
    (C) 7.92
    (D) 6.52

Questions 54–57 refer to the data and graph below.

Four different acid solutions of 0.0100 *M* are prepared, and their pH values are recorded on a laptop computer. One of the solutions contains more than just an acid. At the point designated as "add," all four solutions are diluted with an equal volume of water. The bottom line represents solution 1, and the top line represents solution 4.

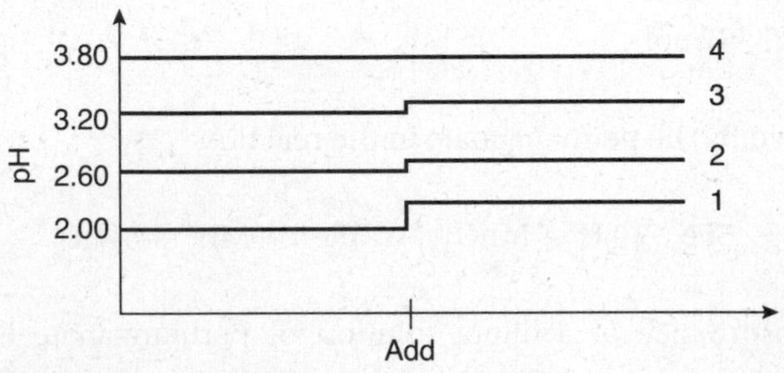

54. Which of the four acids is best described as a strong acid?
    (A) Solution 1 since the initial pH is 2.00
    (B) Solution 2 since there is a small pH change
    (C) Solution 3 since the initial pH is highest and changes when water is added
    (D) Solution 4 since added water doesn't affect the pH

55. Using the data in the graph, which solution seems to be a buffer?
    (A) Solution 1 since the solution is the most dilute
    (B) Solution 2 since water is a base and increases pH
    (C) Solution 3 since the pH is highest with a pH change
    (D) Solution 4 since added water has no effect on buffers

56. Which solution contains only an acid and the acid is the weakest of all?
    (A) Solution 1 since the weakest acid will have the largest pH change
    (B) Solution 2 since water changes the pH the least and therefore is the weakest
    (C) Solution 3 since the initial pH is the largest and the added water changes the pH
    (D) Solution 4 since this acid is so weak that water has no effect on the pH

57. Which acid(s) will require the most 0.0050 M NaOH to neutralize 25.0 mL of a 0.010 M solution of the acid?
(A) Solution 1
(B) Solution 2
(C) Solution 3
(D) They will all require the same volume of NaOH.

---

58. Sulfurous acid is a weak acid, while sulfuric acid is a much stronger acid because
(A) the sulfur in sulfuric acid is more electronegative than the sulfur in sulfurous acid
(B) sulfuric acid has more oxygen atoms in its formula
(C) the O–H bonds in sulfuric acid are much weaker than those in sulfurous acid
(D) sulfurous acid has its hydrogen atoms bound directly to the sulfur atom

59. You cannot prepare a buffer by
(A) mixing a solution of a weak base with a strong acid
(B) mixing a solution of a weak acid with a strong base
(C) mixing a solution of a strong base with a strong acid
(D) mixing a solution of a weak acid with its conjugate base

Photoelectron Spectroscopy Data Table	
Binding Energy (MJ)	**Relative Number of Electrons**
0.58	1
1.09	2
7.19	6
12.1	2
151	2

60. The photoelectron data table for aluminum is shown above. Which of the following statements is not true about these data?
(A) All 13 electrons are listed.
(B) The three valence electrons have the lowest energies at 0.58 MJ/mol and at 1.09 MJ/mol.
(C) The p electrons are all at the same energy of 7.19 MJ/mol.
(D) It takes 0.58 MJ to remove a valence electron.

IF YOU FINISH BEFORE THE 90 MINUTES HAVE ELAPSED, YOU MAY CHECK YOUR WORK IN THIS SECTION. DO NOT GO ON TO SECTION II UNTIL YOU ARE INSTRUCTED TO DO SO.

SECTION II: Free-Response

START ONLY WHEN YOU HAVE 105 MINUTES TO COMPLETE THE WHOLE SECTION.

Time:	1 hour, 45 minutes
Number of Questions:	7, 3 long and 4 short
Percent of Total Score:	50%
Calculator Allowed?	Yes
Pencil or pen with black or dark blue ink	

Instructions

The questions appear on the following pages. You may use the periodic table and the table of equations and constants that were provided with Section I. On the actual exam, you will be directed to write your final answers only in the test booklet. For this practice test, write your answers on separate sheets of lined paper. You will need about three pages for each of questions 1 through 3 and one page for each of questions 4 through 7.

Do not spend too much time on any one question. Budget your time carefully, and answer the easier questions first.

Be sure that your work is well organized, complete, and easy to read. Cross out or erase any mistakes. Erased and crossed-out material will not be scored.

SECTION II

7 Free-Response Questions
(Time: 105 minutes)
CALCULATORS ARE ALLOWED FOR SECTION II

Directions: Questions 1, 2, and 3 are long questions. Each question should take about 20 minutes to answer. Questions 4, 5, 6, and 7 are shorter questions. Each one should take about 5 to 10 minutes to answer. Read the questions carefully, and write your responses on lined paper. Your answers will be graded based on their correctness and relevance to the question asked and on the information cited. Explanations should be well organized and clearly written. Specific answers are always better than broad statements. For calculations, it is to your advantage to show clearly the method used and the steps involved in arriving at your answers since you may receive partial credit.

1. The measurement of the heat of a reaction, q, is measured at constant pressure using a coffee-cup calorimeter. At constant pressure, this measurement can be used to find the $\Delta H°$.

   A student performed an experiment using a coffee-cup calorimeter that contained 0.0525 mol of sodium hydroxide in 75.0 mL of solution. Next the student added 0.0370 mol of hydrochloric acid to this solution using a different 75.0 mL sample. (H_2O has a specific heat of 4.184 J g^{-1} C^{-1}.)

   (a) The figure below shows the data gathered for a coffee-cup calorimeter experiment using an electronic temperature sensor. Fill in the missing information on the two axes. What was the temperature change? Show how the change in temperature for this experiment is determined.

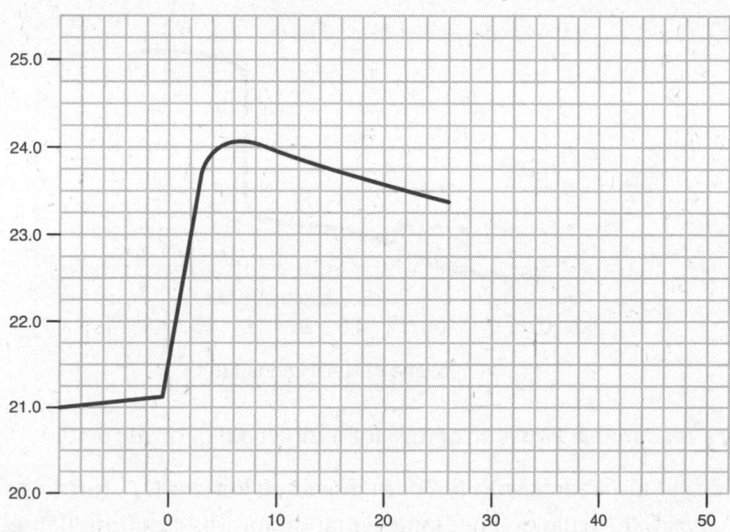

(b) What is the net ionic equation for what occurred in the calorimeter?

(c) Calculate the enthalpy change ($\Delta H°$) of this net ionic equation. (Assume that no significant amount of heat was lost or gained except from the chemical process and that the heat capacity of the system is determined solely by the total volume of water.)

(d) In a similar reaction, the 0.0370 mol of hydrochloric acid was replaced with 75.0 mL of a solution containing 0.0370 mol of acetic acid. A temperature change 0.11°C greater than that in the previous experiment occurred. Write the net ionic equation for acetic acid reacting with sodium hydroxide.

(e) What is the enthalpy change ($\Delta H°$) for the reaction in part (d)?

(f) By combining the data from parts (d) and (e) calculate the heat (enthalpy) of ionization of acetic acid.

2. Oxalic acid is found in relatively high concentrations in parsley, spinach, and rhubarb leaves. Ingesting oxalic acid can cause tissue burns because of its acidic properties. However, its main mode of action is the forming of toxic calcium oxalate in the body. Calcium oxalate precipitates, causing kidney failure. Gallstones are often high in calcium oxalate and are very painful. Small crystals of calcium oxalate precipitate in joints and cause a painful condition. Ethylene glycol ($HOCH_2CH_2OH$, a main ingredient in antifreeze) is poisonous since it is converted into oxalic via biochemical oxidation reactions.

(a) A 2.000 g sample of oxalic acid is burned in pure oxygen, and 1.95 g of CO_2 and 0.400 g of water are collected. What is the empirical formula of oxalic acid? Can this be the molecular formula? Explain your reasoning.

(b) A sample of oxalic acid is purified. It is then dissolved in water and titrated with a sodium hydroxide solution. The titration curve is shown below. What is the molecular formula of oxalic acid?

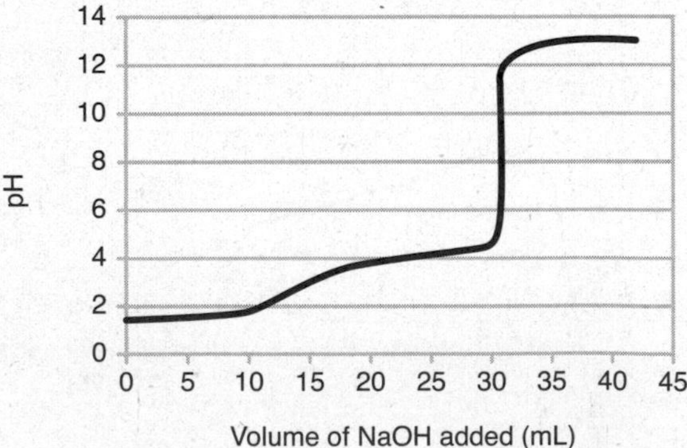

(c) Draw a reasonable Lewis structure for a molecule of oxalic acid.

(d) Oxalic acid also reacts with the permanganate ion, MnO_4^-, forming CO_2 and Mn^{2+}. Write the balanced net ionic equation for this reaction. Remember that permanganate reduces to Mn^{2+} only in a strongly acid solution.

(e) Does the permanganate ion oxidize or reduce the oxalic acid? Justify your answer.

(f) When a weighed mass of pure oxalic acid is titrated with a permanganate solution of known molarity, can the data be used to verify the molecular formula, the empirical formula, or both? Justify your answer.

3. When chlorine gas reacts with ammonia gas, they produce gaseous nitrogen trichloride and hydrogen chloride in equilibrium with the reactants.

(a) Write a balanced chemical equation depicting this equilibrium.

(b) Write the equilibrium expression, K_{eq}, in terms of concentrations for this chemical equation.

(c) What would happen to the concentration of ammonia upon the addition of more chlorine gas to a mixture that is at equilibrium? Explain by adding to the diagram below showing how the ammonia concentration will change with time.

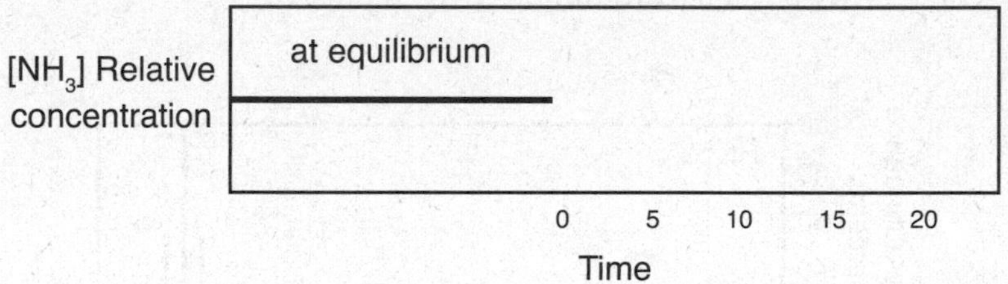

(d) If the value of K_{eq} for this reaction is 2.4×10^{-9}, are there more products or reactants present at equilibrium?

(e) Given initial concentrations of chlorine = 0.10 *M*, ammonia = 0.20 *M*, nitrogen trichloride = 0.20 *M*, and hydrogen chloride = 0.030 *M*, is this system at equilibrium? If not, in which direction will the reaction shift in order to achieve equilibrium? Explain your answer

4. In the early 1800s, John Dalton stated his atomic theory. Based solely on two laws discovered by earlier chemists (alchemists), he reasoned:

i. Matter consists of indestructible particles called atoms.
ii. Atoms of each element are identical in mass and all other properties.
iii. Atoms of different elements differ in mass and all other properties.
iv. Compounds are simple combinations of atoms in fixed ratios.
v. Chemical reactions are simple rearrangements of atoms.

Scientific theories often spawn new insights that can be tested and used. The law of multiple proportions is one such insight. This law states that if two different compounds, made from the same elements, have the same mass of one of the component elements, then the masses of the other elements will be present in a simple whole-number ratio.

	Compound 1	Compound 2
Mass of N	8.40 g	3.64 g
Mass of O	4.80 g	4.16 g

(a) Use the data in the table above to determine if compound 1 is different from compound 2. If the formula for compound 1 is N_2O, what is the formula for compound 2? Does this demonstrate the law of multiple proportions or the law of definite proportions?

(b) Draw the five possible Lewis structures for compound 1, showing all electron pairs. Using formal charges, select the most probable structure. Explain your reasoning.

(c) Below is a mass spectrum of bromine. Bromine has two main isomers with masses of 79 and 81 in virtually equal proportions. Use this figure to explain the atomic mass of bromine listed in your periodic table as 79.90. Then explain what the three peaks clustered around $\frac{m}{e} = 160$ represent.

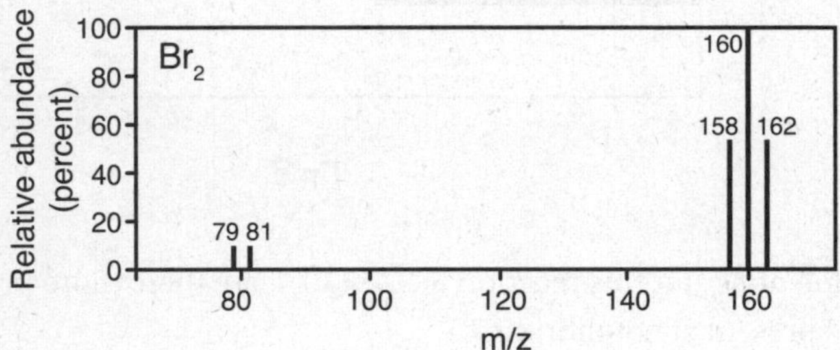

5. Chemists often think at two levels. One is the atomic or molecular level where the interactions of individual atoms and molecules are visualized. The other is often called the laboratory scale where amounts are measured from a few milligrams up to a few kilograms. Traditionally, chemical reactions have been represented with balanced equations. Other representations such as diagrams or simple pictures are often used to illustrate chemical reactions or structures at the atomic level.

(a) Use the following equation to illustrate how it may be interpreted at the atomic level and at the laboratory scale (grams or moles).

$$2OH^- + H_3PO_4 \rightarrow HPO_4^{2-} + 2H_2O$$

(b) In the equation shown above, identify the two pairs of conjugate acids and bases. Be sure to identify the acid and the base within each pair.

(c) The single arrow in the equation shown above generally means that the reaction proceeds very far toward completion. That being said, identify the stronger of the two acids and the stronger of the two bases.

(d) The weak base ethyl amine, $CH_3CH_2NH_2$, is titrated with HBr. Write the formulas for the conjugate acid and conjugate base of ethylamine. There are five boxes associated with five points along the titration curve shown below. In each box, use only ten symbols (e.g., 3CA and 7CB would represent approximately 30% conjugate acid—CA—and 70% conjugate base—CB—in the solution at the indicated point) to show the relative amounts of the conjugate acid (CA) and base (CB) at the designated points along the curve. When there is a very large excess of one ion or another, the box may have all ions of the same kind.

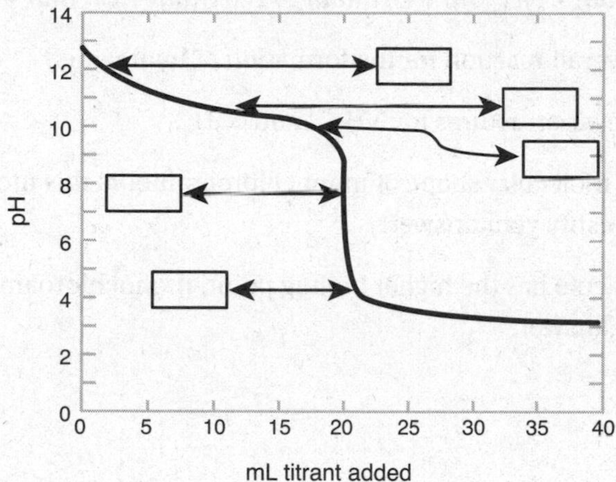

6. The decomposition of dinitrogen pentoxide, N_2O_5, results in the formation of nitrogen dioxide and nitrogen trioxide. Molecules of the general formula NO_x are involved in the depletion of the ozone layer. A decomposition reaction for this molecule is shown below.

$$N_2O_5(g) \rightarrow NO_3(g) + NO_2(g)$$

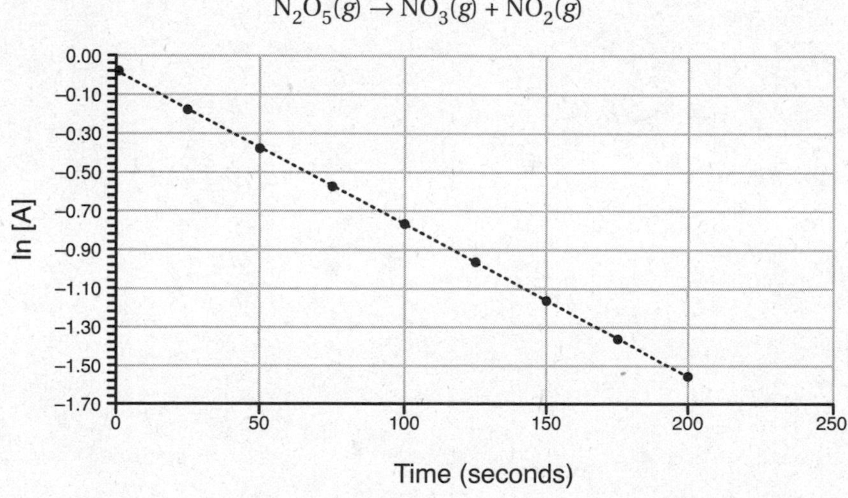

(a) The data shown above illustrate that the reaction is first order. Write the appropriate rate law for this reaction.

(b) Calculate the rate constant for this reaction using the graph above. Justify.

(c) What is the half-life, in seconds, for this reaction?

7. Everyone uses household chemicals to clean. Most do not think about the possible hazards posed by these chemicals. For example, bleach is often used when washing clothes. It is a mixture of chlorine gas and sodium hydroxide that makes NaOCl. Ammonia, NH_3, was used for years to clean windows since it completely evaporates. Mixing bleach and ammonia can form two very toxic compounds, monochloroamine (NH_2Cl) and hydrazine (N_2H_4).

$$NH_3(aq) + OCl^-(aq) \rightarrow NH_2Cl(aq) + OH^-(aq)$$

$$NH_2Cl(aq) + NH_3(aq) + OH^-(aq) \rightarrow N_2H_4(aq) + Cl^-(aq) + H_2O(l)$$

(a) Write the overall reaction for the formation of hydrazine.

(b) Draw the Lewis structures for NH_2Cl and N_2H_4.

(c) What is the molecular shape of monochloroamine? Is this molecule polar or nonpolar? Justify your answer.

(d) Which molecule has the higher boiling point, monochloroamine or hydrazine? Justify your answer.

STOP **END OF EXAM. If you finish before the 90 minutes have elapsed, you may check your work in this section only.**

ANSWER KEY

Practice Test 1

1. **(B)**	11. **(C)**	21. **(C)**	31. **(B)**	41. **(D)**	51. **(A)**
2. **(A)**	12. **(C)**	22. **(D)**	32. **(B)**	42. **(B)**	52. **(A)**
3. **(D)**	13. **(D)**	23. **(D)**	33. **(C)**	43. **(C)**	53. **(D)**
4. **(D)**	14. **(A)**	24. **(D)**	34. **(D)**	44. **(B)**	54. **(A)**
5. **(C)**	15. **(B)**	25. **(D)**	35. **(A)**	45. **(C)**	55. **(D)**
6. **(C)**	16. **(A)**	26. **(D)**	36. **(C)**	46. **(B)**	56. **(C)**
7. **(B)**	17. **(A)**	27. **(C)**	37. **(B)**	47. **(A)**	57. **(D)**
8. **(B)**	18. **(B)**	28. **(A)**	38. **(D)**	48. **(A)**	58. **(C)**
9. **(C)**	19. **(C)**	29. **(A)**	39. **(B)**	49. **(C)**	59. **(C)**
10. **(B)**	20. **(A)**	30. **(A)**	40. **(A)**	50. **(B)**	60. **(C)**

CHAPTER REFERENCES FOR MULTIPLE-CHOICE QUESTIONS

Question	Chapter
1	8, 9, 13
2	4, 7
3	8, 9, 13
4	12, 14
5	4, 7, 8
6	10, 12
7	2, 3, 4
8	5, 6, 7, 14
9	7, 8, 10, 11
10	8, 11
11	8, 13
12	7, 8, 9
13	11
14	10
15	6, 9
16	2
17	1, 2, 5, 6
18	5, 13, 14
19	9, 13
20	5, 13, 14

Question	Chapter
21	13, 14
22	4, 13
23	9, 13
24	13
25	6
26	6, 7, 10
27	7, 8
28	4, 8
29	4, 7, 8
30	4, 6
31	1, 2, 3
32	3, 14
33	9, 11
34	5, 6, 10, 11
35	5, 9, 10
36	5, 6
37	6, 9
38	8, 9, 10
39	7, 11
40	11

Question	Chapter
41	9, 11
42	5, 11
43	5, 14
44	8, 11
45	4, 7
46	4, 7
47	5, 12
48	11, 12
49	3, 8, 12
50	11, 12
51	1, 2
52	5, 10
53	9, 13
54	9, 13
55	9, 13
56	4, 13
57	5, 13
58	4, 13
59	13
60	2

ANSWERS EXPLAINED

Section I—Multiple-Choice

1. **(B)** A higher electronegativity difference between H and the second atom results in a weaker H–*X* bond. The larger diameter of the second atom also weakens the H–*X* bond. With binary compounds, the bond length is more important in weakening this bond. So HI, where iodine is the largest atom, has the weakest H–*X* bond and is the strongest acid. (E.K. 1.C.1; L.O. 1.9–1.11)

2. **(A)** The boiling point indicates the strength of intermolecular forces. Ethane has the lowest intermolecular forces of attraction and thus the lowest surface tension. (E.K. 2.B.1, 2.B.2; L.O. 2.11, 2.16)

3. **(D)** For binary acids in a single group of the periodic table, acid strength increases as bond strength decreases and bond strength decreases as bond length increases. The increasing radius in the table indicates that the shortest and strongest bond belongs to water. So water is the weakest acid. (E.K. 1.C.1; L.O. 1.9–1.11)

4. **(D)** Sulfur and fluorine commonly gain electrons and cause other substances to be oxidized. Cadmium and potassium lose electrons and cause other substances to be reduced. Cadmium is less hazardous and will be preferred in this case. (E.K. 3.B.3; L.O. 3.8)

5. **(C)** Dashed lines traditionally represent attractions that are NOT covalent bonds. In this case, they represent hydrogen bonds. (E.K. 2.B.2; L.O. 2.13)

6. **(C)** When ground to fine particles, most metals will burn; some are even explosive. (E.K. 3.A.1; L.O. 3.1)

7. **(B)** The chemical equation tells us nothing about the position of equilibrium or if the reaction is thermodynamically favored. (E.K. 6.B.2; L.O. 3.1)

8. **(B)** We calculate that the mass of Mg needed to produce 50 mL of H_2 is approximately 0.05 g, so 0.1 g of water left will be a 200 percent error. Recording 25°C rather than 298 K is a 1000 percent error. Since the total gas volume is 50 mL, ±20 mL is a 40 percent error. The vapor pressure error is less than 5 percent and is the smallest error. (E.K. 2.A.2, 3.B.3; L.O. 2.6, 3.8)

9. **(C)** Slightly to the right of the maximum of the distribution curve is the average kinetic energy. (E.K. 2.A.2; L.O. 2.4)

10. **(B)** All the other statements violate the law of conservation of energy. (E.K. 5.B.4; L.O. 5.7)

11. **(C)** Conjugate acid–base pairs always differ by one H^+. Only the ions in (C) meet this criterion. (E.K. 3.B.2; L.O. 3.7)

12. **(C)** Response (C) is the only one that has different phases noted. (E.K. 3.B.1; L.O. 3.6)

***NOTE:** A complete list of the Essential Knowledge (E.K.) needed and the Learning Objectives (L.O.) can be found in the AP Chemistry Course Description:
https://secure-media.collegeboard.org/digitalServices/pdf/ap/ap-chemistry-course-and-exam-description.pdf

13. **(D)** For this reaction, there is no temperature from absolute zero on up where the $\Delta G°$ will not be negative. (E.K. 5.E.2, 5.E.3; L.O. 5.13, 5.14)

14. **(A)** None of the structures has an odd number of electrons, which is necessary for a free radical. (E.K. 4.C.1; L.O. 4.7)

15. **(B)** The number of particles adds up to 18, and they are distributed in 6 liters of volume, giving 3 particles per liter. Since 6 particles per liter represents 633 torr, the pressure should be half of that, or 316 torr. (E.K. 2.A.2; L.O. 2.5)

16. **(A)** In this mass spectrum, we see about 90 percent of the atoms with a mass of 20, about 1 percent with a mass of 21, and 10 percent with a mass of 22. Calculate:

$$(0.9)(20) + (0.01)(21) + (0.1)(22) = 18 + 0.2 + 2.2 = 20.4 \text{ for the average mass}$$

Since each mass was a little less than an integer value, 20.3 is reasonable. (E.K. 1.D.1; L.O. 1.14)

17. **(A)** The carbon and oxygen atoms are balanced. This also represents a limiting reactant process with excess oxygen. (E.K. 1.E.2, 3.A.2; L.O. 1.18, 3.4)

18. **(B)** Point *E* is the inflection point that represents the end point. (E.K. 6.C.1; 6.C.2; L.O. 6.12, 6.13)

19. **(C)** Point *C* is the midpoint of the titration. At this point, the pOH = pK_b. The pOH = 14 – pH. So 14 – pH = pK_b, and this rearranges to the equation shown in the question. (E.K. 6.C.1, 6.C.2; L.O. 6.19)

20. **(A)** This indicator changes mainly within the inflection region of the curve. So the end point will be visually sharp and the result accurate. (E.K. 6.C.1, 6.C.2; L.O. 6.19)

21. **(C)** Points A and D represent the buffer region. Their pH difference is approximately 2 pH units. (E.K. 6.C.1, 6.C.2; L.O. 6.19)

22. **(D)** The equation is M_b(25.0 mL) = (0.075 *M*)(22.5 mL). When solving for M_b, we see that 22.5/25.0 is less than 1. So the molarity of the base must be less than 0.075. (E.K. 6.C.1, 6.C.2; L.O. 6.19)

23. **(D)** We can use the approximation that the square root of the K_b times the initial concentration will give us the [OH^-]. Taking the square root of 4.4×10^{-6} yields approximately 2×10^{-3}. So the pOH will be a little less than 3. Converting to pH results in a range of 10 to 12. (E.K. 6.C.2; L.O. 6.16)

24. **(D)** To calculate the pK_a for the conjugate acid of a weak base, we first calculate the K_a by dividing K_w by K_b for the weak base. Then we take the negative of the logarithm (base 10) of the K_a to get the pK_a. We see that response (D) is log (K_b/K_w), which is mathematically the same as –log (K_w/K_b) = –log K_a = pK_a. (E.K. 6.C.1; L.O. 6.13)

25. **(D)** The pressure times the molar mass, at constant temperature, is related to the mass in the container. The lower the mass, the lower the density is if the containers are the same size. (E.K. 2.A.2; L.O. 2.6)

26. **(D)** Since the temperature is the same in all cases, the average kinetic energy is the same. (E.K. 5.A.1; L.O. 5.2)

27. **(C)** 1-bromopropane should have similar dipole attractive forces as the other molecules due to the bromine. However, it has a longer carbon chain that increases the effect of London forces. (E.K. 2.B.1, 2; L.O. 2.11, 2.13)

28. **(A)** There is no hydrogen bonding in any of these molecules. (E.K. 2.B.3; L.O. 2.16)

29. **(A)** Use the VSEPR theory and the fact that SO_3 has three domains (triangular planar) and that SO_3^{2-} has four domains (tetrahedral domain structure and triangular pyramid molecular shape). (E.K. 2.C.4; L.O. 2.21)

30. **(A)** All three statements in response (A) are true. The other responses are either incomplete or have errors. (E.K. 2.B.2, 2.C.4; L.O. 2.13, 2.21)

31. **(B)** This equation is balanced and also has the compounds that appear in the description of the reaction. (E.K. 6.C.3; L.O. 6.21, 6.22)

32. **(B)** This beaker has the excess KCl along with the correct ratio of K^+ to ClO_4^- ions. (E.K. 2.A.2; L.O. 2.5)

33. **(C)** $K_{sp} = 1.7 \times 10^{-5} = [Pb^{2+}][Cl^-]^2 = (s)(2s)^2 = 4s^3$ (where s = molar solubility). Dividing by 4 gives $s^3 = 4 \times 10^{-6}$. Finally, taking the cube root of 10^{-6} gives 10^{-2}. Actually, the true answer is slightly more than 10^{-2}. (E.K. 6.C.2; L.O. 6.21, 6.22)

34. **(D)** As $N_2O_4(g)$ reacts, we get twice as many molecules, thus increasing the pressure. (E.K. 2.A.2; L.O. 2.6)

35. **(A)** The only way we know of to determine the activation energy is to measure reaction rates at different temperatures. However, no temperatures are mentioned. (E.K. 5.A.1; L.O. 5.1)

36. **(C)** The pressure will increase to 1520 torr, which is 634 torr greater than 886 torr. (E.K. 6.A.1; L.O. 6.1, 6.2)

37. **(B)** If the pressure rises to 886 torr, which means that 126 × 2 torr of NO_2 is produced and 634 torr of N_2O_4 remain. Dividing both by 760 torr gives 0.332 atm NO_2 and 0.834 atm N_2O_4. Since $K_P = (0.332)^2/(0.834)$, we can see that the result must be less than 1. (E.K. 6.A.3; L.O. 6.4)

38. **(D)** This diagram shows the rapid increase (vertical line) for the injection and then the slow reaction (curved line) as the extra NO_2 combines to form N_2O_4. (E.K. 6.B.1; L.O. 5.8)

39. **(B)** All processes that separate molecules or break bonds are endothermic. (E.K. 5.C.2; L.O. 5.8)

40. **(A)** Using general rules, we have a solid and a liquid combining to make one formula unit of solid. A small change in entropy is expected. The enthalpy is very large compared with the entropy values, especially for binary compounds. So this answer is reasonable. (E.K. 5.E.2; L.O. 5.8, 5.13)

41. **(D)** Combining $\Delta H°(3)$ and $2\Delta H°(4)$ will duplicate the reaction and give the proper enthalpy. (E.K. 5.E.3; L.O. 5.8)

42. **(B)** Since the reaction produces only 1/2 mol Br_2, the 0.100 mol is one-fifth of the amount in the reaction. So the heat of reaction is multiplied by 1/5. Since the reaction is effectively reversed, the heat will be absorbed. (E.K. 5.E.3; L.O. 5.8)

43. **(C)** The atomic mass of Rb is 85.47 g/mol and of Br is 79.9 g/mol. With equal masses, we will have more moles of Br than Rb. Since they combine in a 1:1 ratio, we can deduce that Rb will be the limiting reactant and that virtually none will be left when the reaction is complete. (E.K. 1.A.2, 1.A.3, 1.E.1; L.O. 1.2, 4, 17)

44. **(B)** The separation of molecules or the breaking bonds is endothermic. (E.K. 2.A.1, 2.A.3, 5.B.1, 5.B.3; L.O. 2.3, 2.8, 5.5)

45. **(C)** You must know that physical properties often change in a regular manner up and down as a group. This was Mendeleev's famous contribution. So the values for palladium should fall between those of the elements nickel and platinum, which are just above and below palladium, respectively. (E.K. 1.C.1; L.O. 1.9)

46. **(B)** Although London forces tend to be small, molecules that allow many sites for interaction experience stronger attractions. (E.K. 2.B.1; L.O. 2.11)

47. **(A)** The cell diagram can be converted to the following chemical equation:

$$Cu + 2Ag^+ \rightarrow 2Ag + Cu^{2+}$$

Since the standard cell voltage is positive, the reaction proceeds in the direction indicated. Silver ions were reduced to silver atoms at the electrode surface. Two Ag^+ ions (white spheres) have become two silver atoms (dark spheres) plated onto the electrode. (E.K. 3.B.3, 3.C.3; L.O. 3.8, 3.12, 3.13)

48. **(A)** Two Ag^+ ions (white spheres) were reduced to two silver atoms (dark spheres) plated onto the electrode. (E.K. 3.B.3, 3.C.3; L.O. 3.8, 3.12, 3.13)

49. **(C)** Two silver atoms (dark spheres) have left the electrode after losing an electron to form silver ions (white spheres). (E.K. 3.B.3, 3.C.3; L.O. 3.8, 3.12, 3.13)

50. **(B)** When a substance is oxidized, a positive ion enters the solution. A negative charge must come from the salt bridge to keep the positive and negative charges equal. Similarly, when a substance is reduced at the cathode by adding a negative charge from the electrode, a positive charge must come from the salt bridge to balance that charge. (E.K. 3.C.3; L.O. 3.12, 3.13)

51. **(A)** Trends tell us that the atomic radius of sodium should be larger than that of magnesium since atomic radii decrease from left to right across a period. In addition, the first ionization energies increase from left to right across a period. This response has the only reasonable data incorporating those trends. (E.K. 1.C.1; L.O. 1.9, 1.10)

52. **(A)** We need to find a linear change to indicate which plot will be linear. Since the change in time occurs in equal increments, equal changes in one of the columns will give us a straight line. With a few trials, we see that ln A has a uniform change of –0.7 as we go down the table. A plot of ln A vs t is characteristic of a first-order process. (E.K. 1.D.3, 3.B.3, 4.A.3; L.O. 1.15, 1.16, 3.9, 4.12)

53. **(D)** This can be solved without a calculator. The measured pH gives us $[H^+] = [A^-] = 10^{-4.26}$. The initial concentration of the acid, HA, is 0.010 $M = 10^{-2}$:

$$K_a = [H^+][A^-]/[HA] = (10^{-4.26} \times 10^{-4.26})/10^{-2}$$

PRACTICE TEST 1

We can solve this by adding the two exponents in the numerator and subtracting the denominator's exponent: $K_a = 10^{-4.26-4.26-(-2)} = 10^{-6.52}$. Now take the logarithm to get –6.52, and change the sign to get $pK_a = 6.52$. (E.K. 6.C.1; L.O. 6.16)

54. **(A)** A 0.010 *M* solution of a strong acid should have a pH = 2.00. (E.K. 1.E.2, 2.A–2.D; L.O. 1.20, 2.2, 6.12)

55. **(D)** Diluting a buffer with water will have no significant effect on the buffer's pH. (E.K. 6.C.1; 6.16)

56. **(C)** This is the solution with the highest pH that is not a buffer. (E.K. 6.C.1; L.O. 6.17)

57. **(D)** All of the acid solutions, 1, 2, and 3, will require the same amount of base to reach the end point. (E.K. 6.C.1; L.O. 6.17)

58. **(C)** The reason for the difference in strengths is the strength of the –O–H bonds. (A) and (B) are properties that help us determine if that bond is strong or weak. (D) is just wrong. (E.K. 6.C.1; L.O. 6.12)

59. **(C)** Mixing strong acids and strong bases does not result in buffer solutions. (E.K. 6.C.2; L.O. 6.18)

60. **(C)** Aluminum has six $2p$ electrons (7.19 MJ) and one $3p$ electron (0.58 MJ). (E.K. 1.B.2; L.O. 1.7)

Section II—Free-Response

1. (a)

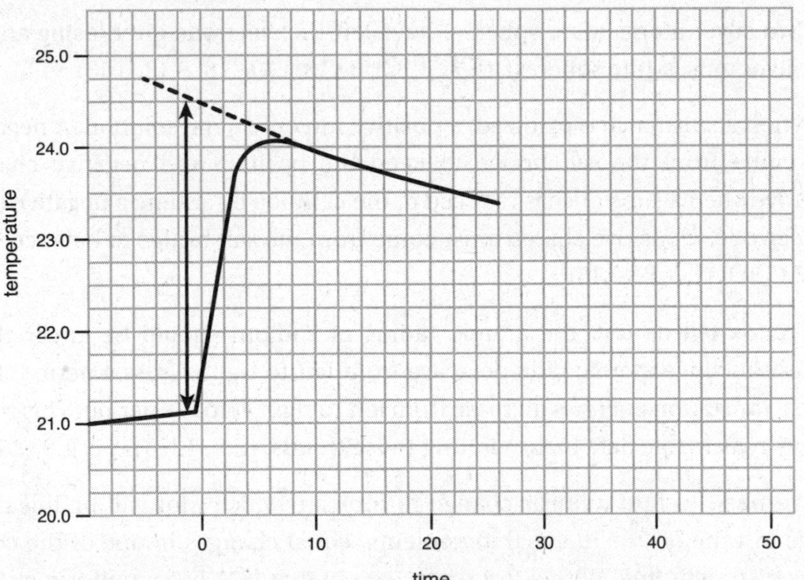

You should extrapolate the cooling portion back to the point of mixing and measure the temperature change at that point. The temperature change is 24.5°C – 21.1°C = 3.40°C in this diagram.

(b) The net ionic equation for the reaction of NaOH with HCl is

$$H^+(aq) + OH^-(aq) \rightarrow H_2O$$

or

$$H_3O^+(aq) + OH^-(aq) \rightarrow 2H_2O$$

(c) Calculate the heat energy evolved as

$$q = (\text{specific heat})(\text{mass})(\Delta T)$$
$$= 4.184\ \text{J g}^{-1}\,°\text{C}^{-1}(150.\ \text{g})(3.40°\text{C})$$
$$= 2134\ \text{J}$$

From the data given, we can see that the sodium hydroxide is the excess reagent and that the hydrochloric acid is the limiting reactant because there are many more moles of NaOH than HCl in the mixture. We deduce that 0.0381 mol of HCl reacts. The enthalpy is assigned a negative sign when heat is evolved (i.e., when the process is exothermic) and we divide by the moles reacted to obtain the molar enthalpy.

$$\Delta H° = \frac{2134\ \text{J}}{0.0370\ \text{mol}} = -57{,}676\ \text{J mol}^{-1} = -57.7\ \text{kJ mol}^{-1}$$

(d)

$$HC_2H_3O_2(aq) + OH^-(aq) \rightarrow C_2H_3O_2^-(aq) + H_2O$$

(e)

$$q = 4.184\ \text{J g}^{-1}\,°\text{C}^{-1}(150.\ \text{g})(3.51°\text{C}) = 2202.9\ \text{J}$$

$$\Delta H° = \frac{2202.9\ \text{J}}{0.0370\ \text{mol}} = -59{,}537\ \text{J mol}^{-1} = -59.5\ \text{kJ mol}^{-1}$$

(f) The ionization reaction is

$$HC_2H_3O_2(aq) + H_2O \rightarrow C_2H_3O_2^-(aq) + H_3O^+(aq)$$

We need to combine

$$HC_2H_3O_2(aq) + OH^-(aq) \rightarrow C_2H_3O_2^-(aq) + H_2O$$

and

$$2H_2O \rightarrow H_3O^+(aq) + OH^-(aq)$$

To get the appropriate heat of reaction, we calculate

$$\Delta H°_{\text{reaction}} = \Delta H°_{\text{acetic acid}} - \Delta H°_{\text{water}}$$
$$= -59.5\ \text{kJ mol}^{-1} - (-57.7\ \text{kJ mol}^{-1}) = -1.8\ \text{kJ mol}^{-1}$$

The heat (enthalpy) of ionization of acetic acid is $-1.8\ \text{kJ mol}^{-1}$.

2. (a) Using the data given, calculate the mass of carbon and hydrogen in the sample:

$$?\text{g C} = 1.95\ \text{g}\ CO_2\left(\frac{1\ \text{mol}\ CO_2}{44.01\ \text{g}\ CO_2}\right)\left(\frac{1\ \text{mol C}}{1\ \text{mol}\ CO_2}\right)\left(\frac{12.01\ \text{g C}}{1\ \text{mol C}}\right) = 0.5321\ \text{g C}$$

$$?\text{g H} = 0.400\ \text{g}\ H_2O\left(\frac{1\ \text{mol}\ H_2O}{18.02\ \text{g}\ H_2O}\right)\left(\frac{2\ \text{mol H}}{1\ \text{mol}\ H_2O}\right)\left(\frac{1.008\ \text{g H}}{1\ \text{mol H}}\right) = 0.0448\ \text{g H}$$

Subtract the masses of C and H from the total 2.000 g sample to get the mass of oxygen.

$$?\text{g O} = 2.000 - 0.5321\ \text{g C} - 0.0448\ \text{g H} = 1.4231\ \text{g O}$$

Convert masses to moles. Then find the simplest ratio of moles of C, H, and O:

$$?\text{mol H} = 0.0448\text{ g H}\left(\frac{1\text{ mol H}}{1.008\text{ g H}_2\text{O}}\right) = 0.0444\text{ mol H}$$

$$?\text{mol C} = 0.5321\text{ g C}\left(\frac{1\text{ mol C}}{12.01\text{ g C}}\right) = 0.0443\text{ mol C}$$

$$?\text{mol O} = 1.4231\text{ g O}\left(\frac{1\text{ mol O}}{16.00\text{ g O}}\right) = 0.0889\text{ mol O}$$

Divide all answers by 0.0443 to obtain the relative number of moles of each:

$$\frac{0.0444\text{ mol H}}{0.0443} = 1.002\text{ mol H}$$

$$\frac{0.0443\text{ mol C}}{0.0443} = 1.000\text{ mol C}$$

$$\frac{0.0889\text{ mol O}}{0.0443} = 2.007\text{ mol O}$$

This gives an empirical formula of CHO_2.

Common organic compounds have an even number of valence electrons so that all electrons are paired. (Otherwise, we will have a free radical with an unpaired electron that is not common.) Carbon and oxygen have 4 and 6 valence electrons respectively. Hydrogen has 1 valence electron. This gives an odd number of valence electrons that will not yield a common compound. So CHO_2 cannot be the molecular formula of oxalic acid.

(b) The titration curve has two end points, indicating a diprotic acid. So there are two empirical formula units in the molecule. The molecular formula is therefore $C_2H_2O_4$. Three empirical units would have an odd number of valence electrons and there is no evidence for larger numbers of empirical units per mol. The empirical formula contains all the atoms needed for an organic acid –COOH (or $-CO_2H$). Two acid groups, back to back, make up a molecule of oxalic acid: HOOCCOOH.

(c) Recall that the organic acid group has a double-bonded oxygen and a singly bonded –OH group connected to the carbon. So the Lewis structure for oxalic acid is

```
       :Ö: :Ö:
        ‖   ‖
  H–Ö–C–C–Ö–H
     ¨       ¨
```

(d) Start by writing the skeleton equations:

$$MnO_4^- \rightarrow Mn^{2+}$$

$$C_2H_2O_4 \rightarrow CO_2$$

Atoms balanced:

$$8H^+ + MnO_4^- \rightarrow Mn^{2+} + 4H_2O$$

$$C_2H_2O_4 \rightarrow 2CO_2 + 2H^+$$

Charges balanced:

$$5e^- + 8H^+ + MnO_4^- \rightarrow Mn^{2+} + 4H_2O$$

$$C_2H_2O_4 \rightarrow 2CO_2 + 2H^+ + 2e^-$$

Electrons equalized:

$$10e^- + 16H^+ + 2MnO_4^- \rightarrow 2Mn^{2+} + 8H_2O$$

$$5C_2H_2O_4 \rightarrow 10CO_2 + 10H^+ + 10e^-$$

Added and canceled:

$$10e^- + 16H^+ + 2MnO_4^- + 5C_2H_2O_4 \rightarrow 10CO_2 + 10H^+ + 2Mn^{2+} + 8H_2O + 10e^-$$

$$6H^+ + 2MnO_4^- + 5C_2H_2O_4 \rightarrow 10CO_2 + 2Mn^{2+} + 8H_2O$$

(e) It is difficult to work with oxidation numbers in organic compounds. Focus on the MnO_4^- reacting to become Mn^{2+}. The oxidation number of Mn in MnO_4^- is +7, while Mn^{2+} has an oxidation number of +2. Manganese is reduced, so the oxalic acid must be oxidized.

(f) The titration data will verify the molar mass of the empirical formula only. However, many empirical formula units may compose the molecular formula and any multiple of the empirical formula can also be verified by the titration data. Therefore, the titration data do not uniquely verify the molar mass and molecular formula.

3. (a) Begin by writing the formulas for the compounds involved in the chemical reaction:

chlorine gas = Cl_2 ammonia = NH_3

nitrogen trichloride = NCl_3 hydrogen chloride = HCl

Write the unbalanced chemical equation:

$$Cl_2(g) + NH_3(g) \rightarrow NCl_3(g) + HCl(g)$$

Now balance the equation by using coefficients:

$$3Cl_2(g) + NH_3(g) \rightarrow NCl_3(g) + 3HCl(g)$$

(b) The equilibrium expression is written as:

$$K_{eq} = \frac{[NCl_3][HCl]^3}{[Cl_2]^3[NH_3]}$$

(c) Using Le Châtelier's principle, you can reason that the addition of more chlorine gas will shift the equilibrium to the right. The concentration of ammonia will decrease since it is reacting with the added chlorine to form more products, thereby reestablishing equilibrium. How the concentration changes with time depends on the kinetics of the reaction. Your sketch should look like the following.

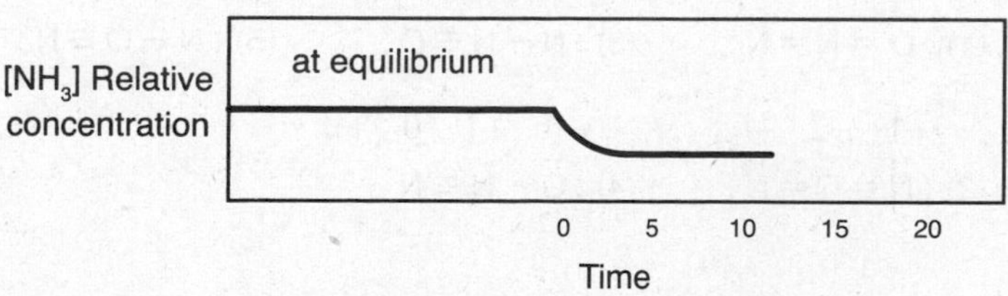

(d) Since the K_{eq} is so small, there are more reactants than products. When the value of K_{eq} is less than 1 the position of the equilibrium lies toward the reactants. If the value of K_{eq} is greater than 1 there would be more products than reactants.

(e) In order to determine whether or not the reaction is at equilibrium, calculate the reaction quotient, Q, using the initial conditions given:

$$Q = \frac{[NCl_3][HCl]^3}{[Cl_2]^3[NH_3]}$$

$$= \frac{(0.20\ M)(0.030\ M)^3}{(0.10\ M)^3(0.20\ M)}$$

$$= 0.027$$

Since K_{eq} is 2.4×10^{-9} (as stated in part (d) of the question), $Q > K$. So the chemical reaction will shift toward the reactants in order to establish equilibrium. So the reaction needs to produce more NH_3 and Cl_2 to reach equilibrium conditions.

4. (a) First we test to see if the two compounds are the same by determining the ratio of the mass of nitrogen divided by the mass of oxygen for the two compounds. The first gives a ratio of 1.75 and the second gives a ratio of 0.875. We conclude that these are two different compounds. Then we test to see if the data obey the law of multiple proportions. The law of multiple proportions predicts that if two different compounds are made from the same two elements and if those compounds have the same mass of one element, then the ratio of masses of the second element will be a ratio of small whole numbers. The data are

	Compound 1	Compound 2
Mass of N	8.40 g	3.64 g
Mass of O	4.80 g	4.16 g

If we presume that the second compound has 8.40 g N (same as compound 1), then we can use the known ratio of 4.16 g O to 3.64 g N to perform the following calculation:

$$?\text{g O} = 8.40\text{ g N}\left(\frac{4.16\text{ g O}}{3.64\text{ g N}}\right) = 9.60\text{ g O}$$

Now we have 8.40 g N in both compounds. We also have 4.80 g O in the first compound and 9.60 g O in the second. We can see that the ratio $\frac{4.80}{9.60} = \frac{1}{2}$ or a ratio of small whole numbers. This is an example of the law of multiple proportions.

(b) The five possibilities for the Lewis structure for N_2O are

(1) $\overset{0}{:\ddot{O}} = \overset{+1}{N} = \overset{-1}{\ddot{N}}:$

(2) $\overset{-1}{:\ddot{N}} = \overset{+2}{O} = \overset{-1}{\ddot{N}}:$

(3) $\overset{-2}{:\underset{\cdot\cdot}{\ddot{N}}} - \overset{+1}{N} \equiv \overset{+1}{O}:$

(4) $\overset{-1}{:\underset{\cdot\cdot}{\ddot{O}}} - \overset{+1}{N} \equiv \overset{0}{N}:$

(5) $\overset{-2}{:\underset{\cdot\cdot}{\ddot{N}}} - \overset{+2}{O} \equiv \overset{0}{N}:$

When the formal charge on each atom is determined as shown, we see that there are no cases with zero formal charge. However, two of the possibilities, choices 1 and 4, have the lowest charges. Among these two, it is more likely that the oxygen will have a negative charge. So the best structure is choice 4.

(c) Almost all elements have more than one isotope. The mass listed in the periodic table is the weighted average of the masses of the isotopes and their natural abundance. Therefore, in a mass spectrum of bromine, which measures the isotopes directly, we see signals for the separate mass 79 and mass 81 isotopes of bromine atoms. In bromine, the two isotopes have almost equal abundances, so the two peaks for 79 and 81 are almost equal. The three peaks around 160 represent three forms of Br_2:

$$^{79}Br - ^{79}Br \text{ total mass} = 158$$

$$^{79}Br - ^{81}Br \text{ total mass} = 160$$

$$^{81}Br - ^{81}Br \text{ total mass} = 162$$

The 1:2:1 relative heights of the peaks occurs because the isotopes can be combined in two different ways to make a molecule of Br_2 with a mass of 160. The Br_2 molecules with a mass of 158 or of 162 can each be made in only one possible way. Therefore the 160 isotope is twice as likely to form in random collisions than either the 158 or 162 isotopes.

5. (a) The chemical equation can be interpreted in different units as indicated below:

	$2OH^-$ +	H_3PO_4	→	HPO_4^{2-} +	$2H_2O$
Molecular scale	2 ions	1 molecule	→	1 ion	2 molecules
Laboratory scale	2 moles	1 mole	→	1 mole	2 moles
Laboratory scale	17 g	49 g	→	48 g	18 g

One major difference between the laboratory scale and the molecular scale is that we cannot use fractions of molecules in the molecular scale. However, fractions of moles are routinely used at the laboratory scale.

(b) Conjugate acids and bases differ from each other by one proton (H^+). In the above equation, the conjugate acid–base pairs are:

Acid	Base
$2H_2O$	$2OH^-$
H_3PO_4	HPO_4^{2-}

Every balanced equation has two conjugate acids and two conjugate bases.

(c) Of the two pairs shown in part (b), the stronger of the two acids will react with the stronger of the two bases. As a consequence, the stronger acid and stronger base are on one side of the arrow and the weaker acid and weaker base are on the other side. In this reaction, the weaker acid and weaker base are on the product side. So the stronger acid and stronger base are H_3PO_4 and OH^-, respectively.

(d) The $CH_3CH_2NH_2$ is the conjugate base, and $CH_3CH_2NH_3^+$ is the conjugate acid. The proportions of the conjugate acid and base are shown in the following figure.

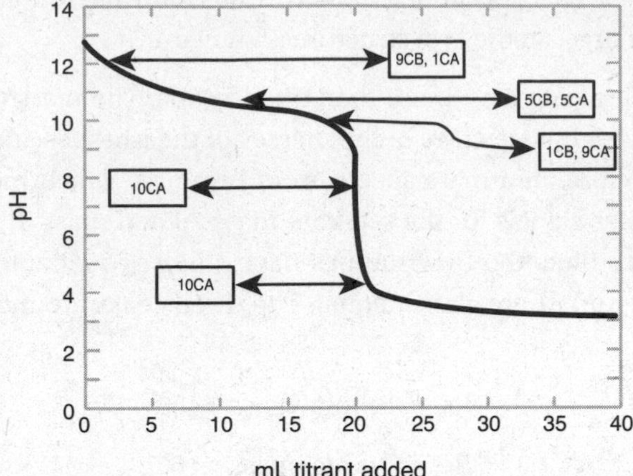

6. (a) Since the reaction exhibits first-order kinetics, the exponents for all concentrations in the rate law add up to 1. Since N_2O_5 is the only reactant:

$$\text{rate} = k\,[N_2O_5]$$

(b) To calculate the rate constant for the reaction, choose any two points on the graph and find the slope of the line. For a first-order reaction, the slope of the line is the rate constant:

$$\ln [A]_0 - \ln [A]_t = kt$$

By rearranging the equation for the line, you can calculate the rate constant. For example, you can use $\ln [A]_0 = -0.40$, $\ln [A]_t = -0.80$, and Δt of 50 seconds:

$$\frac{(\ln [A]_0 \; - \; \ln [A]_t)}{t} = k$$

$$\frac{(-0.40 - (-0.80))}{50 \text{ s}} = k$$

Solving this gives the value for the rate constant = $8.0 \times 10^{-3}\ s^{-1}$.

(c) Now that you have the rate constant, you can calculate the half-life for the reaction:

$$t_{1/2} = \frac{0.693}{8.0 \times 10^{-3} s^{-1}} = 87 \text{ s}$$

7. (a) To write the overall equation, add up the two chemical equations given. Cancel any substance that is the same on both sides of the equation:

$$2NH_3(aq) + OCl^-(aq) \rightarrow N_2H_4(aq) + Cl^-(aq) + H_2O(l)$$

(b) Structure for NH_2Cl:

```
    ..   ..
H — N — Cl:
    |    ..
    H
```

Structure for N_2H_4:

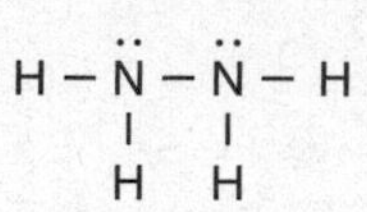

```
    ..   ..
H − N − N − H
    |   |
    H   H
```

(c) The nitrogen central atom has 3 bonding and 1 lone pair of electrons, which forms tetrahedral electron pair (domain) geometry. Due to the lone pair of electrons on the nitrogen, the molecular shape (atoms only) is a triangular pyramid.

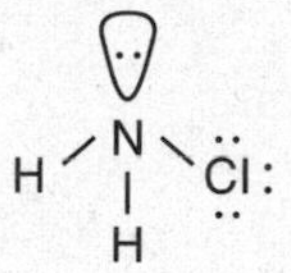

```
     (..)
   ⁄ N ∖ ..
 H   |   Cl :
     H   ..
```

This molecule is polar due the partial charges on N and Cl as shown. A dipole moment is created due to the differences in electronegativity of the atoms involved.

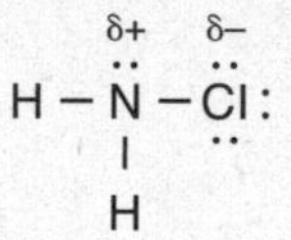

```
     δ+   δ−
     ..   ..
 H − N − Cl :
     |    ..
     H
```

(d) The hydrazine molecule has the higher boiling point due to the presence of more hydrogen bonds as shown in the figure below. (Note that including a figure is not required since it was not requested.)

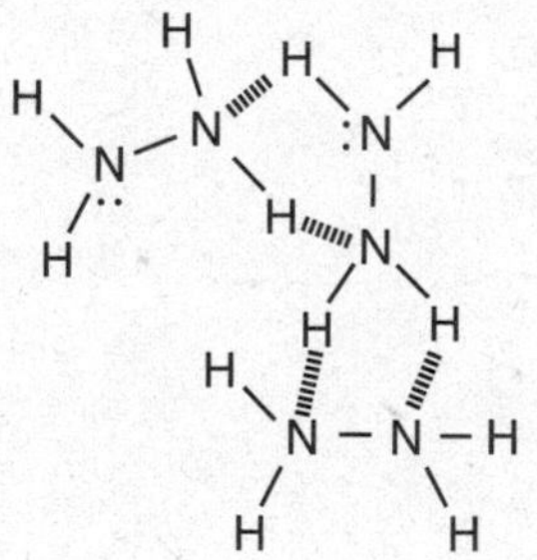

```
          H
  H       \    H    H
   \       N''''  \ /
    N  —  /        :N
   /..       \      |
  H           H''''N
                  /  \
                 H    H
          H      '    '
           \     '    '
            N  —  N  — H
           /       \
          H         H
```

ANSWER SHEET
Practice Test 2

1. Ⓐ Ⓑ Ⓒ Ⓓ	21. Ⓐ Ⓑ Ⓒ Ⓓ	41. Ⓐ Ⓑ Ⓒ Ⓓ
2. Ⓐ Ⓑ Ⓒ Ⓓ	22. Ⓐ Ⓑ Ⓒ Ⓓ	42. Ⓐ Ⓑ Ⓒ Ⓓ
3. Ⓐ Ⓑ Ⓒ Ⓓ	23. Ⓐ Ⓑ Ⓒ Ⓓ	43. Ⓐ Ⓑ Ⓒ Ⓓ
4. Ⓐ Ⓑ Ⓒ Ⓓ	24. Ⓐ Ⓑ Ⓒ Ⓓ	44. Ⓐ Ⓑ Ⓒ Ⓓ
5. Ⓐ Ⓑ Ⓒ Ⓓ	25. Ⓐ Ⓑ Ⓒ Ⓓ	45. Ⓐ Ⓑ Ⓒ Ⓓ
6. Ⓐ Ⓑ Ⓒ Ⓓ	26. Ⓐ Ⓑ Ⓒ Ⓓ	46. Ⓐ Ⓑ Ⓒ Ⓓ
7. Ⓐ Ⓑ Ⓒ Ⓓ	27. Ⓐ Ⓑ Ⓒ Ⓓ	47. Ⓐ Ⓑ Ⓒ Ⓓ
8. Ⓐ Ⓑ Ⓒ Ⓓ	28. Ⓐ Ⓑ Ⓒ Ⓓ	48. Ⓐ Ⓑ Ⓒ Ⓓ
9. Ⓐ Ⓑ Ⓒ Ⓓ	29. Ⓐ Ⓑ Ⓒ Ⓓ	49. Ⓐ Ⓑ Ⓒ Ⓓ
10. Ⓐ Ⓑ Ⓒ Ⓓ	30. Ⓐ Ⓑ Ⓒ Ⓓ	50. Ⓐ Ⓑ Ⓒ Ⓓ
11. Ⓐ Ⓑ Ⓒ Ⓓ	31. Ⓐ Ⓑ Ⓒ Ⓓ	51. Ⓐ Ⓑ Ⓒ Ⓓ
12. Ⓐ Ⓑ Ⓒ Ⓓ	32. Ⓐ Ⓑ Ⓒ Ⓓ	52. Ⓐ Ⓑ Ⓒ Ⓓ
13. Ⓐ Ⓑ Ⓒ Ⓓ	33. Ⓐ Ⓑ Ⓒ Ⓓ	53. Ⓐ Ⓑ Ⓒ Ⓓ
14. Ⓐ Ⓑ Ⓒ Ⓓ	34. Ⓐ Ⓑ Ⓒ Ⓓ	54. Ⓐ Ⓑ Ⓒ Ⓓ
15. Ⓐ Ⓑ Ⓒ Ⓓ	35. Ⓐ Ⓑ Ⓒ Ⓓ	55. Ⓐ Ⓑ Ⓒ Ⓓ
16. Ⓐ Ⓑ Ⓒ Ⓓ	36. Ⓐ Ⓑ Ⓒ Ⓓ	56. Ⓐ Ⓑ Ⓒ Ⓓ
17. Ⓐ Ⓑ Ⓒ Ⓓ	37. Ⓐ Ⓑ Ⓒ Ⓓ	57. Ⓐ Ⓑ Ⓒ Ⓓ
18. Ⓐ Ⓑ Ⓒ Ⓓ	38. Ⓐ Ⓑ Ⓒ Ⓓ	58. Ⓐ Ⓑ Ⓒ Ⓓ
19. Ⓐ Ⓑ Ⓒ Ⓓ	39. Ⓐ Ⓑ Ⓒ Ⓓ	59. Ⓐ Ⓑ Ⓒ Ⓓ
20. Ⓐ Ⓑ Ⓒ Ⓓ	40. Ⓐ Ⓑ Ⓒ Ⓓ	60. Ⓐ Ⓑ Ⓒ Ⓓ

Practice Test 2

SECTION I: Multiple-Choice

START ONLY WHEN YOU HAVE 90 MINUTES TO COMPLETE THE WHOLE SECTION.

Time:	1 hour, 30 minutes
Number of Questions:	60
Percent of Total Score:	50%
Calculator?	None allowed
Pencil required	

Instructions

There are 60 multiple-choice questions for this part of the exam. Enter your answers on the answer sheet provided. On the actual exam, no credit will be given for answers marked on the test itself. For this test, mark your selected answer next to the question and then transfer it to the scoring sheet. (This will allow you to check that all answers are properly transferred to the scoring sheet.) Each question has only one answer. When changing answers, be sure to erase completely.

Useful Hints

Not everyone will know the answers to all of the questions. However, it is to your advantage to provide an answer to all questions. Do not waste time on difficult questions. Answer the easier ones first, and return to the difficult ones you have not answered if time remains.

Your total score is simply the number of questions answered correctly. Wrong answers or blanks on the scoring sheet do not count against you.

You will be allowed to use the following periodic table and the table of equations and constants on this part of the test.

Periodic Table of the Elements

1	2	3	4	5	6	7	8	9	10	11	12	13	14	15	16	17	18
1 **H** 1.008																	2 **He** 4.00
3 **Li** 6.94	4 **Be** 9.01											5 **B** 10.81	6 **C** 12.01	7 **N** 14.01	8 **O** 16.00	9 **F** 19.00	10 **Ne** 20.18
11 **Na** 22.99	12 **Mg** 24.30											13 **Al** 26.98	14 **Si** 28.09	15 **P** 30.97	16 **S** 32.06	17 **Cl** 35.45	18 **Ar** 39.95
19 **K** 39.10	20 **Ca** 40.08	21 **Sc** 44.96	22 **Ti** 47.87	23 **V** 50.94	24 **Cr** 52.00	25 **Mn** 54.94	26 **Fe** 55.85	27 **Co** 58.93	28 **Ni** 58.69	29 **Cu** 63.55	30 **Zn** 65.38	31 **Ga** 69.72	32 **Ge** 72.63	33 **As** 74.92	34 **Se** 78.97	35 **Br** 79.90	36 **Kr** 83.80
37 **Rb** 85.47	38 **Sr** 87.62	39 **Y** 88.91	40 **Zr** 91.22	41 **Nb** 92.91	42 **Mo** 95.95	43 **Tc** (97)	44 **Ru** 101.1	45 **Rh** 102.91	46 **Pd** 106.42	47 **Ag** 107.87	48 **Cd** 112.41	49 **In** 114.82	50 **Sn** 118.71	51 **Sb** 121.76	52 **Te** 127.60	53 **I** 126.90	54 **Xe** 131.29
55 **Cs** 132.91	56 **Ba** 137.33	57 ***La** 138.91	72 **Hf** 178.49	73 **Ta** 180.95	74 **W** 183.84	75 **Re** 186.21	76 **Os** 190.2	77 **Ir** 192.2	78 **Pt** 195.08	79 **Au** 196.97	80 **Hg** 200.59	81 **Tl** 204.38	82 **Pb** 207.2	83 **Bi** 208.98	84 **Po** (209)	85 **At** (210)	86 **Rn** (222)
87 **Fr** (223)	88 **Ra** (226)	89 **†Ac** (227)	104 **Rf** (267)	105 **Db** (270)	106 **Sg** (271)	107 **Bh** (270)	108 **Hs** (277)	109 **Mt** (276)	110 **Ds** (281)	111 **Rg** (282)	112 **Cn** (285)	113 **Nh** (286)	114 **Fl** (289)	115 **Mc** (289)	116 **Lv** (293)	117 **Ts** (294)	118 **Og** (294)

*Lanthanoid Series	58 **Ce** 140.12	59 **Pr** 140.91	60 **Nd** 144.24	61 **Pm** (145)	62 **Sm** 150.4	63 **Eu** 151.97	64 **Gd** 157.25	65 **Tb** 158.93	66 **Dy** 162.50	67 **Ho** 164.93	68 **Er** 167.26	69 **Tm** 168.93	70 **Yb** 173.05	71 **Lu** 174.97
†Actinoid Series	90 **Th** 232.04	91 **Pa** 231.04	92 **U** 238.03	93 **Np** (237)	94 **Pu** (244)	95 **Am** (243)	96 **Cm** (247)	97 **Bk** (247)	98 **Cf** (251)	99 **Es** (252)	100 **Fm** (257)	101 **Md** (258)	102 **No** (259)	103 **Lr** (262)

EQUATIONS AND CONSTANTS

GENERAL INFORMATION

L, mL	= liter(s), milliliter(s)	mm Hg	= millimeters of mercury
g	= gram(s)	J, kJ	= joule(s), kilojoule(s)
nm	= nanometer(s)	V	= volt(s)
atm	= atmosphere(s)	mol	= mole(s)

ATOMIC STRUCTURE

$E = h\nu$

$c = \lambda\nu$

E = energy

ν = frequency

λ = wavelength

Speed of light, $c = 2.998 \times 10^{8}$ m s^{-1}

Planck's constant, $h = 6.626 \times 10^{-34}$ J s

Avogadro's number $= 6.022 \times 10^{23}$ mol^{-1}

Electron charge, $e = -1.602 \times 10^{-19}$ coulomb

EQUILIBRIUM

$$K_a = \frac{[H^+][A^-]}{[HA]}$$

$$K_b = \frac{[OH^-][HB^+]}{[B]}$$

$$K_c = \frac{[C]^c[D]^d}{[A]^a[B]^b}, \text{ where } a\,A + b\,B \rightleftarrows c\,C + d\,D$$

$$K_p = \frac{(P_C)^c(P_D)^d}{(P_A)^a(P_B)^b}$$

$$K_w = [H^+][OH^-] = 1.0 \times 10^{-14} \text{ at } 25°C$$

$$= K_a \times K_b$$

$$pK_a = -\log K_a,\ pK_b = -\log K_b$$

$$pH = -\log[H^+],\ pOH = -\log[OH^-]$$

$$14 = pH + pOH$$

$$pH = pK_a + \log\frac{[A^-]}{[HA]}$$

Equilibrium Constants

K_a (weak acid)

K_b (weak base)

K_c (molar concentrations)

K_p (gas pressures)

K_w (water)

THERMOCHEMISTRY/ELECTROCHEMISTRY AND KINETICS

$\Delta S^\circ = \Sigma S^\circ$ products $- \Sigma S^\circ$ reactants

$\Delta H^\circ = \Sigma \Delta H^\circ_f$ products $- \Sigma \Delta H^\circ_f$ reactants

$\Delta G^\circ = \Sigma \Delta G^\circ_f$ products $- \Sigma \Delta G^\circ_f$ reactants

$\Delta G^\circ = \Delta H^\circ - T\Delta S^\circ$

$= -RT \ln K$

$= -n\mathcal{F}E^\circ$

$q = mc\Delta T$

$I = \dfrac{q}{t}$

$\ln[A]_t - \ln[A]_0 = -kt$

$\dfrac{1}{[A]_t} - \dfrac{1}{[A]_0} = kt$

$t_{1/2} = \dfrac{0.693}{k}$

m = mass
n = number of moles
q = heat
k = rate constant
c = specific heat capacity
S° = standard entropy
H° = standard enthalpy
G° = standard free energy
E° = standard reduction potential
T = temperature
I = current (amperes)
q = charge (coulombs)
t = time (seconds)
$t_{1/2}$ = half-life

Faraday's constant, $\mathcal{F}$ = 96,485 coulombs per mole of electrons

$1 \text{ volt} = \dfrac{1 \text{ joule}}{1 \text{ coulomb}}$

GASES, LIQUIDS, AND SOLUTIONS

$PV = nRT$

$P_A = P_{total} \times X_A$, where $X_A = \dfrac{\text{moles A}}{\text{total moles}}$

$P_{total} = P_A + P_B + P_C + \ldots$

$n = \dfrac{m}{M}$

$K = {}^\circ C + 273$

$D = \dfrac{m}{V}$

KE per molecule $= \dfrac{1}{2}mv^2$

$A = abc$

Molarity, $M = \dfrac{\text{mol solute}}{\text{liters of solution}}$

D = density
P = pressure
T = temperature
m = mass
n = number of moles
v = velocity
V = volume
A = absorbance
a = molar absorptivity
b = path length
c = concentration
KE = kinetic energy
M = molar mass

Gas constant, $R = 8.314 \text{ J mol}^{-1} \text{ K}^{-1}$
$= 0.08206 \text{ L atm mol}^{-1} \text{ K}^{-1}$
$= 62.36 \text{ L torr mol}^{-1} \text{ K}^{-1}$

1 atm = 760 mm Hg = 760 torr

STP = 0.00°C and 1.000 atm

Molar volume of ideal gas = 22.4 L at STP

SECTION I

60 Multiple-Choice Questions
(Time: 90 minutes)
CALCULATORS ARE NOT ALLOWED FOR SECTION I

Note: For all questions, assume that $T = 298$ K, $P = 1.00$ atmosphere, and H_2O is the solvent for all solutions unless the problem indicates a different solvent.

> **Directions:** The questions or incomplete statements that follow are each followed by four suggested answers or completions. Choose the response that best answers the question or completes the statement. Fill in the corresponding circle on the answer sheet.

Questions 1–4 refer to the following information.

Four different acid solutions are prepared, and their pH values are determined and tabulated below.

Concentration (*M*)	Acid 1	Acid 2	Acid 3	Acid 4
0.0010	3.87	3.00	3.17	3.94
0.0050	3.52	2.30	2.82	3.59
0.010	3.37	2.00	2.67	3.44
End point pH for 0.0010 *M* solution	5.79	7.00	7.93	8.84

1. Using the data in the table above, which is the weakest acid?
   (A) Acid 1
   (B) Acid 2
   (C) Acid 3
   (D) Acid 4

2. Which of the four acids is best described as a strong acid?
   (A) Acid 1
   (B) Acid 2
   (C) Acid 3
   (D) Acid 4

3. Which of the following operations will produce an effective buffer solution?
   (A) 50.0 mL of 0.010 *M* acid 3 mixed with 50.0 mL of 0.010 *M* acid 2
   (B) 50.0 mL of 0.010 *M* acid 2 mixed with 25.0 mL of 0.010 *M* NaOH
   (C) 50.0 mL of 0.0050 *M* acid 4 mixed with 0.25 mL of 0.0050 *M* NaOH
   (D) 50.0 mL of 0.0010 *M* acid 2 mixed with 55.0 mL of 0.0010 *M* NaOH

4. Each of the acids were titrated with NaOH to the end point. The end point pH values are shown on the last line of the table. The end point pH values
   (A) suggest that acid 3 is strong and all the rest are weak
   (B) suggest that all the acids are weak
   (C) suggest that acid 1 is most likely polyprotic
   (D) suggest serious problems with the pH meter

---

5. An experiment was performed to determine the moles of hydrogen gas formed (collected over water) when an acid reacts with magnesium metal. To do this, a piece of dry magnesium was weighed. Then 50 mL of hydrogen was collected. Next the Mg was dried to remove about 0.1 mL of water and weighed again to see how much Mg had reacted. The volume of hydrogen was measured and converted into moles of hydrogen. Which mistake will give the largest error in the result?
   (A) Forgetting to dry the magnesium before both weighings
   (B) Failing to take the vapor pressure of water (23 torr at 25°C) into account
   (C) Failing to convert °C to K
   (D) Reading the gas-collecting container to ±20 mL

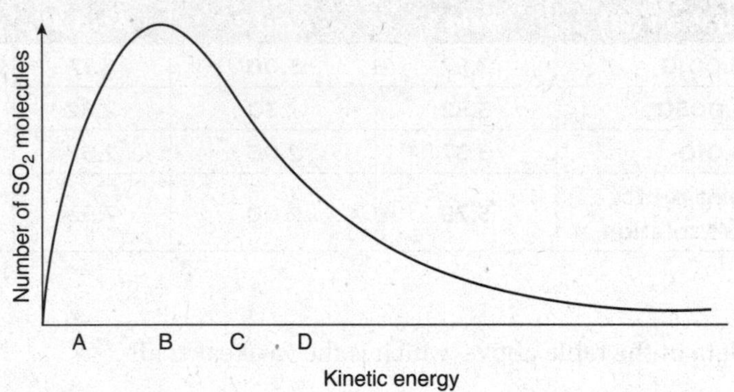

6. The graph above shows the distribution of kinetic energies of a system containing a large number of SO_2 molecules at 300 K. Which letter corresponds to molecules having the largest velocity in this system?
   (A) A
   (B) B
   (C) C
   (D) D

7. A 25 g sample of a liquid was heated to 95°C and then quickly transferred to an insulated container holding 100 g of water at 26°C. The temperature of the mixture rose to reach a final temperature of 37°C. Which of the following can be concluded?
   (A) The sample lost more heat energy than the water gained because the sample temperature changed more than the water temperature did.
   (B) The final temperature is less than the average starting temperatures; therefore the equilibrium constant must be less than 1.
   (C) The sample temperature changed more than the water temperature did; therefore the heat capacity of the sample must be greater than the heat capacity of the water.
   (D) Even though the sample temperature changed more than the water temperature did, the sample lost the same amount of heat energy as the water gained to obey the law of conservation of energy.

8. Three 1-liter flasks are connected to a 3-liter flask by valves. The 3-liter flask is evacuated to start and the entire system is at 585 K. The first flask contains oxygen, the second hydrogen, and the third nitrogen. The pressure of hydrogen is 1.65 atm. The amounts of gas molecules are proportional to their representations in the flasks. If valve 2 is opened first and then the rest of the valves are opened, what will the pressure in the 3-liter flask be after the first valve is opened and after they all are opened? Assume the connections have negligible volume.

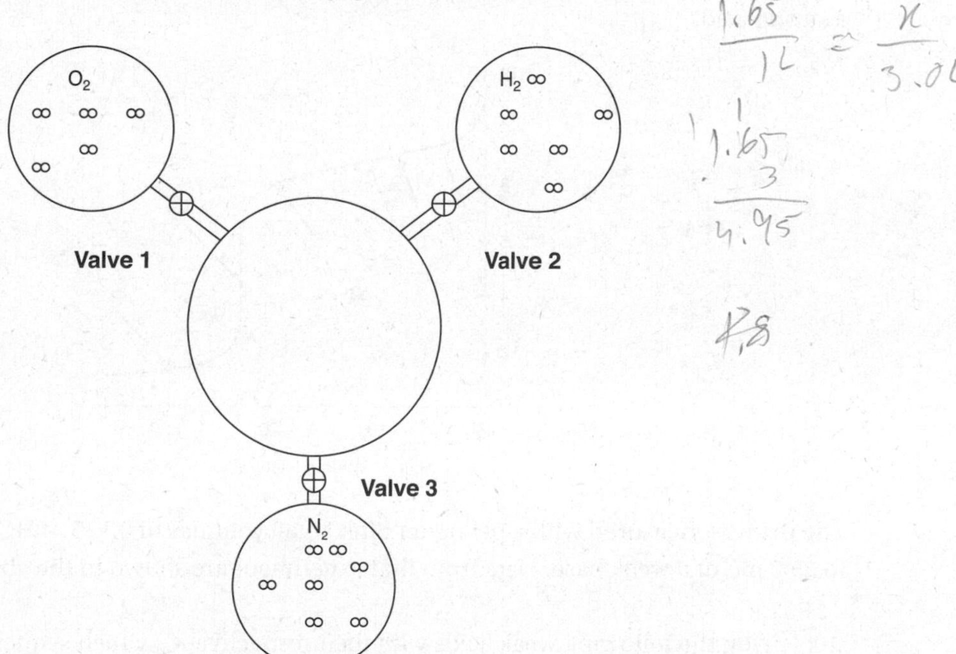

	Valve 2 Opened	All Valves Opened
(A)	1.0 atm	0.5 atm
(B)	0.41 atm	0.82 atm
(C)	0.81 atm	1.65 atm
(D)	2.0 atm	1.0 atm

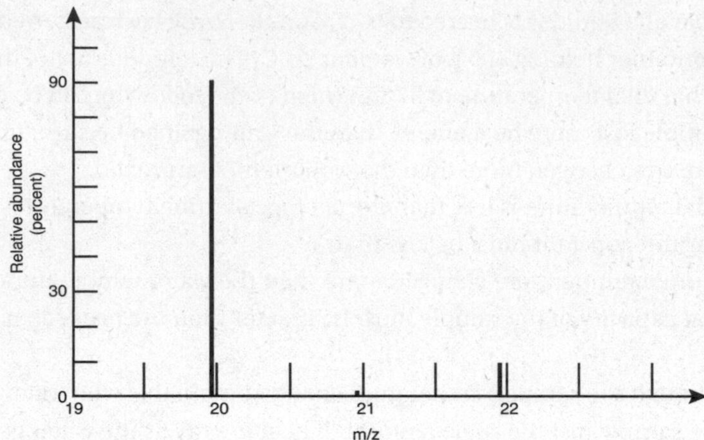

9. A mass spectrum of a naturally occurring sample of an element is shown above. What is the element?
   (A) Ca
   (B) Ne
   (C) K
   (D) Not enough information is provided.

Questions 10–14 refer to the titration curve below, which is of a weak base titrated with a strong acid.

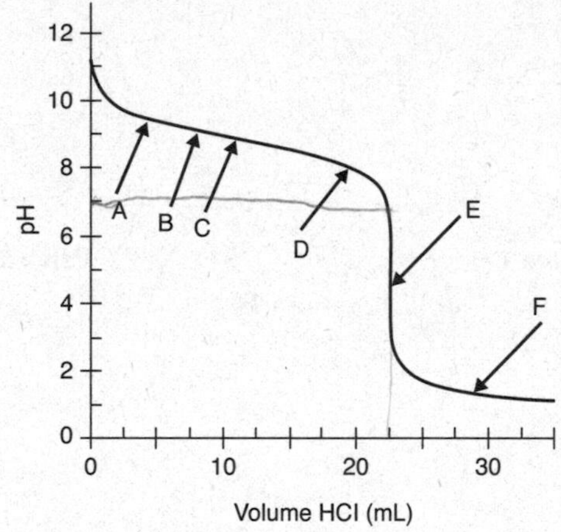

The pH was measured with a pH meter after small volumes of 0.125 *M* HCl were added to 25.0 mL of a weak base. Data from that experiment are shown in the above graph.

10. Given the following weak acids with their respective K_a, which sequence lists the corresponding anions in order of increasing base strength?

$HC_2H_3O_2$	$K_a = 1.8 \times 10^{-5}$
HOCl	$K_a = 3.0 \times 10^{-8}$
HNO_2	$K_a = 7.1 \times 10^{-4}$

   (A) NO_2^-, OCl^-, $C_2H_3O_2^-$
   (B) OCl^-, NO_2^-, $C_2H_3O_2^-$
   (C) $C_2H_3O_2^-$, NO_2^-, OCl^-
   (D) NO_2^-, $C_2H_3O_2^-$, OCl^-

11. Which statement is true about a solution that is made by adding 0.2 mol of sodium fluoride, NaF, to 1.0 L of water?
   (A) The concentration of sodium and fluoride ions will be identical.
   (B) The solution will be neutral.
   (C) The solution will be basic.
   (D) The concentration of sodium ions will be greater than the concentration of fluoride ions.

12. For the best results with a visual indicator (one that changes color), what will optimize the results?
   (A) Add the indicator two drops before the end point.
   (B) Choose an indicator that has a p*K* that is close to the end point pH.
   (C) Choose an indicator that has complementary colors.
   (D) Choose a polyprotic indicator.

13. If the student uses a pH indicator that changes color from pH 6 to 8, which statement best characterizes the expected observations?
   (A) The observed color change will be distinct, and the calculated molarity of the base will be accurate.
   (B) The color will change slowly, and the end point volume will be low.
   (C) The color will change slowly, and the calculated molarity of the base will be low.
   (D) The observed color change will be distinct, but the calculated molarity of the base will be high.

14. A student is performing a titration of a weak acid (0.10 mol L^{-1}) with a strong base of the same concentration. Which of the following is the best representation of the resulting titration curve?

(A)

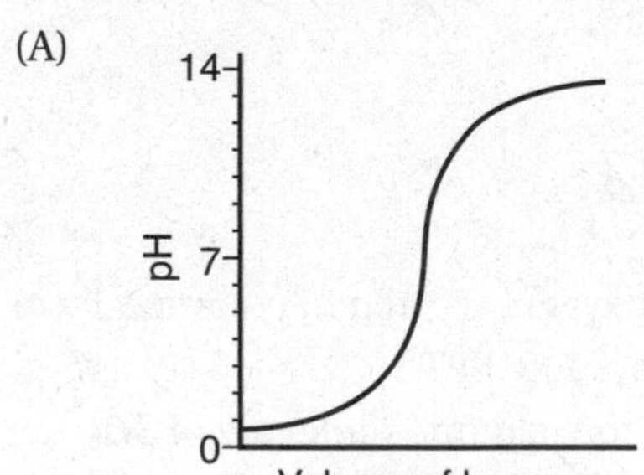

(B)

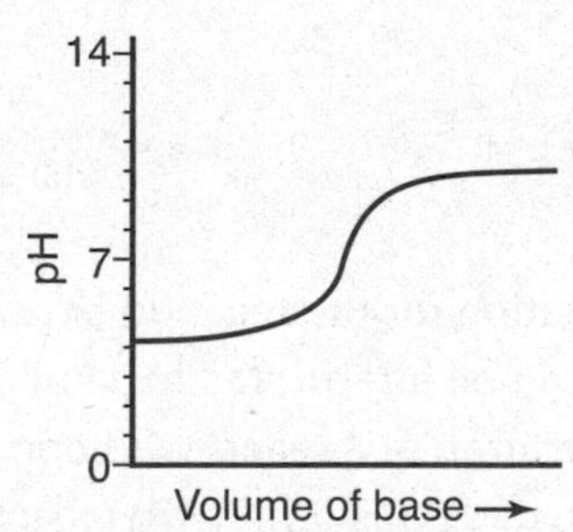

(C)

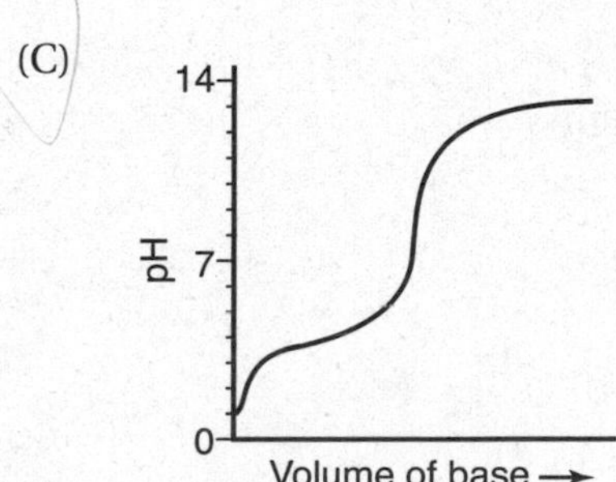

(D)

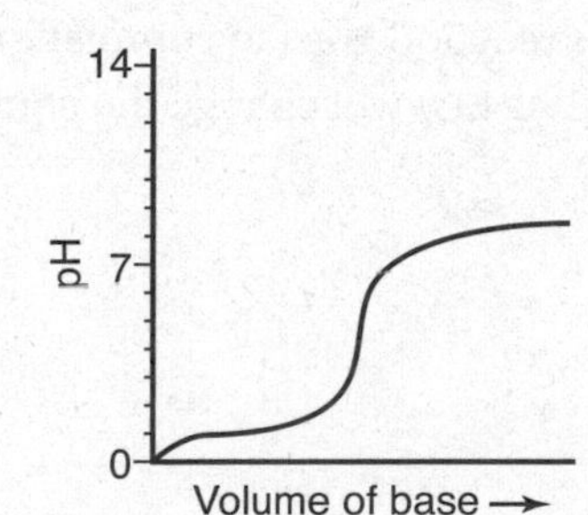

PRACTICE TEST 2

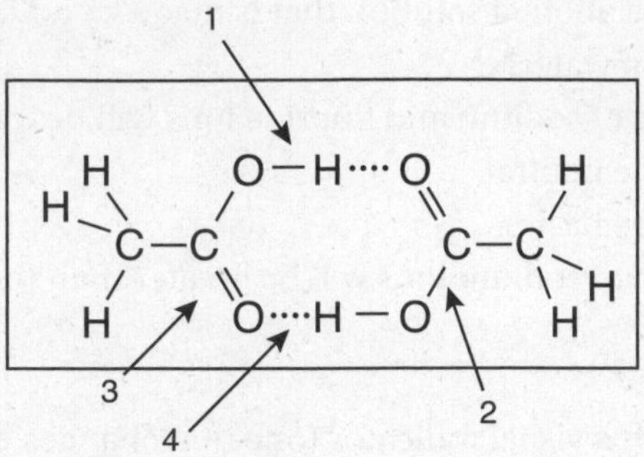

15. In the diagram above, which labeled arrow is pointing toward a covalent bond *and* which is pointing toward a hydrogen bond?

	Covalent Bond	Hydrogen Bond
(A)	1	2
(B)	2	1
(C)	3	4
(D)	4	3

16. Which would be the easiest way to burn a copper penny?
   (A) Hold the copper penny with crucible tongs, and heat strongly in the flame of a Bunsen burner.
   (B) Use the method in (A), but use an oxyacetylene torch to reach a higher temperature.
   (C) Grind the copper penny into very small, dust-sized particles, and spray the particles into a Bunsen burner flame.
   (D) Dissolve the copper penny in acid, precipitate the hydroxide, and heat in a Bunsen burner flame to make the oxide.

$$S(s) + O_2(g) \rightarrow SO_2(g)$$

17. The equation for the reaction between sulfur and oxygen, written in equation form above, can be interpreted in all of the following ways EXCEPT
   (A) one atom of S reacts with one molecule of O_2 to yield one molecule of SO_2
   (B) one mole of sulfur atoms reacts with one mole of oxygen molecules to yield one mole of sulfur dioxide molecules
   (C) this reaction goes to completion
   (D) adding $S(s)$ will change the equilibrium constant

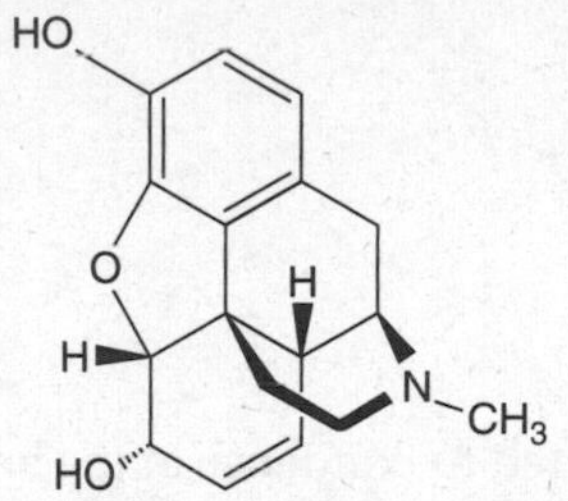

18. Morphine, $C_{17}H_{19}NO_3$ (shown above), has a $K_b = 8.0 \times 10^{-7}$. If a 0.00100 M solution of morphine is prepared, the expected pH will be in which one of the following pH ranges?
(A) 2 to 4
(B) 4 to 6
(C) 8 to 10
(D) 10 to 12

19. What is the pK_a of morphine hydrochloride, $C_{17}H_{20}NO_3{}^+Cl^-$, where $C_{17}H_{20}NO_3{}^+$ is the conjugate acid of the morphine in the previous question?
(A) $\log(8.0 \times 10^{-7})$
(B) $-\log(8.0 \times 10^{-7})$
(C) $\log(1.0 \times 10^{-14}/8.0 \times 10^{-7})$
(D) $-\log(1.0 \times 10^{-14}/8.0 \times 10^{-7})$

Questions 20–23 refer to the following table of information.

Vessel	A	B	C
Gas	Bromomethane	Dibromomethane	Tetrabromomethane
Formula	CH_3Br	CH_2Br_2	CBr_4
Molar mass	95 g/mol	174 g/mol	332 g/mol
Pressure	0.2 atm	0.4 atm	0.2 atm
Temperature	600°C	500°C	500°C

Note that the gases are held in separate, identical, rigid vessels.

20. Which sample has the lowest density?
(A) All have the same density since the temperatures are all the same.
(B) Vessel A
(C) Vessel B
(D) Vessel C

21. The average kinetic energy
   (A) is greatest in vessel A
   (B) is greatest in vessel B
   (C) is greatest in vessel C
   (D) is the same in all vessels

22. Which of these gases is expected to condense at the lowest pressure, assuming that the temperature is held constant?
   (A) Bromomethane
   (B) Dibromomethane
   (C) Tetrabromomethane
   (D) They all condense at the same pressure.

23. Which attractive force is the major cause of condensation?
   (A) Hydrogen bonding
   (B) London forces
   (C) Dipole-dipole attractive forces
   (D) Molar mass

O S O

O

24. Use the octet rule without minimizing formal charges and the arrangement of atoms suggested in the skeleton structure above to construct the Lewis structure for the SO_3 molecule. All of the following statements about this molecule are true EXCEPT
   (A) sulfur trioxide has three resonance structures
   (B) sulfur trioxide has a planar triangle shape
   (C) sulfur trioxide is nonpolar
   (D) the S–O bond order is 5/3

O S O

O

25. Use the octet rule and the arrangement of atoms suggested in the skeleton structure above to construct the Lewis structure for the sulfite ion. Be sure to minimize the formal charges. All of the following statements are true EXCEPT
   (A) the sulfite ion has a double bond
   (B) the sulfite ion has a triangular pyramid shape
   (C) the sulfite ion is nonpolar
   (D) the S–O bond order is 4/3

Questions 26–28 refer to the following information.

$$Pb(ClO_4)_2 + 2KCl \rightarrow PbCl_2(s) + 2KClO_4$$

A chemist mixes a dilute solution of lead perchlorate with a dilute solution of potassium chloride to precipitate lead(II) chloride ($K_{sp} = 1.7 \times 10^{-5}$).

26. Which of the following is the net ionic equation for the experiment described above?
    (A) $2K^+ + ClO_4^{2-} \rightarrow K_2ClO_4(s)$
    (B) $2Cl^- + Pb^{2+} \rightarrow PbCl_2(s)$
    (C) $4Cl^- + Pb^{4+} \rightarrow PbCl_4(s)$
    (D) $2ClO_4^- + Pb^{2+} \rightarrow PbCl_2(s) + 4O_2(g)$

27. If equal volumes of 2.0 millimolar solutions arc mixed, which of the following particulate views represents the experiment after the reactants are mixed thoroughly and the solids are given time to precipitate?

(A)

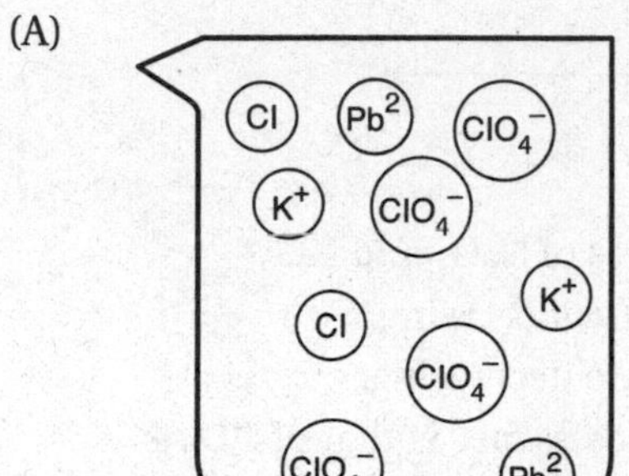

(B)

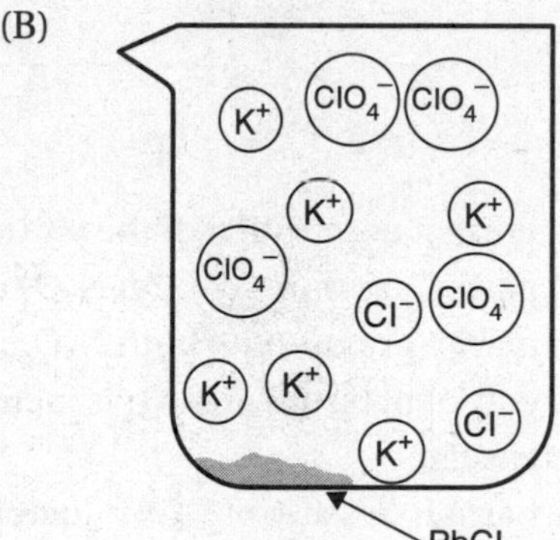

(C)

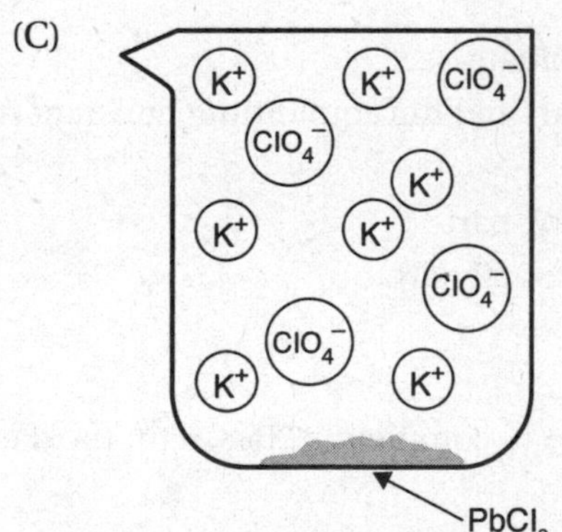

(D)

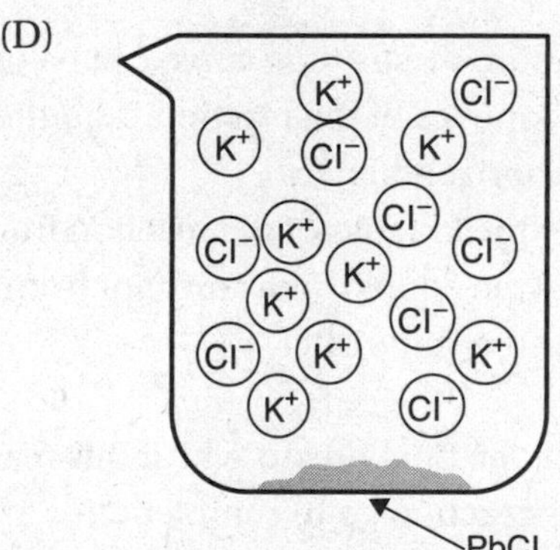

28. Within a factor of 10, what is the approximate molar solubility of $PbCl_2$ in a solution of 0.17 M Pb^{2+} ions?
    (A) 1.7×10^{-5} M
    (B) 5.0×10^{-8} M
    (C) 5.0×10^{-3} M
    (D) 5.0×10^{-15} M

Questions 29–33 refer to the information and diagram below.

$$N_2O_4(g) \rightleftharpoons 2NO_2(g)$$

$N_2O_4(g)$ decomposes into $NO_2(g)$ according to the equation above. A pure sample of $N_2O_4(g)$ is placed into a rigid, evacuated 0.500 L container. The initial pressure of the $N_2O_4(g)$ is 760 torr. The temperature is held constant until the $N_2O_4(g)$ reaches equilibrium with its decomposition products. The figure below shows how the pressure of the system changes while reaching equilibrium.

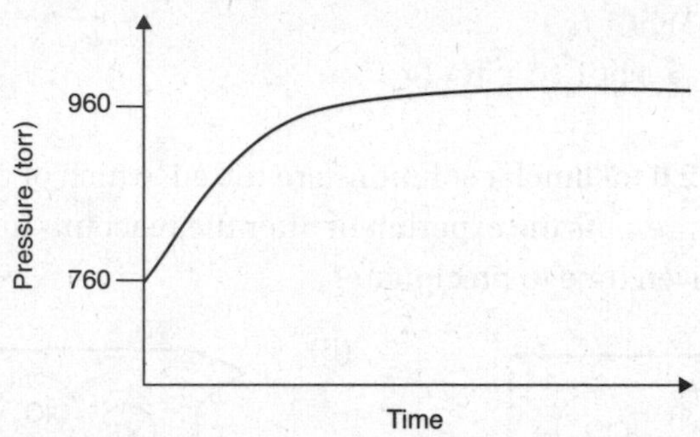

29. Which is a correct description of the partial pressures of N_2O_4 and NO_2?
    (A) The partial pressures of both N_2O_4 and NO_2 are increasing.
    (B) The partial pressures of both N_2O_4 and NO_2 are decreasing.
    (C) The partial pressure of N_2O_4 is increasing twice as fast as that of NO_2 is decreasing.
    (D) The partial pressure of NO_2 is increasing twice as fast as that of N_2O_4 is decreasing.

30. The pressure versus time curve can be used to evaluate
    (A) the kinetics of the reaction from the rising part and the equilibrium constant from the horizontal part
    (B) only the kinetics of the reaction from the rising part
    (C) only the equilibrium constant from the horizontal part
    (D) only the value of Q

31. What will the final pressure be if this reaction goes to completion? Based on the data, does this reaction go to completion?
    (A) 760 torr; yes, the reaction goes to completion because this is the given pressure.
    (B) 900 torr; yes, the reaction goes to completion because it reached equilibrium.
    (C) 1520 torr; no, the reaction does not go to completion because it is not 1520 torr when the reaction stopped.
    (D) 2280 torr; no, the reaction does not go to completion because it is not 2280 torr when the reaction stopped.

32. What is the appropriate combination of numbers for calculating K_p?

(A) $K_p = \dfrac{(400/760)}{(560/760)}$

(B) $K_p = \dfrac{(400/760)^2}{560/760}$

(C) $K_p = \dfrac{(200/760)^2}{560/760}$

(D) $K_p = \dfrac{(400)^2}{560}$

33. Which diagram is an appropriate description of the system if more $NO_2(g)$ is rapidly injected, at time = 0, into the container after equilibrium is established?

(A)

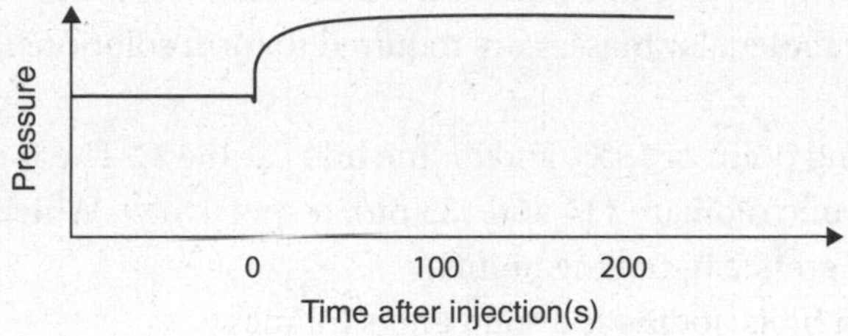

Pressure

0 100 200

Time after injection(s)

(B)

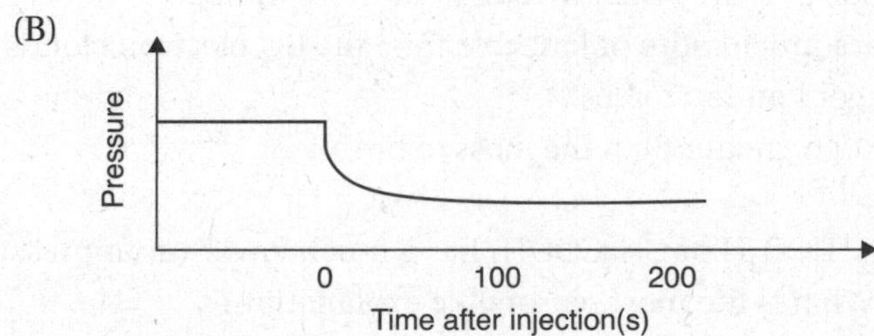

(C)

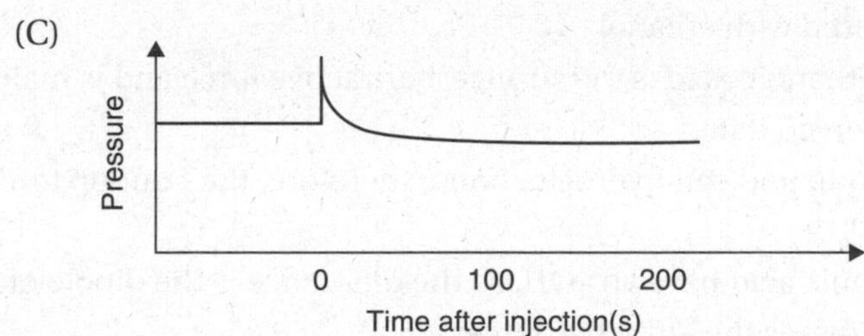

(D)

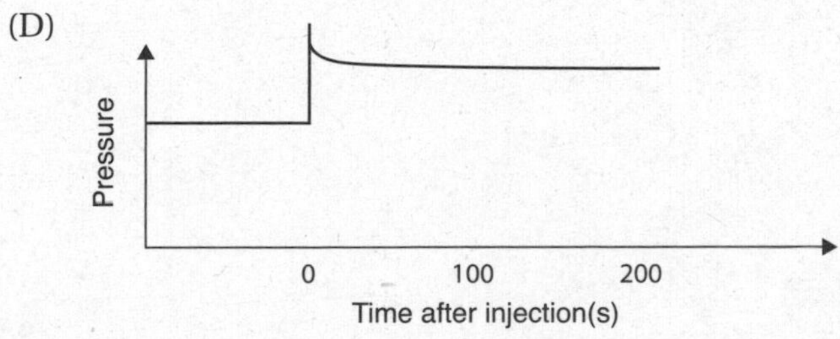

Formation of a solution can often be visualized as a three-step process.

Step 1. Solvent molecules are separated from each other to make space for the solute.

Step 2. Solute molecules are separated so they fit into spaces in the solvent.

Step 3. Separated solute and solvent are brought together filling in the spaces.

34. All of the fundamental principles below are important in understanding the formation of solutions EXCEPT which one?
    (A) Starting at equilibrium, moving particles apart while in the solid or equilibrium phases requires added energy proportional to the attractive forces.
    (B) Bringing particles together releases energy in proportion to their attractive forces.
    (C) The total of the energies in steps 1 to 3 indicates if a solution will form.
    (D) Molecules with similar molecular masses are required to form solutions.

35. Bromine has a normal boiling point of 59°C, and iodine boils at 184°C. The I_2 molecule is much larger than Br_2 (atomic radii are 114 and 133 pm, respectively). Which is the best reason for the large difference in boiling points?
    (A) Bromine is a liquid and boils; iodine is a solid and sublimes.
    (B) The intramolecular bonds in I_2 are much weaker than those in Br_2.
    (C) The I_2 electron clouds are much more polarizable than the Br_2 electron clouds, resulting in much stronger London forces.
    (D) The mass of iodine is much greater than the mass of bromine.

36. Ethanoic acid ($HC_2H_3O_2$ or CH_3CO_2H or CH_3COOH) has a much lower vapor pressure than ethanol (CH_3CH_2OH). What is the most reasonable explanation?
    (A) The polarizability of two oxygen atoms increases the London forces of attraction in ethanoic acid compared with ethanol.
    (B) Hydrogen bonding in ethanoic acid is the strongest attractive force and is mainly responsible for the observed data.
    (C) Ethanol has an –OH group and can hydrogen bond; therefore, the London forces must cause the effect.
    (D) Both ethanol and ethanoic acid have an –OH, so the difference is the dipole of the second oxygen that increases the attractive forces.

37. Which reaction is expected to be endothermic?

(A) $2CO(g) + O_2(g) \rightarrow 2CO_2(g)$

(B) $2C_6H_6(l) + 15O_2(g) \rightarrow 12CO_2(g) + 6H_2O(g)$

(C) $2C_6H_6(l) \rightarrow 2C_6H_6(g)$

(D) $F_2(g) + 2e^- \rightarrow 2F^-(g)$

38. Which of the following is a formation reaction?

(A) $MnO_4^- + 8H^+ + 5e^- \rightarrow Mn^{2+} + 4H_2O$

(B) $2CO(g) + O_2(g) \rightarrow 2CO_2(g)$

(C) $2O_2(g) + N_2(g) \rightarrow 2NO_2(g)$

(D) $2Al(s) + \frac{3}{2}O_2(g) \rightarrow Al_2O_3(s)$

39. Which method cannot be used to determine the enthalpy change of a chemical reaction?

(A) Measure the entropy change for the reaction at 298 K.

(B) Combine the heats of reaction of appropriate reactions using Hess's law.

(C) Use the necessary heats of formation to calculate the heat of reaction.

(D) Use a calorimeter to determine the heat absorbed or evolved in the reaction.

Questions 40–41 refer to the following thermochemical reactions.

$$N_2 + O_2 \rightarrow 2NO \qquad \Delta H° = -180.8 \text{ kJ}$$
$$2NO + O_2 \rightarrow 2NO_2 \qquad \Delta H° = -113.2 \text{ kJ}$$

40. What is the enthalpy of the reaction $N_2 + 2O_2 \rightarrow 2NO_2$?

(A) −67.6 kJ

(B) −294.0 kJ

(C) +294.0 kJ

(D) The enthalpy of the reaction cannot be determined with this information.

41. How much heat is produced or absorbed if 0.200 mol N_2 is reacted with 0.300 mol O_2 to form NO?

(A) 54.2 kJ of heat is absorbed.

(B) 58.8 kJ of heat is evolved.

(C) 22.6 kJ of heat is evolved.

(D) 36.2 kJ of heat is evolved.

Questions 42–45 refer to the cell diagrams and their corresponding $E°_{cell}$ values for the three galvanic cells listed in the table below.

Galvanic Cell	Cell Diagram	$E°_{cell}$
1	$Zn \mid Zn^{2+}(1.0\ M) \parallel Pb^{2+}(1.0\ M) \mid Pb$	0.63
2	$Mg \mid Mg^{2+}(1.0\ M) \parallel Pb^{2+}(1.0\ M) \mid Pb$	2.24
3	$Mg \mid Mg^{2+}(1.0\ M) \parallel Zn^{2+}(1.0\ M) \mid Zn$	?

The chemical reaction occurring in the first galvanic cell is

$$Zn + Pb^{2+} \rightarrow Zn^{2+} + Pb$$

42. What is the standard cell potential for galvanic cell 3 depicted in the table above?
    (A) 1.61 V
    (B) 2.77 V
    (C) −1.61 V
    (D) 0.18 V

43. What is the chemical reaction under study in galvanic cell 3?
    (A) $Pb^{2+} + Mg^{2+} \rightarrow Pb^{2+} + Mg$
    (B) $Zn + Pb^{2+} \rightarrow Zn^{2+} + Pb$
    (C) $Mg + Zn^{2+} \rightarrow Mg^{2+} + Zn$
    (D) $Zn + Mg^{2+} \rightarrow Zn^{2+} + Mg$

44. If the concentration of Mg^{2+} is changed from 1.0 *M* to 0.1 *M* in the galvanic cells above, what will happen to the observed cell voltages in galvanic cells 2 and 3?
    (A) The voltage in both galvanic cells 2 and 3 will increase.
    (B) The voltage will decrease in both galvanic cells 2 and 3.
    (C) The voltage in galvanic cell 2 will increase and will decrease in galvanic cell 3.
    (D) The voltage in galvanic cell 3 will increase and will decrease in galvanic cell 2.

45. Rank Mg, Pb, and Zn from the metal that is easiest to oxidize to the metal that is the most difficult to oxidize.
    (A) Mg > Pb > Zn
    (B) Zn > Pb > Mg
    (C) Mg > Zn > Pb
    (D) Pb > Zn > Mg

Carbon has an atomic radius of 77 pm and a first ionization energy of 1086 kJ/mol.

46. Based on periodic trends and the data given above, what are the most probable values for the ionization energy and atomic radius of nitrogen?
(A) 70 pm, 1402 kJ/mol
(B) 86 pm, 898 kJ/mol
(C) 135 pm, 523 kJ/mol
(D) 40 pm, 995 kJ/mol

47. Dissolving one mole of each of the following compounds in water results in solutions with different pH values. Under those conditions, which of the following acids will have the highest percentage ionization?
(A) HNO_2
(B) $HClO_4$
(C) H_2S
(D) H_3PO_4

Compound	Formula	Normal Boiling Point, K
n-heptane	C_7H_{16}	371
n-octane	C_8H_{18}	398
n-nonane	C_9H_{20}	424
n-decane	$C_{10}H_{22}$	447

48. Based on the data in the table above, which of the following substances has the lowest viscosity?
(A) *n*-heptane
(B) *n*-octane
(C) *n*-nonane
(D) *n*-decane

49. Based on periodic relationships, the concepts related to bond strength, and the concept relating bond strength to acid strength, which of the following correctly predicts the strength of binary acids from strongest to weakest?
(A) $H_2Se > H_2O > H_2S$
(B) $H_2S > H_2Se > H_2O$
(C) $H_2O < H_2S < H_2Se$
(D) $H_2Se > H_2S > H_2O$

50. Chlorine is often used to oxidize other substances. It makes a good bleach because it can oxidize many colored compounds. If chlorine is not available, what other substance can be used for oxidizing substances?
(A) Al
(B) H_2S
(C) Zn
(D) $KMnO_4$

PRACTICE TEST 2

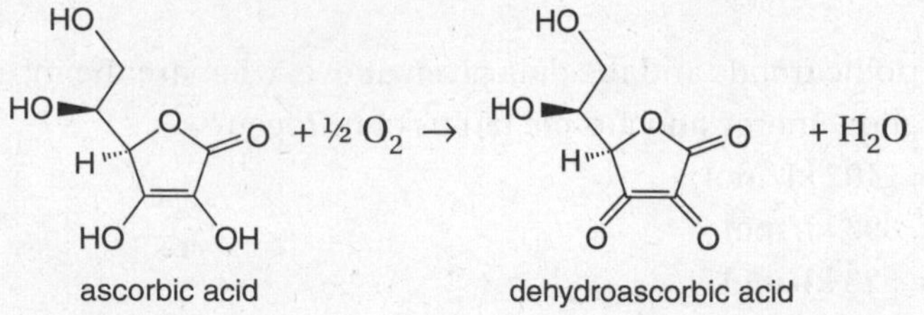

ascorbic acid $+ \frac{1}{2} O_2 \rightarrow$ dehydroascorbic acid $+ H_2O$

Vitamin C is oxidized slowly to dehydroascorbic acid by the oxygen in air. It is catalyzed by ions such as Cu^{2+} and Fe^{3+}. The reaction can be followed by measuring the ultraviolet absorbance at 243 nm.

Time (h)	Absorbance (A)	ln A	1/A	-ln A
0	0.75	-0.29	1.3	0.29
1	0.38	-0.97	2.6	0.97
2	0.19	-1.7	5.3	1.7
3	0.095	-2.4	11	2.4

51. Which of the following is the best interpretation of the data given above?
    (A) The reaction is first order and the rate constant cannot be calculated from these data.
    (B) The reaction is zero order and the rate constant can be calculated from these data.
    (C) The reaction is first order and the rate constant can be calculated from these data.
    (D) The reaction is second order and the rate constant cannot be calculated from these data.

52. A new compound is synthesized and found to be a monoprotic acid with a molar mass of 248 g/mol. When 0.0050 mol of this acid is dissolved in 0.500 L of water, the pH is measured as 3.89. What is the pK_a of this acid?
    (A) 3.89
    (B) 7.78
    (C) 5.78
    (D) 2.33

53. Which of the following particulate diagrams best represents the reaction of gaseous sulfur (black dots) with oxygen (white circles) to form sulfur dioxide?

(A)

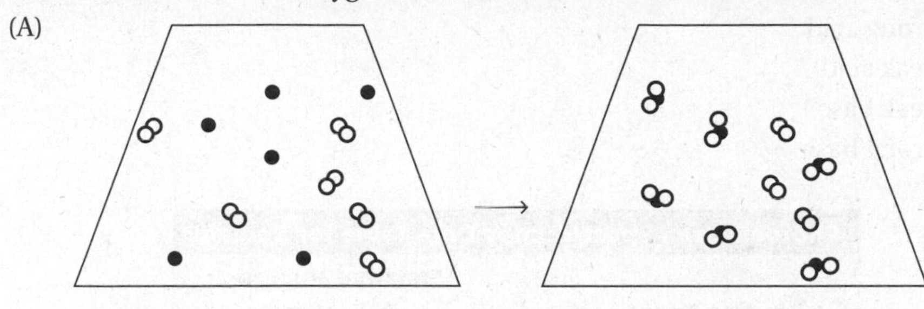

(B)

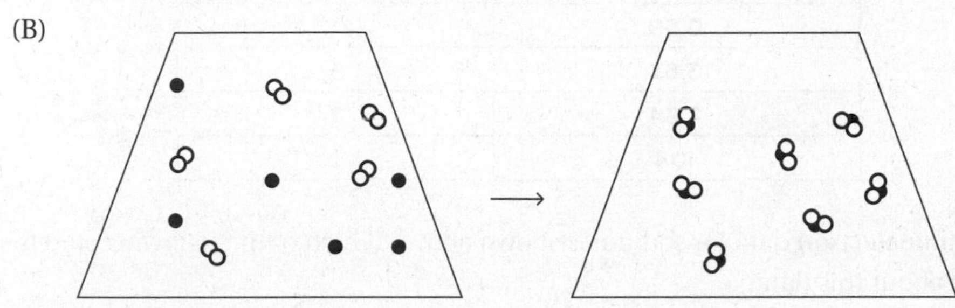

(C)

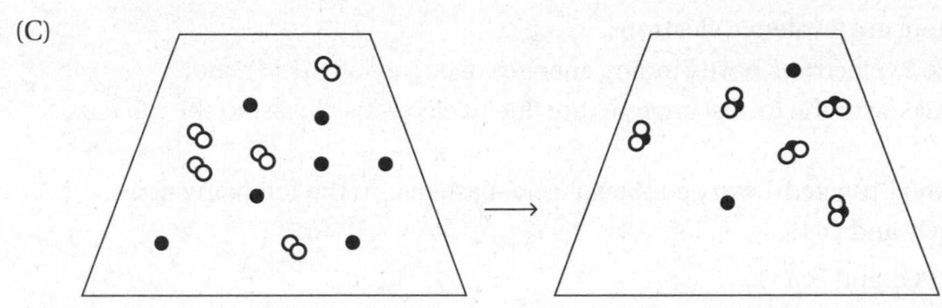

(D)

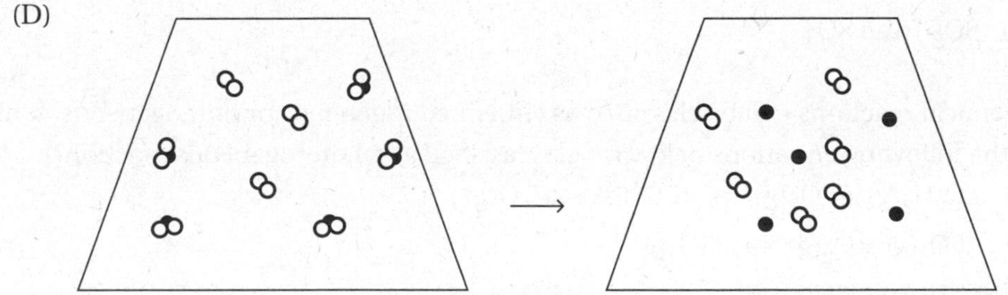

54. Nitrous acid is a weak acid, while nitric acid is a much stronger acid because

(A) the nitrogen in nitric acid is more electronegative than the nitrogen in nitrous acid

(B) nitric acid has more oxygen atoms in its formula

(C) the –O–H bonds in nitric acid are much weaker than in nitrous acid due to the electron withdrawing of more oxygen atoms on nitric acid

(D) nitric acid has the hydrogen atoms bound directly to the nitrogen atom

55. A student has a liter of a 0.100 *M* solution of a strong acid. To prepare a buffer, this should be mixed with

(A) a strong acid
(B) a weak acid
(C) a weak base
(D) a strong base

Photoelectron Spectroscopy Data Table for Sodium	
Binding Energy (MJ)	Relative Number of Electrons
0.50	1
3.67	6
6.84	2
104	2

56. The photoelectron data for sodium is shown above. Which of the following statements is true about this data?

(A) All 11 electrons are shown in the table.
(B) There are 9 valence electrons.
(C) The 2*s* electrons have binding energies of energy of 104 MJ/mol.
(D) Peaks with the lowest energies are due to electrons closest to the nucleus.

57. Identify the Brønsted-Lowry conjugate acid–base pair in the following list.

(A) H_3O^+ and OH^-
(B) H_3PO_4 and H_3PO_3
(C) $HC_2H_3O_2$ and $C_2H_3O_2^-$
(D) SO_3^{2-} and SO_2^{2-}

58. Chemical reactions can be classified as either heterogeneous or homogeneous. Which of the following equations below is best classified as a heterogeneous reaction?

(A) $2C_2H_2(g) + 5O_2(g) \rightarrow 4CO_2(g) + 2H_2O(g)$
(B) $2SO_2(g) + O_2(g) \rightarrow 2SO_3(g)$
(C) $C_2H_2(g) + 5N_2O(g) \rightarrow CO_2(g) + H_2O(g) + 5N_2(g)$
(D) $C(s) + H_2O(g) \rightarrow H_2(g) + CO(g)$

59. Which of the reactions below will not be thermodynamically favored at all temperatures?

	Reaction	$\Delta H°$ (kJ/mol$_{rxn}$)	$\Delta S°$ (J/mol$_{rxn}$K)
(A)	$CO(NH_2)_2(aq) + H_2O(l) \rightarrow CO_2(g) + 2NH_3(g)$	+119.2	+354.8
(B)	$C_2H_5OH(l) + O_2(g) \rightarrow HC_2H_3O_2(l) + H_2O(g)$	–534.3	–131
(C)	$2Fe(s) + 3CO_2(g) \rightarrow Fe_2O_3(s) + 3CO(g)$	+26.7	–15.7
(D)	$C_2H_2(g) + 5N_2O(g) \rightarrow CO_2(g) + H_2O(g) + 5N_2(g)$	–437.4	+272.6

$$CO + NO_2 \rightarrow NO + CO_2$$

60. Consider the following possible mechanism for this gas phase reaction above:

Step 1. $NO_2 + NO_2 \rightarrow NO_3 + NO$ rate-limiting step
Step 2. $NO_3 + CO \rightarrow NO_2 + CO_2$

Which of the following statements is not true?
(A) $NO_3(g)$ is a catalyst for this reaction.
(B) The rate law for this mechanism is Rate = $k\,[NO_2]^2$.
(C) $NO_3(g)$ is an intermediate.
(D) The mechanism adds up to the overall reaction.

IF YOU FINISH BEFORE THE 90 MINUTES HAVE ELAPSED, YOU MAY CHECK YOUR WORK IN THIS SECTION. DO NOT GO ON TO SECTION II UNTIL YOU ARE INSTRUCTED TO DO SO.

SECTION II: Free-Response

START ONLY WHEN YOU HAVE 105 MINUTES TO COMPLETE THE WHOLE SECTION.

Time:	1 hour, 45 minutes
Number of Questions:	7, 3 long and 4 short
Percent of Total Score:	50%
Calculator Allowed?	Yes
Pencil or pen with black or dark blue ink	

Instructions

The questions appear on the following pages. You may use the periodic table and the table of equations and constants that were provided with Section I. On the actual exam, you will be directed to write your final answers only in the test booklet. For this practice test, write your answers on separate sheets of lined paper. You will need about three pages for each of questions 1 through 3 and one page for each of questions 4 through 7.

Do not spend too much time on any one question. Budget your time carefully, and answer the easier questions first.

Be sure that your work is well organized, complete, and easy to read. Cross out or erase any mistakes. Erased and crossed-out material will not be scored.

SECTION II

7 Free-Response Questions
(Time: 105 minutes)
CALCULATORS ARE ALLOWED FOR SECTION II

Directions: Questions 1, 2, and 3 are long questions. Each question should take about 20 minutes to answer. Questions 4, 5, 6, and 7 are shorter questions. Each one should take about 5 to 10 minutes to answer. Read the questions carefully, and write your responses on lined paper. Your answers will be graded based on their correctness and relevance to the question asked and on the information cited. Explanations should be well organized and clearly written. Specific answers are always better than broad statements. For calculations, it is to your advantage to show clearly the method used and the steps involved in arriving at your answers since you may receive partial credit.

1. In communities near the ocean, the overpumping of fresh drinking water from wells drilled into an aquifer (underground reservoir) can lead to saltwater intrusion. The major cations found in seawater are Na^+, K^+, Ca^{2+}, and Mg^{2+}. The major anions are Cl^-, SO_4^{2-}, and Br^-. Generally, humans can taste salt (NaCl) at concentrations above 250 ppm (mg/L). The Environmental Protection Agency (EPA) has set this as a "recommended standard" for the maximum allowable salt concentration.

   (a) What is the molarity of the EPA standard 250 ppm solution?

   (b) The solubility product for silver chloride is 1.8×10^{-10}. If one drop (approximately 0.05 mL) of 0.02 M $AgNO_3$ solution is added to 25 mL of tap water that has a concentration of chloride ions that exactly meets the EPA standard, will a precipitate be observed?

   (c) What concentration of silver nitrate is needed if one drop added to 25.0 mL of tap water will not give a precipitate if the tap water meets the EPA standard but will result in a precipitate if the standard is exceeded?

   (d) The K_{sp} values for AgI and AgBr are 8.5×10^{-17} and 5.4×10^{-13}, respectively. When a silver nitrate solution is added to a mixture that is approximately 0.00663 M in Cl^-, in Br^-, and in I^-, in what sequence will the silver salts precipitate? Justify your answer.

   (e) What special precautions should be taken when working with silver nitrate in the lab?

PRACTICE TEST 2

2. Phosphorus chlorides have been used in the synthesis of various pesticides, flame retardants, and plasticizers. Phosphorus pentachloride has been used to convert carboxylic acids to acyl chlorides. Phosphorus trichloride has been used to convert organic alcohols to alkyl chlorides. For understanding the structural and physical features as well as the physical properties of small molecules, the valence-shell electron-pair repulsion theory works quite well.

(a) Construct the Lewis structures for the molecules PCl_3 and PCl_5. Show all electron pairs.

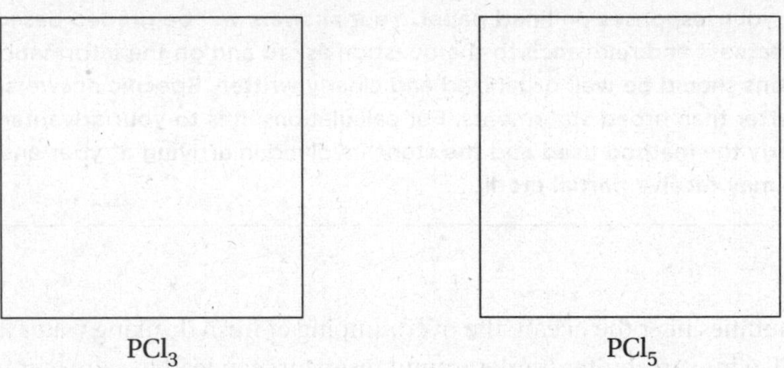

PCl_3 PCl_5

(b) Describe the three-dimensional shape of these molecules.

(c) Is the PCl_3 and/or the PCl_5 a polar molecule? Explain why.

(d) In the appropriate box at the bottom of the question, draw the Lewis structure for propanoic acid ($C_3H_6O_2$).

(e) What is the hybridization of each of the three carbon atoms in propanoic acid?

(f) In the appropriate box, draw a Lewis structure illustrating how two propanoic acid molecules hydrogen-bond to each other.

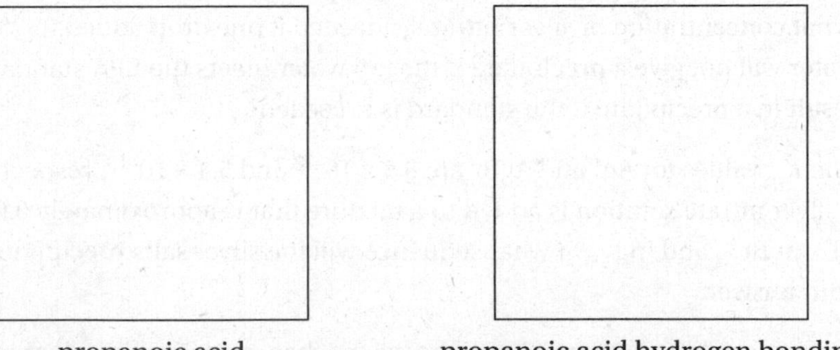

propanoic acid propanoic acid hydrogen bonding

3. A chemist was given the task of determining the identity of a hydrocarbon gas that was isolated by one of his colleagues. The chemist was able to perform several experiments to determine the empirical formula, molecular formula, and possible structures of this compound.

(a) A 0.2880 g sample of the hydrocarbon occupies a volume of 131.0 mL at 24.8°C and 753.0 mm Hg. Determine the molar mass of this hydrocarbon.

(b) The complete combustion of 1.110 g of this same gaseous hydrocarbon yields 3.613 g of CO_2 and 1.109 g of H_2O. Determine the empirical formula of the hydrocarbon.

(c) Based on your answers to parts (a) and (b), what is the molecular formula for this compound?

(d) Draw a possible Lewis structure for this compound.

4. Native Americans were growing corn (maize) long before Europeans "discovered" the New World. For many years, corn was grown for both human consumption and animal food. It was also fermented and distilled to make alcohol for beverages. This alcohol, actually ethanol, is currently being used as an additive to gasoline as a fuel source that burns cleaner than gasoline. The combustion of ethanol is given below:

$$2CH_3CH_2OH(l) + 5O_2(g) \rightarrow 4CO_2(g) + 4H_2O(g)$$

$$2\ H{-}\overset{H}{\underset{H}{C}}{-}\overset{H}{\underset{H}{C}}{-}\ddot{\underset{..}{O}}{-}H + 5\ \ddot{O}{=}\ddot{O} \longrightarrow 4\ \ddot{O}{=}C{=}\ddot{O} + 4\ H{-}\ddot{\underset{..}{O}}{-}H$$

Average Bond Dissociation Energies					
C-H	414 kJ mol^{-1}	C-O	360 kJ mol^{-1}	C=O	799 kJ mol^{-1}
O-H	464 kJ mol^{-1}	C-C	348 kJ mol^{-1}	O=O	498 kJ mol^{-1}

(a) Using bond dissociation energies, estimate the ΔH°_{rxn} for this combustion reaction.

(b) Is this reaction exothermic or endothermic? Justify your answer.

(c) Does this process involve an increase or a decrease in entropy? Explain your answer.

(d) What information would you need to determine if this reaction is thermodynamically favorable? Explain how you would determine this.

5. Propane is widely used as fuel for cooking and heating in remote houses and home barbecue grills. Methane (natural gas) is often piped to residences and businesses for heating and cooking. Switching from one gas to the other is more complicated than just changing connections. The stoichiometry of combustion is an important factor.

(a) Write and balance the combustion equations for methane, CH_4, and for propane, C_3H_8. (Assume complete combustion to carbon dioxide and water with an excess of oxygen.)

(b) A gas burner mixes air with fuel before the fuel burns. Explain whether a methane burner or a propane burner needs larger openings for greater air supply.

(c) If you had to assign a physical state to the substances in the balanced equations in part (a), what would they be? Justify your assignments.

(d) The heat of combustion of methane is –890.31 kJ mol^{-1} methane and for propane is –2219.90 kJ mol^{-1} propane. Which gas produces more heat per mole of CO_2? Show your reasoning clearly. How does this pertain to the current concern about global climate change?

(e) Ethane reacting with oxygen will produce water and three other different substances depending on the oxygen available. Using the symbol E for ethane and the usual formulas for CO_2, O_2, C, CO, and H_2O, draw symbolic representations for the reaction of ethane in excess oxygen, in limited amounts of oxygen, and in very limited amounts of oxygen. In each case, show a starting mixture before reaction and the final mixture after reaction. In each case, start with the 4 molecules of ethane shown and add the appropriate amount of O_2. Draw the correct number of each product and any excess reactant in the product box.

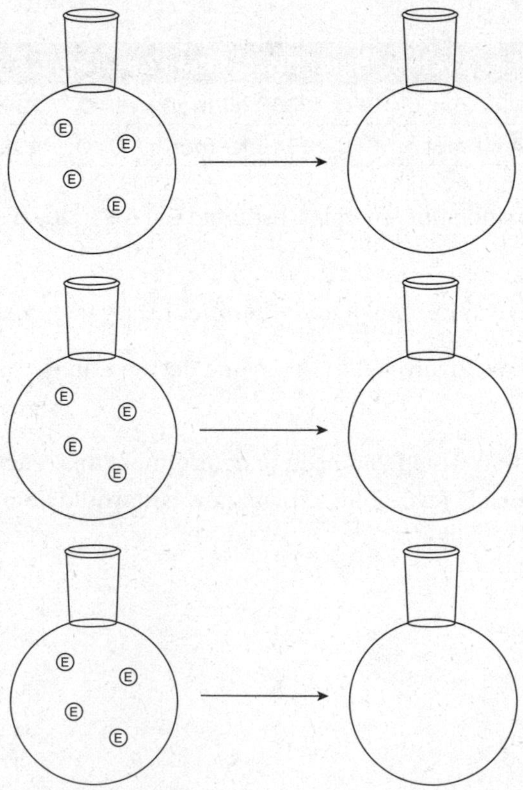

6. Chemical reactions seldom occur in one step according to the balanced chemical equation. Instead they often occur in a series of steps called a mechanism. A reaction mechanism can be deduced by studying the kinetics of a reaction and proposing individual steps, called elementary reactions, which fit the experimental data.

   (a) Nitrogen monoxide, NO, can react with molecular oxygen, O_2, to produce nitrogen dioxide. The following is a possible mechanism:

   $$NO + O_2 \rightarrow NO_3$$

   $$NO_3 + NO \rightarrow 2NO_2$$

   What is the overall reaction that occurs? Do these steps qualify as a reaction mechanism? Justify your response.

   (b) What is the expected rate law if the first step is rate determining? What are the expected units for the rate constant?

   (c) Is NO_3 an intermediate or a catalyst? Justify your answer.

   (d) Draw a potential energy diagram for this exothermic reaction if the second step is rate limiting. Label your diagram fully.

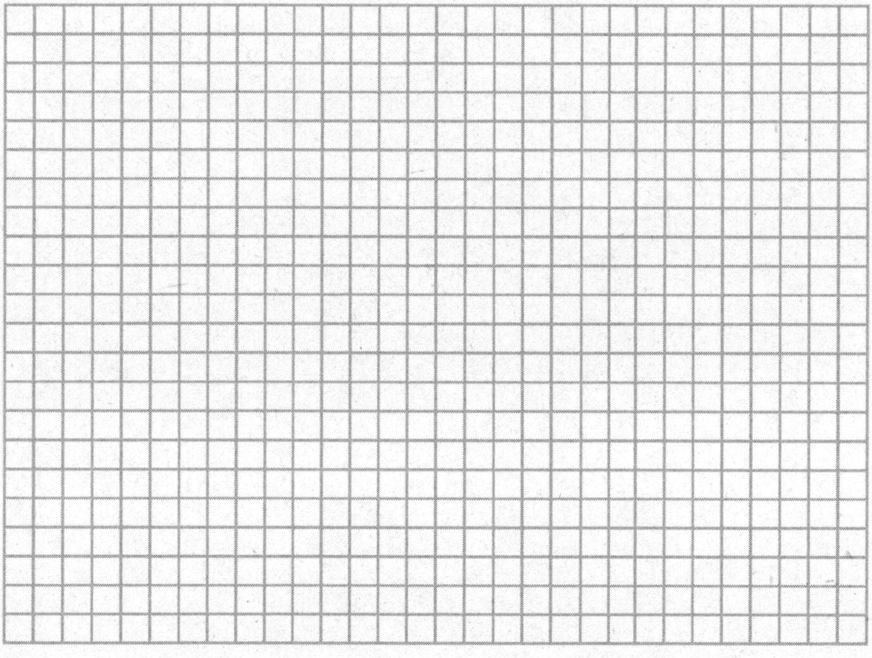

7. The following electrode potentials are for the redox reaction of permanganate ion with iron(III) ions:

$$MnO_4^-(aq) + 8H^+(aq) + 5e^- \rightleftharpoons Mn^{2+}(aq) + 4H_2O(l) \qquad E° = 1.51\text{ V}$$

$$Fe^{3+}(aq) + e^- \rightleftharpoons Fe^{2+}(aq) \qquad E° = 0.77\text{ V}$$

(a) Write the net ionic equation for the reaction using the above set of half-reactions.

(b) Determine the $E°_{cell}$ for this reaction.

(c) Calculate the K_{eq} for this reaction.

(d) Is this reaction thermodynamically favorable as written? Explain.

STOP **END OF EXAM. If you finish before the 90 minutes have elapsed, you may check your work in this section only.**

ANSWER KEY

Practice Test 2

1. **(D)**	11. **(C)**	21. **(A)**	31. **(C)**	41. **(D)**	51. **(C)**
2. **(B)**	12. **(B)**	22. **(C)**	32. **(B)**	42. **(A)**	52. **(C)**
3. **(C)**	13. **(B)**	23. **(B)**	33. **(D)**	43. **(C)**	53. **(C)**
4. **(C)**	14. **(C)**	24. **(D)**	34. **(D)**	44. **(A)**	54. **(C)**
5. **(C)**	15. **(C)**	25. **(C)**	35. **(C)**	45. **(C)**	55. **(C)**
6. **(D)**	16. **(C)**	26. **(B)**	36. **(B)**	46. **(A)**	56. **(A)**
7. **(D)**	17. **(D)**	27. **(A)**	37. **(C)**	47. **(B)**	57. **(C)**
8. **(B)**	18. **(C)**	28. **(C)**	38. **(C)**	48. **(A)**	58. **(D)**
9. **(B)**	19. **(D)**	29. **(D)**	39. **(A)**	49. **(D)**	59. **(C)**
10. **(D)**	20. **(B)**	30. **(A)**	40. **(B)**	50. **(D)**	60. **(A)**

CHAPTER REFERENCES FOR MULTIPLE-CHOICE QUESTIONS

Question	Chapter	Question	Chapter	Question	Chapter
1	9, 13	21	6, 7, 10, 11	41	3, 9, 11
2	9, 13	22	6, 7	42	3, 12, 14
3	9, 13, 14	23	4, 6, 7	43	9, 11, 12
4	13, 14	24	3, 4, 6	44	9, 11, 12
5	6, 7, 14	25	3, 4, 6	45	2, 3, 12
6	6, 7, 10	26	3, 8, 9	46	2, 14
7	7, 11, 14	27	3, 5, 8, 9	47	9, 13, 14
8	6	28	3, 5, 9	48	7, 8
9	2	29	5, 6, 10, 11	49	2, 3, 13, 14
10	5, 9, 13	30	5, 6, 9, 10, 11	50	2, 12, 14
11	8, 13	31	5, 6, 9	51	5, 10, 14
12	5, 13	32	5, 6, 9, 11	52	6, 9, 13
13	5, 13	33	5, 6, 9, 11	53	1, 3, 4, 6
14	5, 9, 13	34	4, 7, 8, 11	54	4, 13
15	4, 7, 8	35	3, 4, 6, 7	55	5, 13, 14
16	2, 10, 12	36	4, 7	56	1, 2
17	1, 2, 3, 4, 5	37	3, 5, 11, 12	57	12, 14
18	9, 13	38	7, 9, 11	58	3, 4, 9, 11
19	9, 13	39	3, 5	59	3, 11
20	5, 6, 14	40	2, 11	60	6, 10

ANSWERS EXPLAINED

Section I—Multiple-Choice

1. **(D)** The weakest acid has the lowest $[H^+]$ and the highest pH. (E.K. 6.C.1; L.O. 6.12, 6.17)

2. **(B)** A strong acid is completely ionized and the molar concentration is equal to the $[H^+]$. (E.K. 6.C.1; L.O. 6.12, 6.17)

3. **(C)** This is the only way to have a significant concentration of both the conjugate acid and conjugate base. (E.K. 6.C.1; L.O. 6.18)

4. **(C)** An end point lower than pH 7 is indicative of a polyprotic acid. (E.K. 6.C.1; L.O. 6.11, 6.13)

5. **(C)** If 25°C is not converted to 298 K, the error is 1000 percent, which is larger than any other error mentioned. (E.K. 2.A.2; L.O. 2.4)

6. **(D)** The kinetic energy is equal to mv^2; therefore, velocity is the highest at point D on this distribution curve. (E.K. 2.A.2; L.O. 2.4)

7. **(D)** All the other responses disobey the law of conservation of energy. (E.K. 5.B.2–5.B.4; L.O. 5.2–5.4)

8. **(B)** Opening valve 2 will increase the volume by 4 times, so the pressure will be 1/4 of the starting pressure. The amount of particles is doubled by opening the second two valves, so the pressure doubles. (E.K. 2.A.2; L.O. 2.4)

9. **(B)** This element is 90 percent of an isotope with a mass of almost 20 and 10 percent of an isotope with a mass close to 22. We calculate $0.90 \times 20 + 0.10 \times 22 = 20.2$, which is closest to the average mass of Ne. (E.K. 1.D.2; L.O. 1.14)

10. **(D)** The weaker the acid is, the stronger is its conjugate base. (E.K. 6.C.1; L.O. 6.16)

11. **(C)** When NaF is dissolved in water, the fluoride ions will react with water to form the weak acid HF. Therefore, the solution will be basic according to the following equation:

$$F^-(aq) + H_2O(l) \rightleftharpoons HF(aq) + OH^-(aq)$$

    (E.K. 6.C.2; L.O. 6.18, 6.20)

12. **(B)** Choosing an indicator with the same pK_a as the end point pH assures the sharpest color change and closest agreement of the end point (experimental) with the equivalence point (theoretical or calculated point). (E.K. 6.C.1; L.O. 6.20)

13. **(B)** Most of the indicator color change is outside the inflection part of the curve. (E.K. 6.C.1; L.O. 6.16, 6.18, 6.20)

14. **(C)** In the titration of a weak acid, the pH of the solution does not start at 1 and has a slight jump at the beginning of the titration that strong acids do not. (E.K. 6.C.1; L.O. 6.13)

***NOTE:** A complete list of the Essential Knowledge (E.K.) needed and the Learning Objectives (L.O.) can be found in the AP Chemistry Course Description:
https://secure-media.collegeboard.org/digitalServices/pdf/ap/ap-chemistry-course-and-exam-description.pdf

15. **(C)** Covalent bonds are shown as solid lines and hydrogen bonds are shown as dashed lines when aligned with the correct atoms. (E.K. 2.B.2; L.O. 2.13)

16. **(C)** In kinetics, the way to increase the reaction rate is to increase the ability of the reactants to meet. Grinding the copper to a dust will make it easy to burn in a flame. (E.K. 4.A.1; L.O. 4.1)

17. **(D)** Only temperature changes result in changes in the equilibrium constant. (E.K. 1.B.2, 6.B.1; L.O. 1.17, 6.8)

18. **(C)** The $[OH^-]$ will be the square root of the $K_b \times$ concentration. We can calculate $(8.0 \times 10^{-7})(0.0010) = 8.0 \times 10^{-10}$, and the square root is approximately 3×10^{-5}. Our answer is between 10^{-5} and 10^{-4}, so the pOH is between 4 and 5. The pH will be between 9 and 10. (E.K. 6.C.1; L.O. 6.15)

19. **(D)** The K_a will be K_w / K_b. The pK_a will be $-\log(K_w / K_b)$. (E.K. 6.C.1; L.O. 6.14)

20. **(B)** The molar mass multiplied by the pressure allows us to compare the amount of material in each flask. Flask *A* has the lowest mass and thus the lowest density. The higher temperature in flask *A* also indicates that it has less mass and lower density. (E.K. 2.A.2; L.O. 2.4)

21. **(A)** The flask with the highest temperature has the gases with the highest average kinetic energy. (E.K. 2.A.2; L.O. 2.4)

22. **(C)** The four bromine atoms with very polarizable electron clouds cause tetrabromomethane to have the strongest intermolecular attractive forces. (E.K. 2.B.A–2.B.1; L.O. 2.12–2.13)

23. **(B)** The London forces from the bromine atoms are most important. No hydrogen bonds are present, and the dipole moments are actually rather small. (E.K. 2.B.1; L.O. 2.11)

24. **(D)** Since the Lewis structure (see page 180) shows three bonding domains and four bonds, three sigma and one pi, the bond order is 4/3, not 5/3. (E.K. 2.C.4; L.O. 2.21)

25. **(C)** The sulfite ion has an unsymmetrical triangular pyramid structure and must be polar. (E.K. 2.C.4; L.O. 2.21)

26. **(B)** This represents the correct net ionic equation. The lead and chloride ions are written correctly with the correct charges. The spectator ions, perchlorate and potassium, have been canceled. (E.K. 1.E.2, 3.A.1; L.O. 1.18, 3.2)

27. **(A)** The K_{sp} calculation indicates that no precipitate should form. (E.K. 6.C.3; L.O. 6.21–6.22)

28. **(C)** This is a common-ion type of calculation. The original concentration of 0.17 $M Pb^{+2}$ suppresses ionization. Solving the K_{sp} equation gives:

$$K_{sp} = [Pb^{2+}][Cl^-]^2 = (0.17)(2x)^2 = 1.7 \times 10^{-5}$$

Dividing by 0.17 gives

$$4x^2 = 1 \times 10^{-4}$$

PRACTICE TEST 2

Then, dividing by 4 produces

$$x^2 = (1 \times 10^{-4})/4$$

Taking the square root gives

$$(1 \times 10^{-2})/2 = 5 \times 10^{-3}$$

(E.K. 6.C.3; L.O. 6.23)

29. **(D)** When the reaction occurs, the N_2O_4 must decrease and the NO_2 must increase. (D) is the only response that expresses this fact. (E.K. 4.B.2; L.O. 4.5)

30. **(A)** All responses are partially correct. (B), (C), and (D) incorrectly state that "only" that parameter can be determined. (E.K. 4.A.1; L.O. 4.1, 6.3)

31. **(C)** The correct pressure is 1520 torr if the reaction goes to completion because the reaction shows that there will be twice the moles of product compared with reactant if all the reactant is consumed. We also conclude the reaction does not go to completion since that pressure is not reached. (E.K. 6.A.3; L.O. 6.4)

32. **(B)** The pressure of N_2O_4 will change by some amount, $-x$. The amount of NO_2 increases by $+2x$. Therefore, the pressure in the vessel will be

$$P = 960 \text{ torr} = 760 \text{ torr} - x + 2x = 760 \text{ torr} + x$$

Calculating for x gives

$$960 - 760 = 200 \text{ torr}$$

Finally, the pressure of N_2O_4 is $760 - x = 560$ torr and the pressure of NO_2 is $2x = 2 \times 200 = 400$ torr. Then we must convert torr to atm by dividing by 760 torr/atm before entering values into the equation. Only (B) has the proper form to fit the chemical equation. (E.K. 6.A.3; L.O. 6.5)

33. **(D)** We expect to see a sharp rise in pressure when more NO_2 is injected into the system. Then there will be a slower decrease due to the slow conversion of NO_2 to N_2O_4. (E.K. 6.A.2; L.O. 6.2)

34. **(D)** Molecular mass is not an important factor by itself in solution formation. (E.K. 2.A.1, 2.B.3; L.O. 2.3, 2.16)

35. **(C)** Polarizability of the electron cloud to form instantaneous dipoles and induced dipole-induced dipole interactions explains the difference in boiling points. (E.K. 1.C.1, 2.C; L.O. 1.9–1.10, 2.17)

36. **(B)** Ethanoic acid molecules strongly hydrogen-bond so that most molecules are part of dimers. (E.K. 2.B.2, 5.D.1–5.D.2; L.O. 2.13, 5.9–5.10)

37. **(C)** When substances are vaporized from the liquid state, heat must be added. This is the definition of an endothermic process. Choices (A) and (B) are combustion reactions, which are expected to be exothermic. The electron affinity for fluorine is also exothermic. (E.K. 5.C.2; L.O. 5.8)

38. **(C)** A formation reaction always has only elements as reactants and one formula unit of the product. Only response (C) meets the definition. (E.K. 5.C.2; L.O. 5.8)

39. **(A)** The entropy change can be used to calculate the heat of reaction, but only if the temperature and Gibbs free energy, standard cell potential, or equilibrium constant to calculate $\Delta G°$ are known. (E.K. 5.E.1; L.O. 5.12, 5.13)

40. **(B)** The desired equation is simply the sum of the two given equations. The heat of reaction is likewise the sum of the two heats of reaction, which is –294.0 kJ. (E.K. 5.C.2, 5.E.4; L.O. 5.8, 5.15)

41. **(D)** If 0.200 mol of N_2 reacts, we would need 0.200 mol of O_2. Therefore, N_2 is the limiting reactant. To calculate the heat evolved, used the equation $q = (-180.8\ \text{kJ/mol}\ N_2)(0.200\ \text{mol}\ N_2) = -36.2\ \text{kJ}$. The minus sign indicates heat is involved. (E.K. 5.C.2; L.O. 5.8)

42. **(A)** The reaction for the first cell is given as

$$Zn + Pb^{2+} \rightarrow Zn^{2+} + Pb$$

For the second cell, the reaction is

$$Mg + Pb^{2+} \rightarrow Mg^{2+} + Pb$$

By reversing the first reaction, adding the two reactions, and canceling the lead, we get the third cell

$$Mg + Zn^{2+} \rightarrow Mg^{2+} + Zn$$

The standard cell potentials are likewise subtracted: $E°_{cell} = 2.24 - 0.63 = +1.61V$. The standard cell potential for this cell, as written, is negative. (E.K. 3.C.3; L.O. 3.12)

43. **(C)** This was shown in answer 42:

$$Mg + Zn^{2+} \rightarrow Mg^{2+} + Zn$$

(E.K. 3.C.3; L.O. 3.12)

44. **(A)** If $E°_{cell}$ is positive and the product concentration is larger than the reactant concentration, the voltage will decrease toward zero. In this case the product concentration is less than the reactant concentration, so the voltage will increase. This will happen in both cells 2 and 3. (E.K. 3.C.3; L.O. 3.13)

45. **(C)** Magnesium will be oxidized by both lead and zinc, so magnesium is easiest to oxidize. Since zinc is oxidized by lead, zinc is the second easiest to oxidize. (E.K. 3.C.3; L.O. 3.13)

46. **(A)** The periodic trends inform us that the radius of nitrogen is expected to be less than that of the preceding carbon. Additionally, the ionization energy increases from left to right across a period, so the ionization energy for nitrogen should be larger than that of carbon. (E.K. 1.C.1; L.O. 1.9–1.10)

47. **(B)** Perchloric acid is a strong acid and is essentially 100 percent ionized. (E.K. 6.C.1; L.O. 6.11–6.12)

48. **(A)** The lowest boiling point points to the lowest attractive forces and thus the lowest viscosity. (E.K. 2.B.1; L.O. 2.11)

49. **(D)** For binary acids, those with longer bond lengths, and thus weaker bonds, are stronger acids. (E.K. 1.C.1, 6.C.1; L.O. 1.11, 6.12)

50. **(D)** The first three substances most commonly get oxidized and cause other substances to be reduced, the opposite of what we want. $KMnO_4$ is a recognized oxidizer and can replace chlorine in many cases. (E.K. 3.A–3.C; L.O. 3.1)

51. **(C)** The reaction is first order since we can see that ln A vs time produces a straight-line graph. Also, we see that at each hour only half of the previous concentration remains, another characteristic of a first-order reaction. We also see that the half-life is $t_{1/2} = 1\text{ h} = 3600\text{ s}$ because it is defined as the time needed for half of the reactant to react. Since $t_{1/2} = \ln 2/k$ for a first-order reaction and the value of $t_{1/2}$ is known, we can calculate $k = \ln 2/t_{1/2}$. (E.K. 4.A.3; L.O. 4.2–4.3)

52. **(C)** We can write $[H^+] = 10^{-3.89}$, which is also equal to [A]

$$K_a = [H^+][A^-]/[HA] = (10^{-3.89})(10^{-3.89})/(10^{-2})$$

Recall that we add exponents in the numerator and subtract the exponents in the denominator. The resulting exponent will be −3.89 + (−3.89) − (−2) = 5.78. So $K_a = 10^{-5.78}$. To get the pK_a, take the negative of the logarithm of $10^{-5.78}$. The result is 5.78. (E.K. 6.C.1–6.C.2; L.O. 6.13, 6.16)

53. **(C)** Only this response is stoichiometrically correct and obeys the law of conservation of mass. We have $5O_2$ and 7S atoms. This gives $5SO_2$ and 2S left over, as shown. (E.K. 2.A.2; L.O. 2.5)

54. **(C)** Weaker bonds to hydrogen mean stronger acids. (E.K. 6.C.1; L.O. 6.16)

55. **(C)** A buffer must contain a conjugate acid–base pair. The only solutions that can possibly produce this combination in this case is the strong acid neutralizing part of a weak base solution. (E.K. 6.C.2; L.O. 6.20)

56. **(A)** All the other statements are misreading the spectroscopy data. (E.K. 1.B.2; L.O. 1.7)

57. **(C)** This pair differs by the one H^+ that is required in the definition of a Brønsted-Lowry conjugate acid–base pair. (E.K. 3.B.2; L.O. 3.7)

58. **(D)** This is the only reaction showing two different phases. (E.K. 3.A–3.C; L.O. 3.1)

59. **(C)** The Gibbs free energy can never be negative using these numbers. (E.K. 5.E.3; L.O. 5.13)

60. **(A)** NO_3 is an intermediate, not a catalyst. (E.K. 4.A; L.O. 4.1)

Section II—Free-Response

1. (a) The information states that 250 ppm is the same as 250 mg L^{-1} of sodium chloride. This is converted to molarity as:

$$? \frac{\text{mol NaCl}}{\text{L}_{\text{solution}}} = \frac{250\text{ mg NaCl}}{\text{L}_{\text{solution}}} \left(\frac{1\text{ g}}{1000\text{ mg}}\right)\left(\frac{1\text{ mol NaCl}}{58.44\text{ g NaCl}}\right) = 4.3\times10^{-3} M_{\text{NaCl}}$$

PRACTICE TEST 2

(b) Start with the solubility product expression:

$$K_{sp} = [Ag^+][Cl^-] = 1.8 \times 10^{-10}$$

The concentration of chloride ions, from part (a), is 4.3×10^{-3} *M*. The concentration of silver ions is calculated from the dilution formula:

$$(0.02\ M\ Ag^+)(0.05\ mL) = [Ag^+](25.05\ mL\ solution)$$

$$[Ag^+] = \frac{(0.02\ M\ Ag^+)(0.05\ mL)}{25.05\ mL\ solution} = 4 \times 10^{-5}\ M\ Ag^+$$

The ion product is $[Ag^+][Cl^-] = (4 \times 10^{-5}\ M)(4.3 \times 10^{-3}\ M) = 2 \times 10^{-7}$.

Since the ion product (2×10^{-7}) is greater than the K_{sp} (1.8×10^{-10}), a precipitate is expected.

(c) Using the K_{sp} and the concentration of chloride ions determined in part (a), the appropriate concentration of silver ions can be determined:

$$K_{sp} = [Ag^+][Cl^-] = 1.8 \times 10^{-10} = [Ag^+](4.3 \times 10^{-3}\ M)$$

$$[Ag^+] = \frac{1.8\times10^{-10}}{4.3\times10^{-3}} = 4.2 \times 10^{-8}\ M\ Ag^+ = 4.2 \times 10^{-8}\ M\ AgNO_3$$

If the salt, specifically the chloride concentration, is above the allowed limit, the drop of silver nitrate will cause a precipitate to form. If the salt concentration is below the limit, no precipitate will form.

(d) As a silver nitrate solution is added, the silver ion concentration slowly increases. Given the same concentrations of anions (Cl^-, Br^-, and I^-), the precipitating cation (Ag^+) will precipitate the anion with the smallest K_{sp} first and the largest K_{sp} last. Therefore, AgI will precipitate first, AgBr second, and AgCl third as silver ions are added to the mixture. If we calculate the silver ion concentration needed to start precipitation, we get:

$$K_{sp}/[Cl^-] = \frac{1.8\times10^{-10}}{0.00663} = 2.7 \times 10^{-8}\ M\ Ag^+$$

$$K_{sp}/[Br^-] = \frac{5.4\times10^{-13}}{0.00663} = 8.1 \times 10^{-11}\ M\ Ag^+$$

$$K_{sp}/[I^-] = \frac{8.5\times10^{-17}}{0.00663} = 1.3 \times 10^{-14}\ M\ Ag^+$$

This leads us to the conclusion that AgI starts to precipitate first, at the lowest $[Ag^+]$. Next AgBr precipitates. AgCl precipitates last, requiring the highest $[Ag^+]$.

(e) Silver nitrate solutions react with the skin and Ag^+ is reduced to colloidal silver, which looks black. Small amounts of black stain on the skin are not harmful, but it does indicate poor laboratory technique. Excessive contact with silver nitrate solutions is not recommended.

2. (a) The Lewis structures of PCl_3 and PCl_5 are shown below.

```
      ..
    : Cl :
 ..   |    ..
: Cl - P - Cl :
 ..   ..   ..
     PCl3
```

```
       ..
      :Cl:    ..
       |    /:Cl:
 ..    |   /
:Cl — P
 ..    |   \  ..
       |    \:Cl:
      :Cl:    ..
       ..
      PCl5
```

(b) The three-dimensional shape of PCl_3 is a triangular pyramid. The three-dimensional shape of PCl_5 is a triangular bipyramid.

(c) The triangular pyramid for PCl_3 is polar. The apex of the pyramid has an electron pair which is expected to be relatively positive compared to the negative end around the three chlorine atoms. The PCl_5 is nonpolar. The two axial chlorines are symmetrically oriented and the P–Cl bond polarities cancel, resulting in no polarity. The three equatorial chlorine atoms are also symmetrically arranged around the "equator" and the polarity of the bonds is canceled because of the symmetry.

(d) The Lewis structure for propanoic acid is shown below.

```
     H   H
     |   |      ..
H —  C — C — C//O:
     |   |     \ ..
     H   H      O—H
                ..
```

propanoic acid

(e) The two carbon atoms on the left have sp^3 hybridization, which also correlates with a tetrahedral shape. The carbon atom on the right has sp^2 hybridization and no lone pairs. The shape is best described as a planar triangle.

(f) The Lewis structure for hydrogen bonding in propanoic acid is shown below.

```
                 O:---- H — O:
                //            \
CH3 — CH2 — C                  C — CH2 — CH3
                \             //
                 :O — H ----:O
```

propanoic acid hydrogen bonding

3. (a) Since this is a gas, use the ideal gas law to obtain the molar mass. First rearrange the formula to find the number of moles of the gas. Next use the values for temperature, pressure, and volume given in the problem. Make sure that everything has the proper units:

$$P = 753.0 \text{ mm Hg}\left(\frac{1 \text{ atm}}{760.0 \text{ mm Hg}}\right) = 0.9908 \text{ atm}$$

$$V = 131.0 \text{ mL}\left(\frac{1 \text{ L}}{1000 \text{ mL}}\right) = 0.1310 \text{ L}$$

$$T = 24.8°\text{C} + 273.15 = 298.0 \text{ K (properly rounded)}$$

$$PV = nRT$$

Rearrange to solve for the number of moles:

$$n = \frac{PV}{RT}$$

$$n = \frac{(0.9908 \text{ atm})(0.1310 \text{ L})}{(0.08206 \text{ L atm mol}^{-1}\text{K}^{-1})(298.0 \text{ K})} = 5.308 \times 10^{-3} \text{ mol}$$

The mass of the sample is given as 0.2880 g. The units for molar mass are g mol^{-1}. The molar mass can be determined by dividing the mass of the gas by the number of moles:

$$\frac{0.2880\text{ g}}{5.308\times10^{-3}\text{ mol}} = 54.3\text{ g mol}^{-1}$$

(b) To find the empirical formula, use the percent composition of each element in a compound. To determine the number of grams of carbon that is in the carbon dioxide, perform a stoichiometric calculation to convert g CO_2 to g C, as shown below. Then use the same procedure to determine the grams of hydrogen in the water:

$$?\text{g C} = 3.613\text{ g }CO_2\left(\frac{1\text{ mol }CO_2}{44.01\text{ g }CO_2}\right)\left(\frac{1\text{ mol C}}{1\text{ mol }CO_2}\right)\left(\frac{12.01\text{ g C}}{1\text{ mol C}}\right) = 0.9860\text{ g C}$$

$$?\text{g H} = 1.109\text{ g }H_2O\left(\frac{1\text{ mol }H_2O}{18.02\text{ g }H_2O}\right)\left(\frac{2\text{ mol H}}{1\text{ mol }H_2O}\right)\left(\frac{1.008\text{ g H}}{1\text{ mol H}}\right) = 0.1241\text{ g H}$$

Now find the number of moles of carbon and of hydrogen, using the atomic mass of each element.

$$0.9860\text{ g C}\left(\frac{1\text{ mol C}}{12.01\text{ g C}}\right) = 0.08210\text{ mol C}$$

$$0.1241\text{ gH}\left(\frac{1\text{ mol H}}{1.008\text{ g H}}\right) = 0.1231\text{ mol H}$$

The next step is to find the lowest whole-number ratio of each element by dividing each by the lowest number of moles calculated above:

$$\frac{0.08210\text{ mol C}}{0.08210} = 1.000\text{ mol C}$$

$$\frac{0.1231\text{ mol H}}{0.08210\text{ mol}} = 1.499\text{ mol H}$$

In relative amounts, we have 1.5 mol H for every mol C. Since it is customary to avoid fractions of a mole in a formula, multiply both numbers by 2 to make whole numbers:

$$1.50\text{ mol H}\times 2 = 3\text{ mol H} = 3\text{ subscript}$$

$$1\text{ mol C}\times 2 = 2\text{ mol C} = 2\text{ subscript}$$

The empirical formula for this compound is C_2H_3.

(c) The empirical mass of the C_2H_3 found in part (b) is 27.03 g mol^{-1}. The molecular formula can be determined by dividing the molar mass by the empirical mass to determine the whole-number multiplier:

$$\frac{54.1\text{ g mol}^{-1}}{27.03\text{ g mol}^{-1}} = 2.00$$

Therefore, the molecular formula can be determined by multiplying the empirical formula's subscripts by 2: $(C_2H_3)\times 2 = C_4H_6$.

(d) To draw the Lewis structure, first find the total number of valence electrons available: 4(4) + 6(1) = 22 valence electrons.

Draw the carbon skeleton by placing a pair of electrons between each of the carbons to represent bonds. This process has used 6 of the valence electrons.

```
C−C−C−C
```

You now have 6 hydrogen atoms to distribute. First try placing them onto the carbons at the ends. Use a pair of electrons for each carbon-hydrogen bond as shown. You have now used 18 electrons.

```
    H           H
    |           |
H − C − C − C − C − H
    |           |
    H           H
```

You still have 4 more electrons to use; place these onto the central carbons.

```
    H           H
    |   ..  ..  |
H − C − C − C − C − H
    |           |
    H           H
```

The two middle carbons do not have full octets. So they must share some electrons to form a triple bond.

```
    H           H
    |           |
H − C − C ≡ C − C − H
    |           |
    H           H
```

In another possible arrangement, put hydrogen atoms onto each carbon as shown (2 on the end and 1 on each carbon in the middle).

```
H − C − C − C − C − H
    |   |   |   |
    H   H   H   H
```

In this case, you still have to place 2 pairs of electrons. So put them onto the two center carbon atoms again.

```
        ..  ..
H − C − C − C − C − H
    |   |   |   |
    H   H   H   H
```

The carbons do not have full octets. So this time, share pairs of electrons between the end (terminal) carbons to form two double bonds.

```
         H
         \       H
          C       /
H −  C  =   \ C = C − H
     |        |
     H        H
```

4. (a) Bond dissociation energies (BDE) can be used to calculate the ΔH°_{rxn}:

$$\Delta H^\circ_{rxn} = \Sigma \text{ BDE (bonds broken)} - \Sigma \text{ BDE (bonds formed)}$$
$$= ((2\text{BDE(C–C)} + (10\text{BDE(C–H)} + (2\text{BDE(C–O))} + (2\text{BDE(O–H))}$$
$$+ (5\text{BDE(O=O))} - ((8\text{BDE(C=O))} + (8\text{BDE(O–H)))}$$
$$= ((2 \times 348 \text{ kJ mol}^{-1}) + (10 \times 414 \text{ kJ mol}^{-1}) + (2 \times 360 \text{ kJ mol}^{-1})$$
$$+ (2 \times 464 \text{ kJ mol}^{-1}) + (5 \times 498 \text{ kJ mol}^{-1}))$$
$$- ((8 \times 799 \text{ kJ mol}^{-1}) + (8 \times 464 \text{ kJ mol}^{-1}))$$
$$= 8974 \text{ kJ} - 10{,}104 \text{ kJ mol}^{-1}$$
$$= -1130 \text{ kJ mol}^{-1}$$

Note: The units for the overal enthalpy are "kilojoules per reaction with the smallest integer coefficients." Sometimes the units are given as kJ_{rxn}.

(b) This process is exothermic due to the negative sign on the calculated value of ΔH°_{rxn}.

(c) The process involves an increase in entropy for the reaction. There are 5 molecules of oxygen gas on the reactant side. However, there are 8 molecules of gas (CO_2 and H_2O) on the product side. The increase in the molecules of a gas provides many more possible energy states for the system. If a reaction has an increase in the possible number of energy states, then the entropy for the reaction increases.

(d) In order for the reaction to be thermodynamically favorable, the value ΔG°_{rxn} must be negative. So ΔG°_{rxn} must be evaluated. We do not know the exact value for the entropy change to solve the following equation:

$$\Delta G^\circ_{rxn} = \Delta H^\circ_{rxn} - T\Delta S^\circ_{rxn}$$

However, we do know that ΔS°_{rxn} is a positive value. In this particular case, we can determine that the free-energy change will be negative at all possible temperatures since the enthalpy will always have a negative value. Additionally, the $-T\Delta S^\circ_{rxn}$ term will always be negative since both T and ΔS°_{rxn} are always positive.

5. (a) The balanced equations are:

$$CH_4(g) + 2O_2(g) \rightarrow CO_2(g) + 2H_2O(l)$$

$$C_3H_8(g) + 5O_2(g) \rightarrow 3CO_2(g) + 4H_2O(l)$$

(b) Since propane needs 5 moles of oxygen and methane needs just 2, the propane burner needs to supply more air than the methane burner.

(c) The usual phase assignments are shown in the answer to part (a). You may make a case for the gas state for water since both reactions occur at elevated temperatures and the question does not specify the temperature.

(d) Burning methane produces –890.31 kJ/mol CO_2.

Burning propane produces –2219.90 kJ/3 mol CO_2 = –740.0 kJ/mol CO_2.

If this is the only consideration, burning methane is environmentally preferable to burning propane.

PRACTICE TEST 2

(e) The following diagrams should be drawn. Observe that the number of each molecule is twice the stoichiometric coefficient since we start with 4 ethane molecules, which is also twice the stoichiometric coefficient. Also, the first set of flasks shows extra oxygen atoms to indicate the excess O_2. Finally, the carbon produced is a solid (soot) and is represented at the bottom of the flask.

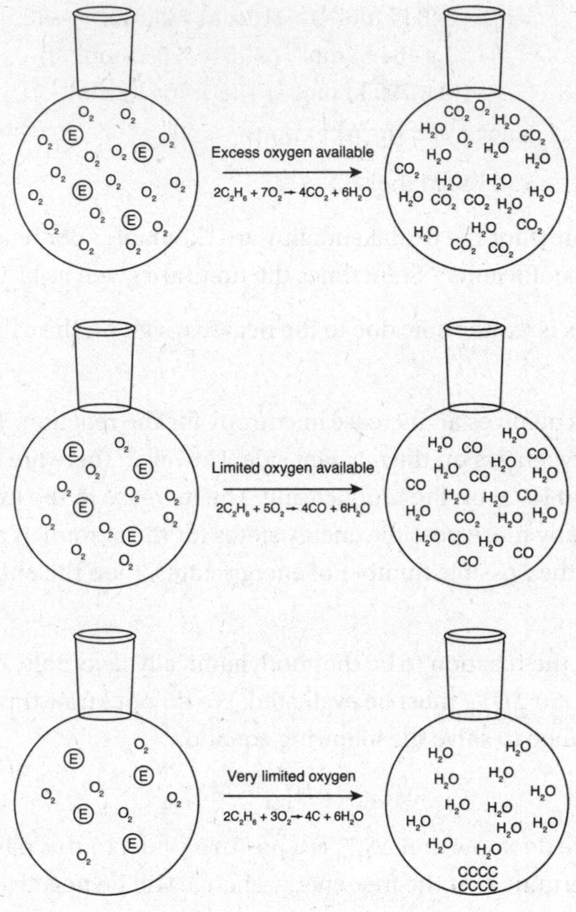

6. (a) The overall equation reads:

$$2NO(g) + O_2(g) \rightarrow 2NO_2(g)$$

Yes, these steps qualify as a reaction mechanism. The two reactions involve only bimolecular collisions and the two equations add up to the overall reaction with reasonable reactants and products.

(b) The rate law for the first elementary reaction involves its reactants with their coefficients as exponents (much like an equilibrium law):

$$\text{Rate} = k[NO][O_2]$$

The units for this rate constant are determined from the rest of the units in the equation:

$$\text{mol L}^{-1}\text{ s}^{-1} = (\text{units for } k)(\text{ mol L}^{-1})(\text{ mol L}^{-1})$$

By rearranging, we get:

$$\frac{\text{mol L}^{-1}\text{ s}^{-1}}{(\text{mol L}^{-1})(\text{mol L}^{-1})} = \text{units for } k$$

Canceling and combining units results in:

$$\frac{s^{-1}}{(\text{mol L}^{-1})} = \text{L mol}^{-1}\text{s}^{-1} = \text{units for } k$$

So L mol^{-1} s^{-1} are the units for a second-order rate constant.

(c) NO_3 is an intermediate that is produced in the first step and is consumed in the second. A catalyst, on the other hand, is present in the reactant mixture and is unchanged in the product mixture. It may or may not be included in the given reactions.

(d) The potential energy diagram should look something like the diagram sketched below. There are two peaks indicating the two steps. The largest peak corresponds to the rate-determining step, which is the second step.

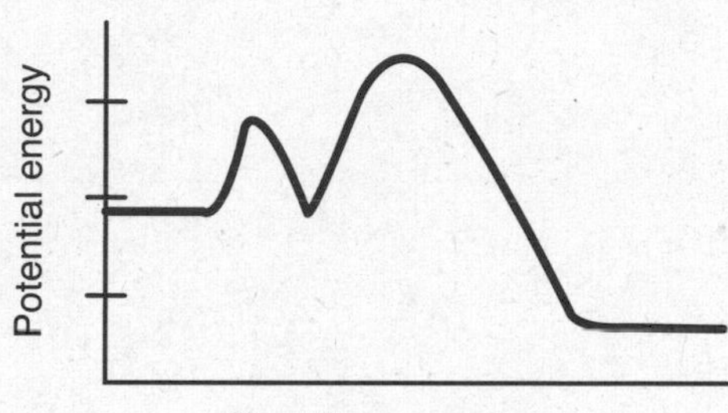

7. (a) In order to get the net ionic equation, the iron half-reaction must be reversed since that half-reaction has a lower potential than the permanganate half-reaction. Iron(II) would be the species that is oxidized. The iron half-reaction must also be multiplied by 5 so that the number of electrons lost equal the number of electrons gained:

$$MnO_4^-(aq) + 8H^+(aq) + 5Fe^{2+}(aq) \rightleftharpoons Mn^{2+}(aq) + 5Fe^{3+} + 4H_2O(l)$$

(b)

$$E^\circ_{cell} = E^\circ_{red} - E^\circ_{ox} = 1.51\text{ V} - (0.77\text{ V})$$

$$= +0.74\text{ V}$$

(c) The equilibrium constant can be found by using the following equation and the values: $T = 298$ K, $R = 8.314$ J mol^{-1} K^{-1}, $n = 5\ e^-$, and $\mathscr{F} = 96{,}485$ C mol^{-1}. Note that all of these values are constants except for n:

$$E^\circ_{cell} = \frac{RT}{n\mathscr{F}} \ln K_{eq}$$

$$\frac{(8.314\text{ J mol}^{-1}\text{ K}^{-1})(298\text{ K})}{(5e^-)(96{,}485\text{ J mol}^{-1})} \ln K_{eq} = 0.00514 \ln K_{eq}$$

Note: For this reaction, all the values are constants except for the temperature, T, which can be changed from experiment to experiment.

Rearrange the equation to find $\ln K_{eq}$:

$$\ln K_{eq} = \frac{0.74}{0.00514} = 144$$

Taking the anti-ln,

$$K_{eq} = 3.45 \times 10^{62}$$

PRACTICE TEST 2

By using either the E°_{cell} or the K_{eq}, the ΔG°_{rxn} can be found:

$$\Delta G^\circ_{rxn} = -RT \ln K_{eq}$$

$$\Delta G^\circ_{rxn} = -(8.314 \text{ J mol}^{-1} \text{ K}^{-1})(298 \text{ K})(\ln 3.45 \times 10^{62})$$

$$= -3.57 \times 10^5 \text{ J mol}^{-1} = -357 \text{ kJ mol}^{-1}$$

or

$$\Delta G^\circ_{rxn} = -(5e^-)(96{,}485 \text{ C mol}^{-1})(0.74 \text{ J C}^{-1})$$

$$= -3.57 \times 10^5 \text{ J mol}^{-1} = -357 \text{ kJ mol}^{-1}$$

(d) This reaction is thermodynamically favorable as written since the ΔG°_{rxn} is a negative value. In order to be thermodynamically favorable, ΔG°_{rxn} must be negative while E°_{cell} must be positive and K_{eq} must be greater than 1.

PRACTICE TEST 2

ANSWER SHEET
Practice Test 3

1. Ⓐ Ⓑ Ⓒ Ⓓ	21. Ⓐ Ⓑ Ⓒ Ⓓ	41. Ⓐ Ⓑ Ⓒ Ⓓ
2. Ⓐ Ⓑ Ⓒ Ⓓ	22. Ⓐ Ⓑ Ⓒ Ⓓ	42. Ⓐ Ⓑ Ⓒ Ⓓ
3. Ⓐ Ⓑ Ⓒ Ⓓ	23. Ⓐ Ⓑ Ⓒ Ⓓ	43. Ⓐ Ⓑ Ⓒ Ⓓ
4. Ⓐ Ⓑ Ⓒ Ⓓ	24. Ⓐ Ⓑ Ⓒ Ⓓ	44. Ⓐ Ⓑ Ⓒ Ⓓ
5. Ⓐ Ⓑ Ⓒ Ⓓ	25. Ⓐ Ⓑ Ⓒ Ⓓ	45. Ⓐ Ⓑ Ⓒ Ⓓ
6. Ⓐ Ⓑ Ⓒ Ⓓ	26. Ⓐ Ⓑ Ⓒ Ⓓ	46. Ⓐ Ⓑ Ⓒ Ⓓ
7. Ⓐ Ⓑ Ⓒ Ⓓ	27. Ⓐ Ⓑ Ⓒ Ⓓ	47. Ⓐ Ⓑ Ⓒ Ⓓ
8. Ⓐ Ⓑ Ⓒ Ⓓ	28. Ⓐ Ⓑ Ⓒ Ⓓ	48. Ⓐ Ⓑ Ⓒ Ⓓ
9. Ⓐ Ⓑ Ⓒ Ⓓ	29. Ⓐ Ⓑ Ⓒ Ⓓ	49. Ⓐ Ⓑ Ⓒ Ⓓ
10. Ⓐ Ⓑ Ⓒ Ⓓ	30. Ⓐ Ⓑ Ⓒ Ⓓ	50. Ⓐ Ⓑ Ⓒ Ⓓ
11. Ⓐ Ⓑ Ⓒ Ⓓ	31. Ⓐ Ⓑ Ⓒ Ⓓ	51. Ⓐ Ⓑ Ⓒ Ⓓ
12. Ⓐ Ⓑ Ⓒ Ⓓ	32. Ⓐ Ⓑ Ⓒ Ⓓ	52. Ⓐ Ⓑ Ⓒ Ⓓ
13. Ⓐ Ⓑ Ⓒ Ⓓ	33. Ⓐ Ⓑ Ⓒ Ⓓ	53. Ⓐ Ⓑ Ⓒ Ⓓ
14. Ⓐ Ⓑ Ⓒ Ⓓ	34. Ⓐ Ⓑ Ⓒ Ⓓ	54. Ⓐ Ⓑ Ⓒ Ⓓ
15. Ⓐ Ⓑ Ⓒ Ⓓ	35. Ⓐ Ⓑ Ⓒ Ⓓ	55. Ⓐ Ⓑ Ⓒ Ⓓ
16. Ⓐ Ⓑ Ⓒ Ⓓ	36. Ⓐ Ⓑ Ⓒ Ⓓ	56. Ⓐ Ⓑ Ⓒ Ⓓ
17. Ⓐ Ⓑ Ⓒ Ⓓ	37. Ⓐ Ⓑ Ⓒ Ⓓ	57. Ⓐ Ⓑ Ⓒ Ⓓ
18. Ⓐ Ⓑ Ⓒ Ⓓ	38. Ⓐ Ⓑ Ⓒ Ⓓ	58. Ⓐ Ⓑ Ⓒ Ⓓ
19. Ⓐ Ⓑ Ⓒ Ⓓ	39. Ⓐ Ⓑ Ⓒ Ⓓ	59. Ⓐ Ⓑ Ⓒ Ⓓ
20. Ⓐ Ⓑ Ⓒ Ⓓ	40. Ⓐ Ⓑ Ⓒ Ⓓ	60. Ⓐ Ⓑ Ⓒ Ⓓ

Practice Test 3

SECTION I: Multiple-Choice

START ONLY WHEN YOU HAVE 90 MINUTES TO COMPLETE THE WHOLE SECTION.

Time:	1 hour, 30 minutes
Number of Questions:	60
Percent of Total Score:	50%
Calculator?	None allowed
Pencil required	

Instructions

There are 60 multiple-choice questions for this part of the exam. Enter your answers on the answer sheet provided. On the actual exam, no credit will be given for answers marked on the test itself. For this test, mark your selected answer next to the question and then transfer it to the scoring sheet. (This will allow you to check that all answers are properly transferred to the scoring sheet.) Each question has only one answer. When changing answers, be sure to erase completely.

Useful Hints

Not everyone will know the answers to all of the questions. However, it is to your advantage to provide an answer to all questions. Do not waste time on difficult questions. Answer the easier ones first, and return to the difficult ones you have not answered if time remains.

Your total score is simply the number of questions answered correctly. Wrong answers or blanks on the scoring sheet do not count against you.

You will be allowed to use the following periodic table and the table of equations and constants on this part of the test.

Periodic Table of the Elements

1 **H** 1.008																	2 **He** 4.00
3 **Li** 6.94	4 **Be** 9.01											5 **B** 10.81	6 **C** 12.01	7 **N** 14.01	8 **O** 16.00	9 **F** 19.00	10 **Ne** 20.18
11 **Na** 22.99	12 **Mg** 24.30											13 **Al** 26.98	14 **Si** 28.09	15 **P** 30.97	16 **S** 32.06	17 **Cl** 35.45	18 **Ar** 39.95
19 **K** 39.10	20 **Ca** 40.08	21 **Sc** 44.96	22 **Ti** 47.87	23 **V** 50.94	24 **Cr** 52.00	25 **Mn** 54.94	26 **Fe** 55.85	27 **Co** 58.93	28 **Ni** 58.69	29 **Cu** 63.55	30 **Zn** 65.38	31 **Ga** 69.72	32 **Ge** 72.63	33 **As** 74.92	34 **Se** 78.97	35 **Br** 79.90	36 **Kr** 83.80
37 **Rb** 85.47	38 **Sr** 87.62	39 **Y** 88.91	40 **Zr** 91.22	41 **Nb** 92.91	42 **Mo** 95.95	43 **Tc** (97)	44 **Ru** 101.1	45 **Rh** 102.91	46 **Pd** 106.42	47 **Ag** 107.87	48 **Cd** 112.41	49 **In** 114.82	50 **Sn** 118.71	51 **Sb** 121.76	52 **Te** 127.60	53 **I** 126.90	54 **Xe** 131.29
55 **Cs** 132.91	56 **Ba** 137.33	57 ***La** 138.91	72 **Hf** 178.49	73 **Ta** 180.95	74 **W** 183.84	75 **Re** 186.21	76 **Os** 190.2	77 **Ir** 192.2	78 **Pt** 195.08	79 **Au** 196.97	80 **Hg** 200.59	81 **Tl** 204.38	82 **Pb** 207.2	83 **Bi** 208.98	84 **Po** (209)	85 **At** (210)	86 **Rn** (222)
87 **Fr** (223)	88 **Ra** (226)	89 **†Ac** (227)	104 **Rf** (267)	105 **Db** (270)	106 **Sg** (271)	107 **Bh** (270)	108 **Hs** (277)	109 **Mt** (276)	110 **Ds** (281)	111 **Rg** (282)	112 **Cn** (285)	113 **Nh** (286)	114 **Fl** (289)	115 **Mc** (289)	116 **Lv** (293)	117 **Ts** (294)	118 **Og** (294)

*Lanthanoid Series	58 **Ce** 140.12	59 **Pr** 140.91	60 **Nd** 144.24	61 **Pm** (145)	62 **Sm** 150.4	63 **Eu** 151.97	64 **Gd** 157.25	65 **Tb** 158.93	66 **Dy** 162.50	67 **Ho** 164.93	68 **Er** 167.26	69 **Tm** 168.93	70 **Yb** 173.05	71 **Lu** 174.97
†Actinoid Series	90 **Th** 232.04	91 **Pa** 231.04	92 **U** 238.03	93 **Np** (237)	94 **Pu** (244)	95 **Am** (243)	96 **Cm** (247)	97 **Bk** (247)	98 **Cf** (251)	99 **Es** (252)	100 **Fm** (257)	101 **Md** (258)	102 **No** (259)	103 **Lr** (262)

EQUATIONS AND CONSTANTS

GENERAL INFORMATION

L, mL	= liter(s), milliliter(s)	mm Hg	= millimeters of mercury
g	= gram(s)	J, kJ	= joule(s), kilojoule(s)
nm	= nanometer(s)	V	= volt(s)
atm	= atmosphere(s)	mol	= mole(s)

ATOMIC STRUCTURE

$E = h\nu$

$c = \lambda\nu$

E = energy
ν = frequency
λ = wavelength

Speed of light, $c = 2.998 \times 10^{8}$ m s^{-1}
Planck's constant, $h = 6.626 \times 10^{-34}$ J s
Avogadro's number = 6.022×10^{23} mol^{-1}
Electron charge, $e = -1.602 \times 10^{-19}$ coulomb

EQUILIBRIUM

$$K_a = \frac{[H^+][A^-]}{[HA]}$$

$$K_b = \frac{[OH^-][HB^+]}{[B]}$$

$$K_c = \frac{[C]^c[D]^d}{[A]^a[B]^b}, \text{ where } aA + bB \rightleftarrows cC + dD$$

$$K_p = \frac{(P_C)^c(P_D)^d}{(P_A)^a(P_B)^b}$$

$$K_w = [H^+][OH^-] = 1.0 \times 10^{-14} \text{ at } 25°C$$

$$= K_a \times K_b$$

$$pK_a = -\log K_a,\ pK_b = -\log K_b$$

$$pH = -\log[H^+],\ pOH = -\log[OH^-]$$

$$14 = pH + pOH$$

$$pH = pK_a + \log\frac{[A^-]}{[HA]}$$

Equilibrium Constants

K_a (weak acid)
K_b (weak base)
K_c (molar concentrations)
K_p (gas pressures)
K_w (water)

THERMOCHEMISTRY/ELECTROCHEMISTRY AND KINETICS

$\Delta S^\circ = \Sigma S^\circ$ products $- \Sigma S^\circ$ reactants

$\Delta H^\circ = \Sigma \Delta H^\circ_f$ products $- \Sigma \Delta H^\circ_f$ reactants

$\Delta G^\circ = \Sigma \Delta G^\circ_f$ products $- \Sigma \Delta G^\circ_f$ reactants

$$\Delta G^\circ = \Delta H^\circ - T\Delta S^\circ$$
$$= -RT \ln K$$
$$= -n\mathcal{F}E^\circ$$

$$q = mc\Delta T$$

$$I = \frac{q}{t}$$

$$\ln[A]_t - \ln[A]_0 = -kt$$

$$\frac{1}{[A]_t} - \frac{1}{[A]_0} = kt$$

$$t_{1/2} = \frac{0.693}{k}$$

m = mass
n = number of moles
q = heat
k = rate constant
c = specific heat capacity
S° = standard entropy
H° = standard enthalpy
G° = standard free energy
E° = standard reduction potential
T = temperature
I = current (amperes)
q = charge (coulombs)
t = time (seconds)
$t_{1/2}$ = half-life

Faraday's constant, $\mathcal{F}$ = 96,485 coulombs per mole of electrons

$$1 \text{ volt} = \frac{1 \text{ joule}}{1 \text{ coulomb}}$$

GASES, LIQUIDS, AND SOLUTIONS

$$PV = nRT$$

$$P_A = P_{total} \times X_A, \text{ where } X_A = \frac{\text{moles A}}{\text{total moles}}$$

$$P_{total} = P_A + P_B + P_C + \ldots$$

$$n = \frac{m}{M}$$

$$K = {}^\circ C + 273$$

$$D = \frac{m}{V}$$

$$\text{KE per molecule} = \frac{1}{2}mv^2$$

$$A = abc$$

$$\text{Molarity, } M = \frac{\text{mol solute}}{\text{liters of solution}}$$

D = density
P = pressure
T = temperature
m = mass
n = number of moles
v = velocity
V = volume
A = absorbance
a = molar absorptivity
b = path length
c = concentration
KE = kinetic energy
M = molar mass

Gas constant, $R = 8.314 \text{ J mol}^{-1} \text{ K}^{-1}$
$= 0.08206 \text{ L atm mol}^{-1} \text{ K}^{-1}$
$= 62.36 \text{ L torr mol}^{-1} \text{ K}^{-1}$

1 atm = 760 mm Hg = 760 torr

STP = 0.00°C and 1.000 atm

Molar volume of ideal gas = 22.4 L at STP

SECTION I

60 Multiple-Choice Questions
(Time: 90 minutes)
CALCULATORS ARE NOT ALLOWED FOR SECTION I

Note: For all questions, assume that $T = 298$ K, $P = 1.00$ atmosphere, and H_2O is the solvent for all solutions unless the problem indicates a different solvent.

Directions: The questions or incomplete statements that follow are each followed by four suggested answers or completions. Choose the response that best answers the question or completes the statement. Fill in the corresponding circle on the answer sheet.

1. An experiment was performed to determine the moles of carbon dioxide gas formed (collected over water) when an acid reacts with limestone. To do this, a piece of dry limestone was weighed. Then carbon dioxide was collected until the limestone disappeared. The atmospheric pressure was measured at 0.965 atm. The temperature was 295 K. The volume of CO_2 was measured to the nearest mL and corrected for the vapor pressure of water. The student continuously got low results from what was expected. Can you suggest why?
   (A) Limestone is never pure $CaCO_3$.
   (B) Limestone is $Ca(HCO_3)_2$.
   (C) Carbon dioxide is rather soluble in water.
   (D) Perhaps there was not enough acid to dissolve all the limestone.

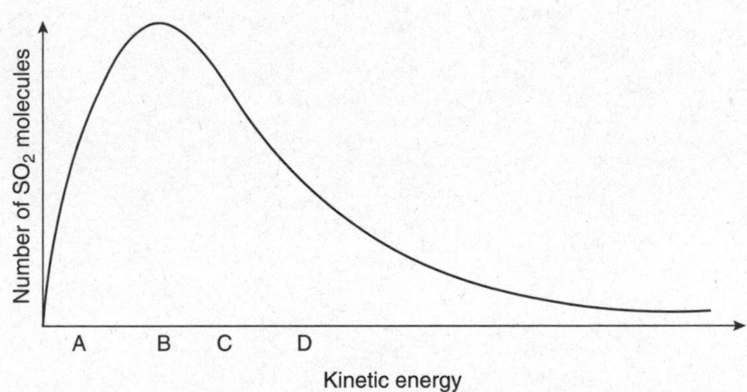

2. Choose the description of the point on the distribution curve that is the most plausible.
   (A) Point *A* represents absolute zero where all motions stops.
   (B) Point *C* is proportional to temperature
   (C) Point *B* is where the transition state occurs.
   (D) At point *D*, the products of the reaction are found.

3. Your supervisor asks you to determine the enthalpy of a certain chemical reaction. Which would you do?
   (A) Measure the ΔS and the ΔG for the reaction, and calculate the ΔH from the Gibbs free energy equation.
   (B) Use a bomb calorimeter to measure the heat of the reaction.
   (C) Use a solution calorimeter such as a coffee-cup calorimeter to measure the heat.
   (D) Use a photoelectron spectrometer to measure the energies of all atoms in the compounds, and use Hess's law to add them.

4. A 25 g sample of a solid was heated to 100°C and then quickly transferred to an insulated container holding 100 g of water at 26°C. The temperature of the mixture rose to reach a final temperature of 37°C. Which of the following can be concluded?
   (A) The sample lost more thermal energy than the water gained because the sample temperature changed more than the water temperature did.
   (B) Even though the sample temperature changed more than the water temperature did, the sample lost the same amount of thermal energy as the water gained.
   (C) The sample temperature changed more than the water temperature did; therefore the heat capacity of the sample must be greater than the heat capacity of the water.
   (D) The final temperature is less than the average starting temperatures; therefore the equilibrium constant must be less than 1.

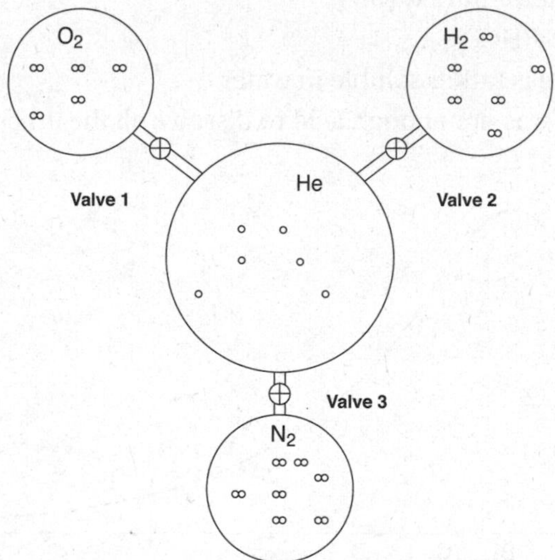

5. Three 1-liter flasks are connected to a 3-liter flask by valves as shown in the diagram above. The 3-liter flask has the relative number of helium atoms as indicated. At the start, the entire system is at 585 K. The first flask contains oxygen, the second contains hydrogen, and the third contains nitrogen. The pressure of hydrogen is 3.00 atm. The number of gas molecules is proportional to their representations in the flasks. If the valves are all opened, what will be the pressure in the system? Assume the connections have negligible volume.
   (A) 1.0 atm
   (B) 2.0 atm
   (C) 3.0 atm
   (D) 4.0 atm

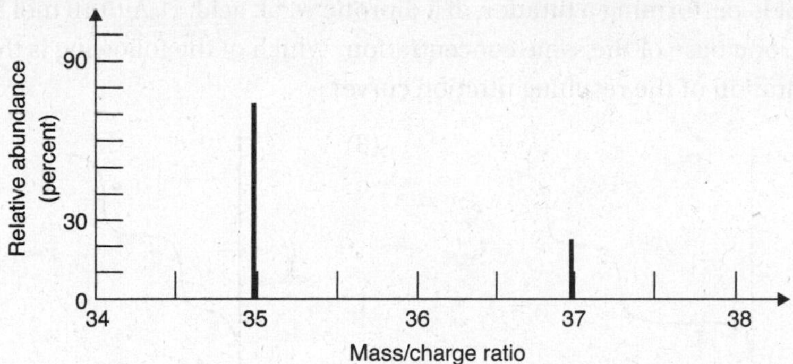

6. A mass spectrum of a naturally occurring sample of an element is shown above. What is the element?
   (A) Cl
   (B) S
   (C) Ar
   (D) There are two peaks, so there must be two compounds.

7. Given the following weak bases with their respective K_b, which sequence lists the corresponding cations in order of increasing acid strength?

NH_3	$K_b = 1.8 \times 10^{-5}$
C_5H_5N	$K_b = 1.5 \times 10^{-9}$
CH_3NH_2	$K_b = 4.4 \times 10^{-4}$

   (A) $C_5H_5NH^+$, $CH_3NH_3^+$, NH_4^+
   (B) NH_4^+, $CH_3NH_3^+$, $C_5H_5NH^+$
   (C) $CH_3NH_3^+$, NH_4^+, $C_5H_5NH^+$
   (D) $C_5H_5NH^+$, $CH_3NH_3^+$, NH_4^+

8. A laboratory assistant was told to create a buffer solution with a pH equal to 5. Which of the following acids would be the best choice?
   (A) HOCl $K_a = 3.0 \times 10^{-8}$
   (B) H_3AsO_4 $K_a = 5.6 \times 10^{-3}$
   (C) $HC_2H_3O_2$ $K_a = 1.8 \times 10^{-5}$
   (D) $H_2C_2O_4$ $K_a = 5.9 \times 10^{-2}$

9. A student is performing a titration of a diprotic weak acid, H_2A (0.10 mol L^{-1}) with a strong base of the same concentration. Which of the following is the best representation of the resulting titration curve?

(A)

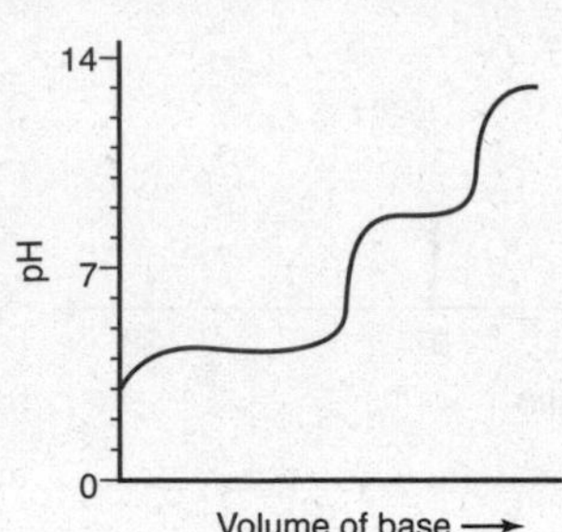

(B)

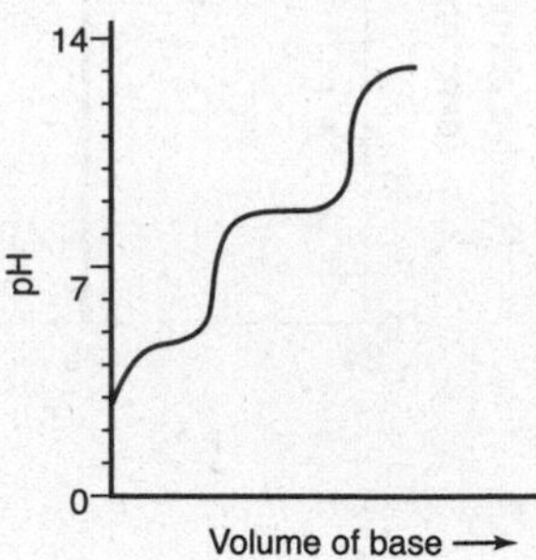

(C)

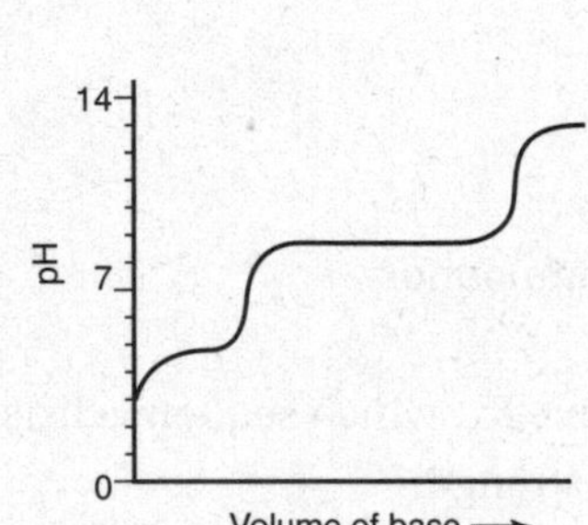

(D)

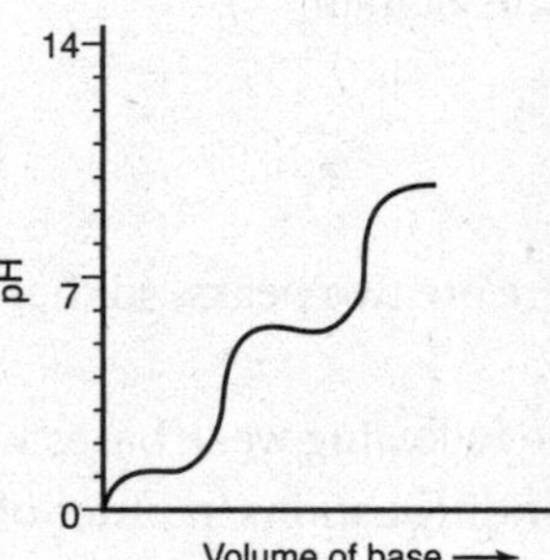

10. A student uses a pH indicator that changes color from pH 8 to 6. Which statement best characterizes the expected observations based on the titration curve above?
   (A) The color will change slowly, and the end point volume will be low.
   (B) The observed color change will be distinct, and the calculated molarity of the base will be accurate.
   (C) The color will change slowly, and the calculated molarity of the base will be low.
   (D) The observed color change will be distinct, but the calculated molarity of the base will be high.

11. When preforming a titration of 0.050 *M* benzoic acid, C_6H_5COOH, ($K_a = 6.6 \times 10^{-5}$) with 0.050 *M* NaOH, which of the following indicators would be the best choice?
   (A) Bromphenol blue, pH range 3.0–4.5
   (B) Phenolphthalein, pH range 8.0–10.1
   (C) Phenol red, pH range 6.9–8.2
   (D) Bromcresol green, pH range 3.8–5.4

Questions 12–14 refer to the information and graph below.

Four different 0.0100 *M* acid solutions were prepared, and their pH values were recorded on a laptop computer. One of the solutions contains more than just an acid. At the point designated as "add," all four solutions are diluted with an equal volume of water. The bottom line represents solution 1, and the top line represents solution 4.

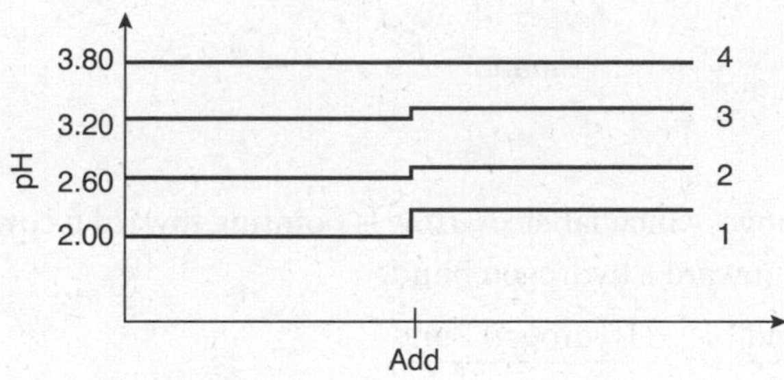

12. Using the data in the graph, which of the solutions does not contain just an acid dissolved in water?
   (A) Solution 4 because the pH does not change upon dilution
   (B) Solution 1 because the pH changes too much
   (C) Solutions 2 and 3 because the pH did not change enough
   (D) Solution 4, which must be pure water since the pH does not change

13. Using the data presented in the graph and the experiment that was performed, which of the weak acids is the weakest?
   (A) Solution 1 because its pH changed the most
   (B) Solution 4 because the pH did not change at all
   (C) Solution 2 because its pK_a appears to be about 3.20
   (D) Solution 3 had the highest pH of all the solutions made with one 0.0100 *M* acid.

14. Which of the acids will react with copper metal?
   (A) Acid 1 and acid 3
   (B) We cannot tell; we need to know what the cations are.
   (C) Acid 2
   (D) We cannot tell; we need to know what the anions are.

15. Chemists often ascribe the macroscopic properties of solids to the underlying microscopic structure. Silicon carbide is almost as hard and brittle as diamond. The solid state structure of silicon carbide is often described as
(A) a molecular crystal
(B) a covalent or network crystal
(C) a metallic crystal
(D) an ionic crystal

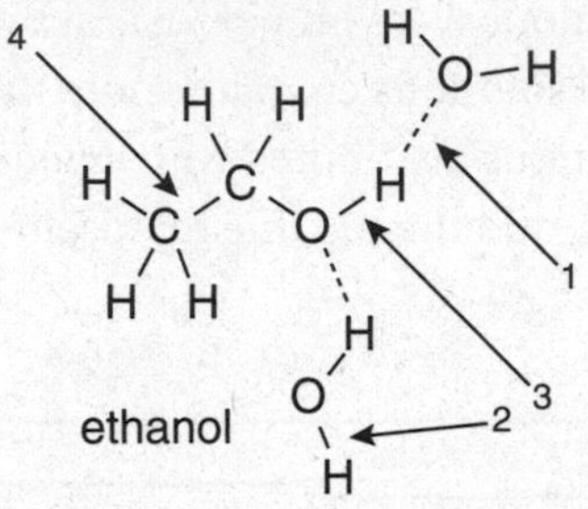

16. In the diagram above, which labeled arrow is pointing toward a covalent bond *and* which is pointing toward a hydrogen bond?

	Covalent Bond	Hydrogen Bond
(A)	1	2
(B)	2	1
(C)	3	4
(D)	4	3

17. Which is the easiest way to burn a silver coin?
(A) Hold the silver coin with crucible tongs, and heat strongly in the flame of a Bunsen burner.
(B) Use the method in (A), but use an oxyacetylene torch to reach a higher temperature.
(C) Grind the silver coin into very small, dust-sized particles, and spray the particles into a Bunsen burner flame.
(D) Dissolve the silver coin in acid, precipitate the hydroxide, and heat in a Bunsen burner flame to make the oxide.

Questions 18–21 refer to the table below.

Vessel	A	B	C
Gas	Argon	Neon	Xenon
Formula	Ar	Ne	Xe
Molar mass	40 g/mol	20 g/mol	132 g/mol
Pressure	0.3 atm	0.4 atm	0.2 atm
Temperature	400°C	500°C	500°C

Note that in addition to the data in the table, the gases are held in separate, identical, rigid vessels.

18. Which sample has the lowest density?
    (A) All have the same density since the temperatures are all the same.
    (B) Vessel *A*
    (C) Vessel *B*
    (D) Vessel *C*

19. The average kinetic energy
    (A) is lowest in vessel *A*
    (B) is lowest in vessel *B*
    (C) is lowest in vessel *C*
    (A) is the same in all vessels

20. Which of these gases is expected to condense at the lowest pressure, assuming that the temperature is held constant?
    (A) Xenon
    (B) Argon
    (C) Neon
    (D) They all condense at the same pressure.

21. Which attractive force is the major cause of condensation of the three compounds in the table above?
    (A) Hydrogen bonding
    (B) London forces
    (C) Dipole-dipole attractive forces
    (D) Molar mass

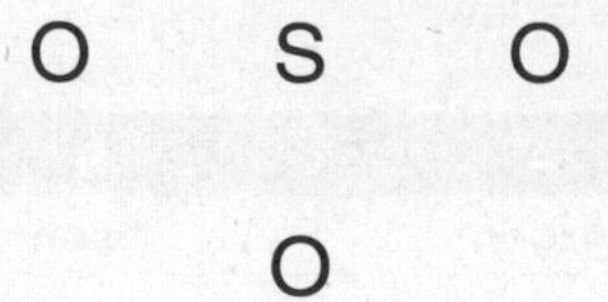

22. Use the arrangement of atoms suggested in the skeleton structure above to construct the Lewis structure for the SO_3 molecule. All of the following statements about this molecule are true EXCEPT
    (A) sulfur trioxide has three resonance structures
    (B) sulfur trioxide has a planar triangle shape
    (C) sulfur trioxide is nonpolar
    (D) the S–O bond order is 5/3

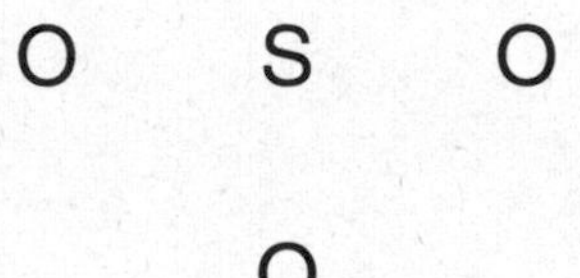

23. Use the arrangement of atoms suggested in the skeleton structure above to construct the Lewis structure for the sulfite ion. Be sure to minimize the formal charges. Which statement is the *least* likely to be true?
    (A) The sulfite ion has a double bond.
    (B) The sulfite ion has a triangular pyramid shape.
    (C) The sulfite ion is nonpolar.
    (D) The S–O bond order is 4/3.

$$S(s) + O_2(g) \rightarrow SO_2(g)$$

24. The reaction between sulfur and oxygen, written in equation form above, can be interpreted in all of the following ways EXCEPT
    (A) one atom of S reacts with one molecule of O_2 to yield one molecule of SO_2
    (B) one mole of sulfur atoms reacts with one mole of oxygen molecules to yield one mole of sulfur dioxide molecules
    (C) the position of equilibrium must be on the product side
    (D) the entropy increase will be large

Questions 25–26 refer to the following diagram.

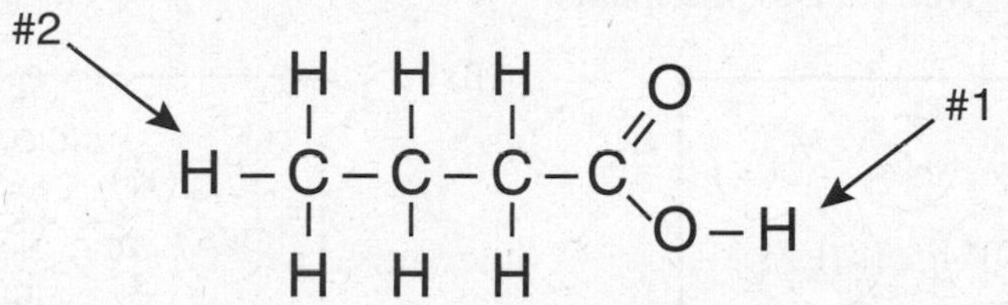

```
#2
        H   H   H   O
        |   |   |  //                #1
    H - C - C - C - C
        |   |   |    \
        H   H   H     O - H
```

25. Which hydrogen in butanoic acid, shown above, will ionize and why?
    (A) Hydrogen #1: it is at the end of the formula.
    (B) Hydrogen #1: its bond to the rest of the molecule is the weakest due to the electronegativity of two nearby oxygen atoms.
    (C) Hydrogen #2: there is a 7-to-1 probability of the hydrogens bound to carbon ionizing.
    (D) Hydrogen #2: this type of H is very electronegative.

26. Butanoic acid $C_4H_8O_2$, shown above, gives rancid butter its foul smell. It has a $K_a = 1.5 \times 10^{-5}$. If a 0.0100 *M* solution of butanoic acid is prepared, the expected pH will be in which one of the following pH ranges?
    (A) 2 to 4
    (B) 4 to 6
    (C) 8 to 10
    (D) 10 to 12

Questions 27–29 refer to the following equation.

$$Tl^{+}(aq) + Cl^{-}(aq) \rightarrow TlCl(s)$$

A chemist mixes a dilute solution of thallium perchlorate with a dilute solution of potassium chloride to precipitate thallium chloride ($K_{sp} = 1.9 \times 10^{-4}$).

27. Which of the following is a molecular equation for the reaction described above?
    (A) $TlClO_3(aq) + KCl(aq) \rightarrow TlCl(s) + KClO_3$
    (B) $Tl(ClO_4)_2(aq) + 2KCl(aq) \rightarrow TlCl(s) + 2KClO_4$
    (C) $TlClO_4(aq) + KCl(aq) \rightarrow TlCl(s) + KClO_4$
    (D) $Tl(ClO_4)_4(aq) + 4KCl(aq) \rightarrow TlCl(s) + 4KClO_4$

28. If equal volumes of 2.0 millimolar solutions are mixed, which of the following particulate views represents the experiment after the reactants are mixed thoroughly and solids are given time to precipitate?

(A)

(B)

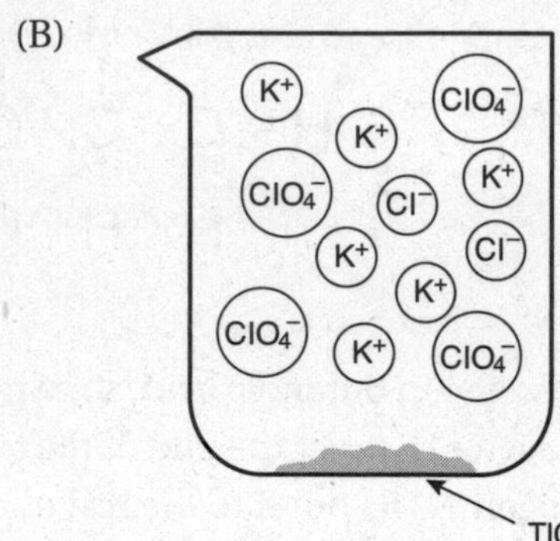

(C)

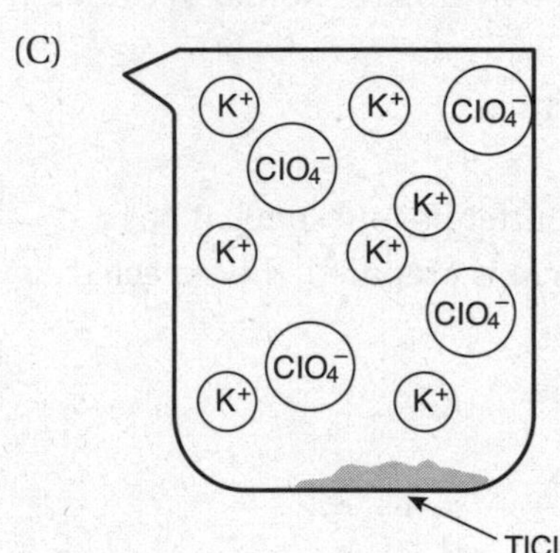

(D)

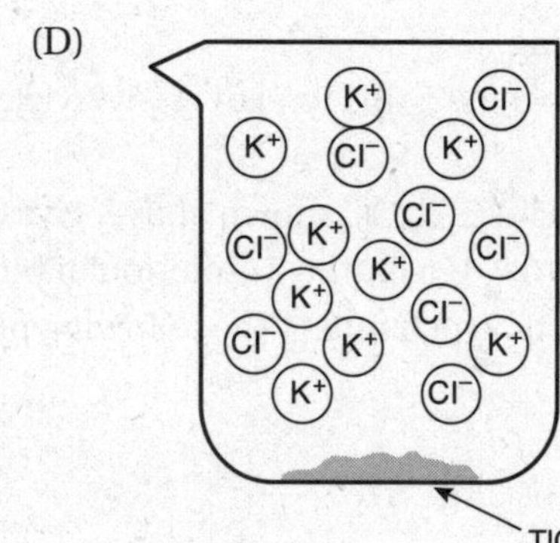

29. Within a factor of 10, what is the approximate molar solubility of thallium chloride in distilled water?

(A) 1.8×10^{-5} mol/L
(B) 10^{-7} mol/L
(C) 10^{-2} mol/L
(D) 10^{-15} mol/L

Questions 30–35 refer to the information and graph below.

$$O_2(g) + 2SO_2(g) \rightleftharpoons 2SO_3(g)$$

$SO_2(g)$ is produced in the combustion of coal, oil, gasoline, and many other natural products. In the atmosphere, it reacts with oxygen to form $SO_3(g)$ via the equation above. A pure sample of $SO_3(g)$ is placed into a rigid, evacuated 0.500 L container. The initial pressure of the $SO_3(g)$ is 760 torr. The temperature is held constant until the system reaches equilibrium. The figure below shows how the pressure of the system changes in reaching equilibrium.

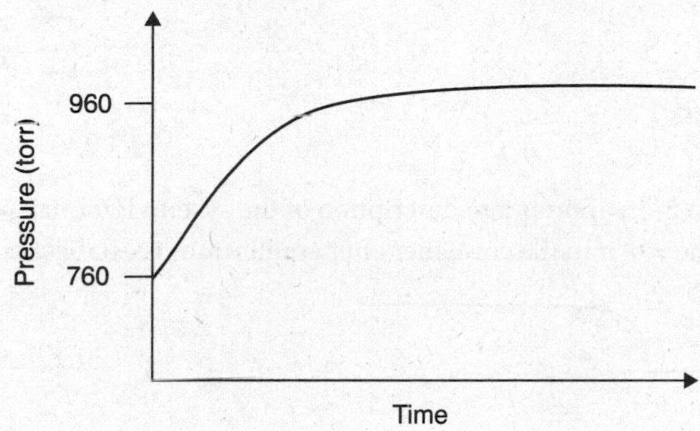

30. Why does the pressure rise in this experiment?
    (A) Actually, the pressure should decrease since three moles of gas are producing two moles in the reaction.
    (B) SO_2 is polar, resulting in more effective collisions with the container walls.
    (C) The nonpolar sulfur trioxide becomes the polar sulfur dioxide, thus increasing the number of collisions per second.
    (D) Two moles of SO_3 are producing three moles of gas, increasing the particles in the container colliding with the walls.

31. The pressure versus time curve can be used to
    (A) determine the rate law for the reverse reaction
    (B) determine the rate law for the forward reaction
    (C) determine the activation energy for the reverse reaction
    (D) determine the enthalpy of this gas phase reaction

32. What will the final pressure be if this reaction goes to completion?
    (A) 760 torr
    (B) 900 torr
    (C) 1140 torr
    (D) 2280 torr

33. Sulfur trioxide decomposed to form some sulfur dioxide and molecular oxygen. What is the proper form for the equilibrium expression for the reaction that actually occurred?

(A) $K_P = \dfrac{P_{O_2} P_{SO_2}}{P_{SO_3}}$

(B) $K_P = \dfrac{P_{O_2} P^2_{SO_2}}{P^2_{SO_3}}$

(C) $K_P = \dfrac{P^2_{SO_3}}{P_{O_2} P^2_{SO_2}}$

(D) $K_P = \dfrac{P_{SO_3}}{P_{O_2} P_{SO_2}}$

34. Which diagram is an appropriate description of the system if more $SO_2(g)$ is rapidly injected, at time = 0, into the container after equilibrium is established?

(A)

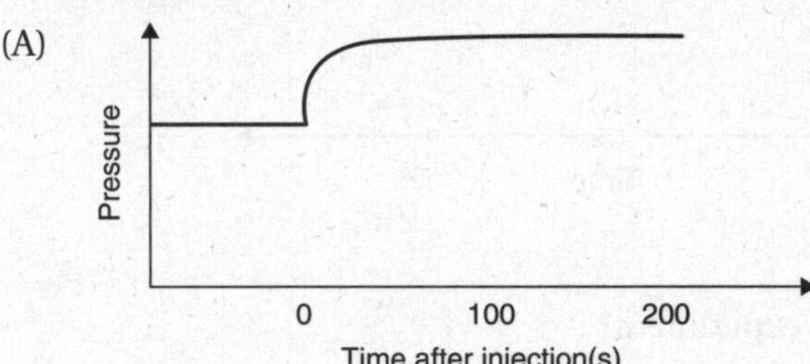

(B)

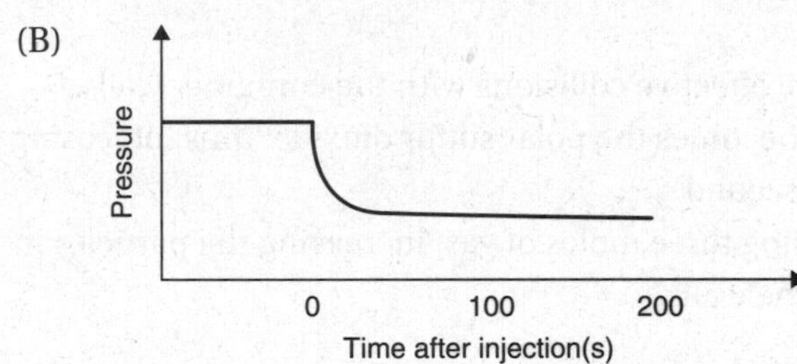

(C)

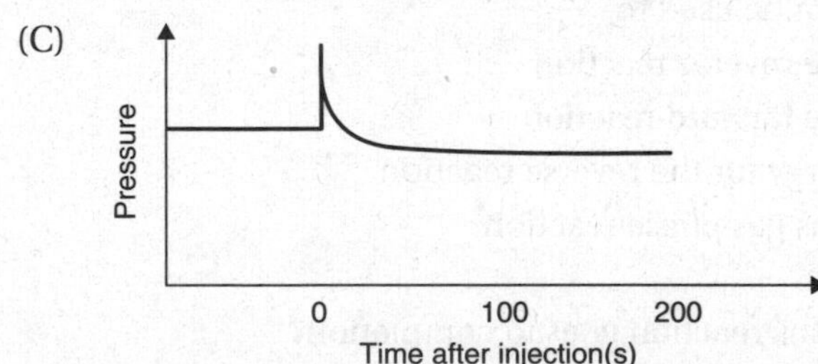

(D)

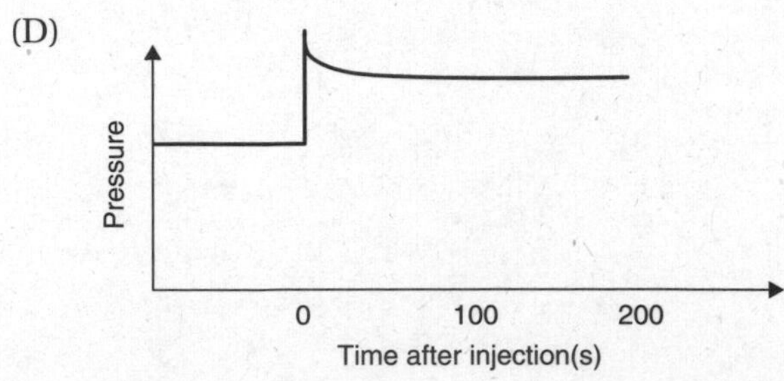

35. Why are $SO_2(g)$ and $SO_3(g)$ pollutants to be concerned about?
(A) They contain yellow sulfur that causes jaundice.
(B) They both are acid anhydrides, forming H_2SO_3 and H_2SO_4, which are a major part of acid rain.
(C) They both have resonance structures that lead to free radicals that cause cancer.
(D) They create the brown layer of smog seen over many cities and destroy O_3, which is essential for life.

---

36. Fluorine has a normal boiling point of 85 K, and chlorine boils at 239 K. The Cl_2 molecule is much larger than F_2 (atomic radius is 99 pm for chlorine and is 64 pm for fluorine). Which is the best reason for the large difference in boiling points?
(A) Chlorine has a higher dipole moment than fluorine.
(B) The intramolecular bonds in Cl_2 are much weaker than those in F_2.
(C) The Cl_2 electron clouds are much more polarizable than the F_2 electron clouds, resulting in much stronger London forces.
(D) The mass of chlorine is much greater than the mass of fluorine.

37. Space-filling representations of carbon tetrachloride and carbon tetrabromide are shown above. The carbon can be seen in the carbon tetrachloride and is hidden by the bromine in carbon tetrabromide. Which of the following is the most reasonable statement? (Assume that the temperatures of CCl_4 and CBr_4 are the same in these comparisons.)
(A) CCl_4 has a higher surface tension compared with CBr_4.
(B) CCl_4 has a higher vapor pressure compared with CBr_4.
(C) CCl_4 has a higher boiling point compared with CBr_4.
(D) CCl_4 has a higher viscosity compared with CBr_4.

Given the following thermochemical reactions, what is the heat of reaction of

$2NO \rightarrow N_2 + O_2$ $\quad \Delta H° = -180.8$ kJ

$2NO + O_2 \rightarrow 2NO_2$ $\quad \Delta H° = -113.2$ kJ

38. What is the enthalpy of the reaction $2NO_2 \rightarrow N_2 + 2O_2$?
    (A) −67.6 kJ
    (B) −294.0 kJ
    (C) +294.0 kJ
    (D) The enthalpy of reaction cannot be determined with this information.

Substance	Heat of formation (kJ/mol)
$H_2O(l)$	-285.9
$H_2O(g)$	-241.8
$CO_2(g)$	-393.5
$C_2H_5OH(l)$	-277.63

39. Select the balanced chemical equation for the combustion of ethyl alcohol, $C_2H_5OH(l)$.
    (A) $C_2H_5OH(l) + 5O_2(g) \rightarrow 2CO_2(g) + 3H_2O(l)$
    (B) $4C_2H_5(l) + 13O_2(g) \rightarrow 8CO_2(g) + 10H_2O(g)$
    (C) $2C_2H_5O(l) + 9O_2(g) \rightarrow 4CO_2(g) + 5H_2O(g)$
    (D) $C_2H_5OH(l) + 3O_2(g) \rightarrow 2CO_2(g) + 3H_2O(g)$

40. Using the table of heats of formation above, determine the heat of combustion for ethyl alcohol, $C_2H_5OH(l)$.
    (A) 2(−277.63) + 7(0.00) − 4(−393.5) − 6(−241.8)
    (B) 2(−393.5) + 3(−241.8) − (−277.63) − 3(0.00)
    (C) 4(−393.5) + 5(−241.8) − 2(−277.63) − 9(0.00)
    (D) 4(−277.63) + 13(0.00) − 8(−393.5) − 10(−241.8)

41. Using the table of heats of formation above, calculate the heat of vaporization of water.
    (A) (−241.8) − (−285.9)
    (B) (−285.9) − (−241.8)
    (C) (241.8) − (−285.9)
    (D) (−285.9) − (241.8)

42. Can the standard entropy change for the vaporization of water at 298 K be determined?
    (A) No, because the equilibrium constant and *Q* are not known
    (B) Yes, it is equal to $1/\Delta H°_{vap}$
    (C) Yes, because at a phase change from liquid to gas or solid to liquid always has $\Delta S° = 0.00$
    (D) Yes, since $\Delta G°$ is zero for water and steam at this temperature, then $\Delta S°$ must be equal to $\Delta H°_{vap}/T$.

43. A chemical system at standard state has all concentrations at 1 *M* and therefore $Q = 1.00$. If the standard cell potential is +0.500 *V*, which of the following is false?
(A) The reaction is thermodynamically favored, and the reaction proceeds in the forward direction.
(B) The reaction is thermodynamically favored, and the concentrations of the products increase.
(C) The reaction is thermodynamically favored, and the concentrations of the reactants decrease.
(D) The reaction is thermodynamically favored, and the reaction proceeds in the reverse direction increasing the concentration of reactants.

Questions 44–46 refer to the galvanic cell shown below. The voltmeter reads 0.658 *V*.

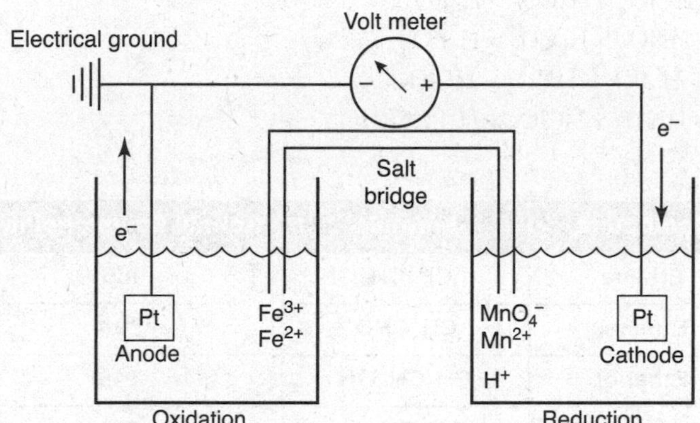

44. What is the cell diagram for the cell above?
(A) Pt | Fe^{2+}, Fe^{3+} || MnO_4^-, H^+, Mn^{2+} | Mn
(B) Pt | Fe^{2+} | Fe^{3+} || MnO_4^-, H^+ | Mn^{2+} | Pt
(C) Pt | Fe^{2+}, Fe^{3+} || MnO_4^-, H^+, Mn^{2+} | Pt
(D) Fe^{2+} | Fe^{3+} || MnO_4^-, H^+

45. Which balanced chemical equation is thermodynamically favored?
(A) $Fe^{2+}(aq) + MnO_4^-(aq) + H^+(aq) \rightarrow Mn^{2+}(aq) + Fe^{3+}(aq) + H_2O$
(B) $5Fe^{3+}(aq) + MnO_4^-(aq) + 8H^+(aq) \rightarrow Mn^{2+}(aq) + 5Fe^{2+}(aq)$
(C) $5Fe^{2+} + MnO_4^- + 8H^+ \rightarrow Mn^{2+} + 5Fe^{3+} + 4H_2O$
(D) $5Fe^{3+} + MnO_4^- + 8H^+ \rightarrow Mn^{2+} + 5Fe^{2+}$

46. For the galvanic cell shown in the diagram, what will be observed if the voltmeter is replaced by a copper wire?
(A) The purple color of the permanganate ion will slowly disappear.
(B) The purple color of the permanganate will increase.
(C) Nothing happens without the voltmeter being present.
(D) The pH of the left cell will not change, but the pH will decrease in the right cell.

The first ionization energy of sodium is 496 kJ/mol, and its atomic radius is 186 pm.

47. Based on the information above and your knowledge of periodic trends, which values are the most reasonable for the radius and first ionization energy of the magnesium atom?
(A) 160 pm, 737 kJ/mol
(B) 86 pm, 398 kJ/mol
(C) 235 pm, 523 kJ/mol
(D) 240 pm, 1200 kJ/mol

48. Dissolving one mole of each of the oxoacids HNO_2, $HClO_4$, H_2CO_3, and H_3PO_4 in 2.0 L of distilled water results in solutions with different pH values. Arrange these acid solutions from the one with the highest pH to the one with the lowest pH.
(A) $HNO_2 > HClO_4 > H_2CO_3 > H_3PO_4$
(B) $HClO_4 > HNO_2 > H_2CO_3 > H_3PO_4$
(C) $H_2CO_3 > H_3PO_4 > HNO_2 > HClO_4$
(D) $H_2CO_3 > HNO_2 > HClO_4 > H_3PO_4$

Compound	Formula	Normal Boiling Point, K
Ethane	CH_3CH_3	185
Ethanal	CH_3CHO	294
Ethanol	CH_3CH_2OH	348
Ethanoic acid	CH_3CO_2H	391

49. Based on the data in the table above, which of the following substances has lowest vapor pressure at any given temperature?
(A) Ethane
(B) Ethanal
(C) Ethanol
(D) Ethanoic acid

50. Based on periodic relationships, the bond strength, and the concept relating bond strength to acid strengths, which of the following correctly predicts the strength of binary acids from strongest to weakest?
(A) $H_2Se > H_2O > H_2S$
(B) $H_2Se > H_2S > H_2O$
(C) $H_2O < H_2S < H_2Se$
(D) $H_2O > H_2S > H_2Se$

51. Silver metal, often amalgamated with mercury, is used to reduce substances to a desired oxidation state. If the silver metal amalgam cannot be used because of concerns about mercury, which of the following would be a reasonable and safe substitute?

(A) $H^+(aq)$
(B) $Na(s)$
(C) $Ca^{2+}(aq)$
(D) $Mg(s)$

52. Which of the following particulate diagrams best represents the reaction of carbon with oxygen to form carbon dioxide?

(A)

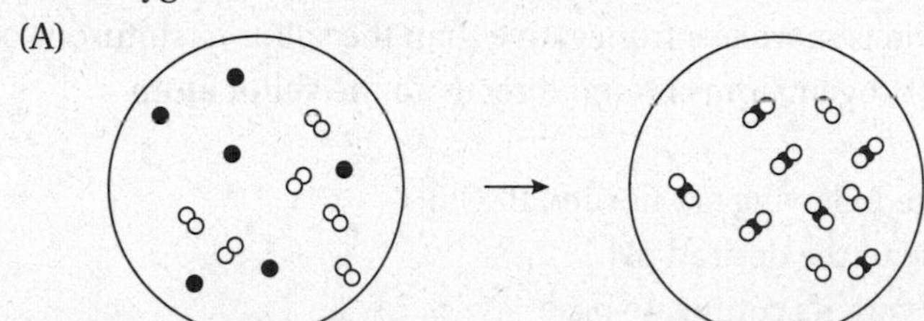

(B)

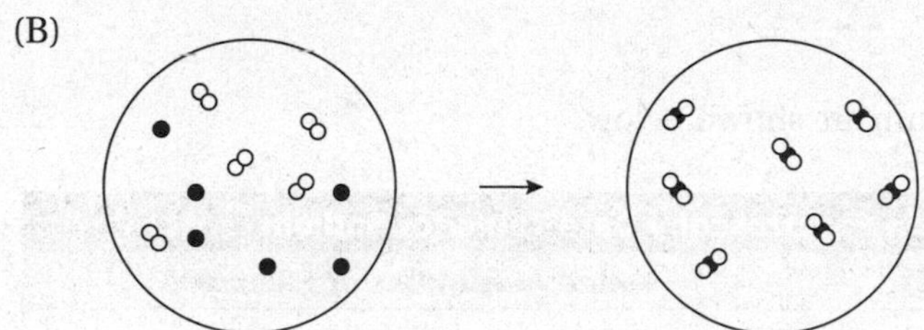

(C)

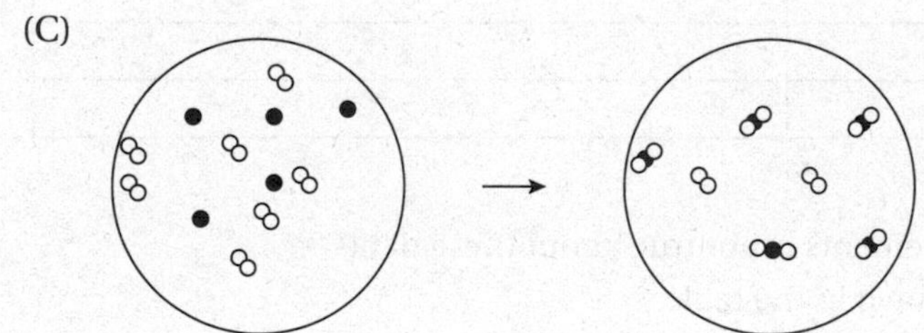

(D)

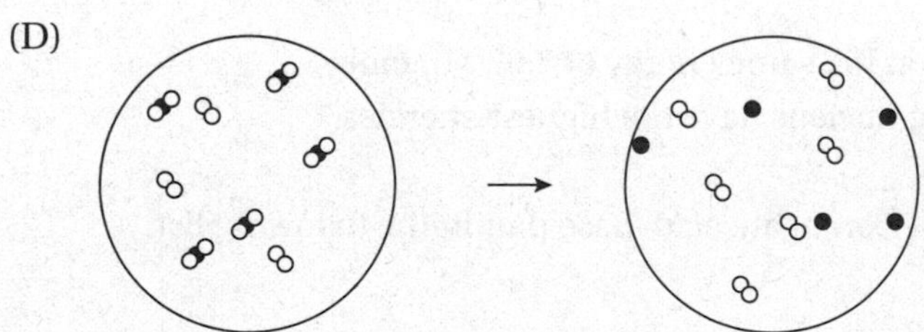

PRACTICE TEST 3

53. 0.25 mol of a weak, monoprotic acid is dissolved in 0.25 L of distilled water. The pH was then measured as 4.26. What is the pK_a of this weak acid?
    (A) 4.26
    (B) 8.52
    (C) 7.52
    (D) 3.66

54. Sulfurous acid is a weak acid, while sulfuric acid is a much stronger acid because
    (A) the O–H bonds in sulfuric acid are much weaker than in sulfurous acid due to the electron withdrawing of the oxygen atoms on sulfuric acid
    (B) sulfuric acid has more oxygen atoms in its formula
    (C) the sulfur in sulfuric acid is more electronegative than the sulfur in sulfurous acid
    (D) sulfuric acid has the hydrogen atoms bound directly to the sulfur atom

55. To prepare a buffer, all of the following are needed EXCEPT
    (A) an acid with a pK_a close to the desired pH
    (B) a conjugate acid along with its conjugate base
    (C) a buffer capacity sufficient to react with added acid or base
    (D) triple-distilled water

The photoelectron data for sodium are shown below.

Photoelectron Spectroscopy Data Table for Sodium	
Binding Energy (MJ)	**Relative Number of Electrons**
0.50	1
3.67	6
6.84	2
104	2

56. Which of the following statements is not true about these data?
    (A) All 11 electrons are shown in the table.
    (B) The 3 valence electrons have the lowest energies at 0.50, 3.67, and 6.84 MJ/mol.
    (C) The *p* electrons are all at the same energy of 3.67 MJ/mol.
    (D) Electrons closest to the nucleus have the highest energies.

57. Identify the Brønsted-Lowry conjugate acid–base pair in the following list.
    (A) H_3O^+ and OH^-
    (B) H_3PO_4 and $H_2PO_4^-$
    (C) $HClO_4$ and ClO_3^-
    (D) SO_3^{2-} and HSO_2^-

58. Chemical reactions can be classified as either heterogeneous or homogeneous. Which of the following equations below is best classified as a heterogeneous reaction?

(A) $2C_2H_2(g) + 5O_2(g) \rightarrow 4CO_2(g) + 2H_2O(g)$

(B) $C_2H_5OH(aq) + O_2(aq) \rightarrow HC_2H_3O_2(aq) + H_2O(aq)$

(C) $C(s) + H_2O(g) \rightarrow H_2(g) + CO(g)$

(D) $C_2H_2(g) + 5N_2O(g) \rightarrow CO_2(g) + H_2O(g) + 5N_2(g)$

59. Which of the following may need to be balanced using the ion-electron method?

(A) $BaCl_2 + Al_2(SO_4)_3 \rightarrow AlCl_3 + BaSO_4$

(B) $H^+ + OH^- \rightarrow H_2O$

(C) $NaOH + H_3PO_4 \rightarrow Na_2HPO_4 + H_2O$

(D) $C_2H_2(g) + N_2O(g) \rightarrow CO_2(g) + H_2O(g) + N_2(g)$

60. Which of the reactions below will become thermodynamically favored only at high temperatures?

	Reaction	$\Delta H°$ (kJ/mol$_{rxn}$)	$\Delta S°$ (J/mol$_{rxn}$K)
(A)	$CO(NH_2)_2(aq) + H_2O(l) \rightarrow CO_2(g) + 2NH_3(g)$	+119.2	+354.8
(B)	$C_2H_5OH(l) + O_2(g) \rightarrow HC_2H_3O_2(l) + H_2O(g)$	–534.3	–131
(C)	$2Fe(s) + 3CO_2(g) \rightarrow Fe_2O_3(s) + 3CO(g)$	+26.7	–15.7
(D)	$C_2H_2(g) + 5N_2O(g) \rightarrow CO_2(g) + H_2O(g) + 5N_2(g)$	–437.4	+272.6

IF YOU FINISH BEFORE THE 90 MINUTES HAVE ELAPSED, YOU MAY CHECK YOUR WORK IN THIS SECTION. DO NOT GO ON TO SECTION II UNTIL YOU ARE INSTRUCTED TO DO SO.

SECTION II: Free-Response

START ONLY WHEN YOU HAVE 105 MINUTES TO COMPLETE THE WHOLE SECTION.

Time:	1 hour, 45 minutes
Number of Questions:	7, 3 long and 4 short
Percent of Total Score:	50%
Calculator Allowed?	Yes
Pencil or pen with black or dark blue ink	

Instructions

The questions appear on the following pages. You may use the periodic table and the table of equations and constants that were provided with Section I. On the actual exam, you will be directed to write your final answers only in the test booklet. For this practice test, write your answers on separate sheets of lined paper. You will need about three pages for each of questions 1 through 3 and one page for each of questions 4 through 7.

Do not spend too much time on any one question. Budget your time carefully, and answer the easier questions first.

Be sure that your work is well organized, complete, and easy to read. Cross out or erase any mistakes. Erased and crossed-out material will not be scored.

SECTION II

7 Free-Response Questions
(Time: 105 minutes)
CALCULATORS ARE ALLOWED FOR SECTION II

Directions: Questions 1, 2, and 3 are long questions. Each question should take about 20 minutes to answer. Questions 4, 5, 6, and 7 are shorter questions. Each one should take about 5 to 10 minutes to answer. Read the questions carefully, and write your responses on lined paper. Your answers will be graded based on their correctness and relevance to the question asked and on the information cited. Explanations should be well organized and clearly written. Specific answers are always better than broad statements. For calculations, it is to your advantage to show clearly the method used and the steps involved in arriving at your answers since you may receive partial credit.

1. Lead(II) is a poisonous heavy metal ion found in rivers and lakes due to runoff from industrial plants. It accumulates in the bones and organs of humans. A common method utilized in determining the amount of lead ion (Pb^{2+}) in an aqueous solution of lead(II) nitrate is to titrate the lead ion with an aqueous solution of sodium iodide. The lead(II) ion is precipitated as lead(II) iodide.

   (a) Write both a balanced molecular and a balanced ionic equation for this reaction. Which reaction is a better representation of what is happening in this solution? Explain.

   (b) What is the concentration of the lead (Pb^{2+}) ion in 75.0 mL of a lead(II) nitrate solution when titrated to the end point with 50.0 mL of a 0.456 *M* sodium iodide solution?

   (c) How many grams of lead(II) iodide would theoretically be formed in the reaction described in part (b)?

   (d) When the student completed the experiment 4.65 g of lead(II) iodide was obtained. What is the percent yield of this experiment?

   (e) If the student passed the end point while performing this titration, what would happen to the calculated value of the percent yield? Explain.

2. Halogenated hydrocarbons have often been used as solvents and dry-cleaning fluids. Misuse of these solvents has led to many cases of environmental pollution costing large sums of money to remediate. One such compound was found and analyzed. A sample of this compound was shown to contain bromine because, upon decomposition, silver nitrate gave a brown precipitate when added to the decomposition products. A 3.856 g sample, now known to contain C, H, and Br, was found by elemental analysis to be composed of 0.8568 g C and 0.1428 g H.

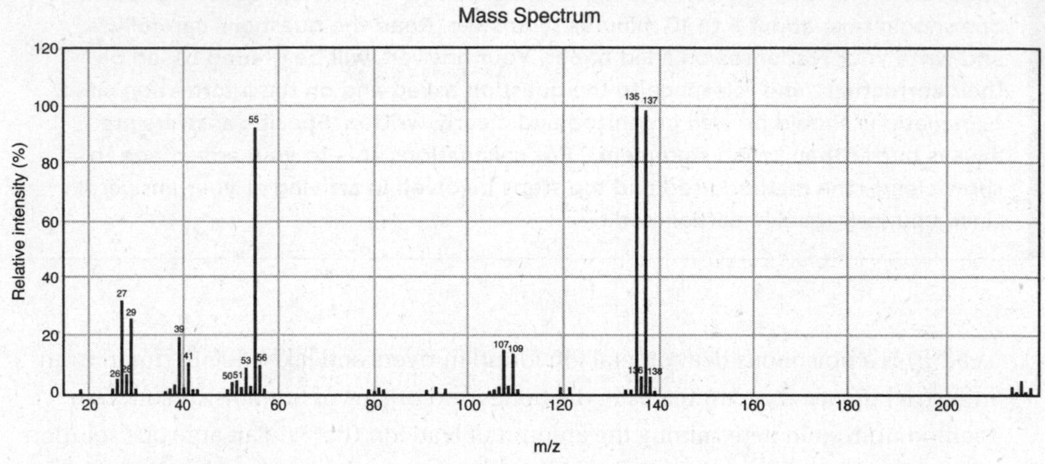

(a) What is the empirical formula for this compound? Explain why this cannot be the molecular formula.

(b) Bromine has two major isotopes, ^{79}Br (50.69%) and ^{81}Br (49.31%). What is the molar mass of this compound based on the mass spectrum above? What is the molecular formula of this compound?

(c) Another compound has the molecular formula $C_2H_4Cl_2$. Write out all the possible Lewis structures (constitutional isomers) for this compound. Show all electron pairs.

(d) What are the intermolecular forces of attraction in each isomer of $C_2H_4Cl_2$? Rank these compounds from most polar to least polar and explain your reasoning.

(e) Rank the isomers of $C_2H_4Cl_2$ based on their expected boiling points. Justify your decision.

3. When iron ore is mined, it comes in two different oxidation states, Fe(II) and Fe(III). A scientist has been given a sample for analysis that contains an unknown percentage of Fe^{2+} and Fe^{3+} ions. The method used to determine the amount of iron in a solution is to perform a redox titration using potassium dichromate. A 0.5580 g sample is dissolved in 100.0 mL of deionized water. A 50.00 mL sample of this solution was titrated with 0.01625 M $K_2Cr_2O_7$, which produced Fe^{3+} and Cr^{3+} ions in acidic solution. This titration took 21.34 mL to reach the end point. The remainder of the solution (50.00 mL) was treated to reduce all of the iron to Fe^{2+} ions. This solution was then titrated with the 0.01625 M $K_2Cr_2O_7$ solution. It took 32.26 mL to reach this end point.

   (a) Explain why one sample was titrated without reducing the Fe(III) and one titration was performed after reducing all Fe(III).

   (b) Write and balance the oxidation and reduction half-reactions in an acidic solution.

   (c) What is the net ionic equation?

   (d) Identify which substance is oxidized and which is reduced.

   (e) What is the mass of Fe(II) ion in the first titration?

   (f) What is the mass of Fe(II) ion in the second titration?

   (g) What is the percent of Fe(II) and Fe(III) ions in the sample? What is the total percent iron?

4. (a) Write the expression for the K_{sp} when solid $Co(OH)_3$ is added to deionized water.

   (b) When solid $Co(OH)_3$ is added to 0.250 L of very pure, triple-distilled water with a pH = 7.00 at 25°C, what is the identity and concentration of the ions of the salt if the K_{sp} at this temperature is 3.1×10^{-45}?

   (c) Should an acid or a base be added to a saturated solution of $Co(OH)_3$ to increase its solubility? Justify your answer with a logical chemical argument.

5. Many metals are present in nature in their oxide forms, such as ZnO, CuO, and Fe_2O_3. One method used to obtain the pure metal from these oxides is to heat them in the presence of carbon. The reduction of iron ore by this process proceeds by the following reaction:

$$2Fe_2O_3(s) + 3C(s) \rightarrow 4Fe(s) + 2CO_2(g)$$

	$Fe_2O_3(s)$	C(s)	Fe(s)	$CO_2(g)$
ΔH°_f (kJ mol^{-1})	-824.2	0	0	-393.5
S°_f (J mol^{-1} K^{-1})	87.4	5.69	27.3	213.6

   (a) Determine the ΔG°_{rxn} for this reaction given the thermodynamic properties at 25°C given in the table above.

   (b) Calculate the value of K_{eq} for this reaction at 25°C.

   (c) Will this reaction be thermodynamically spontaneous as written? Explain.

6. The collision theory for chemical kinetics states that the reaction rate is related to the frequency in which the molecules collide, the fraction of collisions that have sufficient energy for a reaction to occur, and the fraction of the collisions that are effective.

   (a) Illustrate an ineffective and an effective collision of two NOCl molecules according to the following reaction:

   $$2NOCl(g) \rightarrow 2NO(g) + Cl_2(g)$$

   Use the following structure to illustrate your answer:

   (b) Consider the following two-step mechanism for the reaction of NO_2 with Cl_2:

   $NO_2(g) + Cl_2(g) \rightarrow ClNO_2(g) + Cl(g)$ slow

   $NO_2(g) + Cl(g) \rightarrow ClNO_2(g)$ fast

   (i) Write the overall balanced reaction.

   (ii) Write the predicted rate law for this reaction.

7. The strongly acidic environment of the stomach aids in the digestion of food. The pH of the stomach is between 1.6 and 1.8. The stomach acid is comprised of HCl(*aq*) at a concentration between 0.050 and 0.10 *M*. The primary ingredient in some common antacid tablets is the salt $CaCO_3(s)$, which reacts with stomach acid (HCl).

   (a) When a student mixed HCl(*aq*) with $CaCO_3$, she noticed the formation of bubbles. What compound caused the bubbles? Explain.

   (b) Write a balanced molecular and net ionic equation for the neutralization of hydrochloric acid by calcium carbonate. Make sure you include the phases of each compound.

   (c) What mass of this salt would react completely with 35.0 mL of 0.145 *M* HCl?

   (d) A lab assistant is asked to verify that a 0.750 g antacid tablet contains 85.0% calcium carbonate. How should this assistant conduct the analysis to obtain the mass of calcium carbonate and verify the calcium carbonate percentage in the tablet?

END OF EXAM. If you finish before the 90 minutes have elapsed, you may check your work in this section only.

ANSWER KEY

Practice Test 3

1. (C)
2. (B)
3. (C)
4. (B)
5. (B)
6. (A)
7. (C)
8. (C)
9. (B)
10. (A)
11. (C)
12. (A)
13. (D)
14. (D)
15. (B)
16. (B)
17. (C)
18. (C)
19. (A)
20. (A)
21. (B)
22. (D)
23. (C)
24. (D)
25. (B)
26. (A)
27. (C)
28. (A)
29. (C)
30. (D)
31. (A)
32. (C)
33. (B)
34. (D)
35. (B)
36. (C)
37. (B)
38. (C)
39. (D)
40. (B)
41. (A)
42. (D)
43. (D)
44. (C)
45. (C)
46. (A)
47. (A)
48. (C)
49. (D)
50. (B)
51. (D)
52. (C)
53. (B)
54. (A)
55. (D)
56. (B)
57. (B)
58. (C)
59. (D)
60. (A)

CHAPTER REFERENCES FOR MULTIPLE-CHOICE QUESTIONS

Question	Chapter	Question	Chapter	Question	Chapter
1	6, 14	21	4, 6, 7	41	3, 11
2	6, 10	22	4, 6, 7	42	3, 5
3	6, 10	23	4, 6, 7	43	11, 12
4	9, 11, 14	24	1, 2, 3	44	12
5	6, 9, 14	25	4, 13, 14	45	9, 11, 12
6	2	26	3, 5	46	2, 3, 12, 14
7	9, 13	27	8, 11, 14	47	2
8	9, 13, 14	28	8, 9, 11	48	4, 13
9	5, 13, 14	29	5, 6, 14	49	4, 7
10	5, 13, 14	30	9, 10, 11	50	13
11	3, 9, 14	31	5, 10, 11	51	12, 14
12	8, 13, 14	32	5, 9, 11	52	4, 5, 14
13	9, 13	33	9, 11	53	9, 12, 14
14	3, 12, 13	34	5, 6, 9	54	4, 7, 13
15	2, 4, 7	35	4, 6, 14	55	13, 14
16	7, 8	36	6, 7	56	1, 2
17	2, 3, 10	37	4, 7	57	9, 13
18	6, 14	38	3, 11, 12	58	9, 11
19	6, 10	39	3, 5, 11	59	3, 12
20	6, 7	40	7, 8, 11	60	11

ANSWERS EXPLAINED

Section I—Multiple-Choice

1. **(C)** When collecting a gas over a liquid, the solubility of the gas in the liquid must be taken into account. (E.K. 3.A.2; L.O. 3.3)

2. **(B)** At point *C*, which is a little beyond the maximum, the average kinetic energy is directly proportional to temperature. (E.K. 2.A.2, 5.A.1; L.O. 2.4, 5.2)

3. **(C)** Enthalpy is defined as the heat of reaction at constant pressure. The coffee-cup calorimeter operates at constant pressure (equal to the current atmospheric pressure) and is the method of choice. (E.K. 5.B.4; L.0. 5.7)

4. **(B)** This response obeys the law of conservation of energy. (E.K. 5.B.4; L.O. 5.7)

5. **(B)** 6 particles is equivalent to 3 atmospheres. We have 24 particles and 6 liters, or 4 particles per liter. That is 4/6 or 2/3 of the original pressure, or 2.0 atm. (E.K. 1.E.1, 2.A.2; L.O. 1.17, 2.5)

6. **(A)** Mass 35 is about 75 percent and mass 37 is about 25 percent. Calculate

$$(0.75)(35) + (0.25)(37) = 26.25 + 9.25 = 35.5$$

This mass is very close to that of chlorine. (E.K. 1.D.2; L.O. 1.14)

7. **(C)** The stronger the base is, the weaker is its conjugate acid. (E.K. 6.C.2; L.O. 6.19)

8. **(C)** The best, and only, choice for a buffer with a pH of 5 is the acid that has a pK_a within 1 pH unit of the desired pH. (E.K. 6.C.2; L.O. 6.18)

9. **(B)** When titrating a weak diprotic acid with a strong base, there will be two equivalence points. As each equivalence point is reached, there is a sharp rise in pH. Also, the volume of titrant from the start to the first end point is the same as the volume of titrant from the first to the second end point. (E.K. 6.C.1; L.O. 6.13)

10. **(A)** We see that the color will start changing well before the end point, and it will take several mL to change. So the color change will be gradual. This color change also occurs before the end point, so the volume will be low. (E.K. 6.C.2; L.O. 6.18, 6.20)

11. **(C)** When choosing the proper indicator, the color change should happen around the equivalence point of the titration. (E.K. 6.C.1; L.O. 6.13)

12. **(A)** A buffer pH does not change on dilution, and the problem told us that one solution contains more than just an acid. This solution must contain a conjugate acid–base pair. (E.K. 6.C.2; L.O. 6.16–6.18)

13. **(D)** Since we know that solution 4 is a buffer, it is a mixture of the three other solutions. The weakest acid must give the highest pH; therefore, we arrive at solution 3 as the answer. (E.K. 6.C.1; L.O. 6.17)

14. **(D)** Copper reacts with oxidizing acids, and these are characterized by the nature of the anion. (E.K. 3.B.3; L.O. 3 .9)

***NOTE:** A complete list of the Essential Knowledge (E.K.) needed and the Learning Objectives (L.O.) can be found in the AP Chemistry Course Description:
https://secure-media.collegeboard.org/digitalServices/pdf/ap/ap-chemistry-course-and-exam-description.pdf

15. **(B)** All of the atoms are joined in what is essentially one large, covalently bonded crystal. (E.K. 2.D.1; L.O. 2.23)

16. **(B)** Traditionally, solid lines indicate covalent bonds and dashed lines indicate attractive or repulsive interactions. (E.K. 2.B.2; L.O. 2.13)

17. **(C)** Burning of a solid metal is actually a simple kinetics question. When the metal is ground to a powder, it burns easily in a flame. (E.K. 4.A.1; L.O. 4.1)

18. **(C)** The pressure times the molar mass divided by the temperature is directly proportional to the density

$$PV = nRT = (\text{g/molar mass})RT$$

Rearrange to get

$$P(\text{molar mass})/T = (\text{g}/V)R$$

where g/V is the density. (E.K. 2.A.2, 2.B.2; L.O. 2.4, 2.12)

19. **(A)** The average kinetic energy is directly proportional to the temperature. (E.K. 2.A.2; L.O. 2.4)

20. **(A)** Xenon has the largest electron cloud, which should provide stronger London attractive forces. (E.K. 2.B.2; L.O. 2.12, 2.13)

21. **(B)** There are no permanent dipoles or hydrogen-bonding atoms. (E.K. 2.B.1; L.O. 2.11)

22. **(D)** The bond order is 4/3, not 5/3, because there is only one double bond, not two. (E.K. 2.C.4; L.O. 2.21)

23. **(C)** The nonsymmetrical triangular pyramid ensures that the sulfite ion will be polar as well as charged. (E.K. 2.C.4; L.O. 2.21)

24. **(D)** There is one mole of gas on each side of the equation, indicating that the entropy change will not be large. The reaction goes from two particles to one, suggesting a small decrease in entropy. (E.K. 1.E.1, 6.B.1; L.O. 1.17, 6.8)

25. **(B)** The hydrogen with the weakest bond will always be the acidic hydrogen. Sometimes there is more than one acidic hydrogen. (E.K. 2.A–2.D; L.O. 2.2)

26. **(A)** This solution is concentrated enough so that all the assumptions used in calculating pH will work. In particular, $[H^+] = (K_a M_{HA})^{1/2-}$. Solve this equation

$$[H^+] = ((1.5 \times 10^{-5})(0.0100))^{1/2} = (1.5 \times 10^{-7})^{1/2} = (15 \times 10^{-8})^{1/2}$$

The square root of 15 is approximately 4, and the square root of 10^{-8} is 10^{-4}. Since $[H^+] = 4 \times 10^{-4}$ and this number is between 10^{-4} and 10^{-3}, we know the pH must be between pH 3 and 4. (E.K. 6.C.1; L.O. 6.12, 6.13)

27. **(C)** The key here is recognizing that the perchlorate ion is ClO_4^-. Then, the formulas can be written and the equation balanced. (E.K. 1.E.2, 3.A.1; L.O. 6.12, 6.13)

28. **(A)** No precipitate will form. $K_{sp} = 1.9 \times 10^{-4} = [Tl^+][Cl^-]$. Calculating the product of the two ions after dilution is $(0.00200\ M)(0.00200\ M) = 4.0 \times 10^{-6}$. This value is smaller than the K_{sp}, so no precipitate forms. (E.K. 6.C.3; L.O. 6.21, 6.22)

29. **(C)** Using *s* for the molar solubility, the equation becomes

$$K_{sp} = 1.9 \times 10^{-4} = [Tl^+][Cl^-] = s \times s = s^2$$

Since *s* is the square root of 1.9×10^{-4}, *s* is a little more than 10^{-2} molar. (E.K. 6.C.3; L.O. 6.23)

30. **(D)** The number of moles increases while the volume and temperature are held constant. Therefore, the pressure will increase. (E.K. 4.B.2; L.O. 4.5)

31. **(A)** The reaction that is occurring and that the pressure versus time plot is recording is the reverse reaction. Therefore, we can get kinetic information about the reverse reaction. (E.K. 4.A.1, 6.A.3; L.O. 4.1, 6.3)

32. **(C)** The end result will be 50 percent more particles, so the pressure will increase by 50 percent. The math is 760 + (½ × 760) = 1140 torr. (E.K. 6.A.3; L.O. 6.4)

33. **(B)** The reaction occurring is $2SO_3 \rightarrow 2SO_2 + O_2$. The form of the equilibrium expression is correct only in response (B). (E.K. 6.A.3; L.O. 6.5)

34. **(D)** There will be an initial sharp rise in pressure, but the subsequent reaction of SO_2 with O_2 will slowly reduce the pressure. The equilibrium pressure will be higher than before the injection since material was added to the system. (E.K. 6.A.2; L.O. 6.2)

35. **(B)** Acid rain is a major problem, and sulfur oxides are a large part of that problem. (E.K. 6.A.2; L.O. 6.2)

36. **(C)** The only attractive forces in this case are London forces, which depend on a polarizable electron cloud. (E.K. 1.C.1, 2.C; L.O. 1.9, 1.10, 2.17)

37. **(B)** We can deduce that the attractive forces in CBr_4 are stronger than those in CCl_4 because the bromine electron cloud is much more polarizable than the chlorine atom's electron cloud. Response (B) is the only one that agrees with the attractive forces. (E.K. 2.B.2, 5.D.1, 5.D.2; L.O. 2.13, 5.9, 5.10)

38. **(C)** To obtain the desired equation, reverse the second given equation, add, and cancel. Therefore, the heat of the desired reaction is the first heat of reaction plus the second (with its sign changed).

39. **(D)** In a combustion reaction, both the carbon dioxide and water products are gases. Otherwise, the equation is balanced.

40. **(B)** The heats of reaction of the products are multiplied by the stoichiometric coefficients and then added. The signs of the heats of reaction of the reactants are reversed and multiplied by their stoichiometric coefficients and then added. Response (B) is the only one that has the correct coefficients.

41. **(A)** Heats of formation of liquid and gaseous forms of water are given. Start with the heat of formation of the product (i.e., the gaseous H_2O). Then subtract the heat of formation of the reactant (i.e., the liquid form of H_2O). We get an endothermic process (positive heat of reaction) that correlates with the fact that heat is always added to vaporize a liquid.

42. **(D)** At equilibrium ($H_2O(l) \rightleftharpoons H_2O(g)$), since the free energy change is zero and we know the boiling temperature of water is 100°C or 373 K, the entropy change is easily calculated. Since ΔH°_{vap} is positive and T is positive, the entropy change is positive, as expected when water goes from the liquid to the vapor state. (E.K. 5.E.1; L.O. 5.12, 5.13)

43. **(D)** A chemical reaction must be specified to determine if it is thermodynamically favored. If the reaction is thermodynamically favored, it will always proceed in the forward direction. (E.K. 3.B.3, 5.E.1; L.O. 3.8, 5.13)

44. **(C)** This cell diagram gives the correct platinum electrodes. It also has the correct substances for the left cell and the right cell. The diagram has the salt bridge correctly indicated with two vertical lines and phase boundaries between the solid electrodes. The solutions are shown with single vertical lines. (E.K. 3.B.3, 3.C.3; L.O. 3.8, 3.12, 3.13)

45. **(C)** This equation is properly balanced with the correct substances being oxidized and reduced. The states of these substances are obvious from the diagram and do not need to be shown again. (E.K. 3.B.3, 3.C.3; L.O. 3.8, 3.12)

46. **(A)** Permanganate solutions are purple. The copper wire allows electrons to flow freely, and the reaction proceeds in the thermodynamically favored direction. The permanganate is consumed in the reaction, so the purple color decreases. (E.K. 3.C.3; L.O. 3.12, 3.13)

47. **(A)** Magnesium comes right after sodium in the periodic table. Periodic trends tell us that magnesium atoms will have a smaller radius and a higher ionization potential than sodium due to a higher effective nuclear charge in Mg. Response (A) is the only one that meets these criteria. (E.K. 1.C.1; L.O. 1.9, 1.10)

48. **(C)** Perchloric acid is the strongest acid listed and therefore has the lowest pH. The relative strengths of the other acids are based on the effect of oxygen atoms and also on the electronegativity of the central atom. (E.K. 6.C.1.; L.O. 6.11, 6.12)

49. **(D)** Ethanoic acid has the highest boiling point, indicating the strongest intermolecular attractive forces. The stronger the attractive forces are, the lower the vapor pressure. (E.K. 2.B.1; L.O. 2.11)

50. **(B)** The larger the atom is, the longer and weaker the bonds. So H_2Se should be the strongest acid in this list. (E.K. 1.C.1, 6.C.1; L.O. 1.11, 6.12)

51. **(D)** Magnesium will reduce many metal ions. Sodium metal will also work, but it is hazardous to use. (E.K. 3.A–3.C; L.O. 3.1)

52. **(C)** This is the only representation that obeys the law of conservation of mass and also illustrates the limiting reactant situation properly. (E.K. 2.A.2; L.O. 2.5)

53. **(B)** $[H^+] = [A^-] = 10^{-4.26}$. The initial concentration of acid is

$[HA] = 0.25\ \text{mol}/0.25\ \text{L} = 1.0\ M$. Entering these into the K_a equation gives

$$K_a = \frac{(10^{-4.26})(10^{-4.26})}{10^0} = 10^{-4.26-4.26+0} = 10^{-8.52}$$

The pK_a is the negative of the logarithm of $10^{-8.52}$, which is 8.52. (E.K. 6.C.1; L.O. 6.13)

54. **(A)** Weak –O–H bonds make for stronger acids. Electronegativity and lone oxygen atoms just help assess the strength of that bond. (E.K. 6.C.1; L.O. 6.12)

55. **(D)** Quality deionized water is good enough when preparing buffers. (E.K. 6.C.2; L.O. 6.18)

56. **(B)** Sodium has only 1 valence electron, not 3. (E.K. 1.B.2; L.O. 1.7)

57. **(B)** The two phosphate species differ by only one H^+. (E.K. 3.B.2; L.O. 3.7)

58. **(C)** All the other reactions have reactants and products in the same state. (E.K. 3.A.2; L.O. 3.2)

59. **(D)** The ion-electron method is used only to balance redox reactions. Identify a redox reaction by looking for changes in oxidation number. Nitrogen is reduced from an oxidation state of +1 to 0. Carbon is oxidized from –1 to +4. (E.K. 3.B.2; L.O. 3.8)

60. **(A)** At low temperatures, enthalpy dominates and results in a positive value for the Gibbs free-energy. When the temperature is raised, entropy dominates, making the Gibbs free-energy negative. (E.K. 5.E.2; L.O. 5.13)

Section II—Free-Response

1. (a) To write the molecular equation, start by writing the formulas for the two compounds: $Pb(NO_3)_2$ and NaI. Next predict the products of the reaction. One thing to remember is that the positive ions need to be balanced by the negative ions when writing a formula. In this type of reaction (metathesis or double displacement), the positive ions of the two compounds are exchanged. Therefore, the products are PbI_2 and $NaNO_3$:

$$Pb(NO_3)_2(aq) + 2NaI(aq) \rightleftharpoons PbI_2(s) + 2NaNO_3(aq)$$

When writing the ionic equation, start by separating only those compounds that are aqueous:

$$Pb^{2+}(aq) + 2NO_3^-(aq) + 2Na^+(aq) + 2I^-(aq) \rightleftharpoons PbI_2(s) + 2Na^+(aq) + 2NO_3^-(aq)$$

Cancel the spectator ions (NO_3^- and Na^+), and write the net ionic equation:

$$Pb^{2+}(aq) + 2I^-(aq) \rightleftharpoons PbI_2(s)$$

The net ionic equation is a better representation of what is happening in this solution. It illustrates that only the lead(II) ion and the iodide ion are involved in the reaction.

(b) First calculate the number of moles of NaI added in the titration. The definition of molarity is moles of solute per liter of solution. Multiply the molarity by the volume (in L) of the solution to obtain the number of moles of NaI:

$$0.0500\text{ L} \times \frac{0.456\text{ mol}}{1\text{ L}} = 0.0228\text{ mol NaI} = 0.0228\text{ mol I}^-$$

Use the stoichiometric ratio from the equation to find the number of moles of lead ion:

$$0.0228\text{ mol I}^- \times \frac{1\text{ mol Pb}^{2+}}{2\text{ mol I}^-} = 0.0114\text{ mol Pb}^{2+}$$

The question is asking for the concentration of lead ion. So divide the number of moles of lead ions by the volume of the sample in liters:

$$\frac{0.0114 \text{ mol Pb}^{2+}}{0.0750 \text{ L}} = 0.152 \text{ mol L}^{-1}$$

(c) In part (b) the amount of Pb^{2+} ions was calculated to be 0.0114 mol. This answer can be used to calculate the theoretical amount of PbI_2 that would form:

$$0.0114 \text{ mol Pb}^{2+}\left(\frac{461.01 \text{ g PbI}_2}{1 \text{ mol}}\right) = 5.26 \text{ g PbI}_2$$

(d) The percent yield of a reaction is calculated by the following formula

$$\frac{\text{actual yield}}{\text{theoretical yield}} \times 100 = \% \text{ yield}$$

$$\frac{4.65 \text{ g}}{5.26 \text{ g}} \times 100 = 88.4\% \text{ yield}$$

(e) If the student passed the end point during the titration, the amount of iodide ion reported would have been greater. This would have led to an increase in the theoretical yield of PbI_2. Since the percent yield is the actual yield divided by the theoretical yield, the calculated percent yield would have decreased.

2. (a) We know the total mass of the sample and the mass of the carbon and hydrogen in the compound. By subtraction, we can calculate the mass of bromine:

$$3.856 \text{ g total} - 0.8568 \text{ g C} - 0.1428 \text{ g H} = 2.856 \text{ g Br}$$

Calculate the number of moles of each element and then the relative numbers of moles to determine the empirical formula.

$$? \text{ mol C} = 0.8568 \text{ g C}\left(\frac{1 \text{ mol C}}{12.01 \text{ g C}}\right) = 0.07134 \text{ mol C}$$

$$? \text{ mol H} = 0.1428 \text{ g H}\left(\frac{1 \text{ mol H}}{1.008 \text{ g H}}\right) = 0.1417 \text{ mol H}$$

$$? \text{ mol Br} = 2.856 \text{ g Br}\left(\frac{1 \text{ mol Br}}{79.90 \text{ g Br}}\right) = 0.03574 \text{ mol Br}$$

Divide the moles of each element by 0.03574, a process called normalization, to obtain 1 mol Br, 2 mol C, and 4 mol H. The empirical formula is C_2H_4Br. Since there is an odd number of total valence electrons available, it is unlikely that this is the actual molecular formula.

(b) The mass of the empirical formula is 108. We expect the molar mass to be a multiple of 108. The mass spectrum shows large peaks that are not a multiple of 108 and therefore cannot be the molar mass. However, 3 small peaks near 216 are observed. Using that as the molar mass, we find two empirical formula units make up the molecular formula $C_4H_8Br_2$.

(c) This compound has two isomers, 1,2-dichloroethane and 1,1-dichloroethane. Their Lewis structures are

```
                                 ..
      H   H                H   :Cl:
      |   |                |    |
  :Cl−C − C−Cl:          H−C  − C−Cl:
      |   |                |    |
      H   H                H    H
```

(d) Of the three main forces of attraction—hydrogen bonding, dipole-dipole attractions, and London dispersion forces—only the hydrogen bonding is not present in these molecules. The 1,1-dichloroethane has the greatest polarity, and the 1,2-dichloroethane is essentially nonpolar.

(e) Although the 1,1-dichloroethane is more polar, the strong, repeating London dispersion forces strengthen the attraction between many 1,2-dichloroethane molecules. As a result, 1,2-dichloroethane has a higher boiling point. The 1,1-dichloroethane requires alignment of the opposite poles of the molecule. In addition, extensive networks of attraction between molecules do not occur. The result is a lower boiling point for 1,1-dichloroethane.

3. (a) The first titration gave data used to calculate the amount of Fe(II) present but not the Fe(III). After reducing the Fe(III), the second titration had all of the iron in the Fe(II) oxidation state. This titration gave the total amount of iron in the sample.

(b) An oxidation is the loss of electrons to another substance. The oxidation half-reaction involves the Fe^{2+} ion losing one electron to become the Fe^{3+} ion. When writing an oxidation half-reaction, the electrons are placed on the product side:

$$Fe^{2+}(aq) \rightarrow Fe^{3+}(aq) + e^-$$

The reduction half-reaction involves the $Cr_2O_7^{2-}$ gaining electrons to form Cr^{3+}. Since there are two chromium ions in the reactant, write two Cr^{3+} as products. Begin by writing the species involved in the reaction and balancing the metal ions:

$$Cr_2O_7^{2-}(aq) \rightarrow 2Cr^{3+}(aq)$$

This reaction is not yet balanced. In order to balance the oxygen atoms involved, use water molecules and hydrogen ions since the reaction occurs in an acidic solution. Balance the oxygen by adding 1 H_2O for each O you need to balance—in this case, 7 water molecules on the product side:

$$Cr_2O_7^{2-}(aq) \rightarrow 2Cr^{3+}(aq) + 7H_2O(l)$$

Next balance the hydrogen atoms by using 1 H^+ for each H you need to balance. Place those onto the reactant side—in this case, 14 H^+ on the reactant side.

$$Cr_2O_7^{2-}(aq) + 14H^+(aq) \rightarrow 2Cr^{3+}(aq) + 7H_2O(l)$$

Now balance the charge by determining the charges on the reactant side (total of +12) and the charges on the product side (total of +6). Add six electrons to the reactant side so both sides will have a total charge of +6:

$$6e^- + Cr_2O_7^{2-}(aq) + 14H^+(aq) \rightarrow 2Cr^{3+}(aq) + 7H_2O$$

(c) To obtain the net ionic equation, combine the two half-reactions in a manner that cancels all electrons. Therefore, the number of electrons gained and lost must be the same. The oxidation half-reaction needs to be multiplied by 6 to equal the 6 electrons in the reduction half-reaction:

$$6Fe^{2+}(aq) \rightarrow 6Fe^{3+}(aq) + 6e^-$$

Add the two reactions together and cancel any substance that is the same on both sides of the equation to obtain your final overall reaction. Always double-check your equations to make sure that the charges on both sides equal each other.

$$6Fe^{2+}(aq) + Cr_2O_7^{2-}(aq) + 14H^+(aq) \rightarrow 2Cr^{3+}(aq) + 7H_2O(l) + 6Fe^{3+}$$

(d) The species being oxidized is Fe^{2+}, and the species being reduced is $Cr_2O_7^{2-}$.

(e) Use your balanced net ionic equation to determine the amount of iron(II) ion in this first titration. First find the moles of $Cr_2O_7^{2-}$ added using the volume and molarity of the potassium dichromate solution, the mole ratio from the equation, and the atomic mass of iron:

$$\text{mass Fe(II)} = 0.02134\ \text{L}\left(\frac{0.01625\ \text{mol}\ Cr_2O_7^{-2}}{\text{L}}\right)\left(\frac{6\ \text{mol}\ Fe^{2+}}{1\ \text{mol}\ Cr_2O_7^{-2}}\right)\left(\frac{55.85\ \text{g}}{\text{mol}}\right)$$

$$= 0.1162\ \text{g}\ Fe^{2+}$$

This is the amount of Fe^{2+} in the original sample. There was still Fe^{3+} in the sample that was not measured.

(f) The amount of Fe(II) in the second titration can be found the same way:

$$\text{mass Fe(II)} = 0.03226\ \text{L}\left(\frac{0.01625\ \text{mol}\ Cr_2O_7^{-2}}{\text{L}}\right)\left(\frac{6\ \text{mol}\ Fe^{2+}}{1\ \text{mol}\ Cr_2O_7^{-2}}\right)\left(\frac{55.85\ \text{g}}{\text{mol}}\right)$$

$$= 0.1757\ \text{g}\ Fe^{2+}$$

To calculate the mass of Fe^{3+} in the sample, subtract the amount of iron(II) in the first titration from the second titration:

$$0.1757\ \text{g} - 0.1162\ \text{g} = 0.0595\ \text{g}\ Fe^{3+}$$

(g) You were given the mass of the sample containing the iron, so the percentage of each iron ion in the sample can be found. A solution was prepared by dissolving 0.5580 g of sample. Half that amount was titrated in each experiment so the mass of sample titrated was 0.27609.

$$\%\ Fe^{2+} = \frac{\text{mass Fe(II)}}{\text{mass sample}} \times 100 = \frac{0.1162\ \text{g}}{0.2760\ \text{g}} \times 100 = 41.64\%\ Fe^{2+}$$

$$\%\ Fe^{3+} = \frac{\text{mass Fe(III)}}{\text{mass sample}} \times 100 = \frac{0.0595\ \text{g}}{0.2760\ \text{g}} \times 100 = 21.4\%\ Fe^{3+}$$

The total iron content in the sample is the sum of the percentages of Fe^{2+} and Fe^{3+} or 63.0%.

4. (a) The balanced chemical equation for the solubility of $Co(OH)_3(s)$ is shown below:

$$Co(OH)_3(s) \rightleftharpoons Co^{3+}(aq) + 3OH^-(aq)$$

Therefore the K_{sp} can be written as:

$$K_{sp} = [Co^{3+}][OH^-]^3$$

(b) This problem involves the presence of a common ion, in this case OH^-. The solubility of the solid in deionized water at a pH = 7.00 should decrease due to Le Châtelier's principle. Remember that the presence of a common ion will shift the reaction toward the reactant, thereby decreasing the solubility of the salt.

Use an equilibrium table to determine the concentrations upon adding the common ion. Since the pH = 7.00 the $[OH^-] = 1.0 \times 10^{-7}$ *M*. This is greater than the amount of hydroxide present in the solution based on the molar solubility calculated in part (a).

Reaction	$Co(OH)_3(s)$	⇌	$Co^{3+}(aq)$	+	$3OH^-(aq)$
Initial Concentration	Solid		0		1.0×10^{-7} *M*
Change	-*x* dissolves		*x*		+3x
Equilibrium	Some solid		*x*		1.0×10^{-7} *M* + 3x

Plugging into the K_{sp} expression:

$$K_{sp} = [Co^{3+}][OH^-]^3$$

$$3.1 \times 10^{-45} = (x)\ (1.0 \times 10^{-7} + 3x)^3$$

If we assume that the 3*x* is negligible, the equation becomes:

$$3.1 \times 10^{-45} = (x)\ (1.0 \times 10^{-7})^3$$

$$x = 3.1 \times 10^{-24}\ M$$

The concentration of the $Co^{3+} = 3.1 \times 10^{-24}$ *M*, and the $[OH^-] = 1.0 \times 10^{-7}$ *M*.

(c) In order to increase the solubility of $Co(OH)_3$, the addition of an acid will be needed. The acid will react with the hydroxide in the solution, forming water. According to Le Châtelier's principle, any stress to a system will shift in the direction to relieve that stress. In this case, the removal of hydroxide ion by reaction with an acid will cause an increase in the molar solubility of the salt.

5. (a) Given that

$$\Delta G^\circ_{rxn} = \Delta H^\circ_{rxn} - T\Delta S^\circ_{rxn}$$

First calculate the ΔH°_{rxn} and ΔS°_{rxn} for the reaction using the information provided:

$$\Delta H^\circ_{rxn} = \Sigma\Delta H^\circ_f(\text{products}) - \Sigma\Delta H^\circ_f(\text{reactants})$$

$$= ((4 \times 0\ \text{kJ mol}^{-1}) + (2 \times -393.5\ \text{kJ mol}^{-1}) - ((2 \times -824.2\ \text{kJ mol}^{-1}) + (3 \times 0\ \text{kJ mol}^{-1}))$$

$$= 861.4\ \text{kJ}$$

$$\Delta S^\circ_{rxn} = \Sigma\Delta S^\circ_f(\text{products}) - \Sigma\Delta S^\circ_f(\text{reactants})$$

$$= ((4 \times 27.3\ \text{J mol}^{-1}\ \text{K}^{-1}) + (2 \times 213.6\ \text{J mol}^{-1}\ \text{K}^{-1})) - ((2 \times 87.4\ \text{J mol}^{-1}\ \text{K}^{-1}) + (3 \times 5.69\ \text{J mol}^{-1}\ \text{K}^{-1}))$$

$$= 344.53\ \text{J}_{rxn}\ \text{K}^{-1} = 0.345\ \text{kJ}_{rxn}\ \text{K}^{-1}$$

Insert these values into the first equation:

$$\Delta G^\circ_{rxn} = \Delta H^\circ_{rxn} - T\Delta S^\circ_{rxn}$$

$$= 861\ \text{kJ}_{rxn} - (298\ \text{K})(0.345\ \text{kJ}_{rxn}\ \text{K}^{-1})$$

$$= 758\ \text{kJ}_{rxn}$$

(b) To calculate the equilibrium constant, start with:

$$\Delta G^\circ_{rxn} = -RT\ln K_{eq}$$

$$\Delta G^\circ_{rxn} = -(8.314\ \text{J mol}^{-1}\ \text{K}^{-1})(298\ \text{K})\ln K_{eq}$$

$$758\ \text{kJ mol}^{-1} = (-2.48\ \text{kJ mol}^{-1})\ln K_{eq}$$

$$\ln K_{eq} = -306$$

We find that we cannot take the anti-ln of –306, getting an error message on most calculators. It is well known that converting a natural logarithm to a base-10 logarithm is done by dividing the ln by 2.303. We convert the natural logarithm to a base-10 logarithm:

$$\log K_{eq} = \frac{306}{2.303} = -133$$

We can now easily obtain the answer since the antilog is simply 10 raised to the value (–133) we just calculated.

$$K_{eq} = 10^{-133}$$

(c) This reaction is not thermodynamically spontaneous as written since the ΔG°_{rxn} is a positive value and K_{eq} is extremely small. Recall that we make the distinction about thermodynamically spontaneous reactions since some reactions can be spontaneous based on thermodynamics while at the same time being so slow as to appear nonspontaneous. A good example is the case of a mixture of H_2 and O_2 without a spark.

6. (a)

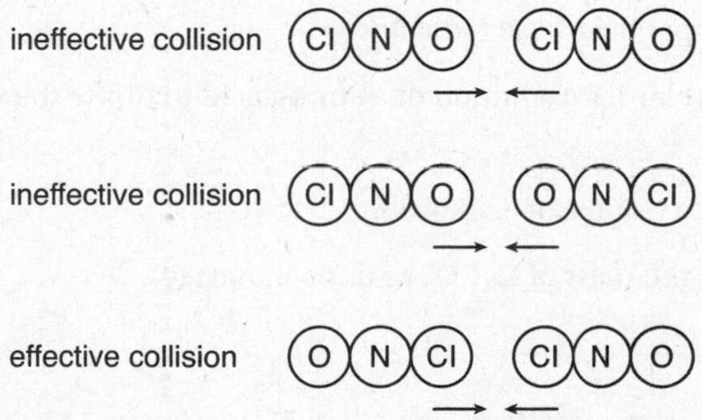

ineffective collision (Cl)(N)(O) → ← (Cl)(N)(O)

ineffective collision (Cl)(N)(O) → ← (O)(N)(Cl)

effective collision (O)(N)(Cl) → ← (Cl)(N)(O)

Using the same diagram as in the problem, the effective collision should have the chlorine atoms coming into contact with each other. All other orientations will be ineffective.

(b) (i) $2NO_2(g) + Cl_2(g) \rightarrow 2ClNO_2(g)$

The Cl is an intermediate in the reaction, so it is not included in the overall reaction.

(ii) Since the first reaction is the slow step, the predicted rate law is as follows:

$$\text{Rate} = k[NO_2][Cl_2]$$

7. (a) The bubbles are carbon dioxide (CO_2). When carbonates react with acids, they form carbonic acid, which decomposes into carbon dioxide gas and liquid water.

(b) The balanced molecular equation is the following:

$$CaCO_3(s) + 2HCl(aq) \rightarrow CaCl_2(aq) + H_2O(l) + CO_2(g)$$

The balanced net ionic equation is the following:

$$CaCO_3(s) + 2H^+(aq) \rightarrow Ca^{2+}(aq) + H_2O(l) + CO_2(g)$$

(c) HCl is a strong acid that ionizes completely in solution, forming hydrogen ions and chloride ions. The hydrogen ions react with the carbonate ion:

$$0.145 \text{ mol/L} \times 0.0350 \text{ L} = 5.08 \times 10^{-3} \text{ mol HCl}$$

Next, use the mole ratio from the net ionic equation to find the moles of calcium carbonate that were neutralized:

$$5.08 \times 10^{-3} \text{ mol H}^+ \left(\frac{1 \text{ mol CaCO}_3}{2 \text{ mol H}^+}\right) = 2.54 \times 10^{-3} \text{ g CaCO}_3$$

Use 100.09 g mol^{-1} as the molar mass of $CaCO_3$:

$$2.54 \times 10^{-3} \text{ mol CaCO}_3 \left(\frac{100.09 \text{ g CaCO}_3}{1 \text{ mol CaCO}_3}\right) = 0.254 \text{ g CaCO}_3$$

(d) The assistant should do the following:

i. Crush the tablet into a fine powder.

ii. Weigh the sample into a beaker, and record the mass.

iii. Dissolve the sample in the appropriate amount of water.

iv. Add a proper acid/base indicator.

v. Use a standardized solution of a strong acid to titrate the solution to the end point.

vi. Record the volume of acid used.

vii. Calculate the mass of $CaCO_3$ as done in part (c).

Appendix 1

Answer Explanations for End-of-Chapter Questions

CHAPTER 1

1. **(B)** The photoelectric effect is the ejection of an electron from a surface by photons with known energy. Measuring the energy of the ejected electron allows the binding energy of that electron to be calculated.

2. **(C)** Response (A) needs paired *s* electrons, response (B) needs *s* electrons. Response (D) shows electrons with the same spins paired, which is not allowed.

3. **(A)** The wave-particle duality of nature applies to everything. At the subatomic level, the photon-electromagnetic wave is the main example. Nuclear reactors and nuclear bombs are evidence of the magnitude of this effect on the macroscopic scale.

4. **(B)** The *p* orbital has the shape shown, which is often called a "dumbbell" shape based on weight-lifting parlance.

5. **(B)** In the ground state, Ca has no *f* electrons. However, its electrons can be excited to the *f* level.

6. **(D)** This answer choice shows the complete electron configuration. Response (A) is the abbreviated version. It is correct for silicon, but the complete configuration was required.

7. **(C)** This wave has the highest frequency (i.e., the most peaks for a given time period).

8. **(C)** The answer is 750 nm. The key to solving this question is the proper conversion of metric units. Solve the equation $\lambda\nu = c$, where $c = 3.00 \times 10^8$ m s^{-1} and $\nu = 4.00 \times 10^{14}$ s^{-1}.

$$\lambda = \frac{3.00\times10^{8}\ \text{m s}^{-1}}{4.00\times10^{14}\ \text{s}^{-1}} = 7.5\times10^{-7}\ \text{m} = 750\ \text{nm}$$

9. **(A)** Arsenic has 15 *p* electrons, 6 in period 2, 6 in Period 3, and 3 in Period 4. All of the other choices have 12 or fewer electrons.

10. **(C)** Antimony has 2 *s* and 3 *p* electrons in the fifth period.

11. **(C)** This region corresponds to the transition elements that have *d* electrons, which are noted for having colored compounds.

12. **(D)** $A = 31$ and $Z = 15$. In a neutral atom the electrons and protons are each equal to Z. The difference between A and Z is the number of neutrons = 16.

13. **(A)** Iron is correct, and the reasoning is correct.

14. **(C)** The electronic configuration of a noble gas always ends with np^6. These gases also have an ns^2 in their structure, where n is the highest principal level.

15. **(A)** Energy is needed to increase the value of n, and energy is released when n decreases. The more levels by which the electron increases, the greater the energy needed.

16. **(D)** Phosphorous is in Group VA and has five valence electrons.

17. **(C)** The Millikan oil drop experiment determined the charge of the electron independent of the mass.

18. **(B)** The law of multiple proportions is a consequence of the atomic theory, not part of it.

19. **(D)** Tungsten comes after the lanthanide elements (atomic numbers 58–71), which have *f* orbitals.

20. **(B)** The symbols *s, p, d,* and *f* in electronic configurations all represent different integer values of the ℓ quantum number.

21. **(D)** 2.36×10^{-34} is the largest common divisor of all five measurements.

22. **(A)** The 4 *d* electrons appear in the fifth period of the periodic table.

23. **(B)** The *f* orbitals in the sixth and seventh periods hold 14 electrons.

24. **(C)** Valence electrons are the electrons in the outermost energy level of the atom. Only the *s* and *p* electrons are in the same energy level as the period in the periodic table.

25. **(D)** $E = h\nu$ is the equation for determining the energy of a photon.

CHAPTER 2

1. **(D)** Bismuth is a metal, and nitrogen is a nonmetal.

2. **(D)** The effective nuclear charge felt by valence electrons increases across a row on the periodic table. Therefore, bromine's electrons feel the most positive charge

3. **(B)** Noble gases have the highest first ionization energies. Ionization energies generally increase as you move from left to right across a period. So you might assume that (C) is correct. However, the first ionization energy for phosphorus is actually higher than that of sulfur. This is because the *p* orbitals are half-filled in phosphorus, and it takes slightly more energy to remove an electron from phosphorus.

4. **(D)** Bromine is a liquid under normal conditions.

5. **(B)** Strontium is one shell (period) larger than Br and Kr. Strontium has a lower effective nuclear charge than tin and thus is larger than tin.

6. **(A)** Elements with high metallic character have low ionization energies. These elements give up electrons more easily than do nonmetals.

7. **(B)** Boiling points vary regularly within groups (columns), and Ni and Pt are just above and below the Pd atom.

8. **(D)** Boiling points decrease for metals and increase for nonmetals from the top to the bottom of a group. Using this generality, barium has a lower boiling point compared to strontium.

9. **(A)** In general, an element close to F has the higher electronegativity of any pair of atoms. Except for response (A), the element closest to F is listed second.

10. **(C)** The anion of sulfur is larger than its corresponding neutral atom since the repulsion of the electrons is greater than the effective nuclear charge for this atom. The more negative the charge is, the larger the ion is. A cation is smaller than its neutral atom since the effective nuclear charge is experienced more by the fewer electrons.

11. **(D)** The effective nuclear charge electrons feel increases from left to right across a period but does not change appreciably going down a group. This effective charge is based on the shielding provided by the core electrons. As you proceed from left to right across a row, the number of electrons in the core does not change but the number of protons does. When proceeding down a group, the core electrons all have similar shielding effects.

12. **(C)** The group C, S, As, H contains either nonmetals or a metalloid. All other choices have at least one or more metals.

13. **(A)** Since Be is in the second group, it is expected to lose 2 electrons easily while the third ionization is very difficult. The other atoms can lose 3 or more electrons with relative ease and have lower-third ionization energies.

14. **(D)** Li and Mg are diagonally related in size and many physical properties stemming from atomic and ionic size. Although Si and C are next to each other in the fourth group, Si is a metalloid and semiconductor, and C is not. There is a larger difference between Si and C than between Mg and Li.

15. **(A)** Most elements are metals.

16. **(D)** We can determine the number of neutrons only if a specific isotope is selected.

17. **(A)** Electrons are the first parts of the atoms to encounter each other in a collision between atoms.

18. **(C)** The atomic mass of an element is a weighted average of all naturally occurring isotopes. The actual abundances are ^{58}Ni (68.2%), ^{60}Ni (26.2%), ^{61}Ni (1.14%), ^{62}Ni (3.63%), ^{64}Ni (0.93%). The average atomic mass would be 59 if Ni-58 and Ni-60 were in equal abundance.

19. **(D)** The definition of electron affinity is the energy necessary to add an electron to an atom in its gaseous state. The acceptance of an electron is an exothermic process for nearly all elements.

20. **(B)** Electron affinity is a measure of an atom's attraction to and affinity for an electron. The more negative (exothermic) the value is, the greater is the tendency to accept an electron. Francium has the lowest electron affinity, and fluorine has the greatest.

21. **(D)** Moving from left to right across a period involves an increase in the number of protons in the nucleus from atom to atom but no increase in the core (or shielding) electrons. Thus the effective nuclear charge increases. This increase in effective charge

attracts the valence electrons closer to the nucleus, resulting in decreasing atomic radius.

22. **(D)** In row *Z*, there is a 10-fold increase in ionization energy from the first to the second electron. This indicates an element with 1 valence electron. The largest increase in ionization energy between the second and third ionization energy is in row *Y*, indicating an element with 2 valence electrons. Similar reasoning is used for the other elements.

CHAPTER 3

1. **(C)** Aluminum has a 3+ charge and the sulfate ion has a 2^- charge. Therefore to have no overall charge, the correct formula must have two Al^{3+} ions and three SO_4^{2-} ions.

2. **(B)** Nitric acid is a strong acid, and sodium hydroxide is a strong base. Therefore, they will dissociate 100% when dissolved in water. Write the molecular, ionic, and net ionic equations:

$$HNO_3(aq) + NaOH(aq) \rightarrow H_2O + NaNO_3(aq)$$
$$H^+(aq) + \mathbf{NO_3^-(aq)} + \mathbf{Na^+(aq)} + OH^-(aq) \rightarrow H_2O + \mathbf{Na^+(aq)} + \mathbf{NO_3^-(aq)}$$
$$H^+(aq) + OH^-(aq) \rightarrow H_2O$$

The hydrogen and hydroxide ions form water. The sodium and nitrate ions in the solution (in bold type) are called spectator ions. These ions were canceled to obtain the net ionic equation.

3. **(D)** The formation of a gas, liquid (water), or covalent compound will all drive a reaction to completion.

4. **(A)** Of the four pairs, only the iron(III) phosphate is insoluble. The +3 and –3 charges provide a very high coulombic attractive force.

5. **(C)** When predicting the products of a double displacement reaction, remember that the cation of one compound will combine with the anion of the other one. The final balanced equation is

$$Pb(C_2H_3O_2)_2 + 2KCl \rightarrow PbCl_2(s) + 2KC_2H_3O_2$$

6. **(C)** Two ammonium ions and one oxalate ion are obtained. The anion does not decompose into other species, and the ammonium cation is enclosed in parentheses with a subscript of 2 indicating that two NH_4^+ ions must result.

7. **(A)** In order to be isoelectronic to a noble gas, the atoms or ions must have the same electron configuration as the noble gas. The electron configurations of neon and of the answer choices are

Ne:	$1s^2\ 2s^2\ 2p^6$
F^-:	$1s^2\ 2s^2\ 2p^6$
Ga^{3+}:	$1s^2\ 2s^2\ 2p^6\ 3s^2\ 3p^6$
P^{3-}:	$1s^2\ 2s^2\ 2p^6\ 3s^2\ 3p^6$

Ne and F^- have identical electron configurations and are therefore isoelectronic, Ga^{3+} and P^{3-} are isoelectronic to argon.

8. **(C)** The ionic radius for sodium is the same in all cases. The ionic radius for the halogen ions increases as the number of orbitals increases. I^- is the largest halogen ion in the list. Therefore, NaI has the largest distance between the sodium ion and iodide ion nuclei.

9. **(C)** Of all the compounds listed first, KBr has the lowest charges and relatively largest distance between the ions. Therefore, the coulolombic attractive forces are the lowest for KBr, which indicates it will be the most soluble.

10. **(C)** Na and F have electronic configurations of their ions that are identical to that of Ne. Each atom in the other pairs is isoelectric with different noble gases.

11. **(D)** Since Al always has a 3+ ion and chloride always has a 1– ion, the name does not need prefixes or roman numerals.

12. **(D)** This is the electronic configuration of the Ar atom. Each of these ions is isoelectronic with Ar.

13. **(A)** A combustion reaction occurs when an organic compound, containing mainly carbon and hydrogen, is reacted with oxygen to form carbon dioxide and water.

14. **(B)** Calcium ions are +2 and oxide ions −2, so the correct formula is CaO.

15. **(A)** The dihydrogen phosphate ion, $H_2PO_4^-$, needs one K^+ for a correct formula. In the other formulas there is an excess of a positive or negative charge.

16. **(C)** The products of this reaction are hydrogen gas and aluminum bromide.

17. **(D)** The solubilities must be known to write correct net ionic equations; these equations do not predict solubility.

18. **(D)** This is the only reaction that has the same number of each atom as reactants and products.

19. **(A)** This diagram shows 4 nitrogen atoms (white circles) on both sides and 12 hydrogen atoms (black circles). Ammonia, NH_3, is properly shown.

20. **(A)** This is the only diagram that has the same number of atoms represented on both sides of the equation.

21. **(C)** Since $PbCl_2$ is the only substance that is insoluble, it will be part of the net ionic equation. The soluble ions Pb^{2+} and Cl^- complete the equation.

CHAPTER 4

1. **(A)** CBr_4 is a tetrahedron. NH_3 is a triangular pyramid but is related to the basic tetrahedron.

2. **(A)** Only H–C ≡ N is a linear structure, with a 180° bond angle.

3. **(A)** There are three resonance forms for the carbonate ion. These indicate that the double and single bonds present are occurring simultaneously and that therefore the bond lengths are equal.

4. **(D)** HCN, being linear, must lie in one plane. The SO_3 is trigonal planar and also is flat.

5. **(B)** A square planar geometry is exhibited when a molecule has an expanded octet with 2 lone pairs of electrons on the central atom. Therefore XeF_4 is labeled correctly. PF_5 has a trigonal bipyramid shape. CCl_4 is a tetrahedron. SF_6 is an octahedron.

6. **(A)** The elements Ba, Zn, C, and Cl are arranged in order from the lower left (Ba) to the upper right corner of the periodic table (Cl).

7. **(C)** SBr_2 has two sets of unpaired electrons on the sulfur atom. PBr_3 has one set of unpaired electrons on the phosphorus. CBr_4 has no unpaired electrons on the carbon. According to VSEPR theory, the presence of lone pairs of electrons on the central atom of a molecule decrease the bond angles between atoms. Lone pairs of electrons take up more space than paired electrons.

8. **(D)** In SF_2Cl_4 the two F atoms may be arranged opposite each other for a nonpolar molecule, or next to each other for a polar molecule.

9. **(C)** The carbonate ion has only one double bond in each resonance structure and therefore one pi bond. Although the entire ion has a charge of –2, it is nonpolar.

10. **(B)** The Lewis structure contains one double bond between the C and O atoms. The structure is as follows:

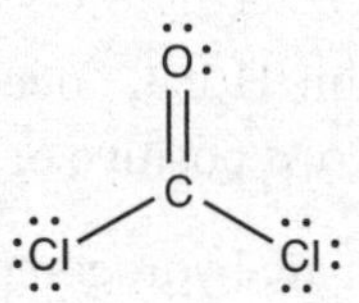

    All the bonds are polar due to differences in electronegativity between carbon and both chlorine and oxygen.

11. **(D)** H_2S is similar in structure to water. With two nonbonding electron pairs, the molecule has a bent structure.

12. **(A)** The cyanide ion, CN^-, is electronically identical to N_2. $:N \equiv N:$, $[:C \equiv N:]^-$

13. **(D)** Square planar structures all have 90° angles, not 120° angles.

14. **(C)** The structures shown are resonance forms of the nitrate ion. All forms are considered to exist simultaneously. The strength and length of all three bonds are equal.

15. **(B)** Ammonia has a nonbonding pair of electrons.

16. **(D)** All of these statements are true.

17. **(C)** Although the B–F bond is polar, the molecule has a trigonal planar shape. The B–F bonds are at 120° from each other, allowing the molecule to be symmetrical. PF_3 has a trigonal pyramid shape due to the lone pair of electrons on the phosphorus.

18. **(B)** All single bonds are sigma bonds. A double bond is composed of a sigma bond and a pi bond. A triple bond consists of one sigma and two pi bonds.

19. **(A)** A 60° angle does not occur in any hybrid structure considered in this text. No atomic orbitals have 60° angles, either.

20. **(A)** The *sp* hybrid requires that two atoms be bound to the central atom with no non-bonding electron pairs. None of the structures named in the question fulfills these requirements. SO_2 and SO_3^{2-} are sp^2, SF_6 is sp^3d^2, SCl_4 is sp^3d, and SCl_2 is sp^3.

21. **(D)** The direction, or orientation, of a bond in space has nothing to do with bond strength.

CHAPTER 5

1. **(D)** Using the molarity as a conversion factor (mol/L), divide the number of moles given by the molarity of the solution:

   Ask: ? L = 3.12 mol NaCl

   Solve: $? \text{ L} = 3.12 \text{ mol NaCl}\left(\frac{1 \text{ L NaCl}}{6.67 \text{ mol NaCl}}\right) = 0.468 \text{ L}$

2. **(A)** Molarity is correct and the reason given is also correct.

3. **(D)** This is a dilution problem so use

$$C_{\text{initial}}V_{\text{initial}} = C_{\text{final}}V_{\text{final}}$$

$$(2.75\ M)(x\text{ L}) = (0.150\ M)(1.25\text{ L})$$

$$x\text{ L} = \frac{(0.150\ M)(1.25\text{ L})}{(2.75\ M)}$$

$$x\text{ L} = 0.0682\text{ L}$$

$$?\text{ mL} = 0.0682\text{ L}\left(\frac{1000\text{ mL}}{\text{L}}\right) = 68.2\text{ mL}$$

4. **(C)** The sequence for this calculation is correctly stated.

5. **(B)** $? \text{ g KClO}_3 = 0.200 \text{ L KClO}_3\left(\frac{0.150 \text{ mol KClO}_3}{1 \text{ L KClO}_3}\right)\left(\frac{122.5 \text{ g KClO}_3}{1 \text{ mol KClO}_3}\right)$

   $= 3.68 \text{ g KClO}_3$

6. **(B)** The reaction is

$$\text{NaOH} + \text{HNO}_3 \rightarrow \text{NaNO}_3 + \text{H}_2\text{O}$$

$$? \text{ mL NaOH} = 35.0 \text{ mL HNO}_3\left(\frac{0.345 \text{ mol HNO}_3}{1000 \text{ mL HNO}_3}\right)\left(\frac{1 \text{ mol NaOH}}{1 \text{ mol HNO}_3}\right)\left(\frac{1000 \text{ mL NaOH}}{0.130 \text{ mol NaOH}}\right)$$

$$= 92.9 \text{ mL NaOH}$$

7. **(A)** When working through the setup, you will convert the volume of CO_2 at STP to moles of CO_2. The moles of CO_2 are then converted to moles of $CaCO_3$ using the mole ratio factor from the equation. Finally, the moles of $CaCO_3$ are converted to grams using its molar mass. The grams of CO_2 are never needed.

8. **(D)** $? \text{ mol H} = 14.3 \text{ g H}\left(\frac{1 \text{ mol H}}{1.0 \text{ g H}}\right) = 14.3 \text{ mol H}$

$$? \text{ mol C} = 85.7 \text{ g C}\left(\frac{1 \text{ mol C}}{12.0 \text{ g C}}\right) = 7.14 \text{ mol C}$$

$$\frac{14.3 \text{ mol H}}{7.14} = 2 \text{ mol H and } \frac{7.14 \text{ mol C}}{7.14} = 1 \text{ mol C.}$$

The empirical formula is CH_2.

9. **(C)** One Al atom, 3 N atoms, and 9 O atoms add up to a mass of 213.01.

10. **(D)** $? \text{ mg } Na_2SO_4$

$$= 0.100 \text{ L Na}^+\left(\frac{0.00100 \text{ mol Na}^+}{1 \text{ L Na}^+}\right)\left(\frac{1 \text{ mol Na}_2\text{SO}_4}{2 \text{ mol Na}^+}\right)\left(\frac{142.04 \text{ g Na}_2\text{SO}_4}{1 \text{ mol Na}_2\text{SO}_4}\right)$$

$$= 7.10 \times 10^{-3} \text{ g Na}_2\text{SO}_4 = 7.10 \text{ mg Na}_2\text{SO}_4$$

11. **(A)** We can determine the mass ratios for the four compounds by writing the total mass of sulfur in the numerator and the molar mass in the denominator. We will use integer numbers for atomic masses:

(A) Al_2S_3 $\quad \frac{(3)(32)}{((2)(27)+(3)(32))} = \frac{96}{150} = 0.64$

(B) $CaSO_4$ $\quad \frac{32}{(40+32+(4)(16))} = \frac{32}{136} = 0.24$

(C) Na_2S $\quad \frac{32}{(2)(23)+32} = \frac{32}{78} = 0.41$

(D) SO_2 $\quad \frac{32}{(32+(2)(16))} = \frac{32}{64} = 0.50$

We can see that the Al_2S_3 has approximately 64% sulfur, while the others obviously have 50% or less.

12. **(D)** $? \text{ mol Al} = 1 \text{ mol Fe}\left(\frac{8 \text{ mol Al}}{9 \text{ mol Fe}}\right) = \frac{8}{9} \text{ mol Al}$

13. **(C)** The reaction is $2KOH + H_2SO_4 \rightarrow K_2SO_4 + 2H_2O$

$$? \frac{\text{mol H}_2\text{SO}_4}{1 \text{ L H}_2\text{SO}_4} = \frac{0.125 \text{ mol KOH}}{1 \text{ L KOH}}\left(\frac{1 \text{ mol H}_2\text{SO}_4}{2 \text{ mol KOH}}\right)\left(\frac{35.4 \text{ mL KOH}}{50.0 \text{ mL H}_2\text{SO}_4}\right)$$
$$= 0.0443 \text{ M H}_2\text{SO}_4$$

14. **(B)** The mass of the empirical formula unit is 14 g unit^{-1}.

$$\frac{83.5\text{ g mol}^{-1}}{14\text{ g unit}^{-1}} = 5.96\text{ unit mol}^{-1}$$

This rounds to 6 empirical formula units per mole and a molecular formula of C_6H_{12}.

15. **(D)**

$$?\text{ g Fe} = 1\text{ atom Fe}\left(\frac{1\text{ mol Fe}}{6.02\times10^{23}\text{ atoms Fe}}\right)\left(\frac{55.85\text{ g Fe}}{1\text{ mol Fe}}\right)$$
$$= 9.28\times10^{-23}\text{ g Fe}$$

16. **(B)** $?\text{ g SO}_2 = 4.00\text{ L SO}_2\left(\frac{1\text{ mol SO}_2}{22.4\text{ L SO}_2}\right)\left(\frac{64.02\text{ g SO}_2}{1\text{ mol SO}_2}\right) = 11.4\text{ g SO}_2$

17. **(D)** Molar mass of K_3PO_4 is 212.3. Mass of potassium in one mole of K_3PO_4 is $3 \times 39.10 = 117.3$.

$$\%K = \frac{117.3\text{ g K}}{212.3\text{ g K}_3\text{PO}_4}\times100 = 55.3\%\text{ K}$$

18. **(C)**

$$?\text{ g C} = 0.357\text{ g CO}_2\left(\frac{1\text{ mol CO}_2}{44.01\text{ g CO}_2}\right)\left(\frac{1\text{ mol C}}{1\text{ mol CO}_2}\right)\left(\frac{12.01\text{ g C}}{1\text{ mol C}}\right)$$
$$= 0.0974\text{ g C}$$
$$\%C = \frac{0.0974\text{ g C}}{0.200\text{ g sample}}\times100 = 48.7\%\text{ C}$$

19. **(B)** There are several ways to answer this question. You can calculate how many liters of O_2 at STP react with 5.50 g of glucose. Based on the answer, you can then decide which is the limiting reactant. Alternatively, you could convert the given amounts to moles of glucose and moles of O_2. Then you could use the mole ratio of the chemical equation to decide which is the limiting reactant. A third method is to calculate how many grams of one of the products can be made with the 5.50 g of glucose and then with the 2.50 L of oxygen. The reactant yielding the *least* product is the limiting reactant. The first method is worked out below.

$$?\text{ L O}_2 = 5.50\text{ g C}_6\text{H}_{12}\text{O}_6\left(\frac{1\text{ mol C}_6\text{H}_{12}\text{O}_6}{180.16\text{ g C}_6\text{H}_{12}\text{O}_6}\right)\left(\frac{6\text{ mol O}_2}{1\text{ mol C}_6\text{H}_{12}\text{O}_6}\right)\left(\frac{22.4\text{ L O}_2}{1\text{ mol O}_2}\right)$$
$$= 4.10\text{ L O}_2$$

Since the given 2.50 L O_2 is less than the needed 4.10 L O_2 calculated above, we must conclude that O_2 is the limiting reactant.

20. **(A)** The question asks for the amount of product. One method is to do the calculations using each reactant and choose the smaller answer.

$$? \text{ g AgCl} = 20.0 \text{ g AgNO}_3\left(\frac{1 \text{ mol AgNO}_3}{170 \text{ g AgNO}_3}\right)\left(\frac{2 \text{ mol AgCl}}{2 \text{ mol AgNO}_3}\right)\left(\frac{143.5 \text{ g AgCl}}{1 \text{ mol AgCl}}\right)$$
$$= 16.9 \text{ AgCl}$$
$$= 15.0 \text{ g CaCl}_2\left(\frac{1 \text{ mol CaCl}_2}{111 \text{ g CaCl}_2}\right)\left(\frac{2 \text{ mol AgCl}}{1 \text{ mol CaCl}_2}\right)\left(\frac{143.5 \text{ g AgCl}}{1 \text{ mol AgCl}}\right)$$
$$= 38.8 \text{ g AgCl}$$

16.9 g AgCl is the correct answer and also defines $AgNO_3$ as the limiting reactant.

21. **(D)** Determine the limiting reactant (this was done in the preceding example, but another method is shown here):

$$? \text{ g AgNO}_3 = 15.0 \text{ g CaCl}_2\left(\frac{1 \text{ mol CaCl}_2}{111 \text{ g CaCl}_2}\right)\left(\frac{2 \text{ mol AgNO}_3}{1 \text{ mol CaCl}_2}\right)\left(\frac{170 \text{ g AgNO}_3}{1 \text{ mol AgNO}_3}\right)$$
$$= 45.9 \text{ g AgNO}_3$$

This calculation shows that we need 45.9 g $AgNO_3$ to react all of the $CaCl_2$. The problem only gives us 20.0 g. Therefore $AgNO_3$ is used up first and is the limiting reactant. Now use the given amount of the limiting reactant to calculate the number of grams of $CaCl_2$ that react.

$$? \text{ g CaCl}_2 = 20.0 \text{ g AgNO}_3\left(\frac{1 \text{ mol AgNO}_3}{170 \text{ AgNO}_3}\right)\left(\frac{1 \text{ mol CaCl}_2}{2 \text{ mol AgNO}_3}\right)\left(\frac{111 \text{ g CaCl}_2}{1 \text{ mol CaCl}_2}\right)$$
$$= 6.53 \text{ g CaCl}_2$$

Since 6.53 g $CaCl_2$ react, $15.0 - 6.53 = 8.47$ g $CaCl_2$ must be left.

22. **(C)** In the correct answer choice, some nitrogen (white circles) are left over and all the hydrogen (black circles) are combined in ammonia molecules. Since all of the hydrogen is bound in ammonia, hydrogen is the limiting reactant.

23. **(D)** $$? \text{ L air} = 1 \text{ mol CH}_4\left(\frac{2 \text{ mol O}_2}{1 \text{ mol CH}_4}\right)\left(\frac{22.4 \text{ L O}_2}{1 \text{ mol O}_2}\right)\left(\frac{1 \text{ L air}}{0.2 \text{ L O}_2}\right)$$
$$= 224 \text{ L air}$$

24. **(C)** Assume a 100-g sample of the compound. Then 25% H = 25 g H and 75% C = 75 g C.

Calculate moles of each:

$$? \text{ mol H} = 25 \text{ g H}\left(\frac{1 \text{ mol H}}{1 \text{ g H}}\right) = 25 \text{ mol H}$$

$$? \text{ mol C} = 75 \text{ g C}\left(\frac{1 \text{ mol C}}{12 \text{ g C}}\right) = 6.25 \text{ mol C}$$

$$\frac{6.25 \text{ mol C}}{6.25} = 1 \text{ mol C}$$

Also,

$$\frac{25 \text{ mol H}}{6.25} = 4 \text{ mol H}$$

The empirical formula is CH_4, and this is the molecular formula too.

25. **(B)** We calculate:

$$? \text{ molecules } C_{12}H_{22}O_{11} =$$

$$2.5\times10^{-6} \text{ g } C_{12}H_{22}O_{11}\left(\frac{1 \text{ mol } C_{12}H_{22}O_{11}}{342 \text{ g } C_{12}H_{22}O_{11}}\right)\left(\frac{6.02\times10^{23} \text{ molecules } C_{12}H_{22}O_{11}}{1 \text{ mol } C_{12}H_{22}O_{11}}\right)$$

$$= 4.4\times10^{15} \text{ molecules of } C_{12}H_{22}O_{11}$$

For the second part of the question, we multiply the number of molecules of sucrose by the fact that there are 45 atoms in each molecule of sucrose

$$4.4 \times 10^{15} \text{ molecules}\left(\frac{45 \text{ atoms}}{\text{molecule}}\right) = 2.0 \times 10^{17} \text{ atoms}$$

There are 2.0×10^{17} atoms in one granule of sucrose.

CHAPTER 6

1. **(A)** Boyle's law states that the change in volume of a gas is inversely proportional to the change in pressure. When graphed, the resulting curve is not linear.
2. **(D)** The effusion rate will increase if the molecular mass is decreased, as it does going from ethane to methane.
3. **(B)** The relationship between temperature and average kinetic energy is a fundamental principle.
4. **(A)** Given that the container of chlorine gas is at 6.0 atm pressure, it can be deduced that each particle represents 1 atm. Therefore, the total pressure in the container of nitrogen should be 3 atm and the container of neon should be 1 atm. The total pressure would be 6.0 + 3.0 + 1.0 = 10 atm if the gases were combined into a 2 L container.

However, the gases were transferred to a 4 L container. So the total pressure would be 5.0 atm based on Dalton's law of partial pressures.

5. **(A)** The more volume per molecule keeps them widely separated, therefore reducing interactions. Increased velocity due to an increased temperature also reduces attractions between gas molecules.

6. **(A)** When the molecules are at moderate pressures, the intermolecular forces among the CO_2 molecules increases, therefore decreasing the total pressure.

7. **(C)** The number of moles of gas can be calculated using the ideal gas equation:

$$n = \frac{PV}{RT} = \frac{(800 \text{ mm Hg}/760 \text{ mm Hg atm}^{-1})(6.45 \text{ L})}{(0.0821 \text{ L atm mol}^{-1} \text{ K}^{-1})(297 \text{ K})}$$
$$= 0.278 \text{ mol}$$

In this example the pressure in millimeters of mercury had to be converted to atmospheres, and the Celsius temperature to Kelvin units.

8. **(D)** At STP 1 mol of an ideal gas occupies 22.4 L. The mass of 22.4 L of gas is the molar mass of the substance. The density is then

$$\text{density} = \frac{\text{mass}}{\text{liters}} = \frac{20.18 \text{ g mol}^{-1}}{22.4 \text{ L mol}^{-1}}$$
$$= 0.901 \text{ g L}^{-1}$$

Equation 6.12 could also be solved to obtain the density:

$$\frac{g}{V} = \frac{(\text{molar mass})\, P}{RT} = \frac{(20.18 \text{ g mol}^{-1})(1.00 \text{ atm})}{(0.0821 \text{ L atm mol}^{-1} \text{ K}^{-1})(273)} = 0.901 \text{ g L}^{-1}$$

9. **(D)** To solve the problem, take the ratio of two ideal gas law equations as follows:

$$\frac{P_i V_i}{P_f V_f} = \frac{n_i R T_i}{n_f R T_f}$$

and cancel variables that are kept constant, *n*, *R*, and *T*. Rearrange the remaining variables.

$$P_f = \frac{P_i V_i}{V_f}$$

$$\frac{(58 \text{ mm Hg})(155 \text{ mL})}{1000 \text{ mL}} = 8.99 \text{ mm Hg}$$

Note that the units of volume must be the same for V_i and V_f, and so 1.00 L was converted to 1000 mL.

10. **(B)** The pressure is directly proportional to the number of moles of gas, and the error in the pressure will be directly reflected in the moles of hydrogen reported. By Dalton's law of partial pressures

$$P_{total} = P_{H_2} + P_{H_2O}$$

The vapor pressure of water at 30 °C is given as 31.82 mm Hg. The correct pressure of hydrogen is

$$\begin{aligned} P_{H_2} &= P_{total} - P_{H_2O} \\ &= 745 \text{ mm Hg} - 31.82 \text{ mm Hg} \\ &= 713.2 \text{ mm Hg} \end{aligned}$$

The error is

$$\begin{aligned} \text{Error} &= \frac{\text{measured} - \text{true}}{\text{true}} \times 100 \\ &= \frac{745 - 713.2}{713.2} \\ &= +4.5\% \end{aligned}$$

11. **(B)** This answer is obtained by using the ideal gas law equation $PV = nRT$ to calculate the pressure of each gas:

$$\begin{aligned} P_{CO_2} &= \frac{nRT}{V} = \frac{(0.016 \text{ mol } CO_2)(0.0821 \text{ L atm mol}^{-1} \text{ K}^{-1})(298 \text{ K})}{2.50 \text{ L}} \\ &= 0.157 \text{ atm} \end{aligned}$$

$$\begin{aligned} P_{CH_4} &= \frac{nRT}{V} = \frac{(0.035 \text{ mol } CH_4)(0.0821 \text{ L atm mol}^{-1} \text{ K}^{-1})(298 \text{ K})}{2.50 \text{ L}} \\ &= 0.343 \text{ atm} \end{aligned}$$

Dalton's law of partial pressures gives

$$P_{total} = 0.157 \text{ atm} + 0.343 \text{ atm} = 0.500 \text{ atm}$$

Conversion of 0.500 atm to mm Hg yields 380 mm Hg.

Another approach to solving this problem would be to add the moles of the two gases and then to use the ideal gas law to calculate the total pressure directly.

$$\begin{aligned} P &= \frac{nRT}{V} = \frac{(0.035 \text{ mol } CH_4 + 0.016 \text{ mol } CO_2)(0.0821 \text{ L atm mol}^{-1} \text{ K}^{-1})(298 \text{ K})}{2.50 \text{ L}} \\ &= 0.500 \text{ atm} \end{aligned}$$

This yields a pressure of 380 mm Hg.

12. **(B)** Carbon dioxide is soluble to some extent in water. To obtain the amount dissolved in the water, the amount actually obtained is subtracted from the theoretical yield. The chemical reaction is

$$2C_2H_6 + 7O_2 \rightarrow 4CO_2 + 6H_2O$$

The theoretical yield is

$$? \text{ g } CO_2 = 1.50 \text{ g } C_2H_6 \left(\frac{1 \text{ mol } C_2H_6}{30.07 \text{ g } C_2H_6}\right)\left(\frac{4 \text{ mol } CO_2}{2 \text{ mol } C_2H_6}\right)\left(\frac{44.01 \text{ g } CO_2}{1 \text{ mol } CO_2}\right)$$
$$= 4.39 \text{ g } CO_2$$

From the gas collection data:

$$n = \frac{PV}{RT} = \frac{(746 \text{ mm Hg}/760 \text{ mm Hg/atm})(2.00 \text{ L})}{(0.0821 \text{ L atm mol}^{-1} \text{ K}^{-1})(298 \text{ K})}$$
$$= 0.0802 \text{ mol } CO_2$$
$$? \text{ g } CO_2 = 0.0802 \text{ mol } CO_2 \left(\frac{44.01 \; CO_2}{1 \text{ mol } CO_2}\right) = 3.53 \text{ g } CO_2$$

The difference between the theoretical yield and actual yield is

$$4.39 \text{ g} - 3.53 \text{ g} = 0.86 \text{ g } CO_2$$

This amount is apparently dissolved in the water of the pneumatic trough.

13. **(B)** If we assume that the volume of the box containing neon is 1 L, then the volume of the box containing argon is 2 L. The density of Ne present is 3(20.2 g) = 60. g/1 L. The density of Ar present is 6(40.0 g) = 240 g/2 L = 120 g/1 L. The density of Ar is greater than that of Ne.

14. **(D)** When combined, the total number of molecules is 9. So the mole fraction of Ne is 3/9, which is 1/3.

15. **(D)** Both an increase in the force of the collisions with the container walls and an increase in the frequency of collisions lead to the increase in pressure with increased temperature.

16. **(D)** The hydrogen gas is collected over water, so the pressure must be corrected for this. The vapor pressure of water must be subtracted from the total pressure (753 torr – 24 torr = 729 torr). The pressure, volume, and temperature need to be converted to the proper units. Using $PV = nRT$, the number of moles is determined to be 0.00764. The mass can then be found using the molar mass of H2 (2.02 g mol^{-1}) to be 0.0154 g.

17. **(A)** The further a gas is from its condensation point (boiling point), the more it behaves as an ideal gas. Condensation can be achieved by cooling a gas and/or by increasing its pressure. Therefore low pressure and high temperature will cause a gas to be as far as possible from its condensation point and therefore to behave most like an ideal gas.

18. **(D)** By applying Graham's law, the molar mass of the compound can be determined. This can then be compared to the molar masses of the compounds listed.

19. **(D)** The rate of diffusion is inversely proportional to the square root of the molecular masses:

$$\sqrt{\frac{\text{molar mass}_1}{\text{molar mass}_2}} = \frac{v_2}{v_1}$$

Entering the given data we have

$$\sqrt{\frac{154 \text{ g CCl}_4 \text{ mol}^{-1}}{44 \text{ g CO}_2 \text{ mol}^{-1}}} = \frac{6.3\times10^{-2} \text{ mol CO}_2 \text{ s}^{-1}}{v_1} = 1.87$$

$$v_1 = \frac{6.3\times10^{-2}}{1.87} = 3.4\times10^{-2} \text{ mol CCl}_4 \text{ s}^{-1}$$

20. **(B)** At STP, 22.4 L of a gas is equivalent to 1 mol. The mass of 22.4 L of this gas is equivalent to

$$22.4 \text{ L mol}^{-1} \times 3.48 \text{ g L}^{-1} = 78 \text{ g mol}^{-1}$$

Of the responses, only C_6H_6 and CaF_2 have molar masses of 78. However, CaF_2 is an ionic solid and highly unlikely to be in the gaseous state; therefore C_6H_6, benzene, is the most appropriate answer.

21. **(A)** The pressure of a gas will not be affected if the number of molecules stays constant. If the average velocity (temperature) of the molecules is changed, the pressure will be changed, since the frequency of collisions with the walls of the container will also change.

22. **(A)**

$$\frac{P_1V_1}{P_2V_2} = \frac{n_1RT_1}{n_2RT_2}$$

R, and *T*, and *V* are constants and cancel, so that $P_1/P_2 = n_1/n_2$. Entering the data from the experiment.

$$\frac{0.800 \text{ atm}}{1.10 \text{ atm}} = \frac{n_1}{n_1 + 0.200}$$

and solving for n_1 yields $n_1 = 0.532$ mol. Two-thirds of n_1 is oxygen, and one-third is nitrogen.

Therefore there is 0.355 mol of O_2 in the flask.

23. **(D)**

$$\frac{P_1V_1}{P_2V_2} = \frac{n_1RT_1}{n_2RT_2}$$

In this problem *n*, *R*, and *T* cancel, and $P_1V_1 = P_2V_2$. Entering data from the problem yields

$$(245 \text{ mm Hg})(1.50 \text{ L}) = (P_2)(0.350 \text{ L}).$$

Solving for P_2 gives 1050 mm Hg.

24. **(D)** The mass of air in the flask = 5.00 L$\left(\frac{1.290\text{ g}}{\text{L}}\right)$ = 6.45 g. The flask must weigh 543.251 – 6.45 = 536.80 g. The mass of the new gas is 566.11 – 536.80 = 29.31 g. The density of the gas is

$$d = \frac{29.31\text{ g}}{5.00\text{ L}} = 5.86\text{ g L}^{-1}.$$

$? \frac{\text{g}}{\text{mol}} = \frac{5.86\text{ g}}{\text{L}}\left(\frac{22.4\text{ L}}{1\text{ mol}}\right) = 131\text{ g mol}^{-1}$, which is the atomic mass of xenon.

25. **(A)**

$$\begin{aligned} ?\text{ L H}_2 &= 0.100\text{ g Mg}\left(\frac{1\text{ mol Mg}}{24.31\text{ g Mg}}\right)\left(\frac{1\text{ mol H}_2}{1\text{ mol Mg}}\right)\left(\frac{22.4\text{ L H}_2}{1\text{ mol H}_2}\right) \\ &= 0.0921\text{ L H}_2 \\ &= 92.1\text{ mL H}_2 \end{aligned}$$

CHAPTER 7

1. **(B)** London forces are the weakest and hydrogen bonds the strongest.

2. **(B)** Salts, because of high forces of attraction (Coulomb forces) have very high melting points. Metals and molecular crystals have a wide range of melting points starting near room temperature. Molecules can be any covalent structure. These can vary from gases with very low boiling points to solids with very high boiling points.

3. **(D)** The substance with the greatest intermolecular forces will have the highest surface tension. Substance (D) has two OH groups that will hydrogen-bond to others. Substance (A) has only one OH group, and so the attractive forces will be much smaller. The other two molecules are attracted by relatively weaker London forces.

4. **(B)** This answer choice is a very basic but true interpretation of this graph. Response (A) is true but does not show the insight of response (B). Response (C) does not make sense, and response (D) suggests experimental error.

5. **(A)** CH_2F_2 is the only compound that does not have an N–H, O–H, or F–H bond in its structure.

6. **(D)** The substance shown in this answer choice has only London forces of attraction (instantaneous dipole–instantaneous dipole attractions). It should have the lowest boiling point because the substance is able to overcome low attractive forces most easily. Response (A) has dipole-dipole attractive forces along with the London forces. Responses (B) and (C) form hydrogen bonds. The attractive forces in the incorrect answers all result in molecules with higher boiling temperatures than the molecule in response (D).

7. **(A)** In order to vaporize, a substance must have enough kinetic energy to overcome the forces of attraction in the matrix. This graph illustrates that a small proportion of the molecules or atoms have sufficient kinetic energy to do so.

8. **(C)** Aluminum is a soft metal that will not even scratch glass.

9. **(C)** Although the instantaneous dipoles are very weak attractive forces, in a large molecule many of them can exceed the strength of a hydrogen bond. In this case, they cause $C_{30}H_{62}$ to be a solid at room temperature.

10. **(A)** The lengths of the horizontal plateaus represent the heats of fusion and vaporization. Although not identical, they are closest to being the same.

11. **(D)** The slope of the curve for the solid, liquid, or gas has units of degree Celsius per joule. Therefore, $\frac{1}{\text{slope} \times \text{mass}}$ is the specific heat with the required units of $J\ g^{-1}\ °C^{-1}$.

12. **(D)** When substances evaporate, the molecules with high kinetic energy leave. Those can also be thought of as the "hot" molecules in the mixture. Thus the remaining molecules are cooler, and we see a temperature drop.

13. **(A)** When the vapor is condensed, a few drops of ethanol ignite easily. The substances in responses (B) and (C) are indicative of hydrogen gas and oxygen gas, respectively. The ethanol burns quietly.

14. **(A)** The hydrogen bonding is expected to give the molecule in the first response the strongest intermolecular forces and thus the lowest vapor pressure.

15. **(C)** The diagram on the right is one representation of dynamic equilibrium, indicating as many molecules evaporating (up arrows) as molecules condensing (down arrows). The other two diagrams indicate that equilibrium has not been attained because the numbers of up and down arrows are different.

16. **(D)** All of the suggested information can be deduced from this graph. Response (C) might not be obvious. However, the tailing of this curve toward the right should suggest more molecules (area under the curve) after the peak than before the peak. Therefore, the average kinetic energy should be after the peak.

CHAPTER 8

1. **(C)** $?\frac{\text{mol CdCl}_2}{\text{L CdCl}_2} = \frac{140\text{ g CdCl}_2}{100\text{ mL CdCl}_2}\left(\frac{1\text{ mol CdCl}_2}{183.3\text{ g CdCl}_2}\right)\left(\frac{1\text{ mL}}{10^{-3}\text{ L}}\right) = 7.64\ M$

2. **(C)** In choice (C), there are 3 solvent particles and 1 solute particle. In order to be a dilute solution, there should be very few solute particles in comparison to the solvent. Response (A) contains only solvent. Response (B) has equal amounts of solute and solvent. Response (D) contains more solute than solvent; therefore, it is supersaturated.

3. **(A)** Molarity $= \frac{\text{mol solute}}{\text{L solution}}$. We know the moles of solute and the mass of the solution. The density of the solution must be measured to convert the mass of solution to the volume. The other values are either not needed or readily available in tables.

4. **(B)** A release of heat energy when substances are mixed indicates stronger attractive forces in the solution than in the pure solvents.

5. **(B)** The mass of a substance does not change with temperature. Therefore, mole fraction and mass percent do not change. Molality by definition is mole of solute per kilograms of solvent, so it also will not change. Molarity is moles of solute per liter of solution. So molarity depends on the volume, which expands and contracts depending on the temperature of the solution.

6. **(C)** Increasing temperature always decreases the solubility of a gas such as O_2.

7. **(B)** Figures (A) and (D) contain only one type of substance while figures (B) and (C) have two. However, the substances illustrated in figure (C) seem to be more ordered. Figure (B) is a mixture of the two substances and is therefore the best representation of a solution.

8. **(B)** In order to make a solution, you need to weigh out the correct number of grams, which in this case is 30.0 g of urea. Transfer the solid to a volumetric flask, and add enough water to make 0.500 L of solution.

9. **(A)** Two driving forces for dissolution are the energy change (which is unfavorable since the solution cools) and the entropy change. We conclude that the entropy change must be great enough to overcome the energy deficit.

CHAPTER 9

1. **(C)** Knowing the enthalpy change of the reaction is needed. Le Châtelier's principle is needed to know how to use that information.

2. **(C)** We calculate the value of Q and if it matches the value of the equilibrium constant, then the system is at equilibrium. If Q does not match the value of the equilibrium constant, the system is not at equilibrium.

3. **(B)** The K_{sp} is the solubility product and is specific for systems that are slightly soluble.

4. **(C)** A system in equilibrium is governed by the equilibrium expression, which specifies a specific ratio of product to reactant concentrations.

5. **(B)** A value that is very small can be dropped only if it is added to or subtracted from a value that is, or is expected to be, very large. The criterion for a number being very large or very small is relative. Usually if a number is 10 times larger than another, the smaller number can be dropped. Also, if a number becomes zero in the rounding process, it can be dropped.

6. **(C)** Nitrogen (N_2) is not involved as either a reactant or a product and therefore has no effect on the reaction given.

7. **(D)** The reaction has been reversed, causing the equilibrium constant to be inverted. In addition, the coefficients are doubled and the equilibrium constant is squared. The new equilibrium constant is

$$K = \frac{1}{(4.5\times10^{+3})^2} = 4.9\times10^{-8}$$

8. **(A)** The reaction is

$$Ag_2CrO_4(s) \rightleftharpoons 2\,Ag^+(aq) + CrO_4^{2-}(aq)$$

and $K_{sp} = [Ag^+]^2[CrO_4^{2-}]$.

9. **(D)** After 6 minutes, all the curves have flattened out and the macroscopic concentration no longer changes.

10. **(D)** Only HCl and CO_2 should appear in the equilibrium expression. Solids and pure liquids have constant concentrations and are included in the equilibrium constant. Response (A) does not have the correct exponent for HCl.

11. **(C)** $K_c = \dfrac{[H_2]\,[I_2]}{[HI]^2} = 0.020$

$$\text{Initial concentration of HI} = \frac{0.200 \text{ mol HI}}{10.0 \text{ L}} = 0.0200\,M$$

Reaction	2HI	⇌	H_2	+	I_2
Initial Conc.	0.0200 *M*		0 *M*		0 *M*
Change	-2*x*		+*x*		+*x*
Equilibrium	0.0200 - 2*x*		+*x*		+*x*
Answer					

$$\frac{[H^2][I^2]}{[HI]^2} = 0.020 = \frac{(x)(x)}{(0.0200-2x)^2}$$

Take the square root of both sides to obtain

$$0.1414 = \frac{x}{0.0200-2x}$$

$$(0.1414)(0.0200-2x) = x$$

$$0.002828 - 0.2828x = x$$

$$0.002828 = x + 0.2828x$$

$$0.002828 = 1.2828x$$

$$\frac{0.002828}{1.2828} = x$$

$$x = 2.2\times10^{-3}$$

Reaction	2HI	⇌	H_2	+	I_2
Initial Conc.	0.0200 *M*		0 *M*		0 *M*
Change	-2*x*		+*x*		+*x*
Equilibrium	0.0200 - 2*x*		+*x*		+*x*
Answer	0.01559 *M*		2.2×10^{-3} *M*		2.2×10^{-3} *M*

Since this is a 10.0-L flask, it will contain 2.2×10^{-2} or 0.022 mol $I_2(g)$.

12. **(B)** For this reaction: $Q = \frac{P_{N_2O_4}}{P^2_{NO_2}}$

The reaction will go in the forward direction if Q is less than K_p. Q must be calculated from pressures in atmospheres. Only response (B) results in a value of Q less than K_p.

13. **(B)** $PbI_2(s) \rightleftharpoons Pb^{2+}(aq) + 2I^-(aq)$ and $K_{sp} = [Pb^{2+}][I^-]^2$

Reaction	PbI_2	$\rightleftharpoons$	Pb^{2+}	+	$2I^-$
Initial Conc.	Solid		0 *M*		0 *M*
Change	$-x$		$+x$		$+2x$
Equilibrium	Solid		$+x$		$+2x$
Answer					

$$7.9 \times 10^{-9} = (x)(2x)^2$$

$$= 4x^3$$

$$1.98 \times 10^{-9} = x^3$$

$$1.25 \times 10^{-3} = x \text{ (This is also the molar solubility.)}$$

14. **(B)** Recall that reactants will be consumed and disappear, so the two top lines are the reactants. A crude estimate is that one reactant changes by 3 units on the concentration scale while the other reactant changes by 2 units. Therefore, *A* and *B* should have coefficients of 3 and 2. Only response (B) has that ratio. For confirmation we look to the products that have increased concentration. We find that one has increased twice as much as the other. Thus, the products should have coefficients of 1 and 2, which they do in response (B).

15. **(D)** We can prove this is true by taking the mass of any solid or liquid compound. Then use its density to determine the volume and calculate its molarity. Taking any other mass of that same compound and doing the same process gives exactly the same molarity.

16. **(B)** We can write the appropriate equation as

$$K = \frac{[A]^2[B]^3}{[X][Y]^2}$$

From the graph we estimate that $[X] = 0.10\ M$, $[Y] = 0.20\ M$, $[A] = 0.32\ M$, and $[B] = 0.22\ M$. Note that these are all estimates. So round the 0.32 to 0.30 and the 0.22 to 0.20. Now solve for K:

$$K = \frac{[0.20]^2[0.30]^3}{[0.10][0.20]^2} = 0.27$$

We may also note that we only rounded the numerator terms down. So the correct answer will be greater than 0.27. If we did not round to one significant figure, the result would have been 0.40. The closest response would still have been (B).

17. **(C)** In an exothermic reaction the amount of product decreases with increasing temperature.

$$\text{Reactants} \rightleftharpoons \text{Products} + \text{Heat}$$

is the general equation for an exothermic reaction. This illustrates that adding heat by increasing the temperature forces the reaction toward the reactant side.

18. **(A)** The reaction rate has nothing to do with the size of the equilibrium constant.

19. **(D)** The iodide precipitates first since it has the smaller K_{sp}. When the silver ion concentration gets large enough, the chloride ions will start precipitating. To find out how much iodide is left when silver starts to precipitate, we first calculate the silver ion concentration when AgCl starts to precipitate.

$$\left[Ag^+\right] = \frac{K_{sp\text{ of }Cl^-}}{\left[Cl^-\right]} = \frac{1.8\times10^{-10}}{0.100} 1.8\times10^{-9}\,M$$

We can now use that concentration of silver ions to calculate the iodide ion concentration as

$$\left[I^-\right] = \frac{K_{sp\text{ of }I^-}}{\left[Ag^+\right]} = \frac{8.3\times10^{-17}}{1.8\times10^{-9}} = 4.6\times10^{-8}\,M$$

20. **(D)** An endothermic reaction places heat on the reactant side of the equation. Increasing the temperature increases the products and decreases the reactants. Therefore the peaks are going downward and they are the reactants. Their sizes are about equal, so their coefficients are 1. For the products, the smaller peak is the same size as that of the reactants, so its coefficient is also 1. The other product must have a coefficient of about 2 since its peak is about twice that of the others. The only equation that has a coefficient of 2 for one of the products is response (D).

21. **(D)** When the initial amount of HI is known, only one: I_2, H_2, or HI needs to be measured to determine the equilibrium constant. Therefore, it is *not* important to measure all three concentrations.

22. **(B)**

Reaction	$2SO_2$	+	O_2	$\rightleftharpoons$	$2SO_3$
Initial Conc.	0.00300 *M*		0.00300 *M*		0.00300 *M*
Change	2*x*		*x*		−2*x*
Equilibrium	0.00300 + 2*x*		0.00300 + *x*		0.00300 − 2*x*
Answer	3.50×10^{-5} *M*				

From the last two lines,

$$3.50 \times 10^{-5} = 0.00300 + 2x$$

Solve for x to get $x = -0.00148$, which yields

$$[O_2] = 0.00152\ M \quad \text{and} \quad [SO_3] = 0.00596\ M$$

on the answer line.

Then

$$K_{eq} = \frac{[SO_3]^2}{[SO_2]^2[O_2]} = \frac{(0.00596)^2}{(3.50\times10^{-5})^2(0.00152)} = 1.9\times10^7$$

23. **(B)** Since the two reactions are added to obtain the overall reaction.

$$K_{overall} = K_{a1}K_{a2} = (2.3\times10^{-4})(4.5\times10^{-7}) = 1.0\times10^{-10}$$

CHAPTER 10

1. **(C)** An increase in temperature increases the number of collisions per second with the proper energy necessary for the reaction to take place.

2. **(C)** Since the rate of the reaction is doubled, the number of molecules with sufficient energy is also doubled.

3. **(C)** An activated complex contains both partially broken bonds in the reactant(s) and partially formed bonds in the product(s) whereas an intermediate has complete bonds and can be isolated occasionally.

4. **(C)** These items all affect the rate of a reaction. The others describe other features of reaction rate theory.

5. **(B)** When we add up the three reactions and cancel out any intermediates that may have formed, we obtain the overall reaction. In this case, the 2Cl and CCl_3 are the intermediates.

6. **(A)** The rate law for a reaction is based on the rate-determining step, which is always the slowest.

7. **(A)** The *x*-axis (*B*) is the reaction coordinate, the *y*-axis (*A*) is the potential energy, and (*C*) represents the activation energy.

8. **(D)** This is an endothermic reaction, with the activation energy larger for the forward reaction than for the reverse reaction.

9. **(C)** Catalysts have an effect on the activation energy.

10. **(D)** If we compare the number of X_2 molecules in the two flasks, we can see that flask *B* has twice the number of molecules as does flask *A*. Therefore, the reaction is twice as fast in flask *B*.

11. **(D)** These are the units for a third-order rate constant.

12. **(A)** Adding an inert gas has no effect on the reaction rate.

13. **(A)** $\ln(2) = kt_{1/2}$

$$k = \frac{\ln(2)}{t_{1/2}} = \frac{0.693}{(36 \text{ min})(60 \text{ s min}^{-1})} = 3.2\times10^{-4} \text{ s}^{-1}$$

14. **(D)**

$$\frac{\text{Rate}_1}{\text{Rate}_2} = \frac{(\text{Conc}_1)^x}{(\text{Conc}_2)}$$

where x is the exponent in the rate law. Consequently, $8 = 2^x$, based on the information in the question, and x must be 3.

15. **(C)** The reaction rate is an exponential function given by the Arrhenius equation. Since the activation energy is not given in this problem, we rely on the rule of thumb that the reaction rate doubles for each 10°C increase in temperature. A 20°C increase in temperature will increase the rate by a factor of approximately 4.

16. **(C)** Integrated rate equations for second-order reactions involve $\frac{1}{[A]}$ terms.

17. **(C)**

$$? \frac{\text{mol O}_2}{\text{L s}} = \frac{2.2\times10^{-2} \text{ mol CO}_2}{\text{L s}}\left(\frac{15 \text{ mol O}_2}{12 \text{ mol CO}_2}\right)$$
$$= 2.8\times10^{-2} \text{ mol O}_2 \text{ L}^{-1}\text{s}^{-1}$$

18. **(C)** The rate law for this reaction is Rate = $k[CH_3Br][OH^-]$. By entering in the concentration values, we find that the rate constant is 35 $M^{-1}s^{-1}$.

19. **(A)** The rate law for a reaction is based on the slowest step of the mechanism.

20. **(C)**

$$\frac{255 \text{ s}}{85\text{s half-life}^{-1}} = 3 \text{ half-lives}$$

$$\left(\frac{1}{2}\right)^3 = \frac{1}{8}$$

21. **(C)** We see that when the $[A]$ is cut in half and the $[B]$ remains the same, the initial rate is decreased by a factor of 4. This means that the exponent for $[A]$ is 2. By comparing trials 1 and 3, the exponent for $[B]$ is found to be 1.

22. **(D)** The constant is found by $k = \text{rate}/[A]^2[B]$ = 15 M^{-2} min^{-1}.

23. **(A)** Using the rate = $k[A]^2[B] = k[A/2]^2[4B]$, the rate will not change at all.

24. **(C)** A plot of $\ln[A]$ versus time that yields a straight line is a first-order reaction. Therefore, the coefficient is 1. The rate constant is always a positive value.

25. **(B)** The catalyst will not alter K_{eq} and therefore will not alter the position of equilibrium. All other effects will be observed.

CHAPTER 11

1. **(A)** A negative value for ΔG° always indicates a thermodynamically favorable reaction.

2. **(B)** The basic idea of a state function is that the path taken to get from the initial state (P, V, T, n) to the final state does not matter. This compares to work, w, where many different paths can be taken and different values for the work are calculated. Work, therefore, is not a state function.

3. **(C)** In thermodynamics, all energy can be classified as either kinetic (the energy of motion where KE = $0.5mv^2$) or potential (the energy of position).

4. **(D)** All items listed in response to this question are valid means of determining if a chemical reaction or physical process is endothermic or exothermic.

5. **(D)** $q_v = -(\text{heat capacity})(\Delta T) = -(3245\text{ J }°\text{C}^{-1})(6.795°\text{C}) = -22.05\text{ kJ}$

   The sign is negative since heat is released in the reaction. Notice that the −22.05 kJ/mol is the heat energy lost by the chemical reaction to the calorimeter. The calorimeter gained that amount of heat.

6. **(B)** At point 1, the ball has reached its highest point and its velocity is zero. So the kinetic energy (KE = $0.5\ mv^2$) is zero. All KE has been converted to potential energy, PE. At point 3, the ball is closest to the center of Earth and the gravitational PE is at its lowest point in this apparatus. The velocity is at its maximum, and so the KE is at its maximum for this device.

7. **(A)** A system where ΔG° is always positive will never be thermodynamically favored. For this to occur, ΔH° must be positive and ΔS° must be negative.

8. **(B)** Condensation occurs when the liquid in the beaker is lower than the atmosphere. Condensation indicates that the temperature has decreased and the process is endothermic. The dissociation equation

   $$KCl(s) \rightarrow K^+(aq) + Cl^-(aq)$$

   yields 2 particles for 1 formula unit and suggests an increase in entropy.

9. **(C)** When CH_4 is burned, the greatest decrease in the moles of gas ($\Delta n_g = -2$) is obtained if liquid water forms.

10. **(D)** During a phase change $\Delta G^\circ = 0$ and $T\Delta S^\circ = \Delta H^\circ$.

    Calculate

    $$\Delta S^\circ = \frac{\Delta H^\circ}{T} = \frac{+43900\text{ J}}{373\text{ K}} = +118\text{ J K}^{-1}$$

11. **(A)** The ΔG represents the slope of the curve. If ΔG is negative, the reaction must proceed in the forward direction to reach equilibrium (the minimum of the curve where $\Delta G = 0$). If ΔG is positive, the reaction needs to proceed in the reverse direction to reach equilibrium.

12. **(B)** The pressure (P_1) on the left side of the diagram is 3.00 atm. The pressure on the right side is P_2 = 1.00 atm, and the volume is 15.0 liters. Using

$$P_1V_1 = P_2V_2$$

we calculate that

$$V_2 = 5.00 \text{ L}$$

The work is

$$w = -P\Delta V = -(1.00 \text{ atm})(15.0 \text{ L} - 5.00 \text{ L})$$
$$= -10.0 \text{ L atm}$$

13. **(D)** Both forms of the universal gas constant are needed. The ratio of the two forms of this constant has units of $\frac{\text{L atm}}{\text{J}}$ and is used as the factor label for the conversion.

14. **(B)** Except for small molecules, the entropy change is a large positive value.

15. **(C)** The entropy of a system is related to the number of possible energy states of that system. Since each particle has an associated energy, we can say that the more particles there are, the higher the entropy of the system is.

16. **(B)** The reaction is

$$2CH_3OH(l) + 3O_2(g) \rightarrow 2CO_2(g) + 4H_2O(g)$$

$$\Delta H° = [(2 \text{ mol})(-393.5 \text{ kJ mol}^{-1}) + (4 \text{ mol})(-241.8 \text{ kJ mol}^{-1})]$$
$$- [(2 \text{ mol})(-238.6 \text{ kJ mol}^{-1})]$$
$$= -1277 \text{ kJ}$$

17. **(D)** The rate of a reaction cannot be determined from thermodynamic quantities.

18. **(C)** The $\Delta H°$ value for first reaction is subtracted from the value for second reaction.

$$\Delta H° = \Delta H° \text{ (second react)} - \Delta H° \text{ (first react)}$$
$$= -113.14 \text{ kJ} - 57.93 \text{ kJ}$$
$$= -171.07 \text{ kJ}$$

19. **(D)** Only changes in temperature change $\Delta G°$. The fact that the temperature change is measured in Celsius degrees is not significant. Thermodynamic calculations involving temperature require conversion to the Kelvin scale.

20. **(D)** The ball bouncing down the stairs is losing potential energy, and the ball never goes back up the stairs to regain the lost energy. Without frictional losses, all others would continue transferring PE and KE forever.

21. **(D)** $\Delta H° = -(2 \text{ mol})(-396 \text{ kJ mol}^{-1})$

$$= +792 \text{ kJ (heats of formation for elements} = 0)$$

$$\Delta S° = (3 \text{ mol } O_2)(205 \text{ J mol}^{-1} \text{ K}^{-1}) + (2 \text{ mol S})(31.8 \text{ J mol}^{-1} \text{ K}^{-1})$$
$$-(2 \text{ mol } SO_3)(256 \text{ J mol}^{-1} \text{ K}^{-1})$$
$$= 167 \text{ J K}^{-1}$$

$$\Delta G° = 792 \text{ kJ} - (298 \text{ K})(0.167 \text{ kJ K}^{-1}) = 742 \text{ kJ}$$

22. **(D)** Heats of combustion of all compounds except one in an equation will allow us to determine the enthalpy of a reaction using Hess's law.

23. **(C)** Heat must be added to evaporate a liquid, and therefore $\Delta H°$ is positive. Since the liquid becomes a gas, $\Delta S°$ is also positive.

24. **(A)** Combustion of organic compounds produces heat; therefore $\Delta H°$ is negative.

CHAPTER 12

1. **(C)** An electrolytic cell uses an external source to provide the energy necessary for a nonspontaneous process to occur. So the cathode in an electrolytic cell is opposite in charge to the cathode in a voltaic cell.

2. **(A)** To determine the $E°$ for the anode, use the equation $E°_{cell} = E°_{cathode} - E°_{anode}$. The $E°_{anode} = -(0.62\ V - 0.34\ V) = -0.28\ V$.

3. **(C)** When writing cell line notation, the anode compartment is written first and the cathode is last. In this case, the Pt is an inert electrode for the H_2 oxidation to H^+.

4. **(D)** The balanced half-reaction is

$$8e^- + 10H^+ + NO_3^- \rightarrow NH_4^+ + 3H_2O$$

5. **(B)** The first two reactions show that Cu and Zn metals are more effective reducing agents than Ag. The last reaction shows that Zn is a better reducing agent than Cu. Consequently, Zn is the strongest reducing agent, and Ag the weakest, of the three.

6. **(B)**

$$E°_{cell} = E°_{reduction} - E°_{oxidation}$$

Since Fe is oxidized and Pb is reduced,

$$E°_{cell} = +\ 1.46 - (+0.77) = +0.69\ V$$

7. **(C)** The more positive cell potential is more likely to occur as written. When comparing the reduction potentials of the two reactions, the Co^{2+} half-reaction is more positive; it is the reaction occurring at the cathode. Therefore, the

$$E°_{cell} = -0.28\ V - (-0.74\ V) = 0.46\ V$$

A positive $E°_{cell}$ is spontaneous as written.

8. **(A)** The reaction at the cathode is

$$Co^{2+}(aq) + 2e^- \rightarrow Co(s)$$

The reaction at the anode is

$$Cr(s) \rightarrow Cr^{3+}(aq) + 3e^-$$

In order for an oxidation-reduction to occur, the same number of electrons must be lost and gained. Therefore, before the two reactions are added, they must be multiplied by the lowest common factor.

$$3Co^{2+}(aq) + 6e^- \rightarrow 3Co(s)$$

$$2Cr(s) \rightarrow 2Cr^{3+}(aq) + 6e^-$$

Adding the two together yields:

$$3Co^{2+}(aq) + 2Cr(s) \rightarrow 3Co(s) + 2Cr^{3+}(aq)$$

9. **(A)** The sulfur in both $H_2S_2O_7$ and H_2SO_4 has an oxidation number of +6.

10. **(D)** H ions are reduced before Na ions.

11. **(B)** Chlorine can have −1, +1, +3, +5, and +7 for its oxidation states.

12. **(A)**

$$\Delta G° = -n\mathcal{F}E°_{cell}$$

and

$$\Delta G° = -RT\ln K$$

Therefore

$$-n\mathcal{F}E°_{cell} = -RT\ln K$$

Calculate

$$\ln K = \frac{n\mathcal{F}E_{cell}}{-RT} = \frac{(2)(96{,}485\text{ C mol}^{-1})(0.82\text{ V})}{(8.314\text{ V C mol}^{-1})(318\text{ K})} = 59.85$$

$$K = 9.8 \times 10^{25}$$

13. **(B)** First determine the number of electrons being transferred in the reaction. The Pb(s) is going from an oxidation number of 0 to +2 in $PbSO_4$ and from +4 to the +2 in $PbSO_4$. Therefore, 2 electrons are transferred. Using the equation $\Delta G° = -nFE°_{cell}$ the value can be determined.

$$\Delta G° = -nFE°_{cell} = -2(96{,}500\text{ J}/V\text{mol})(2.04\ V) = -3.94 \times 10^5\text{ J} = -3.94 \times 102\text{ kJ}$$

14. **(B)**

$$\text{mol} = \frac{It}{n\mathcal{F}} = \frac{(1.23\text{ A})(2.5\text{ h})(60\text{ min h}^{-1})(60\text{ s min}^{-1})}{(2)(96{,}485\text{ C mol}^{-1})} = 0.0574\text{ mol}$$

$$\text{Molar mass} = \frac{\text{g compound}}{\text{mol compound}} = \frac{3.37\text{ g}}{0.0574\text{ mol}} = 58.7\text{ g mol}^{-1}$$

Nickel has a molar mass of 58.7.

15. **(D)** In order to function properly, the overall reaction for this galvanic cell should be

$$Fe(s) + Cu^{2+}(aq) \rightarrow Fe^{2+}(aq) + Cu(s)$$

This will mean that the anode compartment (where oxidation takes place) should contain both Fe(s) and $Fe^{2+}(aq)$.

16. **(C)** If the $E°$ for the reduction half-reaction of X^+ is greater than that of A^{2+}, then A will have a better chance of reducing X^+.

17. **(C)** Chlorine has an oxidation number of +5. The nitrogen in the nitrate ion NO_3^- is also +5.

18. **(B)**

$$Cu^{2+} + 2e^- \rightarrow Cu^0$$

$$\text{mol} = \frac{\text{g Cu}}{\text{molar mass}} = \frac{It}{n\mathcal{F}}$$

Rearranging the equation yields

$$t = \frac{(\text{g Cu})n\mathcal{F}}{(\text{molar mass})} = \frac{(2.00\text{ g})(2)\left(96{,}485\text{ C mol}^{-1}\right)}{\left(63.5\text{ g mol}^{-1}\right)\left(1.25\text{ C s}^{-1}\right)} = 4862\text{ s} = 81\text{ min}$$

19. **(D)** The balanced half-reaction is

$$10e^{-} + 12H^{+} + 2IO_3^{-} \rightarrow I_2 + 6H_2O$$

20. **(D)** The NADH is being oxidized in this reaction, and the oxaloacetate is being reduced. The standard electrode potential can be determined by the following.

$$E^{\circ}_{cell} = E^{\circ}_{red} - E^{\circ}_{ox} = -0.32\ V - (-0.166\ V) = -0.154\ V$$

This reaction is not thermodynamically favored as written since the ΔG°_{rxn} is a positive value.

21. **(D)** Electrolysis does not necessarily produce a gas at either the anode or the cathode.

CHAPTER 13

1. **(C)** In this response, carbonic acid, H_2CO_3 is half neutralized to produce the bicarbonate ion, HCO_3^{-}. Response (A) is meaningless, response (B) is a base that has not reacted with an acid. In response (D) perhaps $Mg(OH)_2$ reacted completely with H_2CO_3.

2. **(D)** The question asks what the solution contains. Therefore, the best answer is to list everything no matter how small the concentrations. Aqueous solutions always contain water along with H^{+} and OH^{-}. The major solutes are the ions K^{+} and F^{-}. A small amount of F^{-} hydrolyzes water to generate HF too.

3. **(A)** When a buffer is mixed with an equal volume of water, the pH has a virtually unnoticeable change. This is because the pH is essentially defined by the ratio of the conjugate acid and the conjugate base and that ratio does not change upon dilution.

4. **(C)** The two salts, $LiHCO_3$ and K_2CO_3, dissociate to give the bicarbonate ion HCO_3^{-} and the carbonate ion CO_3^{2-}. This is the conjugate acid–base pair needed for a buffer

5. **(C)** Carbonic acid is a weaker acid than ethanoic acid, and its salt will be a stronger base, giving a solution with a higher pH. NaCl has no effect on the pH, and A and B are acids with pH values less than 7.

6. **(D)** In solution only conjugate acid–base pairs exist. The oxalate ion and oxalic acid differ by more than one H^{+}. The ions of sodium sulfate (response (C)) can exist in solution together.

7. **(C)** The solution contains 0.100 mol of H_3PO_4, and the addition of 0.250 mol of NaOH represents 2.5 mol base for each mole of phosphoric acid. The first 2 mol of base convert the H_3PO_4 to HPO_4^{2-}. The next 0.5 mol of base converts only half of the HPO_4^{2-} to PO_4^{3-}, and the resulting solution is a mixture of these two ions.

8. **(B)** We need a weak acid with a pK_a within 1 pH unit of 10.00 or a weak base with a pK_b within 1 pH unit of 4.00 (pOH). Ammonia comes closest to these requirements.

9. **(A)** When the concentrations of the conjugate acid and base are equal, $pH = pK_a$.

10. **(C)** The K_b is

$$K_b = \frac{[BH^+][OH^-]}{[B]}$$

The pH is given as 10.45, and so the pOH is 3.55. We can calculate the $[OH^-] = [BH^+] = 2.82 \times 10^{-4}$ *M*. We know that [B] = 0.125 *M*. Enter the data in the equation and get

$$K_b = \frac{[2.82 \times 10^{-4}][2.82 \times 10^{-4}]}{[0.125]}$$

The result is 6.4×10^{-7}.

An alternate estimate is achieved by just using –pOH as the exponent for 10 in the equation:

$$K_b = \frac{[10^{-3.55}][10^{-3.55}]}{[0.125]} = \frac{10^{-7.1}}{0.125} = 8 \times 10^{-7.1}$$

We can see that only response (C) is close to this value, without using a calculator.

11. **(C)** The dinitrogen pentoxide reacts with water:

$$N_2O_5 + H_2O \rightleftharpoons 2HNO_3$$

The other substances formed with water are $Ca(OH)_2$, H_2SO_3, and H_2CO_3.

12. **(C)** When each of these ions is hydrated, it attracts the negative (oxygen) end of water toward the ion. The much larger charge of the Fe^{3+} ion pulls electrons in the water toward the Fe^{3+} to the extent that H^+ can leave the water molecule more easily, making it acidic. The other substances have ions with smaller charges, making them less acidic.

13. **(D)** The greater number of oxygen atoms on HNO_3 weakens the bond with hydrogen, causing it to be a strong acid. Nitrous acid, HNO_2, is a weak acid.

14. **(C)** The anion is amphiprotic and can act as both an acid and a base.

$$pH = \frac{pK_2 + pK_3}{2} = \frac{7.20 + 12.35}{2} = 9.78$$

15. **(D)** Each of these methods results in a solution containing a conjugate acid and its conjugate base in significant amounts, therefore resulting in a buffer solution.

16. **(D)** Starting at a pH greater than 7.0 indicates that the analyte is a base. The curvature of the line indicates it is a weak base. In virtually all acid–base titrations, the titrant is either a strong acid or a strong base, which increases the sharpness of the curve near the end point. So the titrant is a strong acid.

17. **(C)** A horizontal line drawn intersects the *y*-axis at approximately pH = 4.8. A vertical line from the end point to the *x*-axis intersects at 25.0 mL.

18. **(A)** The four points noted are all within the buffer region of the curve. Points 1 and 6 are on parts of the curve where the pH is changing rather rapidly, obviously not what is expected of a buffer.

19. **(B)** The moles are calculated from the end point volume of 25.0 mL. Molarity times volume (in liter units) yields moles.

$$M \times V = (0.0855\ M)(0.025\ \text{L}) = 0.0021 \text{ moles H}^+$$

The curve shows only one end point. So the moles of base must be the same as the moles of acid added.

20. **(B)** Halfway between the start of the titration and the end point is the midpoint where pH = pK_a for weak acids and pOH = pK_b for weak bases. Response (B) applies in this situation.

21. **(B)** The inflection is the point where the slope of the curve is the largest. Although a properly selected indicator will show the end point, response (A) does not indicate that the correct indicator was used. The wrong indicator may change color far from the end point. The first derivative determines the slope, but its maximum is sought for the largest slope. Response (D) is true only for strong acids and strong bases.

22. **(C)** A primary standard is a very pure substance that, in titrations, is used to determine the precise concentrations of standard titrants.

23. **(D)** The equation is

$$\text{NaOH} + \text{HCl} \rightarrow \text{NaCl} + \text{H}_2\text{O}$$

$$?\text{ mmol HCl} = (0.0134\ M)(50.0\text{ mL}) = 0.670\text{ mmol HCl}$$

$$?\text{ mmol NaOH} = (0.0250\ M)(24.0\text{ mL}) = 0.600\text{ mmol NaOH}$$

There is an excess of 0.070 mmol HCl, and the total volume is

$$50.0 + 24.0 = 74.0\text{ mL}$$

$$\text{M}_{\text{acid}} = \frac{0.070}{74} = 9.46\times10^{-4}\ M$$

$$\text{pH} = -\log(9.46 \times 10^{-4}) = 3.02$$

24. **(C)** This is a buffer solution:

$$[\text{HCHO}_2] = \frac{50.0\text{ g}/46.03\text{ g mol}^{-1}}{0.500\text{ L}} = 2.17$$

$$[\text{NaCHO}_2] = \frac{30.0\text{ g}/68.01\text{ g mol}^{-1}}{0.500\text{ L}} = 0.882$$

$$[\text{H}^+] = \frac{K_a[\text{HCHO}_2]}{[\text{NaCHO}_2]} = \frac{1.8\times10^{-4}(2.17)}{0.882} = 4.4\times10^{-4}$$

$$\text{pH} = -\log(4.4\times10^{-4}) = 3.35$$

CHAPTER 14

1. **(A)** The sodium ion gives an orange flame test.

2. **(C)** The bromide ion is oxidized to bromine and then extracted into an organic solvent, producing a distinctive brown color.

3. **(D)** The sulfide ion and ammonium ion are converted to gases that have odors. Acid is added to sulfides to produce $H_2S(g)$ that smells like rotten eggs. Adding base to an ammonium salt produces ammonia that has a characteristic sharp odor.

4. **(B)** Silver ions are precipitated with chloride. It can be confirmed by dissolving the silver chloride in ammonia, forming a soluble $Ag(NH_3)_2^+$ complex ion.

5. **(A)** A buret will measure this volume most accurately.

6. **(B)** Methane does not dissolve in water, so displacement of water is used to collect this compound.

7. **(D)** Centrifugation is best. Fine precipitates clog filters. Drying and distillation are very time-consuming and do not remove any nonvolatile impurities.

8. **(B)** Student 1's data are closer to the true value, while student 2's data are clustered together.

9. **(A)** To prepare a molar solution, the solvent is added to the solute to achieve the total volume of solution desired. By adding the solute to 1 L of water the final volume will not be 1 L.

10. **(B)** Pressure is the dependent variable and is the *y*-axis of the graph.

11. **(D)** To transfer a coarse solid into a test tube, always use a creased piece of paper. A thin-stemmed funnel works only if the solid is very fine.

12. **(C)** A buret is usually graduated in tenths of a millimeter, which means the volume delivered can be estimated to ±0.01 mL.

13. **(C)** $Al(OH)_3$ is amphiprotic. An acid will neutralize the hydroxide ions, and a base will form a hydroxy complex, which is soluble.

14. **(C)** For a monoprotic acid titrated with a monobasic base, $M_aV_a = M_bV_b$. The molarity of the acid is calculated as

$$M_a = \frac{M_bV_b}{V_a}$$

Droplets left in the buret mean that the volume of base reported is larger than the volume that the buret actually delivered. Since the V_b reported is larger than it should be, the M_a calculated using the above equation is too high.

15. **(B)** $C_iV_i = C_fV_f$. When the analysis is started,

$$\begin{aligned} C_i(5.00\text{ mL}) &= (3.0 \times 10^{-3}\ M)(100\text{ mL}) \\ &= 6.0 \times 10^{-2}\ M \end{aligned}$$

The preservative step adds an equal volume of preservative to the sample. Choose any two volumes so that the final volume is twice the initial sample volume. Then

$$\begin{aligned} C_i(50\text{ mL}) &= (6.0 \times 10^{-2}\ M)(100\text{ mL}) \\ &= 1.2 \times 10^{-1}\ M \end{aligned}$$

16. **(D)** Unlike $PbCl_2$, the other chloride precipitates, AgCl and Hg_2Cl_2, are not soluble in hot water. Since all the precipitate dissolves, they are not present. Also, since the unknown was colored, we may deduce that a transition metal ion is present.

17. **(D)** CO_2 and H_2S are the only gases produced when a solution is acidified. CO_2 has no odor, but H_2S smells like rotten eggs. NH_3 is formed when a solution is made alkaline.

18. **(D)** All of these ions are confirmed with flame tests.

19. **(C)** $CrCl_3$ is a soluble salt. Reacting precisely 3 moles of HCl with 1 mole of $Cr(OH)_3$ will produce a solution containing only Cr^{3+} and Cl^- ions, which can be dried to obtain pure $CrCl_3$. The other mixtures contain other ions in addition to Cr^{3+} and Cl^-. This results either in an impure product or in reactions that will not form $CrCl_3$.

20. **(B)** Titration is the most common method. Responses (A) and (D) are possible, but not widely used, methods; response (C) is an imprecise method.

21. **(C)** In order for the molar mass to be too low, either the volume or the concentration of the NaOH solution would be too large. Error III means that a larger volume of NaOH is needed to reach the end point since some of the NaOH has reacted with the CO_2. Error I also means a larger volume of NaOH is necessary to reach the end point since it is not as concentrated as anticipated.

22. **(D)** This is the only choice left because (A), (B), and (C) are all correct.

23. **(A)** The correct approach is to abandon the experiment and start again.

24. **(B)** The data for the experiment should fill most of the graph.

ONLINE

Want more test-taking practice?

Visit *barronsbooks.com/ap/ap-chemistry/* to access the three additional online tests or scan the QR code below:

***Be sure to have a copy of your *AP Chemistry*, 9th edition, on hand to complete the registration process.**

Appendix 2

Electronic Configurations of the Elements

H	$1s^1$				
He	$1s^2$				
Li	$1s^2$	$2s^1$			
Be	$1s^2$	$2s^2$			
B	$1s^2$	$2s^22p^1$			
C	$1s^2$	$2s^22p^2$			
N	$1s^2$	$2s^22p^3$			
O	$1s^2$	$2s^22p^4$			
F	$1s^2$	$2s^22p^5$			
Ne	$1s^2$	$2s^22p^6$			
Na	$1s^2$	$2s^22p^6$	$3s^1$		
Mg	$1s^2$	$2s^22p^6$	$3s^2$		
Al	$1s^2$	$2s^22p^6$	$3s^23p^1$		
Si	$1s^2$	$2s^22p^6$	$3s^23p^2$		
P	$1s^2$	$2s^22p^6$	$3s^23p^3$		
S	$1s^2$	$2s^22p^6$	$3s^23p^4$		
Cl	$1s^2$	$2s^22p^6$	$3s^23p^5$		
Ar	$1s^2$	$2s^22p^6$	$3s^23p^6$		
K	$1s^2$	$2s^22p^6$	$3s^23p^6$	$4s^1$	
Ca	$1s^2$	$2s^22p^6$	$3s^23p^6$	$4s^2$	
Sc	$1s^2$	$2s^22p^6$	$3s^23p^6$	$4s^23d^1$	
Ti	$1s^2$	$2s^22p^6$	$3s^23p^6$	$4s^23d^2$	
V	$1s^2$	$2s^22p^6$	$3s^23p^6$	$4s^23d^3$	
Cr	$1s^2$	$2s^22p^6$	$3s^23p^6$	$4s^13d^5$	
Mn	$1s^2$	$2s^22p^6$	$3s^23p^6$	$4s^23d^5$	
Fe	$1s^2$	$2s^22p^6$	$3s^23p^6$	$4s^23d^6$	
Co	$1s^2$	$2s^22p^6$	$3s^23p^6$	$4s^23d^7$	
Ni	$1s^2$	$2s^22p^6$	$3s^23p^6$	$4s^23d^8$	
Cu	$1s^2$	$2s^22p^6$	$3s^23p^6$	$4s^13d^{10}$	
Zn	$1s^2$	$2s^22p^6$	$3s^23p^6$	$4s^23d^{10}$	
Ga	$1s^2$	$2s^22p^6$	$3s^23p^6$	$4s^23d^{10}4p^1$	
Ge	$1s^2$	$2s^22p^6$	$3s^23p^6$	$4s^23d^{10}4p^2$	
As	$1s^2$	$2s^22p^6$	$3s^23p^6$	$4s^23d^{10}4p^3$	
Se	$1s^2$	$2s^22p^6$	$3s^23p^6$	$4s^23d^{10}4p^4$	
Br	$1s^2$	$2s^22p^6$	$3s^23p^6$	$4s^23d^{10}4p^5$	
Kr	$1s^2$	$2s^22p^6$	$3s^23p^6$	$4s^23d^{10}4p^6$	
Rb	$1s^2$	$2s^22p^6$	$3s^23p^6$	$4s^23d^{10}4p^6$	$5s^1$
Sr	$1s^2$	$2s^22p^6$	$3s^23p^6$	$4s^23d^{10}4p^6$	$5s^2$

Y	$1s^2$	$2s^2 2p^6$	$3s^2 3p^6$	$4s^2 3d^{10} 4p^6$	$5s^2 4d^1$	
Zr	$1s^2$	$2s^2 2p^6$	$3s^2 3p^6$	$4s^2 3d^{10} 4p^6$	$5s^2 4d^2$	
Nb	$1s^2$	$2s^2 2p^6$	$3s^2 3p^6$	$4s^2 3d^{10} 4p^6$	$5s^1 4d^4$	
Mo	$1s^2$	$2s^2 2p^6$	$3s^2 3p^6$	$4s^2 3d^{10} 4p^6$	$5s^1 4d^5$	
Tc	$1s^2$	$2s^2 2p^6$	$3s^2 3p^6$	$4s^2 3d^{10} 4p^6$	$5s^2 4d^5$	
Ru	$1s^2$	$2s^2 2p^6$	$3s^2 3p^6$	$4s^2 3d^{10} 4p^6$	$5s^1 4d^7$	
Rh	$1s^2$	$2s^2 2p^6$	$3s^2 3p^6$	$4s^2 3d^{10} 4p^6$	$5s^1 4d^8$	
Pd	$1s^2$	$2s^2 2p^6$	$3s^2 3p^6$	$4s^2 3d^{10} 4p^6$	$5s^1 4d^{10}$	
Ag	$1s^2$	$2s^2 2p^6$	$3s^2 3p^6$	$4s^2 3d^{10} 4p^6$	$5s^1 4d^{10}$	
Cd	$1s^2$	$2s^2 2p^6$	$3s^2 3p^6$	$4s^2 3d^{10} 4p^6$	$5s^2 4d^{10}$	
In	$1s^2$	$2s^2 2p^6$	$3s^2 3p^6$	$4s^2 3d^{10} 4p^6$	$5s^2 4d^{10} 5p^1$	
Sn	$1s^2$	$2s^2 2p^6$	$3s^2 3p^6$	$4s^2 3d^{10} 4p^6$	$5s^2 4d^{10} 5p^2$	
Sb	$1s^2$	$2s^2 2p^6$	$3s^2 3p^6$	$4s^2 3d^{10} 4p^6$	$5s^2 4d^{10} 5p^3$	
Te	$1s^2$	$2s^2 2p^6$	$3s^2 3p^6$	$4s^2 3d^{10} 4p^6$	$5s^2 4d^{10} 5p^4$	
I	$1s^2$	$2s^2 2p^6$	$3s^2 3p^6$	$4s^2 3d^{10} 4p^6$	$5s^2 4d^{10} 5p^5$	
Xe	$1s^2$	$2s^2 2p^6$	$3s^2 3p^6$	$4s^2 3d^{10} 4p^6$	$5s^2 4d^{10} 5p^6$	
Cs	$1s^2$	$2s^2 2p^6$	$3s^2 3p^6$	$4s^2 3d^{10} 4p^6$	$5s^2 4d^{10} 5p^6$	$6s^2$
Ba	$1s^2$	$2s^2 2p^6$	$3s^2 3p^6$	$4s^2 3d^{10} 4p^6$	$5s^2 4d^{10} 5p^6$	$6s^2$
La	$1s^2$	$2s^2 2p^6$	$3s^2 3p^6$	$4s^2 3d^{10} 4p^6$	$5s^2 4d^{10} 5p^6$	$6s^2 4f^1$
Ce	$1s^2$	$2s^2 2p^6$	$3s^2 3p^6$	$4s^2 3d^{10} 4p^6$	$5s^2 4d^{10} 5p^6$	$6s^2 4f^2$
Pr	$1s^2$	$2s^2 2p^6$	$3s^2 3p^6$	$4s^2 3d^{10} 4p^6$	$5s^2 4d^{10} 5p^6$	$6s^2 4f^3$
Nd	$1s^2$	$2s^2 2p^6$	$3s^2 3p^6$	$4s^2 3d^{10} 4p^6$	$5s^2 4d^{10} 5p^6$	$6s^2 4f^4$
Pm	$1s^2$	$2s^2 2p^6$	$3s^2 3p^6$	$4s^2 3d^{10} 4p^6$	$5s^2 4d^{10} 5p^6$	$6s^2 4f^5$
Sm	$1s^2$	$2s^2 2p^6$	$3s^2 3p^6$	$4s^2 3d^{10} 4p^6$	$5s^2 4d^{10} 5p^6$	$6s^2 4f^6$
Eu	$1s^2$	$2s^2 2p^6$	$3s^2 3p^6$	$4s^2 3d^{10} 4p^6$	$5s^2 4d^{10} 5p^6$	$6s^2 4f^7$
Gd	$1s^2$	$2s^2 2p^6$	$3s^2 3p^6$	$4s^2 3d^{10} 4p^6$	$5s^2 4d^{10} 5p^6$	$6s^2 4f^7 5d^1$
Tb	$1s^2$	$2s^2 2p^6$	$3s^2 3p^6$	$4s^2 3d^{10} 4p^6$	$5s^2 4d^{10} 5p^6$	$6s^2 4f^9$
Dy	$1s^2$	$2s^2 2p^6$	$3s^2 3p^6$	$4s^2 3d^{10} 4p^6$	$5s^2 4d^{10} 5p^6$	$6s^2 4f^{10}$
Ho	$1s^2$	$2s^2 2p^6$	$3s^2 3p^6$	$4s^2 3d^{10} 4p^6$	$5s^2 4d^{10} 5p^6$	$6s^2 4f^{11}$
Er	$1s^2$	$2s^2 2p^6$	$3s^2 3p^6$	$4s^2 3d^{10} 4p^6$	$5s^2 4d^{10} 5p^6$	$6s^2 4f^{12}$
Tm	$1s^2$	$2s^2 2p^6$	$3s^2 3p^6$	$4s^2 3d^{10} 4p^6$	$5s^2 4d^{10} 5p^6$	$6s^2 4f^{13}$
Yb	$1s^2$	$2s^2 2p^6$	$3s^2 3p^6$	$4s^2 3d^{10} 4p^6$	$5s^2 4d^{10} 5p^6$	$6s^2 4f^{14}$
Lu	$1s^2$	$2s^2 2p^6$	$3s^2 3p^6$	$4s^2 3d^{10} 4p^6$	$5s^2 4d^{10} 5p^6$	$6s^2 4f^{14} 5d^1$
Hf	$1s^2$	$2s^2 2p^6$	$3s^2 3p^6$	$4s^2 3d^{10} 4p^6$	$5s^2 4d^{10} 5p^6$	$6s^2 4f^{14} 5d^2$
Ta	$1s^2$	$2s^2 2p^6$	$3s^2 3p^6$	$4s^2 3d^{10} 4p^6$	$5s^2 4d^{10} 5p^6$	$6s^2 4f^{14} 5d^3$
W	$1s^2$	$2s^2 2p^6$	$3s^2 3p^6$	$4s^2 3d^{10} 4p^6$	$5s^2 4d^{10} 5p^6$	$6s^2 4f^{14} 5d^4$
Re	$1s^2$	$2s^2 2p^6$	$3s^2 3p^6$	$4s^2 3d^{10} 4p^6$	$5s^2 4d^{10} 5p^6$	$6s^2 4f^{14} 5d^5$
Os	$1s^2$	$2s^2 2p^6$	$3s^2 3p^6$	$4s^2 3d^{10} 4p^6$	$5s^2 4d^{10} 5p^6$	$6s^2 4f^{14} 5d^6$
Ir	$1s^2$	$2s^2 2p^6$	$3s^2 3p^6$	$4s^2 3d^{10} 4p^6$	$5s^2 4d^{10} 5p^6$	$6s^2 4f^{14} 5d^7$
Pt	$1s^2$	$2s^2 2p^6$	$3s^2 3p^6$	$4s^2 3d^{10} 4p^6$	$5s^2 4d^{10} 5p^6$	$6s^1 4f^{14} 5d^9$
Au	$1s^2$	$2s^2 2p^6$	$3s^2 3p^6$	$4s^2 3d^{10} 4p^6$	$5s^2 4d^{10} 5p^6$	$6s^1 4f^{14} 5d^{10}$
Hg	$1s^2$	$2s^2 2p^6$	$3s^2 3p^6$	$4s^2 3d^{10} 4p^6$	$5s^2 4d^{10} 5p^6$	$6s^2 4f^{14} 5d^{10}$
Tl	$1s^2$	$2s^2 2p^6$	$3s^2 3p^6$	$4s^2 3d^{10} 4p^6$	$5s^2 4d^{10} 5p^6$	$6s^2 4f^{14} 5d^{10} 6p^1$
Pb	$1s^2$	$2s^2 2p^6$	$3s^2 3p^6$	$4s^2 3d^{10} 4p^6$	$5s^2 4d^{10} 5p^6$	$6s^2 4f^{14} 5d^{10} 6p^2$
Bi	$1s^2$	$2s^2 2p^6$	$3s^2 3p^6$	$4s^2 3d^{10} 4p^6$	$5s^2 4d^{10} 5p^6$	$6s^2 4f^{14} 5d^{10} 6p^3$
Po	$1s^2$	$2s^2 2p^6$	$3s^2 3p^6$	$4s^2 3d^{10} 4p^6$	$5s^2 4d^{10} 5p^6$	$6s^2 4f^{14} 5d^{10} 6p^4$
At	$1s^2$	$2s^2 2p^6$	$3s^2 3p^6$	$4s^2 3d^{10} 4p^6$	$5s^2 4d^{10} 5p^6$	$6s^2 4f^{14} 5d^{10} 6p^5$

Rn	$1s^2$	$2s^22p^6$	$3s^23p^6$	$4s^23d^{10}4p^6$	$5s^24d^{10}5p^6$	$6s^24f^{14}5d^{10}6p^6$	
Fr	$1s^2$	$2s^22p^6$	$3s^23p^6$	$4s^23d^{10}4p^6$	$5s^24d^{10}5p^6$	$6s^24f^{14}5d^{10}6p^6$	$7s^1$
Ra	$1s^2$	$2s^22p^6$	$3s^23p^6$	$4s^23d^{10}4p^6$	$5s^24d^{10}5p^6$	$6s^24f^{14}5d^{10}6p^6$	$7s^2$
Ac	$1s^2$	$2s^22p^6$	$3s^23p^6$	$4s^23d^{10}4p^6$	$5s^24d^{10}5p^6$	$6s^24f^{14}5d^{10}6p^6$	$7s^26d^1$
Th	$1s^2$	$2s^22p^6$	$3s^23p^6$	$4s^23d^{10}4p^6$	$5s^24d^{10}5p^6$	$6s^24f^{14}5d^{10}6p^6$	$7s^26d^2$
Pa	$1s^2$	$2s^22p^6$	$3s^23p^6$	$4s^23d^{10}4p^6$	$5s^24d^{10}5p^6$	$6s^24f^{14}5d^{10}6p^6$	$7s^25f^26d^1$
U	$1s^2$	$2s^22p^6$	$3s^23p^6$	$4s^23d^{10}4p^6$	$5s^24d^{10}5p^6$	$6s^24f^{14}5d^{10}6p^6$	$7s^25f^36d^1$
Np	$1s^2$	$2s^22p^6$	$3s^23p^6$	$4s^23d^{10}4p^6$	$5s^24d^{10}5p^6$	$6s^24f^{14}5d^{10}6p^6$	$7s^25f^46d^1$
Pu	$1s^2$	$2s^22p^6$	$3s^23p^6$	$4s^23d^{10}4p^6$	$5s^24d^{10}5p^6$	$6s^24f^{14}5d^{10}6p^6$	$7s^25f^6$
Am	$1s^2$	$2s^22p^6$	$3s^23p^6$	$4s^23d^{10}4p^6$	$5s^24d^{10}5p^6$	$6s^24f^{14}5d^{10}6p^6$	$7s^25f^7$
Cm	$1s^2$	$2s^22p^6$	$3s^23p^6$	$4s^23d^{10}4p^6$	$5s^24d^{10}5p^6$	$6s^24f^{14}5d^{10}6p^6$	$7s^25f^76d^1$
Bk	$1s^2$	$2s^22p^6$	$3s^23p^6$	$4s^23d^{10}4p^6$	$5s^24d^{10}5p^6$	$6s^24f^{14}5d^{10}6p^6$	$7s^25f^9$
Cf	$1s^2$	$2s^22p^6$	$3s^23p^6$	$4s^23d^{10}4p^6$	$5s^24d^{10}5p^6$	$6s^24f^{14}5d^{10}6p^6$	$7s^25f^{10}$
Es	$1s^2$	$2s^22p^6$	$3s^23p^6$	$4s^23d^{10}4p^6$	$5s^24d^{10}5p^6$	$6s^24f^{14}5d^{10}6p^6$	$7s^25f^{11}$
Fm	$1s^2$	$2s^22p^6$	$3s^23p^6$	$4s^23d^{10}4p^6$	$5s^24d^{10}5p^6$	$6s^24f^{14}5d^{10}6p^6$	$7s^25f^{12}$
Md	$1s^2$	$2s^22p^6$	$3s^23p^6$	$4s^23d^{10}4p^6$	$5s^24d^{10}5p^6$	$6s^24f^{14}5d^{10}6p^6$	$7s^25f^{13}$
No	$1s^2$	$2s^22p^6$	$3s^23p^6$	$4s^23d^{10}4p^6$	$5s^24d^{10}5p^6$	$6s^24f^{14}5d^{10}6p^6$	$7s^25f^{14}$
Lr	$1s^2$	$2s^22p^6$	$3s^23p^6$	$4s^23d^{10}4p^6$	$5s^24d^{10}5p^6$	$6s^24f^{14}5d^{10}6p^6$	$7s^25f^{14}6d^1$

Appendix 3

Thermodynamic Data for Selected Elements, Compounds, and Ions (25°C)

Substance	ΔH_f° (kJ mol^{-1})	S° (J mol^{-1} K^{-1})	ΔG_f° (kJ mol^{-1})
aluminum			
$Al(s)$	0	28.3	0
$AlCl_3(s)$	-704	110.7	-629
$Al_2O_3(s)$	-1676	51.0	-1576.4
$Al_2(SO_4)_3(s)$	-3441	239	-3100
barium			
$Ba(s)$	0	66.9	0
$BaCO_3(s)$	-1219	112	-1139
$BaCl_2(s)$	-860.2	125	-810.8
$Ba(OH)_2(s)$	-998.22	-8	-875.3
$Ba(NO_3)_2(s)$	-992	214	-795
$BaSO_4(s)$	-1465	132	-1353
bromine			
$Br_2(l)$	0	152.2	0
$Br_2(g)$	+30.9	245.4	3.11
$HBr(g)$	-36	198.5	53.1
calcium			
$Ca(s)$	0	41.4	0
$CaCO_3(s)$	-1207	92.9	-1128.8
$CaF_2(s)$	-741	80.3	-1166
$CaCl_2(s)$	-795.8	114	-750.2
$CaO(s)$	-635.5	40	-604.2
$Ca(OH)_2(s)$	-986.6	76.1	896.76
$CaSO_4(s)$	-1433	107	-1320.3

Substance	ΔH_f° (kJ mol^{-1})	S° (J mol^{-1} K^{-1})	ΔG_f° (kJ mol^{-1})
carbon			
C(*s*, graphite)	0	5.69	0
C(*s*, diamond)	+1.88	2.4	+2.9
$CCL_4(l)$	-134	214.4	-65.3
$CO(g)$	-110	197.9	-137.3
$CO_2(g)$	-394	213.6	-394.4
$CO_2(aq)$	-413.8	117.6	-385.98
$H_2CO_3(aq)$	-699.65	187.4	-623.08
$HCO_3^-(aq)$	-691.99	91.2	-586.77
$CO^{2-}_3(aq)$	-677.14	-56.9	527.81
$HCN(g)$	+135.1	201.7	+124.7
$CN^-(aq)$	+150.6	94.1	+172.4
$CH_4(g)$	-74.9	186.2	-50.79
$C_2H_2(g)$	+227	200.8	+209
$C_2H_4(g)$	+51.9	219.8	+68.12
$C_2H_6(g)$	-84.5	229.5	-32.9
$C_2H_8(g)$	-104	269.9	-23
$C_2H_{10}(g)$	-126	310.2	-17.0
$C_6H_6(l)$	+49.0	173.3	+124.3
$CH_3OH(l)$	-238	126.8	-166.2
$C_2H_5OH(l)$	-278	161	-174.8
$HCHO_2(g)$	-363	251	+335
$HC_3H_3O_2(l)$	-487.0	160	-392.5
$HCHO(g)$	-108.6	218.8	-102.5
$CH_3CHO(g)$	-167	250	-129
$(CH_3)_3CO(l)$	-248.1	200.4	-155.4
$C_6H_5CO_2H(s)$	-385.1	167.6	-245.3
chlorine			
$Cl_2(g)$	0	223.0	0
$HCl(g)$	-92.5	186.7	-95.27
$HCl(aq)$	-167.2	56.5	-131.2
$HClO(aq)$	-131.3	106.8	-80.21

Substance	ΔH_f° (kJ mol^{-1})	S° (J mol^{-1} K^{-1})	ΔG_f° (kJ mol^{-1})
chromium			
$Cr(s)$	0	23.8	0
$CrCl_2(s)$	-326	115	-282
$CrCl_3(s)$	-563.2	126	-493.7
copper			
$Cu(s)$	0	33.15	0
$CuCl(s)$	-137.2	86.2	-119.87
$CuCl_2(s)$	-172	119	-131
$CuSO_4(s)$	-771.4	109	-661.8
fluorine			
$F_2(g)$	0	202.7	0
$F^-(aq)$	-332.6	13.8	-278.8
$HF(g)$	-271	173.5	-273
hydrogen			
$H_2(g)$	0	130.6	0
$H_2O(l)$	-286	69.96	-237.2
$H_2O(g)$	-242	188.7	-228.6
iron			
$Fe(s)$	0	27	0
$Fe_2O_3(s)$	-822.2	90.0	-741.0
$Fe_3O_4(s)$	-1118.4	146.4	-1015.4
lead			
$Pb(s)$	0	64.8	0
$PbCl_2(s)$	-359.4	136	-314.1
$PbS(s)$	-100	91.2	-98.7
$PbSO_4(s)$	-920.1	149	-811.3
lithium			
$Li(s)$	0	28.4	0
$LiF(s)$	-611.7	35.7	-583.3
$LiCl(s)$	-408	59.29	-383.7

Substance	ΔH_f° (kJ mol^{-1})	S° (J mol^{-1} K^{-1})	ΔG_f° (kJ mol^{-1})
magnesium			
$Mg(s)$	0	32.5	0
$MgCO_3(s)$	-1113	65.7	-1029
$MgF_2(s)$	-1113	79.9	-1056
$MgCl_2(s)$	-641.8	89.5	-592.5
$MgO(s)$	-601.7	26.9	-569.4
$Mg(OH)_2(s)$	-924.7	63.1	-833.9
manganese			
$Mn(s)$	0	32.0	0
$MnO_4^-(aq)$	-542.7	191	-449.4
$KMnO_4(s)$	-813.4	171.71	-713.8
$MnO_2(s)$	-520.9	53.1	-466.1
nitrogen			
$N_2(g)$	0	191.5	0
$NH_3(g)$	-46.0	192.5	-16.7
$NH_4Cl(s)$	-314.4	94.6	-203.9
$NO(g)$	+90.4	210.6	+86.69
$NO_2(g)$	+34	240.5	+51.84
$N_2O(g)$	+81.5	220.0	+103.6
$HNO_3(l)$	-174.1	155.6	-79.9
oxygen			
$O_2(g)$	0	205.0	0
$O_3(g)$	+143	238.8	+163
$OH^-(aq)$	-230.0	-10.75	-157.24
potassium			
$K(s)$	0	64.18	0
$KF(s)$	-567.3	66.6	-537.8
$KCl(s)$	-436.8	82.59	-408.3
$KOH(s)$	-424.8	78.9	-379.1
$K_2SO_4(s)$	-1433.7	176	-1316.4

Substance	ΔH_f° (kJ mol^{-1})	S° (J mol^{-1} K^{-1})	ΔG_f° (kJ mol^{-1})
silver			
$Ag(s)$	0	42.55	0
$AgCl(s)$	-127.1	96.2	-109.8
$AgNO_3(s)$	-124	141	-32
sodium			
$Na(s)$	0	51.0	0
$NaF(s)$	-571	51.5	-545
$NaCl(s)$	-413	72.38	-384.0
$NaOH(s)$	-426.8	64.18	-382
$Na_2SO_4(s)$	-1384.49	149.49	-1266.83
sulfur			
$S(s$, rhombic)	0	31.8	0
$SO_2(g)$	-297	248	-300
$SO_3(g)$	-396	256	-370
$H_2SO_4(aq)$	-909.3	20.1	-744.5
$SF_6(g)$	-1209	292	-1105
tin			
$Sn(s$, white)	0	51.6	0
$SnCl_4(l)$	-511.3	258.6	-440.2
zinc			
$Zn(s)$	0	41.6	0
$ZnCl_2(s)$	-415.1	111	-369.4

Appendix 4

Ionization Constants of Weak Acids

Monoprotic Acid	Name	K_a
HIO_3	iodic acid	1.69×10^{-1}
HNO_2	nitrous acid	7.1×10^{-4}
HF	hydrofluoric acid	6.8×10^{-4}
$HCHO_2$	formic acid	1.8×10^{-4}
$HC_3H_5O_3$	lactic acid	1.38×10^{-4}
$HC_7H_5O_2$	benzoic acid	6.28×10^{-5}
$HC_4H_7O_2$	butanoic acid	1.52×10^{-5}
HN_3	hydrazoic acid	1.8×10^{-5}
$HC_2H_3O_2$	ethanoic acid	1.8×10^{-5}
$HC_3H_5O_2$	propanoic acid	1.34×10^{-5}
HOCl	hypochlorous acid	3.0×10^{-8}
HCN	hydrocyanic acid	6.2×10^{-10}
HC_6H_5O	phenol	1.3×10^{-10}
HOI	hypoiodous acid	2.3×10^{-11}
H_2O_2	hydrogen peroxide	1.8×10^{-12}

Appendix 5

Ionization Constants of Polyprotic Acids

Polyprotic Acid	Name	K_{a1}	K_{a1}	K_{a3}
H_2SO_4	sulfuric acid	Large	1.0×10^{-2}	
H_2CrO_4	chromic acid	5.0	1.5×10^{-6}	
$H_2C_2O_4$	oxalic acid	5.6×10^{-2}	5.4×10^{-5}	
H_3PO_3	phosphorous acid	3×10^{-2}	1.6×10^{-7}	
H_2SO_3	sulfurous acid	1.2×10^{-2}	6.6×10^{-8}	
H_2SeO_3	selenous acid	4.5×10^{-3}	1.1×10^{-8}	
$H_2C_3H_2O_4$	malonic acid	1.4×10^{-3}	2.0×10^{-6}	
$H_2C_8H_4O_4$	phthalic acid	1.1×10^{-3}	3.9×10^{-6}	
$H_2C_4H_4O_6$	tartaric acid	9.2×10^{-4}	4.3×10^{-5}	
H_2CO_3	carbonic acid	4.5×10^{-7}	4.7×10^{-11}	
H_3PO_4	phosphoric acid	7.1×10^{-3}	6.3×10^{-8}	4.5×10^{-13}
H_3AsO_4	arsenic acid	5.6×10^{-3}	1.7×10^{-7}	4.0×10^{-12}
$H_3C_6H_5O_7$	citric acid	7.1×10^{-4}	1.7×10^{-5}	6.3×10^{-6}

Appendix 6

Ionization Constants of Weak Bases

Weak Base	Name	K_b
$(CH_3)_2NH$	dimethylamine	9.6×10^{-4}
CH_3NH_2	methylamine	4.4×10^{-4}
$CH_3CH_2NH_2$	ethylamine	4.3×10^{-4}
$(CH_3)_3N$	trimethylamine	7.4×10^{-5}
NH_3	ammonia	1.8×10^{-5}
N_2H_4	hydrazine	9.6×10^{-7}
C_5H_5N	pyridine	1.5×10^{-9}
$C_6H_5NH_2$	aniline	4.1×10^{-10}

Appendix 7

Solubility Product Constants

Salt	Dissolution Reaction	K_{sp}
fluorides		
MgF_2	$MgF_2(s) \rightleftharpoons Mg^{2+}(aq) + 2F^-(aq)$	6.6×10^{-9}
CaF_2	$CaF_2(s) \rightleftharpoons Ca^{2+}(aq) + 2F^-(aq)$	3.9×10^{-11}
BaF_2	$BaF_2(s) \rightleftharpoons Ba^{2+}(aq) + 2F^-(aq)$	1.7×10^{-6}
PbF_2	$PbF_2(s) \rightleftharpoons Pb^{2+}(aq) + 2F^-(aq)$	3.6×10^{-8}
chlorides		
CuCl	$CuCl(s) \rightleftharpoons Cu^+(aq) + Cl^-(aq)$	1.9×10^{-7}
AgCl	$AgCl(s) \rightleftharpoons Ag^+(aq) + Cl^-(aq)$	1.8×10^{-10}
$PbCl_2$	$PbCl_2(s) \rightleftharpoons Pb^{2+}(aq) + 2Cl^-(aq)$	1.7×10^{-5}
bromides		
CuBr	$CuBr(s) \rightleftharpoons Cu^+(aq) + Br^-(aq)$	5×10^{-9}
AgBr	$AgBr(s) \rightleftharpoons Ag^+(aq) + Br^-(aq)$	5.4×10^{-13}
$PbBr_2$	$PbBr_2(s) \rightleftharpoons Pb^{2+}(aq) + 2Br^-(aq)$	2.1×10^{-6}
iodides		
CuI	$CuI(s) \rightleftharpoons Cu^+(aq) + I^-(aq)$	1×10^{-12}
AgI	$AgI(s) \rightleftharpoons Ag^+(aq) + I^-(aq)$	8.3×10^{-17}
PbI_2	$PbI_2(s) \rightleftharpoons Pb^{2+}(aq) + 2I^-(aq)$	7.9×10^{-9}
hydroxides		
$Mg(OH)_2$	$Mg(OH)_2(s) \rightleftharpoons Mg^{2+}(aq) + 2OH^-(aq)$	7.1×10^{-12}
$Ca(OH)_2$	$Ca(OH)_2(s) \rightleftharpoons Ca^{2+}(aq) + 2OH^-(aq)$	6.5×10^{-6}
$Mn(OH)_2$	$MN(OH)_2(s) \rightleftharpoons Mn^{2+}(aq) + 2OH^-(aq)$	1.6×10^{-13}
$Fe(OH)_2$	$Fe(OH)_2(s) \rightleftharpoons Fe^{2+}(aq) + 2OH^-(aq)$	7.9×10^{-16}
$Fe(OH)_3$	$Fe(OH)_3(s) \rightleftharpoons Fe^{3+}(aq) + 3OH^-(aq)$	1.6×10^{-39}
$Ni(OH)_2$	$Ni(OH)_2(s) \rightleftharpoons Ni^{2+}(aq) + 2OH^-(aq)$	6×10^{-16}
$Cu(OH)_2$	$Cu(OH)_2(s) \rightleftharpoons Cu^{2+}(aq) + 2OH^-(aq)$	4.8×10^{-20}
$Cr(OH)_3$	$Cr(OH)_3(s) \rightleftharpoons Cr^{3+}(aq) + 3OH^-(aq)$	2×10^{-30}
$Zn(OH)_2$	$Zn(OH)_2(s) \rightleftharpoons Zn^{2+}(aq) + 2OH^-(aq)$	3.0×10^{-16}
$Cd(OH)_2$	$Cd(OH)_2(s) \rightleftharpoons Cd^{2+}(aq) + 2OH^-(aq)$	5.0×10^{-15}

Salt	Dissolution Reaction	K_{sp}
sulfites		
$CaSO_3$	$CaSO_3(s) \rightleftharpoons Ca^{2+}(aq) + SO_3^{2-}(aq)$	3×10^{-7}
$BaSO_3$	$BaSO_4(s) \rightleftharpoons Ba^{2+}(aq) + SO_3^{2-}(aq)$	8×10^{-7}
sulfates		
$CaSO_4$	$CaSO_4(s) \rightleftharpoons Ca^{2+}(aq) + SO_4^{2-}(aq)$	2.4×10^{-5}
$BaSO_4$	$BaSO_4(s) \rightleftharpoons Ba^{2+}(aq) + SO_4^{2-}(aq)$	1.1×10^{-10}
Ag_2SO_4	$Ag_2SO_4(s) \rightleftharpoons 2Ag^{+}(aq) + SO_4^{2-}(aq)$	1.5×10^{-5}
$PbSO_4$	$PbSO_4(s) \rightleftharpoons Pb^{2+}(aq) + SO_4^{2-}(aq)$	6.3×10^{-7}
chromates		
Ag_2CrO_4	$Ag_2CrO_4(s) \rightleftharpoons 2Ag^{+}(aq) + CrO_4^{2-}(aq)$	1.2×10^{-12}
Hg_2CrO_4	$Hg_2CrO_4(s) \rightleftharpoons Hg_2^{2+}(aq) + CrO_4^{2-}(aq)$	2.0×10^{-9}
$PbCrO_4$	$PbCrO_4(s) \rightleftharpoons Pb^{2-}(aq) + CrO_4^{2-}(aq)$	1.8×10^{-14}
carbonates		
$MgCO_3$	$MgCO_3(s) \rightleftharpoons Mg^{2+}(aq) + CO_3^{2-}(aq)$	3.5×10^{-8}
$CaCO_3$	$CaCO_3(s) \rightleftharpoons Ca^{2+}(aq) + CO_3^{2-}(aq)$	9.3×10^{-10}
$BaCO_3$	$BaCO_3(s) \rightleftharpoons Ba^{2+}(aq) + CO_3^{2-}(aq)$	5.0×10^{-9}
$MnCO_3$	$MnCO_3(s) \rightleftharpoons Mn^{2+}(aq) + CO_3^{2-}(aq)$	5.0×10^{-10}
$CuCO_3$	$CuCO_3(s) \rightleftharpoons Cu^{2+}(aq) + CO_3^{2-}(aq)$	2.3×10^{-10}
Ag_2CO_3	$Ag_2CO_3(s) \rightleftharpoons 2Ag^{2+}(aq) + CO_3^{2-}(aq)$	8.1×10^{-12}
Hg_2CO_3	$Hg_2CO_3(s) \rightleftharpoons Hg_2{}^{2+}(aq) + CO_3^{2-}(aq)$	8.9×10^{-17}
$ZnCO_3$	$ZnCO_3(s) \rightleftharpoons Zn^{2+}(aq) + CO_3^{2-}(aq)$	1.0×10^{-10}
$PbCO_3$	$PbCO_3(s) \rightleftharpoons Pb^{2+}(aq) + CO_3^{2-}(aq)$	7.4×10^{-14}
sulfides		
MnS	$MnS(s) \rightleftharpoons Mn^{2+}(aq) + S^{2-}(aq)$	3.0×10^{-11}
FeS	$FeS(s) \rightleftharpoons Fe^{2+}(aq) + S^{2-}(aq)$	8.0×10^{-19}
CoS	$CoS(s) \rightleftharpoons Co^{2+}(aq) + S^{2-}(aq)$	5.0×10^{-22}
NiS	$NiS(s) \rightleftharpoons Ni^{2+}(aq) + S^{2-}(aq)$	4.0×10^{-20}
CuS	$CuS(s) \rightleftharpoons Cu^{2+}(aq) + S^{2-}(aq)$	8.0×10^{-37}
Cu_2S	$Cu_2S(s) \rightleftharpoons 2Cu^{2+}(aq) + S^{2-}(aq)$	3.0×10^{-49}
Ag_2S	$Ag_2S(s) \rightleftharpoons 2Ag^{+}(aq) + S^{2-}(aq)$	8.0×10^{-51}
Tl_2S	$Tl_2S(s) \rightleftharpoons 2Tl^{+}(aq) + S^{2-}(aq)$	6.0×10^{-22}
ZnS	$ZnS(s) \rightleftharpoons Zn^{2+}(aq) + S^{2-}(aq)$	2.0×10^{-25}
CdS	$CdS(s) \rightleftharpoons Cd^{2+}(aq) + S^{2-}(aq)$	1.0×10^{-27}
HgS	$HgS(s) \rightleftharpoons Hg^{2+}(aq) + S^{2-}(aq)$	2.0×10^{-53}
SnS	$SnS(s) \rightleftharpoons Sn^{2+}(aq) + S^{2-}(aq)$	1.3×10^{-26}
PbS	$PbS(s) \rightleftharpoons Pb^{2+}(aq) + S^{2-}(aq)$	3.0×10^{-28}
In_2S_3	$In_2S_3(s) \rightleftharpoons 2In^{3+}(aq) + 3S^{2-}(aq)$	4.0×10^{-70}

Glossary

A

absolute uncertainty The uncertainty of ±1 in the last digit of a measurement. If this uncertainty is different from ±1, it is written as part of the number; for example, 23.45 ± 0.05 indicates an uncertainty of ±5 in the last digit.

absolute zero The lowest possible temperature, 0.0 K or –273.16°C.

absorbance, *A* A measure of the amount of light absorbed by a chemical.

absorptivity, *a* A constant the value of which depends on the sample and the wavelength at which the measurement is made in spectroscopy.

accuracy The degree of closeness between a measured value and the true value.

acid Any substance that donates protons, or as Lewis acids are electron-pair acceptors.

acid anhydride The oxide of a nonmetal that forms an acid when dissolved in water.

acid dissociation constant, K_a The value of the equilibrium expression for the dissociation of a weak acid.

activated complex The structures of colliding molecules at the moment of collision, generally thought to be intermediate between the structures of the products and of the reactants.

activation energy The increase in potential energy, due to a molecular collision, necessary to convert a reactant into a product.

activity series A listing of elemental substances in the order of their ability to be oxidized or reduced. This listing makes it possible to predict whether an element will cause the oxidation or the reduction of an ion of another element.

addition reaction The reaction in which a double bond opens to form two additional single bonds.

adhesive force The attractive force between two dissimilar substances.

alcohol An organic compound with an –OH group.

aldehyde An organic compound with a terminal –CHO group.

alkali metals The extremely reactive elements in the first group (column) of the periodic table. They all have ns^1 electrons as valence electrons.

alkaline earth metals The very reactive elements in the second group (column) of the periodic table. They all have ns^2 electrons as valence electrons.

alkanes Organic compounds with the general formula C_nH_{2n+2}.

alkenes Organic compounds with double bonds in their structures.

alkyl group A functional group that is alkane in nature.

alkynes Organic compounds with triple bonds in their structures.

allotrope(s) One or more distinct forms of an element; classification as an allotope is based on structure and physical properties. For example, diamond and graphite are two allotropes of carbon.

alpha particle A helium nucleus.

amines Compounds, related to ammonia, in which one or more of the hydrogen atoms in ammonia have been replaced by organic functional groups.

amino acid An organic acid that contains both an amine and an acid functional group on adjacent carbon atoms.

amorphous A term meaning "without structure."

amphiprotic A term describing a substance that can donate a proton and accept a proton.

amphoteric A term describing a substance that can act as an acid or as a base.

anhydride The oxide of a metal or nonmetal that reacts with water to form an acid or a base, respectively.

anion An ion with a negative charge.

aqueous A term designating a system that involves water or a chemical mixture or solution having water as the solvent.

Arrhenius equation The equation that relates temperature to rate constant. $K = Ae^{-Ea/RT}$.

Arrhenius theory The theory that an acid increases hydrogen ion concentration when dissolved and that a base increases hydroxide ion concentration when dissolved.

aryl group A functional group that is aromatic in nature.

atom A fundamental particle of chemistry. At present, 109 atoms are known and are arranged in an orderly manner in the periodic table.

atomic mass, *A* The relative mass of an element as compared to the mass of the isotope C-12, which is defined as exactly 12.

atomic number, *Z* The number of an element in the periodic table; also, a number representing the number of protons in the nucleus of an atom.

atomic orbital The orbital structure of an element; also, an orbital within an element.

atomic symbol A one- or two-letter abbreviation of an element's name. Some symbols (e.g., Pb for lead) are derived from Latin names of the elements.

autopyrolysis constant of water, K_w The value of the equilibrium expression for the dissociation of water into H^+ and OH^-.

Avogadro's number A quantity equal to 6.02×10^{23}.

Avogadro's principle A statement of the direct relationship between the moles of a gas and the volume of that gas.

axial atom A term used to describe the position of an atom in a covalent molecule of the AX_5 or AX_6 basic structure. The axial atoms are on the vertical axis of the molecule in positions similar to the Earth's North and South Poles.

azimuthal quantum number, ℓ The quantum number that specifies the sublevel in which an electron is located; ℓ may be any number from zero up to $n - 1$. When l is written as its letter equivalent, it represents an orbital where $\ell = 0 = s$; $\ell = 1 = p$; $\ell = 2 = d$; $\ell = 3 = f$.

B

balanced reaction A chemical equation that has the smallest whole-number coefficients for the reactants and products so that there is the same number of atoms of each element on both sides of the arrow.

barometer A closed-end manometer used for measuring atmospheric pressure.

base Any compound that increases the hydroxide concentration of a solution or is a proton acceptor. Lewis bases are electron-pair donors.

base anhydride The oxide of a metal that forms a base when dissolved in water.

base dissociation constant, K_b The value of the equilibrium expression for the dissociation of a weak base.

basic structure One of five basic geometrics—linear, triangular planar, tetrahedral, trigonal planar, or octahedral—that a molecule may take.

battery A galvanic cell used to produce electricity for consumer items such as flashlights, portable radios, and heart pacemakers.

Beer's law The statements that the absorbance of a sample is the product of the absorptivity, optical path length, and sample concentration; *A* = *abc*.

beta particle, β An electron.

bidentate ligand A Lewis base that donates two pairs of electrons.

binary acid An acid that contains hydrogen and one other element in its formula.

body-centered cubic (bcc) A cubic structure in which one atom is at each of the eight corners and one atom is in the center of the unit cell.

Bohr atom The model of the atom developed by Nicls Bohr. This model views electrons as circling the nucleus like a miniature solar system.

boiling point, normal The temperature at which the vapor pressure of a liquid is equal to 760 millimeters of mercury (1.00 atmospheres); also, the temperature at which a gas condenses. Also called condensation point.

boiling-point-elevation constant, k_b The temperature increase of the boiling point per molal noun of solute particles.

bonding electron pair A pair of electrons that participate in a covalent bond.

bond order The average number of bonds per atom covalently bonded to a central atom.

Boyle's law The law that expresses the inverse relationship between the volume and the pressure of a gas; *PV* = constant.

Bragg equation The equation that relates the atomic dimensions in a crystal to the angles at which monochromatic X-rays will undergo constructive reinforcement.

Brønsted-Lowry theory The theory that acids are proton donors and bases are proton acceptors.

buckminsterfullerene The allotrope of carbon that has the formula C_{60}.

buffer capacity The moles of strong acid or strong base required to change the pH of a liter of buffer by l pH unit.

buffer solution An aqueous solution containing a conjugate acid and its conjugate base in a molar ratio greater than 0.1 and less than 10.0.

bumping The violent boiling that occurs when a solution becomes superheated.

buret A tube approximately 1 centimeter in diameter that is used for measuring liquid volumes of 10–100 milliliters.

C

calorimeter An instrument used to determine heat energy.

catalyst A substance that speeds up the rate of a chemical reaction by providing an alternative reaction pathway with a lower activation energy.

cation An ion with a positive charge.

cell voltage, *E* The voltage of a galvanic cell under nonstandard state conditions.

centrifugation The process of separating a solid from a liquid by spinning it rapidly to artificially increase the gravitational force.

chain reaction A nuclear reaction that produces more neutrons than were needed to initiate it, therefore causing more reactions than occurred in the preceding step.

Charles's law The law that expresses the direct relationship between the temperature and volume of a gas: *P*/*V* = constant.

chelate A Lewis base that usually has more than one pair of electrons to donate.

***cis* isomer** An isomer with substituents on the same side of the double bond.

closed system A system in which mass cannot be lost to or gained from the surroundings.

coefficient A number placed in front of a chemical formula to represent the number of molecules of that substance that are included in the equation. This number multiplies the number of atoms in the formula unit.

cohesive force The sum of all the attractive forces in a pure substance.

colligative property Any one of several physical properties of a solution that change depending on the amount of solute particles present in the solution.

collision theory The theory of kinetics, which relates reaction rates to the frequency, energy, and orientation of molecules in collisions.

complex *See* **complex ion**.

complexation reaction A reaction between a Lewis acid and a Lewis base.

complexing agent *See* **Lewis base**.

complex ion A combination of one or more compounds or anions with a metal ion by coordinate covalent bonding. Also called a complex.

compound A combination of two or more elements into a distinct substance with definite physical properties.

concentrated A qualitative term indicating a large amount of solute in a given amount of solvent.

concentration An expression of the amount of solute mixed with a solvent.

concentration vs time curve A graph of reactant or product concentration as a function of time.

condensation The conversion of a gas into a liquid.

condensation point *See* **boiling point**.

condensation reaction A polymerization reaction between an acid and either an alcohol or an amine.

conjugate acid Any substance that has a proton that may be donated to a base.

conjugate acid–base pair A pair of substances whose formulas differ by one H^+ ion.

conjugated double bonds A series of two or more double bonds, each separated by only a single bond in a molecule.

cooling curve A graph showing the changes that occur while a substance is cooling.

coordinate covalent bond The covalent bond between two atoms that is formed when one substance donates both electrons.

covalent bond The bond between two atoms that arises from the sharing of a pair of electrons.

covalent compound A chemical compound in which the atoms are held together with covalent bonds.

covalent crystal A crystal that consists of only one molecule. All atoms are joined to others with covalent bonds. Also called network crystal.

critical mass The minimum mass of a fissile (fissionable) material needed to sustain a nuclear chain reaction.

critical point The temperature and pressure above which a gas cannot be condensed to a liquid.

crystal lattice The arrangement of atoms, ions, or molecules in a crystal structure.

crystallization point *See* **melting point**.

cyclotron A machine for accelerating charged particles, which are usually used to bombard targets in an effort to generate nuclear reactions.

D

Dalton's law of partial pressures The law that the total pressure of a gas is the sum of the individual pressures of all the gases in the mixture; $P_{total} = p_1 + p_2 + \ldots$.

decay The spontaneous emission of a particle in a radioactive event.

ΔE The energy change due to a chemical reaction. It is equal to the heat and work of the reaction or to the change in potential energy of the products as compared to that of the reactants.

$\Delta G°$ The standard free-energy change for a reaction. The temperature is 298 K unless otherwise indicated.

$\Delta G°_f$ The standard free-energy change of formation corresponding to the formation of 1 mole of product from its elements at 298 K.

ΔH The enthalpy change occurring in a chemical process. Without the superscript zero it indicates an extensive property. Often called the heat of reaction.

$\Delta H°$ The standard enthalpy change occurring in a reaction; refers to the heat produced or absorbed when the moles of reactants specified in the chemical reaction react at standard state.

$\Delta H°_f$ The standard heat of formation, which is the heat produced or absorbed when 1 mole of a product is formed.

ΔS The change in entropy between the final state and the initial state in a chemical process. Without the superscript zero it indicates an extensive property.

$\Delta S°$ The standard entropy change of a chemical process; an intensive property based on the moles of substance in the balanced chemical reaction.

dependent variable The variable that an experiment measures; its value depends on the value of the independent variable.

derived structure A molecular structure that is derived from one of the five basic structures.

detergent A chemical substance that has both polar and nonpolar properties and is soluble in both polar and nonpolar solvents.

determinant error An error associated with faulty instruments, calibrations, or techniques.

dextrorotatory A term describing an optical isomer that rotates polarized light to the right.

diagonal relationship A term indicating that two elements in periods 2A and 3A and two groups, x and $x + 1$, respectively, have similar properties; for example, Li and Mg or B and Si.

diagonal trend A term describing the fact that some properties of atoms vary regularly from the lower left corner to the upper right corner of the periodic table. Electronegativity, ionization energy, and electron affinity are some of these diagonal relationships.

diamond An allotrope of carbon in which all carbon atoms have sp^3 hybridization.

diatomic A term describing a molecule that contains only two atoms (e.g., HCl, H_2).

differentiating electron The electron that differentiates one element from an adjacent element in the periodic table.

diffraction The combination of light waves that results in either constructive or destructive reinforcement.

diffusion The movement of molecules from one place to another by random motion.

dilute A qualitative term indicating a small amount of solute in a given amount of solvent.

dimer A substance composed of two identical molecular or ionic units.

dipole A polar molecule. The term *dipole* reminds us that a polar molecule has only two poles, one positive and one negative.

direct relationship A relationship between two variables whereby one must increase when the other increases.

dispersion forces *See* **London forces**.

dissociation The breakup of an ionic compound into ions.

double-replacement reaction A chemical equation in which the cation of one substance replaces the cation of a second substance. At the same time, the cation of the second substance replaces the cation of the first substance.

***dsp*^{3} hybrid orbital** An orbital formed from one *s*, three *p*, and one *d* orbital. The electrons in these orbitals are all equal in energy. Structures are all related to the trigonal bipyramid.

***d*^{2}*sp*^{3} hybrid orbital** An orbital formed from one *s*, three *p*, and two *d* orbitals. The electrons in the hybrid all have the same energy. Structures are all related to the octahedron.

ductile A term describing the property of being able to be drawn into wire forms.

dynamic equilibrium The state in which a chemical process is going in the forward direction at the same rate that it is going in the reverse direction and the concentrations of reactants and products remain constant. Equilibrium follows the kinetic period, in which reaction occurs and the concentrations of reactants and products change.

E

EDTA Ethylenediamine tetraacetic acid, a very useful complexing agent.

effusion The movement of molecules through a small hole from one container to another.

electrode A metal placed in a liquid to transfer electrons in a galvanic or electrolytic cell.

electrolysis The process of using electric current to reduce a chemical substance at the cathode and oxidize a chemical substance at the anode.

electrolyte A substance that dissociates or ionizes completely into ions in solution.

electrolytic cell An arrangement of electrodes and a power source used to force nonspontaneous redox reactions to occur.

electron affinity The energy released or absorbed in adding an electron to an atom.

electron deficient A term describing a Lewis structure that has fewer than an octet of electrons around one or more of its atoms, except hydrogen.

electron 1. The unit of negative charge in the atom. The diffuse electron cloud surrounds the dense nucleus. 2. One of the three particles, along with the proton and neutron, that make up an atom. An electron has a negative charge and virtually no mass in comparison to the neutron and proton. Electrons are arranged in an orderly fashion around the nucleus in a diffuse electron cloud. There are as many electrons as protons in an element.

electronic configuration A listing of the electrons within an atom, based on the sublevels that are filled and the relative energies of these sublevels. For example, the electronic configuration for silicon is $1s^22s^22p^63s^23p^2$.

electronegativity A measure of an atom's tendency to attract electrons. Fluorine has the highest, and francium the lowest, electronegativity in the periodic table.

electroneutrality, law of A statement of the fact that no chemical compound has a net charge. In addition, an element has no net charge.

electrostatic potential energy The energy of attraction of two oppositely charged particles, or the energy of repulsion of two like-charged particles.

element Any one of the 117 distinct particles, known as atoms, that are currently known. Each has distinct chemical and physical properties.

elementary reaction One reaction in a mechanism. It is usually bimolecular or unimolecular, and its coefficients are the exponents in the rate law.

empirical formula The formula that gives only the simplest whole-number ratio of the atoms that make up a compound. *See also* **molecular formula; structural formula**.

endothermic A term describing any process that absorbs heat from the surroundings. Endothermic processes cool the system.

end point The point, where neither reactant is in excess, that marks the end of a titration experiment.

enthalpy The heat content of a chemical substance. Enthalpy is related to the internal potential energy of the substance.

entropy Entropy of a system is proportional to the number of microstates in that system. This is a more precise definition than the randomness of a chemical system.

enzyme One of many naturally occurring catalysts found in biological materials.

enzyme-substrate complex The activated complex formed in an enzyme-catalyzed reaction.

equatorial atom A term used to describe the position of an atom in a covalent molecule of the AX_5 or AX_6 basic structure. The equatorial atoms are around the center of the molecule in positions similar to the Earth's equator.

equilibrium constant, *K* The numerical value of the equilibrium expression. The only variable that has an effect on the equilibrium constant is temperature.

equilibrium expression The basic equation governing chemical equilibrium. Each balanced chemical equation has its own equilibrium expression.

equilibrium table A table of data used to summarize the numerical data and stoichiometric relationships of an equilibrium system.

equivalent A value that is determined for a substance by dividing its mass by the equivalent weight.

equivalent weight The mass of a compound that loses or gains 1 mole of electrons in an oxidation-reduction reaction. In acid-base reactions it is the mass that furnishes or reacts with 1 mole of H^+ ions.

escape energy The minimum kinetic energy of a liquid molecule that is needed for transformation into a gas.

ether An organic compound containing the –C–O–C– functional group.

eudiometer A closed-end manometer.

evaporation The transformation of a liquid into a gas at a temperature below the boiling point.

exact number A number that has no uncertainty. Exact numbers include stoichiometric coefficients, formula subscripts, and most defined quantities.

excess reactant Any reactant that is not completely consumed in a chemical reaction.

exothermic A term describing any process that gives off heat to the surroundings. Exothermic processes heat the system.

extrinsic property A physical or chemical property that varies in proportion to the amount of matter.

F

face-centered cubic (fcc) A cubic structure in which atoms are at the corners and an atom is on each cube face of the unit cell.

factor label A ratio used to convert a number with one set of units into the equivalent number with different units.

Faraday's constant, $\mathscr{F}$ The relationship between the coulomb and moles of electrons; 96,485 C = 1 mol e^-.

filtrate The liquid that passes through a filter.

first law of thermodynamics The law that states that in any chemical or physical process all energy is conserved.

first-order reaction A reaction with a rate law having exponents that add up to 1. The rate law is Rate = k[A].

fissile A term describing a nucleus that is capable of undergoing nuclear fission.

fluorescence A property of some atoms and molecules that allows them to absorb photons of light and reemit them very rapidly, but with a different energy. The emitted light always has a lower energy and longer wavelength than the absorbed light.

formal charge The charge on an atom in a covalent compound, calculated by assuring that all bonding electrons are equally shared.

formation reaction A chemical reaction in which the reactants are elements at standard state and the product is 1 mole of one compound.

formula The representation of a chemical substance using chemical symbols and appropriate subscripts for the numbers of atoms and superscripts to represent charges if the substance is an ion.

free radical A molecule that contains an unpaired electron in its Lewis structure.

freezing-point depression constant, k_f The temperature decrease in the freezing point per molal °C m^{-1} of solute particles.

functional group A group of atoms on an organic compound that represent a characteristic chemical entity.

G

galvanic cell The experimental setup used to convert chemical energy into electric energy. All batteries are galvanic cells.

gamma ray A high-energy photon often emitted in a nuclear reaction.

gas A state of matter characterized by the ability to flow and fill any container completely without regard to the amount of gas in the container.

Gay-Lussac's law The law that expresses the direct relationship between the temperature and pressure of a gas; P/T = constant.

Gibbs free energy The maximum energy from any chemical reaction.

Graham's law of effusion The law that relates the rate at which gases pass through a small hole to the mass of the molecule; $\sqrt{m_1/m_2} = \bar{v}_2/\bar{v}_1$.

gram-atomic mass *See* **atomic mass.**

gram-molar mass *See* **molar mass.**

graph A pictorial method of presenting and evaluating experimental data.

graphite An allotrope of carbon in which carbon has sp^2 hybridization.

gravitational potential energy The attractive energy of any two masses toward each other.

group A column in the periodic table.

H

half-life The time required for half of the reactants to be consumed in a first-order chemical reaction or a radioactive decay.

halide An organic compound with a halogen (e.g., CH_3, CH_2, Cl) in its structure.

halogen An element in the next to last group of the periodic table. Halogens are reactive elements with ns^2, np^5 valence electron structures.

halogenation The addition of a halogen to a double or triple bond.

heat capacity The amount of heat energy that a system needs in order for its temperature to change by 1°C.

heating curve A graph showing the changes that occur while a substance is heated.

heat of combustion The heat released when 1 mole of sample, usually an organic compound, is completely burned in oxygen to form CO_2 and H_2O.

heat of fusion The heat energy needed to convert a solid into a liquid. The units are either joules per gram or joules per mole.

heat of reaction *See* ΔH.

heat of vaporization The heat energy needed to convert a liquid into a gas. The units are either joules per gram or joules per mole.

Henderson-Hasselbalch equation A derivation of the equilibrium expression obtained by taking the negative logarithm of the equilibrium expression; $pH = pK_a + \log([\text{conjugate base}]/[\text{conjugate acid}])$.

Henry's law The law that expresses the relationship between the solubility of a gas and its partial pressure; $c = kp$.

Hess's law The law that states that heats of reaction are additive when chemical reactions are added.

Hund's rule The rule that every orbital in a sublevel must fill with one electron before a second electron of opposite spin can be added to any orbital in that sublevel.

hybrid orbital An orbital constructed by combining electrons, usually froms *s* and *p* orbitals, into a new orbital where all the electrons have the same properties. These orbitals are designated as sp, sp^2, sp^3, dsp^3, and d^2sp^3.

hydrate A substance that contains a fixed number of water molecules. The water molecules are written separately from the formula itself and connected to it with a dot in the center of the line between the chemical formula and the water molecules (e.g., $CuSO_4 \cdot 5\ H_2O$).

hydrogenation The addition of H_2 to a double or triple bond.

hydrogen bonding An extra-strength dipole-dipole attractive force due to a large electronegativity difference between hydrogen and nitrogen, oxygen, or fluorine.

hydrohalogenation The addition of a hydrogen and a halogen to a double or triple bond by using a binary acid such as HF, HCl, HBr, or HI.

hydrolysis reaction The reaction of a substance, usually a conjugate acid or base, with water.

I

ideal gas A gas that obeys the ideal gas law; conceptually, a gas molecule with no volume and no attractive forces with other molecules.

ideal gas law The law that relates temperature, pressure, volume, and moles of gas; $PV = nRT$.

independent variable The variable in an experiment that is under the control of the experimenter.

indeterminate error An error in estimating the uncertain digit in a measurement. Also called random error.

indicator A chemical added to a titration experiment that changes color at the end point.

indicator electrode An electrode placed in a sample in order to measure the concentration of an ion in the sample.

induced dipoles A dipole formed by the interaction of a nonpolar substance and either a polar substance or an instantaneous dipole.

initial conditions A quantitative description of a chemical system at the start of a reaction.

inspection method A method for balancing chemical equations. The inspection involves counting the number of each atom present in the equation and then balancing by adding appropriate coefficients to the reactants or products.

instantaneous dipole A distortion of the electron cloud around an atom or molecule that gives the atom or molecule momentary polarity.

intermediate A substance that appears in the elementary reactions of a mechanism but not in the overall balanced equation.

intermolecular forces The attractive forces—dipole-dipole attractions, London forces, and hydrogen bonding—between molecules and atoms that allows them to condense into liquids and solidify into solids.

internuclear axis An imaginary line connecting the nuclei of two atoms.

intrinsic property A physical or chemical property that does not change with the amount of matter.

inverse relationship A relationship between two variables whereby one must increase if the other decreases.

ion An element that has lost or gained one or more electrons. *See also* **polyatomic ion.**

ion-electron method A method for balancing more complex oxidation-reduction equations. It involves a logical sequence of steps described in Chapter 13.

ionic bond The attraction of a negative anion for a positive cation.

ionic compound A chemical compound composed of negatively charged anions and positively charged cations. The unit is held together by the attraction of the positive charges toward the negative charges.

ionic crystal A crystal formed from cations and anions where the main attractive force is the attraction of positive charges toward negative charges.

ionization The removal or addition of an electron from an atom or molecule. Also the formation of ions when a molecular substance dissolves in water.

ionization energy The energy required to remove an electron completely from an atom.

isoelectronic A term describing any two atoms that have identical electronic configurations. These atoms may be ions or elements.

isolated system A system in which neither mass nor energy is transferred to or from the surroundings.

isomers Distinctly different compounds that have the same elemental composition.

isotope A form of an element with a specified number of protons, neutrons, and electrons.

IUPAC International Union of Pure and Applied Chemistry, which sets nomenclature standards.

K

K The symbol for the equilibrium constant, often written as K_{eq}.

K_a The acid dissociation constant, a special term denoting the equilibrium of a weak acid. A weak acid dissociation is always in the form

$$HA + H_2O \rightleftharpoons A^- + H_3O^+$$

K_b The base dissociation constant, a special term denoting the equilibrium of a weak base. A weak base dissociation is always in the form

$$B + H_2O \rightleftharpoons BH^+ + OH^-$$

K_c The equilibrium constant used when the reactants and products are specified as concentrations.

K_d The dissociation constant, a special term used mainly to describe the dissociation of complex ions.

K_{eq} *See K.*

K_f The formation constant, a special term used to describe the formation of complex ions.

K_p The equilibrium constant used when the reactants and products are specified in terms of partial pressures.

K_{sp} The solubility product, a special term denoting the fact that the equilibrium is between a solid and its solution products.

K_w The autopyrolysis constant of water, equal to 1.0×10^{-14} at 25°C.

ketone An organic compound with a nonterminal –C=O group.

kinetic curve A graph of reactant or product concentration as a function of time.

kinetic energy The energy that matter possesses because of its motion; $KE = \frac{1}{2}mv^2$.

kinetic molecular theory The theory of the motion of molecules in the gas phase, which explains gas pressure, effusion and diffusion rates, and the effect of temperature on the behavior of gases.

L

leveling effect An expression of the fact that the strongest acid in water is the H^+ (H_3O^+) ion and the strongest base is the hydroxide ion, OH^-.

levorotatory A term describing an optical isomer that rotates polarized light to the left.

Lewis acid Any substance that can accept electron pairs.

Lewis base Any substance that can donate electron pairs. Also called ligand; complexing agent; sequestering agent.

Lewis structure A molecular structure based on the concept that all atoms try to achieve the noble gas electronic configuration by sharing electrons.

Lewis theory The theory that acids are electron-pair acceptors and bases are electron-pair donors.

ligand *See* **Lewis base.**

limiting reactant The reactant that is completely consumed in a reaction, causing it to stop. Also called limiting reagent.

limiting reagent *See* **limiting reactant.**

linear A term referring to atoms aligned in a straight line; a three-atom arrangement with a 180° bond angle.

liquid A state of matter characterized by its ability to flow in order to fill any container from the bottom up.

litmus paper A type of pH paper using the indicator litmus, which is pink in acid and blue in base.

lock and key The analogy used to depict how an enzyme recognizes reactants on the basis of physical geometry as well as chemical characteristics.

London forces The attractive forces from instantaneous dipoles. These forces are due to the possibility that the electron clouds around atoms and molecules are not perfectly symmetrical at all times. Also called dispersion forces.

lone pairs Electron pairs in Lewis structures that are not used for bonding.

M

magnetic quantum number, m_ℓ The quantum number that specifies the orbital in which an electron is located and the orientation of the orbital in space; m_ℓ may be any number from -1 to $+1$, including zero.

malleable A term describing the property of being able to be hammered into new shapes.

manometer A device used for measuring gas pressures.

mass A quantity of matter.

mass fraction (wt/wt) A concentration unit defined as the mass of one solute divided by the total mass of the solution.

mass-volume fraction (wt/vol) A concentration unit defined as the mass of one solute in a given total volume of solution.

melting point, normal The temperature at which a solid melts at 1.00 atmosphere of pressure; also, the temperature at which a liquid becomes a solid. Also called crystallization point.

meniscus The curved surface of a liquid in a tube or container.

metal A substance with characteristic properties of high electrical conductivity, malleability, and a metallic silver or yellow luster.

metallic crystal A crystal formed from a metal in the periodic table. Metallic crystals are malleable and ductile and conduct electricity.

metalloid An element that has properties of both metals and nonmetals.

metastable A term describing a physical situation in which a material is stable unless disturbed.

metric base unit One of seven basic units of measurement in the metric system. More complex units are combinations of base units.

metric prefix A prefix (e.g., *milli-*, *pico-*) used with a metric base unit to represent a specific exponential value.

mirror image A description of stereoisomers in which the structure of one isomer is the reflection of the other in a mirror.

molality (*m*) A concentration unit defined as the number of moles of solute dissolved in 1 kilogram of solvent.

molarity (*M*) A concentration unit defined as the number of moles of solute in 1 liter of solution.

molar mass The sum of the gram-atomic masses of all atoms in a chemical formula. Also called molecular mass.

molar volume The volume of 1 mole of gas, usually at standard temperature and pressure (STP).

mole (mol) The quantity of any substance that contains 6.02×10^{23} units of that substance.

mole fraction (χ) A concentration unit defined as the number of moles of solute divided by the total number of moles in a solution.

molecular crystal A crystal formed from a molecule. The attractive forces that hold molecular crystals together are London forces, dipole-dipole attractions, hydrogen bonding, or a combination of these.

molecular formula The formula for a molecular or covalent substance, showing all of the atoms that comprise the molecular unit. A molecular formula may be simplified to an empirical formula if all of the subscripts can be divided by a common number. *See also* **empirical formula; structural formula.**

molecular mass *See* **molar mass.**

molecular orbital An orbital created by the pairing of electrons from different atoms. This orbital encircles the atoms that are bonded together.

molecule A group of atoms bound together by covalent bonds with zero total charge.

molten salt A solid salt that has been heated to a temperature where it becomes a liquid. Also called a fused salt.

monochromatic A term describing light that has a single wavelength.

monodentate ligand A Lewis base that donates one pair of electrons.

monomer One of the individual repeating units of a polymer.

N

natural abundance The percentage of an isotope of an element found in nature.

network crystals *See* **covalent crystal**.

neutralization reactions A chemical reaction of an acid with a base.

neutron One of three particles, along with the electron and proton, that make up an atom. This particle has no charge, but has a mass approximately equal to the proton mass. Neutrons and protons make up the bulk of the mass of the atom.

noble gas An element in the last group in the periodic table. Noble gases are unusually stable elements and all have ns^2, np^6 valence electrons.

nonbending electron pair A pair of electrons in a Lewis structure that is not shared with any other atoms.

nonelectrolyte A substance that does not dissociate at all in solution.

nonmetal An element that is not metallic. Nonmetals do not conduct electricity well and do not have a shiny metallic luster. They are located in the upper right portion of the periodic table.

nonpolar A term describing a bond or molecule that has its charge distributed evenly. Only diatomic elements form truly nonpolar bonds. Symmetrical molecules are nonpolar.

normal A term describing an organic compound in which all carbon atoms are arranged in one straight chain.

normality The number of equivalents of a substance dissolved in 1 liter of solution.

nuclear charge The number of positive charges in the nucleus. This is the same as the number of protons in the nucleus (Z) and is also the atomic number.

nuclear fission A radioactive decay process initiated by the absorption of a neutron; it results in a large nucleus dividing roughly in half.

nuclear fusion The combination of two nuclei to form a new atom.

nuclear mass The total mass of the nucleus. This is the sum of the masses of the protons and neutrons in the nucleus. Since electrons have virtually no mass, it is also the atomic mass (A) of the isotope.

nuclear reactor A device that uses a nuclear reaction to create heat energy for the purpose of generating electricity.

nucleon Either a proton or a neutron, both of which are fundamental particles of the nucleus.

nucleus The center of an atom, which contains the protons and neutrons. The nucleus is extremely dense and comprises a very small fraction of the atom's volume; the rest of the atom is empty space.

O

octahedron A geometric structure of six atoms covalently bound to a central atom. Each atom is 90° from any other.

octet rule A simple but effective rule stating that covalent molecules tend to have octets of electrons around each atom in their structures. These octets simulate the electron configurations of the noble gases.

open system A system in which matter and energy can be exchanged with the surroundings.

optical isomer A stereoisomer that rotates polarized light.

optical path length, *b* The thickness of a sample in a spectroscopic experiment.

orbital A region of space that may be occupied by a maximum of two electrons. The shape of an orbital is defined by the sublevel it is in. The orientation of the orbital depends on its assigned quantum number, m_ℓ. Every orbital in a given sublevel must be filled with one electron before a second electron may fill the orbital.

orbital diagram A diagram in which boxes represent individual valence orbitals. Electrons are represented as arrows to show the spins of the electrons in each orbital.

order of reaction The sum of the exponents in a rate law.

organic acid An acid containing carbon and the –COOH functional group.

organic compound A compound composed of carbon and usually hydrogen.

osmotic pressure The pressure needed to stop the migration of solute through a semipermeable membrane.

oxidation The loss of electrons; also, the increase in oxidation number.

oxidizing agent A substance that causes another to be oxidized; also, a substance that itself is reduced. This term is commonly used by chemists, but it is not on the AP exam.

oxoacids An acid that contains hydrogen, oxygen, and another element in its formula, excluding organic acids.

P

partial pressure The pressure of a single gas in a mixture of gases.

parts per billion (ppb) A unit of measurement similar in concept to percent, obtained by multiplying a fraction by one billion (10^9).

parts per million (ppm) A unit of measurement similar in concept to percent, obtained by multiplying a fraction by one million (10^6).

Pauli exclusion principle The requirement that no two electrons in an atom have the same set of four quantum numbers, n, ℓ, m_ℓ, and m_s.

percent (%) A unit of measurement meaning parts per hundred, obtained by multiplying a fraction by 100.

period A row in the periodic table.

periodic table The table in which the elements are arranged in an orderly fashion that shows the relationships of their chemical and physical characteristics.

pH The negative logarithm of the hydrogen ion concentration in a solution; pH = $-\log H^+$.

phase diagram A graph showing the relationship between temperature and pressure and the conversion of matter among three states: solid, liquid, and gas.

pH indicator A weak acid or a weak base whose conjugate acid and conjugate base have different colors. An indicator changes color indicating the end point of a titration.

pH meter An electronic device used to measure the pH values of solutions.

pH paper Paper with a pH indicator absorbed on it so that it changes color depending on the pH of the solution; pH paper is used to estimate pH. *See also* **litmus paper**.

phosphorescence A property of some atoms and molecules that allows them to absorb photons of light and reemit them seconds to hours later. The emitted light always has a lower energy, longer wavelength than the absorbed light.

pi bond A bond made from the sideways overlap of two *p* orbitals. The electron density of a pi bond lies outside the internuclear axis.

pipet A narrow tube calibrated for precise measurement of small volumes of liquids.

pK_w The negative logarithm of the autopyrolysis constant of water, equal to 14.00.

planar triangle A geometric structure of four atoms, three bonded to a central atom with 120° angles between the atoms, which are all in the same plane.

Planck's constant (*h*) The constant relating the energy of a photon to its frequency.

pneumatic trough An experimental setup for collecting gases by the displacement of water.

pOH The negative logarithm of the hydroxide ion concentration in a solution; pOH= – log OH^-.

polar A term describing the property of a covalent bond or molecule of having one end more positive than the other.

polarizability The tendency for an atom's electron cloud to be deformed so that polarity is created.

polyatomic ion An ion composed of more than one atom covalently bonded together. A polyatomic ion acts as a unit in most chemical reactions.

polyprotic acid An acid with two or more ionizable hydrogen atoms in its formula.

positron A positive electron.

potential energy The energy of matter that may, under appropriate conditions, be converted into work.

precipitate 1. (v.) To cause the formation of a solid by a chemical reaction. 2. (n.) The solid formed as a result of a chemical reaction.

precision The degree of closeness of a group of repeated measurements to each other.

pressure The force per unit area; gas molecules exert this force by collisions with the container walls.

principal quantum number, *n* The quantum number that specifies the energy level of the atom in which an electron is located; *n* may be any integer from 1 to infinity.

product The result of a chemical reaction. Products are placed on the right-hand side of a chemical equation.

proton One of three particles, along with the electron and neutron, that make up an atom. The proton has a positive charge, equal in magnitude (but with the opposite sign) to the charge of the electron. The number of protons is equal to the atomic number, *Z*, of an element. Protons and neutrons make up the bulk of the mass of an atom.

Q

Q *See* **reaction quotient**.

qualitative A term referring to a description of a physical or chemical property without the use of numbers or equations.

qualitative analysis A logical sequence of experiments and observations used to determine the composition of a sample.

quantitative A term referring to a description of a physical or chemical property using numbers or equations.

quantum number One of four numbers (n, ℓ, m_ℓ, m_s) used in the wave-mechanical model of the atom to describe an electron in an atom.

R

racemic mixture An equal molar mixture of L and D optical isomers.

radioactive disintegration series A sequence of radioactive disintegrations from a heavy isotope to a lighter, stable isotope in more than one step.

radioactivity The property that some unstable nuclei have of decaying spontaneously with the emission of a small particle and/or energy.

radioisotope A radioactive isotope of an element.

random error *See* **indeterminate error**.

randomness A qualitative description of the disorder of the molecules in any sample.

Raoult's law The law that expresses the relationship between the vapor pressure of a solution and the mole fraction of solute in that solution.

rate constant, *k* A constant in the direct relationship between the amount of reacting substance and the rate of the reaction.

rate-determining step The slowest reaction in a mechanism, which limits the overall rate of reaction. Also called rate-limiting step.

rate law The mathematical relationship between reactant concentrations and reaction rate. The general form of a rate law is Rate = $k[A]^x[B]^y[C]^z$.

rate-limiting step *See* **rate-determining step**.

reactant One of the starting materials in a chemical reaction. The reactants are placed on the left side of a chemical equation.

reaction mechanism The detailed series of elementary reactions that add up to the overall reaction. *See also* **elementary reaction**.

reaction profile A plot of the potential energy of molecules as they collide in a reaction, illustrating the nature of the activation energy.

reaction quotient, *Q* The value of the ratio of the equilibrium expression when a chemical system is not in a state of equilibrium. The value of Q in comparison to that of K indicates the direction of the reaction.

reaction rate The velocity, in moles per liter per second, at which reactants are converted into products in a chemical reaction.

reagent blank A solution used to set the zero point of a spectrophotometer.

real gas A gas in which there are attractive forces between the molecules and the molecules have a finite volume.

redox A word coined to combine the terms *reduction* and *oxidation*. It indicates that reduction and oxidation always occur together.

reducing agent A substance that causes another substance to be reduced; also, a substance that is oxidized. This term is commonly used by chemists, but it is not on the AP exam.

reduction The gain of electrons; also, the decrease in oxidation number.

reference electrode An electrode in a galvanic cell, which has a constant potential since the concentrations of all reactants are kept constant.

relative mass A term describing the fact that the masses of atoms in the periodic table are relative, without units, as compared to the mass of the carbon-12 isotope.

relative uncertainty The absolute uncertainty of a measurement divided by the value of the measurement.

resonance structure A Lewis structure that can be drawn in more than one equally probable way. The actual structure is a mixture of all possible resonance structures.

reversible process A chemical or physical process that can be changed from one state to another and then back to the original state. A reversible process takes place in infinitesimally small steps.

S

***S*°** The standard entropy of 1 mole of a substance.

saturated A term describing an organic compound in which all carbon-carbon bonds are single bonds. Saturated compounds have the maximum number of hydrogen atoms; that is, they are saturated with hydrogen.

saturated solution A solution that has the maximum amount of solute dissolved in it.

scientific notation A method of writing numbers in which the significant figures are numbers from 1 to 10 and they are multiplied by 10 raised to the appropriate power to indicate the position of the decimal point.

second law of thermodynamics The law that states that in all physical and chemical processes the overall entropy of the universe must increase.

second-order reaction A reaction with a rate law having exponents that add up to 2. The rate law is Rate = k[A][B] or Rate = $[A]^2$.

semipermeable membrane A thin, solid material through which certain molecules can diffuse while others cannot; may be visualized as a barrier with small holes that allow only molecules of a certain size to pass through.

sequestering agent *See* **Lewis base.**

shell The old term for the principal energy level of an electron; the region in space where electrons are located around the nucleus of the atom. Energy levels are numbered starting with the energy level closest to the nucleus. The number of the principal energy level is also known as the principal quantum number.

sigma bond A covalent bond formed by the sharing of a pair of electrons. The electron pair is located along the internuclear axis between the two atoms that share it.

significant figures All the digits in a measurement except preceding zeros.

simple cubic A term describing a cubic structure with one atom in each of the eight corners of a unit cell.

solid A state of matter characterized by a rigid structure that retains its shape without a container.

solubility A property of a solute that refers to the maximum amount of that solute that can be dissolved in a solvent. This term can be a qualitative or quantitative description of a solute.

solute The substance—gas, liquid, or solid—dissolved in the solvent.

solution A uniform mixture of chemicals. In a solution it is impossible to distinguish separate solute and solvent particles.

solvent Typically, the liquid phase in which a gas, another liquid, or a solid is dissolved. In a mixture of two or more liquids the solvent is the liquid present in the largest amount.

***sp* hybrid orbital** An orbital constructed from an *s* and a *p* orbital into one where both have equal energies. The resulting structures related to this orbital are linear.

sp^2 hybrid orbital An orbital constructed from an *s* and two *p* orbitals. The resulting orbitals all have the same energy. Structures related to this orbital are triangular planar.

sp^3 hybrid orbital An orbital constructed from an *s* and three *p* orbitals. The resulting orbitals all have the same energy. Structures related to this orbital are tetrahedral.

specific heat An intrinsic property of matter that describes the quantity of heat energy needed to raise the temperature of 1 gram of substance by 1°C.

spectrophotometer An instrument for determining the amount of light absorbed by a sample.

spin quantum number, m_s The quantum number that specifies the spin of an electron as either $+\frac{1}{2}$ or $-\frac{1}{2}$. Two electrons in the same orbital must have opposite spins.

standard cell voltage, E°_{cell} The voltage of a galvanic cell when the system is at standard state; also, the combination of two standard reduction potentials as $E^\circ_{cell} = E^\circ_{cathode} - E^\circ_{anode}$.

standard reduction potential, *E* The potential (voltage) of a reduction half-reaction at standard state.

standard state Defined temperature, pressure, and concentrations. In electrochemistry the standard state is 1.00 atmosphere pressure, 298 K or 25°C, and 1.00 molar concentrations for all soluble compounds. Solids and pure liquids are also defined as 1.00 molar.

standard temperature and pressure A state defined as having a temperature of 0°C and 1 atmosphere of pressure.

state function A variable whose value depends only on the initial and final states of the system. State functions are ΔH, ΔS, ΔG, and ΔE.

steady-state assumption The assumption that, in evaluating rate constants for elementary reactions, the concentrations of intermediates may be mathematically eliminated by assuming that all prior fast steps are in chemical equilibrium.

stereoisomers Compounds that have the same formula and same bonding but differ in the geometric arrangement of the atoms.

stereospecific A term describing a chemical reaction that produces only one stereoisomer.

stoichiometry Mathematical relationships between chemical substances in a chemical equation.

stopcock The valve on the end of a buret.

STP *See* **standard temperature and pressure.**

strong acid An acid that dissociates completely when dissolved in water.

strong base A base that dissociates completely when dissolved in water.

structural formula A formula that shows the actual arrangement of atoms within the molecule and the bonds between the atoms. *See also* **empirical formula; molecular formula.**

structural isomers Compounds with the same formula but with the atoms bonded in different arrangements.

sublevel A subdivision of an energy level. Electrons in each principal energy level are localized in sublevels. Each sublevel has a distinct shape associated with it. Sublevels are numbered from zero up to one less than the number of the principal energy level. These sublevel numbers are the azimuthal quantum numbers, ℓ. Sublevels are also designated by the letters *s, p, d, f.*

subscript A number placed to the right of, and slightly below, the symbol for an element to represent the number of times that atom is present in the formula unit.

substrate(s) The reactant(s) in an enzyme-catalyzed reaction.

supercritical fluid A gas at a temperature and pressure above the critical point. Such a fluid has properties of both a gas and a liquid.

supersaturated solution A metastable solution that has more than the maximum amount of solute dissolved in it.

supercooling The property of some materials capable of being cooled to temperatures below their melting points without solidifying. Supercooled solutions are also supersaturated and are metastable.

supernatant The liquid remaining above a solid after centrifugation.

surface tension The added attractive force per molecule at the surface of a liquid. Surface tension causes liquids to assume shapes that minimize surface area.

surfactant A substance that lowers the surface tension of liquids.

surroundings All parts of the universe not included in the system being studied.

symmetrical A term describing a geometrical property whereby a structure may be rotated by some angle less than 360° and after rotation the molecule has the same configuration as before.

system The portion of the universe that is under study.

T

TC ("to contain") A label on glassware indicating that the item is calibrated to contain the indicated volume.

TD ("to deliver") A label on glassware indicating that the calibration is based on the volume delivered.

Teflon The addition polymer of CF_2=CF_2 with extraordinary nonstick properties.

tetrahedron A geometric structure with four atoms bound to a central atom by covalent bonds. Each bond is equidistant from any other with a bond angle of 109°.

thermodynamically favored reaction Any reaction that occurs without outside assistance; quantitatively, any reaction that has an equilibrium constant greater than 1.

thermodynamics The study of energy changes in chemical and physical processes.

titration An experimental procedure for reacting two solutions in order to determine the quantity or concentration of one of the solutions.

titration curve A plot of pH versus the volume of titrant added to a sample.

tracer A radioactive element used to detect the movement of materials in a complex system.

***trans* isomer** An isomer with substituents on opposite sides of a double bond.

transition element An element having a *d* electron as the differentiating electron in its electronic configuration.

transition-state theory The reaction-rate theory that details the events and energy changes that occur as two molecules collide.

transuranium element Any of the 17 elements from atomic number 93 to 109.

triangular bipyramid A geometric structure with five atoms covalently bound to a central atom. Three atoms in the equatorial position are 120° from each other. Two additional atoms in the axial positions are 90° from the equatorial atoms.

triple point The temperature and pressure at which all three states of matter—solid, liquid, and gas—are in equilibrium.

U

unit cell The fundamental building block of crystals. An entire crystal is formed by repetitive stacking of the unit cells.

universal gas constant, *R* The constant needed to relate the temperature, pressure, volume, and moles of gas in the ideal gas law, $PV = nRT$.

universe The entirety of all matter and space that exist.

unsaturated A term describing an organic compound that contains one or more double or triple bonds in its structure.

unsaturated solution A solution in which the solute concentration is less than the maximum amount possible.

V

vacuum distillation The laboratory technique of vaporizing and condensing a liquid for the purpose of purification. Vacuum distillation is used to reduce the boiling points of heat-sensitive compounds.

valence electrons The outermost *s* and *p* electrons in an atom. The number and the arrangement of valence electrons define chemical and physical properties.

valence shell electron-pair repulsion (VSEPR) theory A method of evaluating molecular structure by relating the number of bonding and nonbonding electron pairs on an atom to its geometrical structure.

van der Waals forces *See* **London forces.**

vapor pressure The pressure developed by a liquid or solid in a closed container at a constant temperature.

viscosity The ability of a fluid to flow. The more easily a fluid flows, the lower is its viscosity.

volumetric flask A flask calibrated to contain a precise volume of liquid.

volume-volume fraction (vol/vol) A concentration unit defined as the volume of one liquid solute divided by the total volumes of the liquids mixed to prepare a solution.

W

weak acid An acid that dissociates slightly when dissolved in water.

weak electrolyte A substance that partially dissociates into ions in solution.

weight The force developed by the gravitational attraction of two masses.

weighted average An average that depends on the abundance of the objects being averaged.

wetting The spreading of a liquid on a surface that occurs because the adhesive forces overcome the cohesive forces in the liquid.

X

X-ray The high-energy electromagnetic radiation emitted in nuclear decay events or when certain metals are bombarded with energetic electrons.

Z

zero-order reaction A reaction in which the rate is independent of reactant concentration. The rate law is Rate = k.

Index